D1628762

Antidepressants, Antipsychotics, Anxiolytics
Volume 1

Edited by

*Helmut Buschmann, José Luis Díaz,
Jörg Holenz, Antonio Párraga,
Antoni Torrens and José Miguel Vela*

1807–2007 Knowledge for Generations

Each generation has its unique needs and aspirations. When Charles Wiley first opened his small printing shop in lower Manhattan in 1807, it was a generation of boundless potential searching for an identity. And we were there, helping to define a new American literary tradition. Over half a century later, in the midst of the Second Industrial Revolution, it was a generation focused on building the future. Once again, we were there, supplying the critical scientific, technical, and engineering knowledge that helped frame the world. Throughout the 20th Century, and into the new millennium, nations began to reach out beyond their own borders and a new international community was born. Wiley was there, expanding its operations around the world to enable a global exchange of ideas, opinions, and know-how.

For 200 years, Wiley has been an integral part of each generation's journey, enabling the flow of information and understanding necessary to meet their needs and fulfill their aspirations. Today, bold new technologies are changing the way we live and learn. Wiley will be there, providing you the must-have knowledge you need to imagine new worlds, new possibilities, and new opportunities.

Generations come and go, but you can always count on Wiley to provide you the knowledge you need, when and where you need it!

William J. Pesce
President and Chief Executive Officer

Peter Booth Wiley
Chairman of the Board

Antidepressants, Antipsychotics, Anxiolytics

From Chemistry and Pharmacology to Clinical Application

Volume 1

Edited by
Helmut Buschmann, José Luis Díaz, Jörg Holenz,
Antonio Párraga, Antoni Torrens and José Miguel Vela

WILEY-VCH Verlag GmbH & Co. KGaA

The Editors

Helmut Buschmann
Laboratorios Dr. Esteve S.A.
Director R&D
Ave. Marc Deu de Montserrat 5
08041 Barcelona
Spanien

José Luis Díaz
Laboratorios Esteve., S.A.
Medicinal Chemistry
AvMare d.Déu d. Montserrat 221
08041 Barcelona
Spain

Jörg Holenz
Team Leader LI/L01 Chemistry
AstraZeneca R&D Södertälje
15185 Södertälje
Sweden

Antonio Párraga
Laboratorios Dr. Esteve S.A.
Ave. Marc Deu de Montserrat 5
08041 Barcelona
Spain

Antoni Torrens
Laboratorios Dr. Esteve S.A.
Ave. Marc Deu de Montserrat 5
08041 Barcelona
Spain

José Miguel Vela
Laboratorios Dr. Esteve S.A.
Ave. Marc Deu de Montserrat 5
08041 Barcelona
Spain

Coauthors see list of contributors

All books published by Wiley-VCH are carefully produced. Nevertheless, authors, editors, and publisher do not warrant the information contained in these books, including this book, to be free of errors. Readers are advised to keep in mind that statements, data, illustrations, procedural details or other items may inadvertently be inaccurate.

Library of Congress Card No.:
applied for

British Library Cataloguing-in-Publication Data
A catalogue record for this book is available from the British Library.

Bibliographic information published by the Deutsche Nationalbibliothek
Die Deutsche Nationalbibliothek lists this publication in the Deutsche Nationalbibliografie; detailed bibliographic data are available in the Internet at <http://dnb.d-nb.de>.

© 2007 WILEY-VCH Verlag GmbH & Co. KGaA, Weinheim

All rights reserved (including those of translation into other languages). No part of this book may be reproduced in any form – by photoprinting, microfilm, or any other means – nor transmitted or translated into a machine language without written permission from the publishers. Registered names, trademarks, etc. used in this book, even when not specifically marked as such, are not to be considered unprotected by law.

Printing Strauss GmbH, Mörlenbach
Binding Litges & Dopf GmbH, Heppenheim
Cover Design 4T Matthes + Traut GmbH, Darmstadt
Wiley Bicentennial Logo
Richard J. Pacifico

Printed in the Federal Republic of Germany
Printed on acid-free paper

ISBN 978-3-527-31058-6

Contents

Preface 1	XIII
Preface 2	XV
A view from Esteve	XIX
Dedication and Acknowledgements	XXI
List of Contributors	XXIII
Glossary	XXIX

VOLUME 1

1 Depressive Disorders — 1

1.1 Introductory and Basic Aspects — 3

Neurobiology of Mood Disorders

Luz Romero, Ana Montero, Begoña Fernández and José M. Vela

1.1.1 Definition of mood disorders, impact on a global scale and unmet needs	3
1.1.2 Causes and associations in mood disorders: genetics and pharmacogenetics	10
1.1.3 Pathogenesis of mood disorders	14
1.1.4 Concluding remarks	33
References	35

1.2 Clinics — 51

Clinical Aspects of Depressive Disorders

Rosario Pérez-Egea, Victor Pérez, Dolors Puigdemont and Enric Alvarez

1.2.1 Introduction	51
1.2.2 Classification	52
1.2.3 Epidemiology	61
1.2.4 Physiopathology	63
1.2.5 Treatment of affective disorders	68
References	104

1.3 Pharmacology 111
Pharmacotherapy of Depression
Begoña Fernández, Luz Romero and Ana Montero

    11.3.1 Introduction 111
    1.3.2 Current antidepressant treatments 112
    1.3.3 New strategies for antidepressant treatments 116
    1.3.4 Concluding remarks 131
    References 132

1.4 Experimental Research 141
Modeling Human Depression by Animal Models
Ana Montero, Begoña Fernández and Luz Romero

    1.4.1 Introduction 141
    1.4.2 Types of validity 142
    1.4.3 Animal models of depression 145
    1.4.4 Some concluding remarks 160
    References 162

1.5 Chemistry 173
Marketed Drugs and Drugs in Development
Jörg Holenz, José Luis Díaz and Helmut Buschmann

    1.5.1 Summary of drug classes 173
    1.5.2 Tricyclic and tetracyclic antidepressants 173
    1.5.3 Serotonergic agents 193
    1.5.4 Noradrenaline reuptake inhibitors 231
    1.5.5 Monoamine oxidase inhibitors 233
    1.5.6 Miscellaneous agents 239
    1.5.7 Compounds launched in single countries 248
    1.5.8 New opportunities for marketed drugs 249
    1.5.9 Summary of antidepressants in development 249
    References 261

2 Schizophrenia and Other Psychoses — 297

2.1 Introductory and Basic Aspects — 299

Current Status and Challenges in Schizophrenia Research

Francesc Artigas

- 2.1.1 Introduction — 299
- 2.1.2 Clinical diagnosis and assessment of schizophrenia — 300
- 2.1.3 Epidemiology — 306
- 2.1.4 Course of schizophrenia — 307
- 2.1.5 Brain pathology in schizophrenia — 308
- 2.1.6 Pathogenesis and pathophysiology of schizophrenia — 316
- 2.1.7 Concluding remarks — 326
- References — 327

2.2 Clinics — 335

Schizophrenia: A Clinical Review

Salvador Ros and Francisco Javier Arranz

- 2.2.1 Introduction — 335
- 2.2.2 Background — 336
- 2.2.3 Epidemiology — 341
- 2.2.4 General semiology — 341
- 2.2.5 Positive symptoms in schizophrenia — 342
- 2.2.6 Negative symptoms in schizophrenia — 345
- 2.2.7 Cognitive alterations in schizophrenia — 347
- 2.2.8 Characteristics of cognitive deterioration in schizophrenia — 350
- 2.2.9 Methods to evaluate cognitive deterioration in schizophrenia — 356
- 2.2.10 Affective symptoms in schizophrenia — 358
- 2.2.11 Schizophrenia and suicide — 360
- 2.2.12 Onset and states — 365
- 2.2.13 Etiopathogeny — 367
- 2.2.14 Prognosis — 373
- 2.2.15 Schizophrenia therapy — 375
- References — 378

2.3 Pharmacology 389

Pharmacotherapy of Schizophrenia

Analía Bortolozzi, Llorenç Díaz-Mataix and Francesc Artigas

    2.3.1 Antipsychotic drugs: introduction   389

    2.3.2 Atypical Antipsychotics: introduction   401

    2.3.3 Other major investigational approaches   425

    2.3.4 Concluding remarks: challenges in drug discovery   429

    References   432

2.4 Experimental Research 449

Modeling schizophrenia in experimental animals

Pau Celada, Anna Castañé, Albert Adell and Francesc Artigas

    2.4.1 Introduction   449

    2.4.2 Neurophysiology of schizophrenia   450

    2.4.3 Pharmacological models of schizophrenia   453

    2.4.4 Neurodevelopmental models   464

    2.4.5 Genetic models   468

    2.4.6 Other animal models   473

    2.4.7 Concluding remarks   474

    References   475

2.5 Chemistry 491

Marketed Drugs and Drugs in Development

Antonio Párraga, Jörg Holenz and Helmut Buschmann

    2.5.1 Summary of drug classes   491

    2.5.2 Typical antipsychotics   494

    2.5.3 Atypical antipsychotics   549

    2.5.4 Drugs in development   579

    References   586

VOLUME 2

3 Anxiety — 617

3.1 Introductory and Basic Aspects

- 3.1.1 Definition and classification of anxiety disorders — 619

    Francisca López-Ríos, Serafín Gómez-Martín and Antonio Molina-Moreno

    References — 643

- 3.1.2 Epidemiology of anxiety disorders — 645

    Flor Zaldívar-Basurto and José M. García-Montes

    References — 662

- 3.1.3 Etiology of Anxiety — 667

    Margarita Moreno, Ginesa López-Crespo and Pilar Flores

    References — 678

- 3.1.4 Anxiety disorders and drug abuse — 685

    Diana Cardona-Mena, Margarita Moreno and Pilar Flores

    References — 701

- 3.1.5 Neurobiology of anxiety (I): Panic disorder, posttraumatic stress disorder, social phobia and obsessive–compulsive disorder — 709

    Lola Roldán-Tapia, Ginesa López-Crespo, Francisco Antonio Nieto-Escámez and Diana Cardona-Mena

    References — 743

- 3.1.6 Neurobiology of anxiety (II): Childhood anxiety, generalized anxiety disorder and mixed anxiety–depressive disorder — 761

    Francisco Antonio Nieto-Escámez, Ginesa López-Crespo, Lola Roldán-Tapia and Fernando Cañadas-Pérez

    References — 785

- 3.1.7 Consequences of anxiety on memory processes (I): Introductory and clinical aspects — 803

    Fernando Cañadas-Pérez, Francisco Antonio Nieto-Escámez, Lola Roldán-Tapia and Ginesa López-Crespo

    References — 821

- 3.1.8 Consequences of anxiety on memory processes (II): Neurobiological aspects — 831

    Francisco Nieto-Escámez, Fernando Cañadas-Pérez, Lola Roldán-Tapia and Fernando Sánchez-Santed

    References — 864

3.2 Clinics — 881

Clinical Aspects of Anxiety

Blanca Gutiérrez and Jorge Cervilla

- 3.2.1 Conceptual introduction — 881
- 3.2.2 Panic disorder — 882
- 3.2.3 Generalized anxiety disorder — 885
- 3.2.4 Phobic anxiety disorders — 888
- 3.2.5 Obsessive-compulsive disorder — 890
- 3.2.6 Conclusion — 892
- References — 893

3.3 Pharmacology — 895

Pharmacology of Anxiety

José Manuel Baeyens

- 3.3.1 Introduction — 895
- 3.3.2 Drugs used in treatment of anxiety disorders: an overview — 896
- 3.3.3 Generalized anxiety disorder (GAD) — 902
- 3.3.4 Social anxiety disorder (SAD) or social phobia — 905
- 3.3.5 Simple or specific phobia — 908
- 3.3.6 Panic disorder (PD) — 908
- 3.3.7 Posttraumatic stress disorder (PTSD) — 911
- 3.3.8 Obsessive-compulsive disorder (OCD) — 914
- References — 918

3.4 Experimental Research — 923

Modeling Human Anxiety by Animal Models

Enrique Portillo and José Miguel Vela

- 3.4.1 General concepts on animal models — 923
- 3.4.2 General aspects of human anxiety — 925
- 3.4.3 Animal anxiety — 926
- 3.4.4 Potential issues in animal models of anxiety — 937
- 3.4.5 Conclusions: Implication for drug discovery — 940
- References — 942

3.5 Chemistry 951
Marketed Drugs and Drugs in Development
Mónica García-López, Susana Yenes, Helmut Buschmann and Antoni Torrens

- 3.5.1 Summary of drug classes — 951
- 3.5.2 Benzodiazepines — 951
- 3.5.3 Serotonergic drugs — 999
- 3.5.4 GABAergic agents — 1006
- 3.5.5 Dopaminergic agents — 1011
- 3.5.6 Serotonin reuptake stimulants — 1017
- 3.5.7 5-HT2 receptor antagonists — 1019
- 3.5.8 β-adrenoceptor antagonists — 1021
- 3.5.9 Glucocorticoid receptor antagonists — 1026
- 3.5.10 GAT-1 inhibitors — 1029
- 3.5.11 Calcium channel blockers — 1032
- 3.5.12 No-described mechanism of action — 1039
- 3.5.13 Drugs in development — 1051
- References — 1069

4 Attention Deficit and Hyperactivity Disorders — 1089

4.1 Introductory and Basic Aspects — 1091
Neurobiological aspects of Attention Deficit and Hyperactivity Disorders
Javier Burgueño, Rafael Franco and Francisco Ciruela

- 4.1.1 Definition and introductory clues — 1091
- 4.1.2 Etiology of ADHD — 1095
- References — 1100

4.2 Clinics — 1105
Attention Deficit Hyperactivity Disorder: Clinical and Therapeutic Facts
Francisco Javier Arranz and Salvador Ros

- 4.2.1 Introduction — 1105
- 4.2.2 Epidemiology — 1108
- 4.2.3 Etiology — 1114
- 4.2.4 Pathophysiology — 1118
- 4.2.5 Clinical features — 1122
- References — 1146

4.3 Pharmacology 1155

Pharmacology: Targets of Drug Action

Francisco Ciruela, Rafael Franco and Javier Burgueño

    4.3.1 Neurobiological evidence for ADHD as an organic phenomenon   1155

    4.3.2 Neurochemistry of ADHD: Neurotransmitter imbalance   1156

    References   1177

4.4 Chemistry 1183

Marketed Drugs and Drugs in Development

Jörg Holenz, José Luis Díaz and Helmut Buschmann

    4.4.1 Summary of drug classes   1183

    4.4.2 Drugs in development   1192

    References   1194

Index 1197

Preface 1

Psychopharmacology is a recent discipline only initiated fifty years ago that has achieved important advances in the last years. Indeed, the finding of the possible antimaniatic effects of lithium salts in 1949, and the discovery of the sedative effects of chlorpromazine in psychotic patients in 1952 have opened the guidelines of modern psychopharmacology. Since the discovery of these first groups of psychoactive drugs, a huge effort has been devoted to identify new compounds in order to improve their effectiveness and security profile. In spite of this effort, the psychoactive drugs available at the present moment still induce their therapeutic effects based on similar mechanisms of action to those proposed for the first generations of antipsychotic, antidepressant and anxiolytic compounds. Thus, the therapeutic effects of the present antidepressant drugs are yet explained based on the monoaminergic theory of depression. In the same way, antipsychotic drugs available at this moment have the dopamine and the serotonin receptors as therapeutic target, and most of the current anxiolytic compounds primary act by facilitating the activity of the GABAergic system. However, recent studies have started to elucidate new neurobiological mechanisms directly related to the etiopathogenesis of depression, anxiety and psychotic disorders that involve different neurochemical systems, including several neuropeptides, transcription and growth factors. Indeed, the pathophysiological processes underlying these disorders are not limited to the selective changes previously reported on some specific monoaminergic systems. Complex long term changes in neural plasticity and neurogenesis that could even modify the neural morphology seem to play a crucial role in the development of these psychiatric diseases, and represent promising new therapeutic targets. These mechanisms open interesting perspectives for the design of new generations of psychotropic drugs. The modern techniques available in neurobiology from the last decade, such as the different models of genetically modified mice and the leading neuroimaging approaches have provided definitive steps for the advancement in the understanding of these neurobiological disorders.

Psychopharmacology, as well as other medical disciplines, still requires a great research effort for the achievement of new therapeutic strategies for the management of psychiatric diseases. The recent advances in the knowledge of the neurobiological mechanisms underlying the physiopathology of these disorders are providing attractive indications for the design of these new strategies. The possible development of novel generations of psychoactive drugs selectively targeting key components of these emergent mechanisms can probably identify in a near future more effective compounds and with less side effects than the drugs now available in the pharmaceutical market.

This book provides an update of the current state of knowledge on the psychopharmacology field and offers a comprehensive picture of the recent advances made on this exciting discipline. The book is divided in four different chapters devoted to the main psychiatric disorders currently treated with psychotropic drugs: depressive disorders, psychoses and schizophrenia, anxiety, and attention deficit and hyperactive disorders. Each chapter includes a similar strategy to describe the pharmacological approaches to treat these psychiatric disorders. First, an introductory section is included in the different chapters with an exhaustive description of the neurobiological mechanisms and etiopathogenesis of each disorder. A detailed explanation of the main clinical aspects of the psychiatric disorders is also incorporated with a description of the epidemiology, classification and the clinical manifestations of these disorders. The pharmacological aspects include an exhaustive explanation of the different groups of compounds available for the treatment of these diseases as well as the research strategies currently on development for the identification of new therapeutic agents. A comprehensive explanation of the main animal models now available to investigate depression, psychoses and anxiety disorders is also provided. These models are essential tools to further elucidate the physiopathological mechanisms underlying psychiatric disorders and to validate new treatments. However, at the present moment animal models of attention deficit and hyperactive disorders have not been yet well recognized by the scientific community. Finally, a particular emphasis is devoted to the chemical characteristics of the psychotropic drugs with a detailed description of the different groups of compounds including both marketed drugs and drugs under development.

January 2007

Rafael Maldonado, Barcelona
Dept. of Experimental and Health Sciences
Unit of Neuropharmacology
University Pompeu Fabra, Doctor Aiguader, 80.
08003 Barcelona, Spain.

Preface 2

Considerable progress has been made in the treatment of psychiatric diseases for the last decades thanks to the serendipitous discovery of the psychotropic properties of a few drugs about fifty years ago. Antidepressants, antipsychotics, anxiolytics and mood stabilizing agents have all been discovered at the beginning of the second half of the last century, causing a true revolution in the clinical practice of psychiatrists, with the availability of really effective drugs for alleviating patients from major debilitating diseases. Before pharmacotherapy, psychiatry was confined to middle age-like practice, and not really recognized as a clinical discipline when compared to those concerned by organic or infectious diseases.

For the last decades, marked improvement in the phenotypic identification of the major psychiatric diseases such as depression, schizophrenia and related disorders has also been achieved with the consensus description of relevant symptoms in the Diagnostic and Statistical Manual of Mental Disorders, the famous DSM. The 4th revised edition, DSM-IV-R, published in 2000, is still on use, and the fifth edition is currently under way, which will undoubtedly be of great help for a better characterization of the diseases, and, hopefully, the prescription of the most appropriate available drugs for their treatments. This is indeed a tremendous challenge because mental diseases, especially depression and schizophrenia, are extremely heterogeneous, with depressed patients suffering from insomnia or hypersomnia, psychomotor agitation or retardation… and the variable occurrence of opposite negative and positive symptoms in schizophrenic patients. As judiciously emphasized in clinical chapters of this book, these diseases are incredibly complex, and considerable effort has yet to be made for improving their nosographic characterization, a step to be achieved for better physiopathological characterization and, hopefully, clinical treatment.

The first psychotropic drug which allowed psychiatry to enter the modern era of clinical practice is in fact an anti-tuberculosis drug, isoniazid (Rimifon®). Indeed, nurses and physicians, in sanatoriums where patients suffering from tuberculosis were so numerous after the Second World War, noted that those treated with isoniazid were more prone to enjoy daily life, which led to the idea that this drug might be endowed with some antidepressant-like properties. Then iproniazid (Marsilid®), an isoniazid derivative, was synthesized, which revealed to be devoid of anti-tuberculosis effects but still endowed with such antidepressant-like properties. Iproniazid thus became, in 1952, the first of the long series of monoamine oxidase inhibitors (MAOI) that constitutes the first class of antidepressant drugs still on use today. Also in the early fifties of the last century, H. Laborit together with P. Deniker and J. Delay, at Ste Anne hospital in Paris, discovered the antipsychotic properties of an anti-histaminergic compound, chlorpromazine, the first one of the long series of phenothiazine derivatives that were subsequently developed as potential antipsychotics. However, among these anti-histaminergic derivatives, imipramine revealed to be devoid of antipsychotic properties, but endowed with antidepressant properties. Imipramine is actually the first of the long series of tricyclics that have been developed as the second main class of antidepressants, and which were subsequently shown to also act on monoamine neurotransmitters, by blocking their reuptake by nerve terminals.

These few historical elements were just to recall that close relationships really exist between antidepressants and antipsychotics, which fully justify the judicious choice of Helmut Buschmann and his co-editors to group these two classes of psychotropic drugs in this Volume 1 textbook.

Thanks to the subsequent development of animal models of the diseases and availability of these first drugs as pharmacological tools, a very active field of research, Neuropsychopharmacology, has then been set up which provides numerous data on the physiopathological mechanisms associated with depression and schizophrenia, the mechanisms of actions of antidepressant and antipsychotic drugs, and the identification of the molecular targets really involved in their therapeutic actions. As remarkably synthesized in this book, the same neuroactive molecules, mainly monoamines, glutamate, neurotrophins (notably brain derived neurotrophic factor), neuropeptides such as substance P, etc, are most often concerned by the mechanisms of action of both antidepressant and antipsychotic drugs. In addition, modern neuroimaging techniques revealed that depression as well as schizophrenia are associated with functional/structural alterations in the same key areas such as the hippocampus and the frontal lobe. Indeed, a high degree of co-morbidity exists with both depression and psychosis symptoms present in a relatively high proportions of patients, further emphasizing the existence of close relationships between both diseases.

However, in line with the clear-cut distinct properties of antidepressant and antipsychotic drugs, depression and schizophrenia obviously correspond to different disorders with markedly different features. In this regard, epidemiological data are especially informative since the prevalence of schizophrenia appears to be rather stable, with about 1% of population affected worldwide whatever the period considered for the past and coming decades (as estimated by the World Health Organization), while the prevalence of depression is continuously growing in industrial (particularly stressful) societies, up to very probably becoming the first cause of disability in Europe within the next 20 years. These data are indeed in line with evidence showing the major implication of genetic causes for schizophrenia versus environmental causes (stress in particular) for depression, although both genetic and environmental factors very probably interact for each of these diseases, as recently demonstrated in remarkable studies on the impact of candidate genes (such as those encoding the serotonin transporter and neuronal tryptophan hydroxylase) in stress vulnerable subjects. These major differences between depression and schizophrenia are also supported by considerations on animal models aimed at reproducing as best as possible the respective symptoms of each disease, notably in rodents. As thoroughly described in this book, most validated depression models are based on animal exposure to imposed environmental changes that consist of chronic, repeated, stressful situations, and the well established predictive, face and construct validities of these models clearly demonstrate their pertinence for investigating underlying physiopathological mechanisms and discovering new molecular targets for innovative therapeutic strategies. In contrast, so far, validated animal models of schizophrenia, which generate symptoms responsive to antipsychotics, do not rely on environmental alterations but have been obtained by direct action of brain tissues, with lesioning (in the ventral hippocampus for instance), chronic drug treatments (with direct or indirect dopamine receptor agonists, glutamate NMDA receptor antagonists), or gene mutations (knock out mice deficient in NR2 NMDA receptor subunit, STOP protein, dopamine transporter, etc).

These animal models undoubtedly allowed significant progress in the knowledge of the pathophysiological features of the diseases and the molecular and cellular mechanisms of action of antidepressant and antipsychotic drugs. However, notably because of the challenging heterogeneity of both depression and schizophrenia that I mentioned above, none of these models really mimics all the symptoms of the diseases, and there is an absolute, permanent, need of novel, more relevant, models for more thorough investigations. In this regard, the impressive development of molecular genetic technologies, allowing controlled changes of gene expression in a selected cell population at determined life periods under specific environmental conditions in mice will very probably contribute to generate such better models of depression and schizophrenia. Very recent data in the literature lead me to be optimistic in this regard.

To date, available animal models already pointed at novel potential targets for the development of more efficient and better tolerated antidepressant and antipsychotic drugs and one of the greatest merits of this book is to present a very complete overview of these targets (key receptors for the control of the hypothalamo–pituitary–adrenal stress axis, brain mechanisms of circadian rhythm control, cannabinoid receptors, etc) and the various related drugs under development. As also judiciously emphasized in the book, there is indeed an urgent need for really novel drugs, as those based on the « monoaminergic theory of depression » or the « dopaminergic hypothesis of schizophrenia » which are currently available are far from being ideal drugs. Indeed, only 60-70% of patients positively respond to these drugs. In addition, recurrence of illness episodes very often affects patients who have been apparently cured from a first episode. Also, side effects can be very severe and poorly tolerated (in particular extrapyramidal symptoms, tardive dyskinesia, autonomic symptoms caused by antipsychotics; sexual dysfunctions, sleep alterations and gastrointestinal disorders caused by antidepressants), which accounts for a frequently limited compliance to treatments. Significant progress has obviously been achieved with the development of selective serotonin and/or noradrenaline reuptake inhibitors once the molecular targets (i.e. the serotonin and/or noradrenaline transporters) responsible for the antidepressant action of tricyclics were discovered. Similarly, the so-called «atypical» antipsychotics also represent a positive achievement compared to the first «classic» antipsychotics available for treating schizophrenic patients. However, these second generation psychotropic drugs, like those which were empirically discovered half a century ago, only alleviate symptoms of the diseases, and cannot be considered as really curative drugs. Again, drugs really acting on the neurobiological mechanisms causing depression or schizophrenia are eagerly needed for better, really effective and well tolerated, treatments.

All these challenging points raised in the few lines above are thoroughly documented and discussed in relevant chapters in the book. In addition to synthesizing the most relevant data about the clinical and epidemiological features and etiological hypotheses of depression and schizophrenia, the book contains not only the up-to-date established facts concerning the mechanisms of action of antidepressant and antipsychotic drugs, but also a detailed description of their chemical synthesis and metabolism. It is indeed unique compared to previously published books on psychotropic drugs. The same organization for antidepressants on the one hand and antipsychotics on the other hand, with chapters dedicated to (i) basic aspects, (ii) clinical aspects, (iii) pharmacotherapy, (iv) animal models and (v) chemistry and metabolism for each class of drugs makes this book especially convenient, allowing the reader to find rapidly any relevant information. The 17 pages Glossary before starting Chapter 1 is also particularly helpful. For me, who is professionally involved in neuropharmacological research since more than 30 years, this book is obviously a «classic» which I keep out of my library, close to me, for daily consulting. Thanks to its very complete contents, judicious critical assessment of data currently available, and especially convenient organization, I am fully convinced it will also be a superb tool for psychiatrists, pharmacologists, neurobiologists, and students in neurosciences. You all will enjoy reading it.

January 2007

Michel Hamon, Paris
Unit of Neuropsychopharmacology Molecular,
Cellular and Functional - U288 -
Faculty of Medicine Pitié-Salpêtrière,
91-105 Boulevard de l'Hôpital. 75013 Paris, France.

A view from Esteve

Esteve is one of the leading pharmaceutical-chemical companies in Spain and with a strong international presence. The privately owned business group has production facilities in Europe, Asia, and America, subsidiaries in Italy and Portugal and commercial activities in more than 90 countries. The company's segments of business are human medicine, veterinary products and active pharmaceutical ingredients for Esteve own products and the international market. In the human medicine segment, Esteve has a strong focus in analgesia and CNS. A combination of research and development of proprietary products and the marketing of these and licensed products has placed Esteve at the head of the Spanish pharmaceutical sector. For more information please visit: www.esteve.com.

It is the hope of the company that this book will be a useful reference to widespread the knowledge of and to improve psychopharmacology treatment.

Dedicated to the Memory of Lluís Martínez

Acknowledgements

As we publish this edition of *Antidepressants, Antipsychotics, Anxiolytics: From Chemistry and Pharmacology to Clinical Application*, we express our appreciation to those individuals who have made this work possible through their help and advice. We are particularly grateful for the opportunity to have known and worked with Lluís Martínez. His enthusiasm for life continues to be an inspiration to us and to all those who had the opportunity to know him.

The impetus to edit this book grew from our positive experiences with a variety of professionals and colleagues with whom we have collaborated in the context of the research activities developed in Esteve. We wish first to acknowledge this possibility and the support received from Esteve.

We thank the Library Department of Esteve, Begoña Labado, Ana Guerra, Mercè Olivet, Claudia Acosta, Neus Ferrer and Anna Maria Rodríguez for their valuable contribution entering and checking bibliographic references, reviewing the Index Section or copyediting and formatting the manuscript. Their careful attention to detail has made possible the efficient preparation of the manuscript.

Special thanks also go to a number of our colleagues and other excellent contributors for their brilliant work in the elaboration of chapters. We thank them for their positive impact on this book. We are also indebted to Drs. Michel Hamon and Rafael Maldonado, who wrote the Preface Section of this book. Finally, we acknowledge the advice and assistance provided by the staff of Wiley-VCH Verlag GmbH & Co. KGaA; special thanks are due to Waltraud Wuest and Frank Weinreich.

January 2007

Helmut Buschmann
José Luis Díaz
Jörg Holenz
Antonio Párraga
Antoni Torrens
José Miguel Vela

List of Contributors

Albert Adell
Dept. of Neurochemistry and Neuropharmacology, Institut d'Investigacions Biomèdiques de Barcelona, CSIC (IDIBAPS), C/. Roselló, 161. 08036 Barcelona, Spain.

Enric Alvarez
Psychiatry Service, Hospital de la Santa Creu i Sant Pau, Av. Antoni Mª Claret, 175. 08025 Barcelona, Spain.

Francisco Javier Arranz
Medical Dept., CNS Area, Laboratorios Esteve, Av. Mare de Déu de Montserrat, 221. 08041 Barcelona, Spain.

Francesc Artigas
Dept. of Neurochemistry and Neuropharmacology, Institut d'Investigacions Biomèdiques de Barcelona, CSIC (IDIBAPS), C/. Roselló, 161. 08036 Barcelona, Spain.

José Manuel Baeyens
Dept. of Pharmacology and Institute of Neuroscience, School of Medicine, University of Granada, Av. de Madrid, 11. 18012 Granada, Spain.

Analía Bortolozzi
Dept. of Neurochemistry and Neuropharmacology, Institut d'Investigacions Biomèdiques de Barcelona, CSIC (IDIBAPS), C/. Roselló, 161. 08036 Barcelona, Spain.

Javier Burgueño
Dept. of Pharmacology, Laboratorios Esteve, Av. Mare de Déu de Montserrat, 221. 08041 Barcelona, Spain.

Helmut Buschmann
Research Direction, Laboratorios Esteve, Av. Mare de Déu de Montserrat, 221. 08041 Barcelona, Spain.

hbuschmann@esteve.es

Fernando Cañadas-Pérez	Dept. of Neuroscience and Health Sciences, University of Almería, La Cañada. 04120 Almería, Spain.
Diana Cardona-Mena	Dept. of Neuroscience and Health Sciences, University of Almería, La Cañada. 04120 Almería, Spain.
Anna Castañé	Dept. of Neurochemistry and Neuropharmacology, Institut d'Investigacions Biomèdiques de Barcelona, CSIC (IDIBAPS), C/. Roselló, 161. 08036 Barcelona, Spain.
Pau Celada	Dept. of Neurochemistry and Neuropharmacology, Institut d'Investigacions Biomèdiques de Barcelona, CSIC (IDIBAPS), C/. Roselló, 161. 08036 Barcelona, Spain.
Jorge Cervilla	Dept. of Psychiatry and Institute of Neurosciences, School of Medicine, University of Granada, Av. de Madrid, 11. 18012 Granada, Spain.
Francisco Ciruela	Dept. of Biochemistry and Molecular Biology; University of Barcelona, Av. Diagonal, 645. 08028 Barcelona, Spain.
José Luis Díaz	Dept. of Chemistry; Laboratorios Esteve, Av. Mare de Déu de Montserrat, 221. 08041 Barcelona, Spain. jldiaz@esteve.es
Llorenç Díaz-Mataix	Dept. of Neurochemistry and Neuropharmacology, Institut d'Investigacions Biomèdiques de Barcelona, CSIC (IDIBAPS), C/. Roselló, 161. 08036 Barcelona, Spain.
Begoña Fernández	Dept. of Pharmacology, Laboratorios Esteve, Av. Mare de Déu de Montserrat, 221. 08041 Barcelona, Spain.
Pilar Flores	Dept. of Neuroscience and Health Sciences, University of Almería, La Cañada. 04120 Almería, Spain.

Rafael Franco	Dept. of Biochemistry and Molecular Biology; University of Barcelona, Av. Diagonal, 645. 08028 Barcelona, Spain.
Mónica García-López	Dept. of Chemistry, Laboratorios Esteve, Av. Mare de Déu de Montserrat, 221. 08041 Barcelona, Spain.
José M. García-Montes	Dept. of Personality, Assessment and Psychological Treatment, University of Almería, La Cañada. 04120 Almería, Spain.
Serafín Gómez-Martín	Dept. of Personality, Assessment and Psychological Treatment, University of Almería, La Cañada. 04120 Almería, Spain.
Blanca Gutiérrez	Dept. of Psychiatry and Institute of Neurosciences, School of Medicine, University of Granada, Av. de Madrid, 11. 18012 Granada, Spain.
Jörg Holenz	Dept. of Medicinal Chemistry, Laboratorios Esteve, Av. Mare de Déu de Montserrat, 221. 08041 Barcelona, Spain. Current address: Local Discovery Research Area CNS & Pain Control, AstraZeneca R&D Södertälje. 15185 Södertälje, Sweden. jorg.holenz@astrazeneca.com
Ginesa López-Crespo	Dept. of Neuroscience and Health Sciences, University of Almería, La Cañada. 04120 Almería, Spain.
Francisca López-Ríos	Dept. of Personality, Assessment and Psychological Treatment, University of Almería, La Cañada. 04120 Almería, Spain.
Antonio Molina-Moreno	Dept. of Personality, Assessment and Psychological Treatment, University of Almería, La Cañada. 04120 Almería, Spain.
Ana Montero	Dept. of Pharmacology, Laboratorios Esteve, Av. Mare de Déu de Montserrat, 221. 08041 Barcelona, Spain.

Margarita Moreno	Dept. of Neuroscience and Health Sciences, University of Almería, La Cañada. 04120 Almería, Spain.
Francisco Antonio Nieto-Escámez	Dept. of Neuroscience and Health Sciences, University of Almería, La Cañada. 04120 Almería, Spain.
Antonio Párraga	Dept. of Research Coordination, Laboratorios Esteve, Av. Mare de Déu de Montserrat, 221. 08041 Barcelona, Spain. aparraga@esteve.es
Victor Pérez	Psychiatry Service, Hospital de la Santa Creu i Sant Pau, Av. Antoni Mª Claret, 175. 08025 Barcelona, Spain.
Rosario Pérez-Egea	Psychiatry Service, Hospital de la Santa Creu i Sant Pau, Av. Antoni Mª Claret, 175. 08025 Barcelona, Spain.
Enrique Portillo	Dept. of Pharmacology, Laboratorios Esteve, Av. Mare de Déu de Montserrat, 221. 08041 Barcelona, Spain.
Dolors Puigdemont	Psychiatry Service, Hospital de la Santa Creu i Sant Pau, Av. Antoni Mª Claret, 175. 08025 Barcelona, Spain.
Lola Roldán-Tapia	Dept. of Neuroscience and Health Sciences, University of Almería, La Cañada. 04120 Almería, Spain.
Luz Romero	Dept. of Pharmacology, Laboratorios Esteve, Av. Mare de Déu de Montserrat, 221. 08041 Barcelona, Spain.
Salvador Ros	Dept. of Psychiatry, Autonomous University of Barcelona, Hospital del Mar, Pg. Marítim, 25. 08003 Barcelona, Spain.
Fernando Sánchez-Santed	Dept. of Neuroscience and Health Sciences, University of Almería, La Cañada. 04120 Almería, Spain.

Antoni Torrens	Dept. of Chemistry, Laboratorios Esteve, Av. Mare de Déu de Montserrat, 221. 08041 Barcelona, Spain.
	atorrens@esteve.es
José Miguel Vela	Dept. of Pharmacology, Laboratorios Esteve, Av. Mare de Déu de Montserrat, 221. 08041 Barcelona, Spain.
	jvela@esteve.es
Susana Yenes	Dept. of Chemistry, Laboratorios Esteve, Av. Mare de Déu de Montserrat, 221. 08041 Barcelona, Spain.
Flor Zaldívar-Basurto	Dept. of Personality, Assessment and Psychological Treatment, University of Almería, La Cañada. 04120 Almería, Spain.

Glossary

Absorption
Process of taking in. Chemicals can be absorbed into the bloodstream after breathing or swallowing.

Active transport
Active transport is the carriage of a solute across a biological membrane from low to high concentration that requires the expenditure of (metabolic) energy.

Addiction
The compulsive use of drugs for non-medical purpose. It is characterized by a craving for mood-altering drug effects. Addiction refers to a dysfunctional behavior as opposed to the improved function and quality of life.

Address-message concept
Address-message concept refers to compounds in which part of the molecule is required for binding, (address) and part for the biological action (message) (IUPAC).

ADME
Absorption, Distribution, Metabolism, Excretion. See Pharmacokinetics.

Adverse effect
Undesirable and unintended, although not necessarily unexpected, result of therapy or other treatment.

Affinity
Affinity is the tendency of a substance to associate with another. The affinity of a drug is its abillty to bind to its biological target (receptor, enzyme, transport system, etc. For pharmacological receptors it can be thought of as the frequency with which the drug, when brought into the proximity of a receptor by diffusion, will reside at a position of minimum free energy within the force, field of that receptor. For an agonist (or for an antagonist) the numerical representation of affinity is the reciprocal of the equilibrium dissociation constant of the ligand-receptor complex denoted KA, calculated as the rate constant for offset (k_{-1}) divided by the rate constant for onset (k_1).

Agonist
A substance that can stimulate a receptor type to transmit an intracellular message and thus initiate a cellular biochemical change. An agonist is an endogenous substance or a drug that can interact with receptors and initiate a physiological or a pharmacological response (contraction, relaxation, secretion, enzyme activation, etc). An agonist is a drug that binds cellular receptors which are ordinarily stimulated by naturally occurring substances, triggering a response.

Allele	The sequence of nucleotides on a DNA molecule that constitutes the form of a gene at a specific spot or a chromosome. There can be several variations of this sequence, and each of these is called an allele.
Alzheimer's disease	Progressive, neurodegenerative disease characterized (AD) by loss of function and death of nerve cells in several areas of the brain leading to loss of cognitive function such as memory and language.
Analgesia	Absence of pain in response to a stimulation that would normally be painful.
Analgesic	A drug used primarily for relieving pain.
Analogue	An analogue is a drug whose structure is inspired by that of another drug but whose chemical and biological properties may be quite different (IUPAC). See Congener.
Anhedonia	The absence of pleasure or the ability to experience it. A common symptom of depression.
Antagonist	A substance that binds to a receptor type without activating it but which blocks the attachment of agonists to the receptor. An antagonist is according to this definition a drug-or a chemical entity that opposes the physiological effects of another. At the receptor level, it is a chemical entity that opposes the receptor associated responses normally induced by another agent. An antagonist is a Drug that binds a receptor without triggering a response.
Anticonvulsant	A compound commonly used for treating epilepsy but also has applications in treating pain and other pathologies (e.g., phenytoin, carbamezapine, gabapentin and sodium valproate).
Antisense molecule	An antisense molecule is an oligonucleotide or analogue thereof that is complementary to a segment of RNA or DNA and that binds to it and inhibits its normal function. (IUPAC).
Assay	Any combination of targets and compounds which is exposed to a detection device to measure chemical or biological activity.
Basal ganglia	It is a group of nuclei in the brain associated with motor and learning functions. However, no single function can be definitively assigned to the mammalian basal ganglia.
Bioavailability	The percentage of drug that is detected in the systemic circulation after its administration. Losses can be attributed to an inherent lack of absorption/passage into the systemic circulation and/or to metabolic clearance. Detection of drug can be accomplished pharmacodynamically (quantification of a biological response to the drug) or pharmacokinetically

	(quantification of actual drug concentration). Oral bioavailability is associated with orally administered drugs.
Bioinformatics	The use of search programmes (public domain, proprietary or in-house programmes) to analyze DNA and protein sequences to predict the function of a gene sequence.
Bioisostere	A bioisostere is a compound resulting from the exchange of an atom or of a group of atoms with another, broadly similar, atom or group of atoms. The objective of a bioisosteric replacement is to create a new compound presenting similar biological properties to the parent compound. The biolsosteric replacement may be physicochemically or topologically based (IUPAC).
Biotransformation	Biotransformation is the chemical conversion of substances by living, organisms or enzyme preparations derived therefrom (IUPAC).
Brain-derived neurotrophic factor	Neurotrophic grow factor strongly implicated in antidepressant effects.
Catecholamine	It iis a chemical compound derived from the amino acid tyrosine. Some of them are biogenic amines like dopamine, norepinephrine and epinephrine.
Cell	Smallest membrane-bound biological unit capable of replication.
Cell membrane	The phospholipid bilayer that surrounds a cell, forming a selectively-permeable barrier.
Cellular assay	Assay run on whole living cells.
Chemoinformatics	A generic term that encompasses the design, creation, organization, storage, management, retrieval, analysis, dissemination, visualisation and use of chemical information, not only in its own right, but as a surrogate or index for other data, information and knowledge.
Clinical candidate	A compound (small molecule) that has achieved the first ever dose administered to the first human (including patients if they are the first humans to receive the compound).
Clinical trials	Research studies that involve patients.
Clone	Group of identical genes, cells, or organisms derived from a single ancestor.
Cloned DNA	Any DNA fragment that passively replicates in the host organism after it has been joined to a cloning vector.

Cloning	Process of making genetically identical copies.
Coenzyme	A coenzyme is a dissociable, low-molecular weight, non-proteinaceous organic compound (often nucleotide) participating in enzymatic reactions as acceptor or donor of chemical groups or electrons.
Comorbid disorder	Another psychiatric or health problem associated with a given disease.
Computer-aided/structural drug design	Rational drug design – use of high resolution molecular imaging techniques (NMR, x-ray crystallography) to identify the active site of the target molecule and construct an new active substance which binds to this active site.
Conditioned place preference	Paradigm involving repeated pairing of a rewarding stimulus, such as food or drug, with a previously neutral environment. The degree of place preference is considered a measure of reward.
Congener	A congener is a substance literally con- (with) *generated* or synthesized by essentially the same synthetic chemical reactions and the same procedures. Analogues are substances that are analogous in some respect to the prototype and in chemical structure. Clearly congeners may be analogues or vice versa but not necessarily. The term congener, while most often a synonym for homologue, has become somewhat more diffuse in meaning so that the terms congener and analogue are frequently used interchangeably in the literature.
Construct validity	A measure of the model's accuracy with which the test measures that which it is intended to measure.
CRF	Corticotropin-releasing factor is the key regulator of the organism's overall response to stress. CRF has hormone-like effects at the pituitary level secreted by hypothalamic neurons.
Dependence	Inability to do without, in this context, a drug; a problem which can occur in particular with the long-term use of drugs acting on CNS.
Dizygotic	Derived from two separately fertilized eggs. Used especially of fraternal twins.
Docking studies	Docking studies are molecular modeling studies aiming at finding a proper fit between a ligand and its binding site (IUPAC).
Double-blind study	A double-blind study is a clinical study of potential and marketed drugs, where neither the investigators nor the subjects know which subjects will be treated with the active principle and which ones will receive a placebo (IUPAC).

Drug	Any chemical compound that may be used on humans to help in diagnosis, treatment, cure, mitigation, or prevention of disease or other abnormal conditions.
Drug development process	A complex process involving different phases:

• Discovery: Identification of a biological, genetic or protein target linked to a particular disease; subsequent lead identification of a potential drug that interacts with the target to help cure the disease or halt its progression.

• Pre-clinical Phase: Comprehensive *in vitro* and animal testing of the drug candidate to establish its target specificity, toxicity in various doses and pharmacokinetics.

• Clinical Phase I: Human trials conducted to demonstrate safety and effectiveness (efficacy); tests with paid, healthy volunteers to establish dosage, side-effects and pharmacokinetics.

• Clinical Phase II: Trials with small numbers of patients conducted to identify drug performance characteristics (optimal dosing, administration, key indication).

• Clinical Phase III: Pivotal trials conducted with larger patient populations to establish efficacy and provide additional safety information.

• Approval: Data is analyzed and submitted for regulatory review. The U.S. submission to the FDA is called an NDA (New Drug Application) or BLA (Biologic License Application); the European submission to the EMEA (European Medicines Evaluation Agency) is called an MAA (Marketing Authorization Application). After stringent analysis and review of the submission, the regulatory agency provides final approval. |
Drug targeting	Drug targeting is a strategy aiming at the delivery of a compound to a particular tissue of the body (IUPAC).
Dual action drug	A dual action drug is a compound which combines two desired different, pharmacological actions at a similarly efficacious dose (IUPAC).
Efficacy	Efficacy is the property that enables drugs to produce responses. It is convenient to differentiate the properties of drugs into two groups; those which cause them to associate with the receptors (affinity) and those that produce stimulus (efficacy). This term is often used to characterize the level of maximal responses induced by agonists. In fact, not all agonists of a receptor are capable of inducing identical levels of maximal responses. It depends on the efficiency of receptor coupling, i.e., from the cascade of events, which, from the binding of the drug to the receptor, leads to the observed

	biological effect. Efficacy describes the relative intensity with which agonists vary in the response they produce even when they occupy the same number of receptors and with the same affinity. Efficacy is *not* synonymous to intrinsic activity (IUPAC).
Enzyme	Protein that acts as a catalyst, affecting the rate at which chemical reactions occur in cells. An enzyme is any molecular structure that catalyses a physiological chemical reaction.
Epidural	A form of intraspinal analgesia where the agent is injected into the epidural space that surrounds the dura mater, which is the membrane that contains the cerebo-spinal fluid directly outside the spinal cord.
Etiology	The cause or origin of a disease or disorder as determined by medical diagnosis.
Exon	It is the region of DNA within a gene that is not spliced out from the transcribed RNA and are retained in the final messenger RNA (mRNA) molecule.
Extrapyramidal side-effects	The various movement disorders suffered as a result of taking dopamine antagonists, usually antipsychotic (neuroleptic) drugs. These are akathisia (restlessness), dystonia (muscular spasms), drug-induced parkinsonism (muscle stiffness, shuffling gait, drooling, tremor).
Face validity	A measure of the model's ability to reproduce core symptoms of a disease.
fNMR	A method used to visualize the inside of living organisms as well as to detect the composition of geological structures. fNMR is primarily used to demonstrate pathological or other physiological alterations of living tissues and is a commonly used form of medical imaging.
Forced swimming test	Rodent model of depression in which the animals are immersed in a vessel of water. Rodents develop an immobile posture after initial struggling. Most antidepressants, administered acutely before the test, reverse the immobility and promote struggling.
Frontal lobe	It is an area in the brain of vertebrates. Located at the front of each cerebral hemisphere, frontal lobes are positioned in front of (anterior to) the parietal lobes. The temporal lobes are located beneath and behind the frontal lobes.
Gene	Unit of inheritance; a working subunit of DNA containing the code for a specific product, typically, a protein such as an enzyme.

Gene expression	Process by which a gene's coded information is translated into the structures present and operating in the cell (either proteins or RNAs).
Genetics	Scientific study of heredity how particular qualities or traits are transmitted from parents to offspring.
Genome	All the genetic material in the chromosomes of a particular organism; its size is generally given as its total number of base pairs.
Genome projects	Research and technology development efforts aimed at mapping and sequencing some or all of the genorne of human beings and other organisms.
Genome-wide scan	A systematic search of the genome for the location of genes involved in a disease using genetic signposts (markers).
Genomics	Identification and functional characterization of genes; Genomics is the identification of previously unknown human DNA sequences encoding natural human proteins with previously unknown medical use that can be used as targets in Drug Discovery to discover novel therapeutic agents or administered for therapeutic benefit.
Genotype	Genetic constitution of an organism.
Glucocorticoids	A group of the hormones secreted by the pituitary gland, also known as the hypothalamic-anterior pituitary-adrenocortical (HPA) axis.
G-protein coupled receptor (GPCR)	Any cellular macromolecule to which a ligand binds initiating an effect via a G-protein mechanism (Note that it will include only binding activity through binding domain and will not account for any further event that the protein may perform through its effector domain).
Heritability	In genetics, heritability is the proportion of phenotypic variation in a population that is due to genetic variation. Variation among individuals may be due to genetic and/or environmental factors. Heritability analyzes estimate the relative importance of variation in each of these factors.
Heteroreceptor	A heteroreceptor is a receptor regulating the synthesis and/or the release of mediators other than its own ligand (IUPAC).
High-throughput-screening (HTS)	Technique of rapidly searching for molecules with desired biological effects from very large compound libraries.
Hit (compound)	Compound found by screening and having a desired biological effect.

Homologue	The term homologue is used to describe a compound belonging to a series of compounds differing from each other by a repeating unit, usually a methylene group (IUPAC).
Human Genome Project	International research effort aimed at mapping and sequencing all of the genome of the human beings and other organisms.
Hydrophilicity	Hydrophilicity is the tendency of a molecule to be solvated by water (IUPAC).
Hydrophobicity	Hydrophobicity is the association of non polar groups or molecules in an aqueous environment which arises from the tendency of water to exclude non polar molecules (IUPAC). See Lipophilicity.
HPA axis (Hypothalamic–pituitary–adrenal axis)	The hypothalamic–pituitary–adrenal axis (HPA axis) is a major part of the neuroendocrine system that controls reactions to stress.
ICH (International Conference on Harmonization)	The International Conference on Harmonisation of Technical Requirements for Registration of Pharmaceuticals for Human Use (ICH) brings together the regulatory authorities of Europe, Japan and the United States and experts from the pharmaceutical industry in the three regions to harmonise scientific and technical aspects of product registration. They make recommendations which will be adopted by the national / EU authorities after an approval process.
Intracranial self stimulation paradigm	Brief electrical self-stimulation (press a lever) of specific brain areas which is very reinforcing. A change in the threshold current or in the rate of responding is reported to provide a measure of affective state (i.e., an increase in threshold current reflects a depressed affect).
Intrathecal	A form of intraspinal anaesthesia or analgesia in which the agent is injected through the dura mater and arachnoid membrane into the cerebro-spinal fluid which surrounds the spinal cord.
IND	Investigational New Drug. Application must be approved by the Food and Drug Administration (FDA) before a drug can be tested in humans in clinical trials.
Inhibitors	Agents that block or suppress the activity of enzymes such as proteases.
Intrinsic activity	Intrinsic activity is the maximal stimulatory response induced by a compound in relation to that of a criven reference compound. This term has evolved with common use. It was introduced by Ariens as a proportionality factor between tissue response and receptor occupancy. The numerical value of intrinsic activity (alpha) could ran from unity (for full agonists, i.e., agonist inducing the tissue maximal response) to zero (for

	antacyonists). The fractional values within this ran denoting partial agonists. Arien-8 original definition equates the molecular nature of alpha to maximal response only when response is a linear function of receptor occupancy. This function has been verified. Thus, intrinsic activity, which is a drug and tissue parameter, cannot be used as a characteristic drug parameter for classification of drugs or drug receptors. For this purpose, a proportionality factor derived by null methods, namely, relative efficacy, should be used. Finally, "intrinsic activity" should not be used instead of "intrinsic efficacy". A "parcial agonist" should be termed "agonist with intermediate intrinsic efficacy" in a given tissue (IUPAC).
Inverse agonist	An inverse agonist is a drug which acts at the same receptor as that of an agonist, yet produces an opposite effect. Also called negative antagonists.
In vitro	In a test tube.
In vivo	In the living cell or organism as opposed to *in vitro*.
Ion channel	Receptor or carrier proteins which, when activated, allows the passage of ions across cell membranes.
Isosteres	Molecules or ions of similar size containing the same number of atoms and valence electrons.
Lead / Lead compound	As a result of the screening process used during drug discovery, active substances will be identified. Of these active substances, the compound that best fits the desired characteristics profile (pharmacological activity, lack of early toxicity, patentability, etc) will be declared a lead compound. Development activities will then begin to shift from a broad discovery program to a more focused development program centred around the lead compound.
Lead Candidate	A chemical entity (small molecule) or series (a set of structural analogues) that has shown sufficient activity and selectivity for the target, to form the basis for focused medicinal chemistry and optimisation of pharmacological properties.
Lead discovery	Lead discovery is the process of identifying active new chemical entities, which by subsequent modification may be transformed into a clinically useful drug (IUPAC). This phase begins at the initiation of target screening "start of target screening" milestone and concludes with the identification of the first chemical lead compound (or lead series) selected for optimisation – "lead series selected" milestone. It involves the testing of compounds, either *in vitro* or *in vivo*, to determine their target effect.

Lead generation	Lead generation is the term applied to strategies developed to identify compounds which possess a desired but non-optimized biological (IUPAC).
Lead optimization	Lead optimisation is the synthetic modification of a biologically active compound, to fulfill all stereoelectronic, physicochemical, pharmacokinetic and toxicologic required for clinical usefulness (IUPAC). This phase begins with the first chemical lead or lead series selected for optimisation (i.e. the "lead series selected" milestone) and concludes with a decision for an optimized compound to enter preclinical development (i.e. the "pre-clinical candidate selected" milestone). This phase consists of testing of a compound to determine the chemical structure that has the optimum potency and selectivity for the target in question. The phase includes the search for back-up compounds and may also include early ADME and toxicity evaluation.
Learned helplessness	Reduced escape behavior in response to controllable stressors after prior exposures to unavoidable stimuli (usually electric shocks).
Levels of prevalence	The total number of cases of a disease in a given population at a specific time.
Ligand	Chemical messenger, usually released by one cell to communicate with a different cell by binding to specific receptors on the receiving cell's surface.
Ligand Design	The design of ligands using structural information about the target to which they should bind, often by attempting to maximize the energy of the interaction.
Limbic system	The collective name for structures in the human brain involved in emotion, motivation, and emotional association with memory. Includes many different cortical and subcortical brain structures (Amygdala, Cingulate gyrus, Fornicate gyrus, Hippocampus, Hypothalamus, Mammilary body, Nucleus accumbens, Orbitofrontal cortex and Parahippocampal gyrus).
Lipid	A water-insoluble molecule which is soluble in nonpolar solvents such as ether. Divided into two classes: Saponifiable and nonsaponifiable.
Lipophilicity	Lipophilicity represents the affinity of a molecule for a lipophilic environment. It is commonly measured by its distribution behavior in a biphasic system, either liquid-liquid (e.a. partition coefficient in 1-octanol / water) or solid-liquid (retention on reversed-phase high performance liquid chromatography *(RP-HPLC)* or thin-layer chromatography *(TLQ System)* (IUPAC). See Hydrophobicity.

Locus coeruleus	A nucleus in the brain stem responsible for physiological responses to stress and panic.
Macromolecule	A molecule having a molecular weight in the range of a few thousand to many millions: proteins, nucleic acids and polysaccharides.
Medicinal Chemistry	A chemistry-based discipline, also involving aspects of biological, medical and pharmaceutical sciences. It is concerned with the invention, discovery, design, identification and preparation of biologically active compounds, the study of their metabolism, the interpretation of their mode of action at the molecular level and the construction of structure-activity relatioships.
Melanin concentration hormone	Melanin concentration hormone (MCH) is an orexinergic neuropeptide synthesized by neurosecretory cells of the mammalian lateral hypothalamus and the zona incerta, which play an important role in the regulation of energy balance and body weight. The MCH system is also implicated in the regulation of mood and stress responses.
Metabolism	The term metabolism comprises the entire physical and chemical processes involved in the maintenance and reproduction of life in which nutrients are broken down to generate energy and to give simpler molecules (catabolism) which by themselves may be used to form more complex molecules (anabolism). In case of heterotrophic organisms, the energy evolving from catabolic processes is made available for use by the organism (IUPAC).
Me-too drug	A me-too drug is a compound that is structurally very similar to already known drugs, with only minor pharmacological differences (IUPAC).
Micromolar	A concentration representing one millionth of a mole. The amount of pure substance that contains the same number of elementary entities as there are atoms in exactly 12 grams of the isotope carbon-12.
Molecular descriptor	At the root of all methods devised to study the (dis)similarity of compounds; to predict physicochemical properties; to identify compounds with desirable or "drug-like" properties; and to select activity-enriched sets of molecules from "virtual" libraries (i.e., from large numbers of synthetically feasible compounds) are descriptors used to characterize the molecules. These include various types of molecular fingerprints, 3D pharmacophores, physicochemical properties, electrotopological states and connectivity indices. Some represent 2D molecules; some handle 3D ones.
Molecular formula	Shows the actual number of atoms in compound. Ex. C_2H_4.

Molecular modeling	Molecular modeling is a technique for the investigation of molecular structures and properties using computational chemistry and graphical visualization techniques in order to provide a plausible three-dimensional representation under a given set of circumstances (IUPAC).
Monoamine neurotransmitters	Samal molecule neurotransmitters that contain a single amine group. Monoamines include dopamine, serotonin, noradrenaline and adrenaline, and histamine is sometimes also included in this group.
Monoamine oxidase inhibitors	Monoamine oxidase inhibitors are drugs that block degradation of amine transmitters within the cell. The most prominent consequence of monoamine oxidase inhibition is a rapid increase in the intracellular concentrations of monoamines.
Monozygotic	Derived from a single fertilized ovum or embryonic cell mass. Used especially of identical twins.
Ligand	Any atom or molecule attached to a central atom, usually a metallic element, in a co-ordination or complex compound.
Nanomolar	A concentration representing one billionth of a mole.
NCE	New chemical entity. A new chemical entity is a compound not previously described in the literature.
NDA	New drug application. A document that combines all relevant data (with attachments) to allow the US FDA (or an other drug regulatory agency) to review and decide whether to approve marketing of a new drug. Detailed reports of chemistry; pharmacology, toxicology, metabolism, manufacturing, quality controls and clinical data along with proposed labeling are included.
Neuroleptic	Also called antipsychotic. It is applied to a group of drugs used to treat psychosis.
Neuroplasticity	The brain's natural ability to form new connections in order to compensate for injury or changes in one's environment.
Noradrenaline reuptake inhibitors	Norepinephrine reuptake inhibitors (NRIs), also known as noradrenaline reuptake inhibitors (NARIs), are compounds that elevate the extracellular level of the neurotransmitter norepinephrine in the CNS by inhibiting its reuptake from the synaptic cleft into the presynaptic neuronal terminal via the norepinephrine transporter. They act at virtually no other monoamine transporters.
Nuclear receptor	Receptors which are associated to a cell nucleus.
Nucleus accumbens	A collection of neurons located where the head of the caudate and the anterior portion of the putamen meet just lateral to the

	septum pellucidum. The nucleus accumbens, the ventral olfactory tubercle, and ventral caudate and putamen collectively form the ventral striatum. This nucleus is thought to play an important role in reward, pleasure, and addiction.
Odd ratio	A measure of effect size particularly important in Bayesian statistics and logistic regression. It is defined as the ratio of the odds of an event occurring in one group to the odds of it occurring in another group, or to an estimate of that ratio.
Olfactory bulbectomy	The bilateral removal of the olfactory bulbs of rodents with consequent disruption of the limbic-hypothalamic axis. This is associated with several behavioral, neurochemical, neuroendocrine and neuroimmune abnormalities, many of which are comparable with changes seen in depression.
Optimization	The process of synthesizing chemical variations, or analogs, of a lead compound, with the goal of creating those compounds with improved pharmacological properties.
Orphan drug	An orphan drug is a drug for the treatment of a rare disease for which reasonable recovery of the sponsoring firm's research and development expenditure is not expected within a reasonable time. The term is also used to describe substances intended for such uses (IUPAC).
Orphan receptor	Receptor with unknown function binding known ligands.
Partial agonist	A substance which partially (in comparison to an agonist) activates a receptor type to transmit an intracellular message. A partial agonist is an agonist which is unable to induce maximal activation of a receptor population, regardless of the amount of drug applied.
Peptidomimetic	Peptidomimetics are compounds containing non-peptidic structural elements that are capable of mimicking or antagonizing the biological action(s) of a natural parent peptide (IUPAC).
PET	A method that produces cross-sectional x-rays of metabolic processes by means of positron emission tomography.
Pharmacokinetics	Pharmacokinetics refers to the study of absorption, distribution, metabolism and excretion (ADME) of bioactive compounds in a higher organism (IUPAC).
Pharmacology	The science of studying both the mechanisms and the actions of drugs, usually in animal models of disease, to evaluate their potential therapeutic value.
Pharmacophore	A pharmacophore is the ensemble of steric and electronic features that are necessary to ensure the optimal supramolecular interactions with a specific biological taraget

	structure and to trigger (or to block) its biological response. A pharmacophore does not represent a real molecule or a real association of functional groups, but a purely abstract concept that accounts for the common molecular interaction capacities of a group of compounds towards their target structure. The pharmacophore can be considered as the largest common denominator shared by a set of active molecules. This definition discards a misuse often found in the medicinal chemistry literature which consists of naming as a pharmacophore simple chemical functionalities such as guanidines, sulfamides or imidazolines, or typical structural skeletons such as flavones, phenothiazines, prostaglandins or steroids (IUPAC).
Phase I (clinical trial)	The first trials in humans that test a compound for safety, tolerance, and pharmacokinetics. The Phase I trials usually employ healthy volunteers and may expose up to about 50 individuals to the drug. For therapeutic biologics and known toxic compounds, e.g. anticancer agents, only patients with the targeted illness would be used. A Phase I study is a closely monitored clinical trial of a drug or vaccine conducted in a small number of healthy volunteers; used to determine toxicity, pharmacokinetics, preferred route of administration, and safe dosage range of a drug.
Phase II (clinical trial)	The first studies to define efficacy in patients. In general, 100-300 patients would be entered into. Various closely monitored clinical trials during this phase. Dose and dosing regimens are assessed for magnitude and duration of effect during this phase. Some companies further differentiate this phase into Phase 2A and 2B (proof of efficacy and dose finding). A phase II study is a controlled clinical study of a drug or vaccine to identify common short-term side-effects and risks associated with the drug or vaccine, to collect information on its immunogenicity and to demonstrate its efficacy conducted on a limited number of patients with disease.
Phase III (clinical trial)	Expanded controlled and uncontrolled clinical trials intended to gather additional evidence of effectiveness for specific indications and to better understand safety and drug-related adverse effects. Phase III trials are usually large multicenter trials which collect substantial safety experience and may also include specialized studies needed for labeling (e.g., paediatric or elderly, comparative agents). Thousands of patients may be included in the Phase III trials.
Placebo	A placebo is an inert substance or dosage form which is identical in appearance, flavor and odour to the active substance or dosage form. It is used as a active control in a bioassay or in a clinical study (IUPAC).
Physical dependence	Involves the development of a withdrawal syndrome following abrupt discontinuation of treatment or a substantial reduction

	in dose. It is a normal expected response to continuous drug therapy and does not mean that the patient is addicted.
Polymorphism	The occurrence in the same habitat of two or more forms of a trait in such frequencies that the rarer cannot be maintained by recurrent mutation alone.
Potency	Potency is the dose of drug required to produce a specific effect of given intensity as compared to a standard reference. Potency is a comparative rather than an absolute expression of drug activity. Drug potency depends on both affinity and efficacy. Thus, two agonists can be equipotent, but have different intrinsic efficacies with compensating differences in affinity (IUPAC).
Pre-clinical Candidate	An optimised (having sufficient potential as a therapeutic candidate to be tested in humans) compound (small molecule) selected to enter pre-clinical development.
Predictive validity	A measure of the model's ability to identify drugs with clinical efficacy against a disease.
Proband	The first affected individual in a family with a genetic disorder who is manifesting the disease and is diagnosed so.
Prodrug	A prodrug is any compound that undergoes biotransformation before exhibiting its pharmacological effects. Prodrugs can thus be viewed as drugs containing specialized non-toxic protective groups used in a transient manner to alter or to eliminate undesirable properties in the parent molecule (IUPAC).
Protease	Any enzyme that catalyzes the cleavage of a peptide or protein.
Protease inhibitors	Class of drugs designed to inhibit the enzyme protease.
Protein	Large, complex molecule composed of amino-acids. Proteins are essential to the structure, function, and regulation of the body. Examples are hormones, enzymes, and antibodies.
Proteome	Complete profile of all expressed (produced) proteins within a cell, a tissue, or an entire organism at a given time.
Proteomics	Analysis of the functions and interactions of proteins in healthy tissue compared to tissue affected by a disease. Proteomics includes the the separation, identification & characterisation of proteins present in a biological sample and comparison of disease and control samples to identify "disease specific proteins". These proteins may have potential as targets for drugs or as molecular markers of disease.

Receptor	Protein in a cell or on its surface that selectively binds a specific substance (ligand). Upon binding its ligand, the receptor triggers a specific response in the cell.
Scaffold	Core portion of a molecule common to all members of a combinatorial library.
Screening tests	A systematic examination or assessment in order to provide a way to study the effects of potential therapeutics treatments. That kind of model may or may not mimic the psychiatric disorder.
Selective serotonin reuptake inhibitors	A class of antidepressant drugs that act within the brain to increase the amount of the neurotransmitter, serotonin (5-hydroxytryptamine or 5-HT), in the synaptic gap by inhibiting its reuptake. Representative drugs include fluoxetine (Prozac), paroxetine (Paril), sertraline (Zoloft), citalopram (Celexa), estialapram (Lexapro) and fluvoxamine (Luvox).
Serotonin and noradrenalin reuptake inhibitors	A drug treatment for depression. Serotonin and noradrenaline reuptake inhibitors act by blocking reuptake of both neurotransmitters (and also dopamine to some degree).
Simulation models	Paradigms developed to investigate the processes that produce a psychiatric disorder simulating some core symptoms.
Social loss and deprivation	Animal models of depression based on the adverse effects of separation phenomena and social isolation (i.e., neonatal isolation or maternal separation and chronic isolation of adult rats).
Stress models	Presentation of any of several aversive or nociceptive stimuli (physical or psychological stressors) to an animal. Examples include chronic severe or mild stress paradigm and learned helplessness.
Striatum	A subcortical part of the brain consisting of the caudate nucleus and the putamen. It is part of the basal ganglia. The striatum is best known for its role in the planning and modulation of movement pathways but is also involved in a variety of other cognitive processes involving executive function.
Structure-activity relationship	Structure-activity relationship is the relationship between chemical structure and pharmacological activity for a series of compounds (IUPAC).
Substantia nigra	A portion of the midbrain thought to be involved in certain aspects of movement and attention. It consists of two subdivisions, the pars compacta and the pars reticulate.

Substance P	An undecapeptide member of the tackchykinin family of mammalian neuropeptides, wich is the preferred endogenous agonit for neurokin 1 receptors.
Tail suspension test	Rodents are suspended by their tails for several minutes develop an immobile posture after initial struggling. Acute administration of most antidepressant before the tests reverses immobility and promotes struggling.
Tardive dyskinesia	It is a serious neurological disorder caused by the long-term and/or high-dose use of dopamine antagonists, usually antipsychotics and is characterized by repetitive, involuntary, purposeless movements.
Target	Specific biological molecule, such as an enzyme, receptor or ion channel, assumed to be relevant to a certain disease. Most drugs work by binding to a target, thereby affecting its biological function.
Target identification	Identifying a molecule (often a protein) that is instrumental to a disease process (though not necessarily directly involved), with the intention of finding a way to regulate that molecule's activity for therapeutic purposes.
Target validation	Crucial step in the drug discovery process. Following the identification of a potential disease target, target validation verifies that a drug that specifically acts on the target can have a significant therapeutic benefit in the treatment of a given disease.
Teratogen	A teratogen is a substance that produces a malformation in a foetus.
Tolerance	Associated with drug dependence, this phenomenona may occur with chronic administration of a drug. It is characterized by the necessity to progressively increase the dose of the drug to produce its original effect. Tolerance is mainly caused by neuroadaptive changes in the brain.
Toxaemia	An abnormal condition of pregnancy characterized by hypertension and edema and protein in the urine.
Transporters	Carrier proteins which transport molecules across a cell membrane.
Vassopresin	Neuropeptide synthesized in the hypothalamus that modulates HPA axis and improves the effect of CRF on adrenocortico-tropic hormone release.

VOLUME 1

1 Depressive Disorders

1.1 Introductory and Basic Aspects

Neurobiology of Mood Disorders

Luz Romero, Ana Montero, Begoña Fernández and José M. Vela

Dept. of Pharmacology, Laboratorios Esteve, Av. Mare de Déu de Montserrat, 221. 08041 Barcelona, Spain.

1.1.1 Definition of mood disorders, impact on a global scale and unmet needs

The diagnosis of "mood disorders" or "depressive disorders" has evolved over the past 40 years with progressively more precise definitions in each edition of the Diagnostic and Statistical Manual of Mental Disorders (e.g., the "Text Revision" of the Fourth Edition: DSM-IV-TR; American Psychiatric Association, 2000) and International Classification of Diseases and Related Health Problems (ICD-10; WHO, 1993).

The group "depressive disorders" can be divided into three major illnesses:

- Major (unipolar) depressive disorder
- Dysthymic disorder
- Bipolar disorder, also known as Manic-Depressive disorder

Depressive disorders arise from the complex interaction of multiple-susceptibility genes and environmental factors, and disease phenotypes include not only episodic and often profound mood disturbances, but also a range of cognitive, motor, autonomic, endocrine and sleep/wake abnormalities (Manji et al., 2001). Some milestones regarding recognition and treatment of depressive disorders are summarized in Fig. 1.

1.1.1.1 Major (unipolar) depressive disorder

Major depression is a chronic, recurring and potentially life-threatening illness with varied origins, a broad range of symptoms (Box 1), complex genetics and obscure neurobiology that affects up to 20% of the population across the globe (lifetime prevalence), being 2–5 times as high in women as in men (Ayuso-Mateos et al., 2001; Manji et al., 2001; Nestler et al., 2002; Charney, 2004; Lesch, 2004; Gillespie and Nemeroff, 2005; Table 1). This means that in any one year, 19 million people in the United States alone, and 121 million people worldwide, will suffer from some kind of diagnosable depressive

Fig. 1. Milestones of depressive disorders

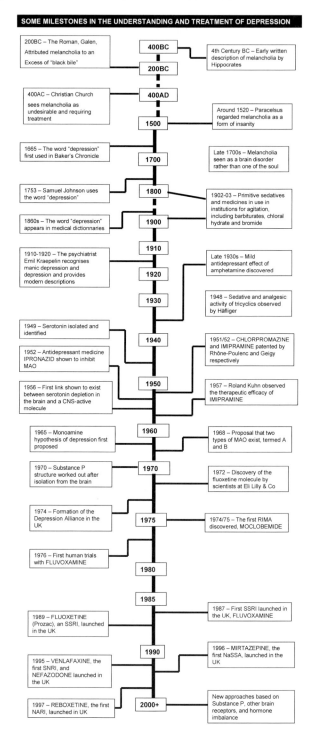

Box. 1. Diagnostic criteria for depression

Diagnostic Statistical Manual of Mental Disorders defines a "major depressive episode" as being characterized by at least five of the following nine symptoms:

- Depressed or irritable mood
- Decreased interest in pleasurable activities and ability to experience pleasure
- Significant weight gain or loss (>5% change in a month)
- Insomnia or hypersomnia; early-morning awakening
- Psychomotor agitation or retardation
- Fatigue or loss of energy
- Feelings of worthlessness or excessive guilt
- Diminished ability to think or concentrate
- Recurrent thoughts of death or suicide

Each symptom must be evident daily or almost every day for at least 2 weeks. The severity of the disorder can range from the very mild to the very severe, and it occurs most commonly in episodes.

"Melancholic" subtype describes particularly severe cases, with prominent circadian variations in symptoms. Manifestations of psychosis (for example, hallucinations or delusions) and anxiety are also seen in many individuals with depression. Individuals with relatively mild but prolonged symptoms, those which persist for at least 2 years, are considered to have "dysthymia".

"Adjustment disorder with depressed mood" describes depressive symptoms that occur after a significant trauma (loss of a job, serious illness, death of a loved one, breakup of a marriage, etc.) and that last for an abnormally long time. If it does not resolve, adjustment disorder can progress to major depression.

"Subsyndromal depression" (or minor depression), is less severe than major depression but carries an increased risk for major depression.

"Seasonal affective disorder" (SAD) strikes during the winter months. SAD is characterized by lethargy, joylessness, hopelessness, anxiety and social withdrawal as well as increased need for sleep and weight gain.

"Premenstrual dysphoric disorder" (PMDD) is a severe distressing and debilitating condition that affects 3-5% of all women. Symptoms – typically depression, anxiety, cognitive and physical manifestations – may occur only during the premenstrual period or may become exacerbated during that time.

Some depressed patients have a variation of the disorder called "bipolar disorder", also known as manic-depressive illness. Bipolar affective disorder refers to patients with depressive "low" episodes alternating with periods of mania, or inappropriate "highs", characterized by an elated mood, increased activity, over-confidence, also known as mania, and impaired concentration.

Other subtypes of depression include "postpartum depression", "psychotic depression" and "atypical depression"

The range of symptoms that comprise depression and the range of diagnostic categories highlight the probable heterogeneity of the illness and the difficulty in establishing any given diagnosis with certainty (Criteria adapted from Diagnostic Statistical Manual of Mental Disorders, 2000; Gruenberg et al., 2005; Berton and Nestler, 2006).

disorders. It is now the leading cause of disability globally and ranks fourth in the top ten leading causes of the global burden of disease based on a survey by the World Health Organization (Figs. 2, 3).

Table 1. Prevalence of major depression.

	Single point	6–12 months	Lifetime
Unipolar depression	3.9% (1.5–6.0)	4.7% (2.6–9.0)	12.6% (4.4–19.5)
Bipolar depression	0.7% (0–2.3)	1.5% (0.1–4.0)	1.3% (0.6–3.3)

Data from 16 epidemiological studies (6 Europe, 2 Canada, 7 USA; Angst, 1995).

Fig. 2. Depression, a major cause of disability worldwide: leading causes of Disability-Adjusted Life Years (DALYs-2000; adapted from WHO Mental Health Report, 2001).

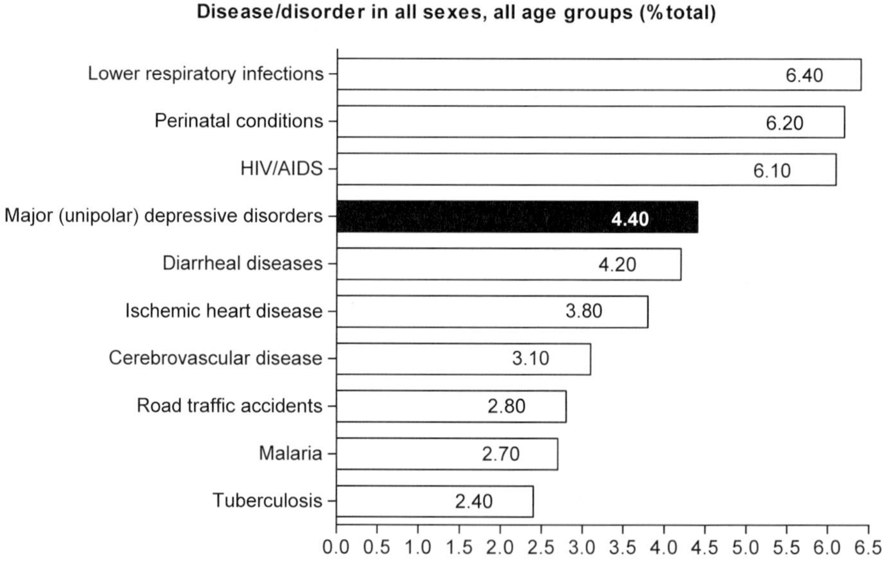

Disease/disorder in all sexes, all age groups (% total)

Condition	%
Lower respiratory infections	6.40
Perinatal conditions	6.20
HIV/AIDS	6.10
Major (unipolar) depressive disorders	**4.40**
Diarrheal diseases	4.20
Ischemic heart disease	3.80
Cerebrovascular disease	3.10
Road traffic accidents	2.80
Malaria	2.70
Tuberculosis	2.40

It has been estimated that, by 2020, unipolar depression will have the dubious distinction of becoming the second cause of the global disease burden (Table 2), a position that the disease already holds for American women.

In 2000, the economic burden (direct and indirect) of depressive disorders in the United States was estimated to be $83.1 billion (Greenberg et al., 2003) and it is estimated that only around 50% of patients suffering from major depressive disorder are assigned the correct diagnosis and go on to receive appropriate treatment. Of those that do receive suitable treatment, approximately 50% reach full remission (Table 3). The recurrence rate for those who recover from the first episode is typically around 35% within 2 years and increases to 60% in 12 years. After three episodes, the likelihood of recurrence is 90%.

Fig. 3. Leading causes of Years of Life Lived with Disability (YLDs-2000; adapted from WHO Mental Health Report 2001).

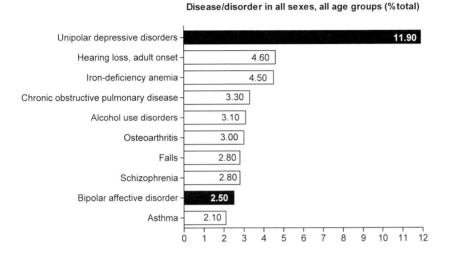

Table 2. Leading causes of Disability-Adjusted Life Years (DALYs-Estimated 2020).

RANK	2020 (Murray and Lopez, 1996)
1	Ischemic heart disease
2	Unipolar major depression
3	Road traffic accidents
4	Cerebrovascular disease
5	Chronic obstructive pulmonary disease

Table 3. Effectiveness of interventions for depression (Mynors-Wallis, 1996; Schulberg et al., 1996).

Intervention	% remission after 3–8 months
Placebo	27
Tricyclics	48–52
Psychotherapy (cognitive or interpersonal)	48–60

During 2002, the market of antidepressants accounted for around $17 billion of the $49 billion market for the Central Nervous System (CNS).

One of the particularly tragic outcomes of a depressive disorder is suicide. Approximately 15–20% of depressive patients end their own lives (Goodwin and Jamison, 1990; Angst et al., 1999; Ebmeier et al., 2006). About 60% of all suicides occur in relation to mood disorders (Mann, 2003). Suicide remains one of the most common and unavoidable outcomes of depression and is now the eighth leading cause of death in the United States.

Several lines of evidence indicate an important contribution of depression to medical morbidity. Depressive disorders often co-occur with anxiety disorders, bulimia, anorexia and substance abuse. Depression is also associated with an increased risk of coronary heart disease (Ferketich et al., 2000) and depressed patients have an increased risk of premature death compared with control subjects (Harris and Barraclough, 1998).

To date, no single antidepressant drug is effective in all patients treated, probably due to the heterogeneity of the disease and to individual differences in the response to the agents used. Moreover, all antidepressant pharmacological treatments need to be administered for several weeks before amelioration signs begin to emerge.

To summarize, depression is a common mental disorder, causing a very high level of disease burden, and is expected to be a rising trend during the coming 15 years. Nowadays, antidepressant drugs of choice are, in the first instance, serotonin selective reuptake inhibitors, such as fluoxetine, paroxetine, fluvoxamine, citalopram and sertraline (NICE, 2004). The main risks associated with serotonin selective reuptake inhibitors are treatment-emergent suicidal behavior and withdrawal symptoms, especially in children and adolescents (Weller et al., 2004). Moreover, existing antidepressant treatments exhibit limited efficacy and a slow onset of action.

The search for an adequate treatment of major depression is one of the main challenges of Neuropharmacology.

1.1.1.2 Dysthymic disorder

Dysthymic disorder (or dysthymia) is a less severe but more prolonged form of depression that lasts for years. A person with dysthymia is unable to obtain any enjoyment from life. Dysthymic adults may be pessimistic, guilt-ridden, irritable, easily hurt by others or withdrawn. According to the American Medical Association, the condition is diagnosed when an individual has a generally depressed mood for most of the day, the majority of the time and for a period of at least two years (one year in a child). Children with the condition may be irritable, cranky, difficult, generally sad or have low self-esteem.

Many patients with dysthymia will at some point experience a superimposed major depressive episode. This condition has been termed *double depression*. Double depression is experienced by as many as 75% of all dysthymic individuals at some point in their lives (Kocsis, 2000).

1.1.1.3 Bipolar disorder (manic depressive disorder)

Bipolar disorder is characterized by cycling mood changes: severe *"lows"* (depression; Box 1) and *"highs"* (mania; Box 2).

Box 2. Diagnostic criteria for maniac phase of bipolar disorders

Symptoms of mania include abnormally and persistently elevated mood accompanied by at least three of the following:

- Abnormal or excessive elation
- Unusual irritability
- Decreased need for sleep
- Grandiose notions
- Increased talkativeness
- Racing thoughts
- Excessive involvement in risky behaviors or activities, including increased sexual behavior, and use of alcohol and illicit drugs
- Increased energy, activity or restlessness
- Poor judgment
- Overly-inflated self-esteem
- Distractibility or irritability
- Inappropriate social behavior (provocative, intrusive or aggressive behavior)
- Denial that anything is wrong

Criteria adapted from Diagnostic Statistical Manual of Mental Disorders, 2000

According to the National Institutes of Mental Health, bipolar disorder affects approximately 2.3 million adult Americans. The lifetime prevalence of bipolar disorder is in the range 1.3–1.6% and does not differ significantly by age, sex, race or ethnicity. There is an exception in rapid cycling, a severe and difficult to treat variant of the disorder, which arises mostly in women (Muller-Oerlinhausen et al., 2002). Bipolar disorder is believed to be widely under-recognized in the primary care setting, which implies that its prevalence could be much higher (Das et al., 2005; Kupfer, 2005).

1.1.2 Causes and associations in mood disorders: genetics and pharmacogenetics

Depression and other mood disorders tend to proliferate as a result of multiple, complex, biological, psychological and social factors, including war and other violent conflicts, natural disasters, poverty and limited access to resources. A combination of genetic and environmental factors join to unchain depression (Sullivan et al., 2000; Wong and Licinio, 2001; Lesch, 2004; Hamet and Tremblay, 2005; Ebmeier et al., 2006). Multiple susceptibility genes of major or small effect, in interaction with each other and in conjunction with environmental events, produce vulnerability to the disorder (Fig. 4).

A significant determinant of depression is the individual's "personal threshold," or vulnerability to depression. Some people are more likely than others to become depressed, but no one is immune. Depression can affect individuals at any stage of the life span, although the incidence is highest during middle age. There is, however, an increasing recognition of depression during adolescence and young adulthood (Lewinsohn et al., 1993; Taylor et al., 2002).

Based on family aggregation and contrasting results from studies in monozygotic and dizygotic twins, major depression has an estimated heritability of 40–50% (major depression) and 40–80% (bipolar disorders), although the specific genes that underlie this risk have not yet been identified. It does not show classic Mendelian inheritance that could be attributable to a single gene. Depression has therefore been classified as a genetically complex disorder, polygenic and epistatic, much like heart disease, hypertension, diabetes and cancer (Lander and Schork, 1994; Sullivan et al., 2000).

Depression with recurrent episodes and possibly early onset may have higher genetic influence and may be thus associated with greater familial aggregation (Sullivan et al., 2000).

Traditional genetic linkage studies and candidate gene methods have been used with fairly limited success in major depression. Genetic models of etiology generally assume a large number of genes with relatively small contributions to liability. Advances in high throughput genotyping and microarray techniques have made it more feasible to identify genes with small effect sizes (Hong and Tsai, 2003). Of special clinical interest are those studies that are based on pathophysiological notions (candidate genes) and in particular those that examine the probability of patients to respond to particular treatments (Ebmeier et al., 2006).

Fig. 4. An integrative approach to depression: genetic and environmental factors (adapted from Wong and Licinio, 2001).

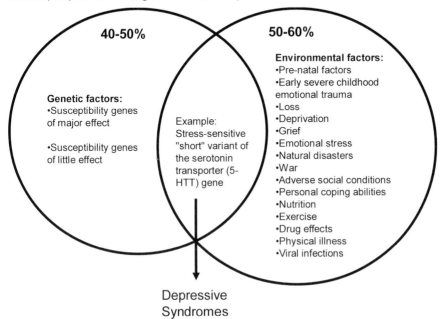

Genetic vulnerability may involve serotonin (5-HT) systems because tryptophan depletion in healthy subjects is reported to induce depressive symptoms only if they have affected relatives (Benkelfat et al., 1994). Allelic variations in the serotonin transporter (5-HTT) gene are of great interest in depression. A polymorphism in the 5' promoter region consisting of a short (s) and a long (l) allele with the s-allele being associated with decreased 5-HT transmembrane transport and reduced transporter mRNA has been described. The gene frequencies in Causasians have been reported to be 41% s-allele and 59% l-allele (Mellerup et al., 2001). The s-allele has been associated with neuroticism in a number of studies (Bellivier et al., 2002), with depressive symptoms, diagnosable depression and suicidality in response to stressful life events in some, but not all, the cases and with the response of patients to selective serotonin reuptake inhibitors (SSRIs; Mann et al., 2000; Caspi et al., 2003). In a study, dietary depletion of circulating tryptophan caused depressive symptoms only in subjects homozygous for the s-allele (Neumeister et al., 2002).

Some patients with depression carry a polymorphism, or genetic variant, in the FKBP5 gene (which encodes a co-chaperone of heat shock protein 90; HSP90) that results in higher affinity of glucocorticoid receptors for cortisol (Binder et al., 2004). These individuals respond much faster to antidepressants and have a higher recurrence of depressive episodes than individuals without this mutation.

Brain-derived neurotrophic factor (BDNF) may have a central role in the effectiveness of antidepressants, but there is no firm evidence of an association of its alleles with major depressive disorder (Hong et al., 2003; Tsai et al., 2003; Schumacher et al., 2005). Certain polymorphisms in the gene promoter of the monoamine oxidase A enzyme (involved in the metabolism of catecholamines and the target of one group of antidepressants, the monoamine oxidase inhibitors) are found in subgroups of patients with major depression or anxiety (Schulze et al., 2000; Du et al., 2004).

A polymorphism in the catechol-O-methyltransferase (COMT; another catecholamine metabolizing enzyme) is associated with treatment response to mirtazepine, but not to paroxetine in major depression (Szegedi et al., 2005).

A single nucleotide polymorphism in the gene of the human tryptophan hydroxylase-2 enzyme (involved in the synthesis of serotonin), with roughly 80% loss of function, is associated with major depression, but not with bipolar illness (Zhang et al., 2005).

A recent search for pharmacokinetic effects of cytochromes *CYP2D6* and *CYP2C19* alleles suggested that, for 14 of 20 investigated antidepressants, at least a doubling of the dose would be needed in extensive metabolizers compared with poor metabolizers. This variation in effects does strengthen the argument for antidepressant plasma monitoring in depression resistant to treatment (Kirchheiner et al., 2001).

No other genes have yet been convincingly linked to major depression (Ebmeier et al., 2006).

In regard to bipolar disorder, molecular genetic studies have reported many linkage loci and candidate genes. However, none of these findings have been consistently replicated. Meta-analyses of linkage studies have also reported conflicting results. Among recently reported candidate genes, BDNF, AKT1, XBP1, G72, GRK3,

GRIN2A, HTR4, IMPA2, GABRA1 and GABRA5 may have some importance (Sklar et al., 2002; Hong et al., 2003; for a review, see Kato et al., 2005).

Among the polymorphisms of monoamine-related genes, some were found to cause functional alteration and to be associated with bipolar disorder in two or more studies. These include monoamine oxidase A (MAO-A; Lim et al., 1995; Rubinsztein et al., 1996; Preisig et al., 2000), serotonin transporter (5-HTT; Collier et al., 1996a, b; Oruc et al., 1997; Furlong et al., 1998) and serotonin 2C receptor (5-HT$_{2C}$; Oruc et al., 1997; Lerer et al., 2001). Catecol-O-methytransferase (COMT) is also included in such genes, although association was found only for ultra-ultra-rapid-cycling bipolar disorder (Kirov et al., 1998; Papolos et al., 1998).

Functional genomics using microarray technology has been used to identify more than 300 genes in animal and human brain affected by antidepressant drug treatments (Sibille and Hen, 2001; Yamada and Higuchi, 2002; Sibille et al., 2004) and this may offer insights for novel therapeutic targets.

Environmental contribution, 50–60%, also remains poorly defined, with suggestions that prenatal factors, early severe childhood emotional trauma, chronic physical illness and even viral infections might be involved (Wong and Licinio, 2001).

Severe obesity may be a causative factor for depression, according to a study that examined depression before and after surgically induced weight loss. Beck Depression Inventory (BDI) questionnaires were completed before and at yearly intervals after gastric-restrictive weight-loss surgery. Results showed that severely obese subjects, especially women with poor body image, were at high risk for depression, while weight loss was associated with reduced BDI scores. The findings also showed that severe obesity might cause or aggravate depression (Dixon et al., 2003). The future identification of genes that confer risk for depression in humans and understanding how specific types of environmental factors interact synergistically with genetic vulnerability will constitute a considerable leap forward in the field of depression. It will be possible to develop more valid animal models of human depression. Important advances will also require the development of evermore penetrating brain imaging methodologies to enable the detection of molecular and cellular biomarkers in living patients.

1.1.3 Pathogenesis of mood disorders

1.1.3.1 The monoamine hypothesis of depression

Historically, the focus in neurobiological studies of mood disorders has been the monoaminergic neurotransmitter systems, which are extensively distributed throughout the network of limbic, striatal and prefrontal cortical neuronal circuits thought to support the behavioral and visceral manifestations of mood disorders (Fig. 5).

One of the first neurochemical theories of depression was the monoamine impairment hypothesis. According to this hypothesis, major depression results from a deficiency/imbalance of available monoamines (the indoleamine serotonin and catecholamines norepinephrine and dopamine) or subnormal monoamine receptors functioning in certain regions of the brain (for a review, see Iversen, 2005).

Some 50 years ago, the monoamine hypothesis of depression initiated a new era of research in "biological psychiatry". The discovery of the first effective antidepressant drugs and the rapid advances in research, which led to the understanding of their mechanisms of action on monoamine systems in the brain, represented the start of the so-called "psychopharmacology revolution" which transformed the practice of psychiatry (for a review, see Iversen, 2005).

The findings first suggesting a link between brain monoamines and depression took place in the 1950s, when it was known that the drug reserpine, used for a while in the treatment of hypertension, tended to cause depression as an unwanted side-effect (Muller et al., 1955). Reserpine causes a profound depletion of the brain stores of serotonin and noradrenaline by blocking vesicular monoamine storage (Shore et al., 1955). Animals and a fraction of humans receiving reserpine develop depression. The marked depression of behavior seen in animals treated with reserpine could be reversed by administering the catecholamine precursor L-DOPA (Carlsson et al., 1957), which also reversed the depressive symptoms induced by reserpine in human subjects.

Moreover, the fortuitous discovery that the drug isoniazid, used for the treatment of tuberculosis exerted a mood-elevating effect (Crane, 1956; Kline, 1961) and the subsequent finding that this drug inhibited monoamine oxidase, one of the enzymes responsible for serotonin and

Fig. 5. Monoaminergic nuclei and principal projection areas in the human brain: support of the behavioral and visceral manifestations of mood disorders

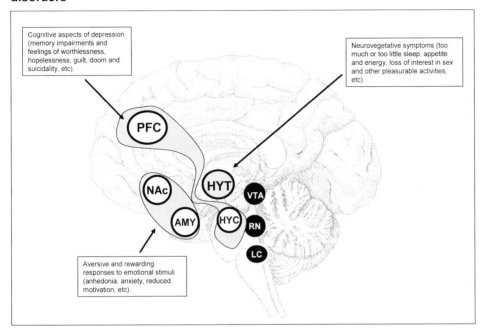

Monoaminergic innervation is extensively distributed throughout the limbic, striatal and prefrontal cortical areas (not shown for simplicity). The ventral tegmental area (VTA) provides dopaminergic input to the nucleus accumbens (NAc), amygdala (AMY), prefrontal cortex (PFC) and other limbic structures. Noradrenaline, from the locus coeruleus (LC), and serotonin, from the raphe nuclei (RN), innervate (not shown) all of the regions shown in the panel. The brain areas illustrated in the panel operate as a series of highly interacting parallel circuits, from which researchers are beginning to formulate the neural circuitry involved in depression (adapted from Berton and Nestler, 2006). HYP: hypothalamus; HYC: hippocampus.

Noradrenaline breakdown (Zeller, 1952) reinforced the monoamine hypothesis of depression.

Further support for the hypothesis was obtained in the 1960s, when it was discovered that tricyclic antidepressants inhibit monoamine transporters (Hertting et al., 1961). This finding led to the first understanding of the mechanism of action of the first generation of antidepressants and triggered the first formulation of the "monoamine hypothesis of depression" as a deficiency of either noradrenaline (Bunney and Davis, 1965; Schildkraut, 1965) or serotonin in the brain (Ashcroft et al., 1966; Coppen et al., 1967; Lapin and Oxenkrug, 1969; see Fig. 6). Initially, the noradrenaline version of this hypothesis was preferred by American researchers and the serotonin deficiency concept found more favor in Europe.

In 1996, impaired serotonergic neurotransmission in the brains of untreated patients with moderate to severe depression was demonstrated *in vivo* for the first time (Mann et al., 1996).

During the past 40 years, following the formulation of the "monoamine hypothesis of depression", there was a lot of research activity aimed at validating the hypothesis. The principal attempts to validate the monoamine hypothesis and results are summarized in Table 4.

Preclinical and clinical evidence has accumulated, mainly during the past two decades, indicating the involvement of the 5-HT system in the therapeutic action of antidepressant drugs (Artigas, 1993; Blier and De Montigny, 1994; Artigas et al., 1996a, b, 2006; Romero et al., 1996a, b, c). In particular, impairment of 5-HT synthesis leads to a transient reappearance of depressive symptoms in patients in remission obtained with various types of antidepressant drugs (Shopsin et al., 1975; Delgado et al., 1990). Conversely, tryptophan and lithium, two compounds that increase 5-HT function (Sharp et al., 1991, 1992) can potentiate the therapeutic effect of antidepressant drugs (De Montigny et al., 1983). Thus, there seems to be a clear association between the antidepressant response and enhanced 5-HT neurotransmission.

Although all currently approved antidepressant drugs appear to act through monoaminergic mechanisms, more than four decades of research have revealed some serious gaps and limitations in the monoamine hypothesis. For example, studies on noradrenaline spillover in plasma and cerebrospinal fluid documented increased noradrenaline output in depression (Veith et al., 1994; Wong et al., 2000). Moreover, the monoamine hypothesis does not fully explain why clinical effects only occur after chronic treatment for at least three weeks, whereas biochemical effects of antidepressants on monoamine systems occur within a few hours of treatment. Preclinical and clinical data support the idea that the acute effects of antidepressants are self-limited by a negative feedback involving the activation of somatodendritic serotonin and noradrenaline autoreceptors, which likely limits their clinical effects(Adell and Artigas, 1991; Artigas et al., 1996a, b, 2006; Romero et al., 1996a, b, c; Perez et al., 1997; Mateo et al., 2001; Adell et al., 2005), and suggest that during chronic treatment with antidepressants adaptive changes occur in pre- and postsynaptic receptors that are responsible for the therapeutic effect. Though theories that postulate

Table 4. Summary of the main lines of evidence implicating monoaminergic neurotransmission in depression (for a review, see Iversen, 2005).

Study approach	Evidence	Selected References
1. Direct measurements of monoamine function in human subjects: depressed vs control		
Monoamine metabolites in body fluids: 5-hydroxyindoleacetic acid (5-HIAA) from 5-HT, 3-methoxy-4-hydroxyphenyl glycol (MHPG) from NA, and homovanillic acid (HVA) from DA	↓ 5-HIAA in CSF: 7 (+) and 8 (−) results	Post and Goodwin, 1976 (review); Gibbons and Davis, 1986; Brown and Linnoila, 1990
	↓ MHPG in CSF: 3 (+) and 3 (−) results	
	↓ HVA in CSF: 2 (+) and 7 (−) results	
	↓ 5-HIAA, MHPG or HVA in urine: (−) results	
Turnover of monoamines in human brain: the probenecid method	↓ Rates of 5-HIAA in CSF: 6 (+) and 3 (−) results	Post and Goodwin, 1976 (review)
	↓ Rates of HVA in CSF: 6 (+) and 2 (−) results	
Post-mortem brain samples and living brain neuroimaging with PET	↓ 5-HT (whole brain, hypothalamus, amygdale)	Iversen, 2005 (review)
	↑ or no changes in 5-HT_2 binding sites (frontal cortex)	Mann et al., 1986; Arango et al., 1990; Hrdina et al., 1993
	↓ 5-HT_2 binding sites (cortex): 2 (+) and 2 (−) results	Yatham et al., 2000
	↑ 5-HT_{1A} binding sites (prefrontal cortex, raphe nuclei)	Stockmeier et al., 1998
	↓ 5-HT_{1A} binding sites (cortical regions). Reduction not altered by treatment with the selective serotonin reuptake inhibitor (SSRI), paroxetine	Sargent et al., 2000

	↓ 5-HTT binding sites *in vivo* using [^{123}I]-β-citalopram	Malison et al., 1998
	↓ 5-HTT binding sites using [^{3}H]-imipramine (post-mortem cortex, hypothalamus, hippocampus)	Stanley et al., 1982; Perry et al., 1983
	↓ 5-HTT binding sites using [^{3}H]-citalopram (cortex, hypothalamus, hippocampus)	Leake et al., 1991
	Any changes in density of 5-HTT binding sites using [^{3}H]-paroxetine (raphe, locus coeruleus)	Klimek et al., 1997
	↓ NAT binding sites using [^{3}H]-nisoxetine (locus coeruleus)	Klimek et al., 1997
	↑ α_2 Adrenoceptor binding sites (frontal cortex, locus coeruleus, other regions)	Meana et al., 1992; Ordway et al., 2003
	MAO-A: 1 (+) and 1 (−) results	Ordway et al., 1999; Du et al., 2002
	↓ 5-HTT binding sites	Briley et al., 1980; Paul et al., 1981; Langer and Galzin, 1988; Nemeroff et al., 1988
	↑ 5-HTT binding sites in patients exhibiting clinical recovery in response to antidepressant drug treatment: 3 (+) and 3 (−) results	Berrettini et al., 1982; Baron et al., 1986, 1987; Langer et al., 1987
	↑ α_2 Adrenoceptor binding sites	García-Sevilla, 1989
	↓ α_2 Adrenoceptor binding sites in patients treated with antidepressant or electroconvulsive therapy (ECT)	García-Sevilla et al., 1990
Monoamine markers in blood platelets	↑ β-Adrenoceptors in depressed patients and their downregulation in response to successful antidepressant treatment	Leonard et al., 1997 (review)

2. Manipulation of monoamines by administration of precursors or by depletion strategies: depressed vs control

		depressed vs control	
Inhibition of monoamine synthesis	Inhibition of NA synthesis with α-methyl-para-tyrosine (AMT)	Depression and sedation in humans	Sjoerdsma et al., 1965
		Reduction in behavior in primates	Redmond et al., 1971a, b
		Any exacerbation of symptoms in depressed patients not taking medication	Miller et al., 1996a
		Clinical relapse in responders to NRIs and not in SSRIs responders	Miller et al., 1996b
	Inhibition of 5-HT synthesis with para-chlorophenylalanine (PCPA)	Reversion of the antidepressant actions of imipramine in depressed patients	Shopsin et al., 1975
	Tryptophan depletion	Recurrence of depression Subjects / 5-HT depletion / NA depletion Healthy controls: − / ± Untreated depressed: − / − Recovered taking SSRI: ++++ / + Recovered taking NRI: − / ++++ Recovered taking NA/5-HT mixed drug: ++++ / ++++	Delgado et al., 1990, 1994, 1999; Benkelfat et al., 1994; Moreno et al., 1999; Delgado, 2004 (review)
Monoamine precursors	5-HTP	Antidepressant effects in a subgroup of depressed patients	van Praag et al., 1983
	L-Tryptophan	Antidepressant effects: 3 (+) and 3 (−) results	Coppen et al., 1967, 1972; Carroll et al., 1970; Bunney et al., 1971; Dunner and Fieve, 1975; Jensen et al., 1975
		Potentiation of tranylcypromine (MAOI): several (+) results	Coppen et al., 1963; Pare et al., 1963; Glassman and Platman, 1969

3. Neuroendocrine markers of monoamine function in depression

Hypothalamus–pituitary–adrenal axis activity	Hypercortisolaemia ("failure to suppress to dexamethasone", 40–50% endogenous depressed patients). Relation between hypercortisolaemia and increased density of α_2 adrenoceptors	Carroll et al., 1976; Checkley, 1980 (review); Arana et al., 1985 (review); Ressler and Nemeroff, 1999 (review)
	5-HT_{1A} and 5-HT_{2C} receptors stimulates cortisol and prolactin secretion in man	Meltzer and Maes, 1994

4. Molecular and genetic approaches

Molecular	Chronic treatment with antidepressants upregulates CREB (cyclic AMP response element protein), BDNF (brain derived neurotrophic factor) and Bcl-2 protein. This may underlie the recent finding that chronic antidepressant treatment causes a proliferation of progenitor cells in the hippocampus	Nibuya et al., 1995, 1996; Duman et al., 1997 (review) Rajkowska, 2000a, b; Chen et al., 2001a, b; Vaidya and Duman, 2001 (review); Nestler et al., 2002 (review)
Genetic	Association of polymorphisms in MAO-A and the SERT in bipolar disorder	Kato, 2001; Schulze et al., 2000; Du et al., 2004
	Association of "short" variant of the SERT gene in depression	Caspi et al., 2003
	Association of polymorphisms in tryptophan hydroxylase-2 enzyme (involved in the synthesis of serotonin) with major depression, but not with bipolar illness	Zhang et al., 2005

these long-term changes in receptor sensitivity have not been unequivocally proved (Siever and Davis, 1985), it is now believed that changes in brain gene expression that are elicited after chronic treatment might underlie the effects of antidepressants (Wong et al., 1996; for a review, see Wong and Licinio, 2001; Nestler et al., 2002; Carlson et al., 2006).

Although many attempts to find evidence for monoamine malfunction in depression proved largely ineffective, it is now accepted that monoamine neurotransmitter systems undoubtedly play an important role in modulating the expression of certain signs/symptoms of mood disorders. Inconsistent results obtained in some studies could be explained by the suggestion that the "monoamine hypothesis" applies only to a subgroup of depressed patients and there is no way of identifying patients in this hypothetical subgroup by clinical criteria (Irvensen, 2005). However, it is worth remembering that the "monoamine hypothesis" rests heavily on the finding that currently approved antidepressant drugs all appear to act through monoaminergic mechanisms – but it is less widely known why these drugs have a number of drawbacks, including delayed clinical response/remission (weeks to months) and, too frequently, incomplete or absent response/remission. A meta-analysis of clinical trials of antidepressants (Fava and Davidson, 1996) suggested that no beneficial response at all was seen in 19–34% of depressed patients treated with antidepressant drugs, whereas there was only a partial response in a further 12–15%. Thus, almost half of all patients treated with antidepressants fail to show a full response. Clearly, there is a great clinical need for the ongoing development of new therapies that are more effective, rapid-acting and easily tolerated than existing ones.

In summary, despite monoamine hypothesis being still controversial, it has guided antidepressant drug discovery in the past 50 years; and the process of discovery continues today (for a review, see Adell et al., 2005). Nevertheless, novel biological approaches beyond the "monoamine hypothesis" are expected to evoke paradigm shifts in the future of depression research.

Fig. 6. The monoamine hypothesis of depression: mechanism of antidepressant action.

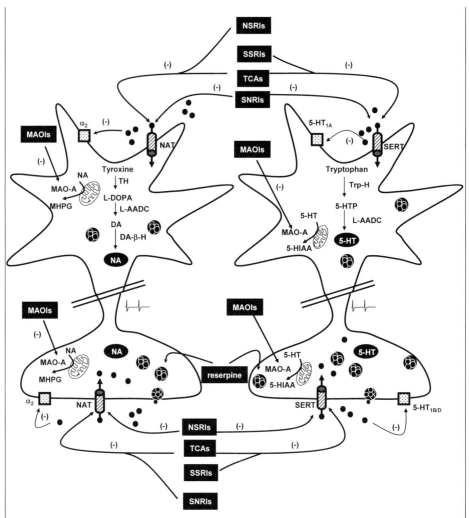

Antidepressants are a heterogeneous group of drugs that act primarily by increasing the availability of monoamines at the synaptic cleft by interfering with their inactivation mechanisms (reuptake or metabolism). Understanding their pharmacology has provided the means for the formulation of the monoamine hypothesis of depression. It has also broadened the approach for developing new drugs, such as the selective serotonin reuptake inhibitors that have less side-effects but are not more efficacious than previously available tricyclic drugs. Acute effects of antidepressants are self-limited by a negative feedback involving the activation of somatodendritic and terminal serotonin and noradrenaline autoreceptors (somatodendritic 5-HT$_{1A}$ and α_2-adrenergic receptors; terminal 5-HT$_{1B/1D}$ and α_2-adrenergic receptors), which likely limits their clinical effects (Adell and Artigas, 1991; Artigas et al., 1996a, b, 2006; Romero et al., 1996a, b, c; Perez et al., 1997; Mateo et al., 2001; Adell et al., 2005). NA: noradrenaline or norepinephrine; 5-HT: serotonin; MAOIs: monoamine oxidase inhibitors; TH: tyroxine hydroxylase; L-AADC: L-aromatic amino acid decarboxylase; DA-β-H: dopamine-betahydroxylase; MHPG: 3-methoxy-4-hydroxyphenylglycol (noradrenaline metabolite); Trp-H: tryptophan hydroxylase; 5-HIAA: 5-hydroxyindole acetic acid (serotonin metabolite); SERT: serotonin transporter; NAT: noradrenaline transporter; NSRIs: noradrenaline and serotonin reuptake inhibitors; SSRIs: selective serotonin reuptake inhibitors; SNRIs: selective noradrenaline reuptake inhibitors; TCAs: tricyclic antidepressants; 5-HT$_{1A}$: serotonin 1A receptor; 5-HT$_{1B}$: serotonin 1B receptor; 5-HT$_{1D}$: serotonin 1D receptor.

1.1.3.2 Neuroendocrine hypothesis of mood disorders: dysregulation of the hippocampus and hypothalamic–pituitary–adrenal axis

The observation that patients with Cushing's syndrome (hypercortisolemia) often experienced severe depression and anxiety, and the increased production and secretion of glucocorticoids such as cortisol in healthy people exposed to stress, in part contributed to the modern stress-diathesis hypothesis of depression in which excess secretion of cortisol is thought to play a significant pathophysiologic role in the etiology of depression and anxiety disorders (Nemeroff, 1996; Pariante and Miller, 2001; Barden, 2004; Gillespie and Nemeroff, 2005).

Currently, several novel approaches to the treatment of depression are being evaluated, based on the present understanding of the hypothalamic–pituitary–adrenal (HPA) axis dysregulation in mood disorders (Zobel et al., 2000; Belanoff et al., 2002).

A prominent mechanism by which the brain reacts to acute and chronic stress is activation of the HPA axis (Fig. 7). This axis is mainly controlled by the neuropeptide corticotrophin-releasing factor (CRF), secreted by neurones whose cell bodies are in the paraventricular nucleus (PVN) of the hypothalamus and other brain areas. CRF, acting in synergy with vasopressin, which is produced in either the same or distinct neurons of the paraventricular nucleus, stimulates the synthesis and release of pro-opiomelanocortin (POMC)-derived peptides (adrenocorticotropin hormone [ACTH], endorphins) from the anterior pituitary. ACTH then stimulates the synthesis and release of glucocorticoids (cortisol in humans, corticosterone in rodents) from the adrenal cortex. Glucocorticoid hormones terminate the stress response by negative feedback action at the level of the pituitary, hypothalamus and limbic brain areas, including the hippocampus (which exerts an inhibitory influence on hypothalamic CRF-containing neurons via a polysynaptic circuit), amygdala (which exerts a direct excitatory influence) and septum (Nestler et al., 2002; Barden, 2004; Berton and Nestler, 2006). This action is mediated by two identified types of corticosteroid receptors (Reul and de Kloet, 1985): the type I or mineralocorticoid receptor (MR) and the type II or glucocorticoid receptor (GR).

Multiple preclinical and clinical lines of evidence point to abnormalities of the axis in depression. Hyperactivity of the HPA axis is observed in approximately half of patients

Fig. 7. Regulation of the hypothalamic–pituitary–adrenal axis.

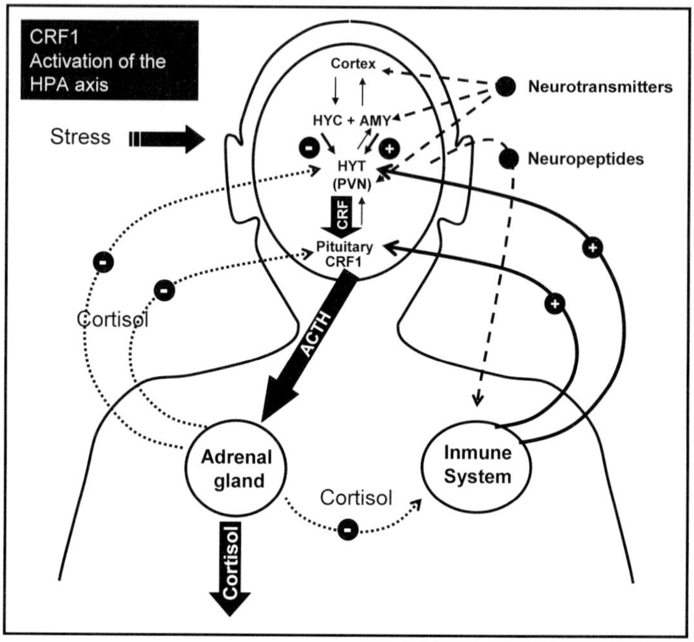

Corticotropin releasing factor (CRF)-containing neurons of the paraventricular nucleus (PVN) of the hypothalamus integrate information relevant to stress. CRF is released into the hypophyseal portal system and triggers the release of corticotrophin (ACTH) from the anterior pituitary via stimulation of CRF1 receptors. ACTH stimulates the secretion of glucocorticoid hormones (cortisol in humans or corticosterone in rodents) from the cortex of the adrenal gland. Increased glucocorticoid levels suppress hypothalamic CRF expression and pituitary ACTH release via negative feedback through hippocampal and hypothalamic glucocorticoid receptors. In this way, glucocorticoids (including synthetic forms such as dexamethasone) repress their own synthesis. The adrenal hypertrophy and consequential hypercortisolemia, associated with some cases of depression, induces a desensitisation of glucocorticoid receptors. This desensitisation of glucocorticoid receptors causes impaired negative feedback, increased activity of macrophages and increased release of proinflammatory cytokines by the immune system, that stimulate HPA axis and disturbed neuropeptide, noradrenaline and serotonin neurotransmission. At high levels, glucocorticoids also impair, and may even damage, the hippocampus, which could initiate and maintain this hypercortisolemic state. Adapted from Berton and Nestler (2006) and CNSforum.com.

with depression, as manifested by increased expression of CRF in the hypothalamus and prefrontal cortex and reduced CRF1 receptor mRNA (probably due to chronic hyperactivity of CRF), increased levels of CRF in the CSF, elevated plasma ACTH, increased cortisol production (urinary free cortisol), reduced feedback inhibition of the axis by CRF and synthetic glucocorticoids ("dexamethasone nonsuppression"), reduced glucocorticoid receptors in hippocampus and increased pituitary and adrenal size (Carroll, 1968; Sapolsky, 2000; Pariante and Miller, 2001; Barden et al., 2004; Merali et al., 2004; Gillespie and Nemeroff, 2005; Korte et al., 2005). Normalization of the overactive HPA system occurs during successful antidepressant pharmacotherapy of depressive illness (Holsboer et al., 1982; Greden et al., 1983; de Kloet et al., 1988; Holsboer-Trachsler et al., 1991; Nemeroff, 1996; Arborelius et al., 1999; Holsboer, 2001). In fact, a failure to normalize HPA activity usually predicts a poor clinical outcome and is associated with early relapse of depression (Banki et al., 1992; Gillespie and Nemeroff, 2005), suggesting that elevated cerebrospinal fluid CRF may be a state marker for depressive vulnerability.

Furthermore, the HPA axis is also involved in neuroplasticity (see also section below). Sustained elevations of glucocorticoids, seen under conditions of prolonged and severe stress, may damage hippocampal CA3 neurons (reduction in dendritic branching, a loss of the highly specialized dendritic spines where the neurons receive their glutamatergic synaptic inputs, even possibly cell death; McEwen, 2000; Sapolsky, 2000; Nestler et al., 2002; Berton and Nestler, 2006). Stress and the resulting hypercortisolemia also reduce neurogenesis in adult hippocampal dentate gyrus (Fuchs and Gould, 2000).

Consistent with these human data are the observations that rodents separated from their mothers early in life show abnormalities in HPA axis function, which resemble those seen in some depressed humans (de Kloet et al., 1988; Francis and Meaney, 1999; Heim and Nemeroff, 2001). These abnormalities can persist into adulthood and be corrected by antidepressant treatments.

Excessive glucocorticoids could be a causative factor for the small reductions in hippocampal volume that have been reported in patients with depression or post-traumatic stress disorder, although this finding remains controversial and it is not known whether these reduced hippocampal volumes are a result of depression or an antecedent cause. It was reported that depression duration correlates with hippocampal volume loss in women with recurrent major depression (Sheline et al., 1999; Bremner et al., 2000; Manji et al., 2001).

In view of the above description, several novel approaches to the treatment of depression are being evaluated based on the present understanding of HPA axis dysregulation in mood disorders (Zobel et al., 2000; Belanoff et al., 2002, see also Chapter 1.3). In this sense, intense attention is being given to antagonists of the CRF1 receptor, the major CRF receptor in brain, although agents directed against CRF2 receptors are also of interest (Arborelius et al., 1999; Holsboer, 2001). The CRF1 antagonist R121919 (Janssen) attenuates behavioral, neuroendocrine and autonomic responses to stress in primates (Habib et al., 2000); and it was effective in the treatment of depressed patients in a limited pilot study of 20 patients (Zobel et al., 2000). Unfortunately, this compound and numerous other CRF1 antagonists were dropped from subsequent studies due to hepatotoxicity and pharmacokinetic issues (Bosker et al., 2004). In the past few years, several pharmaceutical companies have been developing new selective CRF1 antagonists (Integrity database), such as NBI-34041 (GlaxoSmithKline/Neurocrine Biosciences), ONO-2333MS (Ono), SSR-125543A (Sanofi-Aventis) and TS-041 (Janssen/Taisho). These antagonists are currently undergoing clinical trials (Phase I), which should provide crucial information on efficacy, speed of action and safety, which will be necessary to determine whether this class of potential antidepressants has a faster clinical response than compounds based on reuptake blockade (for a review, see Adell et al., 2005).

At the moment, the major disappointment and frustration in this field is the failure to obtain clear proof of concept of the CRF1 antagonist mechanism as either anxiolytic or antidepressant in humans, despite decades of research.

However, clinical evidences suggest that depressive symptoms in patients with psychotic depression or Cushing syndrome might be rapidly ameliorated by glucocorticoid receptor antagonists (Gillespie and Nemeroff, 2005; Flores et al., 2006). The GR antagonist mifepristone (RU486), the use of which is associated with alterations of the HPA axis (Flores et al., 2006), is currently in Phase III clinical trials for psychotic major depression and might be the first nonmonoaminergic-based antidepressant on the market (Adell et al., 2005). Metyrapone, a glucocorticoid synthesis inhibitor, also shows some promise in treating depression when added to a standard antidepressant (Jahn et al., 2004; Berton and Nesler, 2006).

All of these compounds may be effective in treating depression through the interruption of reverberating neuroendocrine loops involving the HPA axis and several areas of the brain (prefrontal cortex, amygdala, hippocampus, hypothalamus) that become excessively activated in response to stress, driven perhaps by hypersecretion of CRF (Gold et al., 2002; Gillespie and Nemeroff, 2005).

Although the majority of the literature has reported HPA overactivity in major depression, recent data provide evidence that ACTH, CRH and cortisol activity may only be elevated in some subtypes of major depression and that some depressed patients may actually have low HPA activity (Heim et al., 2000; Posener et al., 2000) Hence, both decreased and elevated HPA axis activity may be found in specific depressive subtypes. In many ways, this parallels the findings in catecholamine activity in depressed patients (for a review, see Schatzberg et al., 2002).

1.1.3.3 Chronobiological hypothesis of depression: depressed have a phase shift

Abnormalities in the circadian regulation of sleep, temperature and activity cycles have been described in major depression (Kupfer et al., 1982). Reduced latency to the onset of rapid-eye-movement (REM) sleep, reduced slow-wave sleep (stages 3, 4), as well as early-morning awakening led to the formulation of the hypothesis that depressed patients have a phase shift of the biological clock (for a review, see Lewy, 2002). Moreover, sleep deprivation and exposure to bright white light ameliorates depressive symptoms by correcting this phase shift. Sleep deprivation is an effective but short-lasting non-pharmacological treatment of depression and light therapy is an effective treatment for seasonal affective illness (Gerner et al., 1979). REM sleep deprivation is more effective than is total sleep deprivation (the effects last longer). In fact, antidepressant drugs suppress REM sleep and increase slow-wave sleep (for a review, see Mayers and Baldwin, 2005; Argyropoulos and Wilson, 2005; Lam, 2006).

Many patients with depression report their most serious symptoms in the morning, with some improvement as the day progresses. This might represent an exaggeration of the diurnal fluctuations in mood, motivation, energy level and responses to rewarding stimuli that are commonly seen in the healthy population. However, the molecular basis for these rhythms seen under normal and pathological conditions is poorly understood.

Most research on circadian rhythms focuses on the suprachiasmatic nucleus (SCN) of the hypothalamus, which is considered the master circadian pacemaker of the brain (Reppert and Weaver, 2002; Takahashi, 2004). An increasing number of reports shows that circadian genes (Clk, Per, Bmal, Cry, neuronal Pas domain protein 2 or NPAS2, glycogen synthase kinase 3β or GSK3β, etc.) may act outside the SCN, including limbic regions implicated in mood regulation (hippocampus, striatum, ventral tegmental area-nucleus accumbens pathway, etc.). Results of early preclinical studies support the hypothesis that abnormalities in circadian gene function could contribute to certain symptoms of depression and other mood disorders and suggest that these circadian transcription factors could be potential targets for possible new treatment drugs for depression (Li et al., 2002; Reppert and Weaver, 2002; Manji et al., 2003; Takahashi, 2004; McClung et al., 2005; Uz et al., 2005; Berton and Nestler, 2006; Manev and Uz, 2006).

1.1.3.4 Infectious hypothesis of mood disorders

The Borna-disease virus (BDV) hypothesis of depression was proposed in the 1990s, based on the fact that this infection causes disturbances in behavior and cognitive functions that can even lead to a fatal neurological disease. BDV is a neurotrophic, single-stranded enveloped RNA virus that persistently infects birds, rodents and primates (Lipkin et al., 1990). Several positive reports support the hypothesis: (1) presence of BDV antibodies and viral genomic transcripts in patients with recurrent depression (Rott et al., 1985; Bode, 1995; Bode et al., 1995), (2) isolation of infectious BDV from mononuclear cells of patients (Bode et al., 1996) and (3) detection of BDV antigen and RNA in human autopsy brain samples from patients with depression (de la Torre et al., 1996). However, there are also several negative reports that did not show a clear association of BDV with depression (Kim et al., 1999; Tsuji et al., 2000). Therefore, it is currently accepted that BDV infect a small, but significant portion of patients with depression; but only detailed, carefully controlled, prospective studies will determine whether there is a causal relation between some forms of depression and BDV (Wong and Licinio, 2001).

1.1.3.5 Neural circuitry and neuroplasticity in mood disorders: beyond the monoamine hypothesis. the most recent approaches for novel therapeutic targets

1.1.3.5.1 Neural circuitry of depression: neuroimaging and neuropathological studies

The broad range of symptoms of depression suggests that many brain regions might be involved. This is supported by recent human functional brain imaging (positron emission tomography; PET), which show disturbed blood flow and glucose metabolism in several brain areas of depressive patients (Fig. 8), including regions of the prefrontal and cingulated cortex (orbital cortex, subgenual anterior cingulated cortex, dorsomedial prefrontal cortex, dorsolateral prefrontal cortex, dorsal anterior cingulated cortex), hippocampus, ventral striatum, amygdala (Amy) and mediodorsal thalamus, that modulate emotional behaviors (Drevets, 2001; Mayberg, 2003). In the same way, morphometric magnetic resonance imaging (MRI) and post-mortem histopathological investigations have also reported abnormalities in many of these same brain regions (for a review, see Carlson et al., 2006).

Although disagreement exists regarding the specific locations and the direction of some of these abnormalities, regional cerebral blood flow (CBF) and glucose metabolism are consistently increased in the amygdala, orbital cortex and medial thalamus, while they are decreased in the dorsomedial/dorsal anterolateral PFC and dorsal anterior cingulate cortex (subgenual PFC) in unmedicated subjects with familial major depression relative to healthy controls (Drevets, 2000a, b, 2001; Manji et al., 2001; Carlson et al., 2006; Fig. 8). It has been postulated that imbalance within these circuits, rather than an increase or decrease in any single region of the circuit, seems to predispose to and mediate the expression of the affective disorder (Carlson et al., 2006).

Structural imaging studies have reported reduced gray matter volumes in specific key areas of the orbital and medial prefrontal cortex, ventral striatum and hippocampus, and enlargement of third ventricles (suggesting volumetric reduction of the thalamus and hypothalamus) in mood disorders relative to healthy control subjects (for a review, see Drevets, 2000a, b; Manji et al., 2001, 2003; Brambilla et al., 2002; Bremner, 2002; Bremner et al., 2002; Fossati et al., 2004; Kanner, 2004; Lacerda et al.,

2004, 2005; Dickstein et al., 2005; Rosso, 2005; Carlson et al., 2006). The most prominent reduction is reported in the left (but not right) subgenual prefrontal cortex. Complementary postmortem neuropathological studies have shown abnormal reductions in cortex volume, glial cell counts and/or neuron size in the subgenual PFC, orbital cortex, dorsal anterolateral PFC, amygdala, basal ganglia and dorsal raphe nuclei (Öngür et al., 1998; Rajkowska et al., 1999, Rajkowska 2000a, b; Cotter et al., 2001; Manji and Duman, 2001; Carlson et al., 2006).

These findings led to the identification of two presumptive circuits involved in the pathophysiology of mood disorders (Drevets, 2000a, b, 2001; Carlson, 2006): (1) the limbic–thalamic–cortical (LTC) circuit (amygdala, mediodorsal thalamus, orbital and medial PFC; all regions interconnected by excitatory glutamatergic projections) and (2) the limbic–cortical–striatal–pallidal–thalamic (LCSPT) circuit (components of the LTC circuit as well as related areas of the striatum and pallidum). The involvement of these circuits in mood disorders is supported not only by imaging findings in primary depressive patients, but also by the increased frequency of mood disturbance in various neurological conditions (for example, Huntington's and Parkinson's diseases, strokes, and tumours) that involve lesions in these areas (Mayeux, 1982; Starkstein and Robinson, 1989; Starkstein et al., 1989; Folstein, 1991; Folstein et al., 1991; Carlson et al., 2006) and based upon evidence from electrophysiological lesion analysis and brain-mapping studies of humans and experimental animals (LeDoux, 2000; Drevets, 2000b).

All of this knowledge lead to the consideration that mood disorders are associated with activation of regions that putatively mediate emotional and stress responses (such as the amygdala). However, areas that appear to inhibit emotional expression (such as the posterior orbital cortex) show histological abnormalities that might interfere with the modulation of emotional or stress responses (Drevets, 2000a, b, 2001; Manji et al., 2001). In this sense, depression severity positively correlates with the elevation of CBF and glucose metabolism in the amygdala (Drevets, 2001). Many studies have demonstrated that effective antidepressant treatment decreases to normative levels CBF and metabolism in the amygdala and orbital/insular cortex (for a review, see Carlson et al., 2006). This fact is compatible with evidence that chronic antidepressant drug treatment has inhibitory effects on amygdala function in experimental animals (Nestler et al., 2002).

Fig. 8. Brain areas where neuropathological/morphometric changes have been reported in mood disorders.

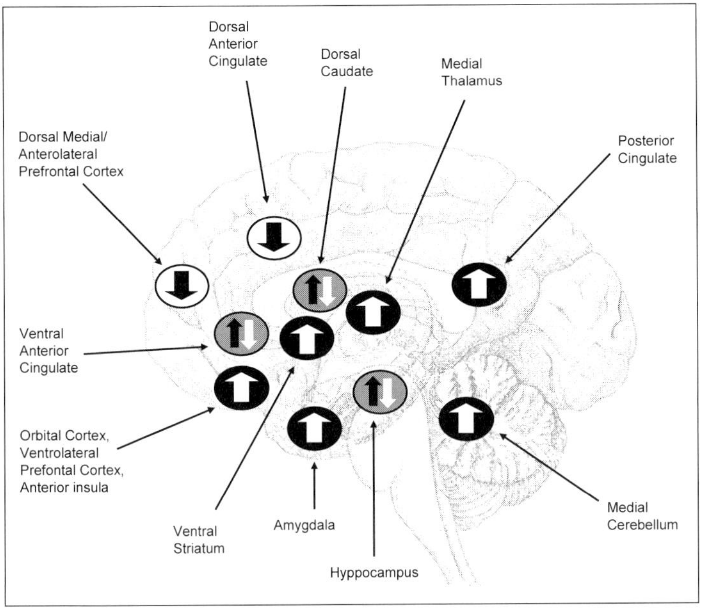

Arrows indicate direction of abnormalities relative to control. Adapted from Carlson et al. (2006).

In summary, functional brain imaging studies have identified key brain circuits and abnormalities that mediate the behavioral, cognitive and somatic manifestations of mood disorders; but it is not known yet whether these abnormalities confer vulnerability to abnormal mood episodes, compensatory changes to other pathogenic processes, or the sequelae of recurrent mood episodes. The increasing ability to monitor and modulate activity in these circuits is beginning to yield greater insight into the neurobiological basis of mood disorders (Carlson et al., 2006).

1.1.3.5.2 Neuroplasticity and cellular resilience in mood disorders

Growing recent evidence suggests that mechanisms of neural plasticity and cellular resilience, including impairments of neurotrophic signaling cascades, mitochondrial function as well as altered glutamatergic and glucocorticoid signaling, might underlie the pathophysiology of mood disorders and that antidepressant treatments exert major effects on signaling pathways that regulate neuroplasticity and cell survival (Duman et al., 1997; Altar,

1999; Manji et al., 2001; Kato and Kato, 2000; Nestler et al., 2002; Carlson et al., 2006; Berton and Nestler, 2006). These disturbances in structural plasticity and cellular resilience seem to be critical factors in precipitating and perpetuating disruption of the affective circuits discussed above (see Fig. 8). In this sense, although not universal, a reduction in the hippocampal volume and cortical neuronal death/atrophy have been reported in depressed patients (Bremner et al., 2000, 2002; Bremner, 2002); and chronic, but not acute antidepressant treatments and electroconvulsive shock can increase neurogenesis in the hippocampal dentate gyrus (Altar, 1999; Malberg et al., 2000). Together, preclinical and clinical results seem to indicate that in mood disorders there are impairments of neuroplasticity and cellular resilience, although it remains unclear whether they are a primary etiological factor or just an epiphenomenon.

Stress and depression likely contribute to impairments of cell-survival pathways by a variety of mechanisms, including cyclic AMP responsive element binding protein (CREB), brain-derived neurotrophic factor (BDNF), the protein Bcl-2 and mitogen-activated protein (MAP) kinases (CREB/BDNF/Bcl-2 cascade), facilitating glutamatergic transmission via NMDA and non-NMDA receptors and reducing the energy capacity of cells (Manji et al., 2001; Nestler el al., 2002; Berton and Nestler, 2006; Carlson et al., 2006).

In rodents, acute and chronic stress decreases levels of BDNF expression in the dentate gyrus and pyramidal cell layer of the hippocampus (Smith et al., 1995). This reduction appears to be mediated via stress-induced glucocorticoids and via other mechanisms, such as stress-induced increases in serotonergic transmission (Smith et al., 1995; Vaidya et al., 1997). Conversely, chronic (but not acute) administration of a variety of antidepressants increases the expression of BDNF and its receptor TrkB (Nibuya et al., 1995) and can prevent the stress-induced decreases in BDNF levels. Lithium and valproic acic (VPA) robustly upregulate the cytoprotective protein Bcl-2. Lithium and VPA also inhibit GSK-3β, biochemical effects shown to have neuroprotective effects. VPA also activates the ERK-MAP-kinase pathway, effects which may play a major role in neurotrophic effects and neurite outgrowth (for a review, see Manji et al., 2001). Antidepressant treatments also increase CREB expression in the hippocampus (Thome et al., 2000; Chen et al., 2001a), while CREB overexpression in the dentate gyrus induces antidepressant effects.

Importantly, on autopsy, reduced BDNF levels in the hippocampus have been reported in some patients with depression; and there is also evidence that antidepressants increase hippocampal BDNF levels in humans (Chen et al., 2001b).

Likewise, aminergic antidepressants have been shown to increase neurogenesis in the hippocampus, mainly due to the well known involvement of 5-HT in neural development (Altar, 1999). HPA axis hyperactivation and consequential sustained release of glucocorticoids also play a role in cell survival in the hippocampus; and the subtypes of depression most frequently associated with HPA activation are those most likely to be associated with hippocampal volume reductions (Sapolsky, 2000; Watson et al., 2004; Carlson et al., 2006). Substantial evidence suggests that excessive glutamate transmission (which may be precipitated by stress) plays a key role in atrophy of hippocampal pyramidal neurons. Conversely, NMDA receptor blockade have been found to have a positive effect on neuroplasticity in this area (Brown et al., 1999; Sapolsky, 2000).

Overall, these findings support the possibility that drugs activating the CREB/BDNF/Bcl-2 cascade or inhibiting glucocorticoid or glutamate effects could help to repair and prevent some stress-induced damages in neural circuits involved in mood disorders and therefore exert an antidepressant effect. These new therapeutic targets are discussed in Chapter 1.3. These findings could also explain why the response to commonly used antidepressants is delayed: they would require sufficient time for levels of neurotrophic factors to gradually rise and exert their neuroplastic/neuroprotective effects (Nestler et al., 2002; Berton and Nestler, 2006).

1.1.4 Concluding remarks

Mood disorders, with major depression in the first instance, are the main challenge in contemporary medicine. Depression is a heterogeneous syndrome, very common, incapacitating, causing profound suffering and occasionally lethal. It also extends across a wide range of severity and requires a large choice of treatments. The Global Burden of Disease Study has identified major depression among the leading causes of disability worldwide and it is expected that these illnesses would likely represent an increasingly greater health, societal and economic problem in future years (Murray and López, 1996). Despite the global impact of this disease, little is known about the etiology and pathophysiology. Available evidence consistently indicates that neurobiological substrates underlying depression phenotypes are the outcome of a combination of multiple-susceptibility (and likely protective) genes and environmental factors, although the specific genes that underlie this risk have not yet been clearly identified. Depression phenotypes include not only episodic and often profound mood disturbances, but also a range of cognitive, motor, autonomic, endocrine and sleep/wake abnormalities.

Traditionally, mood disorders have been considered to have a neurochemical basis (essentially monoaminergic neurotransmission), but recent studies have associated these complex disorders with pathological abnormalities, such as regional reductions in central nervous system volume and in the numbers and/or sizes of glia and neurons in discrete brain areas. Although the precise cellular mechanisms underlying these morphometric changes are unknown, the data indicate that mood disorders are associated with impairments of neuroplasticity and cellular resilience. It remains unclear whether these impairments correlate with the magnitude or duration of the biochemical perturbations present in mood disorders, whether they reflect an enhanced vulnerability to the deleterious effects of these perturbations (for example, due to genetic factors and/or early adverse life events), or whether they represent the fundamental etiological process in mood disorders (Manji et al., 2001; Nestler et al., 2002; Carlson et al., 2006; Berton and Nestler, 2006; Ebmeier et al., 2006). A number of preclinical and clinical studies have shown that signaling pathways involved in regulating cell survival and cell death are long-term targets for the actions of current antidepressant treatments.

General problems in all current aminergic antidepressant drugs are limited efficacy, a slow onset of action and the existence of neurobiological differences among depressed patients, which plays a role in treatment outcome. Today, there is a great clinical need for novel therapeutic strategies in addressing these illnesses (Adell et al., 2005).

Future advances in genetics, genomics and functional imaging techniques may facilitate the molecular dissection of depression phenotypes (etiologically based disease subtypes, favorable drug responses, antidepressant resistance, etc.) leading to a better understanding of the neurobiology of mood disorders and helping to identify new targets for drug development – leading ultimately to the discovery of definitive, more refined and specifically targeted therapies, eventually cures and preventive measures.

References

Adell A, Artigas F. *Differential effects of clomipramine given locally or systemically on extracellular 5-hydroxytryptamine in raphe nuclei and frontal cortex. An in vivo brain microdialysis study.* Naunyn Schmiedebergs Arch Pharmacol. **1991**, 343:237-244.

Adell A. *Antidepressant properties of substance P antagonists: relationship to monoaminergic mechanisms?* Curr Drug Targets CNS Neurol Disord. **2004**, 3:113-121.

Adell A, Castro E, Celada P, Bortolozzi A, Pazos A, Artigas F. *Strategies for producing faster acting antidepressants.* Drug Discov Today. **2005**, 10:578-585.

Altar CA. *Neurotrophins and depression.* Trends Pharmacol Sci. **1999**, 20:59-61.

American Psychiatric Association. *Diagnostic Statistical Manual of Mental Disorders.* 4th ed. American Psychiatric Press; Washington DC, **2000**.

Anden N, Grabowska M. *Pharmacological evidence for a stimulation of dopamine neurons by noradrenaline neurons in the brain.* Eur J Pharmacol. **1976**, 39:275-282.

Angst J. *The epidemiology of depressive disorders.* Eur Neuropsychopharmacol. **1995**, 5 Suppl:95-98.

Angst J, Angst F, Stassen HH. *Suicide risk in patients with major depressive disorder.* J Clin Psychiatry. **1999**, 60 Suppl 2:57-62.

Arana GW, Baldessarini RJ, Ornsteen M. *The dexamethasone suppression test for diagnosis and prognosis in psychiatry. Commentary and review.* Arch Gen Psychiatry. **1985**, 42:1193-1204.

Arango V, Ernsberger P, Marzuk PM, Chen JS, Tierney H, Stanley M, Reis DJ, Mann JJ. *Autoradiographic demonstration of increased serotonin 5-HT2 and beta-adrenergic receptor binding sites in the brain of suicide victims.* Arch Gen Psychiatry. **1990**, 47:1038-1047.

Arborelius L, Owens MJ, Plotsky PM, Nemeroff CB. *The role of corticotropin-releasing factor in depression and anxiety disorders.* J Endocrinol. **1999**, 160:1-12.

Argyropoulos SV, Wilson SJ. *Sleep disturbances in depression and the effects of antidepressants.* Int Rev Psychiatry. **2005**, 17:237-245.

Artigas F. *5-HT and antidepressants: new views from microdialysis studies.* Trends Pharmacol Sci. **1993**, 14:262.

Artigas F, Bel N, Casanovas JM, Romero L. *Adaptative changes of the serotonergic system after antidepressant treatments.* Adv Exp Med Biol. **1996a**, 398:51-59.

Artigas F, Romero L, De Montigny C, Blier P. *Acceleration of the effect of selected antidepressant drugs in major depression by 5-HT1A antagonists.* Trends Neurosci. **1996b**, 19:378-383.

Artigas F, Adell A, Celada P. *Pindolol augmentation of antidepressant response.* Curr Drug Targets. **2006**, 7:139-147.

Ashcroft GW, Crawford TB, Eccleston D, Sharman DF, MacDougall EJ, Stanton JB, Binns JK. *5-hydroxyindole compounds in the cerebrospinal fluid of patients with psychiatric or neurological diseases.* Lancet. **1966**, 2:1049-1052.

Ayuso-Mateos JL, Vazquez-Barquero JL, Dowrick C, Lehtinen V, Dalgard OS, Casey P, Wilkinson C, Lasa L, Page H, Dunn G, Wilkinson G. *Depressive disorders in Europe: prevalence figures from the ODIN study.* Br J Psychiatry. **2001**, 179:308-316.

Banki CM, Karmacsi L, Bissette G, Nemeroff CB. *CSF corticotropin-releasing hormone and somatostatin in major depression: response to antidepressant treatment and relapse.* Eur Neuropsychopharmacol. **1992**, 2:107-113.

Barden N. *Implication of the hypothalamic-pituitary-adrenal axis in the physiopathology of depression.* J Psychiatry Neurosci. **2004**, 29:185-193.

Baron M, Barkai A, Gruen R, Peselow E, Fieve RR, Quitkin F. *Platelet [3H]imipramine binding in affective disorders: trait versus state characteristics.* Am J Psychiatry. **1986**, 143:711-717.

Baron M, Barkai A, Gruen R, Peselow E, Fieve RR, Quitkin F. *Platelet 3H-imipramine binding and familial transmission of affective disorders.* Neuropsychobiology. **1987**, 17:182-186.

Belanoff JK, Rothschild AJ, Cassidy F, DeBattista C, Baulieu EE, Schold C, Schatzberg AF. *An open label trial of C-1073 (mifepristone) for psychotic major depression.* Biol Psychiatry. **2002**, 52:386-392.

Bellivier F, Leroux M, Henry C, Rayah F, Rouillon F, Laplanche JL, Leboyer M. *Serotonin transporter gene polymorphism influences age at onset in patients with bipolar affective disorder.* Neurosci Lett. **2002**, 334:17-20.

Benkelfat C, Ellenbogen MA, Dean P, Palmour RM, Young SN. *Mood-lowering effect of tryptophan depletion. Enhanced susceptibility in young men at genetic risk for major affective disorders.* Arch Gen Psychiatry. **1994**, 51:687-697.

Berrettini WH, Nurnberger JI, Jr., Post RM, Gershon ES. *Platelet 3H-imipramine binding in euthymic bipolar patients.* Psychiatry Res. **1982**, 7:215-219.

Berton O, Nestler EJ. *New approaches to antidepressant drug discovery: beyond monoamines.* Nat Rev Neurosci. **2006**, 7:137-151.

Binder EB, Salyakina D, Lichtner P, Wochnik GM, Ising M, Putz B, Papiol S, Seaman S, Lucae S, Kohli MA, Nickel T, Kunzel HE, Fuchs B, Majer M, Pfennig A, Kern N, Brunner J, Modell S, Baghai T, Deiml T, Zill P, Bondy B, Rupprecht R, Messer T, Kohnlein O, Dabitz H, Bruckl T, Muller N, Pfister H, Lieb R, Mueller JC, Lohmussaar E, Strom TM, Bettecken T, Meitinger T, Uhr M, Rein T, Holsboer F, Muller-Myhsok B. *Polymorphisms in FKBP5 are associated with increased recurrence of depressive episodes and rapid response to antidepressant treatment.* Nat Genet. **2004**, 36:1319-1325.

Blier P, De Montigny C. *Current advances and trends in the treatment of depression.* Trends Pharmacol Sci. **1994**, 15:220-226.

Bode L, Zimmermann W, Ferszt R, Steinbach F, Ludwig H. *Borna disease virus genome transcribed and expressed in psychiatric patients.* Nat Med. **1995**, 1:232-236.

Bode L. *Human infections with Borna disease virus and potential pathogenic implications.* Curr Top Microbiol Immunol. **1995**, 190:103-130.

Bode L, Durrwald R, Rantam FA, Ferszt R, Ludwig H. *First isolates of infectious human Borna disease virus from patients with mood disorders.* Mol Psychiatry. **1996**, 1:200-212.

Bosker FJ, Westerink BH, Cremers TI, Gerrits M, van der Hart MG, Kuipers SD, van der PG, ter Horst GJ, den Boer JA, Korf J. *Future antidepressants: what is in the pipeline and what is missing?* CNS Drugs. **2004**, 18:705-732.

Brambilla P, Nicoletti MA, Harenski K, Sassi RB, Mallinger AG, Frank E, Kupfer DJ, Keshavan MS, Soares JC. *Anatomical MRI study of subgenual prefrontal cortex in bipolar and unipolar subjects.* Neuropsychopharmacology. **2002**, 27:792-799.

Bremner JD, Narayan M, Anderson ER, Staib LH, Miller HL, Charney DS. *Hippocampal volume reduction in major depression.* Am J Psychiatry. **2000**, 157:115-118.

Bremner JD. *Structural changes in the brain in depression and relationship to symptom recurrence.* CNS Spectr. **2002**, 7:129.

Bremner JD, Vythilingam M, Vermetten E, Nazeer A, Adil J, Khan S, Staib LH, Charney DS. *Reduced volume of orbitofrontal cortex in major depression.* Biol Psychiatry. **2002**, 51:273-279.

Briley MS, Langer SZ, Raisman R, Sechter D, Zarifian E. *Tritiated imipramine binding sites are decreased in platelets of untreated depressed patients.* Science. **1980**, 209:303-305.

Brown ES, Rush AJ, McEwen BS. *Hippocampal remodeling and damage by corticosteroids: implications for mood disorders.* Neuropsychopharmacology. **1999**, 21:474-484.

Brown GL, Linnoila MI. *CSF serotonin metabolite (5-HIAA) studies in depression, impulsivity, and violence.* J Clin Psychiatry. **1990**, 51 Suppl:31-41.

Bunney WE, Jr., Davis JM. *Norepinephrine in depressive reactions. A review.* Arch Gen Psychiatry. **1965**, 13:483-494.

Bunney WE, Jr., Brodie HK, Murphy DL, Goodwin FK. *Studies of alpha-methyl-para-tyrosine, L-dopa, and L-tryptophan in depression and mania.* Am J Psychiatry. **1971**, 127:872-881.

Carlson PJ, Singh JB, Zarate CA, Jr., Drevets WC, Manji HK. *Neural circuitry and neuroplasticity in mood disorders: insights for novel therapeutic targets.* NeuroRx. **2006**, 3:22-41.

Carlsson A, Lindqvist M, Magnusson T. *3,4-Dihydroxyphenylalanine and 5-hydroxytryptophan as reserpine antagonists.* Nature. **1957**, 180:1200.

Carroll BJ, Martin FI, Davies B. *Pituitary-adrenal function in depression.* Lancet. **1968**, 1:1373-1374.

Carroll BJ, Mowbray RM, Davies B. *Sequential comparison of L-tryptophan with E.C.T. in severe depression.* Lancet. **1970**, 1:967-969.

Carroll BJ, Curtis GC, Mendels J. *Neuroendocrine regulation in depression. II. Discrimination of depressed from nondepressed patients.* Arch Gen Psychiatry. **1976**, 33:1051-1058.

Caspi A, Sugden K, Moffitt TE, Taylor A, Craig IW, Harrington H, McClay J, Mill J, Martin J, Braithwaite A, Poulton R. *Influence of life stress on depression: moderation by a polymorphism in the 5-HTT gene.* Science. **2003**, 301:386-389.

Charney DS. *Psychobiological mechanisms of resilience and vulnerability: implications for successful adaptation to extreme stress.* Am J Psychiatry. **2004**, 161:195-216.

Checkley SA. *Neuroendocrine tests of monoamine function in man: a review of basic theory and its application to the study of depressive illness.* Psychol Med. **1980**, 10:35-53.

Chen AC, Shirayama Y, Shin KH, Neve RL, Duman RS. *Expression of the cAMP response element binding protein (CREB) in hippocampus produces an antidepressant effect.* Biol Psychiatry. **2001a**, 49:753-762.

Chen B, Dowlatshahi D, MacQueen GM, Wang JF, Young LT. *Increased hippocampal BDNF immunoreactivity in subjects treated with antidepressant medication.* Biol Psychiatry. **2001b**, 50:260-265.

Collier DA, Arranz MJ, Sham P, Battersby S, Vallada H, Gill P, Aitchison KJ, Sodhi M, Li T, Roberts GW, Smith B, Morton J, Murray RM, Smith D, Kirov G. *The serotonin transporter is a potential susceptibility factor for bipolar affective disorder.* Neuroreport. **1996a**, 7:1675-1679.

Collier DA, Stober G, Li T, Heils A, Catalano M, Di Bella D, Arranz MJ, Murray RM, Vallada HP, Bengel D, Muller CR, Roberts GW, Smeraldi E, Kirov G, Sham P, Lesch KP. *A novel functional polymorphism within the promoter of the serotonin transporter gene: possible role in susceptibility to affective disorders.* Mol Psychiatry. **1996b**, 1:453-460.

Coppen A, Shaw DM, Farrell JP. *Potentiation of the antidepressive effect of a monoamine-oxidase inhibitor by tryptophan.* Lancet. **1963**, 1:79-81.

Coppen A, Shaw DM, Herzberg B, Maggs R. *Tryptophan in the treatment of depression.* Lancet. **1967**, 2:1178-1180.

Coppen A, Brooksbank BW, Peet M. *Tryptophan concentration in the cerebrospinal fluid of depressive patients.* Lancet. **1972**, 1:1393.

Cotter D, Mackay D, Landau S, Kerwin R, Everall I. *Reduced glial cell density and neuronal size in the anterior cingulate cortex in major depressive disorder.* Arch Gen Psychiatry. **2001**, 58:545-553.

Crane GE. *The psychiatric side-effects of iproniazid.* Am J Psychiatry. **1956**, 112:494-501.

Das AK, Olfson M, Gameroff MJ, Pilowsky DJ, Blanco C, Feder A, Gross R, Neria Y, Lantigua R, Shea S, Weissman MM. *Screening for bipolar disorder in a primary care practice.* JAMA. **2005**, 293:956-963.

de Kloet ER, Rosenfeld P, Van Eekelen JA, Sutanto W, Levine S. *Stress, glucocorticoids and development.* Prog Brain Res. **1988**, 73:101-120.

de la Torre JC, Gonzalez-Dunia D, Cubitt B, Mallory M, Mueller-Lantzsch N, Grasser FA, Hansen LA, Masliah E. *Detection of borna disease virus antigen and RNA in human autopsy brain samples from neuropsychiatric patients.* Virology. **1996**, 223:272-282.

De Montigny C, Cournoyer G, Morissette R, Langlois R, Caille G. *Lithium carbonate addition in tricyclic antidepressant-resistant unipolar depression. Correlations with the neurobiologic actions of tricyclic antidepressant drugs and lithium ion on the serotonin system.* Arch Gen Psychiatry. **1983**, 40:1327-1334.

Delgado PL, Charney DS, Price LH, Aghajanian GK, Landis H, Heninger GR. *Serotonin function and the mechanism of antidepressant action. Reversal of antidepressant-induced remission by rapid depletion of plasma tryptophan*. Arch Gen Psychiatry. **1990**, 47:411-418.

Delgado PL, Price LH, Miller HL, Salomon RM, Aghajanian GK, Heninger GR, Charney DS. *Serotonin and the neurobiology of depression. Effects of tryptophan depletion in drug-free depressed patients*. Arch Gen Psychiatry. **1994**, 51:865-874.

Delgado PL, Miller HL, Salomon RM, Licinio J, Krystal JH, Moreno FA, Heninger GR, Charney DS. *Tryptophan-depletion challenge in depressed patients treated with desipramine or fluoxetine: implications for the role of serotonin in the mechanism of antidepressant action*. Biol Psychiatry. **1999**, 46:212-220.

Delgado PL. *How antidepressants help depression: mechanisms of action and clinical response*. J Clin Psychiatry. **2004**, 65 Suppl 4:25-30.

Dickstein DP, Milham MP, Nugent AC, Drevets WC, Charney DS, Pine DS, Leibenluft E. *Frontotemporal alterations in pediatric bipolar disorder: results of a voxel-based morphometry study*. Arch Gen Psychiatry. **2005**, 62:734-741.

Dixon JB, Dixon ME, O'Brien PE. *Depression in association with severe obesity: changes with weight loss*. Arch Intern Med. **2003**, 163:2058-2065.

Drevets WC. *Functional anatomical abnormalities in limbic and prefrontal cortical structures in major depression*. Prog Brain Res. **2000a**, 126:413-431.

Drevets WC. *Neuroimaging studies of mood disorders*. Biol Psychiatry. **2000b**, 48:813-829.

Drevets WC. *Neuroimaging and neuropathological studies of depression: implications for the cognitive-emotional features of mood disorders*. Curr Opin Neurobiol. **2001**, 11:240-249.

Du L, Faludi G, Palkovits M, Sotonyi P, Bakish D, Hrdina PD. *High activity-related allele of MAO-A gene associated with depressed suicide in males*. Neuroreport. **2002**, 13:1195-1198.

Du L, Bakish D, Ravindran A, Hrdina PD. *MAO-A gene polymorphisms are associated with major depression and sleep disturbance in males*. Neuroreport. **2004**, 15:2097-2101.

Duman RS, Heninger GR, Nestler EJ. *A molecular and cellular theory of depression*. Arch Gen Psychiatry. **1997**, 54:597-606.

Dunner DL, Fieve RR. *Affective disorder: studies with amine precursors*. Am J Psychiatry. **1975**, 132:180-183.

Ebmeier KP, Donaghey C, Steele JD. *Recent developments and current controversies in depression*. Lancet. **2006**, 367:153-167.

Fava M, Davidson KG. *Definition and epidemiology of treatment-resistant depression*. Psychiatr Clin North Am. **1996**, 19:179-200.

Ferketich AK, Schwartzbaum JA, Frid DJ, Moeschberger ML. *Depression as an antecedent to heart disease among women and men in the NHANES I study. National Health and Nutrition Examination Survey*. Arch Intern Med. **2000**, 160:1261-1268.

Flores BH, Kenna H, Keller J, Solvason HB, Schatzberg AF. *Clinical and biological effects of mifepristone treatment for psychotic depression*. Neuropsychopharmacology. **2006**, 31:628-636.

Folstein SE. *The psychopathology of Huntington's disease.* Res Publ Assoc Res Nerv Ment Dis. **1991**, 69:181-191.

Folstein SE, Peyser CE, Starkstein SE, Folstein MF. Subcortical triad of Huntington's disease-a model for a neuropathology of depression, dementia, and dyskinesia. In: Carrol BJ, Barrett JE, eds. *Psychopathology and the brain.* Raven Press, New York, **1991**, pp 65-75.

Fossati P, Radtchenko A, Boyer P. *Neuroplasticity: from MRI to depressive symptoms.* Eur Neuropsychopharmacol. **2004**, 14 Suppl 5:S503-S510.

Francis DD, Meaney MJ. *Maternal care and the development of stress responses.* Curr Opin Neurobiol. **1999**, 9:128-134.

Fuchs E, Gould E. *Mini-review: in vivo neurogenesis in the adult brain: regulation and functional implications.* Eur J Neurosci. **2000**, 12:2211-2214.

Furlong RA, Ho L, Walsh C, Rubinsztein JS, Jain S, Paykel ES, Easton DF, Rubinsztein DC. *Analysis and meta-analysis of two serotonin transporter gene polymorphisms in bipolar and unipolar affective disorders.* Am J Med Genet. **1998**, 81:58-63.

Garcia-Sevilla JA. *The platelet alpha 2-adrenoceptor as a potential biological marker in depression.* Br J Psychiatry Suppl. **1989**, 67-72.

Garcia-Sevilla JA, Padro D, Giralt MT, Guimon J, Areso P. *Alpha 2-adrenoceptor-mediated inhibition of platelet adenylate cyclase and induction of aggregation in major depression. Effect of long-term cyclic antidepressant drug treatment.* Arch Gen Psychiatry. **1990**, 47:125-132.

Gerner RH, Post RM, Gillin JC, Bunney WE, Jr. *Biological and behavioral effects of one night's sleep deprivation in depressed patients and normals.* J Psychiatr Res. **1979**, 15:21-40.

Gibbons RD, Davis JM. *Consistent evidence for a biological subtype of depression characterized by low CSF monoamine levels.* Acta Psychiatr Scand. **1986**, 74:8-12.

Gillespie CF, Nemeroff CB. *Hypercortisolemia and depression.* Psychosom Med. **2005**, 67 Suppl 1:S26-S28.

Glassman AH, Platman SR. *Potentiation of a monoamine oxidase inhibitor by tryptophan.* J Psychiatr Res. **1969**, 7:83-88.

Gold PW, Drevets WC, Charney DS. *New insights into the role of cortisol and the glucocorticoid receptor in severe depression.* Biol Psychiatry. **2002**, 52:381-385.

Goodwin FK, Jamison KR. Suicide. In: Goodwin FK, Jamison KR, eds. *Manic-depressive illness.* Oxford University Press, New York, **1990**, pp 227-244.

Greden JF, Gardner R, King D, Grunhaus L, Carroll BJ, Kronfol Z. *Dexamethasone suppression tests in antidepressant treatment of melancholia. The process of normalization and test-retest reproducibility.* Arch Gen Psychiatry. **1983**, 40:493-500.

Greenberg PE, Kessler RC, Birnbaum HG, Leong SA, Lowe SW, Berglund PA, Corey-Lisle PK. *The economic burden of depression in the United States: how did it change between 1990 and 2000?* J Clin Psychiatry. **2003**, 64:1465-1475.

Gruenberg AM, Goldstein RD, Pincus HA. Classification of Depression : Research and Diagnostic Criteria: DMS-IV and ICD-10. In: Licinio J, Wong ML, eds. *Biology of Depression. From*

Novel Insights to Therapeutic Strategies. Wiley-VCH Verlag GmbH & Co. KGaA, Weinheim, **2005**.

Habib KE, Weld KP, Rice KC, Pushkas J, Champoux M, Listwak S, Webster EL, Atkinson AJ, Schulkin J, Contoreggi C, Chrousos GP, McCann SM, Suomi SJ, Higley JD, Gold PW. *Oral administration of a corticotropin-releasing hormone receptor antagonist significantly attenuates behavioral, neuroendocrine, and autonomic responses to stress in primates.* Proc Natl Acad Sci U S A. **2000**, 97:6079-6084.

Hamet P, Tremblay J. *Genetics and genomics of depression.* Metabolism. **2005**, 54:10-15.

Harris EC, Barraclough B. *Excess mortality of mental disorder.* Br J Psychiatry. **1998**, 173:11-53.

Heim C, Newport DJ, Heit S, Graham YP, Wilcox M, Bonsall R, Miller AH, Nemeroff CB. *Pituitary-adrenal and autonomic responses to stress in women after sexual and physical abuse in childhood.* JAMA. **2000**, 284:592-597.

Heim C, Nemeroff CB. *The role of childhood trauma in the neurobiology of mood and anxiety disorders: preclinical and clinical studies.* Biol Psychiatry. **2001**, 49:1023-1039.

Herting G, Axelrod J, Whitby LG. *Effect of drugs on the uptake and metabolism of H3-norepinephrine.* J Pharmacol Exp Ther. **1961**, 134:146-153.

Holsboer-Trachsler E, Stohler R, Hatzinger M. *Repeated administration of the combined dexamethasone-human corticotropin releasing hormone stimulation test during treatment of depression.* Psychiatry Res. **1991**, 38:163-171.

Holsboer F, Liebl R, Hofschuster E. *Repeated dexamethasone suppression test during depressive illness. Normalisation of test result compared with clinical improvement.* J Affect Disord. **1982**, 4:93-101.

Holsboer F. *Stress, hypercortisolism and corticosteroid receptors in depression: implications for therapy.* J Affect Disord. **2001**, 62:77-91.

Hong CJ, Huo SJ, Yen FC, Tung CL, Pan GM, Tsai SJ. *Association study of a brain-derived neurotrophic-factor genetic polymorphism and mood disorders, age of onset and suicidal behavior.* Neuropsychobiology. **2003**, 48:186-189.

Hong CJ, Tsai SJ. *The genomic approaches to major depression.* Current Pharmacogenomics. **2003**, 1:67-74.

Hrdina PD, Demeter E, Vu TB, Sotonyi P, Palkovits M. *5-HT uptake sites and 5-HT2 receptors in brain of antidepressant-free suicide victims/depressives: increase in 5-HT2 sites in cortex and amygdala.* Brain Res. **1993**, 614:37-44.

Iversen L. The Monoamine Hypothesis of Depression. In: Licinio J, Wong ML, eds. *Biology of Depression. From Novel Insights to Therapeutic Strategies.* Wiley-VCH Verlag GmbH & Co. KGaA, Weinheim, **2005**, pp 71-86.

Jahn H, Schick M, Kiefer F, Kellner M, Yassouridis A, Wiedemann K. *Metyrapone as additive treatment in major depression: a double-blind and placebo-controlled trial.* Arch Gen Psychiatry. **2004**, 61:1235-1244.

Jensen K, Fruensgaard K, Ahlfors UG, Pihkanen TA, Tuomikoski S, Ose E, Dencker SJ, Lindberg D, Nagy A. *Letter: Tryptophan/imipramine in depression.* Lancet. **1975**, 2:920.

Kanner AM. *Structural MRI changes of the brain in depression*. Clin EEG Neurosci. **2004**, 35:46-52.

Kato T, Kato N. *Mitochondrial dysfunction in bipolar disorder*. Bipolar Disord. **2000**, 2:180-190.

Kato T. *Molecular genetics of bipolar disorder*. Neurosci Res. **2001**, 40:105-113.

Kato T, Kuratomi G, Kato N. *Genetics of bipolar disorder*. Drugs Today (Barc). **2005**, 41:335-344.

Kim YK, Kim SH, Choi SH, Ko YH, Kim L, Lee MS, Suh KY, Kwak DI, Song KJ, Lee YJ, Yanagihara R, Song JW. *Failure to demonstrate Borna disease virus genome in peripheral blood mononuclear cells from psychiatric patients in Korea*. J Neurovirol. **1999**, 5:196-199.

Kirchheiner J, Brosen K, Dahl ML, Gram LF, Kasper S, Roots I, Sjoqvist F, Spina E, Brockmoller J. *CYP2D6 and CYP2C19 genotype-based dose recommendations for antidepressants: a first step towards subpopulation-specific dosages*. Acta Psychiatr Scand. **2001**, 104:173-192.

Kirov G, Murphy KC, Arranz MJ, Jones I, McCandles F, Kunugi H, Murray RM, McGuffin P, Collier DA, Owen MJ, Craddock N. *Low activity allele of catechol-O-methyltransferase gene associated with rapid cycling bipolar disorder*. Mol Psychiatry. **1998**, 3:342-345.

Klimek V, Stockmeier C, Overholser J, Meltzer HY, Kalka S, Dilley G, Ordway GA. *Reduced levels of norepinephrine transporters in the locus coeruleus in major depression*. J Neurosci. **1997**, 17:8451-8458.

Kline NS. *Comprehensive therapy of depressions*. J Neuropsychiatr. **1961**, 2(Suppl 1):15-26.

Kocsis JH. *New strategies for treating chronic depression*. J Clin Psychiatry. **2000**, 61 Suppl 11:42-45.

Korte SM, Koolhaas JM, Wingfield JC, McEwen BS. *The Darwinian concept of stress: benefits of allostasis and costs of allostatic load and the trade-offs in health and disease*. Neurosci Biobehav Rev. **2005**, 29:3-38.

Kupfer DJ, Shaw DH, Ulrich R, Coble PA, Spiker DG. *Application of automated REM analysis in depression*. Arch Gen Psychiatry. **1982**, 39:569-573.

Kupfer DJ. *The increasing medical burden in bipolar disorder*. JAMA. **2005**, 293:2528-2530.

Lacerda AL, Keshavan MS, Hardan AY, Yorbik O, Brambilla P, Sassi RB, Nicoletti M, Mallinger AG, Frank E, Kupfer DJ, Soares JC. *Anatomic evaluation of the orbitofrontal cortex in major depressive disorder*. Biol Psychiatry. **2004**, 55:353-358.

Lacerda AL, Brambilla P, Sassi RB, Nicoletti MA, Mallinger AG, Frank E, Kupfer DJ, Keshavan MS, Soares JC. *Anatomical MRI study of corpus callosum in unipolar depression*. J Psychiatr Res. **2005**, 39:347-354.

Lam RW. *Sleep disturbances and depression: a challenge for antidepressants*. Int Clin Psychopharmacol. **2006**, 21 Suppl 1:S25-S29.

Lander ES, Schork NJ. *Genetic dissection of complex traits*. Science. **1994**, 265:2037-2048.

Langer SZ, Galzin AM, Poirier MF, Loo H, Sechter D, Zarifian E. *Association of [3H]-imipramine and [3H]-paroxetine binding with the 5HT transporter in brain and platelets: relevance to studies in depression* . J Recept Res. **1987**, 7:499-521.

Langer SZ, Galzin AM. *Studies on the serotonin transporter in platelets*. Experientia. **1988**, 44:127-130.

Lapin IP, Oxenkrug GF. *Intensification of the central serotoninergic processes as a possible determinant of the thymoleptic effect*. Lancet. **1969**, 1:132-136.

Leake A, Fairbairn AF, McKeith IG, Ferrier IN. *Studies on the serotonin uptake binding site in major depressive disorder and control post-mortem brain: neurochemical and clinical correlates*. Psychiatry Res. **1991**, 39:155-165.

LeDoux JE. *Emotion circuits in the brain*. Annu Rev Neurosci. **2000**, 23:155-184.

Leonard BE. *The role of noradrenaline in depression: a review*. J Psychopharmacol. **1997**, 11:S39-S47.

Lerer B, Macciardi F, Segman RH, Adolfsson R, Blackwood D, Blairy S, Del Favero J, Dikeos DG, Kaneva R, Lilli R, Massat I, Milanova V, Muir W, Noethen M, Oruc L, Petrova T, Papadimitriou GN, Rietschel M, Serretti A, Souery D, Van Gestel S, Van Broeckhoven C, Mendlewicz J. *Variability of 5-HT2C receptor cys23ser polymorphism among European populations and vulnerability to affective disorder*. Mol Psychiatry. **2001**, 6:579-585.

Lesch KP. *Gene-environment interaction and the genetics of depression*. J Psychiatry Neurosci. **2004**, 29:174-184.

Lewinsohn PM, Rohde P, Seeley JR, Fischer SA. *Age-cohort changes in the lifetime occurrence of depression and other mental disorders*. J Abnorm Psychol. **1993**, 102:110-120.

Lewy AJ. *Circadian phase sleep and mood disorders*. In: Davis KL, Charney D, Coyle JT, Nemeroff C, eds. *Neuropsychopharmacology: The Fifth Generation of Progress*. 5th ed. Lippincott Williams & Wilkins, Philadelphia, **2002**, pp 1879-1894.

Li X, Bijur GN, Jope RS. *Glycogen synthase kinase-3beta, mood stabilizers, and neuroprotection*. Bipolar Disord. **2002**, 4:137-144.

Licinio J, Wong ML. *The pharmacogenomics of depression*. Pharmacogenomics J. **2001**, 1:175-177.

Lim LC, Powell J, Sham P, Castle D, Hunt N, Murray R, Gill M. *Evidence for a genetic association between alleles of monoamine oxidase A gene and bipolar affective disorder*. Am J Med Genet. **1995**, 60:325-331.

Lipkin WI, Travis GH, Carbone KM, Wilson MC. *Isolation and characterization of Borna disease agent cDNA clones*. Proc Natl Acad Sci U S A. **1990**, 87:4184-4188.

Malberg JE, Eisch AJ, Nestler EJ, Duman RS. *Chronic antidepressant treatment increases neurogenesis in adult rat hippocampus*. J Neurosci. **2000**, 20:9104-9110.

Malison RT, Price LH, Berman R, van Dyck CH, Pelton GH, Carpenter L, Sanacora G, Owens MJ, Nemeroff CB, Rajeevan N, Baldwin RM, Seibyl JP, Innis RB, Charney DS. *Reduced brain serotonin transporter availability in major depression as measured by [123I]-2 beta-carbomethoxy-3 beta-(4-iodophenyl)tropane and single photon emission computed tomography*. Biol Psychiatry. **1998**, 44:1090-1098.

Manev H, Uz T. *Clock genes: influencing and being influenced by psychoactive drugs*. Trends Pharmacol Sci. **2006**, 27:186-189.

Manji HK, Drevets WC, Charney DS. *The cellular neurobiology of depression*. Nat Med. **2001**, 7:541-547.

Manji HK, Duman RS. *Impairments of neuroplasticity and cellular resilience in severe mood disorders: implications for the development of novel therapeutics*. Psychopharmacol Bull. **2001**, 35:5-49.

Manji HK, Quiroz JA, Sporn J, Payne JL, Denicoff K, Gray A, Zarate CA, Jr., Charney DS. *Enhancing neuronal plasticity and cellular resilience to develop novel, improved therapeutics for difficult-to-treat depression*. Biol Psychiatry. **2003**, 53:707-742.

Mann JJ, Stanley M, McBride PA, McEwen BS. *Increased serotonin2 and beta-adrenergic receptor binding in the frontal cortices of suicide victims*. Arch Gen Psychiatry. **1986**, 43:954-959.

Mann JJ, Malone KM, Diehl DJ, Perel J, Cooper TB, Mintun MA. *Demonstration in vivo of reduced serotonin responsivity in the brain of untreated depressed patients*. Am J Psychiatry. **1996**, 153:174-182.

Mann JJ, Huang YY, Underwood MD, Kassir SA, Oppenheim S, Kelly TM, Dwork AJ, Arango V. *A serotonin transporter gene promoter polymorphism (5-HTTLPR) and prefrontal cortical binding in major depression and suicide*. Arch Gen Psychiatry. **2000**, 57:729-738.

Mann JJ. *Neurobiology of suicidal behaviour*. Nat Rev Neurosci. **2003**, 4:819-828.

Mateo Y, Fernandez-Pastor B, Meana JJ. *Acute and chronic effects of desipramine and clorgyline on alpha(2)-adrenoceptors regulating noradrenergic transmission in the rat brain: a dual-probe microdialysis study*. Br J Pharmacol. **2001**, 133:1362-1370.

Mayberg HS. *Positron emission tomography imaging in depression: a neural systems perspective*. Neuroimaging Clin N Am. **2003**, 13:805-815.

Mayers AG, Baldwin DS. *Antidepressants and their effect on sleep*. Hum Psychopharmacol. **2005**, 20:533-559.

Mayeux R. Depression and dementia in Parkinson's disease. In: Marsden CO, Fahn S, eds. *Movement disorders*. Butterworth, London, **1982**, pp 75-95.

McClung CA, Sidiropoulou K, Vitaterna M, Takahashi JS, White FJ, Cooper DC, Nestler EJ. *Regulation of dopaminergic transmission and cocaine reward by the Clock gene*. Proc Natl Acad Sci U S A. **2005**, 102:9377-9381.

McEwen BS. *Allostasis and allostatic load: implications for neuropsychopharmacology*. Neuropsychopharmacology. **2000**, 22:108-124.

Meana JJ, Barturen F, Garcia-Sevilla JA. *Alpha 2-adrenoceptors in the brain of suicide victims: increased receptor density associated with major depression*. Biol Psychiatry. **1992**, 31:471-490.

Mellerup E, Bennike B, Bolwig T, Dam H, Hasholt L, Jorgensen MB, Plenge P, Sorensen SA. *Platelet serotonin transporters and the transporter gene in control subjects, unipolar patients and bipolar patients*. Acta Psychiatr Scand. **2001**, 103:229-233.

Meltzer HY, Maes M. *Effects of buspirone on plasma prolactin and cortisol levels in major depressed and normal subjects*. Biol Psychiatry. **1994**, 35:316-323.

Merali Z, Du L, Hrdina P, Palkovits M, Faludi G, Poulter MO, Anisman H. *Dysregulation in the suicide brain: mRNA expression of corticotropin-releasing hormone receptors and GABA(A) receptor subunits in frontal cortical brain region*. J Neurosci. **2004**, 24:1478-1485.

Miller HL, Delgado PL, Salomon RM, Berman R, Krystal JH, Heninger GR, Charney DS. *Clinical and biochemical effects of catecholamine depletion on antidepressant-induced remission of depression*. Arch Gen Psychiatry. **1996a**, 53:117-128.

Miller HL, Delgado PL, Salomon RM, Heninger GR, Charney DS. *Effects of alpha-methyl-para-tyrosine (AMPT) in drug-free depressed patients*. Neuropsychopharmacology. **1996b**, 14:151-157.

Moreno FA, Gelenberg AJ, Heninger GR, Potter RL, McKnight KM, Allen J, Phillips AP, Delgado PL. *Tryptophan depletion and depressive vulnerability*. Biol Psychiatry. **1999**, 46:498-505.

Muller-Oerlinghausen B, Berghofer A, Bauer M. *Bipolar disorder*. Lancet. **2002**, 359:241-247.

Muller JC, Pryor WW, Gibbons JE, Orgain ES. *Depression and anxiety occurring during Rauwolfia therapy*. J Am Med Assoc. **1955**, 159:836-839.

Murray CJL, Lopez AD. *The Global Burden of Disease: a comprehensive assessment of mortality and disability from diseases, injuries, and risk factors in 1990 and projected to 2020*. Harvard University Press; Cambridge, **1996**.

Mynors-Wallis L. *Problem-solving treatment: evidence for effectiveness and feasibility in primary care*. Int J Psychiatry Med. **1996**, 26:249-262.

Nemeroff CB, Knight DL, Krishnan RR, Slotkin TA, Bissette G, Melville ML, Blazer DG. *Marked reduction in the number of platelet-tritiated imipramine binding sites in geriatric depression*. Arch Gen Psychiatry. **1988**, 45:919-923.

Nemeroff CB. *The corticotropin-releasing factor (CRF) hypothesis of depression: new findings and new directions*. Mol Psychiatry. **1996**, 1:336-342.

Nestler EJ, Barrot M, DiLeone RJ, Eisch AJ, Gold SJ, Monteggia LM. *Neurobiology of depression*. Neuron. **2002**, 34:13-25.

Neumeister A, Konstantinidis A, Stastny J, Schwarz MJ, Vitouch O, Willeit M, Praschak-Rieder N, Zach J, de Zwaan M, Bondy B, Ackenheil M, Kasper S. *Association between serotonin transporter gene promoter polymorphism (5HTTLPR) and behavioral responses to tryptophan depletion in healthy women with and without family history of depression*. Arch Gen Psychiatry. **2002**, 59:613-620.

Nibuya M, Morinobu S, Duman RS. *Regulation of BDNF and trkB mRNA in rat brain by chronic electroconvulsive seizure and antidepressant drug treatments*. J Neurosci. **1995**, 15:7539-7547.

Nibuya M, Nestler EJ, Duman RS. *Chronic antidepressant administration increases the expression of cAMP response element binding protein (CREB) in rat hippocampus*. J Neurosci. **1996**, 16:2365-2372.

NICE. *Depression: management of depression in primary and secondary care*. National Institute for Clinical Excellence; London, **2004**.

Ongur D, Drevets WC, Price JL. *Glial reduction in the subgenual prefrontal cortex in mood disorders*. Proc Natl Acad Sci U S A. **1998**, 95:13290-13295.

Ordway GA, Farley JT, Dilley GE, Overholser JC, Meltzer HY, Balraj EK, Stockmeier CA, Klimek V. *Quantitative distribution of monoamine oxidase A in brainstem monoamine nuclei is normal in major depression.* Brain Res. **1999**, 847:71-79.

Ordway GA, Schenk J, Stockmeier CA, May W, Klimek V. *Elevated agonist binding to alpha2-adrenoceptors in the locus coeruleus in major depression.* Biol Psychiatry. **2003**, 53:315-323.

Oruc L, Verheyen GR, Furac I, Jakovljevic M, Ivezic S, Raeymaekers P, Van Broeckhoven C. *Association analysis of the 5-HT2C receptor and 5-HT transporter genes in bipolar disorder.* Am J Med Genet. **1997**, 74:504-506.

Papolos DF, Veit S, Faedda GL, Saito T, Lachman HM. *Ultra-ultra rapid cycling bipolar disorder is associated with the low activity catecholamine-O-methyltransferase allele.* Mol Psychiatry. **1998**, 3:346-349.

Pare CM. *Potentation of monoamine-oxidase inhibitors by tryptophan.* Lancet. **1963**, 35:527-528.

Pariante CM, Miller AH. *Glucocorticoid receptors in major depression: relevance to pathophysiology and treatment.* Biol Psychiatry. **2001**, 49:391-404.

Paul SM, Rehavi M, Skolnick P, Ballenger JC, Goodwin FK. *Depressed patients have decreased binding of tritiated imipramine to platelet serotonin "transporter".* Arch Gen Psychiatry. **1981**, 38:1315-1317.

Perez V, Gilaberte I, Faries D, Alvarez E, Artigas F. *Randomised, double-blind, placebo-controlled trial of pindolol in combination with fluoxetine antidepressant treatment.* Lancet. **1997**, 349:1594-1597.

Perry EK, Marshall EF, Blessed G, Tomlinson BE, Perry RH. *Decreased imipramine binding in the brains of patients with depressive illness.* Br J Psychiatry. **1983**, 142:188-192.

Posener JA, DeBattista C, Williams GH, Chmura KH, Kalehzan BM, Schatzberg AF. *24-Hour monitoring of cortisol and corticotropin secretion in psychotic and nonpsychotic major depression.* Arch Gen Psychiatry. **2000**, 57:755-760.

Post RM, Goodwin FK. *Handbook of psychopharmacology.* Plenum Press; New York, **1976**, pp 137-186.

Preisig M, Bellivier F, Fenton BT, Baud P, Berney A, Courtet P, Hardy P, Golaz J, Leboyer M, Mallet J, Matthey ML, Mouthon D, Neidhart E, Nosten-Bertrand M, Stadelmann-Dubuis E, Guimon J, Ferrero F, Buresi C, Malafosse A. *Association between bipolar disorder and monoamine oxidase A gene polymorphisms: results of a multicenter study.* Am J Psychiatry. **2000**, 157:948-955.

Rajkowska G, Miguel-Hidalgo JJ, Wei J, Dilley G, Pittman SD, Meltzer HY, Overholser JC, Roth BL, Stockmeier CA. *Morphometric evidence for neuronal and glial prefrontal cell pathology in major depression.* Biol Psychiatry. **1999**, 45:1085-1098.

Rajkowska G. *Histopathology of the prefrontal cortex in major depression: what does it tell us about dysfunctional monoaminergic circuits?* Prog Brain Res. **2000a**, 126:397-412.

Rajkowska G. *Postmortem studies in mood disorders indicate altered numbers of neurons and glial cells.* Biol Psychiatry. **2000b**, 48:766-777.

Redmond DE, Jr., Maas JW, Kling A, Dekirmenjian H. *Changes in primate social behavior after treatment with alpha-methyl-para-tyrosine.* Psychosom Med. **1971a**, 33:97-113.

Redmond DE, Jr., Maas JW, Kling A, Graham CW, Dekirmenjian H. *Social behavior of monkeys selectively depleted of monoamines*. Science. **1971b**, 174:428-431.

Reppert SM, Weaver DR. *Coordination of circadian timing in mammals*. Nature. **2002**, 418:935-941.

Ressler KJ, Nemeroff CB. *Role of norepinephrine in the pathophysiology and treatment of mood disorders*. Biol Psychiatry. **1999**, 46:1219-1233.

Reul JM, de Kloet ER. *Two receptor systems for corticosterone in rat brain: microdistribution and differential occupation*. Endocrinology. **1985**, 117:2505-2511.

Romero L, Bel N, Artigas F, De Montigny C, Blier P. *Effect of pindolol on the function of pre- and postsynaptic 5-HT1A receptors: in vivo microdialysis and electrophysiological studies in the rat brain*. Neuropsychopharmacology. **1996a**, 15:349-360.

Romero L, Bel N, Casanovas JM, Artigas F. *Two actions are better than one: avoiding self-inhibition of serotonergic neurones enhances the effects of serotonin uptake inhibitors*. Int Clin Psychopharmacol. **1996b**, 11 Suppl 4:1-8.

Romero L, Hervas I, Artigas F. *The 5-HT1A antagonist WAY-100635 selectively potentiates the presynaptic effects of serotonergic antidepressants in rat brain*. Neurosci Lett. **1996c**, 219:123-126.

Rosso IM. *Review: hippocampal volume is reduced in people with unipolar depression*. Evid Based Ment Health. **2005**, 8:45.

Rott R, Herzog S, Fleischer B, Winokur A, Amsterdam J, Dyson W, Koprowski H. *Detection of serum antibodies to Borna disease virus in patients with psychiatric disorders*. Science. **1985**, 228:755-756.

Rubinsztein DC, Leggo J, Goodburn S, Walsh C, Jain S, Paykel ES. *Genetic association between monoamine oxidase A microsatellite and RFLP alleles and bipolar affective disorder: analysis and meta-analysis*. Hum Mol Genet. **1996**, 5:779-782.

Sapolsky RM. *Glucocorticoids and hippocampal atrophy in neuropsychiatric disorders*. Arch Gen Psychiatry. **2000**, 57:925-935.

Sargent PA, Kjaer KH, Bench CJ, Rabiner EA, Messa C, Meyer J, Gunn RN, Grasby PM, Cowen PJ. *Brain serotonin1A receptor binding measured by positron emission tomography with [11C]WAY-100635: effects of depression and antidepressant treatment*. Arch Gen Psychiatry. **2000**, 57:174-180.

Schatzberg AF, Garlow SJ, Nemeroff CB. Molecular and cellular mechanisms in depression. In: Davis KL, Charney D, Coyle JT, Nemeroff C, eds. *Neuropsychopharmacology: The Fifth Generation of Progress*. 5th ed. Lippincott Williams & Wilkins, Philadelphia, **2002**, pp 1039-1050.

Schildkraut JJ. *The catecholamine hypothesis of affective disorders: a review of supporting evidence*. Am J Psychiatry. **1965**, 122:509-522.

Schulberg HC, Block MR, Madonia MJ, Scott CP, Rodriguez E, Imber SD, Perel J, Lave J, Houck PR, Coulehan JL. *Treating major depression in primary care practice. Eight-month clinical outcomes*. Arch Gen Psychiatry. **1996**, 53:913-919.

Schulze TG, Muller DJ, Krauss H, Scherk H, Ohlraun S, Syagailo YV, Windemuth C, Neidt H, Grassle M, Papassotiropoulos A, Heun R, Nothen MM, Maier W, Lesch KP, Rietschel M.

Association between a functional polymorphism in the monoamine oxidase A gene promoter and major depressive disorder. Am J Med Genet. **2000**, 96:801-803.

Schumacher J, Jamra RA, Becker T, Ohlraun S, Klopp N, Binder EB, Schulze TG, Deschner M, Schmal C, Hofels S, Zobel A, Illig T, Propping P, Holsboer F, Rietschel M, Nothen MM, Cichon S. *Evidence for a relationship between genetic variants at the brain-derived neurotrophic factor (BDNF) locus and major depression.* Biol Psychiatry. **2005**, 58:307-314.

Sharp T, Bramwell SR, Lambert P, Grahame-Smith DG. *Effect of short- and long-term administration of lithium on the release of endogenous 5-HT in the hippocampus of the rat in vivo and in vitro.* Neuropharmacology. **1991**, 30:977-984.

Sharp T, Bramwell SR, Grahame-Smith DG. *Effect of acute administration of L-tryptophan on the release of 5-HT in rat hippocampus in relation to serotoninergic neuronal activity: an in vivo microdialysis study.* Life Sci. **1992**, 50:1215-1223.

Sheline YI, Sanghavi M, Mintun MA, Gado MH. *Depression duration but not age predicts hippocampal volume loss in medically healthy women with recurrent major depression.* J Neurosci. **1999**, 19:5034-5043.

Shopsin B, Gershon S, Goldstein M, Friedman E, Wilk S. *Use of synthesis inhibitors in defining a role for biogenic amines during imipramine treatment in depressed patients.* Psychopharmacol Commun. **1975**, 1:239-249.

Shore PA, Silver SL, Brodie BB. *Interaction of reserpine, serotonin, and lysergic acid diethylamide in brain.* Science. **1955**, 122:284-285.

Sibille E, Hen R. *Combining genetic and genomic approaches to study mood disorders.* Eur Neuropsychopharmacol. **2001**, 11:413-421.

Sibille E, Arango V, Galfalvy HC, Pavlidis P, Erraji-Benchekroun L, Ellis SP, John MJ. *Gene expression profiling of depression and suicide in human prefrontal cortex.* Neuropsychopharmacology. **2004**, 29:351-361.

Siever LJ, Davis KL. *Overview: toward a dysregulation hypothesis of depression.* Am J Psychiatry. **1985**, 142:1017-1031.

Sjoerdsma A, Engelman K, Spector S, Udenfriend S. *Inhibition of catecholamine synthesis in man with alpha-methyl-tyrosine, an inhibitor of tyrosine hydroxylase.* Lancet. **1965**, 2:1092-1094.

Sklar P, Gabriel SB, McInnis MG, Bennett P, Lim YM, Tsan G, Schaffner S, Kirov G, Jones I, Owen M, Craddock N, DePaulo JR, Lander ES. *Family-based association study of 76 candidate genes in bipolar disorder: BDNF is a potential risk locus. Brain-derived neutrophic factor.* Mol Psychiatry. **2002**, 7:579-593.

Smith MA, Makino S, Kvetnansky R, Post RM. *Stress and glucocorticoids affect the expression of brain-derived neurotrophic factor and neurotrophin-3 mRNAs in the hippocampus.* J Neurosci. **1995**, 15:1768-1777.

Stanley M, Virgilio J, Gershon S. *Tritiated imipramine binding sites are decreased in the frontal cortex of suicides.* Science. **1982**, 216:1337-1339.

Starkstein SE, Preziosi TJ, Berthier ML, Bolduc PL, Mayberg HS, Robinson RG. *Depression and cognitive impairment in Parkinson's disease.* Brain. **1989**, 112 (Pt 5):1141-1153.

Starkstein SE, Robinson RG. *Affective disorders and cerebral vascular disease.* Br J Psychiatry. **1989**, 154:170-182.

Stockmeier CA, Shapiro LA, Dilley GE, Kolli TN, Friedman L, Rajkowska G. *Increase in serotonin-1A autoreceptors in the midbrain of suicide victims with major depression-postmortem evidence for decreased serotonin activity.* J Neurosci. **1998**, 18:7394-7401.

Sullivan PF, Neale MC, Kendler KS. *Genetic epidemiology of major depression: review and meta-analysis.* Am J Psychiatry. **2000**, 157:1552-1562.

Szegedi A, Rujescu D, Tadic A, Muller MJ, Kohnen R, Stassen HH, Dahmen N. *The catechol-O-methyltransferase Val108/158Met polymorphism affects short-term treatment response to mirtazapine, but not to paroxetine in major depression.* Pharmacogenomics J. **2005**, 5:49-53.

Takahashi JS. *Finding new clock components: past and future.* J Biol Rhythms. **2004**, 19:339-347.

Taylor L, Faraone SV, Tsuang MT. *Family, twin, and adoption studies of bipolar disease.* Curr Psychiatry Rep. **2002**, 4:130-133.

Thome J, Sakai N, Shin K, Steffen C, Zhang YJ, Impey S, Storm D, Duman RS. *cAMP response element-mediated gene transcription is upregulated by chronic antidepressant treatment.* J Neurosci. **2000**, 20:4030-4036.

Tsai SJ, Cheng CY, Yu YW, Chen TJ, Hong CJ. *Association study of a brain-derived neurotrophic-factor genetic polymorphism and major depressive disorders, symptomatology, and antidepressant response.* Am J Med Genet B Neuropsychiatr Genet. **2003**, 123:19-22.

Tsuji K, Toyomasu K, Imamura Y, Maeda H, Toyoda T. *No association of borna disease virus with psychiatric disorders among patients in northern Kyushu, Japan.* J Med Virol. **2000**, 61:336-340.

Uz T, Ahmed R, Akhisaroglu M, Kurtuncu M, Imbesi M, Dirim AA, Manev H. *Effect of fluoxetine and cocaine on the expression of clock genes in the mouse hippocampus and striatum.* Neuroscience. **2005**, 134:1309-1316.

Vaidya VA, Marek GJ, Aghajanian GK, Duman RS. *5-HT2A receptor-mediated regulation of brain-derived neurotrophic factor mRNA in the hippocampus and the neocortex.* J Neurosci. **1997**, 17:2785-2795.

Vaidya VA, Duman RS. *Depresssion--emerging insights from neurobiology.* Br Med Bull. **2001**, 57:61-79.

van Praag HM. *In search of the mode of action of antidepressants. 5-HTP/tyrosine mixtures in depressions.* Neuropharmacology. **1983**, 22:433-440.

Veith RC, Lewis N, Linares OA, Barnes RF, Raskind MA, Villacres EC, Murburg MM, Ashleigh EA, Castillo S, Peskind ER, . *Sympathetic nervous system activity in major depression. Basal and desipramine-induced alterations in plasma norepinephrine kinetics.* Arch Gen Psychiatry. **1994**, 51:411-422.

Watson S, Gallagher P, Ritchie JC, Ferrier IN, Young AH. *Hypothalamic-pituitary-adrenal axis function in patients with bipolar disorder.* Br J Psychiatry. **2004**, 184:496-502.

Weller IVD, Ashby D, Brook R, Chambers M.G.A., Chick JD, Drummond DC, Ebmeier KP, Gunnell DJ. *Report of the Committee on Safety of Medicines expert working group on the safety of selective serotonin reuptake inhibitor antidepressants.* Medicines and Healthcare products Regulatory Agency; London, **2004**.

WHO. *The ICD-10 classification of mental and behavioural disorders: diagnostic criteria for research.* World Health Organization; Geneva, **1993**.

WHO Mental Health report. *Mental Health: New Understanding, New Hope.* (http://www.who.int/whr/2001/en/) 2001.

Wong ML, Khatri P, Licinio J, Esposito A, Gold PW. *Identification of hypothalamic transcripts upregulated by antidepressants.* Biochem Biophys Res Commun. **1996**, 229:275-279.

Wong ML, Kling MA, Munson PJ, Listwak S, Licinio J, Prolo P, Karp B, McCutcheon IE, Geracioti TD, Jr., DeBellis MD, Rice KC, Goldstein DS, Veldhuis JD, Chrousos GP, Oldfield EH, McCann SM, Gold PW. *Pronounced and sustained central hypernoradrenergic function in major depression with melancholic features: relation to hypercortisolism and corticotropin-releasing hormone.* Proc Natl Acad Sci U S A. **2000**, 97:325-330.

Wong ML, Licinio J. *Research and treatment approaches to depression.* Nat Rev Neurosci. **2001**, 2:343-351.

Yamada M, Higuchi T. *Functional genomics and depression research. Beyond the monoamine hypothesis.* Eur Neuropsychopharmacol. **2002**, 12:235-244.

Yatham LN, Liddle PF, Shiah IS, Scarrow G, Lam RW, Adam MJ, Zis AP, Ruth TJ. *Brain serotonin2 receptors in major depression: a positron emission tomography study.* Arch Gen Psychiatry. **2000**, 57:850-858.

Zeller EA, Barsky J. *In vivo inhibition of liver and brain monoamine oxidase by 1-Isonicotinyl-2-isopropyl hydrazine.* Proc Soc Exp Biol Med. **1952**, 81:459-461.

Zhang X, Gainetdinov RR, Beaulieu JM, Sotnikova TD, Burch LH, Williams RB, Schwartz DA, Krishnan KR, Caron MG. *Loss-of-function mutation in tryptophan hydroxylase-2 identified in unipolar major depression.* Neuron. **2005**, 45:11-16.

Zobel AW, Nickel T, Kunzel HE, Ackl N, Sonntag A, Ising M, Holsboer F. *Effects of the high-affinity corticotropin-releasing hormone receptor 1 antagonist R121919 in major depression: the first 20 patients treated.* J Psychiatr Res. **2000**, 34:171-181.

1.2 Clinics

Clinical Aspects of Depressive Disorders

Rosario Pérez-Egea, Victor Pérez, Dolors Puigdemont and Enric Álvarez

Psychiatry Service, Hospital de la Santa Creu i Sant Pau, Av. Antoni Mª Claret, 175. 08025 Barcelona, Spain.

1.2.1 Introduction

Achilles' sudden mood swings and Belafonte's sadness, described by Homer in the Iliad, are probably among the first descriptions of affective disorders. Hippocrates (460–380 BC) raised the first "biologistic" hypothesis, which denied any divine intervention upon depressive symptoms and focused on the possible "corruption" of the brain – the organ where he located emotions – due to an excess or alteration of "black bile", the definition at the origin of the word "melancholy". During the same period Aristotle (384–322 BC) wrote a monograph on this issue, stating that there is a tendency to melancholy in all those who have reached eminence in philosophy, politics or poetry'. In the 1st century, Areteus the Cappadocian gave one of the best descriptions of melancholy; in the 2nd century, Galen of Pergamon attempted to explain its biochemistry through his temperament theory.

Medieval obscurantism attributed depressive states to demonic causes, and many patients were burnt in the Inquisition fires. These patients were treated differently in the Eastern and Arab world, although far from Hippocratic considerations.

In the Renaissance, Vesalius followed his classic predecessors and located the cause of depression in the brain. That is how the Western culture rejected definitely any divine intervention upon the onset of affective diseases. In those times, the definition of melancholy was somewhat loose and was considered – when relatively mild – as a common maladjustment in people with keenly developed sensitivity, intelligence or perception. *The Anatomy of Melancholy* by Robert Burton (1577–1640) is a good example of this type of analysis.

In the 19th and 20th centuries, Esquirol's conception according to a clinical–anatomical model (1820), Billod's taxonomic skills (1856), Griensinger's introduction of the term "depression" as presently conceptualized (1861) and Freud's psychologistic conceptions (1917) all crystallized in Kraepelin's integrating work (1921), an essential tool for understanding the present classification of mental diseases. The German psychiatrists Kahlbaum and Kraepelin are usually acknowledged for having developed an overall, modern systematization for the classification of mental disorders. Kahlbaum provided the terms "dysthymia" – referring to the chronic state of melancholy – and "cyclothymia" – referring to a disorder characterized by emotional ups and downs. Concepts developed by his predecessors led Kraepelin to propose a general system for the classification of psychoses. The term "melancholy" was preserved, but only to designate depressive disorders occurring in senility, and the term "depression" was given importance for the first time. Because it evoked more physiological grounds of the disease, this term steadily gained acceptance and could be found in medical dictionaries already in 1860. In 1896, Kraepelin defined affective disorders as a unique, recurrent, good-prognosis disease with a tendency to interepisode recovery, thereby differentiating it from psychotic disorders. In the 1970s, the distinction between unipolar and bipolar depression for endogenous affective psychoses, proposed by Leonhard in 1957, was accepted even by American psychiatry. The validity of bipolar disorders is further strengthened by family and genetic studies, as well as by the efficacy of lithium salts in its treatment. The conceptualization of unipolar depression, however, remains controversial, as shown in the next section.

1.2.2 Classification

The present classification systems originate from the need to gather epidemiological information on mental diseases. This goal has been decisive in determining classifications and diagnostic criteria, thereby limiting, according to many authors, its usefulness in research. A clear example of this limitation is the concept of Major Depressive Disorder. Indeed, since its inclusion in the *Diagnostic Criteria for Research* (RCD; Spitzer et al., 1978), this concept has been of great epidemiological usefulness by contributing a greater detection and treatment of these patients; conversely, it has hampered the ability to screen homogenous patient groups – a must in clinical research. In the 1980s, Major Depressive Disorder was renamed Major Depressive Episode in the *DSM III* (1980) so as to eliminate references to its possible etiology. Likewise, the general term "Affective Disorders" was replaced by that of "Mood Disorders" in the *DSM III-R* (1987) and in the *DSM IV* (1994). In addition, while little change is found in the diagnostic criteria for Major Depressive Episodes, its specifications are modified and therefore the ability to subclassify unipolar depressive disorders is progressively lost.

The classification of affective disorders has been controversial and is not yet resolved. Key issues have been raised, including:

1. *Unitary, dichotomic or multiple classifications?* Lewis (1938) was the first author to raise a convincing unitary approach, suggesting that depression subtypes were but attempts to differentiate between acute-chronic, mild-severe forms. Kendell (Kendell and Gourlay, 1970) was the main defender of this position and proposed the idea of a depression continuum, with severe or psychotic forms at one end and chronic or mild forms at the other. According to these authors, while the continuum ends may differ in key aspects (e.g., response to treatment or long-term outcome), precise limits between them cannot be drawn. Most present-day authors, however, propose a dichotomic subdivision of depression: endogenous vs reactive, bipolar vs unipolar, "pure" depression vs depression spectrum disorder, agitated vs retarded, and so forth. The majority of these dichotomic systems originated from clinical experience in order to delimitate subtypes likely to predict the evolution or the therapeutic response.

2. *What are the limits between depression and normality or other psychiatric disorders?* The limits of affective disorders are quite diffused and doubt is cast on their nature as a disease, a syndrome, a symptom, or even a normal reaction of human nature to the adversities of life. Ever since Aristotle, depression or sadness have been considered a normal response to unfortunate events, a mood occurring more frequently in extremely sensitive and reflexive individuals. Conversely, clinical psychiatry has always conceptualized depression as a disease and not as a normal response, an entity differing completely from a mere feeling of discomfort. Presently, no consensus has been reached on the boundaries between depression and normal mood fluctuation. Thus, while some authors vaguely define depression as a set of characteristic symptoms to be considered even if they are of short duration or a reaction to recent events, others state that these symptoms should be categorized as demoralization or feeling of discomfort (Clayton et al., 1972; Frank, 1973; Wing et al., 1974). These authors would rather exclude relatively mild, short or reactive disorders from the concept of affective disorder, with stricter criteria being required for

its diagnosis. In the past two decades, the emphasis was placed on diagnostic criteria and structured interviews, in an effort to differentiate depression from dejection, discomfort or grief. Examples of this are Feighner's criteria (Feighner et al., 1972), the Diagnostic Criteria for Research (Spitzer et al., 1978) and, more recently, DSM (APA, 1994) or ICD 10 criteria (WHO, 1992). Even under these operative criteria, however, many researchers believe that the definition of depression is still highly imprecise, overinclusive and deprived of clinical usefulness to predict outstanding factors such as evolution or response to treatment. The border between affective disorders and other alterations such as schizophrenia or anxiety disorders is also not clear. Hence, the clinical features of some patients combine affective symptoms with delusional symptomatology of extravagant content or first-order symptoms. In an attempt to classify these patients, Kasanin proposed in 1933 the concept of schizoaffective disorder, defined as acute psychosis with a mixture of schizophrenic and affective symptoms, usually with a relatively favorable prognosis. While the schizoaffective disorder has traditionally been classified together with schizophrenias, some authors consider that it should be conceptualized as a subtype of affective disorder, an independent disease or a continuum between affective disorder and schizophrenia. Because depression syndromes are frequently associated with anxiety symptoms, the boundaries between depressive and anxious states are equally problematic. Indeed, severely depressed patients may present with marked, anxious agitation, while mildly depressed ones usually complain of stress, nervousness and apprehension. As with the schizoaffective disorder, the issue of classifying these patients into affective disorders, anxiety disorders or into an independent group is hard to solve.

Mood disorders are mainly characterized by mood changes, either in the form of low spirits (depression) or high spirits (euphoria). In the latter case, an exalted mood is usually accompanied, among other symptoms, by exaggerated self-esteem, insomnia, hyperactivity, verbal diarrhoea, distractibility and accelerated thinking, conforming what is known as a manic episode. Although the bipolar disorder is not the object of the present chapter, it should be outlined that patients presenting with a depressive episode but with a background of manic episodes will be diagnosed with bipolar disorder – present depressive episode.

The ubiquity of depressive symptoms has them appear in highly diverse psychiatric and organic conditions. Before a depressive disorder can be diagnosed, one must discard the presence of physical diseases or the influence of substance consumption, as well as the presence of other primary psychiatric disorders (anxiety disorders, psychotic disorders, somatoform disorders) likely to develop with depression symptomatology.

Sadness and its related symptoms, their duration, intensity and degree of interference in the individual's life, will conform the different nosological entities into which depression is classified today. The diagnostic criteria of depressive disorders will be analyzed according to the *DSM-IV-TR* classification (American Psychiatric Association, 2000), with some nuances being added when comparing it with the WHO classification, i.e. *ICD-10* (World Health Organization, 1993). Importantly, the two classification systems are not substantially different, and accordingly they share similar pros and cons (pros: specification of operative diagnostic criteria from an atheoretical, descriptive view; cons: scarce ability to differentiate and hierarchize the qualitative aspects of the different forms of depression, with further loss of aetiopathogenetic involvement in diagnosis). In fact, one could say that present classifications diagnose the different entities based on a sum of symptoms and their duration, thus implying high reliability and low validity altogether and thereby leading to poor diagnostic stability.

1.2.2.1 DSM-IV-TR criteria for major depressive episode

These criteria include the presence of a minimum of five of the symptoms listed below, for a period of 2 weeks. These symptoms produce clinically significant discomfort and include social and occupational impairment, as well as impairment of other areas important to the individual's activity. One of the symptoms must be depressed mood, or loss of interest, or loss of ability to feel pleasure.

ICD-10 criteria add lack of vitality or increased fatigability to these two main DSM-IV-TR symptoms (sadness, anhedonia), and two of the three symptoms must be present. In addition, ICD-10 emphasizes the value of symptoms like loss of confidence, low self-esteem and feelings of inferiority by means of a particular section differentiated from that described by DSM-IV-TR on self-reproach and feelings of guilt.

A. Depressed mood verbalized by the subject him/herself (feelings of sadness or emptiness) or observed by others (weeping...), most part of the day and practically every day.

B. Important decrease in interest or in the ability to feel pleasure in most activities, most part of the day, almost every day (according to the subject or observed by others).

C. Important weight loss without being on a diet, or weight gain (higher than 5% of bodyweight in 1 month), or loss or increase of appetite almost every day.

D. Insomnia or hypersomnia almost every day.

E. Agitation or psychomotor slowdown almost every day.

F. Fatigue or loss of energy almost every day.

G. Excessive or inappropriate (even delusional) feelings of uselessness or guilt, also on a continuous basis.

H. Decreased ability to think or concentrate, or indecision.

I. Recurring thoughts of death, self-harm ideation with or without a specific plan, or self-harm attempt.

1.2.2.1.1 Severity, psychosis and remission specifications

- *Mild depressive episode:* few or no symptoms, other than those indispensable for diagnosis. The symptoms only lead to a slight disability at an occupational or interpersonal level.

- *Moderate depressive episode:* slight to severe functional disability.

- *Severe depressive episode without psychotic symptoms:* several symptoms in addition to the indispensable ones. Symptoms interfere notably on occupational activity and relationships with others.

- *Severe depressive episode with psychotic symptoms:* the subject has delusional ideas or hallucinations. This depressive subtype must be differentiated from the onset of delirium (acute confusional state) in the course of depression, where delirium and hallucinations (particularly visual ones) may occur, but whose course is fluctuant, predominantly nocturnal and accompanied by impaired level of conscience. These clinical features are indicative of an organic cause or the effect of substances or drugs (for example, anticholinergic side-effects of the antidepressant treatment itself, particularly with tricyclic antidepressants).

- *Mood-consistent psychotic symptoms:* delusional ideas and hallucinations dealing with typical topics of depression (uselessness, guilt/punishment deserving, disease, nihilism).
- *Mood-inconsistent psychotic symptoms:* delusional ideas and hallucinations dealing with topics other than the above, e.g. delusional ideas of persecution or control, thought insertion/broadcast.

- *Partially remitting depressive episode:* the criteria are not fully met despite a few symptoms of major depressive episode still persisting, or there is a period without significant symptoms of less than 2 months after the major depressive episode.
- *Totally remitting depressive episode:* no significant depressive signs or symptoms in the last 2 months.

1.2.2.1.2 Course specifications

Major depressive disorder is a single episode defined as the subject having had only one major depressive disorder in his/her life, or if it is the first episode. *Major recidivant depressive disorder* is diagnosed when two or more major depressive episodes have been observed within an interval of at least 2 months.

The specifications of the longitudinal course of a depressive disorder refer to the achievement or non-achievement of total remission between the two latest depressive episodes, leading to the diagnosis of *major depressive disorder with or without full interepisodic recovery.*

A depressive episode is considered to be *chronic* when all major depressive episode criteria have been met continuously for the past 2 years.

A *seasonal affective disorder* (SAD) is a sustained time-related relationship between the onset and the remission of depressive episodes and specific times of year, e.g. episodes occur regularly in fall and disappear in spring.

In *postpartum depressive episodes*, depression signs appear during the first 4 weeks after delivery.

1.2.2.1.3 Melancholic symptom specifications

- *Depressive episode with melancholic symptoms:*

    A. One of the following symptoms must be present during the most severe period of the present episode:
    - Loss of pleasure in practically all activities.
    - Absence of reactivity to usually pleasant stimuli.

    B. Three or more of the following symptoms:
    - One distinctive quality of depressed mood (for example, depression is experienced differently than the type of feeling experienced after the death of a loved one).
    - Depression is worse in the morning.
    - Early waking (at least 2 hours before usual time).
    - Psychomotor slowdown or agitation.
    - Significant anorexia or weight loss.
    - Excessive or inappropriate feelings of guilt.

The ICD-10 classification uses the term *somatic syndrome* as a synonym for melancholic, biological or endogenomorphic symptoms, plus a marked loss of libido.

Despite the difficulty of describing or grasping the distinctive mood quality of melancholic forms of depression, the key features contained in the concept of *endogenous depression* include anhedonia and areactivity, as well as motor alterations (particularly inhibition), weight loss and early waking. The Newcastle group established an index to differentiate endogenous depression from neurotic depression (Carney et al., 1965). Together with some of the previous symptoms, the presence of an adapted previous personality and the absence of psychogenetic elements (stress or psychological difficulties at the core of the depressive disorder) would be elements supporting the episode's endogenicity.

1.2.2.1.4 Atypical symptom specifications

- *Depressive episode with atypical symptoms:*

    A. Reactive mood: it improves in response to real or potentially positive situations.

    B. Two or more of the following symptoms:
    - Increased appetite or weight.

- Hypersomnia.
- Dejection.
- Long-lasting hypersensitivity to interpersonal rejection, not limited to the depressive episode and with significant repercussion at a social or occupational level.

In the literature, the concept of atypical depression gathers a somewhat confusing ensemble of symptoms. The core of this type of depressions, which are not only moderate forms of depression and which can also benefit from specific therapeutic approaches, would be the presence of reactive depressive mood, together with specific guiding symptoms such as hyperphagia, hypersomnia and lethargy (Vallejo and Urretavizcaya, 2000). While there are other diagnosis-orienting symptoms such as predominant anxiety, asthenia, mood worsening in the evening, or sleep-onset insomnia, they may also be found in other clinical features (anxiety disorders, reactive and adaptative disorders, personality disorders, etc.).

1.2.2.1.5 Catatonic symptom specifications

- *Depressive episode with catatonic symptoms:*

    Presence of two of the following symptoms:

    A. Motor immobility, manifested either by *catalepsy* (decreased reactivity, with same attitude or position maintained for a prolonged period of time, including wax-like flexibility (flexibilitas cerea), where the subject's limbs can be "molded" and maintained into any position) or by *stupor* (highly decreased reactivity to the environment, decreased activity and decreased spontaneous movements).

    B. Excessive motor activity (purposeless agitation not influenced by external stimuli).

    C. Extreme negativism (purposeless resistance to all instructions or maintenance of a rigid posture against attempts to be moved).

    D. Peculiar voluntary movements: the subject adopts strange or inappropriate postures, exhibits stereotyped movements, mannerisms or exaggerated gesticulation.

    E. Echolalia (repeating other's words or sentences) or echopraxia (repeating or imitating somebody else's movements, gestures or posture).

1.2.2.2 DSM-IV-TR criteria for dysthymic disorder

A. Depressed mood most part of the day and most days, for at least 2 years, reported by the subject or observed by others.

B. Presence of two or more of the following symptoms:
   - Appetite loss or increase.
   - Insomnia or hypersomnia.
   - Lack of energy or fatigue.
   - Low self-esteem.
   - Difficulty in concentrating or decision-making.
   - Feelings of despair.

Depending on the onset of the disorder, dysthymia may be classified as of *early onset* (before the age of 21) or *late onset* (21 and older). The presence or absence of atypical symptoms, described in the previous section, can also be specified.

There must be an absence of a previous manic, hypomanic or mixed episode, which would lead to diagnose a bipolar disorder. Conversely, a major depressive episode can indeed be superimposed on dysthymia, with both diagnoses being mandatory. This is called *double depression*.

The appearance of the concept of dysthymic disorder in the DSM-III (APA, 1980) attempted to replace that of neurotic depression, which in addition covered excessively heterogeneous entities. It was accepted that the main categories to include the previous diagnoses of neurotic depression would be major depression without melancholy, dysthymic disorder (gathering most of these cases) and adaptative disorder with depressed mood (Vallejo and Menchón, 2000). The main characteristics of dysthymia, together with its chronicity, would be the mild to moderate intensity of depressive symptoms, the coexistence of neurotic symptoms (hypochondriac worries, anxious-phobic symptoms) and the concurrence of characterological aspects suggesting that, in many cases, depressive symptoms are expressions of an underlying personality disorder.

In connection with this, the issues of whether personality disorders predispose to dysthymia, whether both entities are linked by another type of primary disorder or whether dysthymia is another type of personality disorder by itself, are still under discussion. Many authors support the hypothesis that the genesis of dysthymia is a personality disorder – not only from group B – that configures a type of existence marked by a chronic conflict and that is clinically expressed through anxious and depressive symptoms (Vallejo and Menchón, 2000).

The high comorbidity of these patients is to be emphasized. Indeed, dysthymic disorders are 65–100% associated with other Axis-I or Axis-II diagnoses of DSM-IV-TR (Mezzich,1987; Markowitz, 1994). The most frequent would be major depressive disorder (the abovementioned "double depression"), substance use disorder (alcohol, stimulants) and personality disorders, which will influence the approach to these patients. The most important problem of present classifications is the conceptual overlap between major depression and dysthymia; indeed, practically all items conforming the diagnostic criteria for dysthymia are also part of the criteria for major depression.

1.2.3 Epidemiology

The past two decades have shown depression to be one of the main causes of disability in the world today. According to WHO data, depression is the fourth cause of disability and death in the world, only surpassed by respiratory tract diseases, partum/puerperium-related diseases and diarrhoea in children. According to these data, unipolar depression will be the world's second cause of disability in the year 2020, only surpassed by cardiovascular diseases. In addition, depressive patients have higher mortality rates than the general population, even if suicide-related mortality is excluded (Vázquez-Barquero, 1999). This is probably due to the negative influence of depression on the evolution of physical conditions. WHO data also confirm that depression is not a disease exclusive to developed societies, although these allow faster diagnose and treatment.

Epidemiological studies have provided essential information on the prevalence, comorbidity, distribution, risk factors and course of mental disorders, as well as data on quality of life, social cost, demand and use of related healthcare services. Depressive disorders in the community reach a steady annual prevalence of around 5%, while lifetime prevalence is approximately 15% (Murphy et al., 2000). Incidence rates range between 500 and 1,000 new depressive episodes for each 100,000 inhabitants and year (Vázquez-Barquero, 1999).

1.2.3.1 Epidemiology of major depressive disorder

The prevalence of unipolar major depression ranges over 5–9% in women and 2–4% in men (Boyd and Weissman, 1982; Vázquez-Barquero, 1999). Most recent studies using diagnostic criteria and structured interviews like Diagnostic Interview Schedule (DIS) or Composite International Diagnostic Interview (CIDI) have shown a prevalence of major depressive episode ranging between 0.6% in Taiwan and 10.3% in the USA. Lifetime prevalence rates range between 0.9% in Taiwan and 17.1% in the USA, and between 4% and 12% in most studies. A recent study conducted by the Depression Research in European Society (DEPRES), using the Mini International Neuropsychiatric Interview (MINI) in 78,463 subjects, found a 6-month prevalence of 6.9%, ranging between 3.8% in Germany and 9.9% in the United Kingdom (Lépine et al., 1997).

A field of great interest in the epidemiology of depression is that of the sociodemographic factors related to affective disorders. The most relevant ones associated to major depression are the following (Smith and Weissman, 1992):

- *Sex*. The prevalence of major depressive disorder is two-fold in women between 12 and 55 years of age (sex-related prevalence is similar at younger and older ages). The age of onset of the first depressive episode is similar in both sexes, with a higher risk of chronicity also in women (Kessler et al., 1994).

- *Age*. The risk of suffering a major depressive episode increases with age, with a maximum peak between 45 and 64 years in women and between 35 and 64 years in men.

- *Marital status*. Its influence in the rate of depressive disorders is clear and seems to be sex-related. Single women have lower rates than married ones, while single men have higher rates than married ones.

- *Socioeconomic status*. A reverse relationship has been described between this factor and the prevalence of depression.

- *Social stress*. A direct relationship has been observed between severe loss events and depression (Brown, 1994). It is commonly said that, although less that 10% of the subjects experiencing a relevant vital event will further develop depression, the risk of suffering it is 6 times higher if stressing events have been endured.

- *Social support*. Social support seems to have a protective effect at the onset of depression, and may modify its course and influence its prognosis. The presence of a social support network makes the development of psychopathologies less likely and will improve their course if these conditions appear.

1.2.3.2 Epidemiology of dysthymia

Dysthymia has a lifetime prevalence of around 3% (Horwarth and Weissman, 1995). Importantly, there is a possibility that the dysthymic patient presents with a major depressive episode comorbidity, or overlapping his/her chronic disorder. Hence, 40% of patients with major depression are estimated to also meet the criteria for dysthymic disorder: the so-called "double depression" (Kaplan and Saddock, 1999).

Regarding dysthymia-related sociodemographic factors, sex has a relationship similar to that of major depression (approximately two women for every man). Although of early onset (frequently before age 21), the prevalence of dysthymia tends to increase with age (Horwarth and Weissman, 1995). Regarding *marital status*, married people have lower dysthymia prevalence rates as compared to single, divorced or widowed people.

1.2.4 Physiopathology

1.2.4.1 Biochemistry

A great part of the research in the field of depression has focused on the search of biological alterations. One objective is to seek physiopathological explanations to help understand the underlying biological mechanisms likely to be present in depression, with further therapeutic application.

In the 1960s and 1970s, research on the biology of depression rested on the knowledge of the mechanisms of action of the antidepressants known at that time. In 1965, Schildkraut formulated the *catecholaminergic hypothesis*, which suggested that depression was caused by a lack of noradrenaline (NA) in certain brain circuits. Later on, in 1972, Alec Coppen et al. emphasized the etiological role of serotonin deficit (5-HT) in their *indoleamine hypothesis* of depression. The *monoaminergic hypothesis of depression* was thus established, where an essential factor in antidepressant treatment was considered to be the increase of monoamines into the synaptic cleft.

According to the present state of knowledge, there is considerable evidence involving the serotoninergic system in the physiopathology of depression. Most of the presently used antidepressants act on the serotoninergic system: tricyclic antidepressants (TCA) and selective serotonin reuptake inhibitors (SSRI) reduce serotonin reuptake (5-HT), thereby increasing its levels at the synapse, while monoamine oxidase inhibitors (MAOI) act by inhibiting the metabolism of serotonin and other neurotransmitters, causing the same effect at a synaptic level (Menchón and Urretavizcaya, 1998; Gastó, 1999).

With time, several hypotheses have been raised regarding the biological basis of depression (see Table 1). The majority of biological data have failed to be integrated in a general theory and, importantly, the exact physiopathological mechanism of depression is still unknown. Despite its limitations, the classic monoaminergic hypothesis remains valid and continues to be investigated (Vallejo, 1999).

Table 1. Hypotheses on the biological bases for depression and on the mechanism of action of antidepressants

Year	Hypothesis on depression	Antidepressant mechanisms of action
1960	Monoamine deficiency	Inhibition of monoamine reuptake
1970	Receptor supersensitivity	Subregulation
	Postsynaptic receptor supersensitivity	Subsensitization of β receptor stimulating the production of cAMP
1980	Second messenger dysregulation	Effect of AD on second messengers
1990	Alteration in G protein function	Action of AD on G proteins

cAMP, cyclic adenosinmonophosphate; AD, antidepressants (source: Gastó and Vieta, 2000)

The aminergic neurotransmitter systems share a number of characteristics (Artigas, 1999). Because these systems are formed by long-axon, extensively branched neurons, very few neurons innervate vast cerebral territories. The cell bodies of these neurons are primarily found in the midbrain and brainstem areas. The noradrenergic neurons of the locus coeruleus and the serotoninergic neurons of the nucleus raphe dorsalis supply the noradrenergic and serotoninergic innervation of most part of the encephalon. Serotoninergic and noradrenergic neurons are also part of the brainstem activating reticular system, involved in the regulation of sleep-awake cycles. Both neurotransmitter systems have a wide diversity of membrane receptors coupled to different intracellular signal-transducing systems, which allows them to produce highly diverse biological responses. They may act as *neurotransmitters* (by increasing conductance to given ions at the postsynaptic neuron) or as *neuromodulators* (by indirect influence on the postsynaptic response to neurotransmitters such as GABA or glutamate, by means of processes such as intracellular calcium mobilization or protein phosphorylation). Also important is the existence of presynaptic receptors, called autoreceptors, which can be activated by the neurotransmitter synthesizing and releasing the neuron. Autoreceptors are part of the neurotransmitter release regulating system and are responsible for self-inhibition mechanisms: when activated, they usually regulate by reducing electric and metabolic neuronal activity, as well as neurotransmitter synthesis and release rate. Autoreceptors may be found at somatodendritic or terminal level. At the terminal level, they may be sensitive to neurotransmitters other than those synthesized by the neuron, and are then called heteroreceptors.

The *serotoninergic system* is involved in a great variety of functions, such as responses to anxiety, mood and perception regulation, motor activity regulation, sleep, appetite, aggressive behavior, sex drive (inhibited with increased serotoninergic tone), circadian rhythm, hormone secretion (by stimulating the production of growth hormone, prolactin, cortisol, corticotropin-releasing factor), and thermoregulation. Descending fibres modulate the sensitivity to painful stimuli and sexual function. Several subtypes of serotoninergic receptors have been identified. $5-HT_{1A}$ and $5-HT_{1B}$ receptors act as autoreceptors of nerve cell bodies and terminals, respectively. When occupied by 5-HT or specific agonists, serotoninergic neurons undergo a decreased release of 5-HT. The remaining serotoninergic receptors are basically located postsynaptically (Montes and Gurpegui, 2000).

The *noradrenergic system* regulates mechanisms of attention, learning, response, behavior maintenance, affective states and anxiety. Neocortex, hippocampus and amygdala (noradrenergic innervation areas) are activated in stress situations, thereby increasing NA release and thus facilitating attention and alertness. Because these central mechanisms and the peripheral ones are coordinated, sympathetic nervous system activation and catecholamine release by suprarenal glands also prepare the body for these situations. There are two types of adrenergic receptors, α and β, all belonging to the family of G protein-coupled receptors. These receptors have been further subdivided in up to ten different subtypes: three of the β type (β_1, β_2, β_3), four of the α_1 type (α_{1A}, α_{1B}, α_{1C}, α_{1D}) and three of the α_2 type (α_{2A}, α_{2B}, α_{2C}). α_2 receptors act as autoreceptors and are located in the somatodendritic region and at the synaptic terminals of noradrenergic neurons (Álamo et al., 1999). The occupation of these autoreceptors leads to inhibit the release of the neurotransmitter excreted by the neurons where they are located.

Importantly, the noradrenergic and serotoninergic systems interact at different levels and function coordinately. Specifically, the serotoninergic terminals in several areas of the brain (hippocampus, frontal cortex) do have α_2 adrenergic heteroreceptors that diminish serotonin release (Artigas, 1999).

Apart from the noradrenergic and indolaminergic hypotheses of depression, Janowsky et al. suggested in 1972 that there might be a central cholinergic-adrenergic imbalance favoring acetylcholine in depressed patients, and the contrary in manic patients, thereby postulating the cholinergic hypersensitivity hypothesis of depression. In contrast, Randrup (Randrup and Braestrup, 1977) suggested that dopamine also had a role in affective disorders by defining a central dopamine decrease in depression.

In short, it is accepted that depression involves an alteration of the catecholaminergic systems that may include both serotonin and noradrenaline, as well as dopamine and acetylcholine, and that all these systems are interrelated. This proves the difficulty of establishing a simplistic hypothesis on the etiology of depressive disorders.

1.2.4.2 Neuroendocrinology

While it is accepted, as commented above, that depression involves alterations of the noradrenergic and serotoninergic systems and, to a lesser extent, of the dopaminergic and cholinergic systems, it should be emphasized that many studies postulate the existence of imbalances at other levels, which might account for the lack of response observed in some patients to therapies with drugs acting solely at a monoaminergic level.

The hypothalamus, hormone secretion regulator par excellence, synthesizes and releases peptides which act on the hypophysis and stimulate or inhibit the release of certain hormones into the blood. These substances, called neuropeptides, seem to play an important neurotransmitting/neuromodulating role in the CNS (Gibert et al., 1999). The development of drugs likely to affect these systems might be useful for treating disorders like depression.

It has also been shown that a great number of depressive patients show hyperactivity of certain hormone systems. In the context of affective disorders, the most investigated hormone systems have been the connections between hypothalamus and hypophysis with the thyroidal, adrenocortical and somatotropin-release systems. The evidence that patients with alterations in these hormone systems (hypo- and hyperthyroidism, acromegalia, Cushing syndrome) frequently suffer from affective disorders has led to attempt to relate alterations of these systems with hypothetic causes of affective disorders.

The most significant alterations found in these systems in depressed patients are briefly described below.

Hypothalamo-pituitary-adrenocortical axis

Many studies in depressed patients have shown a high level of glucocorticoids in plasma, urine and cerebrospinal fluid (CSF), as well as abnormalities in the circadian secretion of cortisol. High concentrations of corticotrophin-releasing factor (CRF) have been found in the CSF of depressed patients, and also a decreased sensitivity of CSF in the frontal cortex of suicidals. This suggests an excessive activity of CRF-secreting neurons accounting for the hyperstimulation of this axis.

The biological marker most frequently used in depressed patients to identify an alteration in corticotrope axis regulation is the lack of suppression of cortisol secretion after a single dose of exogenous dexamethasone (abnormality of the dexamethasone suppression test, or DST). Although a specificity of 95% in melancholic depressive patients was described in the earlier tests performed in the eighties, their actual sensitivity is 45%. Also importantly, DST alterations have been found in other psychic disorders (schizophrenia, alcoholism, obsessive–compulsive disorder, etc.) as well as in various situations (weight loss, lack of sleep and hospitalization stress, among others). DST seems to go back to normal with clinical improvement and be altered again in case of relapse.

Hypothalamo–pituitary–thyroid axis

It is generally stated that 20% of depressed subjects have some kind of thyroidal impairment. Most of them show a flat TSH response to TRH stimulation. One hypothesis to explain this response is chronic hypersecretion of TRH in depressed patients. This finding is barely specific, given that a similar response is found in other mental disorders such as anxiety disorders, schizophrenia, obsessive-compulsive disorder and alcoholism, and also according to certain variables such as sex, age and the presence of renal or hepatic disorders.

There are studies which suggest effects of thyroid hormones on brain neurotransmission, even postulating a direct neurotransmitting effect of thyroid hormones and TRH.

Hypothalamo–pituitary–somatotropin axis

Several studies have found a flat GH response to clonidine (α-2-agonist) in endogenous depressive patients versus non-endogenous and controls. Other studies found a frequent correlation of this test with a flat GH response to the stimulation with growth regulating factor (GRF).

1.2.5 Treatment of affective disorders

1.2.5.1 Mechanism of action of antidepressant drugs

Noradrenergic and serotoninergic neurons constitute the therapeutic target of almost 99% of all antidepressant treatments. These act on the control mechanisms of the extracellular synaptic concentration of 5-HT and NA. The blockage of 5-HT and/or NA neuronal transportation by means of reuptake inhibitors (tricyclic antidepressants, selective serotonin and/or noradrenalin reuptake inhibitors) increases the synaptic concentration of amines, which activate the postsynaptic receptors.

In addition, continuous treatment with antidepressant drugs leads to a great number of changes in the density and sensitivity of the aminergic receptors. However, because none of these changes is common to all of them, they cannot be associated to the clinical effect.

While these neurotransmitter systems are the first step in the mechanism of action of antidepressants in the different mental diseases, most of the present data are oriented towards intracellular signal transduction processes as the final target of antidepressants. Some studies have observed that antidepressant treatments induce cAMP and protein kinase A (PKA) activation. In turn, this increase translates into a greater expression of transcription factor CREB (cyclic AMP response element binding protein), which is involved in the regulation of the expression of several genes, particularly BDNF (brain-derived neurotrophic factor). The role of this neurotrophic factor in neuronal survival has been widely demonstrated. Several studies have provided evidence that stress decreases BDNF expression in the hippocampus and the cerebral cortex, while antidepressant treatment produces the opposite effect (Duman et al., 1997).

Two issues on the mechanism of action of antidepressant drugs are still unresolved today and constitute one of the major challenges in the treatment of depression. One is the gap between their pharmacological action and their clinical action, which is found in all existing antidepressant types. The other is the reason for their limited response.

5-HT and NA are still key elements in the treatment of depression. Consequently, until the genes involved in clinical improvement are known and until molecules able to selectively modulate their expression are discovered, the pharmacological treatment of depression will rest on the regulation of aminergic neural activity.

1.2.5.2 General aspects of pharmacological treatment

1.2.5.2.1 Basic aspects

According to Preskorn (1994), and as in any other field of medicine, the appropriate treatment of depression involves four aspects.

A. Diagnosis

- The patient's disease may benefit from a pharmacological treatment.
- The disease lasts longer without pharmacological treatment.
- The disease may worsen if no treatment is given.
- The lack of treatment may lead to chronicity problems or to the patient's death.

B. Drug
- The selection of the best drug is based on safety, efficacy and easy-to-use criteria.

C. Dose
- First-choice dosage is that warranting best performance and less untoward effects.

D. Duration
- Each condition requires treatment for a given time. Some diseases, and specifically depression, require prolonged treatment that may vary according to the stage of the condition (Álvarez, 1998).

1.2.5.2.2 Appropriate information to the patient regarding treatment

It is important to remember that the first cause of the lack of response to antidepressant treatment is patient non-compliance. Therefore, in order to fully warrant treatment compliance, it is necessary to identify the reasons for non-compliance by the patient, including:

- Duration and complexity of treatment.
- Tolerance to treatment
- Lack of "faith" in the physician
- Doubts on medication efficacy
- Perceived control of disease
- Severity of disease
- Lack of follow-up and support
- Lack of social support
- Low education
- Concomitant organic alterations
- Toxic substance abuse

Compliance may be improved if, at the beginning of treatment, the patient and his/her family are informed in a clear, understandable manner about the following aspects: dosage, time of onset of therapeutic activity (with particular insistence on the latency period, explained later on), need to continue treatment after improvement, comments on possible side-effects without dramatizing, etc.

1.2.5.2.3 Treatment efficacy

It is important to remember that the onset of therapeutic improvement does not usually occur before the third week after the establishment of the complete dose, and that the remission of the majority of symptoms is usually not achieved before the sixth or eighth week. The first symptoms that improve are nocturnal rest and anorexia, while sadness is one of the last.

While there is increasing consensus on the therapeutic efficacy, the concepts of therapeutic response and therapeutic remission need to be ascertained. In the case of depression, most authors define *response* as a 50% reduction of the symptomatology presented by the patient at the beginning of treatment. *Remission* refers to the reduction of the symptomatology below the threshold by which most psychiatrists would consider that there is no disease (final score in Hamilton's scale below 7) (Hamilton, 1967; Hendlund and Vieweg, 1979). Depressed patients need complete treatments guaranteeing the remission of the disease and not only a good therapeutic response. Indeed, residual symptoms, as well as relapses and recurrences, may appear. The treatment of depression calls for a drug with maximum potential for remission induction.

A patient is considered refractory when no response has been obtained eight weeks after the establishment of the complete dose. The strategy that most suits the characteristics of the patient should then be indicated (see later on). The therapeutic efficacy of antidepressants in monotherapy is approximately 65%, and therefore one in every three patients receiving correct treatment is resistant to the initial antidepressant therapy. Treatment efficacy can be increased up to 90% by means of the different potentiation strategies.

1.2.5.2.4 Maintenance therapy

The treatment of depression needs to be continued upon the remission of the initial symptoms, in order to consolidate recovery and help prevent possible relapses. The clinical practice shows that many patients do not undergo this phase of treatment. It is therefore necessary to remind the patient, as early as during the first visit, of the need and importance of completing the whole treatment. An early withdrawal or even a dose reduction significantly increase the risk of relapse.

The general recommendations for maintenance therapy (Servei Català de la Salut, 2000) include:

- *Single moderate depressive episode:* maintain treatment with the antidepressant at the effective dose from 6 to 12 months after the remission of the acute episode.

- *Single severe episode or 2 episodes in less than one year:* maintain treatment between 2 and 3 years (as indicated by the specialist physician).

- *Three or more episodes in less than 5 years:* maintain treatment at least 5 years (as indicated by the specialist physician). In these cases, and given a higher-than-90% risk of suffering a new episode, indefinite treatment according to episode severity, patient age and family history could be licitly considered.

1.2.5.2.5 Treatment withdrawal

In order to prevent possible effects related to abrupt treatment discontinuation, a gradual dose reduction is recommended. As shown in several studies, abrupt discontinuation of TCA and MAOI administration may cause withdrawal symptoms in 20–80% of patients, particularly: physical discomfort, anxiety or agitation, vivid dreams or insomnia, motor restlessness, tremors and behavioral alterations. The probabilities for the onset of withdrawal symptoms can be significantly reduced by suitable preparation of the patient and gradual dose reduction.

A discontinuation syndrome has also been described with SSRI (Paddin et al., 1997) and venlafaxine (Fava et al., 1997). The incidence of this syndrome ranges between 0.06–0.9%, and paroxetine is so far the most involved SSRI, with an incidence of up to 5%. This syndrome may develop with balance impairment (vertigo is apparently the most frequent symptom), gastrointestinal symptoms (nausea, vomiting, diarrhoea), pseudoinfluenza symptoms (fatigue, lethargy, myalgia, rhinorrhea, headache), sleep disturbances (somnolence, insomnia, nightmares) and sensorial impairment (paresthesia) (Pacheco et al., 1998). In general, symptoms are mild and transient (1–2 weeks), and their incidence can be minimized by gradual dose reduction.

1.2.5.2.6 Overview of antidepressant drugs

Antidepressants cover a wide range of indications, including *depression, anxiety, obsessive–compulsive disorders, eating disorders, panic* or *social phobia.*

These drugs are characterized in that they increase synaptic activity by blocking two types of structures: serotonin transporters (SERT), noradrenaline transporters (NAT) and several receptors. Unnecessary blockade of a given receptor, e.g. dopaminergic, adrenergic or cholinergic receptors leads to side-effects without increasing efficacy. Selective drugs are those where unnecessary receptor blockade is scarce (Gibert et al., 1999).

A great number of drugs with proven antidepressant activity are currently available in the Spanish market. An overview of these drugs and recommended doses, classified according to their mechanism of action, is shown in Table 2. The important differences between the different antidepressants must be taken into account for prescribing purposes.

1.2.5.2.7 Antidepressant drug selection

While the majority of studies suggest a similar efficacy of the various antidepressants, the differences in the pharmacological characteristics may favor a given drug. Choice will also depend on factors relating to the patient (age, other diseases, drugs taken, history of good personal or family response to a given antidepressant) or the physician him/herself (information or experience available with the different drugs). The criteria for antidepressant selection are summarized in Table 3. The ideal drug is the safest, the most effective and the easiest to use.

Table 2. Antidepressant drugs available in the Spanish market, classified according to their mechanism of action

Mechanism of action	Drug	Selective	Dose (mg/day)
SERT and NAT blockade	Imipramine	No	100–300
	Amitriptyline	No	100–200
	Clomipramine	No	5–125
	Venlafaxine	Yes	75–300
	Duloxetine	Yes	60–120
NAT blockade	Nortryptiline	No	100–175
	Maprotiline	No	75–150
	Lofepramine	No	70–140
	Reboxetine	Yes	4–12
SERT blockade	Fluvoxamine	Yes	100–200
	Fluoxetine	Yes	20–40
	Paroxetine	Yes	20–40
	Sertraline	Yes	50–200
	Citalopram	Yes	20–60
α_2 adrenoreceptor blockade	Mianserin	No	60–120
	Mirtazapine	Yes	30–60
MAO inhibition	Phenelzine	No	30–60
	Tranylcypromine	No	10–40
MAO-A reversible inhibition	Moclobemide	Yes	450–900

5-HT, serotonin; MAO, monoaminooxidase; NAT, noradrenaline transporter; SERT, serotonin transporter

1.2.5.2.8 Side-effects of antidepressant drugs

A key point when choosing an antidepressant treatment is its side-effect profile. This profile is shaped by the action on the different neurotransmitter systems. The affinity of various antidepressants for the different receptors is shown in Table 4. The side-effects of antidepressants according to their pharmacodynamic action are summarized in Table 5, and possible therapeutic approaches are also suggested.

Table 3. Criteria for antidepressant drug selection (Pérez-Solà et al., 2000)

Safety
- Toxicity
- Tolerance
- Pharmacokinetic interactions
- Pharmacodynamic interactions
- Previous experience

Efficacy
- Probability of response
- Spectrum of activity
- Maintenance of response
- Preventive efficacy

Simplicity
- Easy administration
- Easy monitorization

Cost-effectiveness

Table 4. Affinity of various antidepressants for different receptors (Ayuso-Mateos and Ayuso-Gutiérrez, 1998)

	Muscarinic	Histaminic	Adrenergic	Serotoninergic
Amitriptyline	+++	+++	+++	+
Imipramine	++	++	+++	++
Clomipramine	+++	+	+++	++++
Nortryptiline	++	++	++	0
Dotiepine	+	++++	++	0
Citalopram	0	0	0	+++
Fluoxetine	0	0	0	+++
Fluvoxamine	0	0	0	+++
Sertraline	0	0	0	+++
Paroxetine	+	0	0	+++
Trazodone	0	+	+++	++
Mianserin	0	+++	+++	0
Moclobemide	0	0	++	++
Phenelzine	+	0	+++	++

Emphasis will be placed on the most characteristic side-effects when the different pharmacological groups are described.

Table 5. Side-effects of antidepressants according to their pharmacodynamic action
(Pérez-Solá et al., 2000)

Pharmacological action	Adverse event and therapeutic approach (between brackets)
M_1 acetylcholine receptors (atropinic effects)	Dry mucosa (chewing-gum, hydro-buccal), constipation with risk of fecalomas, even intestinal obstruction (fibre-based diet: fibre tables 4 t/meal with 2 glasses of water), sinus tachycardia (propranolol, 15-30 mg/day), blurred sight (correction lenses), urinary retention (risk of acute urine retention, betanecol), delayed ejaculation, cognitive impairment (reduce dose or switch to SSRI), decreased convulsive threshold (reduce dose), precipitation of closed-angle glaucoma (intraocular pressure control and, if required, switch to SSRI)
Antihistaminic (H1)	Sedation (caffeine), weight gain (poor result of hypocaloric diets, walking at least 1 hour a day is recommended)
α_1 -adrenergic receptor blockade	Orthostatic hypotension: risk of falls and fractures, stroke (stand up from bed or chair very slowly and always leaning; when crouching, bend legs instead of trunk; increase sodium salt in diet), dizziness, cognitive impairment (reduce dose, switch to SSRI)
NAT blockade (increased NA activity)	Anxiety, hypertension, tachycardia, dry mouth, constipation, decreased convulsive threshold (reduce dose and add β-blockers such as propranolol at doses of 20-40 mg/day)
SERT blockade (increased 5-HT activity)	Restlessness, insomnia, akathisia (reduce dose or more gradual start; indicates intolerance if maintained), nausea (metoclopramide, clebopride), anorexia and weight loss, decreased sex drive (reduce dose, sildenafil), diarrhoea (astringent diet), headache (paracetamol).

NAT, noradrenaline transporter; SERT, serotonin transporter

1.2.5.2.9 Drug interactions

Many patients eligible for antidepressant therapy are already taking drugs concomitantly for other diseases. Importantly, the antidepressant may lead to pharmacological interactions with these drugs. For instance, it may compete for protein binding and significantly increase the plasma levels of any of the other drugs.

The most important pharmacokinetic factor to be considered is that most antidepressant drugs may be substrates and/or inhibitors of some of the isoenzymes of cytochrome P450 (Ereshefsky et al., 1996; Cuenca, 1998). This is particularly important because these enzymes are known to be responsible for the hepatic metabolism of approximately 90% of the drugs used in the clinical field.

Whether the antidepressant to be prescribed has or has not an important inhibitory activity on any of these isoenzymes must be ascertained. If it does have inhibitory activity, the metabolization of other drugs requiring this isoenzyme may decrease, thereby increasing their plasma levels and even their toxic effects. In practice, however, many theoretical interactions only affect patients at risk, like those suffering from severe cardiovascular, metabolic, hepatic or renal diseases (interactions with anticoagulants, digitalis, antiarhythmics, etc.).

Table 6. Antidepressant risk potential (*in vitro* and *in vivo*)

Relative category	CYP1A2	CYP2C	CYP2D6	CYP 3A3/4
High	Fluvoxamine	Fluvoxamine Fluoxetine	Paroxetine Fluoxetine	Fluvoxamine Nefazodone Tertiary TCA
Moderate to minimal	Tertiary TCA Fluoxetine	Tertiary TCA	Secondary TCA Sertraline	Sertraline Fluoxetine
Low to minimal	Paroxetine		Nefazodone Fluvoxamine Venlafaxine (*in vitro*)	Venlafaxine (*in vitro*)

Table 7. Examples of drugs likely to interact with an antidepressant (Kaplan and Sadock, 1999)

CYP1A2	CYP2C	CYP2D6	CYP 3A3/4
Theophylline	Mephenytoin	Desipramine, secondary TCA	Terfenadine
Imipramine (minor)	Diazepam	Flecainide/encainide	Astemizole
Caffeine	Hexobarbital	Risperidone	Ketoconazole
Phenacetin	Imipramine	Phenothiazines	Alprazolam
Acetaminofen	Phenacetin	Haloperidol (minor)	Triazolam
Warfarin (minor)	Warfarin	Codeine	Erythromycin
Phenothiazines	Propranolol	Propranolol (minor)	Nifedipine
	Tertiary TCA	Quinidine[a]	Cyclosporine
			Corticosteroids

[a] Inhibits at 2D6, is not a substrate.

The degree of inhibition, and therefore the risk potential of the main antidepressants on the different isoenzymes of cytochrome P450 (Nemeroff et al., 1996), are shown in Table 6. Examples of drugs likely to interact with an antidepressant, as a result of being substrates and/or inhibitors of the same enzyme system, are shown in Table 7. For example, fluvoxamine is a powerful inhibitor of CYP1A2 and is involved in the development of interactions with drugs such as theophylline and phenothiazines (Cuenca, 1990).

1.2.5.3 Antidepressant drugs

1.2.5.3.1 Tricyclic and tetracyclic antidepressants

The tricyclic and tetracyclic antidepressant group is usually abbreviated as TCA. This group shares many pharmacodynamic and pharmacokinetic properties, as well as a similar side-effect profile. TCAs have a long, proven history of efficacy in the treatment of depression. They have been extensively studied both in controlled and follow-up studies in large series of patients, thus providing thorough experience on their use. As with the rest of antidepressant drugs, their efficacy has been shown to range between 60% and 80% in correctly treated depressed patients (suitable doses and treatment duration), with a latent period (response time) ranging between 2 and 6 weeks.

All tricyclic drugs have a similar structure and include three intertwined rings to which a side-chain is bound. The binding of a fourth ring conforms a tetracyclic drug. Tertiary amines (two methyl groups in the nitrogen atom of the side-chain) include: imipramine, amitriptyline, clomipramine, trimipramine and *doxepine*. Secondary amines (with only one methyl group in that position) are *desipramine, nortryptiline* and *protryptiline*, while tetracyclic drugs include *amoxapine, maprotiline* and *mianserin*.

Mechanism of action

These drugs are known for their central biogenic amine potentiation capacity, on account of NA, 5-HT and DA neuronal reuptake blockade. In general, tertiary amines are more powerful in blocking 5-HT reuptake, while secondary amines are more powerful in blocking NA reuptake. This explanation, however, cannot suffice: while amine transportation blockade occurs as treatment is started, symptom remission is not apparent until 10–15 days after (Pérez-Solá et al., 2000).

Lately, research has focused on adaptive phenomena occurring after continuous antidepressant administration:

hyposensitization of β postsynaptic and $α_2$ presynaptic receptors, hypersensitivity of $α_1$ presynaptic receptors and possible subsensitivity of serotoninergic receptors.

Pharmacokinetics

TCAs are well absorbed by oral route and metabolized by hepatic microsomal enzymes. They are strongly bound to plasma proteins (80–95%) and are highly liposoluble. The metabolism of these drugs shows a high interindividual variability, depending on genetic factors, age and administration of other drugs. The half-life of TCAs and their respective doses are listed in Table 8 below.

Table 8. Pharmacokinetic characteristics of TCAs (Vázquez, 1997)

	Half-life ($t½$; h)	Starting dose (mg/day)	Mean maintenance dose (mg/day)	Dose range (mg/day)
Imipramine	11–25	25–50	75–150	50–300
Amitriptyline	16–26	50–75	100–150	50–300
Clorimipramine	17–28	25–50	75–150	50–150
Doxepine	11–23	25–50	75–150	50–300
Desipramine	17–27			50–300
Nortryptiline	18–44	30–50	75–150	30–100
Protryptiline	67–89			50–60
Amoxapine	8–30	50–100	100–300	50–600
Maprotiline		25–50	75–150	50–250
Mianserin		20–30	30–90	30–200
Amineptine		25–50	100–200	25–300
Viloxacine	2–5	50–100	150–300	150–300

Precautions

Not recommended during pregnancy and lactation. Recent acute myocardial infarction is an absolute contraindication. Relative contraindications include: prostatic hypertrophy, glaucoma, epilepsy, untreated hypertension, alcoholism, arrhythmias and other heart diseases, certain liver and kidney diseases, severe confusional states, hyperthyroidism, history of AMI (Carrasco, 1996).

Indications

- Major depressive disorders.
- Depression associated with medical condition.
- Anxiety disorder with agoraphobia: imipramine is the most studied one, usually administering the normal antidepressant dose.
- Obsessive-compulsive disorder: clomipramine seems to be the most effective drug.
- Eating disorders: imipramine and desipramine have shown to be effective both in nervous anorexia and nervous bulimia studies.
- Pain disorder: amitriptyline and others are frequently used for chronic pain treatment, including stress headache (at lower doses than those used for antidepressant treatment).
- Other disorders: narcolepsy and posttraumatic stress disorder; sometimes used in attention deficit hyperactivity disorder and enuresis in children (imipramine).

Prescription

Because effective TCA doses are usually accompanied by side-effects, treatment must be started at low doses and further modified according to the observed clinical response and toxicity, until the appropriate dose is reached. Some authors recommend follow-up by means of periodic controls of the blood drug concentration in non-responding patients, in case of toxicity or when pharmacological interactions are suspected. The appropriate dose is reached when therapeutic response is obtained without severe adverse events (López-Ibor, 1999). In the elderly, dose increase must be performed even slower, and therapeutic responses with lower final doses are frequently obtained.

Adverse events

TCA-related adverse events are usually of early onset and little importance. However, even the mildest ones may endanger patient compliance, particularly if they appear before clinical response is experienced. Between 5% and 40% of patients taking TCAs report side-effects, which sometimes are severe enough to lead to dose variation or drug withdrawal (5–10%).

- *Anticholinergic effects:* dry mouth or mucosa, sweating, constipation, dizziness, blurred sight, tachycardia, mydriasis, glaucoma, impaired ejaculation.
- *Cardiovascular effects:* postural hypotension, tachycardia, ECG alterations, ventricular arrhythmias and risk of sudden death.
- *Central effects:* sedation, agitation, insomnia, confusion, extrapyramidal effects, decreased convulsive threshold, sexual dysfunctions, precipitation of manic episodes, headache.
- *Neuroendocrine effects:* bodyweight increase, decreased libido, amenorrhea, galactorrhea.
- *Allergic or hypersensitivity reactions:* exanthema, photosensitization, anemia, agranulocytosis, leukopenia, eosinophilia and leukocytosis.

Intoxication

TCA-related overdose is a severe, life-threatening condition on account of cardiac arrhythmias, severe hypotension or convulsive picture. Symptoms appear within 24 hours and include agitation, delirium, convulsions, exacerbated tendinous reflexes, intestinal and urinary paralysis, blood pressure and temperature dysregulation, and mydriasis. The patient slips into a coma and may present with respiratory depression. Because the patient is at risk of cardiac arrhythmias during the following 3 or 4 days, he/she must be monitored in an intensive care unit.

1.2.5.3.2 Monoamine oxidase inhibitors

While the monoamine oxidase inhibitors (MAOI) group of drugs is presently not considered of first choice, all physicians ought to know the dietetic rules and the precautions to be considered when prescribing any other drug to MAOI-treated patients.

According to the inhibitory action on one of the two isoenzymes of MAO (A and B) and to the reversibility of their action, MAOIs are classified as:

- Non-selective and irreversible (classic MAOIs): *phenelzine, tranylcypromine.*
- MAO-A selective and reversible: *moclobemide.*

- MAO-B irreversible: *selegiline* and others. Because dopamine is the preferred substrate of this isoenzyme, it is used in the treatment of Parkinson's disease.

Mechanism of action

Their basic action consists in the inhibition of the MAO enzyme, involved in monoamine breakdown, thereby leading to an increased number of neurotransmitters (DA, NA, 5-HT) at the synaptic space. Some adaptation mechanisms also take place further on, like β postsynaptic and $α_2$ presynaptic receptor hyposensitization (Pérez and Álvarez, 2000).

Pharmacokinetics

Good absorption by oral route. Almost 50% binding to plasma proteins. Metabolization takes place in the liver. Importantly, the time required for enzyme resynthesis in the case of irreversible inhibitors is 1–2 weeks, and therefore the action of the drug is maintained even after its discontinuation.

Precautions

MAOI should not be administered together with other antidepressants. In the case of TCA, allow at least 10 days after treatment suspension, and up to 5 weeks in the case of fluoxetine.

MAOIs are contraindicated in liver disease, kidney disease, severe cardiovascular disease, asthma, pheochromocytoma, alcoholism and in patients unable to comply with all necessary dietetic restrictions. Relative contraindications include HBP, epilepsy and Parkinson's disease.

Interactions

MAOIs potentiate the toxicity of levodopa, opiates and metoprolol. Tranylcypromine and phenelzine potentiate the effect of sympathicomimetics, amphetamines and oral antidiabetics.

Indications

Similar to those of TCAs. MAOIs are considered to be particularly effective in atypical depression, characterized by hyperphagia, hypersomnia and anxiety. The most frequent reason for MAOI indication is therapeutic failure secondary to SSRI or TCA treatment. MAOIs are also effective in anxiety disorders with agoraphobia, posttraumatic stress disorder, eating disorders, social phobia and pain disorders.

Adverse effects

MAOIs have few anticholinergic effects, and in this sense are better tolerated than TCAs. They also produce orthostatic hypotension, digestive alterations (nausea, vomiting, constipation), edemas and, less frequently, anxiety, sleep disturbances, increased transaminases. They may cause severe *hypertensive crises* due to excess of amines not broken down by MAO if dietetic rules (aimed at eliminating tyramine- or dopamine-rich foods) or pharmacological restrictions (prohibition of drugs with sympathomimetic activity) are not complied with (Table 9).

Table 9. Instructions for treatment with MAOIs (modified from: Ayuso-Mateos and Ayuso-Gutiérrez, 1998)

- *Restriction of tyramine-containing foods:* alcoholic beverages (particularly red wine), fermented cheese and sausage, canned fish or meat, smoked and salt-cured foods, pâté, chicken liver, mushroom, snails, broad beans, bananas, yogurt, chocolate, giblets, caviar, figs, avocados, coffee, tea.
- *Relative pharmacological restrictions:* TCAs, oral antidiabetics and anticoagulants, insulin, sulfamides, clonidine, reserpine, guanetidine, β-blockers, anesthetics.
- *Absolute pharmacological restrictions:* SSRIs, ephedrine, caffeine, methylphenidate, levodopa, amphetamines and anorexigens.
- *Also important is to consider the existence of common OTC drug products likely to interact with MAOI:* antiinfluenza drugs, nasal decongestives, antipyretics and local anesthetics.

Intoxication

Intoxication develops with an asymptomatic period of 1–6 hours, after which hyperthermia, delirium, mydriasis and osteotendinous hyperreflexia develop. This may evolve to metabolic acidosis, convulsive crisis and cardiovascular collapse.

1.2.5.3.3 Selective serotonin reuptake inhibitors

Because the selective activity of selective serotonin reuptake inhibitors (SSRIs) leads to less side-effects than TCAs and offers a more favorable profile, the great contribution of this drug group is tolerance. This allows treating patients with organic diseases, where tricyclics have to be used cautiously or avoided. The toxicity of SSRIs is also much lower in case of overdose – a key circumstance in depressive patients.

Main SSRIs include *fluoxetine*, *citalopram*, *fluvoxamine*, *paroxetine* and *sertraline* (see section 1.5.3.1). All SSRIs strongly and selectively inhibit 5-HT reuptake by the presynaptic neuron, thus increasing 5-HT concentration at the synapse. With time, a decreased receptor density and/or sensitivity is also observed (Warrington, 1992; Richelson, 1994).

Pharmacokinetics

SSRIs are well absorbed by oral route and largely bound to proteins (with less affinity shown by citalopram and fluvoxamine). All SSRIs undergo metabolization in the liver and are basically eliminated in the urine.

When choosing a SSRI, consideration should be given to their different chemical structures, which lead to significant changes in their pharmacokinetic parameters.

Management

The management of these drugs is much easier than that of TCAs. The therapeutic dose may be administered from the beginning and maintained – if effective – throughout the treatment period. In general, elderly patients show better tolerance to SSRI-related side-effects as compared to TCAs. However, because the usual hepatic and renal impairment found in elderly patients favor the accumulation of these drugs in the body, lower starting doses of SSRI are advisable.

Since all SSRIs inhibit cytochrome P450, interactions may occur with drugs thus metabolized – as discussed in the Overview section. Caution must be exerted regarding possible interactions with other psychodrugs. SSRIs must never be administered with MAOIs, and severe side-effects may appear with tryptophan, cimetidine and barbiturates.

Indications

- Major depressive disorder and treatment of depressive episodes in bipolar disorder.
- Mild to moderate depression.
- Obsessive-compulsive disorder: higher doses than those used in depression.
- Anguish disorder and other anxiety disorders: slow, progressive dose escalation.
- Bulimia.
- Social phobia.

Adverse events

SSRIs have fewer side-effects than TCAs, and a different profile. They have less sedative and anticholinergic effects, which are also less cardiotoxic. Adverse events include:

- *Gastrointestinal:* nausea, dyspepsia, diarrhoea, flatulence, anorexia. These are the most frequent and usually subside with time.
- *Extrapyramidal:* akatysia and parkinsonism (Steur, 1993).
- *Neuropsychiatric:* restlessness, anxiety, irritability, tremors, agitation, insomnia, diurnal somnolence, headache.
- *Sexual dysfunction:* decreased libido, impotence, anorgasmia.
- Onset of *hypomanic* or *manic conditions*, particularly in bipolar patients.

Serotoninergic syndrome

Initially described in patients treated with a SSRI-MAOI combination (Sternbach, 1988; Brasseur, 1989), the serotoninergic syndrome has also been described with the above combination plus venlafaxine (Hodgman et al., 1997). This syndrome may include, confusion, hypomania, agitation, nervousness, myoclonia, profuse sweating, tremors, shuddering, diarrhoea, poor motor coordination, fever, coma and risk of death.

1.2.5.3.4 Serotonin-noradrenaline reuptake inhibitors: venlafaxine and duloxetine

Serotonin-noradrenaline reuptake inhibitors (SNRIs) act by blocking 5-HT and NA reuptake in the CNS. This double blockade could account for a faster downregulation of β-adrenergic receptors as compared to other antidepressants, and thus have a shorter latency period (Rudolph et al., 1991).

Because of their practically null affinity for muscarinic, histaminergic and α-adrenergic receptors, they are deprived of the side-effects secondary to the blockade of these receptors.

Indications

SNRIs are indicated for the treatment of all types and intensities of depression. They have shown to be effective in patients with treatment-resistant depression (Nierenberg et al., 1994; Holliday and Benfield, 1995). Venlafaxine was approved by the FDA in 1999 for treating generalized anxiety disorders. As yet, Duloxetine does not have this indication.

Consistent data are available on the efficacy of Venlafaxine in treating depression with associated anxiety symptoms, panic disorders, obsessive-compulsive disorders, social phobia, attention deficit hyperactivity disorder, personality disorders and chronic pain (Gelenberg et al., 2000; Hackett, 2000). Duloxetine has an indication in neuropathic pain.

Adverse events

The most frequent adverse events include: nausea, headache, somnolence, ejaculatory dysfunction, sweating, dry mouth and tremors. These effects tend to subside as from the second or third week.

In the case of Venlafaxine, a small but significant increase in blood pressure has been reported in a minority of patients, particularly at doses higher than 200 mg/day (Feighner, 1994). The existence of previous blood hypertension does not necessarily imply an increase in blood pressure.

A lower toxicity as compared to classic antidepressants has been observed in case of overdose (Feighner, 1994).

1.2.5.3.5 Noradrenaline reuptake inhibitors: reboxetine

Reboxetine is a selective NA reuptake inhibitor with low affinity for α-noradrenergic, muscarinic, dopaminergic and histaminergic receptors (Dostert et al., 1997).

The initial dose (4 mg/day) must be increased progressively according to the response, up to a maximum of 12 mg/day. The dose should be fractioned into one morning intake and one noon/afternoon intake (to prevent insomnia).

Reboxetine is not sedative and does not lead to motor slowdown. While few, mild side-effects occur, these can be more marked in the elderly. Adverse events include dry mouth, constipation, vertigo, insomnia, headache, tachycardia, postural hypotension.

Reboxetine is contraindicated in case of recent acute myocardial infarction.

1.2.5.3.6 Antidepressants with action on α-adrenergic receptor

Mianserin

Mianserin is a tetracyclic antidepressant with important noradrenergic activity and antihistaminergic action. The average maintenance dose is 30–90 mg/day. This drug causes somnolence. There is a (rare) possibility of bone marrow depression, particularly during the first weeks of treatment and in the elderly.

Mirtazapine

Also of tetracyclic structure, mirtazapine is a $α_2$, $5-HT_2$ and $5-HT_3$ receptor antagonist and also a powerful antihistaminergic. It shows little anticholinergic action. Its half-life (20–40 hours) allows one single daily intake. The initial dose (15 mg/day) may be increased up to 45 mg/day. It is best taken before going to bed. Its most frequent side-effects include somnolence, sedation, dizziness, dry mouth, and increased appetite and bodyweight. These are usually transient and moderate. Occasionally it may cause orthostatic hypotension and – infrequently – agranulocytosis (Montgomery, 1995).

1.2.5.3.7 Antidepressants acting on serotoninergic receptors: trazodone

Trazodone inhibits 5-HT reuptake and, due to its main metabolite, has a postsynaptic serotoninergic agonistic effect. Therapeutic range: 200–600 mg/day. Its most frequent side-effects include orthostatic hypotension, nausea, vomiting, sedation and priapism. One of its metabolites (CPP) has anxiogenic, migrainous and pro-obsessive effects, and therefore paradoxal effects have been described. Trazodone must not be used in combination with cysapride, MAOIs (except as a sedative at low doses of 50 mg to 10 mg) or with certain antihistaminics (like terphenadine or astemizole).

1.2.5.4 Complementary drug treatments in depression

The association of anxiolytics and/or hypnotics in the treatment of depression is recommended when the depressive episode is accompanied by anxiety or insomnia. Today's anxiolytic drugs are very useful and broadly used, given their clear benefits and little risk. These drugs should be prescribed for specific goals, without placing high hopes on them nor disregarding all necessary psychotherapeutic support. Importantly, these drugs should not be withdrawn abruptly for fear of dependence.

1.2.5.4.1 Benzodiazepines

Benzodiazepines (BDZ) are the treatment of choice for anxiety. They are easy to use, effective and have a good safety margin. They act by improving the subjective and objective symptoms of anxiety (by relieving vegetative symptomatology and muscle stress, as well as by diminishing the state of alertness or hypervigilance usually seen in anxiety conditions).

Because their pharmacodynamic actions are similar, BZD have the following effects: anxiolytic, myorelaxant, anticonvulsive (diazepam, clonazepam, clobazam), hypnotic, antiaggressive, avoidance response inhibitor. It may also produce anterograde amnesia (only at very high doses).

Pharmacokinetics

Well absorbed by oral route, the intramuscular route is rather erratic. BZDs differentiate themselves by their pharmacokinetic properties. Roughly, short half-life BZDs are useful as sleep inducers, while BZDs with longer half-lives are more used to treat generalized anxiety.

BZD classification, pharmacological properties and doses are shown in Table 10.

Table 10. Classification and pharmacological properties of benzodiazepines

BZD	Onset of action	Active metabolites	Commonest indication	Equivalent dose^a (mg)	Usual dose in adults (mg/day)
Long half-life (> 30 h)					
Clobazam	Intermediate	Yes	Anxiolytic	10	10–40
Chlorazepate	Fast	Yes	Anxiolytic	7,5	10–60
Chlordiazepoxide	Intermediate	Yes	Anxiolytic	10	15–100
Diazepam	Fast	Yes	Anxiolytic	5	2–60
Flurazepam	Fast	Yes	Hypnotic	5	15–30
Ketazolam	Fast	Yes	Hypnotic	7,5	15–75
Quazepam	Fast	Yes	Hypnotic	5	7.5–30
Intermediate half-life (= 30 h)					
Bromazepam	Slow	Yes	Anxiolytic	3	3–6
Flunitrazepam	Slow	No	Hypnotic	0,5	1
Nitrazepam	Slow	No	Hypnotic	2,5	5–10
Short half-life (< 10–24 h)					
Alprazolam	Intermediate	No	Anxiolytic	0,5	0.5–3
Bentazepam	Intermediate	-	Anxiolytic	50	50–100
Lorazepam	Intermediate	No	Anxiolytic	1	2–6
Lormetazepam	Slow	-	Hypnotic	0,5	0.5–1
Oxazepam	Slow	No	Anxiolytic	15	30–120
Temazepam	Intermediate	Yes	Hypnotic	5	15–30
Ultrashort half-life (< 5 h)					
Triazolam	Fast	No	Hypnotic	0.1–0.003	0.125–0.25
Midazolam	Fast	Yes	Hypnotic	1.2–1.7	7.5–10

^{a} Most powerful drugs have equivalent doses < 1, medium potency 1–10 and low potency > 10. BZD, benzodiazepines.

Adverse events

Adverse events are generally mild, well tolerated and transient. The most important events are: feeling of imbalance, excessive somnolence or sedation, dry mouth, diarrhoea or constipation, decreased libido, morning headaches (infrequent) and hypomnesia (mainly in the elderly). Paradoxal reactions, such as irritability or excitation, occur very rarely.

Management

Excessive doses may lead to important sedation and inappropriately low doses may be ineffective and create even more anxiety in the patient – due to apparent lack of effect. Also, a short half-life BZD must be prescribed for a fast, strong anxiolytic effect. Conversely, a long half-life BZD should be prescribed to achieve a maintained effect. Also to be remembered is that BZD discontinuation must be slow and gradual.

Interactions

BZDs have few incompatibilities with other drugs. If any, these incompatibilities are quite irrelevant clinically. Pharmacological interactions may appear when used with alcohol, other CNS depressors and oral contraceptives.

Risk of dependence

Only when very high doses (almost always higher than those prescribed) are taken for a long time. The risk is higher with short-acting BZDs, particularly if used with alcohol or other drugs.

Overdose

BZDs are very safe drugs in case of overdose relating to self-injury or simple error. Unless other drugs were added, almost no deaths by suicide have been described with BZDs. Importantly, an intravenously administered antagonist – flumazenil – is available in all emergency care units in case of overdose.

1.2.5.5 Management of various forms of depression

1.2.5.5.1 General aspects of management

Three key issues must be considered in the pharmacological handling of depressive disorders where insufficiencies are still being detected: (1) underuse of antidepressant drugs and overuse of benzodiazepines, (2) insufficient or subtherapeutic dose prescription, and (3) duration of treatments shorter than necessary. A common denominator for these factors might be a lack of specific training or updated knowledge in the field of antidepressant treatment, combined with difficulties or fears relating to side-effects or toxicity. This would condition both the prescription of sufficient doses for the right period of time by the physician and the "adherence" to treatment (or minimization of medication dropout) by the patient. In this sense, a concept gaining importance in the frame of a suitable doctor/patient relationship is that of psychoeducation, i.e. providing appropriate information on the effects of medication and monitoring the degree of compliance and the problems likely to be encountered during treatment, particularly in its early phases (Lin et al., 1995; Hirschfeld et al., 1997).

At the beginning of treatment, it is essential to provide clear, understandable information on the following aspects:

- The symptoms presented by the subject are a disease altogether, and any feeling of guilt regarding his/her situation or functioning deficits should be removed.

- The idea that improvement with appropriate treatment is the rule (2 out of 3 cases will respond to the first antidepressant) and not the exception must be conveyed (Ayuso Mateos and Ayuso Gutiérrez, 1998).

- Medication administration (dosage, schedules, gradual escalation).

- Characteristics, importance and highest probabilities of onset and remission of side-effects (without dramatizing them).

- Period necessary for the onset of the therapeutic effect and the need to maintain treatment after achieving improvement.

It is also important to check the extent to which the patient has understood the basics of treatment. After a clinical interview, the subject usually understands part of the information (according to variables like cultural level and lack of familiarization with medical terminology) and remembers small fragments (particularly the end of the interview). Besides insisting on the key aspects with clear, repetitive messages, the patient (or the accompanying person if the patient's condition cannot guarantee understanding and compliance) may be invited to repeat the most relevant aspects or to decipher the hieroglyphic into which a prescription is frequently turned, before requesting an interpretation by third parties or the pharmacist.

Before selecting the optimal treatment for each case, the following key issues should be considered in the management of major depression (Spanish Society of Psychiatry, 1997).

- Pharmacological treatment with antidepressants is of paramount importance to obtain a good therapeutic response of the episode. The effectiveness of antidepressants in monotherapy, that is, their therapeutic response (equal to or higher than 50% decrease in the Hamilton Depression Rating Scale) ranges between 60% and 70%.
- We should opt for monotherapy drugs that obtain high percentages of remission and may prevent disease recurrence.
- It is advised to start treatment at low doses that are increased progressively until the minimum effective dose is achieved as from the first week. This dosing schedule – valid for tricyclics and MAOIs and useful to minimize side-effects – may be obviated in other pharmacological groups with better tolerance like SSRIs, which can be initiated at therapeutic doses (to be further increased according to the clinical response).
- In the presence of significant anxiety or insomnia, concomitant administration of benzodiazepines during the first weeks is advised, with further gradual withdrawal.
- Most side-effects disappear or subside after the first week of treatment.
- The latency period of the antidepressant action is 2–6 weeks (significant changes are usually observed around week 3–4).
- Symptoms like sleep or appetite disturbances usually improve before sadness does. The risk of suicide is maintained until complete remission, and is sometimes increased at the beginning of improvement.
- The remission of most symptoms is not achieved before 6–8 weeks. If no improvement is obtained until that time with full doses of antidepressants, switching to another antidepressant with a different pharmacological profile should be considered, or the aspects approached in resistant forms of depression should be addressed.

- Upon the consecution of complete improvement or episode remission with treatment given in acute phase, such treatment is to be maintained for 6–12 months at the effective dose in order to prevent relapses (relapse is the reactivation or reappearance of symptoms of the same episode, the evolution or natural history of which is estimated at approximately 6 months). This is the treatment continuation phase. After that, maintenance or prophylactic treatment aims at preventing disease recurrences in cases of recurrent depressive disorder (Horwarth et al., 1992), as commented above. At any rate, antidepressant medication will be withdrawn progressively to prevent discontinuation symptoms or cholinergic or serotoninergic rebound.

1.2.5.5.2 Response-predicting factors according to clinical variables, depressive subtypes and psychosocial variables

According to the clinical information presently available, antidepressants are equipotent at equivalent doses. However, there is a number of response-predicting factors related to certain subtypes, clinical characteristics and psychosocial factors (Goodnick, 1997). While all antidepressants – classic and recent – are effective in all types of depression and no specific indications can be attributed to any of them, certain aspects do influence the choice of the antidepressant (Herrán et al., 1998; Ayuso-Mateos and Ayuso-Gutiérrez, 1998; Crespo, 1999; Theobald et al., 2000):

- The existence of a *comorbid personality disorder* is predictive of worse response to antidepressant treatment.

- The presence of *low social support* is predictive of worse response, even in melancholic depressions. The occurrence of *negative vital events* during treatment is also associated with poor response.

- A *history of good response* to an antidepressant, as determined by the patient (or even a relative) favors the administration of the same drug in case of recurrence.

- Treatment should preferably be started with a drug of the *SSRI group, venlafaxine or other new antidepressants* in patients where poor compliance is foreseen, in those suffering cardiovascular diseases or in those at risk of drug intoxication.

- Classic treatment with *tricyclic and tetracyclic antidepressants* offers proven efficacy and lower cost, but have more important anticholinergic, antiadrenergic and antihistaminic side-effects.
- *Endogenicity* elements are predictive of good response to tricyclic antidepressants and to electroconvulsive therapy (ECT).
- *Psychotic depression* seems to respond better to ECT than the combination of more antipsychotic antidepressants or antidepressants alone (Kroessler, 1985), although a combination of antidepressants and antipsychotics has also shown high levels of efficacy. The predictor of best response to ECT is the combination of motor inhibition and psychotic symptoms.
- *Atypical depressions* respond well to MAOIs. Recent studies also report a good response to cognitive therapy, which focuses on the management of the hypersensitivity to interpersonal rejection or the magnified, distorted perception of it shown by these patients (Jarrett et al., 1999).
- The presence of *family history of bipolar disorder*, as well as *cyclothymic personality*, are indicative of a possibly good response to lithium. Lithium is more effective in bipolar depressions.

1.2.5.6 Approach to treatment-resistant depressions

Treatment-resistant depression (TRD) is a depressive episode that has failed to show sufficient improvement after treatment with a drug of proven antidepressant activity, at sufficient doses and for a suitable period of time (Álvarez et al., 1999). For example, the use of a 200 mg/day dose of imipramine for a minimum of 6 weeks under this maximum dose, without including the period of time used to reach it. Between 30% and 40% of the patients suitably treated with antidepressant monotherapy will show null or insufficient response after a first therapeutic attempt, while 5–10% will not respond to more aggressive therapies (Klein and Davis, 1970).

The most valid method for determining whether a "therapeutic response" has or has not been achieved is the consensus between patient, family and physician. The quantitation of this aspect, however, usually requires a 50% decrease in the baseline score of the Hamilton Depression Rating Scale and a final score below 7 (therapeutic remission).

In these cases, the following strategies will be useful (adapted from Álvarez et al., 1999):

1.2.5.6.1 Optimization

- *Reconsider diagnose* by discarding conditions likely to be misdiagnosed, such as: nonaffective psychiatric disorders, comorbidity with other psychiatric disorders, medical diseases and drugs likely to induce depressive states (e.g., antihypertensives, immunosupressors or corticoids).

- *Check whether the therapeutic indication is correct*, bearing in mind the subtypes of depression for which specific treatments are available (atypical depressions, bipolar depression or psychotic depressions).

- *Monitor antidepressant plasma levels*. While most antidepressants show no clear, proven relationship between plasma levels and efficacy, plasma level determination allows detecting medication noncompliance or pharmacokinetic problems in the absorption and/or metabolization of the drug. The evaluation of nonadherence to treatment is essential, as it is estimated to generate up to 20% of resistant cases (Souery and Mendlewicz, 1998).

- *Force initial treatment*: Incorrect treatment is the main cause for lack of response in major depression. Dose recommendation and treatment duration must be based on the literature. The optimal dose must be used for a minimum of four weeks, with adequate duration estimated at 6–8 weeks. If no response is obtained, the dose must be increased until the maximum dose for that antidepressant or the tolerance limit are reached, and/or treatment duration must be extended to 8–10 weeks.

1.2.5.6.2 Potentiation

Without altering the current antidepressant treatment, potentiation consists in *adding substances* without antidepressant activity of their own but likely to increase the potency of the drug they are added to. This is considered to be a good option in patients having obtained partial response to treatment because it allows to maintain the achieved improvement. The most frequently used substances are: *lithium*, *tryptophan* (an amine precursor amino acid), *triiodothyronine* (T3, particularly in patients with subclinic hypothyroidism) or *pindolol* (presynaptic serotoninergic blockade).

Potentiation strategies with or without sufficient scientific evidence are described overleaf.

Strategies supported by scientific evidence

Lithium salts: This is the most investigated and documented strategy. Lithium must be added without discontinuing the antidepressant treatment, and should be maintained for the same time period as the antidepressant drug throughout the initial and continuation treatments. If prophylactic treatment is required, the use of lithium alone may be considered on account of its efficacy in preventing recurrences. Efficacy estimates are 40–60%. While improvement may start within 24–48 hours, the available data advise to extend the potentiating attempt up to the 3rd or 4th week before deciding on its uselessness (Álvarez et al., 1997). The metaanalysis published by Bauer (Bauer and Döpfmer, 1999) concludes that lithium potentiation is the treatment of choice in depressed patients not responding to monotherapy, with a minimum waiting time of 7 days with lithemias ≥ 0.5 mmol/l.

Triiodothyronine (T3): T3 improves central NA activity by increasing postsynaptic receptor sensitivity. Recommended doses are 25–50 µg/day, with response assessment after 3 weeks. Treatments longer than 8–12 weeks must be avoided due to a withdrawal-related risk of hypothyroidism. The most frequent side-effects are nervousness and insomnia. A controlled clinical trial versus placebo and lithium (Joffe et al., 1993) found no differences between the two procedures, both being higher than placebo.

Atypical antipsychotics: While more data are needed, this is a safe, well tolerated combination, particularly indicated to cover the hyperarousal symptoms of depression (insomnia, weight loss, anxiety, agitation). Although efficacy data in treatment-resistant depression are available for most atypicals, the most studied and proven drug for this indication is olanzapine (Thase, 2002).

Strategies supported by less scientific evidence

Psychostimulants: Although only a few, non-controlled, small-sample studies are available, good results have been obtained. *Dextroamphetamine* 5–20 mg/day, *methylphenidate* 5–40 mg/day and *modaphinil* 200–400 mg/day have been used. Response latency is very short and is occasionally seen few hours after the first intake. These drugs are more useful in the elderly and in patients with dysthymic disorder and double depression. High blood pressure, anxiety, irritability and insomnia have been reported. This, together with the potential risk of abuse of these substances and their short half-life, limits their use. The introduction of drugs with a more prolonged half-life may help prevent these limitations.

Serotonin precursors (tryptophan): While adding a 5-HT precursor would seem a logical way of potentiating the antidepressant effect, the use of serotonin precursors is limited by its poor gastric tolerance and the high doses required by their pharmacokinetics, where only a small amount of the orally administered dose reaches its target in the CNS. Maximum daily dose of 1 g may be administered, always with food.

Anticonvulsivant drugs: All published studies were open-labeled, performed with small populations and offering contradictory results. Some efficacy has been suggested for valproate, carbamazepine, lamotrigine and gabapentin.

Hormone therapy: Decreased estrogen levels, particularly in postmenopausal women, have been suggested in the etiopathogenesis of refractoriness in some depressive conditions. Dihydroepiandrosterone (DHEA) is a precursor to testosterone and estrogens; its metabolite DHEA-S has been involved in mood regulation and feeling of well-being. A small double-blind preliminary study (Wolkowitz et al., 1999) suggests its use as AD potentiator in refractory depressions.

A preliminary placebo-controlled study (Pope et al., 2003) reported significant improvement in men aged 30–65 with resistant depression and low or borderline testosterone levels by potentiating AD treatment with 1% testosterone transdermal gel.

Antiglucocorticoid drugs (ketoconazole, aminoglutetimide and metirapone) have been involved in the treatment of depression (Murphy et al., 1991; O'Dwyer et al., 1995; Wolkowitz and Reus, 1999), on account of the well known relationship between depression and the activation of the hypothalamic–pituitary–adrenal axis.

Buspirone: This is a 5-HT$_{1A}$ receptor partial agonist whose efficacy as a SSRI potentiator has been described in five open-labeled studies at doses of 10–50 mg/day. The two placebo-controlled studies reported negative, albeit not conclusive results (Appelberg et al., 2001).

Pindolol: Because this β-adrenergic blocker antagonizes 5-HT$_{1A}$ presynaptic autoreceptors, a faster response to antidepressants might be produced. Several controlled studies in resistant depression (Moreno et al., 1997; Pérez et al., 1999) are available where no significant differences were found between pindolol and placebo as a potentiator in resistant depression at doses of 2.5 mg t.i.d.

Inositol: Some authors have suggested that inositol deficit is a factor related to resistant depression. Preclinical data suggest that postsynaptic modulation of the second-messenger system (phosphatidilinositol and cyclic AMP) might be useful in the treatment of depression. Two double-blind, controlled studies (Nemets et al., 1999; Levine et al., 1999) do not support its use in SSRI-resistant patients.

Dopaminergic agents: No controlled studies have been performed. Open-labeled studies suggest good results, although very small samples were used. Potentiations have been described with the following agents: *pergolide*, 0.25–2.0 mg/day; *amantadine*, 200–400 mg/day; *pramipexol*, 0.375–1.0 mg/day, and *bromocryptine*, 5 mg/day.

Typical antipsychotics: While their usefulness has been suggested in open-labeled studies with reserpine (Zohar et al., 1991), the results obtained in controlled studies were inconsistent (Price et al., 1987). Typical antipsychotics have been used empirically in a subtype of anxious depression considered as more frequently resistant, apparently with good results. Their use has been limited by the higher risk of dyskinesia in these patients.

Other substances used in the treatment of TRD. Opiates have shown their efficacy in small non-controlled studies, suggesting their use as AD potentiator. Their potential risk of abuse, however, has limited their usefulness. Positive cases with oxycodeine and oxymorphine (Stoll and Rueter, 1999), and buprenorphine (Bodkin et al., 1995) have been described.

Folate and *S-adenosyl-methionine* are substances involved in brain methylation processes for which antidepressant properties have been described (Alpert et al., 2000; Coppen et al., 1963, 2000).

A controlled study seemed to demonstrate that *omega-3 fatty acids* are effective in resistant depression versus placebo. These data, however, need to be replicated by independent studies.

1.2.5.6.3 Combination

Combination refers to using *two antidepressants simultaneously*, with different mechanisms of action that make them complementary to one another. For example, TCAs with SSRIs, noradrenergics with serotoninergics. The overall response is estimated at 60%, and combinations are usually well tolerated. Roughly, combination treatment should be maintained for 6–9 months after remission, upon which gradual discontinuation of one of the two antidepressants should be attempted.

Antidepressant combination should be justified by meeting a few basic principles: mechanisms of action – not mere drugs – are to be combined; synergic combinations – not mathematically exact ones – are to be sought. A successful combination is that which multiplies the antidepressant action (in terms of efficacy) and whose mechanisms of action minimize side-effects (in terms of tolerability).

Because few controlled, very small sample studies are available (Davidson et al., 1978; Fava et al., 1994; Medhus et al., 1990; Ferreri et al., 2001), the combination of antidepressants should not be used as first-line treatment of resistant depression. In practice, however, it is one of the most commonly used strategies.

Examples of accepted, safe combinations are:

- Combination of selective serotoninergic drugs (e.g., clomipramine, SSRI) and noradrenergic drugs (e.g., maprotiline, reboxetine).
- Reuptake inhibitors (SSRI, venlafaxine) with mirtazapine.

1.2.5.6.4 Switching

Switching is *changing from one antidepressant to another* in a different chemical group. While few data on controlled studies are available, a review of the literature suggests a response rate of approximately 50% when switching from one antidepressant to another. This would be a reasonable option in patients without therapeutic response after 6 weeks of treatment with correct doses, particularly if side-effects appear. No conclusive data are available on switching to a same-class or different-class antidepressant. Nonetheless, it is generally recommended to switch to a different-class antidepressant if two of the same class have proved ineffective. The most documented strategy is switching from TCAs to MAOIs.

1.2.5.6.5 Non-pharmacological strategies

Electroconvulsive Therapy (ECT) is a biological treatment widely used in modern psychiatry. In severe depression, ECT has shown an efficacy similar to TCAs and MAOIs. Fifty percent of drug-resistant patients do respond to ECT, which is the most effective treatment in resistant depression. A course of ECT must be followed by prophylactic treatment – usually with antidepressant drugs – to prevent relapse. Even in case of previous pharmacological resistance, maintenance treatment with antidepressants may be appropriate. Indeed, it has been hypothesized that ECT might modify this resistance. Possible maintenance treatment with ECT will be considered in patients with a history of multiple depression recurrence and/or resistance, or drug intolerance.

Repetitive Transcranial Magnetic Stimulation (rTMS) is a non-invasive technique which has been proposed as an alternative therapeutic approach to affective disorders refractory to drug treatment. For the time being, however, not sufficient evidence has been produced to endorse rTMS in the treatment of depression. Nonetheless, given the poor quality of the data, a possible beneficial effect of the technique cannot be ruled out (Martin et al., 2002).

The studies performed with *Vagus Nerve Stimulation* (VNS) suggest antidepressant efficacy in otherwise resistant patients, with responses likely to be maintained for months in a high percentage of cases (Rush et al., 2000; Marangell et al., 2002). VNS is a proven, well tolerated, safe technique, with mild side-effects according to the patients.

Photostimulation could be considered as a coadjuvant strategy in refractory patients with a history of seasonal affective disorder or with increased seasonal symptoms (Levitt et al., 1991).

While *sleep deprivation* accounts for a 40% response in patients with refractory depression (Álvarez, 1997), it involves an unsustained response that calls for prospective, controlled studies.

Psychosurgery should only be considered in certain untreatable depressions where the only real possibility of improvement is this therapeutic route. Bilateral cingulotomy would be the technique of choice (Bouckoms, 1991).

Psychosocial treatments, such as psychotherapeutic interventions, may be of dramatic importance in candidate patients. Some authors have suggested that resistance cannot be confirmed until psychotherapeutic interventions have been performed. While the results of the published studies do not endorse the use of psychotherapy as a consistent strategy in resistant depression, the combination of depression-specific psychotherapy and drug therapy under certain circumstances has yielded better results than drug therapy alone (Thase, 2001).

In summary, and given the frequent lack of direct comparative data to assess the most effective strategy, the empirical data available suggest that the decision to wait, potentiate, associate or switch will frequently depend on the physician's own experience, the severity of the disease, the adverse effects of the first medication, the patient's predisposition to take more than one drug, or the response. ECT has proved to be the most effective treatment in resistant depression. Although not complication-free, it should be used at any time during treatment, as required by the patient's evolution. Finally, it should be borne in mind that the goal to be achieved when treating depression must be complete remission or the return to previous psychosocial functioning.

1.2.5.7 Treatment of dysthymia

The treatment of dysthymia is a controversial issue on account of (previously approached) nosological and conceptual difficulties, as well as due to the great heterogeneity of the concept. This leads to major variations in the treatment responses reported by the different studies, ranging from placebo-like responses to responses close to those obtained with endogenous depressions (Howland, 1991).

To some extent, the therapeutic approach of dysthymic disorders shares the problem of non-melancholic depressions, much more confusing than melancholy problems on account of being a more heterogeneous condition. Two subgroups of dysthymic patients, involving different responses to the different therapeutic strategies, can be established (Menchón and Vallejo, 1999):

- One subgroup of dysthymic patients where a better response to drug treatment should be expected, presenting with a larger biological base and corresponding to what Prof. Akiskal called "*subaffective dysthymias*" (Akiskal, 1996), supporting the view that dysthymia is a constitutional, genetically attenuated variant of endogenous affective disorders. These dysthymias include family history of unipolar and bipolar affective disorders, diurnal mood variations, tendency to asthenia and hypersomnia, obsessive and narcissistic personality traits, major depression episodes during dysthymia (inhibition-hypersomnia type), as well as spontaneous or secondary-to-antidepressants hypomanic episodes.
- Patients where psychotherapy would play a more relevant role, relating to personalities with predominance of low feeling tone, insecurity and obsessivity, or relating to *personality disorders* in cluster B of DSM-IV-TR (immature or "dramatic" personalities including borderline, narcissistic, histrionic and antisocial disorders).

1.2.5.7.1 Pharmacological treatment

The long duration of the disorder and its supposedly basically psychological origin have led some clinicians to view drugs as a second-choice option. However, and despite the abovementioned heterogeneity and the fact that response rates vary considerably, many controlled studies on dysthymia (Howland, 1991; Lapierre, 1994; Invernizzi et al., 1997; Gorman and Kent, 1999, Menchón and Vallejo, 1999) provide increasing credibility to the pharmacological treatment of these conditions. Some studies suggest that *SSRIs* may be the drugs of choice. However, the superiority of these treatments as compared to classic *MAOIs* has yet to be proven; indeed, some authors consider MAOIs as the most resolutive drugs in dysthymic patients (Howland, 1991; Vallejo and Menchón, 1996).

Because it is increasingly evident that dysthymic patients are particularly sensitive to the side-effects of antidepressants (Invernizzi et al., 1997; Vallejo and Menchón, 2000), the use of the new molecules, featuring a more selective mechanism of action and better tolerance, is advised. The alternatives to the failure of a first pharmacological attempt in dysthymia would be the same as those described in the approach of resistant depression (Kaplan and Sadock, 1999). Despite all this, 50% of dysthymic patients fail to respond to psychopharmacology (Vallejo and Menchón, 2000), and only 10–15% of the conditions have abated one year after the initial diagnose (Kaplan and Sadock, 1999). Also important is the fact that the poor diagnostic stability of the dysthymic disorder is coupled with a likewise poor tendency to spontaneous remission.

The future psychopharmacology of dysthymic disorders should aim to define treatment-respondent subtypes according to clinical variables (presence or absence of comorbidity with MDD or anxiety), personality, sociodemographic variables (age, duration of the condition, social-family environment, etc.) and biological substrate (detecting markers that identify dysthymic patients with an aetiopathogenic biological component).

1.2.5.7.2 Psychological treatment

Psychotherapy is particularly important in dysthymia, given that almost 50% of patients do not respond to psychodrugs and others are not susceptible for a number of reasons (untoward effects, medication rejection, concomitant medical diseases…).

Cognitive therapy, widely accepted in depressive disorders, may be useful in dysthymic patients, and data support the efficacy of short cognitive therapies in some patients (Markowitz, 1994). However, more studies on its use are required. With regard to interpersonal therapy – of proven efficacy in non-melancholic depression – few publications with cognitive therapy are available in dysthymic patients. Because no studies are available, there is no evidence that dysthymia might benefit from analytical therapy; in fact, it is even postulated that this orientation might have a negative effect in these patients.

1.2.5.7.3 Combined treatment

Generally speaking, the combination of *cognitive or behavioral therapy and pharmacotherapy* is probably the most effective treatment of this disorder (Kaplan and Sadock, 1999). Because no studies clearly showing the efficacy of these combined approaches are available, both the empirical perspective and common sense advise providing guidelines for a better adaptation to the consequences of chronic depressive symptomatology. Also to be implemented is the patient's orientation, advice and training in skills designed to face the difficulties – particularly interpersonal ones – generated by the disorder, on the basis of a symptom-abating, complication-preventing pharmacological intervention.

References

Akiskal HS. *Dysthymia as a temperamental variant of affective disorder.* Eur Psychiatry. **1996**, 11:117S-122S.

Alpert JE, Mischoulon D, Nierenberg AA, Fava M. *Nutrition and depression: focus on folate.* Nutrition. **2000**, 16:544-546.

American Psychiatric Association. *DSM-IV-TR. Diagnostic and Statistical Manual of Mental Disorders.* 4th ed. American Psychiatric Publishing; Washington, DC, **2000**, pp 1-943.

Appelberg BG, Syvalahti EK, Koskinen TE, Mehtonen OP, Muhonen TT, Naukkarinen HH. *Patients with severe depression may benefit from buspirone augmentation of selective serotonin reuptake inhibitors: results from a placebo-controlled, randomized, double-blind, placebo wash-in study.* J Clin Psychiatry. **2001**, 62:448-452.

Artigas F. Neurotransmisión aminérgica y su modulación por fármacos antidepresivos. In: Roca M, ed. Médica Panamericana S.A., Madrid, **1999**.

Ayuso Mateos JL, Ayuso Gutiérrez JL. Uso clínico de los fármacos antidepresivos. In: Vázquez-Barquero JL, ed. *Psiquiatría en atención primaria.* Grupo Aula Médica, Madrid, **1998**.

Álamo C, Cuenca E, López-Muñoz F. Neurotransmisión noradrenérgica en el sistema nervioso central. In: Vallejo C, Cuenca E, eds. *Depresión y noradrenalina.* Ed.Doyma, Barcelona, **1999**, pp 1-15.

Álvarez E. Utilización de fármacos antidepresivos y "activación". *Temas clave en el manejo de la depresión.* Edimsa, Madrid, **1998**.

Álvarez E, Pérez-Solá V, Pérez-Blanco J. Tratamiento de la depresión resistente. In: Roca M, ed. *Trastornos del humor.* Médica Panamericana S.A., Madrid, **1999**.

Álvarez P. *Deprivación del sueño.* Psiquiatría Biológica. **1997**, 4:50-54.

Bauer M, Dopfmer S. *Lithium augmentation in treatment-resistant depression: meta-analysis of placebo-controlled studies.* J Clin Psychopharmacol. **1999**, 19:427-434.

Bodkin JA, Zornberg GL, Lukas SE, Cole JO. *Buprenorphine treatment of refractory depression.* J Clin Psychopharmacol. **1995**, 15:49-57.

Bouckoms AJ. The role of stereotactic cingulotomy in the treatment of intractable depression. In: Amsterdam JD, ed. *Refractory Depression.* Raven Press, New York, **1991**, pp 233-242.

Boyd J, Weissman MM. Epidemiology. *Handbook of affective disorders.* Churchill Livingstone, London, **1982**, pp 109-125.

Brasseur R. *A multicentre open trial of fluoxetine in depressed out-patients in Belgium.* Int Clin Psychopharmacol. **1989**, 4 Suppl 1:107-111.

Brown GW, Harris TO, Hepworth C. *Life events and endogenous depression. A puzzle reexamined.* Arch Gen Psychiatry. **1994**, 51:525-534.

Carney MW, Roth M, Garside RF. *The diagnosis of depressive syndromes and the prediction of E.C.T. response.* Br J Psychiatry. **1965**, 111:659-674.

Carrasco JL. Empleo de fármacos antidepresivos en el paciente con patologías médicas. In: Roca M, Bernardo M, eds. *Trastornos depresivos en patologías médicas.* Masson, Barcelona, **1996**.

Clayton PJ, Halikas JA, Maurice WL. *The depression of widowhood.* Br J Psychiatry. **1972**, 120:71-77.

Coppen A, Shaw DM, Farrell JP. *Potentiation of the antidepressive effect of a monoamine-oxidase inhibitor by tryptophan.* Lancet. **1963**, 1:79-81.

Coppen A, Prange AJ, Jr., Whybrow PC, Noguera R. *Abnormalities of indoleamines in affective disorders.* Arch Gen Psychiatry. **1972**, 26:474-478.

Coppen A, Bailey J. *Enhancement of the antidepressant action of fluoxetine by folic acid: a randomised, placebo controlled trial.* J Affect Disord. **2000**, 60:121-130.

Crespo JM. Depresión mayor. In: Roca M, ed. *Trastornos del humor.* Médica Panamericana, Madrid, **1999**.

Cuenca E, Cuenca-Söderberg O. *Fluvoxamina. Aspectos farmacológicos y clínicos.* An Psiquiatría. **1990**.

Cuenca E. Interacciones farmacológicas de los agentes antidepresivos y citocromo P-450. *Temas clave en el manejo de la depresión.* Edimsa, Madrid, **1998**, pp 13-25.

Dostert P, Benedetti MS, Poggesi I. *Review of the pharmacokinetics and metabolism of reboxetine, a selective noradrenaline reuptake inhibitor.* Eur Neuropsychopharmacol. **1997**, 7 Suppl 1:S23-S35.

Duman RS, Heninger GR, Nestler EJ. *A molecular and cellular theory of depression.* Arch Gen Psychiatry. **1997**, 54:597-606.

Ereshefsky L, Riesenman C, Lam YW. *Serotonin selective reuptake inhibitor drug interactions and the cytochrome P450 system.* J Clin Psychiatry. **1996**, 57 Suppl 8:17-24.

Fava M, Bouffides E, Pava JA, McCarthy MK, Steingard RJ, Rosenbaum JF. *Personality disorder comorbidity with major depression and response to fluoxetine treatment.* Psychother Psychosom. **1994**, 62:160-167.

Fava M, Mulroy R, Alpert J, Nierenberg AA, Rosenbaum JF. *Emergence of adverse events following discontinuation of treatment with extended-release venlafaxine.* Am J Psychiatry. **1997**, 154:1760-1762.

Feighner JP, Robins E, Guze SB, Woodruff RA, Jr., Winokur G, Munoz R. *Diagnostic criteria for use in psychiatric research.* Arch Gen Psychiatry. **1972**, 26:57-63.

Feighner JP. *The role of venlafaxine in rational antidepressant therapy.* J Clin Psychiatry. **1994**, 55 Suppl A: 62-68.

Ferreri M, Lavergne F, Berlin I, Payan C, Puech AJ. *Benefits from mianserin augmentation of fluoxetine in patients with major depression non-responders to fluoxetine alone*. Acta Psychiatr Scand. **2001**, 103:66-72.

Frank JD. *Persuasion and healing. A comparative study of psychotherapy.* 2nd ed. Johns Hopkins University Press; Baltimore, **1973**, pp 1-378.

Gastó C, Vieta E. Trastornos afectivos. In: Cervilla J, Gracía-Ribera C, eds. *Fundamentos biológicos en psiquiatría*. Masson, Barcelona, **2000**.

Gastó P. Antidepresivos en los trastornos de personalidad. In: Vallejo J, Gastó C, eds. *Antidepresivos en la clínica psiquiátrica*. Mosby-Doyma, Barcelona, **1996**, pp 79-89.

Gelenberg AJ, Lydiard RB, Rudolph RL, Aguiar L, Haskins JT, Salinas E. *Efficacy of venlafaxine extended-release capsules in nondepressed outpatients with generalized anxiety disorder: A 6-month randomized controlled trial*. JAMA. **2000**, 283:3082-3088.

Gibert J, Moreno M, Ignacio J. Antidepresivos. In: Roca M, ed. *Trastornos del humor*. Médica Panamericana, Madrid, **1999**, pp 793-812.

Gibert J, Rojas M, Micó J. Neurotransmisión peptidérgica y depresión. In: Roca M, ed. Médica Panamericana, Madrid, **1999**.

Goodnick PJ. Predictores de la respuesta al tratamiento en trastornos del estado de ánimo. *Predictores Biológicos*. Edika Med, Barcelona, **1997**, pp 187-193.

Gorman JM, Kent JM. *SSRIs and SMRIs: broad spectrum of efficacy beyond major depression*. J Clin Psychiatry. **1999**, 60 Suppl 4:33-38.

Hackett D. *Venlafaxine XR in the treatment of anxiety*. Acta Psychiatr Scand Suppl. **2000**, 30-35.

Hamilton M. *A rating scale for depression*. J Neurol Neurosurg Psychiatry. **1960**, 23:56-62.

Hamilton M. *Development of a rating scale for primary depressive illness*. Br J Soc Clin Psychol. **1967**, 6:278-296.

Herrán A, Núñez MJ, Vázquez-Barquero JL. Trastornos del estado de ánimo. In: Vázquez-Barquero JL, ed. *Psiquiatría en Atención Primaria*. Grupo Aula Médica, Madrid, **1998**.

Hirschfeld RM, Keller MB, Panico S, Arons BS, Barlow D, Davidoff F, Endicott J, Froom J, Goldstein M, Gorman JM, Marek RG, Maurer TA, Meyer R, Phillips K, Ross J, Schwenk TL, Sharfstein SS, Thase ME, Wyatt RJ. *The National Depressive and Manic-Depressive Association consensus statement on the undertreatment of depression*. JAMA. **1997**, 277:333-340.

Hodgman MJ, Martin TG, Krenzelok EP. *Serotonin syndrome due to venlafaxine and maintenance tranylcypromine therapy*. Hum Exp Toxicol. **1997**, 16:14-17.

Holliday SM, Benfield P. *Venlafaxine. A review of its pharmacology and therapeutic potential in depression*. Drugs. **1995**, 49:280-294.

Howland RH. *Pharmacotherapy of dysthymia: a review*. J Clin Psychopharmacol. **1991**, 11:83-92.

Invernizzi G, Mauri MC, Waintraub L. *Antidepressant efficacy in the treatment of dysthymia*. Eur Neuropsychopharmacol. **1997**, 7 Suppl 3:S329-S336.

Janowsky DS, el Yousef MK, Davis JM, Sekerke HJ. *A cholinergic-adrenergic hypothesis of mania and depression.* Lancet. **1972**, 2:632-635.

Jarrett RB, Schaffer M, McIntire D, Witt-Browder A, Kraft D, Risser RC. *Treatment of atypical depression with cognitive therapy or phenelzine: a double-blind, placebo-controlled trial.* Arch Gen Psychiatry. **1999**, 56:431-437.

Joffe RT, Singer W, Levitt AJ, MacDonald C. *A placebo-controlled comparison of lithium and triiodothyronine augmentation of tricyclic antidepressants in unipolar refractory depression.* Arch Gen Psychiatry. **1993**, 50:387-393.

Kaplan HJ, Sadock BJ. *Sinopsis de psiquiatría.* 8th ed. Panamericana; Madrid, **1999**.

Kendell RE, Gourlay J. *The clinical distinction between psychotic and neurotic depressions.* Br J Psychiatry. **1970**, 117:257-260.

Kessler RC, McGonagle KA, Nelson CB, Hughes M, Swartz M, Blazer DG. *Sex and depression in the National Comorbidity Survey. II: Cohort effects.* J Affect Disord. **1994**, 30:15-26.

Klein DF, Davis JM. *The drug treatment of depression.* JAMA. **1970**, 212:1962-1963.

Kroessler D. *Relative Efficacy Rates for Therapies of Delusional Depression.* Convuls Ther. **1985**, 1:173-182.

Lapierre YD. *Pharmacological therapy of dysthymia.* Acta Psychiatr Scand Suppl. **1994**, 383:42-48.

Levine J, Mishori A, Susnosky M, Martin M, Belmaker RH. *Combination of inositol and serotonin reuptake inhibitors in the treatment of depression.* Biol Psychiatry. **1999**, 45:270-273.

Levitt AJ, Joffe RT, Kennedy SH. *Bright light augmentation in antidepressant nonresponders.* J Clin Psychiatry. **1991**, 52:336-337.

Lewis A. *Melancholia: a clincial survey of depressive states.* J Ment Sci. **1934**, 80:277-378.

Lin EH, Von Korff M, Katon W, Bush T, Simon GE, Walker E, Robinson P. *The role of the primary care physician in patients' adherence to antidepressant therapy.* Med Care. **1995**, 33:67-74.

Lin EH, Katon WJ, Simon GE, Von Korff M, Bush TM, Rutter CM, Saunders KW, Walker EA. *Achieving guidelines for the treatment of depression in primary care: is physician education enough?* Med Care. **1997**, 35:831-842.

López-Ibor Aliño JJ. In: Roca M, ed. *Trastornos del humor.* Médica Panamericana, Madrid, **1999**.

Marangell LB, Rush AJ, George MS, Sackeim HA, Johnson CR, Husain MM, Nahas Z, Lisanby SH. *Vagus nerve stimulation (VNS) for major depressive episodes: one year outcomes.* Biol Psychiatry. **2002**, 51:280-287.

Markowitz JC. *Psychotherapy of dysthymia.* Am J Psychiatry. **1994**, 151:1114-1121.

Martin JL, Barbanoj MJ, Schlaepfer TE, Clos S, Perez V, Kulisevsky J, Gironell A. *Transcranial magnetic stimulation for treating depression.* Cochrane Database Syst Rev. **2002**, 2, CD003493.

Medhus A, Heskestad S, Thue JF. *[A combination effect of mianserin (Tolvon) and tricyclic antidepressive agents. Improved antidepressive therapy].* Tidsskr Nor Laegeforen. **1990**, 110:3527-3528.

Menchón JM, Urretavizcaya M. Aspectos biológicos de los trastornos afectivos. In: Menchón JM, Urretavizcaya M, eds. *Trastornos afectivos.* 1ª ed. Médica Panamericana, Madrid, **1998**.

Menchón JM, Vallejo J. Distimia. In: Roca M, ed. *Trastornos del humor.* Médica Panamericana, Madrid, **1999**, pp 409-443.

Mezzich JE, Fabrega H, Jr., Coffman GA. *Multiaxial characterization of depressive patients.* J Nerv Ment Dis. **1987**, 175:339-346.

Montes MR, Gurpegui M. Neuroquímica psiquiátrica. In: Cervilla J, García-Ribera C, eds. *Fundamentos biológicos en psiquiatría.* Masson, Barcelona, **2000**.

Montgomery SA. *Safety of mirtazapine: a review.* Int Clin Psychopharmacol. **1995**, 10 Suppl 4:37-45.

Moreno FA, Gelenberg AJ, Bachar K, Delgado PL. *Pindolol augmentation of treatment-resistant depressed patients.* J Clin Psychiatry. **1997**, 58:437-439.

Murphy BE, Dhar V, Ghadirian AM, Chouinard G, Keller R. *Response to steroid suppression in major depression resistant to antidepressant therapy.* J Clin Psychopharmacol. **1991**, 11:121-126.

Murphy JM, Monson RR, Laird NM, Sobol AM, Leighton AH. *A comparison of diagnostic interviews for depression in the Stirling County study: challenges for psychiatric epidemiology.* Arch Gen Psychiatry. **2000**, 57:230-236.

Nemeroff CB, DeVane CL, Pollock BG. *Newer antidepressants and the cytochrome P450 system.* Am J Psychiatry. **1996**, 153:311-320.

Nemets B, Mishory A, Levine J, Belmaker RH. *Inositol addition does not improve depression in SSRI treatment failures.* J Neural Transm. **1999**, 106:795-798.

Nierenberg AA, Feighner JP, Rudolph R, Cole JO, Sullivan J. *Venlafaxine for treatment-resistant unipolar depression.* J Clin Psychopharmacol. **1994**, 14:419-423.

O'Dwyer AM, Lightman SL, Marks MN, Checkley SA. *Treatment of major depression with metyrapone and hydrocortisone.* J Affect Disord. **1995**, 33:123-128.

Pacheco L, Malo P, Aragüés E. Síndrome de discontinuación con los ISRS. Aplicaciones prácticas a la clínica diaria. *Temas clave en el manejo de la depresión.* Edimsa, Madrid, **1998**.

Paddin JJ, Martínez E, Arias F. *Fenómenos de abstinencia tras la discontinuación del tratamiento con los inhibidores de recaptación de serotonina.* Psiquiatría Biológica. **1997**, 4:162-166.

Perez V, Soler J, Puigdemont D, Alvarez E, Artigas F. *A double-blind, randomized, placebo-controlled trial of pindolol augmentation in depressive patients resistant to serotonin reuptake inhibitors. Grup de Recerca en Trastorns Afectius.* Arch Gen Psychiatry. **1999**, 56:375-379.

Pérez-Solá V, Álvarez E, Pérez-Blanco J. Fármacos antidepresivos. In: Cervilla J, García-Ribera C, eds. *Fundamentos biológocos en psiquiatría.* Masson, Barcelona, **2000**, pp 315-325.

Pérez Retuerto M, Vázquez-Bourgon ME, Gaite L, Vázquez-Barquero JL. La exploración del estado mental. In: Vázquez-Barquero JL, ed. *Psiquiatría en atención primaria.* 2ª ed. Grupo Aula Médica, Madrid, **1998**.

Pope HG, Jr., Cohane GH, Kanayama G, Siegel AJ, Hudson Jl. *Testosterone gel supplementation for men with refractory depression: a randomized, placebo-controlled trial*. Am J Psychiatry. **2003**, 160:105-111.

Preskorn, S. *Tratamiento ambulatorio de la depresión*. Professional Communications, **1994**.

Price LH, Charney DS, Heninger GR. *Reserpine augmentation of desipramine in refractory depression: clinical and neurobiological effects*. Psychopharmacology (Berl). **1987**, 92:431-437.

Randrup A, Braestrup C. *Uptake inhibition of biogenic amines by newer antidepressant drugs: relevance to the dopamine hypothesis of depression*. Psychopharmacology (Berl). **1977**, 53:309-314.

Richelson E. *The pharmacology of antidepressants at the synapse: focus on newer compounds*. J Clin Psychiatry. **1994**, 55 Suppl A:34-39.

Rudolph R, Entsuah R, Derivan A. *Early clinical response in depression to venlafaxine hydrochloride*. Biol Psychiatry. **1991**, 29:630S.

Rush AJ, George MS, Sackeim HA, Marangell LB, Husain MM, Giller C, Nahas Z, Haines S, Simpson RK, Jr., Goodman R. *Vagus nerve stimulation (VNS) for treatment-resistant depressions: a multicenter study*. Biol Psychiatry. **2000**, 47:276-286.

Schildkraut JJ. *The catecholamine hypothesis of affective disorders: a review of supporting evidence*. Am J Psychiatry. **1965**, 122:509-522.

Servei Català de Salut, Brugulat P, Casaus P, Fernández de Sanmamed, MJ, Foz, G, Mercader, M, Molina C, Palao D, Pastor C, Pérez Arnau F, Pérez Simó R, Saenz I. *Recomanacions per a l'atenció als problemes de salut mental més freqüents en l'atenció primària de salut*. Servei Català deSalut. Barcelona, **2000**.

Smith M, Weissman MM. Epidemiology. In: Paykel ES, ed. *Handbook of the Affective Disorders*. Churchill Livingstone, Edinburgh, **1992**, pp 111-129.

Souery D, Mendlewicz J. *Compliance and therapeutic issues in resistant depression*. Int Clin Psychopharmacol. **1998**, 13 Suppl 2:S13-S18.

Spitzer RL, Endicott J, Robins E. *Research diagnostic criteria: rationale and reliability*. Arch Gen Psychiatry. **1978**, 35:773-782.

Sternbach H. *Danger of MAOI therapy after fluoxetine withdrawal*. Lancet. **1988**, 2:850-851.

Steur EN. *Increase of Parkinson disability after fluoxetine medication*. Neurology. **1993**, 43:211-213.

Stoll AL, Rueter S. *Treatment augmentation with opiates in severe and refractory major depression*. Am J Psychiatry. **1999**, 156:2017.

Thase ME, Friedman ES, Howland RH. *Management of treatment-resistant depression: psychotherapeutic perspectives*. J Clin Psychiatry. **2001**, 62 Suppl 18:18-24.

Thase ME. *What role do atypical antipsychotic drugs have in treatment-resistant depression?* J Clin Psychiatry. **2002**, 63:95-103.

Theobald DE, Kasper M, Nick-Kresl CA, Rader M, Passik SD. *Documentation of indicators for antidepressant treatment and response in an HMO primary care population.* J Managed Care. **2000**, 6:494-498.

Vallejo J, Menchón JM. Tratamiento de las depresiones no melancólicas. In: Vallejo J, Gastó C, eds. *Antidepresivos en la clínica psiquiátrica.* Mosby-Doyma, Madrid, **1996**.

Vallejo J. *Árboles de decisión en psiquiatría.* 2ª ed. Editorial Médica JIMS, Barcelona, **1999**.

Vallejo J. Teorías bioquímicas clásicas de la depresión. In: Vallejo J, Cuenca E, eds. *Depresión y noradrenalina.* Doyma, Barcelona, **1999**.

Vallejo J, Menchón JM. Distimia y otras depresiones no melancólicas. In: Vallejo J, Gastó C, eds. *Trastornos afectivos: ansiedad y depresión.* 2ª ed. Masson, Barcelona, **2000**, pp 261-288.

Vallejo J, Urretavizcaya M. Depresión atípica. In: Vallejo J, Gastó C, eds. *Trastornos afectivos: ansiedad y depresión.* 2ª ed. Masson, Barcelona, **2000**, pp 308-327.

Vallejo J. Clasificación de los trastornos afectivos. In: Vallejo J, Gastó C, eds. *Trastornos afectivos: ansiedad y depresión.* 2ª ed. Masson, Barcelona, **2000**, pp 192-216.

Vázquez-Barquero JL. Epidemiología de los trastornos del humor. In: Roca M, ed. *Trastornos del humor.* Médica Panamericana, Madrid, **1999**.

Warrington SJ. *Clinical implications of the pharmacology of serotonin reuptake inhibitors.* Int Clin Psychopharmacol. **1992**, 7 Suppl 2:13-19.

Wing JK, Cooper JE, Sartorius N. *The measurement and classification of psychiatric symptoms.* Cambridge University Press; Cambridge, **1974**.

Wolkowitz OM, Reus VI. *Treatment of depression with antiglucocorticoid drugs.* Psychosom Med. 999, 61:698-711.

Wolkowitz OM, Reus VI, Keebler A, Nelson N, Friedland M, Brizendine L, Roberts E. *Double-blind treatment of major depression with dehydroepiandrosterone.* Am J Psychiatry. **1999**, 156:646-649.

World Health Organization. *The ICD-10 classification of mental and behavioural disorders: diagnostic criteria for research.* World Health Organization; Geneva, **1993**.

Zohar J, Kaplan Z, Amsterdam JD. Reserpine augmentation in resistant depression: a review. In: Amsterdam JD, ed. *Refractory Depression.* Raven Press, New York, **1991**, pp 219-222.

1.3 Pharmacology

Pharmacotherapy of Depression

Begoña Fernández, Luz Romero and Ana Montero

Dept. of Pharmacology, Laboratorios Esteve, Av. Mare de Déu de Montserrat, 221. 08041 Barcelona, Spain.

1.3.1 Introduction

Depression is a chronic, recurring and potentially life-threatening illness that affects up to 20% of population across the world. It is one of the top ten causes of morbidity and mortality worldwide, based on a survey by the World Health Organization.

All available antidepressants are based on serendipitous discoveries of the clinical efficacy of two classes of antidepressants more than 50 years ago: tricyclic antidepressants (TCAs) and monoamine oxidase inhibitors (MAOIs). Since the introduction of these drugs, the treatment of depression has been dominated by monoamine hypothesis, which postulates that symptoms of depression results from perturbations in serotonergic and/or noradrenergic transmission. In this context, the well documented clinical efficacy of TCAs and MAOIs is due, at least in part, to the enhancement of noradrenergic and/or serotonergic transmission. Unfortunately, their very broad mechanisms of action also include many undesired effects related to their potent activity on cholinergic, adrenergic and histaminergic systems.

The introduction of selective serotonin reuptake inhibitors (SSRIs) over 20 years ago was the next major step in the evolution of antidepressants to develop drugs as effective as TCAs but with a higher safety and tolerability profile. During the past two decades, SSRIs have become the most widely prescribed medication in psychiatric practice. The evolution of antidepressants continued, resulting in the introduction of dual noradrenalin and serotonin reuptake inhibitors, selective noradrenalin reuptake inhibitors and atypical antidepressants that bind to serotonergic and noradrenergic receptors but with a mechanism of action not well known. However, these newer drugs are neither more efficacious nor more rapid acting than their predecessors and approximately 30% of population does not respond to current therapies.

In recent years, new strategies based on the better understanding of pathophysiology of depression are being developed. The goal of this chapter is to give a brief overview of the current antidepressant treatments available today, the major advances from monoamine-based treatment strategies and the new emerging approaches in the treatment of depression.

1.3.2 Current antidepressant treatments

Almost all the available medications for depression are based on the tryciclic antidepressants. These drugs provided a template for the development of newer classes of antidepressants including selective serotonin reuptake inhibitors (SSRIs) characterized by their ability to preferentially increase serotonin release, a more recent class of novel antidepressants that block both serotonin and noradrenalin reuptake (SNRIs) and noradrenalin reuptake inhibitors (NRIs) (Table 1). All these newer class of antidepressant show the same therapeutic efficacy than TCAs and reduced side-effects.

Table 1. Currently available antidepressant treatments.

Treatment	Mechanism of action
Tricyclics	Inhibition of mixed noradrenaline and serotonin reuptake
Selective serotonin reuptake inhibitors (SSRIs)	Inhibition of serotonin-selective reuptake
Noradrenaline reuptake inhibitors (NRIs)	Inhibition of noradrenaline-selective reuptake
Serotonin and noradrenaline reuptake inhibitors (SNRIs)	Inhibition of mixed noradrenaline and serotonin reuptake
Monoamine oxidase inhibitors (MAOIs)	Inhibition of monoamine oxidase A (MAO_A)
Lithium	Unknown
Atypical antidepressants	Unknown

1.3.2.1 Tricyclics

Tricyclic antidepressants (TCA) are a class of antidepressant drugs first used in the 1950s. They are named after the drugs' molecular structure, which contains three rings of atoms. The exact mechanism of action is not well understood, but it is generally thought that tricyclic antidepressants work by inhibiting the reuptake of the neurotransmitters noradrenalin and serotonin by nerve cells. Although this pharmacologic effect occurs immediately, the patient's symptoms often do not ameliorate until several weeks.

For many years, they were the first choice for the pharmacological treatment of depression. Although still considered effective, they have been increasingly replaced by newer antidepressants that show the same therapeutic efficacy as TCAs and reduced side-effects. The main side-effects of TCAs are related to their anti-muscarinic properties, including dry mouth, blurred vision and decreased gastrointestinal motility and secretion, besides others. However, TCAs are sometimes still used to treat refractory depression that has failed to respond to newer therapies.

The first TCA discovered was imipramine, which was discovered accidentally in a search for a new antipsychotic in the late 1950s. Antidepressant drugs in the tricyclic drug group include:

- Amitriptyline (Elavil®, Endep®, Tryptanol®)
- Clomipramine (Anafranil®)
- Desipramine (Norpramin®, Pertofrane®)
- Doxepin (Adapin®, Sinequan®)
- Imipramine (Tofranil®)
- Lofepramine (Gamanil®, Lomont®)
- Nortriptyline (Pamelor®)
- Protriptyline (Vivactil®)
- Trimipramine (Surmontil®)

1.3.2.2 Monoamine oxidase inhibitors

Monoamine oxidase inhibitors (MAOIs) are a class of antidepressant drugs prescribed for the treatment of depression. MAOIs act by inhibiting the activity of monoamine oxidase, preventing the breakdown of monoamine neurotransmitters and so increasing the available stores. There are two isoforms: MAO-A and MAO-B. MAO-A preferentially deaminates serotonin, melatonin, adrenaline and noradrenalin. MAO-B preferentially deaminates phenylethylamine and trace amines. Dopamine is equally deaminated by both types.

Due to their serious side-effects, MAOIs are used less frequently than other classes of antidepressant drugs. However, they are tried in some cases where patients are unresponsive to other treatments, often with a marked success. They are particularly effective in treating atypical depression.

Monoamine oxidase inhibitors include:

- Isocarboxazid (Marplan)
- Moclobemide (Aurorix, Manerix, Moclodura®)
- Phenelzine (Nardil)
- Tranylcypromine (Parnate)
- Selegiline (Selegiline Eldepryl)
- Harmala

1.3.2.3 Selective serotonin reuptake inhibitors

Selective serotonin reuptake inhibitors (SSRIs) are a newer class of antidepressants for treating depression. These drugs are designed to allow the available neurotransmitter serotonin to be utilized more efficiently. SSRIs inhibit the reuptake of neurotransmitter serotonin in the synaptic cell, increasing levels of serotonin within the synaptic cleft.

SSRIs are described as selective because they affect only the reuptake pumps responsible for serotonin, as opposed to earlier antidepressants, which also affect other monoamine neurotransmitters. Because of this, SSRIs lack some of the side-effects of the more general drugs showing the same clinical efficacy.

There are many drugs of this class, including the following:

- Citalopram (Celexa, Cipramil, Emocal. Sepram)
- Escitalopram oxalate (Lexapro, Cipralex, Esertia)
- Fluoxetine (Prozac, Fontex, Seromex, Seronil, Sarafem, Fluctin)
- Fluvoxamine maleate (Luvox, Faverin)
- Paroxetine (Paxil, Seroxat, Aropax, Deroxat)
- Sertraline (Zoloft, Lustral, Serlain)
- Trazodone (Desyrel)

1.3.2.4 Noradrenaline reuptake inhibitors

Noradrenalin reuptake inhibitors (NRIs) are compounds that elevate extracellular noradrenalin in the central nervous system by inhibiting its reuptake into the synapse via the noradrenalin transporter. For the past decade, the role of noradrenalin in depression has been somewhat neglected in favor of serotonin because of the advent of the SSRIs. However, the recent development of reboxetine, the first selective noradrenalin reuptake inhibitor, has allowed clinical investigation of the role of noradrenergic system in different aspects of depressive disorders.

Adverse events predicted by the neuroanatomy of the noradrenergic system, such as tremor and cardiovascular effects, occur less frequently than expected. Selective noradrenalin reuptake inhibition therefore offers a significant improvement in antidepressant pharmacotherapy and an opportunity to increase our understanding of the role of noradrenalin in depression (for a review, see Montgomery, 1999).

1.3.2.5 Serotonin and noradrenaline reuptake inhibitors

Although progress toward more specific medications over the years has generally yielded the benefit of enhanced tolerability, a similar trend has not been observed toward greater efficacy in depression. In fact, the SSRIs are no more effective than TCAs in the treatment of depression; and as yet none of the SSRIs has distinguished itself clearly as the best of the class. In recent years, interest has turned to medications that specifically target both the serotonin and the noradrenalin systems, the serotonin-noradrenalin reuptake inhibitors (SNRIs). In a sense, this new focus represents a step away from the specificity or selectivity associated with SSRIs but still embraces some concept of selectivity, at less in comparison with TCAs, which also exert an effect on other neurotransmitter systems. There is evidence suggesting that dual reuptake blockers are more rapid and/or efficacious than SSRIs (Danish University Antidepressant Group, 1986, 1990; Anderson and Tomenson, 1994; Thase et al., 2001). However, they do not yet appear to meet the requirements of a fast-acting antidepressant.

1.3.2.6 Lithium

Lithium has various molecular actions (for example, inhibition of phosphatidylinositol phosphatases, adenyl cyclases, glycogen synthase kinase 3β and G proteins) but it is unknown which of these actions is responsible for its antidepressant action.

1.3.2.7 Atypical antidepressants

In this group, we could include drugs like bupropion, mirtazapine or tieneptine. Although these drugs have reported monoamine-based mechanisms (bupropion inhibits dopamine uptake, mirtazapine is an α_2 adrenergic receptor antagonist with an affinity for $5-HT_3$ and $5-HT_2$ receptors and tianeptine is an activator of monoamine reuptake), these actions are not necessarily the mechanism that underlie the drugs' therapeutic benefit.

1.3.3 New strategies for antidepressant treatments

1.3.3.1 Monoaminergic strategies

Despite the available treatments for depression being safe and effective, fewer than 50% of all patients with depression show full remission with optimized treatment. In addition to the need to administer the drugs for weeks to see clinical effects, side-effects are still a serious problem even with the newer antidepressants. Therefore, there is still a great need for faster-acting, safer and more effective treatments for depression.

1.3.3.1.1 SSRI/Aminergic antagonism

In order to accelerate the onset of antidepressant actions and to limit side-effects, one of the current drug development strategies focuses on designing new antidepressants with dual modes of action. The delayed clinical efficacy is thought to reflect the time required for desensitization of the receptors regulating monoamine release (e.g., $5-HT_{1A}$, $5-HT_{2C}$ and α_2 adrenergic receptors between others). Thus, one of the new strategies for antidepressant development looks for SSRI compounds that also block these monoaminergic receptors (Schechter et al., 2005).

1.3.3.1.1.1 SSRI/ $5-HT_{1A}$ antagonists

$5-HT_{1A}$ are somatodendritic receptors expressed at the presynaptic level, in the midbrain raphe nuclei, and postsynaptically to serotonin nerve terminals, mainly in cortico–limbic areas. Activation of $5-HT_{1A}$ receptors hyperpolarizes the neuronal membrane, reducing the firing rate of raphe nuclei serotonergic neurons (Sprouse and Aghajanian, 1987) and as a consequence reducing 5-HT release in serotonergic terminal areas (Artigas et al.,

1996). Most antidepressants increase serotonin release in the brain. However, this increase is offset by the negative feedback mediated by 5-HT$_{1A}$ autoreceptor activation. The delayed clinical efficacy of antidepressant drugs is believed to result from the indirect activation of presynaptic 5-HT$_{1A}$ receptors in the raphe nuclei. In this context, acute administration of SSRIs inhibits the serotonergic neuron firing rate and subsequent release of serotonin in terminal regions. However, following long-term SSRI treatment (14–21 days), 5-HT$_{1A}$ autoreceptors desensitize, resulting in a more pronounced elevation in extracellular serotonin concentration compared to acute treatment (Blier and Montigny, 1994; Perez et al., 2001).

In 1993, Artigas et al. proposed that 5-HT$_{1A}$ receptor antagonists could accelerate the clinical effects of antidepressants by preventing this negative feedback. In this context, clinical data using this combination strategy demonstrate that the antidepressant activity of SSRIs is accelerated and/or enhanced when combined with the mixed 5-HT$_{1A/\beta}$-adrenoceptor antagonist pindolol (Blier and Bergeron, 1998). While the results of these clinical studies remain controversial, there is enough collective evidence suggesting that combination of SSRIs and 5-HT$_{1A}$ antagonists could be a useful strategy in the treatment of depression (Celada et al., 2004). In this way, several companies recently published on the synthesis of such dual-acting compounds, showing both SSRI and full/partial 5-HT$_{1A}$ receptor antagonism (Mewshaw et al., 2004; Hughes et al., 2005; Table 2).

1.3.3.1.1.2 SSRI/ 5-HT$_{2A}$ antagonists

5-HT$_{2A}$ serotonin receptors are positively coupled to phospholipase C (PLC) and mobilize intracellular calcium. There are mainly expressed in GABAergic interneurons in the cortex (Sheldon and Aghajanian, 1991; Francis et al., 1992; Morilak et al., 1993; Burnet et al., 1995) but also in projections of pyramidal glutamatergic neurons (Burnet et al., 1995; Wright et al., 1995).

In recent years, there were many studies suggesting that atypical antipsychotic drugs and some antidepressants (mianserine and mirtazapine) augment the clinical response to SSRIs in treatment-resistant patients (Ostrof and Nelson, 1999; Shelton et al., 2001; Marangell et al., 2002; Carpenter et al., 2002). All these drugs occupy 5-HT$_2$ receptors in the brain at clinical doses and block 5-HT$_2$-mediated responses, mainly 5-HT$_{2A}$ responses

(Marek et al., 2003). To determine whether 5-HT$_{2A}$ receptor antagonism could possibly account for the additional benefits of combining an SSRI with an atypical antipsychotic, studies have been developed with YM992, a SSRI/5-HT$_{2A}$ antagonist. With regard to its effect on serotonin transmission, there were no differences between the actions of this drug when compared with the actions produced by SSRIs (Canuso et al., 2004). However, the effect of YM992 on neuronal noradrenergic activity was drastically different from that of SSRIs and even NRIs or MAOIs (Blier and de Montigny, 1985; Szabo and Blier, 2001, 2002). Behind SSRIs, NRIs or MAOI chronic treatment inhibits spontaneous noradrenergic neuronal activity. YM992 would not dampen noradrenergic transmission (Blier and Szabo, 2005). There are studies showing an inhibition of noradrenalin release after 5-HT$_{2A}$ activation (Done and Sharp, 1992, 1994). Therefore, it is possible that the beneficial effect of blocking 5-HT$_{2A}$ receptors when combined with SSRIs may be due to the action on noradrenalin neurons.

1.3.3.1.1.3 SSRI/5-HT$_{2C}$ antagonists

In vivo microdialysis studies report that the effect of SSRIs on serotonin release in cortex and hippocampus is enhanced by 5-HT$_{2C}$ receptor antagonist administration (Cremers et al., 2003; Mørk and Hogg, 2003). These observations are in agreement with the fact that fluoxetine increases cortical extracellular 5-HT levels more in 5-HT$_{2C}$ receptor knockout mice than in wild-type mice (Cremers et al., 2004). Using behavioral models of depression, a marked increase in the antidepressant effects of SSRIs is shown when combined with 5-HT$_{2C}$ receptor antagonists (Cremers et al., 2003, 2004). Altogether, this supports the combination of SSRI and 5-HT$_{2C}$ block as a good strategy in the treatment of depression.

1.3.3.1.1.4 SSRI/α$_2$ antagonists

The improvement in the antidepressant effect of SNRIs versus SSRIs and the success of therapies with NRIs show the importance of elevated noradrenalin in the treatment of depression. α$_2$ Adrenergic receptors placed in terminal areas (autoreceptors) regulate noradrenalin release in the synaptic cleft through a feedback inhibitory mechanism (Langer, 1974; Dubocovich, 1984; Starke, 1987; Miller, 1998). There are also α$_2$ adrenergic receptors placed on somatodendritic areas regulating the firing rate of locus coeruleus noradrenergic neurons (Cederbaun and Aghajanian, 1976). In contrast, there are α_2 adrenoceptors placed on non-noradrenergic terminals (heteroreceptors) regulating the release of serotonin (Göthert et al., 1981; Feuerstein et al., 1993) and dopamine (Andén and Grabouska, 1976). Activation of α_2 adrenoceptors can inhibit the release of the three monoamines: noradrenalin, serotonin and dopamine. Thus, blocking the α_2 adrenoceptors in combination with monoamine reuptake inhibition could be a good strategy in order to improve the antidepressant effect. In this context, neurochemical data show that co-administration of α_2 adrenergic receptors antagonists with antidepressants improves the ability of these drugs to increase the release of noradrenalin, serotonin and dopamine in the rat cortex (Gobert et al., 1997). According with this hypothesis, clinical studies emphasize that combining non-specific α_2 adrenoceptors antagonists with SSRIs accelerate the onset of antidepressant actions (Cappiello et al., 1995; Sanacora et al., 2004). At the experimental level, α_2 adrenoceptor antagonists have been used to improve the effects of noradrenalin reuptake inhibitors (Invernizzi and Garattini, 2004).

1.3.3.1.2 Serotonin noradrenaline and dopamine reuptake inhibition

Although antidepressant treatments are mainly focused on noradrenergic and serotonergic modulation, the dopaminergic mesolimbic and mesocortical systems are fundamental in hedonia and motivation. Therefore, an important role of dopamine in understanding depression treatment should be taken into account. Recently, a single molecule that inhibits noradrenalin, serotonin and dopamine transporters has shown antidepressant activity in animal models of depression (Skolnick et al., 2003). In this context, certain companies are developing molecules with this mechanism of action for the treatment of depression (Table 2).

1.3.3.1.3 Selective serotonin approaches

It seems that the antidepressant effect of SSRIs is mediated by increased serotonin and the subsequent activation of certain serotonin receptors. However, it is probably not necessary to implicate the 14 subtypes of serotonin receptors to achieve the antidepressant effect. Even more, side-effects induced by SSRIs could be mediated, at least in part, by activation of some of the serotonin receptors. Thus, a new approach in order to improve antidepressant treatments consists in selectively targeting serotonergic postsynaptic receptors involved in the antidepressant effect of SSRIs.

1.3.3.1.3.1 5-HT$_{2C}$

The 5-HT$_{2C}$ receptor warrants consideration in the development of novel treatment strategies for depression. Recent findings indicate that RNA editing of these receptors is regulated by SSRI treatment and altered in the prefrontal cortex of suicide victims (Gurevich et al., 2002a, 2002b). However, the role of these receptors in depression is controversial. In the previous section, we talked about a new approach to improve antidepressant treatment consisting of the combination of SSRs and 5-HT$_{2C}$ receptor antagonists (see Section 1.3.3.1.1.3, *SSRI/5-HT$_{2C}$ antagonists*). However, other studies have reported that at least part of the effects elicited by SSRIs seems to be mediated by 5-HT$_{2C}$ receptor activation (Berendsen and Broekkamp, 1994), suggesting that some of the therapeutic effects of SSRIs may be mediated at least in part by 5-HT$_{2C}$ receptor agonism. Accordingly, there are a number of common effects in rodents that have been noted for SSRIs and 5-HT$_{2C}$ agonists (Broekkamp and Berendsen, 1992). Supporting this hypothesis, recent data report that compounds acting like 5-HT$_{2C}$ agonists have shown antidepressant effects in depression animal models (Dunlop et al., 2005). The discovery of 5-HT$_{2C}$ receptor agonists is one of the new strategies to improve antidepressant treatments.

1.3.3.1.3.2 5-HT$_7$

There is evidence that implicates 5-HT$_7$ receptors in the physiopathology of depression and suggests that its modulation could exert antidepressant effects (Schwartz, 1993; Barnes and Sharp, 1999; Mullins et al., 1999; Wood et al., 2002). In situ hybridisation studies revealed CNS expression in amygdala, cortex, hippocampus, thalamus, septum, hypothalamus and suprachiasmatic nucleus (Gustafson et al., 1996). 5-HT$_7$ receptors are thus highly expressed in cortical and limbic areas (Thomas et al., 2000; Martín-Cora and Pazos, 2004), both regions implicated in depressive disorders. As 5-HT$_7$ receptors are also expressed in the suprachiasmatic nucleus, it is important to note that 5-HT$_7$ receptors modulate REM sleep and play a role in circadian rhythm regulation. In this context, a number of experimental and clinical observations support a relationship between disturbances in circadian rhythms and sleep and in the processes underlying unipolar depression (Boivin et al., 1997). This CNS distribution, coupled with the receptor's relatively high affinity for several psychoactive drugs, including antidepressants (e.g., mianserin, maprotiline, imipramine and amitriptyline) (Monsma et al.,1993; Shen et al., 1993; Lucchelli et al., 2000) implicated the 5-HT_7 receptor as a potential therapeutic target in depression. In fact, acute antidepressant treatment increases c-Fos expression in the suprachiasmatic nucleus and this effect is blocked by a non-selective 5-HT_7 receptor antagonist, ritanserine (Mullins et al., 1999). Pharmacological adrenalectomy increases 5-HT_7 receptor mRNA expression in the hippocampus, effect partly reversed by corticosterone replacement. 5-HT_7 receptor expression may thus explain, at least partially, the therapeutic actions of adrenal steroid synthesis inhibitors in resistant depression (Yau et al.,1997).

In spite of the finding of 5-HT_7 ligands with antidepressant effects as a common strategy in the pharmaceutical industry, the mechanism of action by which 5-HT_7 receptors exert their antidepressant action remains unclear (Thomas and Hagan, 2004). A reduced expression of glucorticoid receptors is an important issue in the physiopathology of depression (Pariante and Miller, 2001). In primary cultures of hippocampus, the activation of 5-HT_7 receptors increases the expression of glucocorticoid receptors (Laplante et al., 2002). Thus, it seems that 5-HT_7 agonists could be a good tool for the treatment of depression. However, after chronic antidepressant treatment, it has been described that a desensitization of 5-HT_7 receptors occurs in the hypothalamus (Sleight et al., 1995; Mullins et al., 1999), suggesting that 5-HT_7 receptor antagonists could accelerate the antidepressant effect or exert antidepressant effects by themselves. More studies will be necessary in order to clarify the role of 5-HT_7 receptors in depression and their contribution in the antidepressant treatment. In this way, antidepressants belonging to different chemical classes behave as antagonists at the 5-HT_7 receptor (Lucchelli et al., 2000) and 5-HT_7 receptor knockout mice display an "antidepressant-like" phenotype, as shown by a significant decrease in immobility compared to controls in the forced swim and tail suspension tests (Guscott et al., 2005; Hedlund et al., 2005), preclinical assays routinely used to assess antidepressant potencial. This phenotype of knockout mice correlates well with the antidepressant activity found in the same assays when treating mice with selective 5-HT_7 receptor antagonists (Guscott et al., 2005; Hedlund et al., 2005). Additional studies investigating the role of 5-HT_7 ligands in depression will be required to further elucidate the role of this receptor and their contribution in the antidepressant treatment.

1.3.3.1.3.3 5-HT_6

Several tricyclic antidepressant compounds such as amitriptyline and atypical antidepressant compounds as mianserine display a high affinity for the 5-HT_6 receptor (Monsma et al., 1993; Kohen et al., 2001). Interestingly, extracts of the plant *Hypericum perforatum* (St John's Wort), used in the treatment of mild depression, also had a high affinity for the human 5-HT_6 receptor (Gobbi et al., 2001; Simmen et al., 1999). This finding led to the implication of a role for the receptor in the pathogenesis and/or treatment of affective disorders. Furthermore, 5-HT_6 receptor expression appears to be regulated by glucocorticoids. In particular, Yau et al. (1997) demonstrated that adrenalectomy increases the expression of 5-HT_6 receptors in specific hippocampal subfields. Since elevated levels of glucocorticoids are observed in some depressive subjects and glucocorticoid synthesis blockers have been used as an antidepressant therapy, it is possible that 5-HT_6 receptor modulation by glucocorticoids may have clinical significance. However, some antidepressants have been shown to increase the expression of growth factors (i.e. BDNF) in the hippocampus, suggesting that their therapeutic effects actually depend on an increased growth and synaptic plasticity in this region (Russo-Neustadt et al., 2005). One candidate 5-HT receptor for mediating these changes in synaptic plasticity is the 5-HT_6 receptor as activation of this receptor upregulates BDNF mRNA in the hippocampus following either acute or short-term (4 days) treatment, an effect blocked by 5-HT_6 antagonism (De Foubert et al., 2004). A polymorphism in the 5-HT_6 receptor gene has also been associated with the response to antidepressant treatment in patients with major depressive disorder (Lee et al., 2005). However, 5-HT_6 blockade is not effective in modulating the neurochemical effects induced by NA and/or 5-HT reuptake inhibitor antidepressants (Dawson and Li, 2003) and at the moment there are no studies reporting antidepressant activity in animal models of depression using 5-HT$_6$ ligands. More studies will be necessary in order to elucidate the role of 5-HT$_6$ receptors in depression.

1.3.3.2 Non-monoaminergic strategies

1.3.3.2.1 HPA-based strategies

Excessive stimulation of the hypothalamic–pituitary–adrenal (HPA) axis is implicated in depression. Indeed, depressed patients show hyperactivity of the HPA (Barden, 2004). The result of numerous clinical studies suggests that normalization of the HPA axis might be necessary for stable remission of symptoms of depression; and a failure of normalize HPA activity usually predicts a poor antidepressant action. Based on these data, new strategies to improve antidepressant treatment are being developed.

1.3.3.2.1.1 CRF Antagonists

HPA is mainly controlled by corticotrophin-releasing factor (CRF), secreted by hypothalamic neurons present in the paraventricular nucleus (PVN). There are two CRF receptor subtypes: CRF1 and CRF2. CRF1 receptors are placed in corticolimbic areas and in the pituitary, while CRF2 receptors show mainly a peripheral location. The expression of CRF is increased in the hypothalamus of patients with depression. These patients also show high levels of CRF in the cerebrospinal fluid (CSF) and reduced feedback inhibition of the axis by CRF and glucocorticoids (Barden, 2004).

CFR modulates the HPA axis mainly by activating CRF1 receptors placed in the pituitary. Therefore, the CRF1 receptor has emerged as a target of interest for antidepressant development and CRF antagonists might represent a novel class of antidepressant drugs. There are lots of studies indicating that several synthetic antagonists exhibit antidepressant/anxiolitic activity in preclinical animal models, including learned helplessness (a model with documented sensitivity to drugs showing antidepressant activity in humans; Mansbach et al., 1997), the forced swimming test, chronic mild stress and olfactory bulbectomized rats (Ducottet et al., 2003). At the clinical level, R121919 (Janssen) has demonstrated antidepressant efficacy (Zobel et al., 2000). Unfortunately, clinical development of this compound was discontinued, believed to be because of hepatotoxicity. In the past few years, several pharmaceutical companies have been developing new selective CRF1 receptor antagonists (Table 2). Clinical studies with these new compounds will help us to know more about the efficacy and safety of these compounds.

1.3.3.2.1.2 Vasopressin receptor antagonists

The neuropeptide vasopressin (V), which is synthesized in the PVN and supraoptic hypothalamic nuclei, modulates the HPA axis and improves the effect of CRF on adrenocorticotropic hormone (ACTH) release. The central vasopressinergic system has been examined as a platform for psychiatric drug development, including depression. V is also found outside the hypothalamus, notably in the amygdala, and is believed to exert effects throughout the limbic system through activation of V1a and V1b receptors. V levels are increased in some patients with depression and might contribute to HPA axis abnormalities observed in these patients. Postmortem studies indicate that SSRI treatment normalizes V levels (Holmes et al., 2003). Together, this leads many to hypothesize the utility of a central vasopressinergic receptor antagonism as a potentially novel antidepressant strategy. Non-peptide V1b antagonists show antidepressant-like effects in rodents (Holmes et al., 2003). This is in contrast to V1b knockout mice, which show normal stress responses (Wersinger et al., 2002; Winslow and Insel, 2004). V antagonists have yet to be evaluated in clinical studies.

1.3.3.2.1.3 Glucocorticoids

The functionality of the HPA axis is feedback-regulated by glucocorticoid receptors (GR) which are located in the brain and periphery. As mentioned before, insufficient feedback suppression of the HPA axis by CRF and glucocorticoids is seen in a large subset of patients with depression. This neuroendocrine abnormality was reproduced in adult mice with selective deletion of GR2 in the forebrain. Interestingly, this mutation resulted in a depression-like phenotype and many of these abnormalities were corrected after antidepressant treatment. Moreover, transgenic mice over-expressing GR2 in the forebrain are more sensitive to the acute effects of antidepressants (Wei et al., 2004). Chronic treatment with TCAs and NRIs increased GR expression (Pariante and Miller, 2001;, Calfa et al., 2003; Adell et al., 2005). Based on these finding, a new strategy is the use of GR antagonists as targets for antidepressant drugs. In this context, the GR antagonist mifepristone is currently in Phase III clinical trials for psychotic major depression and might be the first non-monoaminergic-based antidepressant on the market (Adell et al., 2005). However, the clinical use of existing drugs is limited by their severe side-effects.

1.3.3.2.2 NK1 antagonists

Substance P (SP) is an undecapeptide member of the tackchykinin family of mammalian neuropeptides, which is the preferred endogenous agonist for neurokin 1 (NK1) receptors. The relationship between SP and depression is based in several lines of evidence. First, there is a colocalization of SP with serotonergic and noradrenergic systems, strongly implicated in mood regulation. In particular, there is a high density of NK1 receptors in brain areas related with the control of emotional responses (e.g., amygdala, hippocampus, frontal cortex, raphe nuclei and locus coeruleus; Maeno et al., 1993; Saffroy et al., 2003). Second, acute and chronic stressors increase SP (Takayama et al., 1986) and administration of SP or NK1 agonists induces a stress response in animal models (Helke et al., 1990; Kramer et al., 1998), which is blocked by NK1 receptor antagonists (Culman et al., 1997). Elevated levels of SP have been reported in CSF and plasma samples of depressive patients (Rimon et al., 1984); and antidepressant treatment decreases biosynthesis of SP in rat forebrain (Shirayama et al., 1996). Collectively, all these data suggest that the NK1 receptor antagonist could show an antidepressant-like profile. In this context, numerous NK1 antagonists have been developed by the pharmaceutical industry showing antidepressant activity in preclinical animal models (Herpfer et al., 2005). In 1998, Kramer et al. published the first evidence that chronic treatment with a NK1 receptor antagonist might be antidepressant in humans. However, their results were replicated in some studies, but not in others, and the validity of NK1 receptor antagonists as effective antidepressants remains thus unclear. Nevertheless, pharmaceutical companies have not slowed interest in the continued clinical development of NK1 antagonists and at this moment there are numerous NK1 receptor antagonists in clinical development (Table 2).

Although NK1 receptor antagonism was initially claimed as a novel and unique mechanism of action, further studies have suggested that its therapeutic action could be associated with changes in monoaminergic systems (Adell, 2004; Blier et al., 2004). In this way, chronic administration of NK1 receptor antagonists increased the firing of serotonergic neurons in the dorsal raphe nucleus (Haddjeri and Blier, 2001; Conley et al., 2002) and it has been shown that the NK1 receptor antagonists improve the neurochemical effects of SSRIs (Guiard et al., 2004). These data suggest the possibility of a new therapeutic approach to the treatment of depression using NK1 receptor antagonists as augmentation agents in combination with traditional antidepressants in order to improve antidepressant response (Ryckmans et al., 2002).

The investigations into NK2 antagonists are mainly focused on inflammation. However, support from preclinical data in animal models of depression suggests antidepressant-like effects of these compounds (Griebel et al., 2001; Schechter et al., 2005). In this context, some companies are developing NK2 receptor antagonists for the treatment of depression.

1.3.3.2.3 MCH antagonists

Melanin concentration hormone (MCH) is a 19-amino-acid orexinergic neuropeptide synthesized by neurosecretory cells of the mammalian lateral hypothalamus and the zona incerta (Bittencourt et al., 1992), which plays an important role in the complex regulation of energy balance and body weight. MCH exerts its action through activation of two G protein-coupled receptors (GPCRs): MCH1-R and MCH2-R. MCH1-R is the only subtype identified in rats. Evidence implicates the MCH system in the regulation of mood and stress responses. MCH antagonists produce antidepressant-like effects in preclinical antidepressant models similar to those observed with the SSRI fluoxetine (Georgescu et al., 2005). These data support a rationale for a novel mechanism of action for depression. The clinical evaluation of these compounds awaits evaluation.

1.3.3.2.4 Neurotrophins

An alternate hypothesis of depression postulates that neurotrophic growth factors (NGF) are involved or mediate the mechanism of action of antidepressant drugs. The NGF neurotrophins family activates two classes of receptors; a low-affinity p75 receptor, which is common to all neurotrophins, and the high-affinity tyrosine kinases (trk) receptors, which are associated with specific neurotrophins and encode transmembrane receptor trks that mediate multiple signaling pathways. It has been shown that trkB is the subtype receptor for BDNF.

Antidepressant research has focused on BDNF and trkB. Antidepressant treatment increases mRNA and protein levels of BDNF in the rat hippocampus and cortex, indicating that up-regulation of this factor could be related with the antidepressant effect. The time-course necessary for this modulation corresponds with the time necessary for clinical efficacy. Acute or chronic stress decreases expression of BDNF in the hippocampus and this is prevented by chronic antidepressant treatment. At the clinical level, depressed patients taking antidepressants have high BDNF levels in the hippocampus, while those untreated show low BDNF levels in serum (Chen et al., 2001). Postmortem studies have reported decreased hippocampal trkB and BDNF mRNA in suicides compared with controls (Dwivedi et al., 2003). BDNF acutely administered into the lateral ventricle or hippocampus produces antidepressant-like effects in the forced swimming and learned helplessness paradigms (Siuciak et al., 1997; Shirayama et al., 2002). Inducible knockout of BDNF from the hippocampus and other forebrain regions prevents the antidepressant effects of reuptake-inhibitor antidepressants in these paradigms (Monteggia et al., 2004).

Together, these data support the possibility that drugs that activate BDNF signaling in the hippocampus might be antidepressant. However, BDNF is not an easy drug target. It binds trkB receptor as a dimer, it does not cross the blood-brain barrier and it is difficult to develop small molecule agonists of trkB. BDNF activation of trkB leads to diverse physiological effects by regulating a complex cascade of post-receptor pathways, which involve Ras-Raf-ERK (extracellular-signal regulated kinase), phosphatidylinositol 3-kinase (PI3K)-Akt (v-akt murine thymoma viral oncogene homologue) and PLCγ. Another strategy in the development of new antidepressants consists of the regulation of BDNF actions at post-receptor level. However, it is not well known which of these signaling proteins are most important for the antidepressant actions of BDNF. Another challenge is that, although BDNF exerts antidepressant actions at the level of hippocampus, it induces pro-depression-like effects at the reward circuit level. (Eisch et al., 2003; Berton et al., 2006). These findings raise caution about the goal of developing antidepressants based on BDNF. In consequence, in spite of the multiple evidence that relates depression and antidepressant actions with neurotrophic mechanisms, it is difficult to translate these discoveries in new treatment approaches.

1.3.3.2.5 Phosphodiesterase inhibitors

Phosphodiesterase-IV (PDE-IV) catalyzes the degradation of cAMP. Activation of the cAMP pathway leads to the activation of the transcription factor cAMP response element-binding protein (CREB), which is an important regulator of BDNF gene expression. Induction of CREB in the hippocampus exerts antidepressant-like effects in the forced swimming test (Chen et al., 2001). Therefore, PDE-IV inhibition could be a valid strategy in the treatment of depression. Rilopram, a PDE-IV inhibitor that increases BDNF, shows an antidepressant effect in animal models (Takahashi et al., 1992).

The development of PDE-IV inhibitors as antidepressants shows two main problems. First, PDE-IV inhibitors induce nausea and vomiting. Second, stimulation of the cAMP pathway and CREB in the nucleus accumbens might be pro-depressant.

1.3.3.2.6 Glutamatergic targets

The N-methyl-D-aspartate (NMDA) receptor is an ionotropic glutamate receptor distributed with highest density in the cortico–limbic regions of the brain. Chronic antidepressant administration can modify NMDA receptor function (Skolnick et al., 1996; Petrie et al., 2000) and antagonism of these receptors induces antidepressant-like effects in animal models (Papp and Moryl, 1994). A single injection of ketamine, an NMDA receptor antagonist, showed an antidepressant effect in a placebo-controlled trial. The application of ketamine and related drugs is limited by their severe spycomimetic actions. However, preclinical research suggests that NMDA receptor antagonists show antidepressant-like effects in animal models of depression, while chronic treatment with antidepressants down-regulates NMDA receptors and reduces glutamate release through presynaptic mechanism (Korte et al., 2005; Sanacora et al., 2003; Paul and Skolnick, 2003). Deletion of a novel NMDA receptor subunit induces an anxiolytic- and antidepressant-like profile (Miyamoto et al., 2002).

α-Amino-3-hydroxy-5-methyl-4-isoxazole propionic acid (AMPA) receptors have been also related with antidepressant action. It has been reported that AMPA activation increases BDNF expression and stimulates neurogenesis in the hippocampus (Duman et al., 2004). Therefore, another glutamatergic strategy in the development of new antidepressants has been the evaluation of AMPA receptors potentiators in animal models of depression (Sanacora et al., 2003; Paul and Skolnick, 2003; Alt et al., 2005). In this context, positive alosteric modulators of NMDA receptors were reported to have the same antidepressant activity than TCAs and SSRIs in the forced swim and tail suspension tests.

The metabotropic glutamate receptors (mGluRs) are GPCR and it seems that they could regulate specific populations of NMDA or AMPA receptors. Specific modulations of mGluRs can give us a new strategy more selective and safer to find antidepressants based on glutamatergic modulations.

Table 2. Antidepressants in drug discovery

Drug name	Pharmacological action	Company	Developmental phase
EMSAM® (selegiline)	MAO-B inhibitor, weak MAO-A inhibitor	Bristol-Mayers Squibb, Somerset (Mylan/Watson)	Approved
Valdoxan (agomelatine, S-200989	5-HT_{2C} antagonist, 5-HT_{2B} antagonist, melatonin M1/M2 receptor agonist	Servier	Submitted in the European Union; decision expected early to mid-2006
DVS-233 SR (desvenlafaxine)	Metabolite of Effexor® (venalfaxine)	Wyeth	Pre-registration. NDA to be submitted in early 2006
Gepirone ER	5-HT_{1A} partial agonist	Fabre-Kramer	Amended NDA scheduled to be submitted in March
SR 58611	β-3-adrenoceptor agonist	Sanofi-Aventis	Phase III
Saredutant (SR 48968)	NK2 antagonist	Sanofi-Aventis	Phase III
PRX-00023	5-HT_{1A} agonist, sigma receptor antagonist, reuptake inhibitor	Predix	Phase II

Radafaxine, GW353162	Noradrenaline reuptake inhibitor and weak dopamine reuptake inhibitor (metabolite of bupropion)	GSK	Phase III (beginning in 2006)
DOV 216,303	Dopamine/serotonin/noradrenaline reuptake inhibitor	DOV/Merck	Phase II – complete
DOV 21, 947	Dopamine/serotonin/noradrenaline reuptake inhibitor	DOV/Merck	Phase II
Miraxion, LAX-101	"Purified" omega 3	Amarin	Phase II
GW372475	Dopamine, serotonin and noradrenaline reuptake inhibitor	GSK, NeuroSearch	Phase II
Nemifitide (INN 00835)	Pentapeptide analog of melanocyte-inhibiting factor (MIF-1) administered intravenously (mechanism unknown)	Innapharma	Phase II
ORG 34517/34850	GR antagonist	Organon	Phase II
Vestipitant, GW597599	NK1 antagonist	GSK	Phase II
CP-122,721	NK1 antagonist	Pfizer	Phase II
VPI-013, OPC-14523	$5-HT_{1A}$ agonist, sigma receptor agonist (also a serotonin reuptake inhibitor at higher doses)	Vela	Phase II
Casopitant, GW679769	NK1 antagonist	GSK	Phase II
YKP-10A, R228060	Phenylalanine derivate	Janssen (Johnson & Johnson)/SK Pharmaceuticals	Phase II
SSR149415	V1B antagonist	Sanofi-Aventis	Phase II
Elzasonan, CP 448,187	$5-HT_{1B}$ and $5-HT_{1D}$ receptor antagonist	Pfizer	Phase II
Delucemine, NPS 1506	NMDA antagonist	NPS	Phase I
Lu AA21004	5-HT reuptake inhibitor	Lundbeck	Phase I
Lu AA24530		Lundbeck	Phase I
SEP-225289	Dopamin/serotonin/noradrenaline reuptake inhibitor	Sepracor	Phase I
DMP904	CRF1 antagonist	Novartis	Phase I
DMP696	CRF1 antagonist	Bristol-Meyers Squibb	Phase I
CP-316,311	CRF1 antagonist	Bristol-Meyers Squibb	Phase I
GW876008	CRF1 antagonist	Pfizer	Phase I
ONO-2333Ms	CRF1 antagonist	Neurocrine/GSK	Phase I

JNJ-19567470 or TS-041	CRF1 antagonist	Ono Pharmaceuticals	Phase I
SSR 125543	CRF1 antagonist	Janssen (Johnson & Johnson), Taisho	Phase I
ND7001	PDE2 inhibitor	Neuro3d	Phase I
GW823296	NK1 antagonist	GSK	Phase I
R1576	GPCR	Roche	Phase I

1.3.4 Concluding remarks

To date, all available antidepressant drugs used clinically have monoamine-based mechanisms. Current treatments present three main problems: (1) the lag between drug administration and clinical efficacy, (2) depressed patients not responding to treatment and (3) side-effects. In recent years, a great effort has been expended in order to increase the safety and efficacy of antidepressant treatments and to decrease their side-effects. In this context, SSRIs, NRIs and SNRIs are safer than TCAs or MAOIs, but still need chronic administration to get clinical efficacy; and a relatively high percentage of depressed patients do not respond to these treatments. Pharmacological approaches based on monoaminergic systems have tried to solve these problems by combining reuptake blockers with aminergic receptor blockers or by the use of selective aminergic receptor ligands. Unfortunately, these new approaches have not solved the problem of efficacy and non-monoamine-based strategies are being tested. In this context, the use of neurpeptides and glucocorticoids as new targets for depression treatment is a very common strategy in the pharmaceutical industry (Table 2). At the same time, the relationship between mood disorders and neuroplasticity and cell survival in brain has suggested numerous biomarkers for depression.

In summary, it is hoped that novel approaches in depression drug discovery will be useful in the improvement of antidepressant treatment. To date, preclinical data are exciting and clinical testing will determine the advantages of these strategies over existing therapies.

References

Adell A. *Antidepressant properties of substance P antagonists: relationship to monoaminergic mechanisms?* Curr Drug Targets CNS Neurol Disord. **2004**, 3:113-121.

Adell A, Castro E, Celada P, Bortolozzi A, Pazos A, Artigas F. *Strategies for producing faster acting antidepressants.* Drug Discov Today. **2005**, 10:578-585.

Alt A, Witkin JM, Bleakman D. *AMPA receptor potentiators as novel antidepressants.* Curr Pharm Des. **2005**, 11:1511-1527.

Anden N, Grabowska M. *Pharmacological evidence for a stimulation of dopamine neurons by noradrenaline neurons in the brain.* Eur J Pharmacol. **1976**, 39:275-282.

Anderson IM, Tomenson BM. *The efficacy of selective serotonin re-uptake inhibitors in depression: A meta-analysis of studies against tricyclic antidepressants.* Journal of Psychopharmacology. **1994**, 8:238-249.

Artigas F. *5-HT and antidepressants: new views from microdialysis studies.* Trends Pharmacol Sci. **1993**, 14:262.

Artigas F, Romero L, de Montigny C, Blier P. *Acceleration of the effect of selected antidepressant drugs in major depression by 5-HT1A antagonists.* Trends Neurosci. **1996**, 19:378-383.

Barden N. *Implication of the hypothalamic-pituitary-adrenal axis in the physiopathology of depression.* J Psychiatry Neurosci. **2004**, 29:185-193.

Barnes NM, Sharp T. *A review of central 5-HT receptors and their function.* Neuropharmacology. **1999**, 38:1083-1152.

Bel N, Artigas F. *Chronic treatment with fluvoxamine increases extracellular serotonin in frontal cortex but not in raphe nuclei.* Synapse. **1993**, 15:243-245.

Berendsen HH, Broekkamp CL. *Comparison of stimulus properties of fluoxetine and 5-HT receptor agonists in a conditioned taste aversion procedure.* Eur J Pharmacol. **1994**, 253:83-89.

Berton O, McClung CA, Dileone RJ, Krishnan V, Renthal W, Russo SJ, Graham D, Tsankova NM, Bolanos CA, Rios M, Monteggia LM, Self DW, Nestler EJ. *Essential role of BDNF in the mesolimbic dopamine pathway in social defeat stress.* Science. **2006**, 311:864-868.

Bittencourt JC, Presse F, Arias C, Peto C, Vaughan J, Nahon JL, Vale W, Sawchenko PE. *The melanin-concentrating hormone system of the rat brain: an immuno- and hybridization histochemical characterization.* J Comp Neurol. **1992**, 319:218-245.

Blier P, de Montigny C. *Serotoninergic but not noradrenergic neurons in rat central nervous system adapt to long-term treatment with monoamine oxidase inhibitors.* Neuroscience. **1985**, 16:949-955.

Blier P, de Montigny C. *Current advances and trends in the treatment of depression.* Trends Pharmacol Sci. **1994**, 15:220-226.

Blier P, Bergeron R. *The use of pindolol to potentiate antidepressant medication.* J Clin Psychiatry. **1998**, 59 Suppl 5:16-23.

Blier P, Gobbi G, Haddjeri N, Santarelli L, Mathew G, Hen R. *Impact of substance P receptor antagonism on the serotonin and norepinephrine systems: relevance to the antidepressant/anxiolytic response.* J Psychiatry Neurosci. **2004**, 29:208-218.

Blier P, Szabo ST. *Potential mechanisms of action of atypical antipsychotic medications in treatment-resistant depression and anxiety.* J Clin Psychiatry. **2005**, 66 Suppl 8:30-40.

Boivin DB, Czeisler CA, Dijk DJ, Duffy JF, Folkard S, Minors DS, Totterdell P, Waterhouse JM. *Complex interaction of the sleep-wake cycle and circadian phase modulates mood in healthy subjects.* Arch Gen Psychiatry. **1997**, 54:145-152.

Broekkamp CL, Berendsen HH. *The importance of 5-HT1C receptors for anti-depressant effects.* Pol J Pharmacol Pharm. **1992**, 44:20.

Burnet PW, Eastwood SL, Lacey K, Harrison PJ. *The distribution of 5-HT1A and 5-HT2A receptor mRNA in human brain.* Brain Res. **1995**, 676:157-168.

Calfa G, Kademian S, Ceschin D, Vega G, Rabinovich GA, Volosin M. *Characterization and functional significance of glucocorticoid receptors in patients with major depression: modulation by antidepressant treatment.* Psychoneuroendocrinology. **2003**, 28:687-701.

Canuso C, Gharabawi G, Bouhours Peal. *Results from open-label phase of Arise-RD (augmentation with risperidone in resistant depression) [abstract].* Int J Psychopharmacol. **2004**, 7:S345.

Cappiello A, McDougle CJ, Malison RT, Heninger GR, Price LH. *Yohimbine augmentation of fluvoxamine in refractory depression: a single-blind study.* Biol Psychiatry. **1995**, 38:765-767.

Carpenter LL, Yasmin S, Price LH. *A double-blind, placebo-controlled study of antidepressant augmentation with mirtazapine.* Biol Psychiatry. **2002**, 51:183-188.

Cedarbaum JM, Aghajanian GK. *Noradrenergic neurons of the locus coeruleus: inhibition by epinephrine and activation by the alpha-antagonist piperoxane.* Brain Res. **1976**, 112:413-419.

Celada P, Puig M, Amargos-Bosch M, Adell A, Artigas F. *The therapeutic role of 5-HT1A and 5-HT2A receptors in depression.* J Psychiatry Neurosci. **2004**, 29:252-265.

Chen AC, Shirayama Y, Shin KH, Neve RL, Duman RS. *Expression of the cAMP response element binding protein (CREB) in hippocampus produces an antidepressant effect.* Biol Psychiatry. **2001**, 49:753-762.

Conley RK, Cumberbatch MJ, Mason GS, Williamson DJ, Harrison T, Locker K, Swain C, Maubach K, O'Donnell R, Rigby M, Hewson L, Smith D, Rupniak NM. *Substance P (neurokinin 1) receptor antagonists enhance dorsal raphe neuronal activity.* J Neurosci. **2002**, 22:7730-7736.

Cremers TI, Bosker,FJ, Hogg S, Arnt J, Mork A, et al. *5-HT2C antagonists augment antidepressant effects of SSRIs.* Proceedings of the 10th International Conference on in Vivo Methods. Sweden, **2003**.

Cremers TI, Giorgetti M, Bosker FJ, Hogg S, Arnt J, Mork A, Honig G, Bogeso KP, Westerink BH, den Boer H, Wikstrom HV, Tecott LH. *Inactivation of 5-HT(2C) receptors potentiates consequences of serotonin reuptake blockade.* Neuropsychopharmacology. **2004**, 29:1782-1789.

Culman J, Klee S, Ohlendorf C, Unger T. *Effect of tachykinin receptor inhibition in the brain on cardiovascular and behavioral responses to stress.* J Pharmacol Exp Ther. **1997**, 280:238-246.

Danish University Antidepressant Group. *Citalopram: clinical effect profile in comparison with clomipramine. A controlled multicenter study.* Psychopharmacology (Berl). **1986**, 90:131-138.

Danish University Antidepressant Group. *Paroxetine: a selective serotonin reuptake inhibitor showing better tolerance, but weaker antidepressant effect than clomipramine in a controlled multicenter study.* J Affect Disord. **1990**, 18:289-299.

Dawson LA, Li P. *Effects of 5-HT(6) receptor blockade on the neurochemical outcome of antidepressant treatment in the frontal cortex of the rat.* J Neural Transm. **2003**, 110:577-590.

De Foubert G, Murray TK, O'Neill MJ, Zetterström TS. *5-HT6 receptor-selective upregulation of brain-derived neurotrophic factor mRNA in the rat brain.* Federation of European Neuroscience Societies (FENS). **2004**.

Done CJ, Sharp T. *Evidence that 5-HT2 receptor activation decreases noradrenaline release in rat hippocampus in vivo.* Br J Pharmacol. **1992**, 107:240-245.

Done CJ, Sharp T. *Biochemical evidence for the regulation of central noradrenergic activity by 5-HT1A and 5-HT2 receptors: microdialysis studies in the awake and anaesthetized rat.* Neuropharmacology. **1994**, 33:411-421.

Dubocovich ML. *Presynaptic alpha-adrenoceptors in the central nervous system.* Ann N Y Acad Sci. **1984**, 430:7-25.

Ducottet C, Griebel G, Belzung C. *Effects of the selective nonpeptide corticotropin-releasing factor receptor 1 antagonist antalarmin in the chronic mild stress model of depression in mice.* Prog Neuropsychopharmacol Biol Psychiatry. **2003**, 27:625-631.

Duman RS. *Role of neurotrophic factors in the etiology and treatment of mood disorders.* Neuromolecular Med. **2004**, 5:11-25.

Dunlop J, Sabb AL, Mazandarani H, Zhang J, Kalgaonker S, Shukhina E, Sukoff S, Vogel RL, Stack G, Schechter L, Harrison BL, Rosenzweig-Lipson S. *WAY-163909 [(7bR, 10aR)-1,2,3,4,8,9,10,10a-octahydro-7bH-cyclopenta-[b][1,4]diazepino[6,7,1h i]indole], a novel 5-hydroxytryptamine 2C receptor-selective agonist with anorectic activity.* J Pharmacol Exp Ther. **2005**, 313:862-869.

Dwivedi Y, Rizavi HS, Conley RR, Roberts RC, Tamminga CA, Pandey GN. *Altered gene expression of brain-derived neurotrophic factor and receptor tyrosine kinase B in postmortem brain of suicide subjects.* Arch Gen Psychiatry. **2003**, 60:804-815.

Eisch AJ, Bolanos CA, de Wit J, Simonak RD, Pudiak CM, Barrot M, Verhaagen J, Nestler EJ. *Brain-derived neurotrophic factor in the ventral midbrain-nucleus accumbens pathway: a role in depression.* Biol Psychiatry. **2003**, 54:994-1005.

Feuerstein TJ, Mutschler A, Lupp A, Van V, V, Schlicker E, Gothert M. *Endogenous noradrenaline activates alpha 2-adrenoceptors on serotonergic nerve endings in human and rat neocortex.* J Neurochem. **1993**, 61:474-480.

Francis PT, Pangalos MN, Pearson RC, Middlemiss DN, Stratmann GC, Bowen DM. *5-Hydroxytryptamine1A but not 5-hydroxytryptamine2 receptors are enriched on neocortical pyramidal neurones destroyed by intrastriatal volkensin.* J Pharmacol Exp Ther. **1992**, 261:1273-1281.

Georgescu D, Sears RM, Hommel JD, Barrot M, Bolanos CA, Marsh DJ, Bednarek MA, Bibb JA, Maratos-Flier E, Nestler EJ, Dileone RJ. *The hypothalamic neuropeptide melanin-concentrating hormone acts in the nucleus accumbens to modulate feeding behavior and forced-swim performance.* J Neurosci. **2005**, 25:2933-2940.

Gobbi M, Moia M, Pirona L, Morizzoni P, Mennini T. *In vitro binding studies with two hypericum perforatum extracts--hyperforin, hypericin and biapigenin--on 5-HT6, 5-HT7, GABA(A)/benzodiazepine, sigma, NPY-Y1/Y2 receptors and dopamine transporters.* Pharmacopsychiatry. **2001**, 34 Suppl 1:S45-S48.

Gobert A, Rivet JM, Cistarelli L, Melon C, Millan MJ. *Alpha2-adrenergic receptor blockade markedly potentiates duloxetine- and fluoxetine-induced increases in noradrenaline, dopamine, and serotonin levels in the frontal cortex of freely moving rats.* J Neurochem. **1997**, 69:2616-2619.

Göthert M, Huth H, Schlicker E. *Characterization of the receptor subtype involved in alpha-adrenoceptor-mediated modulation of serotonin release from rat brain cortex slices.* Naunyn Schmiedebergs Arch Pharmacol. **1981**, 317:199-203.

Griebel G, Perrault G, Soubrie P. *Effects of SR48968, a selective non-peptide NK2 receptor antagonist on emotional processes in rodents.* Psychopharmacology (Berl). **2001**, 158:241-251.

Guiard BP, Przybylski C, Guilloux JP, Seif I, Froger N, De Felipe C, Hunt SP, Lanfumey L, Gardier AM. *Blockade of substance P (neurokinin 1) receptors enhances extracellular serotonin when combined with a selective serotonin reuptake inhibitor: an in vivo microdialysis study in mice.* J Neurochem. **2004**, 89:54-63.

Gurevich I, Englander MT, Adlersberg M, Siegal NB, Schmauss C. *Modulation of serotonin 2C receptor editing by sustained changes in serotonergic neurotransmission.* J Neurosci. **2002a**, 22:10529-10532.

Gurevich I, Tamir H, Arango V, Dwork AJ, Mann JJ, Schmauss C. *Altered editing of serotonin 2C receptor pre-mRNA in the prefrontal cortex of depressed suicide victims.* Neuron. **2002b**, 34:349-356.

Guscott M, Bristow LJ, Hadingham K, Rosahl TW, Beer MS, Stanton JA, Bromidge F, Owens AP, Huscroft I, Myers J, Rupniak NM, Patel S, Whiting PJ, Hutson PH, Fone KC, Biello SM, Kulagowski JJ, McAllister G. *Genetic knockout and pharmacological blockade studies of the 5-HT7 receptor suggest therapeutic potential in depression.* Neuropharmacology. **2005**, 48:492-502.

Gustafson EL, Durkin MM, Bard JA, Zgombick J, Branchek TA. *A receptor autoradiographic and in situ hybridization analysis of the distribution of the 5-ht7 receptor in rat brain.* Br J Pharmacol. **1996**, 117:657-666.

Haddjeri N, Blier P. *Sustained blockade of neurokinin-1 receptors enhances serotonin neurotransmission.* Biol Psychiatry. **2001**, 50:191-199.

Hedlund PB, Huitron-Resendiz S, Henriksen SJ, Sutcliffe JG. *5-HT7 receptor inhibition and inactivation induce antidepressantlike behavior and sleep pattern.* Biol Psychiatry. **2005**, 58:831-837.

Helke CJ, Krause JE, Mantyh PW, Couture R, Bannon MJ. *Diversity in mammalian tachykinin peptidergic neurons: multiple peptides, receptors, and regulatory mechanisms.* FASEB J. **1990**, 4:1606-1615.

Herpfer I, Lieb K. *Substance P receptor antagonists in psychiatry: rationale for development and therapeutic potential.* CNS Drugs. **2005**, 19:275-293.

Holmes A, Heilig M, Rupniak NM, Steckler T, Griebel G. *Neuropeptide systems as novel therapeutic targets for depression and anxiety disorders.* Trends Pharmacol Sci. **2003**, 24:580-588.

Hughes ZA, Starr KR, Langmead CJ, Hill M, Bartoszyk GD, Hagan JJ, Middlemiss DN, Dawson LA. *Neurochemical evaluation of the novel 5-HT1A receptor partial agonist/serotonin reuptake inhibitor, vilazodone.* Eur J Pharmacol. **2005**, 510:49-57.

Invernizzi RW, Garattini S. *Role of presynaptic alpha2-adrenoceptors in antidepressant action: recent findings from microdialysis studies.* Prog Neuropsychopharmacol Biol Psychiatry. **2004**, 28:819-827.

Kohen R, Fashingbauer LA, Heidmann DE, Guthrie CR, Hamblin MW. *Cloning of the mouse 5-HT6 serotonin receptor and mutagenesis studies of the third cytoplasmic loop.* Brain Res Mol Brain Res. **2001**, 90:110-117.

Korte SM, Koolhaas JM, Wingfield JC, McEwen BS. *The Darwinian concept of stress: benefits of allostasis and costs of allostatic load and the trade-offs in health and disease.* Neurosci Biobehav Rev. **2005**, 29:3-38.

Kramer MS, Cutler N, Feighner J, Shrivastava R, Carman J, Sramek JJ, Reines SA, Liu G, Snavely D, Wyatt-Knowles E, Hale JJ, Mills SG, MacCoss M, Swain CJ, Harrison T, Hill RG, Hefti F, Scolnick EM, Cascieri MA, Chicchi GG, Sadowski S, Williams AR, Hewson L, Smith D, Carlson EJ, Hargreaves RJ, Rupniak NM. *Distinct mechanism for antidepressant activity by blockade of central substance P receptors.* Science. **1998**, 281:1640-1645.

Langer SZ. *Presynaptic regulation of catecholamine release.* Biochem Pharmacol. **1974**, 23:1793-1800.

Laplante P, Diorio J, Meaney MJ. *Serotonin regulates hippocampal glucocorticoid receptor expression via a 5-HT7 receptor.* Brain Res Dev Brain Res. **2002**, 139:199-203.

Lee SH, Lee KJ, Lee HJ, Ham BJ, Ryu SH, Lee MS. *Association between the 5-HT6 receptor C267T polymorphism and response to antidepressant treatment in major depressive disorder.* Psychiatry Clin Neurosci. **2005**, 59:140-145.

Lucchelli A, Santagostino-Barbone MG, D'Agostino G, Masoero E, Tonini M. *The interaction of antidepressant drugs with enteric 5-HT7 receptors.* Naunyn Schmiedebergs Arch Pharmacol. **2000**, 362:284-289.

Maeno H, Kiyama H, Tohyama M. *Distribution of the substance P receptor (NK-1 receptor) in the central nervous system.* Brain Res Mol Brain Res. **1993**, 18:43-58.

Mansbach RS, Brooks EN, Chen YL. *Antidepressant-like effects of CP-154,526, a selective CRF1 receptor antagonist.* Eur J Pharmacol. **1997**, 323:21-26.

Marangell LB, Johnson CR, Kertz B, Zboyan HA, Martinez JM. *Olanzapine in the treatment of apathy in previously depressed participants maintained with selective serotonin reuptake inhibitors: an open-label, flexible-dose study.* J Clin Psychiatry. **2002**, 63:391-395.

Marek GJ, Carpenter LL, McDougle CJ, Price LH. *Synergistic action of 5-HT2A antagonists and selective serotonin reuptake inhibitors in neuropsychiatric disorders.* Neuropsychopharmacology. **2003**, 28:402-412.

Martin-Cora FJ, Pazos A. *Autoradiographic distribution of 5-HT7 receptors in the human brain using [3H]mesulergine: comparison to other mammalian species*. Br J Pharmacol. **2004**, 141:92-104.

Mewshaw RE, Zhou D, Zhou P, Shi X, Hornby G, Spangler T, Scerni R, Smith D, Schechter LE, Andree TH. *Studies toward the discovery of the next generation of antidepressants. 3. Dual 5-HT1A and serotonin transporter affinity within a class of N-aryloxyethylindolylalkylamines*. J Med Chem. **2004**, 47:3823-3842.

Miller RJ. *Presynaptic receptors*. Annu Rev Pharmacol Toxicol. **1998**, 38:201-227.

Miyamoto Y, Yamada K, Noda Y, Mori H, Mishina M, Nabeshima T. *Lower sensitivity to stress and altered monoaminergic neuronal function in mice lacking the NMDA receptor epsilon 4 subunit*. J Neurosci. **2002**, 22:2335-2342.

Monsma FJ, Jr., Shen Y, Ward RP, Hamblin MW, Sibley DR. *Cloning and expression of a novel serotonin receptor with high affinity for tricyclic psychotropic drugs*. Mol Pharmacol. **1993**, 43:320-327.

Monteggia LM, Barrot M, Powell CM, Berton O, Galanis V, Gemelli T, Meuth S, Nagy A, Greene RW, Nestler EJ. *Essential role of brain-derived neurotrophic factor in adult hippocampal function*. Proc Natl Acad Sci U S A. **2004**, 101:10827-10832.

Montgomery SA. *Predicting response: noradrenaline reuptake inhibition*. Int Clin Psychopharmacol. **1999**, 14 Suppl 1:S21-S26.

Morilak DA, Garlow SJ, Ciaranello RD. *Immunocytochemical localization and description of neurons expressing serotonin2 receptors in the rat brain*. Neuroscience. **1993**, 54:701-717.

Mork, A. and Hogg, S. Augmentation of paroxetine by the 5-HT2 antagonist, irindalone: evidence for increased efficacy. Proceedings of the 10th International Conference on in Vivo Methods. Stockholm, Sweden. 2003.

Mullins UL, Gianutsos G, Eison AS. *Effects of antidepressants on 5-HT7 receptor regulation in the rat hypothalamus*. Neuropsychopharmacology. **1999**, 21:352-367.

Ostroff RB, Nelson JC. *Risperidone augmentation of selective serotonin reuptake inhibitors in major depression*. J Clin Psychiatry. **1999**, 60:256-259.

Papp M, Moryl E. *Antidepressant activity of non-competitive and competitive NMDA receptor antagonists in a chronic mild stress model of depression*. Eur J Pharmacol. **1994**, 263:1-7.

Pariante CM, Miller AH. *Glucocorticoid receptors in major depression: relevance to pathophysiology and treatment*. Biol Psychiatry. **2001**, 49:391-404.

Paul IA, Skolnick P. *Glutamate and depression: clinical and preclinical studies*. Ann N Y Acad Sci. **2003**, 1003:250-272.

Perez V, Puiigdemont D, Gilaberte I, Alvarez E, Artigas F. *Augmentation of fluoxetine's antidepressant action by pindolol: analysis of clinical, pharmacokinetic, and methodologic factors*. J Clin Psychopharmacol. **2001**, 21:36-45.

Petrie RX, Reid IC, Stewart CA. *The N-methyl-D-aspartate receptor, synaptic plasticity, and depressive disorder. A critical review*. Pharmacol Ther. **2000**, 87:11-25.

Rimon R, Le Greves P, Nyberg F, Heikkila L, Salmela L, Terenius L. *Elevation of substance P-like peptides in the CSF of psychiatric patients*. Biol Psychiatry. **1984**, 19:509-516.

Russo-Neustadt AA, Chen MJ. *Brain-derived neurotrophic factor and antidepressant activity.* Curr Pharm Des. **2005**, 11:1495-1510.

Ryckmans T, Balancon L, Berton O, Genicot C, Lamberty Y, Lallemand B, Pasau P, Pirlot N, Quere L, Talaga P. *First dual NK(1) antagonists-serotonin reuptake inhibitors: synthesis and SAR of a new class of potential antidepressants.* Bioorg Med Chem Lett. **2002**, 12:261-264.

Saffroy M, Torrens Y, Glowinski J, Beaujouan JC. *Autoradiographic distribution of tachykinin NK2 binding sites in the rat brain: comparison with NK1 and NK3 binding sites.* Neuroscience. **2003**, 116:761-773.

Sanacora G, Rothman DL, Mason G, Krystal JH. *Clinical studies implementing glutamate neurotransmission in mood disorders.* Ann N Y Acad Sci. **2003**, 1003:292-308.

Sanacora G, Berman RM, Cappiello A, Oren DA, Kugaya A, Liu N, Gueorguieva R, Fasula D, Charney DS. *Addition of the alpha2-antagonist yohimbine to fluoxetine: effects on rate of antidepressant response.* Neuropsychopharmacology. **2004**, 29:1166-1171.

Schechter LE, Ring RH, Beyer CE, Hughes ZA, Khawaja X, Malberg JE, Rosenzweig-Lipson S. *Innovative approaches for the development of antidepressant drugs: current and future strategies.* NeuroRx. **2005**, 2:590-611.

Schwartz WJ. *A clinician's primer on the circadian clock: its localization, function, and resetting.* Adv Intern Med. **1993**, 38:81-106.

Sheldon PW, Aghajanian GK. *Excitatory responses to serotonin (5-HT) in neurons of the rat piriform cortex: evidence for mediation by 5-HT1C receptors in pyramidal cells and 5-HT2 receptors in interneurons.* Synapse. **1991**, 9:208-218.

Shelton RC, Tollefson GD, Tohen M, Stahl S, Gannon KS, Jacobs TG, Buras WR, Bymaster FP, Zhang W, Spencer KA, Feldman PD, Meltzer HY. *A novel augmentation strategy for treating resistant major depression.* Am J Psychiatry. **2001**, 158:131-134.

Shen Y, Monsma FJ, Jr., Metcalf MA, Jose PA, Hamblin MW, Sibley DR. *Molecular cloning and expression of a 5-hydroxytryptamine7 serotonin receptor subtype.* J Biol Chem. **1993**, 268:18200-18204.

Shirayama Y, Mitsushio H, Takashima M, Ichikawa H, Takahashi K. *Reduction of substance P after chronic antidepressants treatment in the striatum, substantia nigra and amygdala of the rat.* Brain Res. **1996**, 739:70-78.

Shirayama Y, Chen AC, Nakagawa S, Russell DS, Duman RS. *Brain-derived neurotrophic factor produces antidepressant effects in behavioral models of depression.* J Neurosci. **2002**, 22:3251-3261.

Simmen U, Burkard W, Berger K, Schaffner W, Lundstrom K. *Extracts and constituents of Hypericum perforatum inhibit the binding of various ligands to recombinant receptors expressed with the Semliki Forest virus system.* J Recept Signal Transduct Res. **1999**, 19:59-74.

Siuciak JA, Lewis DR, Wiegand SJ, Lindsay RM. *Antidepressant-like effect of brain-derived neurotrophic factor (BDNF).* Pharmacol Biochem Behav. **1997**, 56:131-137.

Skolnick P, Layer RT, Popik P, Nowak G, Paul IA, Trullas R. *Adaptation of N-methyl-D-aspartate (NMDA) receptors following antidepressant treatment: implications for the pharmacotherapy of depression.* Pharmacopsychiatry. **1996**, 29:23-26.

Skolnick P, Popik P, Janowsky A, Beer B, Lippa AS. *Antidepressant-like actions of DOV 21,947: a "triple" reuptake inhibitor.* Eur J Pharmacol. **2003**, 461:99-104.

Sleight AJ, Carolo C, Petit N, Zwingelstein C, Bourson A. *Identification of 5-hydroxytryptamine7 receptor binding sites in rat hypothalamus: sensitivity to chronic antidepressant treatment.* Mol Pharmacol. **1995**, 47:99-103.

Sprouse JS, Aghajanian GK. *Electrophysiological responses of serotoninergic dorsal raphe neurons to 5-HT1A and 5-HT1B agonists.* Synapse. **1987**, 1:3-9.

Starke K. *Presynaptic alpha-autoreceptors.* Rev Physiol Biochem Pharmacol. **1987**, 107:73-146.

Szabo ST, Blier P. *Effect of the selective noradrenergic reuptake inhibitor reboxetine on the firing activity of noradrenaline and serotonin neurons.* Eur J Neurosci. **2001**, 13:2077-2087.

Szabo ST, Blier P. *Effects of serotonin (5-hydroxytryptamine, 5-HT) reuptake inhibition plus 5-HT(2A) receptor antagonism on the firing activity of norepinephrine neurons.* J Pharmacol Exp Ther. **2002**, 302:983-991.

Takahashi T, Nowakowski RS, Caviness VS, Jr. *BUdR as an S-phase marker for quantitative studies of cytokinetic behaviour in the murine cerebral ventricular zone.* J Neurocytol. **1992**, 21:185-197.

Takayama H, Ota Z, Ogawa N. *Effect of immobilization stress on neuropeptides and their receptors in rat central nervous system.* Regul Pept. **1986**, 15:239-248.

Thase ME, Entsuah AR, Rudolph RL. *Remission rates during treatment with venlafaxine or selective serotonin reuptake inhibitors.* Br J Psychiatry. **2001**, 178:234-241.

Thomas DR, Atkinson PJ, Ho M, Bromidge SM, Lovell PJ, Villani AJ, Hagan JJ, Middlemiss DN, Price GW. *[(3)H]-SB-269970--A selective antagonist radioligand for 5-HT(7) receptors.* Br J Pharmacol. **2000**, 130:409-417.

Thomas DR, Hagan JJ. *5-HT7 receptors.* Curr Drug Targets CNS Neurol Disord. **2004**, 3:81-90.

Wei Q, Lu XY, Liu L, Schafer G, Shieh KR, Burke S, Robinson TE, Watson SJ, Seasholtz AF, Akil H. *Glucocorticoid receptor overexpression in forebrain: a mouse model of increased emotional lability.* Proc Natl Acad Sci U S A. **2004**, 101:11851-11856.

Wersinger SR, Ginns EI, O'Carroll AM, Lolait SJ, Young WS, III. *Vasopressin V1b receptor knockout reduces aggressive behavior in male mice.* Mol Psychiatry. **2002**, 7:975-984.

Winslow JT, Insel TR. *Neuroendocrine basis of social recognition.* Curr Opin Neurobiol. **2004**, 14:248-253.

Wood MD, Thomas DR, Watson JM. *Therapeutic potential of serotonin antagonists in depressive disorders.* Expert Opin Investig Drugs. **2002**, 11:457-467.

Wright DE, Seroogy KB, Lundgren KH, Davis BM, Jennes L. *Comparative localization of serotonin1A, 1C, and 2 receptor subtype mRNAs in rat brain.* J Comp Neurol. **1995**, 351:357-373.

Yau JL, Noble J, Widdowson J, Seckl JR. *Impact of adrenalectomy on 5-HT6 and 5-HT7 receptor gene expression in the rat hippocampus.* Brain Res Mol Brain Res. **1997**, 45:182-186.

Zobel AW, Nickel T, Kunzel HE, Ackl N, Sonntag A, Ising M, Holsboer F. *Effects of the high-affinity corticotropin-releasing hormone receptor 1 antagonist R121919 in major depression: the first 20 patients treated.* J Psychiatr Res. **2000**, 34:171-181.

1.4 Experimental research

Modeling Human Depression by Animal Models

Ana Montero, Begoña Fernández and Luz Romero

Dept. of Pharmacology, Laboratorios Esteve, Av. Mare de Déu de Montserrat, 221. 08041 Barcelona, Spain.

1.4.1 Introduction

Animal models that simulate some aspects of the human diseases provide a powerful methodology for investigating these problems and developing relevant therapies. However, especially in neuroscience research, homologous models are very rare. Isomorphic models are more common but, although they display similar symptoms, the condition is not provoked by the same events as the human condition. Most common are partial models which focus only on limited aspects (Willner, 1991a; De Deyn et al., 2000). The two most important applications of animal models in neurosciences are, on the one hand, the development and testing of hypothesis about neurological and psychiatric disorders and their neural substrates and, on the other hand, the screening and identification of new therapies, usually drugs (De Deyn et al., 2000; Geyer and Markou, 2002). Nevertheless, currently available animal models do not fulfil these two needs (Mitchell and Redfern, 2005; Wong, 2005).

Depression is a very complex psychological disorder. Diagnosis Statistical Manual of Mental Disorders IV (DSM-IV) defines two core symptoms in the diagnosis of a depressive episode: depressed mood (a subjective feeling impossible to simulate in animals) and loss of interest or pleasure (anhedonia). Some symptoms can easily be modelled in animals, for instance, body weight change, psychomotor retardation or anhedonia, whereas others cannot. Therefore, probably one of the most important limitations of the animal models of depression emerges from the impossibility to translate into the non-verbal behavior of animals symptoms, such as suicidal ideation, negativism, or feelings of worthlessness or guilt. Thus, animal models can only reproduce some features of depression (Willner, 1984, 1990; Geyer and Markou, 2002; Moreau, 2002; Frazer and Morilak, 2005; Wong, 2005).

Depression is a heterogeneous syndrome not only in terms of symptomatology but also in terms of its etiology. It is highly heritable, although the specific genes that underlie the risk of depression have not yet been identified. The non-genetic risk also remains poorly defined, with suggestions that early childhood trauma, emotional stress or physical illness might be involved (Berton and Nestler, 2006).

Given the heterogeneity of depression, it is not surprising to find that this disorder may be simulated in a variety of different ways and with a variety of purposes. A wide diversity of animal models has been used as screening tests to discover and develop novel antidepressant drug therapies, as simulations for investigating aspects of the neurobiology of depressive illness and as experimental models for examining the neuropharmacological mechanisms associated with antidepressant treatments (Willner, 1984, 1990, 1991a; Jesberger and Richardson, 1986; Mitchell and Redfern, 2005).

This chapter will review briefly the criteria for the development of animal models and provide an overview of the currently available models of depression.

1.4.2 Types of validity

In general, the validity of a model refers to the extent to which a model is useful for a given purpose (Geyer and Markou, 1995, 2002). Usually, not only the utility but also the quality of a model depends upon its validity (Cryan and Holmes, 2005). For this reason, researchers have proposed specific criteria for evaluating whether an experimental procedure in an animal has validity as a model of psychiatric disease. Some of the most widely cited criteria were developed by McKinney and Bunney (1969). They suggested that the minimum requirements for a valid animal model of depression are that it was reasonably analogous to the human disorder in its manifestations or symptomatology, that there was a behavioral change that can be monitored objectively, that the behavioral changes observed should be reversed by the same treatment modalities that are effective in humans and that it should be reproducible between investigators. In contrast, Geyer and Markou (1995) proposed that the only criteria that are necessary and sufficient for initial use are that the paradigm had strong predictive validity and that the behavioral readout was reliable and robust, both in the same laboratory and between laboratories.

Ideally, a valid model should resemble the pathology it simulates in terms of its symptomatology, etiology and background (i.e., neuropathological, neurophysiological or electrophysiological features) and should display concordant effects of therapy as the condition it is supposed to imitate (McKinney, 1984; Moreau, 2002). Nevertheless, current animal models of depression do not meet every requirement. Thus, depending on the desired purpose of the test that one wishes to validate, different types of validity are relevant (Geyer and Markou, 1995, 2002). Usually, any assessment of the validity of animal models of depression addresses three dimensions: predictive, face and construct validity (Willner, 1984, 1991b; Moreau, 2002; Mitchell and Redfern, 2005).

1.4.2.1 Predictive validity

Predictive validity is determined by appropriate response of the animal model to therapeutic agents (Moreau, 2002). Thus, the predictive validity of an animal model of depression is a measure of that model's ability to identify drugs with clinical efficacy against depression (Mitchell and Redfern, 2005).

However, a positive response to antidepressant drugs is insufficient to define an animal model of depression. Some antidepressants are active in animal models of anxiety and panic, following chronic treatment (Bodnoff et al., 1988; Fontana et al., 1989). Therefore, a valid test should respond to effective antidepressant treatment, including electroconvulsive shock therapy (ECS), and should fail to respond to ineffective agents. Furthermore, it should minimize the identification of false negatives and false positives (e.g., psychomotor stimulants, anticholinergics, opiates; Mitchell and Redfern, 2005).

1.4.2.2 Face validity

Face validity refers to phenomenological similarity between the behavior exhibited by the animal model and the specific symptoms of the human condition (Geyer and Markou, 1995; Moreau, 2002; Lyons, 2004). The face validity of an animal model of depression is a measure of the model's ability to reproduce core symptoms of the disease (Mitchell and Redfern, 2005). However, as mentioned in the Introduction, not all of the clinical symptoms of depression can be modelled in animals; symptoms such as excessive guilt, feelings of worthlessness, suicidal ideation or decreased self-esteem are necessarily excluded (Willner, 1984, 1990; Moreau, 2002; Wong, 2005).

A DSM-IV diagnosis of major depression requires the presence of a least one of two core symptoms: anhedonia and depressed mood. Anhedonia can be modelled in animals (as detailed in Section 2.4.3.3, *Reward Models*) but a depressed mood cannot. In assessing the face validity of animal models of depression, anhedonia assumes therefore a central position. The subsidiary symptoms of depression in DSM-IV that are amenable to modeling in animals include phychomotor changes, fatigue or loss of energy and disturbances of sleep or food intake. Interestingly, psychomotor activity, sleep and appetite may be increased or decreased in depression (Willner, 1991a; Mitchell and Redfern, 2005).

1.4.2.3 Construct validity

Construct validity refers to the theoretical rationale for linking a process in the model to a process hypothesized to produce a key symptom of a given disorder in humans (Lyons, 2004). In other words, construct validity of a test is the accuracy with which the test measures that which it is intended to measure (Geyer and Markou, 1995; Moreau, 2002).

Evaluation of this type of validity requires a comparison of the causal pathology of the disease state and the cellular and neurochemical mechanisms underlying the model. There is little in the literature about biochemical markers or neurochemical abnormalities associated with depression that can be employed to provide a standard against which to validate animal models (Mitchell and Redfern, 2005). Therefore, the assessment of the construct validity of animal models of depression is limited by this lack of knowledge of causal pathology. Moreover, there are a variety of psychological factors implicated in the etiology of depression: chronic mild stress, adverse childhood experiences, undesirable life events, some personality traits (introversion and impulsiveness), etc. Biological factors including genetic influences and a variety of physical illnesses and medications should also be considered (Akiskal, 1985, 1986). Furthermore, in most cases, the immediate precipitant of a depression cannot be clearly identified (except, for instance, in post-partum depression or in the seasonal affective disorder). For all these reasons, depression is usually better understood as resulting from an accumulation of a number of different risk factors (Aneshensel and Stone, 1982; Akiskal, 1985; Mitchell and Redfern, 2005).

Despite these limitations, a number of generalizations are possible. The major group of animal models of depression is based on responses to stressors, but there are a wide variety of animal models modeling other kind of factors in the etiology of depression, as will be reviewed in this chapter.

1.4.3 Animal models of depression

The models commonly used are diverse and were developed mainly based on the behavioral consequences of lesion, stress or genetic manipulations. Most of these models aim to mimic or simulate some core symptoms of the clinical situation. Another type of animal models, valuable as screening tests, aim to predict accurately antidepressant activity (Cryan et al., 2002; Cryan and Mombereau, 2004; Mitchell and Redfern, 2005).

1.4.3.1 Lesion model: olfactory bulbectomy

The bilateral removal of the olfactory bulbs of rodents, with consequent disruption of the limbic–hypothalamic axis, results in a complex constellation of behavioral, neurochemical, neuroendocrine and neuroimmune alterations, many of which are comparable with changes seen in depression (Kelly et al., 1997; Harkin et al., 2003; Song and Leonard, 2005).

Regarding the behavioral changes, the most consistent response is the hyperactivity shown in a novel, brightly lit open-field apparatus, which seems to be related to increases in defensive behavior (Kelly et al., 1997; Cryan et al., 1998; Harkin et al., 2003; Zueger et al., 2005). Other behavioral alterations include irritability, an impairment of passive avoidance learning, heightened acoustic startle response to stress and anhedonia-like behaviors, such as decreased sucrose preference and sexual behavior (Fontana et al., 1989; Harkin et al., 2003; Song and Leonard, 2005).

Bulbectomized animals also show changes in the immune system and in the noradrenalin (NA), dopamine (DA), serotonin (5-HT), gamma-aminobutyric acid (GABA), acetylcholine (ACh) and glutamatergic neurotransmitter systems (Kelly et al., 1997). They also show an elevation of circulating corticosteroid levels (as do stressed animals), which appears to be an increased corticosteroid response to stress (Broekkamp et al., 1986; Harkin et al., 2003).

These deficits are reversed by chronic, but not acute, treatment with antidepressants (Kelly et al., 1997; Cryan et al., 1998; Harkin et al., 2003; Song and Leonard, 2005), including agonists at the 5-HT$_{1A}$ receptor (Borsini et al., 1997; McGrath and Norman, 1999), 5-HT-NA reuptake inhibitors (SNRIs; McGrath and Norman, 1998) and selective NA reuptake inhibitors (Harkin et al., 1999). However, monoamine oxidase (MAO) inhibitors are not effective in this test (Jesberger and Richardson, 1986).

1.4.3.2 Stress models

Exposure to trauma and stress has been shown to be one of the main predisposing factors to major depression (Lloyd, 1980; Anisman and Zacharko, 1990; Kessler, 1997; Sullivan et al., 2000; Berton and Nestler, 2006). Specifically, uncontrollable stressful events can generate symptoms of a major depressive episode, resulting in an inability to react, for instance, to normally pleasant events (Nelson and Charney, 1981; Moreau, 2002). Therefore, many models and tests for assessing depression-related behavior in rodents involve exposure to stressful situations (Willner, 1991a; Cryan and Holmes, 2005).

1.4.3.2.1 Learned helplessness

Learned helplessness is one of the earliest and best studied models of depression. It emerged from classic studies of Seligman (Overmier and Seligman, 1967; Seligman and Maier, 1967). The learned helplessness paradigm is based on the observation that animals exposed to uncontrollable stress (usually electric shocks) are subsequently less able to learn to escape shock than animals exposed to comparable, or indeed, identical, patterns of controllable shock (Maier and Seligman, 1976; Mitchell and Redfern, 2005).

The learned helplessness paradigm has good face validity because there is similarity between the behavioral characteristics of learned-helpless animals and signs of depression in humans (Willner, 1984, 1986; Mitchell and Redfern, 2005). For example, learned-helpless animals exhibit loss of appetite and weight, decreased locomotor activity, poor performance in both appetite- and adversity-motivated tasks, loss of response for rewarding brain stimulation, altered sleep patterns and early morning awakening, social impairments and striking deficits in learning appropriate avoidance/escape behavior (Geyer and Markou, 1995; Weiss and Kilts, 1998; Maier, 2001). Uncontrollable stress also induces significant changes in noradrenergic, serotonergic, GABAergic and adenosinergic brain systems (Petty and Sherman, 1981, 1983; Maier and Watkins, 2005) that could mediate the behavioral abnormalities observed in the learned-helpless animals.

Moreover, Lachman and collaborators (1993) reported that rats can be selectively bred for helplessness, suggesting a genetic component. Specifically, Vollmayr and Henn (2001) recently described a procedure in which mild shocks induce learned helplessness in only some of the subjects, which may mimic the variable human predisposition for depressive illness. This procedure was used as the basis for a selective breeding programme, which produced a "congenital learned helplessness" and a "congenital non-learned helplessness" strain.

The predictive validity of the model is indicated by the fact that pharmacological treatments clinically effective in depression, such as tricyclic antidepressants (TCAs), MAO inhibitors, atypical antidepressants and also ECS are effective in reducing the behavioral and physical abnormalities seen in animals exposed to uncontrollable stress (Sherman et al., 1982; Willner, 1984, 1986).

Taken together, these findings demonstrate that learned helplessness models have a high degree of validity in terms of etiology, symptoms, physiology and recovery responses to known antidepressant drugs (Lyons, 2004; Mitchell and Redfern, 2005). Nevertheless, this model has some limitations. A well known problem with this paradigm is that the behavioral manifestations in rats persist for only several days, while episodes of depression in humans following negative life-events may last for several months (Lyons, 2004). One explanation for the discrepancy in time is that the rats do not spontaneously generate memories or "ruminations" about their experiences outside of the context in which the experiences occurred. Maier (2001) recently demonstrated that, by "reminding" rats of a previously stressful experience through repeated exposures to contextual cues, the learned helplessness can be clearly prolonged. In addition, it is important to note that inescapable shock has a variety of other simpler effects that could also explain many of the behavioral impairments, such as decreased locomotor activity (Glazer and Weiss, 1976; Anisman et al., 1979) and analgesia (Lewis et al., 1980). Regarding the predictive validity, some false positives have also been reported (Wilner, 1984; Geyer and Markou, 1995).

1.4.3.2.2 Chronic severe stress

Repeated presentation of the same stressor usually leads to adaptation. However, adaptation can be prevented by presenting a variety of stressors in an unpredictable sequence. Katz and collaborators (1981) developed a procedure whereby rats were submitted to a variety of chronic, unpredictable stressors such as electric shocks, immersion in cold water, tail pinch, reversal of the light/dark cycle, etc. After three weeks of exposure to a variety of stressors, rats failed to show the typical increase in open-field locomotor activity. Subsequent experiments showed that animals exhibit other behavioral deficits. For example, chronically stressed animals do not increase drinking when saccharine or sucrose is added to their drinking water (Katz, 1982; Katz and Sibel, 1982). This behavior may reflect the development of an anhedonic state in animals. Therefore, a chronic stress regimen is able to induce dysfunctioning of the reward systems.

These behavioral deficits can be prevented by the administration of different antidepressants drugs, as well as ECS (Katz, 1982; Willner, 1990), but not by some drugs such as the MAOI tranylcypromine (Mitchell and Redfern, 2005).

1.4.3.2.3 Chronic mild stress

Willner adapted the chronic severe stress procedure by using less severe stressors which were supposed to provide a better analogy with mild unpredictable stressors encountered in daily life (Willner et al., 1987). The procedure involves relatively continuous exposure of rats or mice to a variety of mild stressors, such as periods of food and water deprivation, small temperature reductions, changes of cages mates and other similarly individually innocuous, but unpredictable, manipulations (Mitchell and Redfern, 2005).

The chronic mild stress (CMS) procedure induces various long-term behavioral, neurochemical, neuroimmune and neuroendocrine alterations that resemble those observed in depressed patients. Over a period of weeks of chronic exposure to the mild stress regime, rats gradually reduce their consumption of a preferred dilute sucrose solution, and in untreated animals, this deficit persists for several weeks following cessation of stress (Kupfer and Thase, 1983; Willner et al., 1987; Willner, 1997; Moreau, 2002). These effects reflect a generalized insensitivity to reward, i.e., an anhedonic state. In particular, CMS also impairs responsiveness to reward as assessed by different methods, including suppression of place preference conditioning (Papp et al., 1991, 1992, 1993; Willner, 1997; Strekalova et al., 2004) and increased threshold for intracranial self-stimulation (ICSS; Moreau et al., 1992; see Section 1.4.3.3, *Reward Models*). In fact, anhedonia has been chosen as an essential characteristic of this model since it provides this simulation with remarkable face validity (Moreau, 2002).

In addition to decreasing responsiveness to rewards, CMS also causes the appearance of many other symptoms of major depressive disorder including decreases in sexual, aggressive and investigative behaviors and decreases in locomotor activity. Animals exposed to CMS also show an advanced phase shift of diurnal rhythms, diurnal variation in symptom severity and a variety of sleep disorders characteristic of depression. They also gain weight more slowly, leading to a relative loss of body weight, and show signs of increased activity in the hypothalamic–pituitary–adrenal (HPA) axis, including adrenal hypertrophy and corticosterone hypersecretion. Abnormalities are also detected in the immune system (D'aquila et al., 1994; Gorka et al., 1994; Ayensu et al., 1995; Moreau et al., 1995, 1998; Mitchell and Redfern, 2005).

Normal behavior can be gradually restored by chronic, but not acute, treatment with a wide variety of antidepressants, including TCAs, selective 5-HT reuptake inhibitors (SSRIs), a specific NA reuptake inhibitor (maprotiline), MAO-A inhibitors, atypical antidepressants such as mianserin, buspirone and amisulpride, and ECS (Willner, 1997; Moreau, 2002; Mitchell and Redfern, 2005). Regarding stress-induced anhedonia, medications effective in antagonizing this behavioral alteration include representatives of the TCAs (Willner et al., 1987; Moreau et al., 1992), monoamine reuptake inhibitors such as fluoxetine and maprotiline (Muscat et al., 1992), MAO inhibitors such as moclobemide and brofaromine (Moreau et al., 1993; Papp et al., 1996), and atypical antidepressants such as mianserin (Cheeta et al., 1994; Moreau et al., 1994). ECS (Moreau et al., 1995) and lithium (Sluzewska and Nowakowska, 1994) are also active in this model. Interestingly, the antagonism of stress-induced anhedonia requires two to four weeks of treatment, similar to the time-course of antidepressant drugs in humans (Moreau, 2002).

Therefore, this simulation appears to be specific and selective in its response to all categories of clinically used antidepressant treatments and also exhibits good face validity. However, while the CMS has a great many positive features, a major drawback is that the model has proved extremely difficult to implement reliably (Mitchell and Redfern, 2005).

1.4.3.2.4 Withdrawal from chronic psychomotor stimulants

It is frequently assumed that stimulant drug treatment is a form of stress, since in many respects stimulant treatment and stress appear to be interchangeable (Antelman et al., 1980). Like stimulant drugs, stressors activate the mesolimbic and mesocortical DA projections (Blanc et al., 1980). However, the relationship of drug-induced depressions to major depressive disorder is uncertain; and the validity of this model is therefore questionable (Mitchell and Redfern, 2005).

In humans, withdrawal from several drugs of abuse is characterized by the symptom of anhedonia (Markou et al., 1998). In animals, withdrawal from drugs of abuse, such as amphetamine, also produces deficits in reward-related behaviors (sucrose, ICSS; Barr et al., 2002; Cryan et al., 2003). For example, a number of studies have reported that response to ICSS is reduced during the days following withdrawal from chronic amphetamine treatment (Barrett and White, 1980; Kokkinidis and Zacharko, 1980). Other studies have reported that the threshold for ICSS is elevated following amphetamine withdrawal (Leith and Barrett, 1980; Cassens et al., 1981). This effect can be alleviated by two days of imipramine or amitriptyline treatment; and with continued treatment, a normal response can be restored (Kokkinidis et al., 1980). Amphetamine withdrawal also increases immobility in both the rat and mouse forced swimming test (FST) and the mouse tail suspension test (TST; see below), induces escape deficits in the mouse learned-helplessness model and leads to decrements in sexual motivation (Barr et al., 2002; Cryan et al., 2003).

The self-administration of cocaine also produces similar alterations. Following 24 h of cocaine self-administration, ICSS thresholds are elevated for several hours (Koob, 1989), indicating that cocaine withdrawal induces a state of anhedonia. Acute administration of the DA receptor agonist, bromocriptine, restored ICSS thresholds to normal (Markou and Koob, 1989). Repeated administration of the TCA desipramine is reported to shorten the duration of post-cocaine anhedonia (Markou et al., 1992).

1.4.3.2.5 Despair paradigm: forced swimming test

The FST is probably the most extensively used rodent model of depression (Porsolt et al., 1997; Cryan et al., 2002). In the FST, which was introduced by Porsolt and collaborators in 1977, mice or rats are forced to swim in a confined environment. The animals initially swim around and attempt to escape, and eventually assume an immobile posture. The immobility is thought to reflect either a failure of persistence in escape-directed behavior, i.e., behavioral despair, or the development of passive behavior that disengages the animals from active forms of coping with stressful stimuli (Lucki, 1997). This behavioral despair paradigm is conceptually similar to the learned helplessness paradigm in assuming that, after uncontrollable stress, animals have learned to "despair" (Willner, 1991a; Geyer and Markou, 1995).

The FST is currently a popular model, due to the low cost of the experiments and because it is arguably the most reliable model available even across laboratories (Borsini and Meli, 1988; Holmes, 2003; Cryan and Mombereau, 2004). The behavioral despair model has one of the highest degrees of predictive validity in terms of identifying major classes of antidepressants including TCAs, MAO inhibitors, atypical antidepressants, etc. (Boursini and Meli, 1988; West, 1990; Willner, 1991a; Detke et al., 1997; Petit-Demouliere et al., 2005; Berton and Nestler, 2006). However, in spite of this high degree of predictive validity, the FST has some limitations. The major drawback is that it is unreliable in the detection of the effects of SSRIs (Lucki, 1997), which are the most widely prescribed antidepressant drugs today. Another limitation is the false positives. For this reason, the FST requires additional tests to distinguish antidepressants from some other drug classes, such as psychomotor stimulants (Dalvi and Lucki, 1999; Petit-Demouliere et al., 2005). The ability of acute antidepressants treatments to reverse the immobility when the same compounds are effective in humans only after chronic treatment has also been considered by some authors as a partial failure to demonstrate good predictive validity (Geyer and Markou, 1995).

Regarding the construct validity of this model, the stress-induced alterations in the monoamine neurotransmitters believed to be involved in the actions of antidepressants indicate some etiological and theoretical validity for this model (Geyer and Markou, 1995). For example, FST has been shown to alter dialysate 5-HT levels in several brain regions (Kirby et al., 1997), while chronic treatment with antidepressants has been consistently shown to increase serotonergic neurotransmission (Markou et al., 1998). Nevertheless, the construct validity of this model is difficult to establish and is thus questionable (Petit-Demouliere et al., 2005).

Several studies have reported that inbred and outbred mouse strains display marked differences in baseline immobility and variability in the FST, demonstrating the presence of a genetic component (Dalvi and Lucki, 1999; Lucki et al., 2001; David et al., 2003). Moreover, some data also exist with transgenic mice in this model. For example, transgenic mice overexpressing human growth factor alpha show lengthened immobility (Hilakivi-Clarke and Goldberg, 1993). Another group of investigators have shown that transgenic mice with impaired glucocorticoid receptor function are less immobile than non-transgenic mice (Montkowski et al., 1995). Other data come from knockout mice studies. For instance, the decrease in immobility observed after paroxetine administration in wild-type mice is absent in $5-HT_{1B}$ receptor knockout in the test (Gardier et al., 2001).

Taken together, the FST provides a useful model not only to study antidepressant activity but also to study neurobiological and genetic mechanisms underlying stress and antidepressant response (Porsolt, 2000; Lucki et al., 2001; Nestler et al., 2002; Petit-Demouliere et al., 2005).

1.4.3.2.6 Tail suspension test

The TST shares a common theoretical basis and behavioral measure with the FST (Steru et al., 1985). In the TST, mice are suspended by their tails from an elevated bar or a hook for several minutes. Typically, mice immediately engage in several "agitation- or escape-like" behaviors, followed temporally by increasing bouts of immobility. The TST avoids problems of hypothermia or motor dysfunction that could interfere with performance in a FST (Steru et al., 1985; Ripoll et al., 2003; Cryan and Mombereau, 2004).

Administration of most antidepressants before the test reverses immobility and promotes struggling. The TST has been shown to be sensitive to various antidepressants (Steru et al., 1985; Perrault et al., 1992). SSRIs are very active in this test, in contrast to the behavioral despair test, in which SSRIs show less marked effects (Porsolt and Lenègre, 1992). These differential pharmacological effects suggest that the two models, while conceptually similar, might model different aspects of depression (Porsolt, 2000; Renard et al., 2003; Cryan and Mombereau, 2004). As in the FST, acute antidepressant administration, which is not effective in human depression, is effective in this test (Berton and Nestler, 2006). This may represent a partial disadvantage in demonstrating the good predictive validity of the model.

Some inter-strain variation in behavioral responses to antidepressants has been reported in the TST, in the same way as in the FST (Van der Heyden et al., 1987; Porsolt, 2000; Ripoll et al., 2003; Cryan and Mombereau, 2004). Liu and Gershenfeld (2001) examined the baseline and imipramine-induced behaviors of 11 strains of mice in the TST. Clear inter-strain differences were observed for baseline scores in the TST; and only three strains (DBA/2J, NMRI, FVB/NJ) responded to the antidepressant imipramine. The TST can be a useful alternative to the FST when examining antidepressants with some mouse strains (Dalvi and Lucki, 1999). Vaugeois and collaborators (1996) selectively bred mice for high and low immobility scores in the TST and showed that imipramine is only active in the high immobility time.

Since FST and TST have many characteristics in common, it is important to emphasize some of the much-discussed issues about both models.

A key issue regarding the validity of the FST and TST as models of depression is the finding that the administration of various classes of clinically effective antidepressant treatments before either test causes mice to actively and persistently engage in escape-directed behaviors for longer periods of time than after vehicle treatment. So, both tests have gained enormous popularity as rapid screening assays for novel antidepressants (Cryan et al., 2005a, 2005b; Berton and Nestler, 2006). However, some concerns have emerged about the validity of the FST and TST as models of depression. For example, these tests are sensitive to acute antidepressant administration, whereas chronic treatment is required for full clinical efficacy, suggesting that they might not be tapping into the same long-term adaptative changes in neuronal circuitry that underlie antidepressant effects in humans (Geyer and Markou, 2002; Cryan and Holmes, 2005).

In contrast, as an important contribution, FST and TST are frequently used as phenotypic screens for depression-related behaviors in mutant mice, with decreases in basal immobility interpreted as an antidepressant-like phenotype and, conversely, increased immobility taken as evidence of increased depression-related behavior in the mutant (Cryan and Mombereau, 2004; Park et al., 2005). Therefore, in these tests, immobility seems to be the result of an inability or reluctance to maintain effort rather than a generalized hypoactivity. This provides an interesting correlate with clinical observations that depressed patients show important psychomotor impairments, particularly in those tests requiring sustained expenditure of effort (Willner, 1990).

1.4.3.3 Reward models

These paradigms have been proposed to model a specific psychological process, i.e., reward, which appears to be altered in depressed individuals (Geyer and Markou, 1995). Thus, these paradigms are not considered models of an entire syndrome, but rather provide operational measures of anhedonia (Geyer and Markou, 1995; Anisman and Matheson, 2005). Anhedonia, the loss of interest in normally pleasurable, rewarding activities, is a core symptom of depression. Pleasure-seeking in rodents can be assessed by simple preference for a highly palatable solution, such as sucrose or saccharine, over water, by measuring the preference for a place associated to a hedonic stimulus, or through ICSS (Moreau, 2002; Cryan and Holmes, 2005). Probably the most important contribution of the experimental data coming from these three behavioral paradigms is the converging evidence that stress induces abnormalities in reward processes (Geyer and Markou, 1995).

1.4.3.3.1 Intracranial self stimulation paradigm

The intracranial self-stimulation (ICSS) paradigm involves brief electrical self-stimulation of specific brain sites, which is very reinforcing, as indicated by the fact that animals will work for it. Both the rate of responding for ICSS and the psychophysically defined threshold(s) for ICSS have been used as measures of the reward value of the stimulation (Markou and Koob, 1991, 1992). The lowering of thresholds is interpreted as an increase in the reward value of the stimulation, whereas elevated thresholds is interpreted as a decrease in the reward value of the stimulation (Geyer and Markou, 1995; Moreau, 2002).

Two manipulations have been used to produce an anhedonic state in animals, i.e., a decrease in ICSS response rates or elevated thresholds. On the one hand, exposure to uncontrollable stress produces decreases in the reward value of ICSS. These stress-induced alterations in ICSS behavior could be reversed by repeated antidepressant treatment (e.g., Zacharko et al., 1984; Moreau et al., 1992), indicating the predictive validity of the model in terms of pharmacological isomorphism. On the other hand, the withdrawal from long-term exposure drugs of abuse such as cocaine, amphetamine, nicotine, morphine, or ethanol can also produce changes in the thresholds for ICSS reward (e.g., Markou and Koob, 1991; Epping-Jordan et al., 1998; Markou et al., 1998; Lin et al., 1999). For example, when rats are withdrawn after being allowed to self-administer cocaine for prolonged periods of time, their thresholds for ICSS reward are elevated (Markou and Koob, 1991). This post-cocaine elevation in thresholds can be reversed by administration of some antidepressants (Markou et al., 1998).

1.4.3.3.2 Sucrose preference

In this paradigm, rats are usually exposed chronically to a series of mild stressors and their consumption of a sweet saccharin or sucrose solution is monitored. Stressed animals tend to consume less sweet solution than controls, suggesting an induction of a mild anhedonic state by stress (Willner et al., 1987; Geyer and Markou, 1995; Moreau, 2002). This effect of stress is reversed by antidepressant treatment (Willner, 1997), indicating good predictive validity in terms of pharmacological isomorphism.

Nevertheless, the validity of volumetric measures of sucrose consumption as a hedonic measure has been seriously questioned. Specifically, the amount of sucrose consumption is highly correlated with body weight, which is often lower in stressed animals compared to controls, especially when food deprivation is one of the stress manipulations. Therefore, some authors suggest that sucrose preference, in contrast to total sucrose consumption, may be a more appropriate reward measure (Geyer and Markou, 1995).

1.4.3.3.3 Conditioned place preference

This paradigm involves repeated pairing of a rewarding stimulus, such as food or drug, with a previously neutral environment (Geyer and Markou, 1995). The degree of place preference is considered a measure of reward. Exposure to CMS prevents or attenuates the formation of an association between rewarding stimuli and a previously neutral environment, indicating decreased sensitivity to rewards after stress manipulations (Papp et al., 1991; Willner, 1997).

1.4.3.4 Animal models of social loss and deprivation

An inadequate socialization, i.e., a low level of social support, or the experience of stressful events mainly during childhood have long been known to confer a long-lasting vulnerability to depression (Aneshensel and Stone, 1982; Brown, 1989; Willner, 1989b, 1991a; Heim and Nemeroff, 2001; Penza et al., 2005). The presumed etiological role in depression of social loss and deprivation events has led to the development of a number of animal models of depression based on the adverse effects of separation phenomena and social isolation (Mitchell and Redfern, 2005).

Probably, the most familiar of these models is the neonatal isolation or maternal separation. This model involves non-human primates, either infants isolated from their parents, or juveniles isolated from their peer group. In an initial stage, the separation response consists in protest-like behaviors such as agitation, sleeplessness and distress calls. This initial stage is followed by despair behaviors characterized by a decrease in activity, appetite, play, social interaction and a facial expression of sadness (Suomi, 1976; Henn and McKinney, 1987). These depressive-like behaviors have been associated with neurochemical changes, for instance, decreased concentrations of NA, with relatively little effect on DA or 5-HT (Kraemer et al., 1989) that are normalized after the isolation (Kraemer et al., 1984a, 1984b). Chronic treatment with desipramine (Hrdina et al., 1979), imipramine (Suomi et al., 1978), oxaprotiline (McKinney and Kraemer, 1989) and ECS (Lewis and McKinney, 1976) have been reported to reverse some, but not all, of the effects of separation in monkeys.

In rats, behavioral changes associated with postnatal maternal separation include slowed learning and less stable memory sleep disturbances, increased activity and stress (Hofer and Shair, 1987; Penza et al., 2005). The development of anhedonia, measured by a reduction in sucrose solution ingestion, has also been observed (Ladd et al., 2000). Maternal separation can also produce lasting alterations in neurotransmitter systems that are implicated in emotionality, including corticotropin-releasing factor, 5-HT, NA and glutamate (Meaney, 2001). Alterations in HPA axis responses to stress are also well established (Arborelius et al., 1999; Ladd et al., 2000). Antidepressant medication such as fluoxetine serves a protective function to maternal separation in animals (Lee et al., 2001).

Not only has isolation during childhood been studied, but also adult isolation. Chronic isolation of adult rats has been found to cause a disruption of cooperative social behavior (Berger and Schuster, 1982) reminiscent of the poor social performance of depressed people (Lewinsohn, 1974). The loss of valued social companionship also induces a well studied form of chronic hypercortisolism in squirrel monkeys (Parker et al., 2003). When squirrel monkeys are separated from social companions, they respond with prolonged increases in plasma levels of cortisol (Lyons, 2004).

In rats, a model of social defeat has been used. Social defeat is a potent stressor which is associated with a decrease in aggressive behavior (Albonetti and Farabollini, 1994). A single social defeat has been reported to produce a gradual, but long-lasting, increase in immobility in the FST, which is prevented by chronic treatment with clomipramine (Koolhaas et al., 1990). A similar model has been developed in submissive C57BL/6J mice (Korte et al., 1991). In a modified rat model, defeat of dominant pair-housed rats by rats of a different, more aggressive strain results in the loss of dominant status relative to their previously submissive partners, which is restored by chronic imipramine treatment (Willner et al., 1995).

Nevertheless, in spite of the development of these significant number of animal models based on social isolation, with the exception of some of the primate studies, these models have largely ignored both the complexity of childhood social deprivation phenomena and the mediation of their effects through later social relationships (Mitchell and Redfern, 2005).

1.4.3.5 Genetic models

Major depression is a heritable disorder that probably involves multiple genes, each with small effects (Wong and Licinio, 2001). Advances in genomic research and modern technologies capable of producing targeted gene deletions and gene transfers have opened a vast field to help us to elucidate the functional significance of potentially relevant genes in animal models of complex disorders, such as depression (Porsolt, 2000; Cryan and Mombereau, 2004; Lyons, 2004; Wong, 2005).

Mouse strain differences provide a principal resource for the study of genetic factors that underlie depression-related behaviors (Cryan and Holmes, 2005). That issue is important not only because it demonstrates the presence of a genetic component in depression-related behaviors, but also because it shows the significance of the rational selection of the best mouse strain for studying models of depression (Lucki, 2001; Ripoll et al., 2003).

A promising genetic approach involves selective breeding for clinically relevant, well defined behavioral traits in rodents, followed by genome-wide scans to identify specific predisposing gene candidates (Lyons, 2004). To discriminate genetic influences on depressive-like behavior, several investigators have undertaken selective breeding programs of animals, based on individual responsiveness in animal models of depression (Cryan and Mombereau, 2004). An example is the swim-test susceptible rat, selectively bred for exaggerated passivity in response to uncontrollable stress (Weiss and Kilts, 1998; Petit-Demouliere et al., 2005). Other breeding efforts include animals bred for spontaneous high or low immobility scores in the TST (Vaugeois et al., 1996) or animals susceptible to learned helplessness (Vollmayr et al., 2001).

The Flinders sensitive line (FSL) rat is the result of selective breeding for sensitivity to the hypothermic effect of ACh receptor agonists and is based on the hypothesis that central cholinergic system are important in depression, since increased cholinergic sensitivity has been reported in depressed patients (Henn and Vollmayr, 2005; Mitchell and Redfern, 2005; Overstreet et al., 2005). FSL animals have a number of behavioral alterations, including increased rapid eye movement sleep, reduced locomotor activity, a greater immobility in the FST and a greater vulnerability to the suppressive effect of CMS on responsiveness to sweet reward (Overstreet and

Janowsky, 1991; Pucilowski et al., 1991, 1993; Overstreet, 1993; Overstreet et al., 2005). The depressive-like behaviors may be normalized by chronic treatment with several antidepressants, such as imipramine, desipramine or sertraline (Overstreet, 1991; Schiller et al., 1992; Overstreet et al., 1995). FLS animals also exhibit markedly elevated levels of 5-HT, NA and DA in specific brain areas, which are normalized during chronic treatment with desipramine (Zangen et al., 1997, 1999).

Another example of a rat strain resulting from the selective breeding is the Fawn-hooded (FH) rat. The FH rat strain exhibits hypercortisolaemia and a blunted response to desamethasone-induced suppression of cortisol secretion (Owens and Nemeroff, 1991), consistent with the hyperactivity of the HPA axis which is observed in depressive illness. FH rats exhibit increased immobility in the FST in some studies (but not all; see Hall et al., 1998; Lahmame et al., 1996), increased ethanol intake and preference (Overstreet et al., 1992), higher levels of anxiety (Hall et al., 2000; Kantor et al., 2001) and also impaired social behavior (Kantor et al., 2000). These observations suggest that the FH strain could be a suitable model of comorbid depressive illness, alcoholism and anxiety (Mitchell and Redfern, 2005). The elevated levels of plasma cortisol are reduced by chronic treatment with the TCAs, imipramine and clomipramine, and with the MAO-A inhibitor, clorgyline (Aulakh et al., 1993).

The Rouen depressed or helpless mice are another promising murine selective breeding program to depression. Mice with high and low immobility in the TST are selected in this program (El Yacoubi et al., 2003; Vaugeois et al., 1996). In addition to increased immobility in the TST, helpless animals also have an increased immobility in the FST in advanced generations. These mice also show alterations in sleep, such as decreased wakefulness and decreased rapid eye movement sleep latency. In addition, the animals have an elevated basal corticosterone, which is suggestive of a disturbance in HPA axis function (Cryan and Mombereau, 2004).

Other genetic models have been developed, based on an underlying alteration (i.e., knockout mice) in the function of a selective neurotransmitter system involved in antidepressant action, be it ACh receptor-mediated (Yadid et al., 2000), 5-HT$_{1A}$ receptor-mediated (Knapp et al., 2000) or NA transporter-mediated responses (Cryan et al., 2002). There are also models of transgenic mouse strains, for instance, expressing glucocorticoid receptor antisense

(Pepin et al., 1992; Montkowski et al., 1995) or overexpressing corticotropin-releasing factor (Bale and Vale, 2004; Barden, 2004; Charmandari et al., 2005; de Kloet et al., 2005; Gillespie and Nemeroff, 2005; Keck et al., 2005).

1.4.4 Some concluding remarks

Animal models are indispensable tools to provide insights into the neurophatology that underlies depression and in the search to identify new antidepressant drugs. Animal models of depression, however, are not exempt from drawbacks such as the impossibility to translate every symptom of this human affective disorder into relevant and valid animal models. Despite the difficulties, numerous attempts have been made to create animal models of depression. On the one hand, various paradigms have been developed to investigate the processes that produce the psychiatric disorder simulating some core symptoms (i.e., simulation models). On the other hand, some other models have been developed with a more limited purpose: to provide a way to study the effects of potential therapeutics treatments (i.e., screening tests). In such cases, that kind of model may or may not mimic the psychiatric disorder. Rather, these models are only intended to reflect the efficacy of known therapeutic agents and thus lead to the discovery of new pharmacotherapies (Geyer and Markou, 1995, 2002).

The two approaches show inherent conflicts. To mimic a psychiatric syndrome such as depression in its entirety would require a highly complex model. However, reliable and reproducible data is more likely to be generated from a single model that perhaps mimics only one key symptom of the disease (Geyer and Markou, 2002; Mitchell and Redfern, 2005). Moreover, only a few symptoms of depression can be modelled, for instance, psychomotor and body weight changes, but not other symptoms such as suicidal ideation or feelings of guilt. The screening tests also have significant drawbacks. The requirements for any antidepressant screening test are that it is cheap, robust, reliable and easy to use and that it accurately predicts antidepressant activity (Danysz et al., 1991; Willner, 1991b). Therefore, the success of such screening tests and its consequent continued use often relies on acute treatment. However, one of the direct consequences of this approach is that such tests are incapable, by virtue of their design, of responding to the major current challenge of discovering new antidepressants that have a shorter onset of action (Mitchell and Redfern, 2005). Animal models sensitive to chronic antidepressant treatments have the capacity to detect a rapidly acting antidepressant treatment. Furthermore, the pharmacotherapy of depression typically requires chronic drug treatment; thus, the validity of an animal model could be called into question by an acute response to conventional antidepressant treatment (Geyer and Markou, 2002; Mitchell and Redfern, 2005). In contrast to these arguments, it can also be argued that, if a test responds acutely to antidepressants, the response would be potentiated by chronic treatment (Willner, 1989a). Another much-discussed issue related to the screening tests is that, generally, the use of these models has simply resulted in the identification of further "me too" compounds (novel compounds whose acute pharmacological profile is similar to those already available to the clinician) because usually such models have been developed and validated by reference to the effects of known therapeutic drugs. This fact clearly limits their ability to identify novel mechanisms and targets for future drug discovery (Geyer and Markou, 1995, 2002; Mitchell and Redfern, 2005). In the light of this current situation, one must acknowledge the absence of a perfect animal model for studies of depression or antidepressant action (Berton and Nestler, 2006). Thus, probably the selection of the best animal model of depression depends on the specific research purposes.

Despite these and other limitations, animal models continue to offer essential opportunities for generating, testing and extending current theories and pharmacotherapies for depression (Lyons, 2004). It is also clear that current models need to be refined continuously or new models developed to enhance the models' utility for both detecting novel targets for antidepressant activity and contributing to a better understanding of the underlying pathophysiology of depression (Cryan et al., 2002). One of the possible ways to achieve this improvement may be the information obtained from clinical science. In fact, the parallel development of animal models of depression and the clinic would involve important advances and mutual benefits (Geyer and Markou, 1995, 2002; Matthews et al., 2005; Willner, 1991a).

References

Akiskal HS. *Interaction of biologic and psychologic factors in the origin of depressive disorders.* Acta Psychiatr Scand Suppl. **1985**, 319:131-139.

Akiskal HS. *A developmental perspective on recurrent mood disorders: a review of studies in man.* Psychopharmacol Bull. **1986**, 22:579-586.

Albonetti ME, Farabollini F. *Social stress by repeated defeat: effects on social behaviour and emotionality.* Behav Brain Res. **1994**, 62:187-193.

Aneshensel CS, Stone JD. *Stress and depression: a test of the buffering model of social support.* Arch Gen Psychiatry. **1982**, 39:1392-1396.

Anisman H, Irwin J, Sklar LS. *Deficits of escape performance following catecholamine depletion: implications for behavioral deficits induced by uncontrollable stress.* Psychopharmacology (Berl). **1979**, 64:163-170.

Anisman H, Zacharko RM. *Multiple neurochemical and behavioral consequences of stressors: implications for depression.* Pharmacol Ther. **1990**, 46:119-136.

Anisman H, Matheson K. *Stress, depression, and anhedonia: caveats concerning animal models.* Neurosci Biobehav Rev. **2005**, 29:525-546.

Antelman SM, Eichler AJ, Black CA, Kocan D. *Interchangeability of stress and amphetamine in sensitization.* Science. **1980**, 207:329-331.

Arborelius L, Owens MJ, Plotsky PM, Nemeroff CB. *The role of corticotropin-releasing factor in depression and anxiety disorders.* J Endocrinol. **1999**, 160:1-12.

Aulakh CS, Hill JL, Murphy DL. *Attenuation of hypercortisolemia in fawn-hooded rats by antidepressant drugs.* Eur J Pharmacol. **1993**, 240:85-88.

Ayensu WK, Pucilowski O, Mason GA, Overstreet DH, Rezvani AH, Janowsky DS. *Effects of chronic mild stress on serum complement activity, saccharin preference, and corticosterone levels in Flinders lines of rats.* Physiol Behav. **1995**, 57:165-169.

Bale TL, Vale WW. *CRF and CRF receptors: role in stress responsivity and other behaviors.* Annu Rev Pharmacol Toxicol. **2004**, 44:525-557.

Barden N. *Implication of the hypothalamic-pituitary-adrenal axis in the physiopathology of depression.* J Psychiatry Neurosci. **2004**, 29:185-193.

Barr AM, Markou A, Phillips AG. *A 'crash' course on psychostimulant withdrawal as a model of depression.* Trends Pharmacol Sci. **2002**, 23:475-482.

Barrett RJ, White DK. *Reward system depression following chronic amphetamine: antagonism by haloperidol.* Pharmacol Biochem Behav. **1980**, 13:555-559.

Berger BD, Schuster R. An animal model of social interaction: Implications for the analysis of drug action. In: Spiegelstein M, Levy A, eds. *Behavioral Models and the Analysis of Drug Action.* Elsevier, Amsterdam, **1982**, pp 415-428.

Berton O, Nestler EJ. *New approaches to antidepressant drug discovery: beyond monoamines.* Nat Rev Neurosci. **2006**, 7:137-151.

Blanc G, Herve D, Simon H, Lisoprawski A, Glowinski J, Tassin JP. *Response to stress of mesocortico-frontal dopaminergic neurones in rats after long-term isolation.* Nature. **1980**, 284:265-267.

Bodnoff SR, Suranyi-Cadotte B, Aitken DH, Quirion R, Meaney MJ. *The effects of chronic antidepressant treatment in an animal model of anxiety.* Psychopharmacology (Berl). **1988**, 95:298-302.

Borsini F, Meli A. *Is the forced swimming test a suitable model for revealing antidepressant activity?* Psychopharmacology (Berl). **1988**, 94:147-160.

Borsini F, Cesana R, Kelly J, Leonard BE, McNamara M, Richards J, Seiden L. *BIMT 17: a putative antidepressant with a fast onset of action?* Psychopharmacology (Berl). **1997**, 134:378-386.

Broekkamp CL, O'Connor WT, Tonnaer JA, Rijk HW, Van Delft AM. *Corticosterone, choline acetyltransferase and noradrenaline levels in olfactory bulbectomized rats in relation to changes in passive avoidance acquisition and open field activity.* Physiol Behav. **1986**, 37:429-434.

Brown GW. A psychological view of depression. In: Bennett D, Freeman H, eds. *Community Psychiatry: The Scientific Background.* Churchill-Livingstone, London, **1991**, pp 71-114.

Cassens G, Actor C, Kling M, Schildkraut JJ. *Amphetamine withdrawal: effects on threshold of intracranial reinforcement.* Psychopharmacology (Berl). **1981**, 73:318-322.

Charmandari E, Tsigos C, Chrousos G. *Endocrinology of the stress response.* Annu Rev Physiol. **2005**, 67:259-284.

Cheeta S, Broekkamp C, Willner P. *Stereospecific reversal of stress-induced anhedonia by mianserin and its (+)-enantiomer.* Psychopharmacology (Berl). **1994**, 116:523-528.

Cryan JF, McGrath C, Leonard BE, Norman TR. *Combining pindolol and paroxetine in an animal model of chronic antidepressant action--can early onset of action be detected?* Eur J Pharmacol. **1998**, 352:23-28.

Cryan JF, Markou A, Lucki I. *Assessing antidepressant activity in rodents: recent developments and future needs.* Trends Pharmacol Sci. **2002**, 23:238-245.

Cryan JF, Hoyer D, Markou A. *Withdrawal from chronic amphetamine induces depressive-like behavioral effects in rodents.* Biol Psychiatry. **2003**, 54:49-58.

Cryan JF, Mombereau C. *In search of a depressed mouse: utility of models for studying depression-related behavior in genetically modified mice.* Mol Psychiatry. **2004**, 9:326-357.

Cryan JF, Holmes A. *The ascent of mouse: advances in modelling human depression and anxiety.* Nat Rev Drug Discov. **2005**, 4:775-790.

Cryan JF, Mombereau C, Vassout A. *The tail suspension test as a model for assessing antidepressant activity: review of pharmacological and genetic studies in mice.* Neurosci Biobehav Rev. **2005a**, 29:571-625.

Cryan JF, Valentino RJ, Lucki I. *Assessing substrates underlying the behavioral effects of antidepressants using the modified rat forced swimming test.* Neurosci Biobehav Rev. **2005b**, 29:547-569.

D'Aquila PS, Brain P, Willner P. *Effects of chronic mild stress on performance in behavioural tests relevant to anxiety and depression.* Physiol Behav. **1994**, 56:861-867.

Dalvi A, Lucki I. *Murine models of depression.* Psychopharmacology (Berl). **1999**, 147:14-16.

Danysz W, Archer T, Fowler CJ. Screening for new antidepressant compounds. In: Willner P, ed. *Behavioural models in psychopharmacology: Theoretical, industrial and clinical perspective.* Cambridge University Press, Cambridge, **1991**, pp 126-156.

David DJ, Renard CE, Jolliet P, Hascoet M, Bourin M. *Antidepressant-like effects in various mice strains in the forced swimming test.* Psychopharmacology (Berl). **2003**, 166:373-382.

De Deyn PP, D'Hooge R, van Zutphen LFM. *Animal models of human disorders - general aspects.* Neuroscience Research Communications. **2000**, 26:141-148.

de Kloet ER, Joels M, Holsboer F. *Stress and the brain: from adaptation to disease.* Nat Rev Neurosci. **2005**, 6:463-475.

Detke MJ, Johnson J, Lucki I. *Acute and chronic antidepressant drug treatment in the rat forced swimming test model of depression.* Exp Clin Psychopharmacol. **1997**, 5:107-112.

El Yacoubi M, Bouali S, Popa D, Naudon L, Leroux-Nicollet I, Hamon M, Costentin J, Adrien J, Vaugeois JM. *Behavioral, neurochemical, and electrophysiological characterization of a genetic mouse model of depression.* Proc Natl Acad Sci U S A. **2003**, 100:6227-6232.

Epping-Jordan MP, Watkins SS, Koob GF, Markou A. *Dramatic decreases in brain reward function during nicotine withdrawal.* Nature. **1998**, 393:76-79.

Fontana DJ, Carbary TJ, Commissaris RL. *Effects of acute and chronic anti-panic drug administration on conflict behavior in the rat.* Psychopharmacology (Berl). **1989**, 98:157-162.

Frazer A, Morilak DA. *What should animal models of depression model?* Neurosci Biobehav Rev. **2005**, 29:515-523.

Gardier AM, Trillat AC, Malagie I, David D, Hascoet M, Colombel MC, Jolliet P, Jacquot C, Hen R, Bourin M. *Récepteurs 5-HT1B de la sérotonine et effets antidépresseurs des inhibiteurs de recapture sélectifs de la sérotonine.* C R Acad Sci III. **2001**, 324:433-441.

Geyer MA, Markou A. Animal models of psychiatric disorders. In: Bloom BS, Kupfer DJ, Bunney BS, Ciaranello RD, Davis KL, Koob GF, Meltzer HY, Schuster CR, Shader RI, Watson SJ, eds. *Psychopharmacology: The fourth generation of progress.* Raven Press, New York, **1995**, pp 787-798.

Geyer MA, Markou A. The role of preclinical models in the development of psychotropic drugs. In: Davis KL, Charney D, Coyle JT, Nemeroff C, eds. *Neuropsychopharmacology: The fifth generation of progress.* Lipincott Williams & Wilkins, Philadelphia, **2002**, pp 445-454.

Gillespie CF, Nemeroff CB. *Hypercortisolemia and depression.* Psychosom Med. **2005**, 67 Suppl 1:S26-S28.

Glazer HI, Weiss JM. *Long-term interference effect: An alternative to "learned helplessness".* Journal of Experimental Psychology: Animal Behavior Processes. **1976**, 2:202-213.

Gorka Z, Moryl E, Papp M. *Chronic mild stress influences the diurnal activity rhythms in rats.* Behav Pharmacol. **1994**, 5:88.

Hall FS, Huang S, Fong GF, Pert A. *The effects of social isolation on the forced swim test in Fawn hooded and Wistar rats*. J Neurosci Methods. **1998**, 79:47-51.

Hall FS, Huang S, Fong GW, Sundstrom JM, Pert A. *Differential basis of strain and rearing effects on open-field behavior in Fawn Hooded and Wistar rats*. Physiol Behav. **2000**, 71:525-532.

Harkin A, Kelly JP, McNamara M, Connor TJ, Dredge K, Redmond A, Leonard BE. *Activity and onset of action of reboxetine and effect of combination with sertraline in an animal model of depression*. Eur J Pharmacol. **1999**, 364:123-132.

Harkin A, Kelly JP, Leonard BE. *A review of the relevance and validity of olfactory bulbectomy as a model of depression*. Clinical Neuroscience Research. **2003**, 3:253-262.

Heim C, Nemeroff CB. *The role of childhood trauma in the neurobiology of mood and anxiety disorders: preclinical and clinical studies*. Biol Psychiatry. **2001**, 49:1023-1039.

Henn FA, McKinney WT. Animal models in psychiatry. In: Meltzer H, ed. *Psychopharmacology: The Third Generation of Progress*. Raven Press, New York, **1987**, pp 697-704.

Henn FA, Vollmayr B. *Stress models of depression: forming genetically vulnerable strains*. Neurosci Biobehav Rev. **2005**, 29:799-804.

Hilakivi-Clarke LA, Goldberg R. *Effects of tryptophan and serotonin uptake inhibitors on behavior in male transgenic transforming growth factor alpha mice*. Eur J Pharmacol. **1993**, 237:101-108.

Hofer MA, Shair HN. *Isolation distress in two-week-old rats: influence of home cage, social companions, and prior experience with littermates*. Dev. Psychobiol. **1987**, 20:465-476.

Holmes PV. *Rodent models of depression: reexamining validity without anthropomorphic inference*. Crit Rev Neurobiol. **2003**, 15:143-174.

Hrdina PD, von Kulmiz P, Stretch R. *Pharmacological modification of experimental depression in infant macaques*. Psychopharmacology (Berl). **1979**, 64:89-93.

Jesberger JA, Richardson JS. *Effects of antidepressant drugs on the behavior of olfactory bulbectomized and sham-operated rats*. Behav Neurosci. **1986**, 100:256-274.

Kantor S, Anheuer ZE, Bagdy G. *High social anxiety and low aggression in Fawn-Hooded rats*. Physiol Behav. **2000**, 71:551-557.

Kantor S, Graf M, Anheuer ZE, Bagdy G. *Rapid desensitization of 5-HT(1A) receptors in Fawn-Hooded rats after chronic fluoxetine treatment*. Eur Neuropsychopharmacol. **2001**, 11:15-24.

Katz RJ, Roth KA, Carroll BJ. *Acute and chronic stress effects on open field activity in the rat: implications for a model of depression*. Neurosci Biobehav Rev. **1981**, 5:247-251.

Katz RJ, Sibel M. *Animal model of depression: tests of three structurally and pharmacologically novel antidepressant compounds*. Pharmacol Biochem Behav. **1982**, 16:973-977.

Katz RJ. *Animal model of depression: pharmacological sensitivity of a hedonic deficit*. Pharmacol Biochem Behav. **1982**, 16:965-968.

Keck ME, Ohl F, Holsboer F, Muller MB. *Listening to mutant mice: a spotlight on the role of CRF/CRF receptor systems in affective disorders*. Neurosci Biobehav Rev. **2005**, 29:867-889.

Kelly JP, Wrynn AS, Leonard BE. *The olfactory bulbectomized rat as a model of depression: an update.* Pharmacol Ther. **1997**, 74:299-316.

Kessler RC. *The effects of stressful life events on depression.* Annu Rev Psychol. **1997**, 48:191-214.

Kirby LG, Chou-Green JM, Davis K, Lucki I. *The effects of different stressors on extracellular 5-hydroxytryptamine and 5-hydroxyindoleacetic acid.* Brain Res. **1997**, 760:218-230.

Knapp DJ, Sim-Selley LJ, Breese GR, Overstreet DH. *Selective breeding of 5-HT(1A) receptor-mediated responses: application to emotion and receptor action.* Pharmacol Biochem Behav. **2000**, 67:701-708.

Kokkinidis L, Zacharko RM, Predy PA. *Post-amphetamine depression of self-stimulation responding from the substantia nigra: reversal by tricyclic antidepressants.* Pharmacol Biochem Behav. **1980**, 13:379-383.

Kokkinidis L, Zacharko RM. *Response sensitization and depression following long-term amphetamine treatment in a self-stimulation paradigm.* Psychopharmacology (Berl). **1980**, 68:73-76.

Koob GF. Anhedonia as an animal model of depression. In: Koob GF, Ehlers CL, Kupfer DJ, eds. *Animal Models of Depression.* Birkhauser, Boston, **1989**, pp 162-183.

Koolhaas JM, Hermann PM, Kemperman C, Bohus B, van den Hoofdakker RH, Beersma DGM. *Single social defeat in male rats induces a gradual but long lasting behavioural change: a model of depression?* Neuroscience Research Communications. **1990**, 7:35-41.

Korte SM, Smit J, Bouws GAH, Koolhaas JM, Bohus B. Neuroendocrine evidence for hypersensitivity in serotonergic neuronal system after psychosocial stress of defeat. In: Olivier B, Mos J, Slangen J, eds. *Animal Models in Psychopharmacology.* Birkhauser, Basel, **1991**, pp 199-203.

Kraemer GW, Ebert MH, Lake CR, McKinney WT. *Cerebrospinal fluid measures of neurotransmitter changes associated with pharmacological alteration of the despair response to social separation in rhesus monkeys.* Psychiatry Res. **1984a**, 11:303-315.

Kraemer GW, Ebert MH, Lake CR, McKinney WT. *Hypersensitivity to d-amphetamine several years after early social deprivation in rhesus monkeys.* Psychopharmacology (Berl). **1984b**, 82:266-271.

Kraemer GW, Ebert MH, Schmidt DE, McKinney WT. *A longitudinal study of the effect of different social rearing conditions on cerebrospinal fluid norepinephrine and biogenic amine metabolites in rhesus monkeys.* Neuropsychopharmacology. **1989**, 2:175-189.

Kupfer DJ, Thase ME. *The use of the sleep laboratory in the diagnosis of affective disorders.* Psychiatr Clin North Am. **1983**, 6:3-25.

Lachman HM, Papolos DF, Boyle A, Sheftel G, Juthani M, Edwards E, Henn FA. *Alterations in glucocorticoid inducible RNAs in the limbic system of learned helpless rats.* Brain Res. **1993**, 609:110-116.

Ladd CO, Huot RL, Thrivikraman KV, Nemeroff CB, Meaney MJ, Plotsky PM. *Long-term behavioral and neuroendocrine adaptations to adverse early experience.* Prog Brain Res. **2000**, 122:81-103.

Lahmame A, Gomez F, Armario A. *Fawn-hooded rats show enhanced active behaviour in the forced swimming test, with no evidence for pituitary-adrenal axis hyperactivity*. Psychopharmacology (Berl). **1996**, 125:74-78.

Lee HJ, Kim JW, Yim SV, Kim MJ, Kim SA, Kim YJ, Kim CJ, Chung JH. *Fluoxetine enhances cell proliferation and prevents apoptosis in dentate gyrus of maternally separated rats*. Mol Psychiatry. **2001**, 6:610, 725-610, 728.

Leith NJ, Barrett RJ. *Effects of chronic amphetamine or reserpine on self-stimulation responding: animal model of depression?* Psychopharmacology (Berl). **1980**, 72:9-15.

Lewinsohn PM. A behavioural approach to depression. In: Friedman RJ, Katz MM, eds. *The psychology of depression: Contemporary theory and research*. John Wiley & Sons Inc, New York, **1974**, pp 157-185.

Lewis JK, McKinney WT. *Effects of electroconvulsive shock on the behaviour of normal and abnormal rhesus monkeys*. Behav Psychiatr. **1976**, 37:687-693.

Lewis JW, Cannon JT, Liebeskind JC. *Opioid and nonopioid mechanisms of stress analgesia*. Science. **1980**, 208:623-625.

Lin D, Koob GF, Markou A. *Differential effects of withdrawal from chronic amphetamine or fluoxetine administration on brain stimulation reward in the rat--interactions between the two drugs*. Psychopharmacology (Berl). **1999**, 145:283-294.

Liu X, Gershenfeld HK. *Genetic differences in the tail-suspension test and its relationship to imipramine response among 11 inbred strains of mice*. Biol Psychiatry. **2001**, 49:575-581.

Lloyd C. *Life events and depressive disorder reviewed. II. Events as precipitating factors*. Arch Gen Psychiatry. **1980**, 37:541-548.

Lucki I. *The forced swimming test as a model for core and component behavioral effects of antidepressant drugs*. Behav Pharmacol. **1997**, 8:523-532.

Lucki I, Dalvi A, Mayorga AJ. *Sensitivity to the effects of pharmacologically selective antidepressants in different strains of mice*. Psychopharmacology (Berl). **2001**, 155:315-322.

Lucki I. *A prescription to resist proscriptions for murine models of depression*. Psychopharmacology (Berl). **2001**, 153:395-398.

Lyons DM. Animal models. In: Nemeroff CB, Schatzberg AF, eds. *Textbook of psychopharmacology*. 3rd ed. American Psychiatric Press, Arlington, **2004**, pp 105-114.

Maier SF, Seligman MEP. *Learned helplessness: Theory and evidence*. Journal of Experimental Psychology: General. **1976**, 1:3-46.

Maier SF. *Exposure to the stressor environment prevents the temporal dissipation of behavioral depression/learned helplessness*. Biol Psychiatry. **2001**, 49:763-773.

Maier SF, Watkins LR. *Stressor controllability and learned helplessness: the roles of the dorsal raphe nucleus, serotonin, and corticotropin-releasing factor*. Neurosci. Biobehav. Rev. **2005**, 29:829-841

Markou A, Koob GF. *Bromcriptine reverses post-cocaine anhedonia in a rat model of cocaine withdrawal*. American College of Neuropsychopharmacology. **1989**, Abstracts: 157.

Markou A, Koob GF. *Postcocaine anhedonia. An animal model of cocaine withdrawal.* Neuropsychopharmacology. **1991**, 4:17-26.

Markou A, Hauger RL, Koob GF. *Desmethylimipramine attenuates cocaine withdrawal in rats.* Psychopharmacology (Berl). **1992**, 109:305-314.

Markou A, Koob GF. *Construct validity of a self-stimulation threshold paradigm: effects of reward and performance manipulations* . Physiol Behav. **1992**, 51:111-119.

Markou A, Kosten TR, Koob GF. *Neurobiological similarities in depression and drug dependence: a self-medication hypothesis.* Neuropsychopharmacology. **1998**, 18:135-174.

Matthews K, Christmas D, Swan J, Sorrell E. *Animal models of depression: navigating through the clinical fog.* Neurosci Biobehav Rev. **2005**, 29:503-513.

McGrath C, Norman TR. *The effect of venlafaxine treatment on the behavioural and neurochemical changes in the olfactory bulbectomised rat.* Psychopharmacology (Berl). **1998**, 136:394-401.

McGrath C, Norman TR. *(+)-S-20499 -- a potential antidepressant? A behavioural and neurochemical investigation in the olfactory bulbectomised rat.* Eur Neuropsychopharmacol. **1999**, 9:21-27.

McKinney WT. *Animal models of depression: an overview.* Psychiatr Dev. **1984**, 2:77-96.

McKinney WT, Kraemer GW. *Effects of oxaprotiline on the response to peer separation in rhesus monkeys.* Biol Psychiatry. **1989**, 25:818-821.

McKinney WT, Jr., Bunney WE, Jr. *Animal model of depression. I. Review of evidence: implications for research.* Arch Gen Psychiatry. **1969**, 21:240-248.

Meaney MJ. *Maternal care, gene expression, and the transmission of individual differences in stress reactivity across generations.* Annu Rev Neurosci. **2001**, 24:1161-1192.

Mitchell PJ, Redfern PH. *Animal models of depressive illness: the importance of chronic drug treatment.* Curr Pharm Des. **2005**, 11:171-203.

Montkowski A, Barden N, Wotjak C, Stec I, Ganster J, Meaney M, Engelmann M, Reul JM, Landgraf R, Holsboer F. *Long-term antidepressant treatment reduces behavioural deficits in transgenic mice with impaired glucocorticoid receptor function.* J Neuroendocrinol. **1995**, 7:841-845.

Moreau JL, Jenck F, Martin JR, Mortas P, Haefely WE. *Antidepressant treatment prevents chronic unpredictable mild stress-induced anhedonia as assessed by ventral tegmentum self-stimulation behavior in rats.* Eur Neuropsychopharmacol. **1992**, 2:43-49.

Moreau JL, Jenck F, Martin JR, Mortas P, Haefely W. *Effects of moclobemide, a new generation reversible Mao-A inhibitor, in a novel animal model of depression.* Pharmacopsychiatry. **1993**, 26:30-33.

Moreau JL, Bourson A, Jenck F, Martin JR, Mortas P. *Curative effects of the atypical antidepressant mianserin in the chronic mild stress-induced anhedonia model of depression.* J Psychiatry Neurosci. **1994**, 19:51-56.

Moreau JL, Scherschlicht R, Jenck F, Martin JR. *Chronic mild stress-induced anhedonia model of depression; sleep abnormalities and curative effects of electroshock treatment.* Behav Pharmacol. **1995**, 6:682-687.

Moreau JL. *Simulating the anhedonia symptom of depression in animals.* Dialogues in clinical neuroscience: Drug development. **2002**, 4:351-360.

Muscat R, Papp M, Willner P. *Reversal of stress-induced anhedonia by the atypical antidepressants, fluoxetine and maprotiline.* Psychopharmacology (Berl). **1992**, 109:433-438.

Nelson JC, Charney DS. *The symptoms of major depressive illness.* Am J Psychiatry. **1981**, 138:1-13.

Nestler EJ, Gould E, Manji H, Buncan M, Duman RS, Greshenfeld HK, Hen R, Koester S, Lederhendler I, Meaney M, Robbins T, Winsky L, Zalcman S. *Preclinical models: status of basic research in depression.* Biol Psychiatry. **2002**, 52:503-528.

Overmier JB, Seligman ME. *Effects of inescapable shock upon subsequent escape and avoidance responding.* J Comp Physiol Psychol. **1967**, 63:28-33.

Overstreet DH. *Commentary: a behavioral, psychopharmacological, and neurochemical update on the Flinders Sensitive Line rat, a potential genetic animal model of depression.* Behav Genet. **1991**, 21:67-74.

Overstreet DH, Janowsky DS. *A cholinergic supersensitivity model of depression.* In: Boulton AA, Baker GB, Martin-Iverson MT, eds. *Animal models in psychiatry, II.* Humana Press, Totowa, **1991**, pp 81-114.

Overstreet DH, Rezvani AH, Janowsky DS. *Genetic animal models of depression and ethanol preference provide support for cholinergic and serotonergic involvement in depression and alcoholism.* Biol Psychiatry. **1992**, 31:919-936.

Overstreet DH. *The Flinders sensitive line rats: a genetic animal model of depression.* Neurosci Biobehav Rev. **1993**, 17:51-68.

Overstreet DH, Pucilowski O, Rezvani AH, Janowsky DS. *Administration of antidepressants, diazepam and psychomotor stimulants further confirms the utility of Flinders Sensitive Line rats as an animal model of depression.* Psychopharmacology (Berl). **1995**, 121:27-37.

Overstreet DH, Friedman E, Mathe AA, Yadid G. *The Flinders Sensitive Line rat: a selectively bred putative animal model of depression.* Neurosci Biobehav Rev. **2005**, 29:739-759.

Owens MJ, Nemeroff CB. *Physiology and pharmacology of corticotropin-releasing factor.* Pharmacol Rev. **1991**, 43:425-473.

Papp M, Willner P, Muscat R. *An animal model of anhedonia: attenuation of sucrose consumption and place preference conditioning by chronic unpredictable mild stress.* Psychopharmacology (Berl). **1991**, 104:255-259.

Papp M, Lappas S, Muscat R, Willner P. *Attenuation of place preference conditioning but not place aversion conditioning by chronic mild stress.* J Psychopharmacol. **1992**, 6:352-356.

Papp M, Muscat R, Willner P. *Subsensitivity to rewarding and locomotor stimulant effects of a dopamine agonist following chronic mild stress.* Psychopharmacology (Berl). **1993**, 110:152-158.

Papp M, Moryl E, Willner P. *Pharmacological validation of the chronic mild stress model of depression.* Eur J Pharmacol. **1996**, 296:129-136.

Park SK, Nguyen MD, Fischer A, Luke MP, Affar eB, Dieffenbach PB, Tseng HC, Shi Y, Tsai LH. *Par-4 links dopamine signaling and depression.* Cell. **2005**, 122:275-287.

Parker KJ, Schatzberg AF, Lyons DM. *Neuroendocrine aspects of hypercortisolism in major depression.* Horm Behav. **2003**, 43:60-66.

Penza KM, Heim C, Nemeroff CB. Loss and deprivation: From animal models to clinical presentation. In: Licinio J, Wong ML, eds. *Biology of depression: From novel insights to therapeutic strategies.* Wiley-VCH Verlag GmbH & Co. KgaA, Weinheim, **2005**, pp 689-714.

Pepin MC, Pothier F, Barden N. *Impaired type II glucocorticoid-receptor function in mice bearing antisense RNA transgene.* Nature. **1992**, 355:725-728.

Perrault G, Morel E, Zivkovic B, Sanger DJ. *Activity of litoxetine and other serotonin uptake inhibitors in the tail suspension test in mice.* Pharmacol Biochem Behav. **1992**, 42:45-47.

Petit-Demouliere B, Chenu F, Bourin M. *Forced swimming test in mice: a review of antidepressant activity.* Psychopharmacology (Berl). **2005**, 177:245-255.

Petty F, Sherman AD. *GABAergic modulation of learned helplessness.* Pharmacol. Biochem. Behav. **1981**, 15:567-570.

Petty F, Sherman AD. *Learned helplessness induction decreases in vivo cortical serotonin release.* Pharmacol. Biochem. Behav. **1983**, 18:649-650.

Porsolt RD, Bertin A, Jalfre M. *Behavioral despair in mice: a primary screening test for antidepressants.* Arch Int Pharmacodyn Ther. **1977a**, 229:327-336.

Porsolt RD, Le Pichon M, Jalfre M. *Depression: a new animal model sensitive to antidepressant treatments.* Nature. **1977b**, 266:730-732.

Porsolt RD, Lenègre A. Behavioural models of depression. In: Elliott JM, Heal DJ, Marsden CA, eds. *Experimental approaches to anxiety and depression.* John Wiley, Chichester, **1992**, pp 73-87.

Porsolt RD. *Animal models of depression: utility for transgenic research.* Rev Neurosci. **2000**, 11:53-58.

Pucilowski O, Danysz W, Overstreet DH, Rezvani AH, Eichelman B, Janowsky DS. *Decreased hyperthermic effect of MK-801 in selectively bred hypercholinergic rats.* Brain Res Bull. **1991**, 26:621-625.

Pucilowski O, Overstreet DH, Rezvani AH, Janowsky DS. *Chronic mild stress-induced anhedonia: greater effect in a genetic rat model of depression.* Physiol Behav. **1993**, 54:1215-1220.

Renard CE, Dailly E, David DJ, Hascoet M, Bourin M. *Monoamine metabolism changes following the mouse forced swimming test but not the tail suspension test.* Fundam Clin Pharmacol. **2003**, 17:449-455.

Ripoll N, David DJ, Dailly E, Hascoet M, Bourin M. *Antidepressant-like effects in various mice strains in the tail suspension test.* Behav Brain Res. **2003**, 143:193-200.

Schiller GD, Pucilowski O, Wienicke C, Overstreet DH. *Immobility-reducing effects of antidepressants in a genetic animal model of depression.* Brain Res Bull. **1992**, 28:821-823.

Seligman ME, Maier SF. *Failure to escape traumatic shock.* J Exp Psychol. **1967**, 74:1-9.

Sherman AD, Sacquitne JL, Petty F. *Specificity of the learned helplessness model of depression.* Pharmacol Biochem Behav. **1982**, 16:449-454.

Sluzewska A, Nowakowska E. *The effects of carbamazepine, lithium and ketoconazole in chronic mild stress model of depression in rats.* Behav Pharmacol. **1994**, 5:86.

Song C, Leonard BE. *The olfactory bulbectomised rat as a model of depression.* Neurosci Biobehav Rev. **2005**, 29:627-647.

Steru L, Chermat R, Thierry B, Simon P. *The tail suspension test: a new method for screening antidepressants in mice.* Psychopharmacology (Berl). **1985**, 85:367-370.

Strekalova T, Spanagel R, Bartsch D, Henn FA, Gass P. *Stress-induced anhedonia in mice is associated with deficits in forced swimming and exploration.* Neuropsychopharmacology. **2004**, 29:2007-2017.

Sullivan PF, Neale MC, Kendler KS. *Genetic epidemiology of major depression: review and meta-analysis.* Am J Psychiatry. **2000**, 157:1552-1562.

Suomi SJ. Factors affecting responses to social separation in rhesus monkeys. In: Serban G, Kling A, eds. *Animal models in human psychobiology.* Plenum Press, New York, **1976**, pp 9-26.

Suomi SJ, Seaman SF, Lewis JK, DeLizio RD, McKinney WT, Jr. *Effects of imipramine treatment of separation-induced social disorders in rhesus monkeys.* Arch Gen Psychiatry. **1978**, 35:321-325.

van der Heyden JA, Molewijk E, Olivier B. *Strain differences in response to drugs in the tail suspension test for antidepressant activity.* Psychopharmacology (Berl). **1987**, 92:127-130.

Vaugeois JM, Odievre C, Loisel L, Costentin J. *A genetic mouse model of helplessness sensitive to imipramine.* Eur J Pharmacol. **1996**, 316:R1-R2.

Vollmayr B, Faust H, Lewicka S, Henn FA. *Brain-derived-neurotrophic-factor (BDNF) stress response in rats bred for learned helplessness.* Mol Psychiatry. **2001**, 6:471-474, 358.

Vollmayr B, Henn FA. *Learned helplessness in the rat: improvements in validity and reliability.* Brain Res Brain Res Protoc. **2001**, 8:1-7.

Weiss JM, Kilts CD. Animal models of depression and schizophrenia. In: Nemeroff CB, Schatzberg AF, eds. *Textbook of psychopharmacology.* 2nd ed. American Psychiatric Press, Washington DC, **1998**, pp 88-123.

West AP. *Neurobehavioral studies of forced swimming: the role of learning and memory in the forced swim test.* Prog Neuropsychopharmacol Biol Psychiatry. **1990**, 14:863-877.

Willner P. *The validity of animal models of depression.* Psychopharmacology (Berl). **1984**, 83:1-16.

Willner P. *Validation criteria for animal models of human mental disorders: learned helplessness as a paradigm case.* Prog Neuropsychopharmacol Biol Psychiatry. **1986**, 10:677-690.

Willner P, Towell A, Sampson D, Sophokleous S, Muscat R. *Reduction of sucrose preference by chronic unpredictable mild stress, and its restoration by a tricyclic antidepressant.* Psychopharmacology (Berl). **1987**, 93:358-364.

Willner P. Sensitization to the actions of antidepressant drugs. In: Emmett-Oglesby MW, Goudie AJ, eds. *Psychoactive Drugs: Tolerance and Sensitization.* Humana Press, New York, **1989a**, pp 407-460.

Willner P. Towards a theory of serotonergic dysfunction in depression . In: Archer T, Bevan P, Cools A, eds. *Behavioural Pharmacology of 5-HT*. Lawrence Erlbaum Associates, Hillsdale, NJ., **1989b**, pp 157-178.

Willner P. *Animal models of depression: an overview*. Pharmacol Ther. **1990**, 45:425-455.

Willner P. Animal models of depression. In: Willner P, ed. *Behavioural models in psychopharmacology: Theoretical, industrial and clinical perspective*. Cambridge University Press, Cambridge, **1991a**, pp 91-125.

Willner P. Behavioral models in psychopharmacology. In: Willner P, ed. *Behavioural models in psychopharmacology: Theoretical, industrial and clinical perspective*. Cambridge University Press, Cambridge, **1991b**, pp 3-18.

Willner P, D'Aquila PS, Coventry T, Brain P. *Loss of social status: preliminary evaluation of a novel animal model of depression*. J Psychopharmacol. **1995**, 9:207-213.

Willner P. *Validity, reliability and utility of the chronic mild stress model of depression: a 10-year review and evaluation*. Psychopharmacology (Berl). **1997**, 134:319-329.

Wong ML, Licinio J. *Research and treatment approaches to depression*. Nat Rev Neurosci. **2001**, 2:343-351.

Wong ML. Major Depression and Animals Models. In: Licinio J, Wong ML, eds. *Biology of Depression: From Novel Insights to Therapeutic Strategies*. Wiley-VCH Verlag GmbH & Co. KgaA, Weinheim, **2005**, pp 669-688.

Yadid G, Nakash R, Deri I, Tamar G, Kinor N, Gispan I, Zangen A. *Elucidation of the neurobiology of depression: insights from a novel genetic animal model*. Prog Neurobiol. **2000**, 62:353-378.

Zacharko RM, Bowers WJ, Kelley MS, Anisman H. *Prevention of stressor-induced disturbances of self-stimulation by desmethylimipramine*. Brain Res. **1984**, 321:175-179.

Zangen A, Overstreet DH, Yadid G. *High serotonin and 5-hydroxyindoleacetic acid levels in limbic brain regions in a rat model of depression: normalization by chronic antidepressant treatment*. J Neurochem. **1997**, 69:2477-2483.

Zangen A, Overstreet DH, Yadid G. *Increased catecholamine levels in specific brain regions of a rat model of depression: normalization by chronic antidepressant treatment*. Brain Res. **1999**, 824:243-250.

Zueger M, Urani A, Chourbaji S, Zacher C, Roche M, Harkin A, Gass P. *Olfactory bulbectomy in mice induces alterations in exploratory behavior*. Neurosci Lett. **2005**, 374:142-146.

1.5 Chemistry

Marketed Drugs and Drugs in Development

Jörg Holenz, José Luis Díaz and Helmut Buschmann

Laboratorios Esteve, Av. Mare de Déu de Montserrat, 221. 08041 Barcelona, Spain.

1.5.1 Summary of drug classes

According to their chemical structural class as well as their mode of action, the antidepressants currently available on the market are clustered into the following groups: tricyclic and tetracyclic antidepressants (TCAs; Section 1.5.2), serotoninergic agents like e.g. selective serotonin reuptake inhibitors (SSRIs; Section 1.5.3), noradrenaline reuptake inhibitors (NARIs; Section 1.5.4), monoamine oxidase inhibitors (MAOs; Section 1.5.5) and miscellaneous agents (Section 1.5.6). Section 1.5.7 presents compounds launched in single countries, whereas Section 1.5.8 lists some drugs originally marketed for other indications, but now also used against depression. Finally, Section 1.5.9 gives a summary on drugs in development. In the redaction of this chapter and particularly in order to search for the primary literature, the authors used, i.a., Prous Integrity database (http://integrity.prous.com) and "Martindale – The Complete Drug Reference" (Sweetman, 2004).

1.5.2 Tricyclic and tetracyclic antidepressants

Historically, the first available class of clinically potent antidepressants are the tricyclic and tetracyclic antidepressants (TCAs). TCAs are a class of antidepressant drugs first used in the 1950s. They are named after the drugs' common molecular structure, which contains three (four) rings of atoms. The exact mechanism of action is not well understood, however it is generally thought that tricylic antidepressants work by inhibiting the reuptake of the neurotransmitters norepinephrine, dopamine or serotonin in nerve cells. Although this pharmacologic effect occurs immediately, often the patient's symptoms do not respond for several weeks. Even though with the market introduction of e.g. the selective serotonin reuptake inhibitors newer classes of potent antidepressants are available, tricyclic antidepressants such as especially amitriptyline and imipramine still play a major role in the therapy of depressants, despite of their pronounced, mainly anti-muscarinic side-effects and the potential TCA poisoning. In the following, the TCAs on the market are

Amitriptyline

Amitriptyline

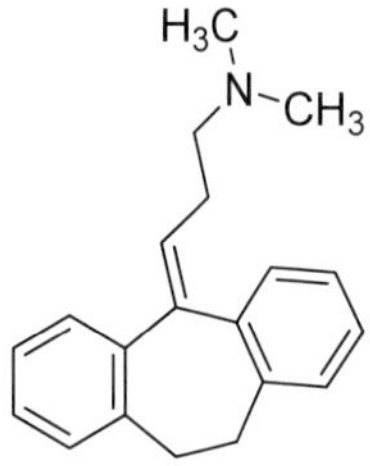

[50-48-6], N,N-Dimethyl-3-(10,11-dihydro-5H-dibenzo[a,d]cyclohepten-5-ylidene)-1-propanamine, $C_{20}H_{23}N$, M_r 277.40; [549-18-8] (hydrochloride salt), $C_{20}H_{24}ClN$, M_r 313.86

presented summarizing chemical properties, synthesis, important patents, clinical use as well as pharmacokinetics and metabolism.

- Patented by Hoffmann-La Roche in 1961 (Rey-Bellet and Spiegelberg, 1961)
- Launched in 1961 by AstraZeneca in the United States and by Roche and Merck Sharp and Dohme in Europe
- Compound is in phase III trials for interstitial cystitis, fibromyalgia and pain

Synthesis (Rey-Bellet and Spiegelberg, 1961): Starting from 5-oxodibenzo[a,d]cyclohepta-1,4-diene, addition of magnesium and 3-chloropropyl-dimethylamine in tetrahydrofuran (Winthrop et al., 1962) affords 5-(3-dimethylamino-propyl)-10,11-dihydro-5H-dibenzo[a,d]cyclohepten-5-ol, which is dehydrated with H_2SO_4 to give, after basic treatment, N,N-dimethyl-3-(10,11-dihydro-5H-dibenzo[a,d]cyclohepten-5-ylidene)-1-propanamine.

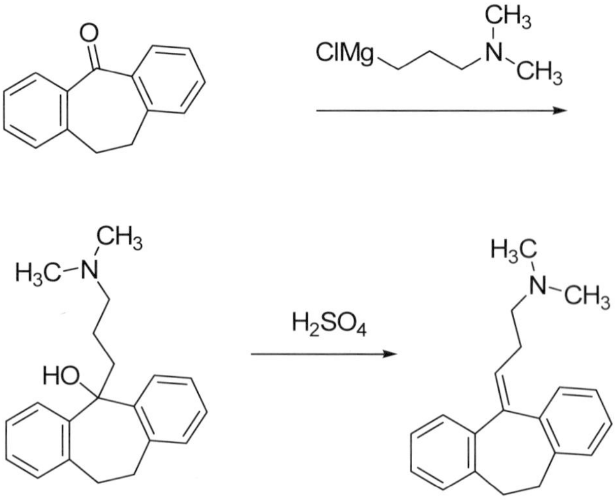

Scheme 1: Synthesis of amitriptyline

An alternative synthesis using cyclopropyl magnesium bromide was published in 1962 (Hoffsommer et al., 1962).

Clinical Use, Pharmacokinetics and Metabolism: Amitriptyline can be seen as the prototype of the classic

TCA antidepressants. It is widely used in melancholic depression as well as in some classes of atypical depression and represents, even though in the meanwhile there are many newer classes of antidepressants available, the main standard for first therapy, and also the one by which others are measured and rated. Amitriptyline has marked antimuscarinic and sedative properties and prevents the reuptake (inactivation) of noradrenaline and serotonin in nerve terminals. Its complete mode of action in Depression however, as for the vast majority of its class, is not completely understood. While the sedative effect of amitriptyline is immediately observed, it may take up to four weeks until an antidepressant effect is observed clinically. Most common side-effects include drowsiness (sedation), orthostatic hypotension and "muscarinic" side-effects like dry mouth, constipation, urinary retention or blurred vision and disturbances in accommodation. After longer treatment, in most cases, tolerance is achieved with respect to many of the above mentioned side-effects. Amitriptyline is readily absorbed from the GI tract, leading to peak plasma concentrations within a few hours. During extensive first-pass metabolism, amitriptyline is cleaved (*N*-demethylation) to its main active metabolite nortriptyline (Caccia and Garattini, 1990). Other metabolic pathways include *N*-oxidation and hydroxylation steps. Amitriptyline as well as noramitriptyline show extensive plasma and tissue protein binding (Schulz et al., 1985), so that interindividual free plasma concentrations as well as elimination half-lifes [between 9–36 (!) hours] and thus therapeutic effects differ widely (Brøsen and Gram, 1989; Llerena et al., 1993; Wood and Zhou, 1991).

Imipramine

- Imipramine was the first tricyclic antidepressant, developed at Geigy (Ciba-Geigy since 1970; Ciba since 1992 and Novartis since 1996 after fusion with Sandoz) in the early 1950s (Haefliger and Schindler, 1951; Kuhn, 1957)

- Launched in 1958 for the treatment of depression by Novartis

- In 2001 Novartis sold the US rights of imipramine (Tofranil®) to Mallinckrodt, affiliate of Tyco International Ltd

Imipramine

[*50-49-7*], 5-[3-(Dimethylamino)propyl]-10,11-dihydro-5*H*-dibenz[*b,f*]azepine, $C_{19}H_{24}N_2$, M_r 280.41; [*113-52-0*] (hydrochloride salt), $C_{19}H_{25}ClN_2$, M_r 316.87

Synthesis. Typical *N*-alkylation affords imipramine from commercially available 10,11-dihydro-5*H*-dibenz[*b*,*f*]azepine (for instance, see: Schmolka and Zimmer, 1984).

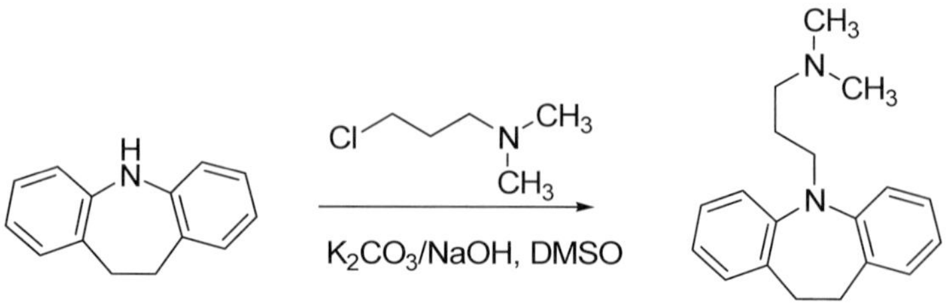

Scheme 2: Synthesis of imipramine

Clinical Use, Pharmacokinetics and Metabolism: Imipramine represents, besides amitriptyline, the main prototype and most important representative of the class of TCAs. It is widely used in melancholic depression and – in some cases – in atypical depression. Next to its application in depression treatment, like many other TCAs, it finds a use in the treatment of anxiety disorders (Lepola et al., 1993; Clark et al., 1994), hyperactivity (ADHD), micturition disorders, narcoleptic syndrome and pain (Cannon et al., 1994; Kvinesdal et al., 1984; Walsh, 1986; Hummel et al., 1994). In terms of side-effects, imipramine has a very similar spectrum as amitriptyline, but with less pronounced sedation. Apart from this, vasospastic episodes (Appelbaum and Kapoor, 1983; Anderson and Morris, 1988) and TCA-typical drug interactions have been reported. As other members of the TCA family, imipramine is extensively distributed throughout the body and bound to plasma and tissue protein, resulting in elimination half-lifes of 9–28 hours, with significant interindividual differences (Sallee and Pollock, 1990). The metabolism and kinetics of imipramine in humans are well documented (Gram and Christiansen, 1975, Gram et al., 1983) and have been extensively studied experimentally and clinically. It was found, that in humans imipramine was metabolized mainly by *N*-dealkylation to desmethylimipramine (desipramine) and by hydroxylation to 2-hydroxyimipramine.

The studies revealed substantial interindividual differences in plasma concentration. Experimental studies have shown that 2-hydroxyimipramine and 2-hydroxydesipramine exert cardiac effects and reuptake inhibition of adrenalin and serotonin much like those of imipramine and desipramine.

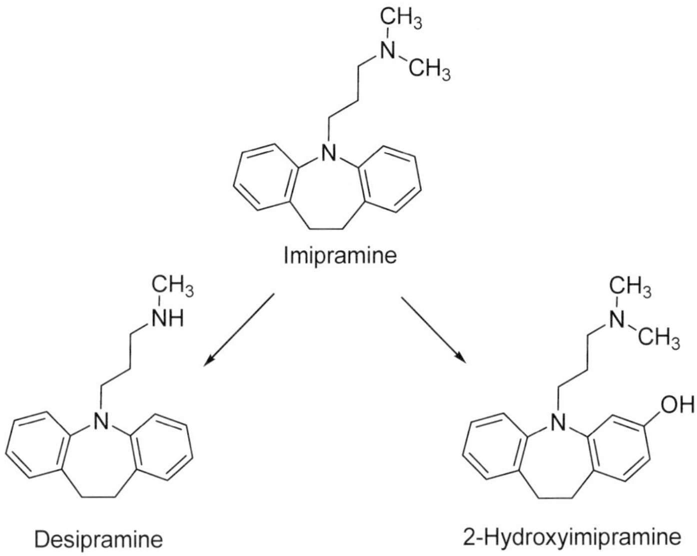

Scheme 3: Metabolism of imipramine

The *N*-oxide and 10-hydroxyimipramine (Coutts and Su, 1993) have only been detected in traces using *in vitro* methods. The N^+-glucuronide (Qian and Zheng, 2006) of imipramine is also reported.

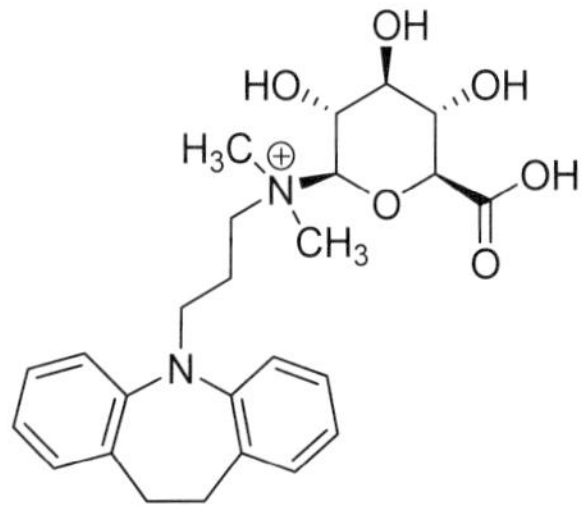

Imipramine-N^+-glucoronide

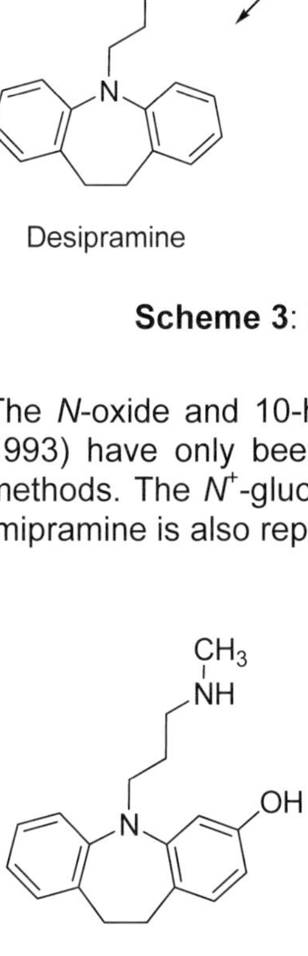

2-Hydroxydesipramine

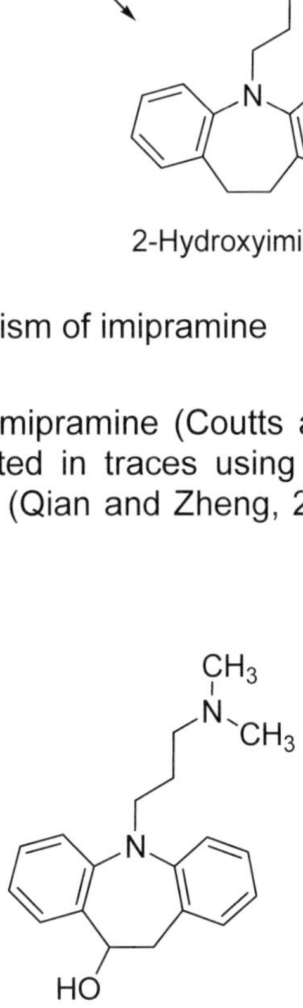

10-Hydroxyimipramine

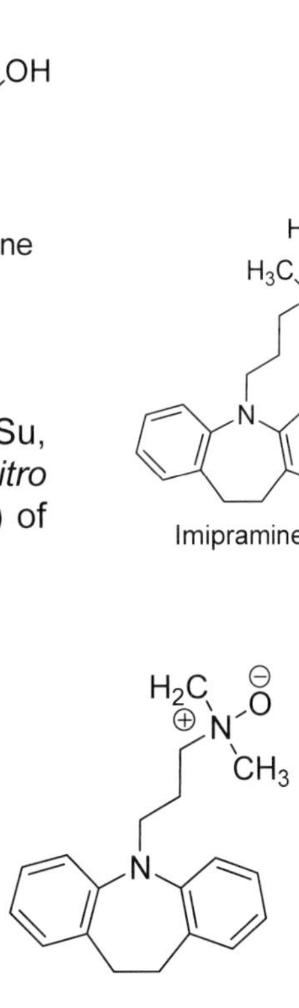

Imipramine-*N*-oxide

Trimipramine

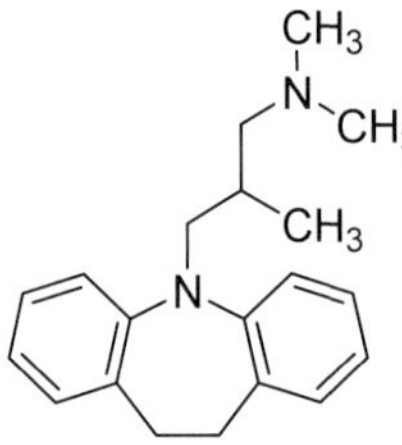

[739-71-9], (±)-5-[3-(Dimethylamino)-2-methylpropyl]-10,11-dihydro-5H-dibenz[b,f]azepine, $C_{20}H_{26}N_2$, M_r 294.43; [521-78-8] (maleate salt), $C_{24}H_{30}N_2O_4$, M_r 410.51

Trimipramine

- Patented in the 1950s (Rhone-Poulenc, 1959)
- Launched in 1961 for the treatment of depression by Rhone Poulenc, now part of Sanofi-Aventis

Synthesis (Rhone-Poulenc, 1959): The synthesis of the racemate was described starting from iminobibenzyl-5-carboxylic acid which is treated with phosgene and with 3-(dimethylamino)-2-methylpropan-1-ol. The carboxylate was heated until the evolution of carbon dioxide ceased to afford trimipramine base.

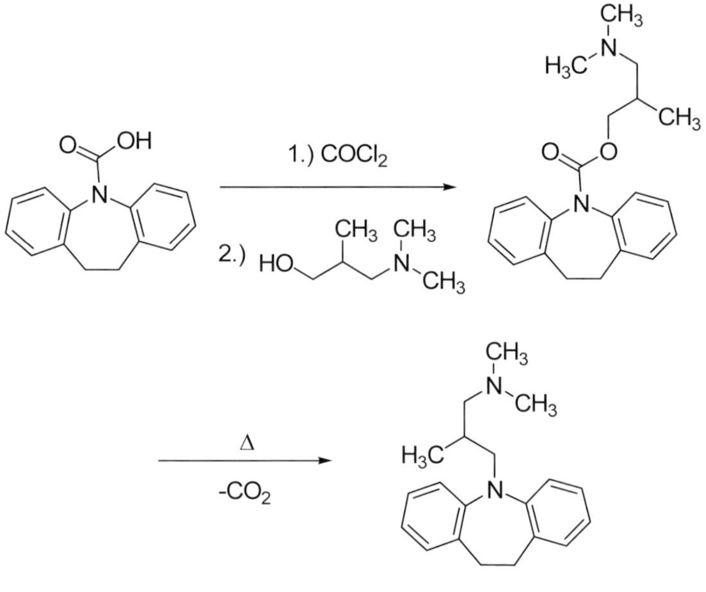

Scheme 4: Synthesis of trimipramine

The enantiomers of trimipramine are obtained first by separation of diastereomeric salts formed by treatment of racemic 5-(3-hydroxy-2-methylpropyl)-iminodibenzyl phthalate with strychnine (Rhone-Poulenc, 1964). Decomposition of the corresponding salt with hydrocloric acid, followed by treatment with *p*-toluenesulfonic acid chloride and finally S_N2 type substitution with dimethylamine in benzene yields the corresponding trimipramine enantiomers.

Clinical Use and Pharmacokinetics: Trimipramine possesses similar clinical actions as the other TCAs, e.g.

like amitriptyline, thus providing efficient treatment for depression. It has – similar to amitriptyline – marked sedative properties. Next to its use in treatment of depression, trimipramine is used – like other TCAs – against pain (Macfarlane et al., 1986) and against peptic ulcer therapy. Trimipramine is readily absorbed after oral administration, peak plasma concentrations are observed already after 2 hours. It shows extensive plasma protein binding and half-lifes of 9–11 hours (Abernethy et al., 1984; Maurer, 1989; Musa, 1989).

Opipramol

- Patented and launched by Geigy (Novartis) in 1961 (Geigy, 1961)

Opipramol

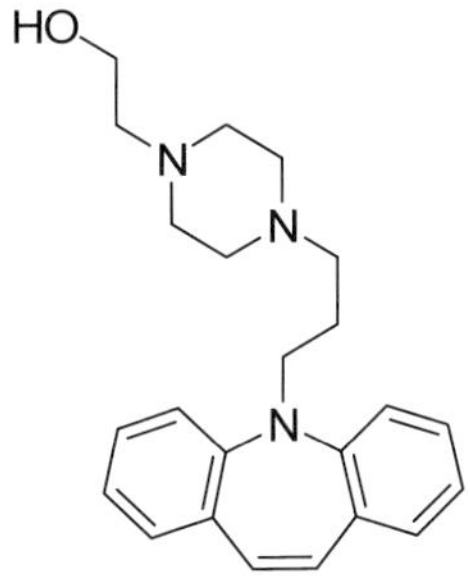

Synthesis (Schindler, 1962): To N-(3-chloropropyl)dibenz[b,f]azepine (from dibenz[b,f]azepine, 1-chloro-3-bromopropane, using sodium amide as base) N-(2-acetoxyethyl)piperazine and sodium iodide are added to afford N-(4-acetoxyethylpiperazinopropyl)dibenz[b,f]azepine which is next saponified to give opipramol.

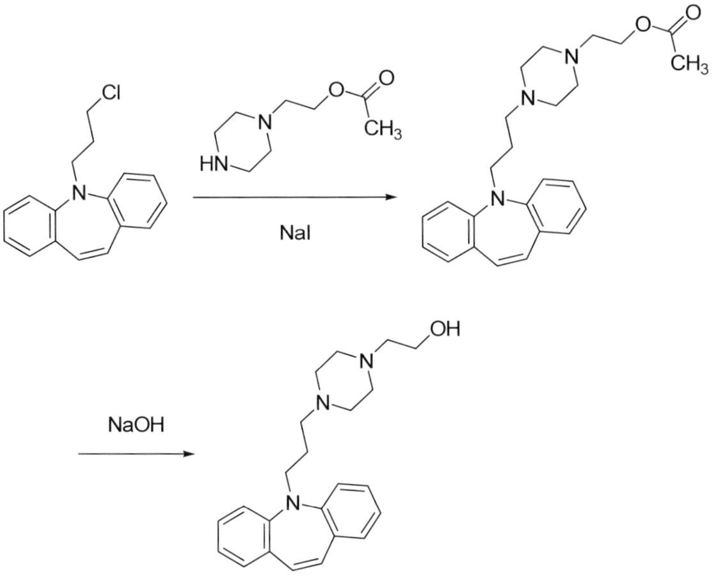

[315-72-0], 4-[3-(5H-Dibenz[b,f]azepin-5-yl)propyl]piperazine-1-ethanol, $C_{23}H_{29}N_3O$, M_r 363.50;
[909-39-7] (dihydrochloride salt), $C_{23}H_{31}Cl_2N_3O$, M_r 436.42

Scheme 5: Synthesis of opipramol

Clinical Use: Opipramol is a classic TCA with a clinical spectrum similar to amitriptyline, providing effective

Butriptyline

CH3
H3C-N
CH3

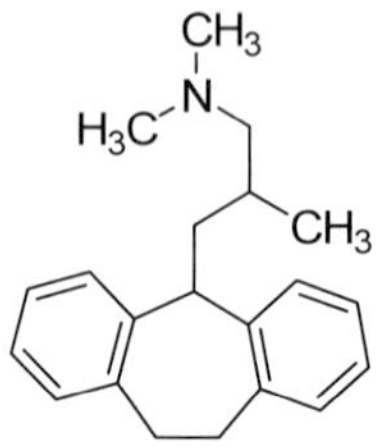

[35941-65-2], (±)-N,N, beta-Trimethyl-10,11-dihydro-5H-dibenzo[a,d]cycloheptene-5-propanamine, $C_{21}H_{27}N$, M_r 293.45;
[5585-73-9] (hydrochloride salt), $C_{21}H_{28}ClN$, M_r 329.91

treatment of depression and having similar, mainly antimuscarinic-like side-effects.

Butriptyline

- Described (Winthrop et al., 1962) and patented in the 1960s (Davis and Winthrop, 1962) and launched in 1974 by Ayerst, McKenna & Harrison, a Canadian company founded in 1925 and acquired in 1943 by American Home Products Corporation (AHP), Wyeth-Ayerst Canada since 1993

Synthesis (Davis, 1972): 2-Phenethylbenzoic acid is converted to the acid chloride, which is treated with hydrogen in the presence of palladium on barium sulfate carrier poisoned with quinoline-S. The aldehyde is reacted with a Grignard reagent and the resulting o-phenetylbenzyl alcohol is finally cyclodehydrated by refluxing with phosphorous pentoxide on celite in toluene to afford the racemic product.

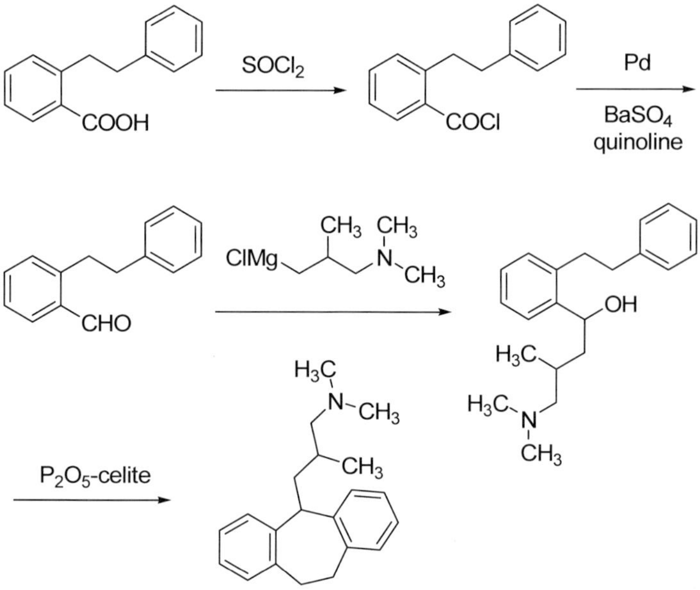

Scheme 6: Synthesis of butriptyline

Clinical Use: Butriptyline possesses – as a member of the TCA family – similar clinical actions and uses as well as similar side-effects as e.g. amitriptyline. However, in depression therapy, it is claimed to have less marked sedative effects than amitriptyline. In addition, drug

Desipramine

- Patented in 1962 by Geigy (Geigy, 1962)
- Launched in 1966 for the treatment of depression by Geigy (today Novartis)
- It is the active *in vivo* metabolite of imipramine

Synthesis: Described by Geigy in the 1960s (Geigy, 1962): 10,11-Dihydro-5*H*-dibenzo[b,f]azepine (iminodibenzyl) is reacted with 1-chloro-3-bromopropane and sodium amide. The reaction of the chloropropyl product with methylamine gives the desired target compound.

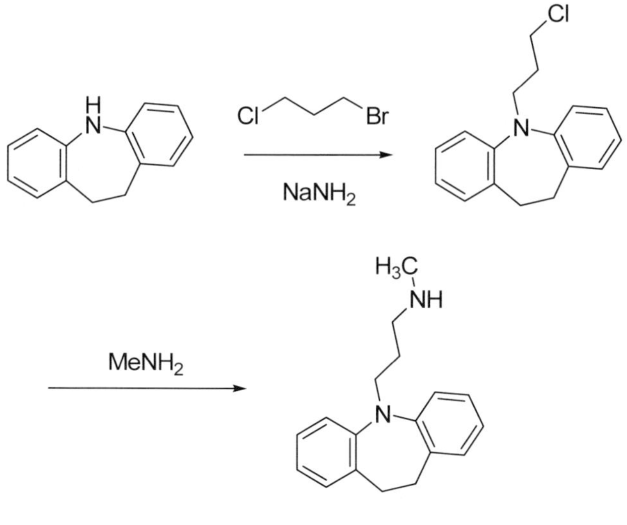

Desipramine

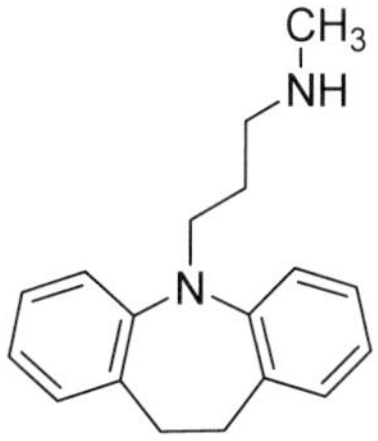

[50-47-5], 5-[3-(Methylamino)propyl]-10,11-dihydro-5*H*-dibenz[b,f]azepine, $C_{18}H_{22}N_2$, M_r 266.38; [58-28-6] (hydrochloride salt), $C_{18}H_{23}ClN_2$, M_r 302.84

Scheme 7: Synthesis of desipramine

Clinical Use, Pharmacokinetics and Metabolism: Desipramine has a similar clinical profile to Amitriptyline and provides as the other TCAs effective treatment for depression. Chemically, desipramine represents the principal active metabolite of imipramine and has, like imipramine, less sedation as compared to amitriptyline. Besides the management of depression, desipramine, like other TCAs, has been used as a treatment for ADHD (Rapport et al., 1993; Pataki et al., 1993; Singer et al., 1995), as an analgesic e.g. in neuropathy (Kishore-Kumar

---

interactions with direct-acting sympathomimetic amines and with adrenergic neurone blocking antihypertensive drugs may occur less likely (Ghose et al., 1977).

et al., 1990; Max et al., 1992; Coquoz et al., 1993; Gordon et al., 1993) or in the management of withdrawal symptoms (Giannini et al., 1986; Weiss, 1988; Gawin et al., 1989; Fischman et al., 1990; Clarck, 1989). In terms of its side-effects, desipramine shows less pronounced antimuscarinic and sedative actions as compared with amitriptyline. The metabolism of desipramine is similar to the one of its parent compound imipramine. A linear relationship between human plasma levels and clinical outcome could be established in two studies (Task Force on the Use of Laboratory Tests in Psychiatry, 1985).

Nortriptyline

Nortriptyline

[72-69-5], N-Methyl-3-(10,11-dihydro-5H-dibenz[a,d]cyclohepten-5-ylidene)-1-propanamine, $C_{19}H_{21}N$, M_r 263.38; [894-71-3] (hydrochloride salt), $C_{19}H_{22}ClN$, M_r 299.84

- Described in patents of different companies published in the 1960s: Hoffmann-La Roche A.G. (Roche) (Hoffmann-La Roche, 1962), Merck & Co. (MSD) (Engelhardt and Christy, 1962), Geigy A.G. (Novartis) (Schindler, 1962), and Eli Lilly & Co. (Peters and Hennion, 1963)

- Launched in 1964 for the treatment of depression by Lilly

Synthesis (Peters and Hennion, 1963): the synthesis of nortriptyline can be performed from amitriptylin by demethylation (Schindler, 1962) or for instance, starting from 5-oxodibenzo[a,d]cyclohepta-1,4-diene, which is treated with the sodium salt of N-methylpropargylamine (from sodium amide/ammonia treatment of 2-chloro-3-methylamino-1-propene). The resulting 5-hydroxy-5-(3-methylamino-1-propynyl)dibenzo[a,d]cyclohepta-1,4-diene is hydrogenated with 5% palladium on aluminum oxide to give 5-hydroxy-5-(3-methylaminopropyl)dibenzo[a,d]cyclo-hepta-1,4-diene, which is dehydrated at 195 C to give N-methyl-3-(10,11-dihydro-5H-dibenz[a,d]cyclohepten-5-ylidene)-1-propanamine (Nortriptyline).

Clinical Use, Pharmacokinetics and Metabolism: Nortriptyline is a typical TCA with similar clinical use and side-effects as e.g. amitriptyline. Being the principal active metabolite of amitriptyline, it provides – like many other TCAs – effective treatment of depression and enuresis. Besides these indications, nortriptyline has been used for skin disorders and against tinnitus (Sullivan et al., 1993). In terms of side-effects, nortiptyline is less sedating than amitriptyline. Nortriptyline has a longer plasma half-life than amitriptyline and is subject to first-pass metabolism to

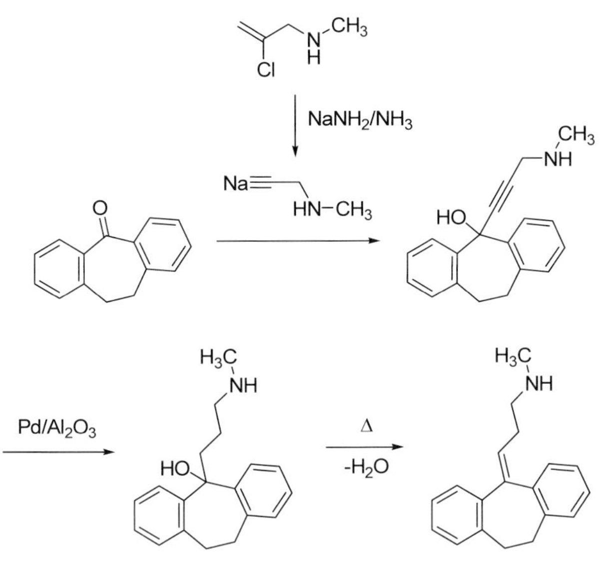

Scheme 8: Synthesis of nortriptyline

10-hydroxynortriptyline, the principal active metabolite of nortriptyline (Park and Kitteringham, 1987; Nordin et al., 1985; Nordin and Bertilsson, 1995; Jerling and Alván, 1994; Bertilsson et al., 1982).

Dosulepin

- Described in the 1960s by the Czech company Sdruzeni Podniku pro Zdravotnickon Vyrobu; SPOFA (Spofa, 1962)

Synthesis (Spofa, 1962): *S*-Benzylthiosalicylic acid is treated with polyphosphoric acid and the resulting cyclic ketone is reacted with 3-dimethylaminopropyl magnesium chloride to afford an alcohol which is dehydrated with sulfuric acid.

Clinical Use, Pharmacokinetics and Metabolism: Dosulepin (also referred to as dothiepin) is a member of the TCA family and has similar clinical uses as amitriptyline, thus providing effective treatment of depression (Goldstein and Claghorn, 1980; Lancaster and Gonzales, 1989; Donovan et al., 1991) and also against pain (Feinmann et al., 1984; Caruso et al., 1987) and tinnitus. Like amitriptyline it has sedative properties, however its anti-muscarinic side-effects are less pronounced. Dosulepin is readily absorbed from the GI tract and

Dosulepin

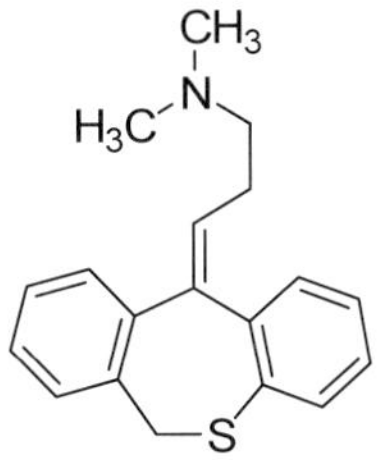

[113-53-1], *N,N*-Dimethyl-3-dibenzo[*b,e*]thiepin-11(6*H*)-ylidene-1-propanamine, $C_{19}H_{21}NS$, M_r 295.44; [897-15-4] (hydrochloride salt), $C_{19}H_{22}ClNS$, M_r 331.90

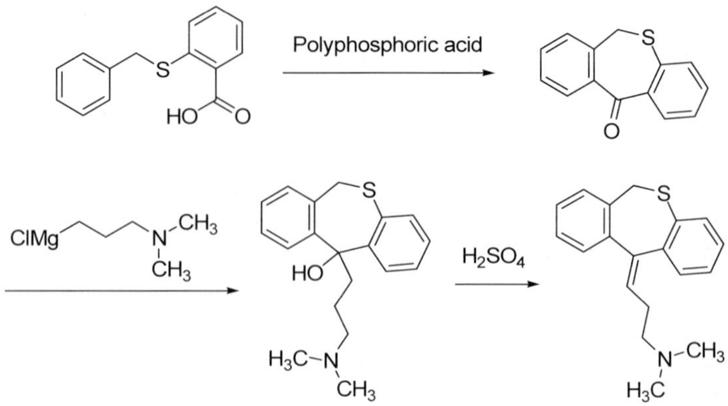

Scheme 9: Synthesis of dosulepin

extensively demethylated while undergoing a first-pass effect. Metabolic pathways include next to hydroxylation and N-oxidation steps, S-oxidation. Elimination half-lifes vary from 14 to 24 hours interindividually (Maguire et al., 1982; Yu et al., 1986; Ilett et al., 1993).

Clomipramine

- Patented by Geigy AG (Novartis) in 1963 (Schindler and Dietrich, 1963)
- Launched in 1967 for the treatment of depression, anxiety and obsessive-compulsive disorder by Novartis

Synthesis (Schindler and Dietrich, 1963): 3-Chloroiminodibenzyl is treated with sodium amide and 3-dimethylaminopropyl chloride in toluene to afford clomipramine.

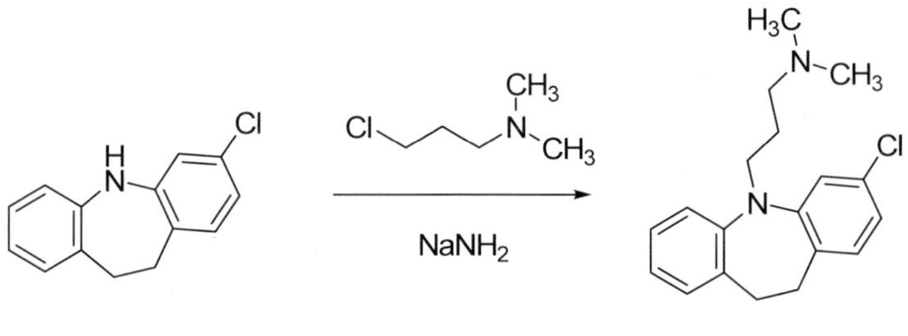

Clomipramine

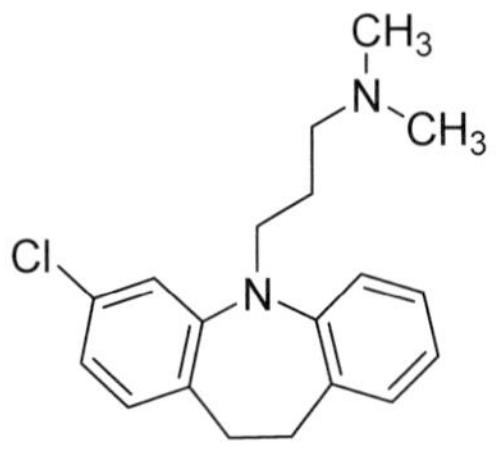

[303-49-1], 3-Chloro-5-[3-(dimethylamino)propyl]-10,11-dihydro-5H-dibenz[b,f]azepine, $C_{19}H_{23}ClN_2$, M_r 314.85; [17321-77-6] (hydrochloride salt), $C_{19}H_{24}Cl_2N_2$, M_r 351.31

Scheme 10: Synthesis of clomipramine

Clinical Use, Pharmacokinetics and Metabolism: Clomipramine is member of the class of tricyclic antidepressants and exacts similar clinical and side-effects as Amitriptyline. It is especially used for the effective treatment of depression if sedative effects are required as well. Next to the typical TCA pharmacological effects, clomipramine possesses a serotonin reuptake inhibitory component leading to its use in the treatment of obsessive–compulsive disorders (Insel et al., 1983; Marks et al., 1988; Jenike et al., 1989; Flament et al., 1985; Leonhard et al., 1988; Mc Tavish and Benfield, 1990; Kelly and Myers, 1990). Like other TCAs, it may be used in the treatment of anxiety-related disorders such as panic disorders (Mc Tavish and Benfield, 1990) and phobias (Yanchick et al., 1994; Swedo et al., 1989, 1993).

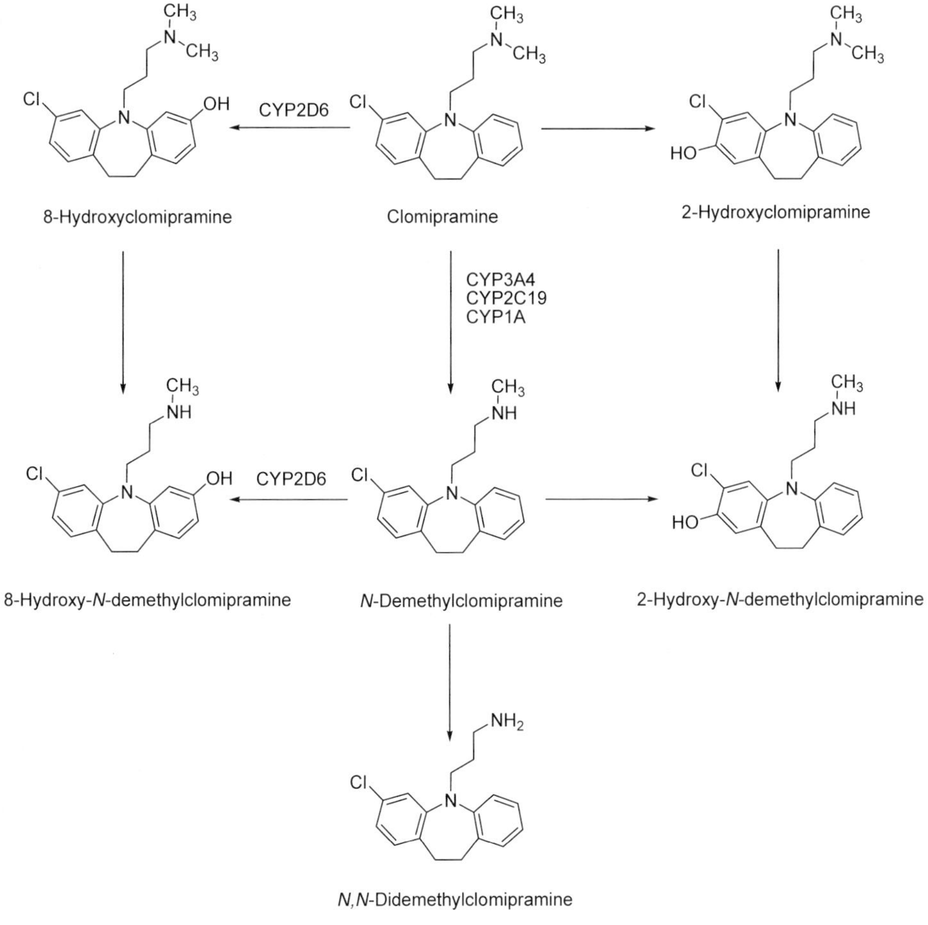

Scheme 11: Metabolism of clomipramine

Clomipramine is most widely used for primary treatment of catalepsy and sleep paralysis associated with narcolepsy. Clinical antidepressant effects with clomipramine usually are displayed some weeks after the start of treatment, although a faster achievement of the antidepressant effect has been reported (Pollock et al., 1985). Adverse effects and drug interactions are similar to those reported for other TCA members, e.g. for amitriptyline.

The major metabolic pathways of clomipramine are N-demethylation and hydroxylation. Orally administered clomipramine undergoes an important first-pass metabolism to yield the pharmacologically active N-demethylclomipramine. Both clomipramine and N-demethylclomipramine are further hydroxylated to their 8-hydroxy derivatives, namely 8-hydroxyclomipramine and 8-hydroxy-N-demethylclomipramine, which have been detected in human plasma. In human liver and yeast microsomes, more metabolites are identified, as e.g. 2-hydroxyclomipramine, 2-hydroxy-N-demethylclomipramine and didemethylclomipramine. N-demethylation of clomipramine was catalyzed by CYP3A4, CYP2C9 and CYP1A2, while the 8-hydroxylation was mediated by CYP2D6. In rat serum or plasma, hydroxylated metabolites were not measured, although they were produced *in vitro* using rat liver microsomes (Nielsen et al., 1996; Weigmann et al., 2000).

Protriptyline

- Patented in 1966 by Merck & Co. (Tishler et al., 1966)
- Launched in 1967 for the treatment of depression by Merck & Co.

Synthesis (Tishler et al., 1966): N-Methyl-5H-dibenzo[a,d]cycloheptene-5-propanamine is prepared by refluxing the ethylcarbamate derivative obtained from reaction of potassium amide and ethyl 3-chloropropyl(methyl)carbamate with 5H-dibenzo[a,d]cycloheptatriene in potassium hydroxide.

Clinical Use, Pharmacokinetics and Metabolism: As a member of the TCA family, protriptyline displays a similar clinical spectrum as e.g. amitriptyline, providing effective treatment for depression. It has considerable less sedative properties than other TCAs and stimulant effects, especially useful in the treatment of apathic and withdrawn

Protriptyline

[438-60-8], N-Methyl-5H-dibenz[a,d]cycloheptene-5-propanamine, $C_{19}H_{21}N$, M_r 263.38;
[1225-55-4] (hydrochloride salt), $C_{19}H_{22}ClN$, M_r 299.84

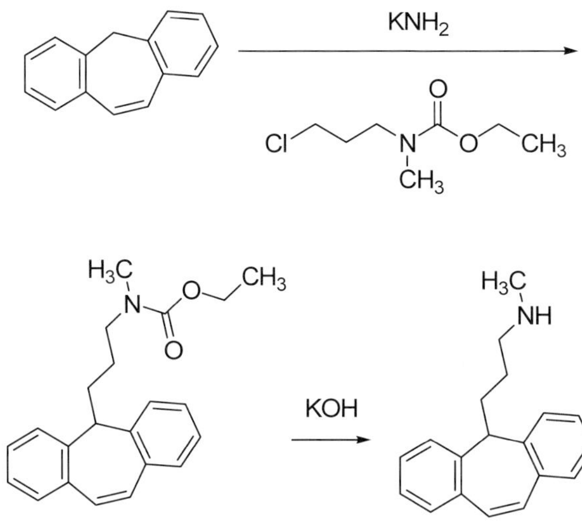

Scheme 12: Synthesis of protriptyline

patients, but also leading to side-effects like insomnia, anxiety or agitation as well as tachycardia and hypotension. Besides depression, protriptyline (as well as clomipramine and imipramine) is used in the management of narcoleptic syndrome (Schmidt et al., 1977). Protriptyline is slow, but completely absorbed in the GI tract, peak plasma concentrations being displayed some hours after ingestion. Metabolic pathways include hydroxylation and *N*-oxidation, followed by conjugate addition. Protriptyline shows a high level of plasma and tissue protein binding and has extremely long elimination half-lifes of between 55 and 198 (!) hours.

Doxepin

- Patented by Pfizer in 1967 (Tretter, 1967)
- Histamine H_1 and H_2 receptor antagonist launched in 1969 for the treatment of depression by Pfizer
- In 1998, Bioglan launched doxepin for the treatment of eczema-associated pruritus

Synthesis (Nakanishi, 1972): Doxepine was prepared from 6,11-dihydrodibenz[*b,e*]oxepin-11-one by treatment with a mixture of Me$_2$N(CH$_2$)$_3$P(Ph$_3$Br) and butyl lithium.

Doxepin

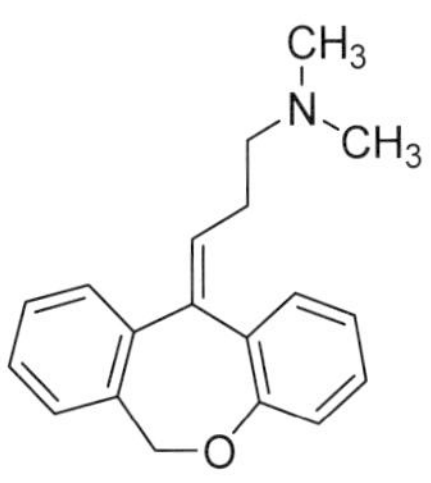

[*1668-19-5*], *N,N*-Dimethyl-3-dibenz[*b,e*]oxepin-11(6*H*)-ylidene-1-propanamine, C$_{19}$H$_{21}$NO, M_r 279.38; [*1229-29-4*] (hydrochloride salt), C$_{19}$H$_{22}$ClNO, M_r 315.84

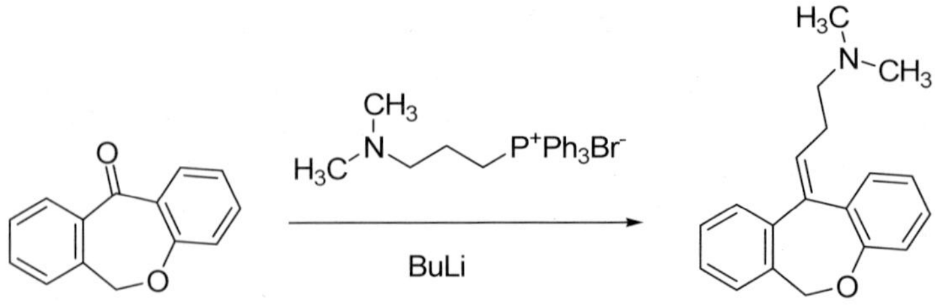

Scheme 13: Synthesis of doxepin

The synthesis of the starting material 6,11-dihydrodibenz[b,e]oxepin-11-one is described in a Pfizer patent (Bloom and Tretter, 1969): starting from ethyl o-bromomethylbenzoate which – after reaction with phenol in the presence of sodium hydroxide – gives 2-phenoxymethylbenzoic acid. Cyclization with trifluoroacetic anhydride yields the expected ketone.

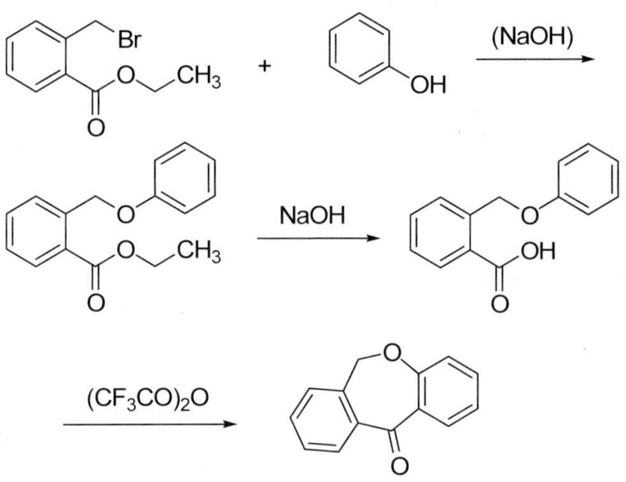

Scheme 14: Synthesis of 6,11-dihydrodibenz[b,e]oxepin-11-one

Clinical Use, Pharmacokinetics and Metabolism: With regard to its clinical use, doxepine has a similar spectrum as amitriptyline and other TCAs, thus representing an efficient treatment for depression. It has marked sedation and other typical TCA side-effects, even after topical application. Doxepine is readily absorbed from the GI tract and is extensivly demethylated while undergoing first-pass effect. Further metabolic steps comprise, hydroxylation and *N*-oxidation, followed by phase II metabolism (conjugate addition). Due to extense distribution throughout the body and plasma and tissue protein binding, long elimination half-lifes are observed (8–24 hours). As doxepine possesses – besides the typical TCA receptor binding profile – pronounced binding to histaminic H_1 and H_2 receptors, it is used in topical form for the treatment of severe pruritus and other forms of skin diseases.

Maprotiline

- Patented by Ciba (Novartis) in the 1960s (Wilhelm et al., 1969) and launched in 1975 for the treatment of depression by Novartis

Synthesis (Wilhelm et al., 1969): 2-(9-Anthryl)propionic acid is reduced with lithium aluminum hydride to yield 9-(3-hydroxypropyl)anthracene, which is refluxed with thionyl chloride and then reacted in an autoclave with methylamine. 9-(3-Methylaminopropyl)anthracene is heated in toluene with ethylene in an autoclave at 50 atm./150°C to afford maprotiline.

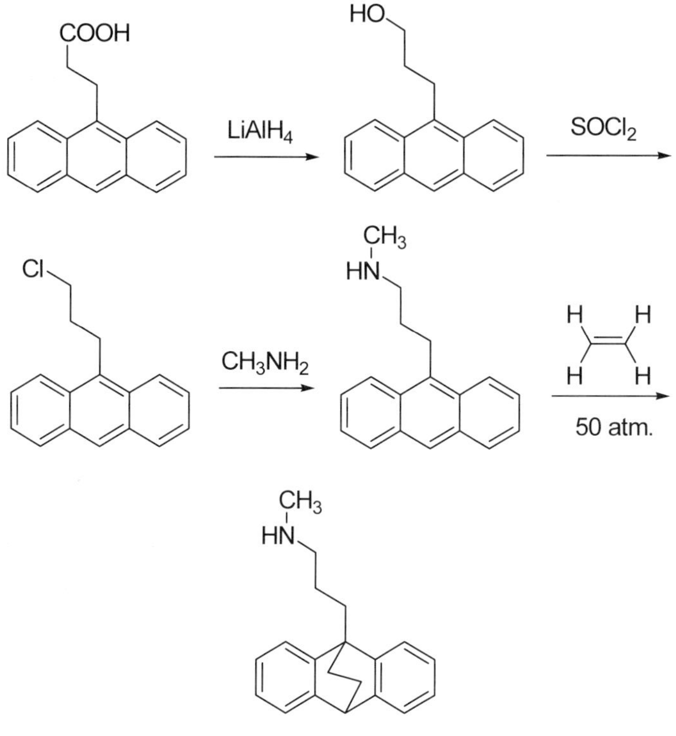

Maprotiline

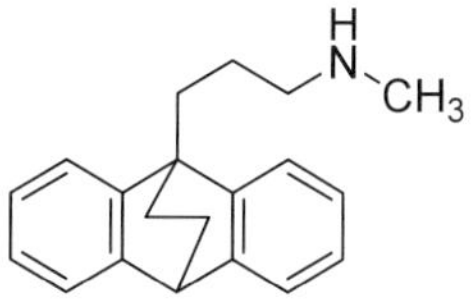

[10262-69-8], *N*-Methyl-9,10-ethanoanthracene-9(10*H*)-propylamine, $C_{20}H_{23}N$, M_r 277.40; [10347-81-6] (hydrochloride salt), $C_{20}H_{24}ClN$, M_r 313.86

Scheme 15: Synthesis of maprotiline

Clinical Use, Pharmacokinetics and Metabolism: Structurally, maprotiline is a tetracyclic antidepressant, however, clinically, it behaves very similar to most of the TCAs, e.g. amitriptyline, providing effective treatment for depression. Antimuscarinic side-effects occur less frequently as compared to amitriptyline, except for skin rashes (Oakley and Hodge, 1985) and seizures (Committee on Safety of Medicines, 1985; Jabbari et al., 1985) so that maprotiline should not be given to patients with a history of seizure disorders, as e.g. epilepsy. After complete but slow absorption from the GI tract, maprotiline undergoes extensive demethylation while the hepatic first-pass effect, resulting in its main active metabolite. Aromatic and aliphatic hydroxylation, followed by phase II conjugate addition, are further metabolic steps. High plasma protein binding and extremely long elimination half-lifes (50–90 hours!) characterize the pharmacokinetics of maprotiline and its metabolites (Maguire et al., 1980; Alkalay et al., 1980; Firkusny and Gleiter, 1994).

Amoxapine

Amoxapine

[14028-44-5], 2-Chloro-11-(1-piperainyl)dibenzo[b,f][1,4]oxazepine, $C_{17}H_{16}ClN_3O$, M_r 313.78

- Tetracyclic antidepressant patented in 1969 (Howell et al., 1969) by American Cyanamid Co., acquired by Wyeth in 1995

- Launched in 1980 by American Cyanamid

Synthesis (Castaner and Playle, 1976): Reaction of O-(p-chlorophenoxy)aniline with ethyl chlorocarbonate gives ethyl O-(p-chlorophenoxy)phenylcarbanilate, which is next condensed with N-carbetoxypiperazine by means of sodium methoxide to afford ethyl 4-[(O-(p-chlorophenoxy)phenyl)carbamoyl]-1-piperazine carboxylate, which is decarboxylated and cyclized with phosphorous pentoxide in refluxing phosphorus (III) oxychloride to afford amoxapine.

Clinical Use, Pharmacokinetics and Metabolism: Pharmacologically, amoxapine belongs to the class of TCA (although chemically tetracyclic). Its clinical actions and uses are similar to those of amitriptylin, thus providing effective treatment for depression. Chemically, amoxapine is the N-desmethyl derivative of the antipsychotic loxapine and has been discussed controversially to have a faster onset of action in depression therapy (claimed within 4–7 days) than amoxapine (Jue et al., 1982). Side-effects comprise the same ones as described for amitriptyline,

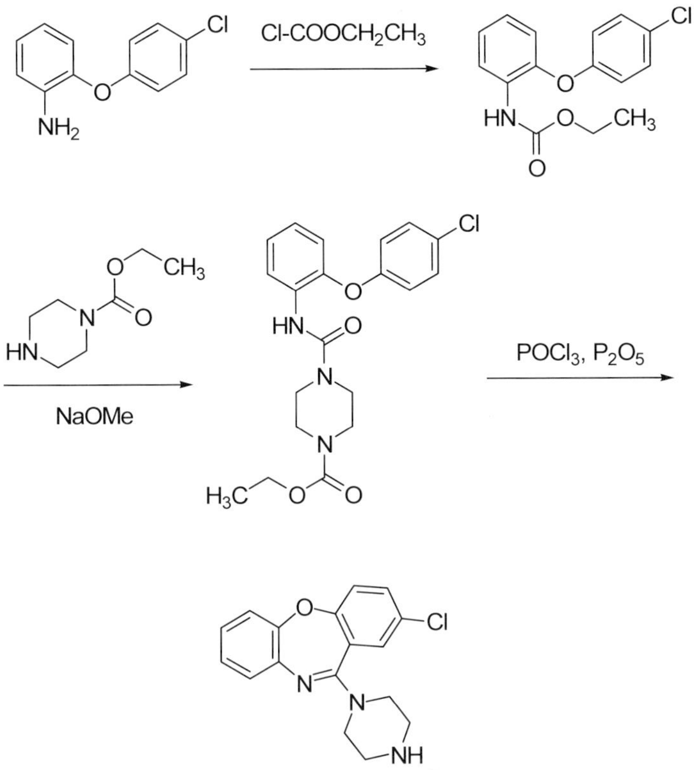

Scheme 16: Synthesis of amoxapine

and, due to the structural similarity with loxapine, antidopaminergic side-effects like tardive dyskinesias (Tao et al., 1985), neuroleptic malignant syndrome (Devarajan, 1989), chorea (Patterson, 1983) and oculogyric crisis (Hunt-Fugate et al., 1984). In terms of antimuscarinic side-effects, it has been claimed that amoxapine possesses significantly less of those side-effects (e.g., constipation, blurred vision or dry mouth) than amitriptyline, given the lower *in vitro* dopamine sites binding affinity of amoxapine as compared to amitriptyline (Bourne et al., 1993). Pharmakokinetics include ready absorption in the GI tract, and – like loxapine – hydroxylation and conjugation as principal metabolic pathways. Amoxapine and its principal metabolites show extensive plasma protein binding and have long half-lifes (between 6 and 30 hours). The metabolites also show pharmacological activity.

Lofepramine

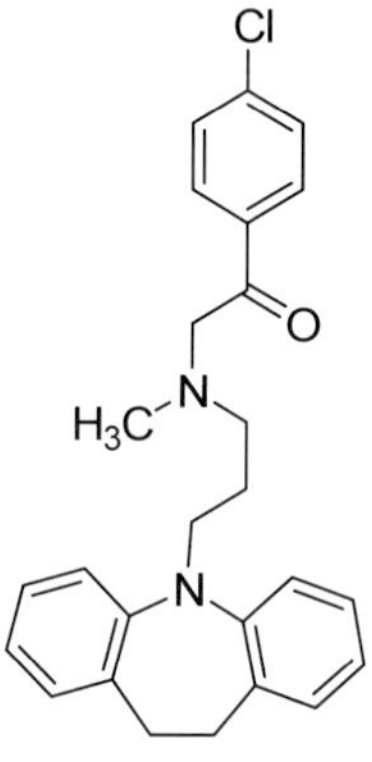

[23047-25-8], 1-(4-Chlorophenyl)-2-[3-(10,11-dihydro-5H-dibenz[b,f]azepin-5-yl)propylmethylamino]ethanone, $C_{26}H_{27}ClN_2O$, M_r 418.96;
[26786-32-3] (hydrochloride salt), $C_{26}H_{28}Cl_2N_2O$, M_r 455.42

Amineptine

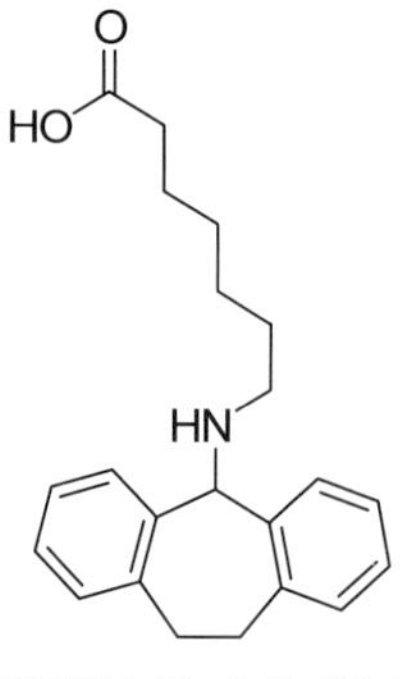

[57574-09-1], 7-(10,11-Dihydro-5H-dibenzo[a,d]cyclohepten-5-ylamino)heptanoic acid, $C_{22}H_{27}NO_2$, M_r 337.45

Lofepramine

- Tricyclic antidepressant developed by Leo (Sweden) (Eriksoo et al., 1972) and launched in 1980 by Merck

Synthesis (Castaner and Arrigoni-Martelli, 1976): from desipramine (N-methyl-3-(10,11-dihydro-5H-dibenz[b,f]azepin-5-yl)propylamine) and 4-chlorophenacyl bromide by reaction with sodium hydrogencarbonate.

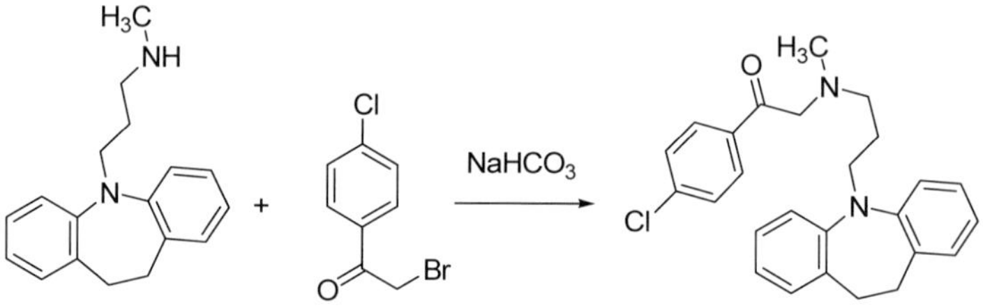

Scheme 17: Synthesis of lofepramine

Clinical Use, Pharmacokinetics and Metabolism: Lofepramine is a TCA similar to amitriptyline in terms of its clinical use and typical side-effects. One of the metabolites of lofepramine is desipramine. Like imipramine and desipramine, it has less sedative properties than amitriptyline (Lancaster and Gonzales, 1989). Lofepramine has – in comparison to amitriptyline – less pronounced antimuscarinic side-effects and less cardiotoxic effects. However, it should not be used in patients with renal impairment. In comparison with other TCAs, lofepramine is considered as especially safe in repect of overdosage or attempted suicide, leading to moderate cardiotoxicity, CNS toxicity and CNS depression, only (Heath, 1984; Dormann, 1985; Crome and Ali, 1986). Lofepramine is readily absorbed from the GI tract and extensively demethylated while undergoing first-pass effect. Further metabolic pathways include hydroxylation, N-oxidation and conjugate addition steps. Lofepramine shows high plasma protein binding.

Amineptine

- Patented in the 1970s (Malen et al., 1973) by Science Union et Cie., Societe Francaise de Recherche Medicale (in partnership with Laboratories Servier) and launched in 1983 by Servier and Pfizer

Synthesis (Malen et al., 1973): Reaction of 5-chloro-10,11-dihydro-5H-dibenzo[a,d]-cycloheptene with ethyl 7-aminoheptanoate in nitrometane gives the corresponding ethyl 7-(dibenzo[a,d]cycloheptadiene-5-yl)aminoheptanoate which is treated with hydrochloric acid to afford amineptine.

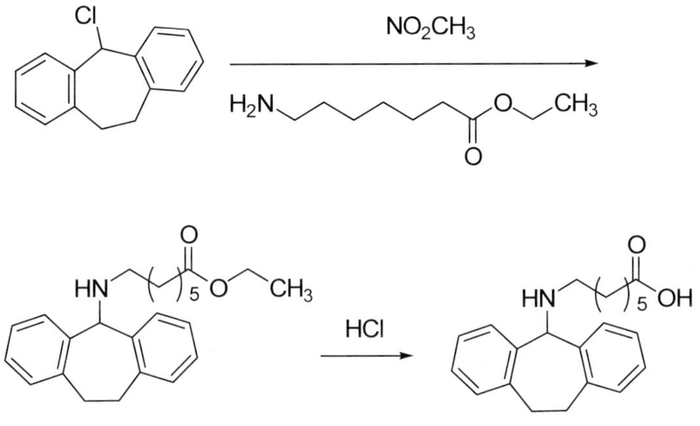

Scheme 18: Synthesis of amineptine

Clinical Use and Metabolism: Amineptine is a classic TCA with clinical antidepressant properties similar to its prototype amitriptyline. Hepatic adverse events seem to be more common with amineptine than with other TCAs. Like with other TCAs, abuse and prolonged withdrawal has been reported. Amineptine metabolism leads to a lactam form, which was correlated in some cases with very severe acne-type lesions after administration of high doses of amineptine (Vexiau et al., 1988).

1.5.3 Serotonergic agents

Given the pronounced side-effects of the tricyclic antidepressants (TCAs), substantial research efforts were started in the early 1970s in order to design more selective agents focussing on influencing brain 5-HT levels. A low level of utilization of serotonin is currently seen as one among several neurochemical symptoms of depression.

1.5.3.1 Selective serotonin reuptake inhibitors

Selective serotonin reuptake inhibitors like e.g. fluoxetine and sertraline represent the most successful class of currently used antidepressants, evolving their effects at the serotonin transporter, increasing extracellular 5-HT levels by inhibiting its reuptake into the presynaptic cell. They have no or only weak effects on other monoamine transporters, thus having little direct influence on the level of other neurotransmitters. SSRIs are considered to be considerably safer than TCAs, since the toxic dose is much higher and they possess a more favorable side efect profile as compared with the TCAs. However, like with the TCAs, usually it can take some weeks from the start of the therapy until a clinically relevant effect can be observed (induction period). Structurally, they all are derived from phenoxyphenylpropylamines (PPPAs), by employing Medicinal Chemistry guided methods like scaffold hopping. Given the wide use of especially fluoxetine (Prozac©), also in children, the term "Prozac© nation" was formed in connection with a controverse public discussion on the risk benefit ratio of the SSRIs. In the following, the SSRIs on the market are presented summarizing chemical properties, synthesis, important patents, clinical use as well as pharmacokinetics and metabolism.

Fluoxetine

Fluoxetine

[54910-89-3], 3-(p-trifluoromethylphenoxy)-N-methyl-3-phenylpropylamine, $C_{17}H_{18}F_3NO$, M_r 309.33; [59333-67-4] (hydrochloride salt), $C_{17}H_{19}ClF_3NO$, M_r 345.79

- In 1974 first patent filed by Eli Lilly (Molloy and Schmiegel, 1982)

- First publication appeared in 1974 describing the discovery of Fluoxetine as a selective serotonin-reuptake inhibitor (Wong et al., 1974)

- Approval for the treatment of depression by the United States FDA on 1987

- Approval in the United States for the treatment of obsessive-compulsive disorder (OCD) on 1994

- Approval for the treatment of bulimia nervosa in the United States in 1996

- Market exclusivity expired in the United States in 2001 opening the generic prescription market

Synthesis (Castaner and Paton, 1977): Reduction of β-dimethylaminopropiophenone with diborane gives *N,N*-dimethyl-3-phenyl-3-hydroxypropylamine, which is converted into *N,N*-dimethyl-3-phenyl-3-chloropropylamine with refluxing thionyl chloride. Condensation with 4-trifluoromethylphenol, treatment with BrCN in benzene followed by hydrolysis with KOH at 130°C affords fluoxetine.

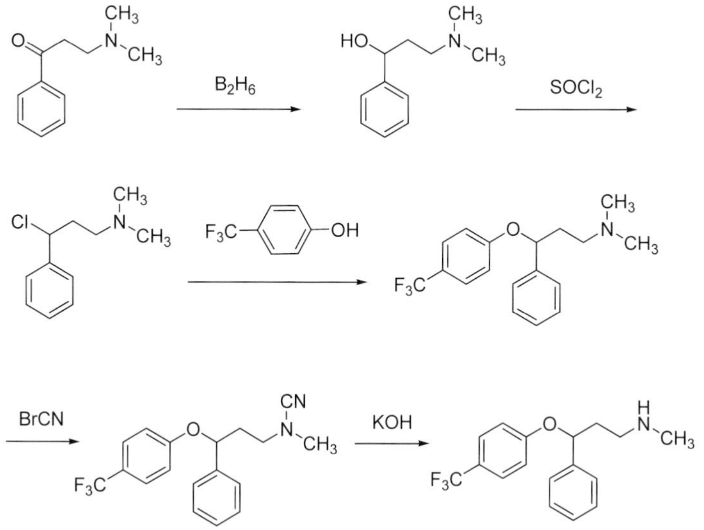

Scheme 19: Synthesis of fluoxetine

The enantiomers of fluoxetine and its main metabolite norfluoxetine possess different clinical efficacies. (S)-Fluoxetine was first synthesized starting from (S)-(−)-3-chloro-1-phenylpropanol (Robertson et al., 1988). At the same time, a versatile enantioselective synthesis was developed by Sharpless (Gao and Sharpless, 1988).

In 2000, Sepracor Inc. filed a patent on a synthetic process starting from benzoylacetonitrile (Hilborn and Jurgens, 2000; Hilborn et al., 2001), leading to the individual enantiomers of fluoxetine.

Since then, a range of other alternative enantioselective processes has been described in the literature, employing asymmetric synthesis (Corey et al., 1989; Panunzio et al., 2004; Wang et al., 2005), chemoenzymatic (Kumar et al., 1991; Bracher and Litz, 1996; Kamal et al., 2002) or microbiological (Chenevert et al., 1992) methods.

Based on the structure of fluoxetine, by application of "scaffold-hopping", medicinal chemistry-driven approaches to new and potent SSRIs were conducted and the resulting candidates successfully marketed. Scaffold hopping is a well-known Medicinal Chemistry concept using 2D or 3D modeling tools in order to substitute a core scaffold known to produce high-affinity ligands by another

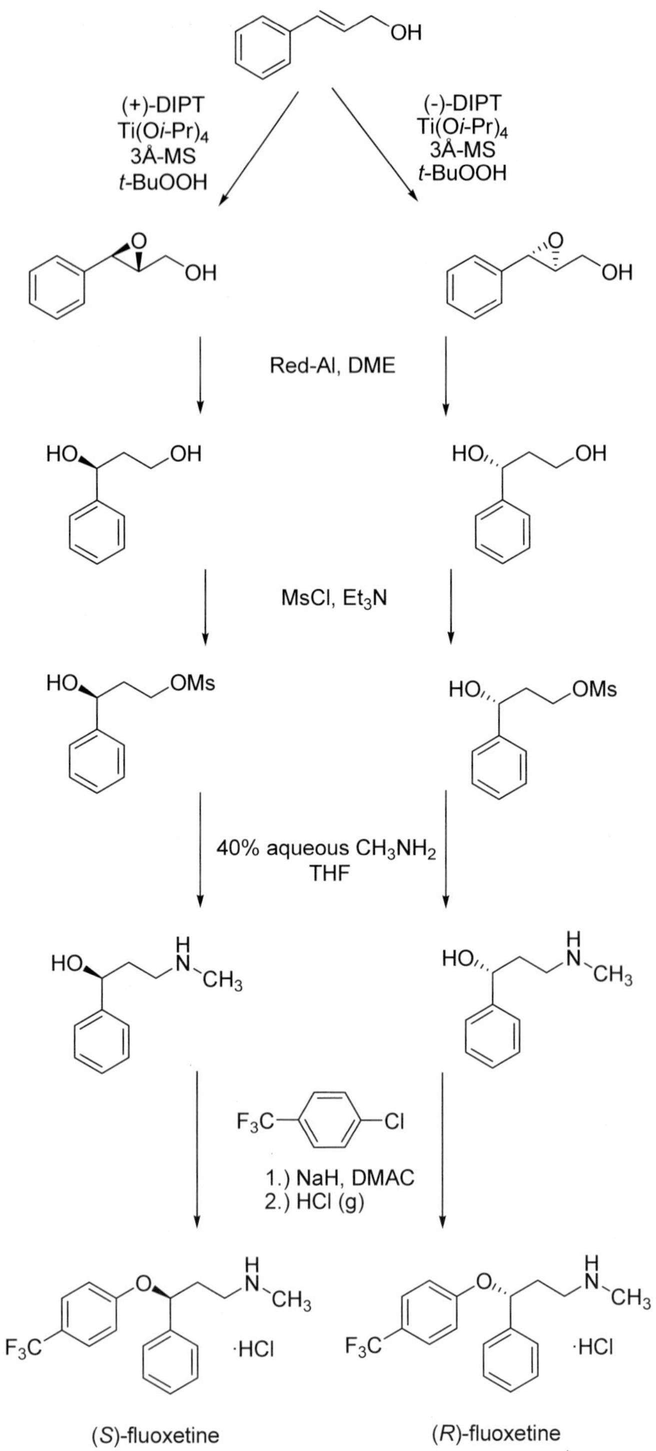

Scheme 20: Sharpless' asymmetric synthesis of enantiomers of fluoxetine

one leading to similar geographical orientation (topographical) of the substituents (being pharmacophore key motifs), thus resulting in unprecedented biomimetic compounds with potentially high binding affinities as well.

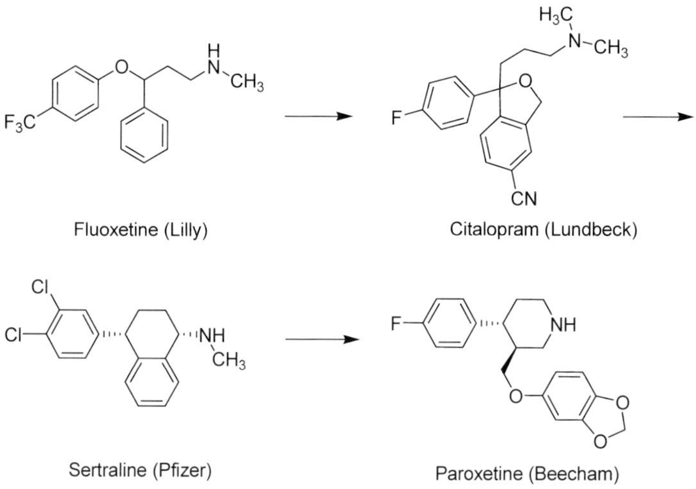

Scheme 21: Scaffold hopping for selective serotonin reuptake inhibitors (Böhm et al., 2004)

These three drugs will be discussed in the following sections.

Clinical Use, Pharmacokinetics and Metabolism: Fluoxetine is the prototype of the class of selective serotonin reuptake inhibitors, effective in the treatment of depression. It has little effect on noradrenaline uptake and causes much less antimuscarinic side-effects and less cardiotoxicity as e.g. TCAs like amitriptyline. Its mode of action in depression is not fully understood. It provides a valuable alternative in depression treatment to the TCAs, even though, as with the TCAs, it may take some weeks until a clinically observable effect is seen (Gram, 1994; Finley, 1994). Besides treatment of depression (Song et al., 1993; Edwards, 1994; Montgomery et al., 1994; Roose et al., 1994; Fava et al., 1994; Fava et al., 1995, Anderson and Tomenson, 1995; Goodnick et al., 1995), fluoxetine is useful for treatment of alcoholism (Gorelick, 1989, Naranjo et al., 1990; Slywka and Hart, 1993), anxiety (Turner et al., 1985; Fontaine and Choinard, 1989; Montgomery et al; 1993; Wood et al., 1993; Pigott et al., 1990; Yanchick et al., 1994; Schneier et al., 1990; van Ameringen et al.,

1993; van der Kolk et al., 1994), disturbed behavior (Coccaro et al., 1990; Cornelius et al., 1990; Richer and Crismon, 1993), eating disorders (Beumont et al., 1993; Fairburn and Peveler; 1990; Goldstein et al., 1995; Goldbloom and Olmsted, 1993; Levine et al., 1989; Bray, 1993; Anonymous, 1994; Fichtner and Braun, 1994; Bross and Hoffer, 1995), hyperactivity (Gammon and Brown, 1993; Bussing and Levin, 1993), hypochondriasis (Fallon et al., 1993), narcoleptic syndrome (Langdon et al., 1986), orthostatic hypotension (Grubb et al., 1994) and pain (Max et al., 1992; Geller, 1989; Cantini et al., 1994; Sosin, 1993; Sasper et al., 1994), as well as in pathological crying or laughing (Seliger et al., 1992; Hanger, 1993), peripheral vascular disease (Bolte and Avery, 1993; Jaffe, 1995) and premenstrual syndrome (Stone et al., 1991; Menkes et al., 1992; Wood et al., 1992; Steiner et al., 1995; Mortola, 1994; Pearlstein and Stone, 1994). Most common side-effects include the GI tract as well as the nervous system. Selective serotonin reuptake inhibitors such as fluoxetine are less sedative than TCAs and have few antimuscarinic or cardiotoxic effects. Other reported side-effects include effects on blood (Alderman et al., 1992), body weight, the endocrine system, the eyes (Ahmad, 1991), hair loss (Ogilvie, 1993) and sexual function (Hollander, 1994; Balon, 1995; Hopkins and Gelenberg, 1995; Hollander, 1995; Norden, 1994), as well as epileptogenic (Weber, 1989; Ware and Stwart, 1989) and extrapyramidal effects (Eisenhauer and Germain, 1993; Lipinski et al., 1989; Rothschild and Locke, 1991). Recently, effects on mental state, especially with regard to suicides, have been discussed controversially. However, metaanalyses have not confirmed an increased risk (Rothschild and Locke, 1991; Fichtner et al., 1991; Beasley et al., 1991; Goldstein et al., 1993; Li Wan Po, 1993; Nakielny, 1994; Healy, 1994; Power and Cowen, 1992; Anonymous, 1992). Fluoxetine should be used with care in patients with impaired hepatic or renal function and it interacts with a range of drugs due to inhibition of their metabolism (Committee on Safety of Medicines, 1989; Brannan et al., 1994; Graber et al., 1994; Kline et al., 1989; Dursun et al., 1993; Neuvonen et al., 1993). Fluoxetine is readily absorbed from the GI tract after oral administration with peak plasma concentrations after 6–8 hours. While undergoing a hepatic first-pass effect, it is extensively metabolized into its desmethyl, main active metabolite norfluoxetine. Another main metabolite is *p*-trifluoromethylphenol, produced by oxidative *O*-dealkylation with high contribution of CYP2C19. In addition to CYP2D6 and 2C9, which produce the greatest amount of fluoxetine N-demethylation, other cytochrome P450 isoforms are involved as well: 3A4, 2C19, 1A2, 2B6 and 3A5. For a detailed review on the pharmacokinetics and metabolism of fluoxetine, also taking into account stereoselective aspects, see Mandrioli et al. (2006) and references cited in this review. Fluoxetine is used as a racemate. However, the fact that the S-enantiomer of norfluoxetine is more active as the R-enantiomer of norfluoxetine may possibly provide some rationale for the clinical development of S-fluoxetine. Fluoxetine is widely distributed throughout the body and has a long elimination half-life of about 2–3 days, the one of its metabolite norfluoxetine even being much higher [7–9 days (!)], having clinical implications, as steady state is reached only after weeks of treatment (Altamura et al., 1994).

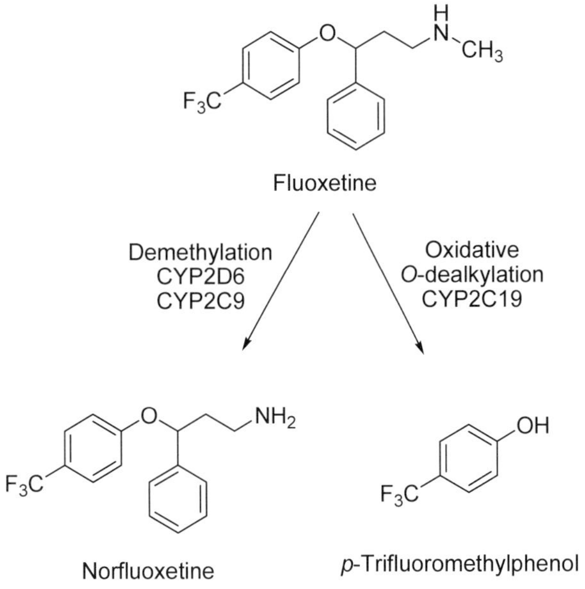

Scheme 22: Main metabolites of fluoxetine

The discovery of fluoxetine has been described in an excellent and highly enjoyable review by Wong et al. (2005): after the initial publication of fluoxetine it took more than 16 years and a lot of scientific as well as regulatory and marketing "challenges" to bring this drug to its approval on 29 December 1987.

Citalopram

Citalopram

[59729-33-8], 1-(3-Dimethylaminopropyl)-1-(4-fluorophenyl)-5-phthalancarbonitrile, $C_{20}H_{21}FN_2O$, M_r 324.39; [59729-32-7] (hydrobromide salt), $C_{20}H_{22}BrFN_2O$, M_r 405.30

- First described (patent application) in 1976 by Kefalas A/S (now Lundbeck A/S; Boegesoe and Toft, 1979) for the treatment of depression
- Launched in 1989 for the treatment of depression by Lundbeck
- Launched in 1995 for the treatment of anxiety by Lundbeck/Recordati
- (S)-(+) Enantiomer (named: escitalopram) is greater than two orders of magnitude more potent than (R)-(−) enantiomer *in vitro*
- Escitalopram has been developed by Lundbeck and Forest and launched in Europe in 2002 (Cipralex®); Forest received an approvable letter from the United States FDA in 2002 for the treatment of major depressive disorder in adults (Lexapro)

Synthesis (Castaner and Roberts, 1979): The first process described by Lundbeck starts with the reaction of 5-bromophthalide and 4-fluorophenylmagnesium bromide to give 4-bromo-4'-fluoro-2-(hydroxymethyl)benzophenone, which is then reduced to afford 4-bromo-4'-fluoro-2-(hydroxymethyl)benzhydrol. Cyclization with phosphoric, toluene sulfonic or sulfuric acid yields 5-bromo-1-(4-fluorophenyl)phtalan, which is then converted into 1-(4-fluorophenyl)-5-phtalancarbonitrile by means of cuprous cyanide. To afford citalopram, a nucleophilic substitution reaction (S_N2 type) using 3-(dimethylamino)propyl chloride in the presence of NaH is performed in the final step.

Since then, new and more efficient methods have been developed by Lundbeck using different starting materials instead of 5-bromophthalide, for instance 1-oxo-1,3-dihydroisobenzofuran-5-carboxylic acid (Dall'asta et al., 2000), 1-oxo-1,3-dihydroisobenzofuran-5-carbonitrile (Rock et al., 2000), or 5-aminophthalide (Petersen et al., 1998). Some further improvements or modifications of these synthetic routes have also been described by Lundbeck (Petersen 2001; Petersen and Dancer, 2001).

Escitalopram was first prepared by isolating the (S)-enantiomer from the racemate citalopram (Boegesoe and Perregaard, 1989). An asymmetric synthesis was published in 2001 for escitalopram oxalate (Sorbera et al., 2001). The 1-oxo-1,3-dihydroisobenzofuran-5-carboxylic acid is converted with 2-amino-2-methyl-1-propanol to the

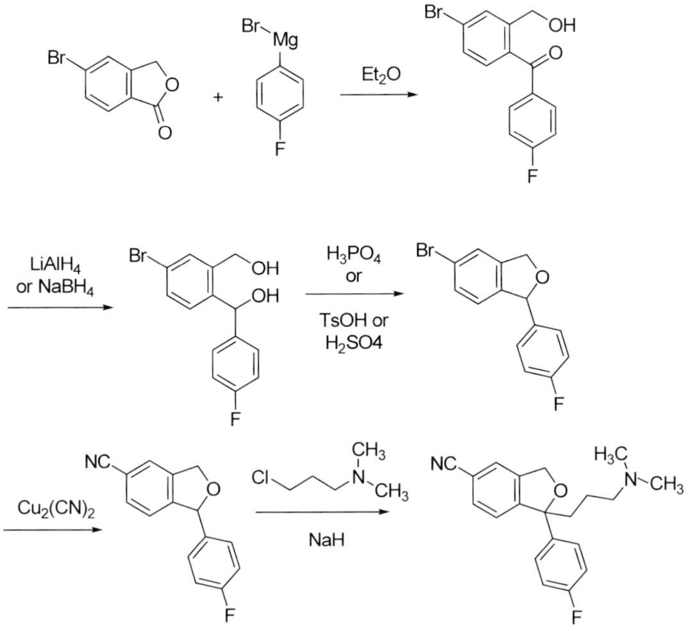

Scheme 23: Synthesis of citalopram

corresponding amide. The amide is cyclized to afford the corresponding oxazoline, which is then treated with 4-fluorophenylmagnesium bromide to give the respective benzophenone. After treatment with 3-(dimethylamino)propylmagnesium chloride to the corresponding carbinol, optical resolution by diastereomeric salt formation is performed with tartaric acid or camphor-10-sulfonic acid (CSA) to give the desired (S)-enantiomer. Final cyclization with methanesulfonylchloride/triethylamine and treatment with phosphorus(III) oxychloride yields escitalopram.

Clinical Use, Pharmacokinetics and Metabolism: Citalopram is a selective serotonin reuptake inhibitor (SSRI) with a clinical profile similar to fluoxetine. It offers, as the other SSRI members an efficient alternative treatment of depression to the TCAs (Milne and Gloa, 1991; Andersen et al., 1994). In addition, it is useful in treating alcoholism (Naranjo et al., 1987; Naranjo et al., 1992), pain (Sindrup et al., 1992) and pathological crying and laughing (Andersen et al., 1993). Citalopram is readily absorbed form the GI tract after ingestion, reaching peak plasma concentrations after 2–4 hours. Its protein binding is low. Metabolism occurs with demethylation, deamination and oxidation steps into inactive metabolites (Milne and

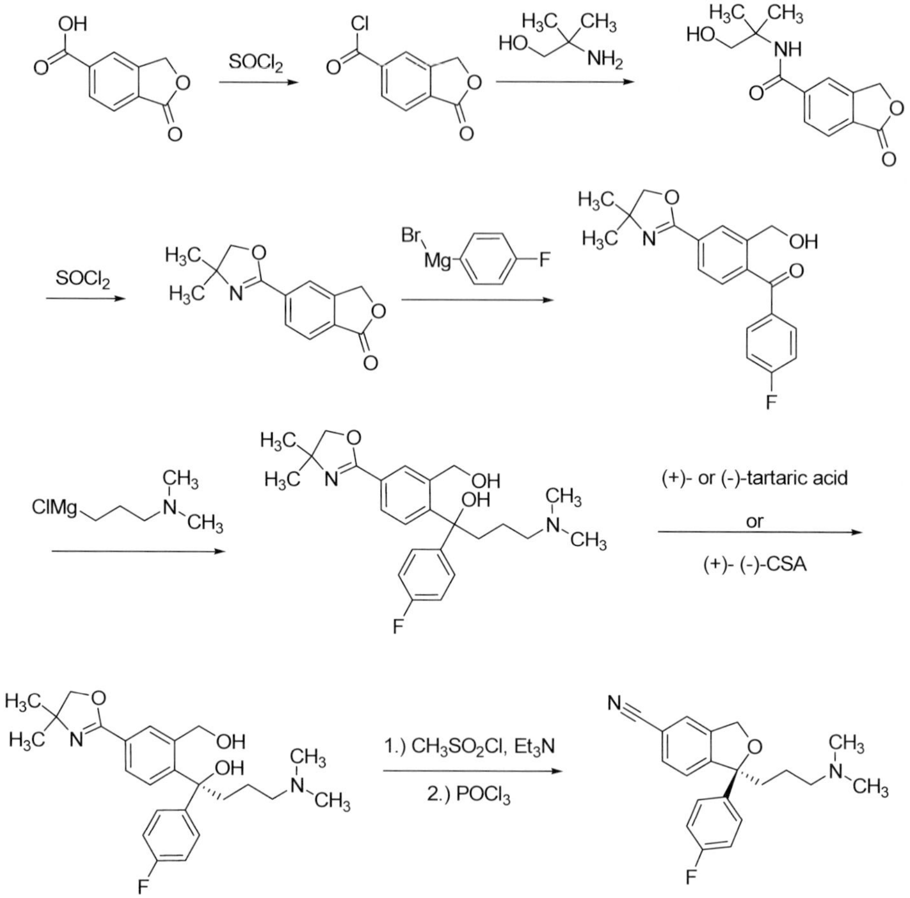

Scheme 24: Synthesis of escitalopram

Goa, 1991). Citalopram is a racemate, the more active S-enantiomer was already launched (as escitralopram) in some countries as a follow-up drug.

Fluvoxamine

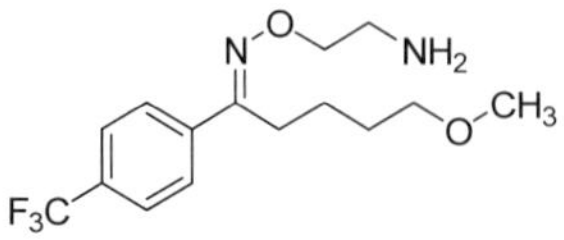

Fluvoxamine

[54739-18-3], 5-Methoxy-1-[4-(trifluoromethyl)phenyl]-1-pentanone (E)-O-(2-aminoethyl)oxime, $C_{15}H_{21}F_3N_2O_2$, M_r 318.33; [61718-82-9] (maleate salt), $C_{19}H_{25}F_3N_2O_6$, M_r 434.41

- First patent filed in 1975 by Solvay Pharmaceuticals (Welle and Claassen, 1978)
- Launched in 1985 for the treatment of depression and in 1986 for obsessive compulsive disorder
- Registered in 2005 for the treatment of social phobia

Synthesis (Castaner and Thorpe, 1978): Fluvoxamine can be prepared in three different ways all starting from p-trifluoromethyl-5-methoxy-valerophenone: (a) by condensation with 2-aminooxyethylamine dihydrochloride, (b) by reaction with hydroxylamine to yield the corresponding oxime, which is then condensed with 2-chloroethylamine, and (c) by reaction of the oxime with lithium and ethylene oxide to afford the O-(2-hydroxyethyl)oxime, which is esterified with mesyl chloride to the O-(2-mesyloxyethyl)oxime. This intermediate is finally treated with ammonia in methanol at 100°C in a pressure vessel to give the target compound.

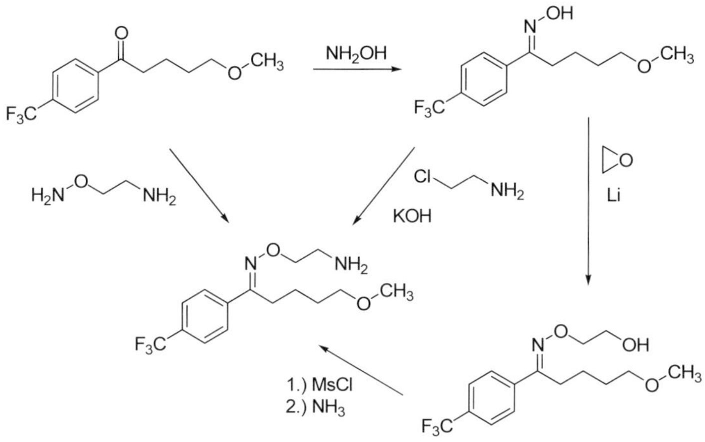

Scheme 25: Synthesis of fluvoxamine

Clinical Use, Pharmacokinetics and Metabolism: Fluvoxamine is a selective serotonin reuptake inhibitor (SSRI) with a clinical profile similar to fluoxetine. It offers, as the other SSRI members an efficient alternative treatment of depression to the TCAs, causing less antimuscarinic side-effects and less cardiotoxicity as e.g. amitriptyline (Grimsley and Jann, 1992; Mendlewicz, 1992; Wilde et al., 1993; Palmer and Benfield, 1994). As with the TCAs, it may take some weeks until a clinical benefit can be seen in depression therapy (Harris et al., 1991; Ottevanger, 1991; Rahman et al., 1991; Bougerol et al., 1992; Franchini et al., 1994). In addition, fluvoxamine is useful in the treatment of anxiety disorders (Jenike et al., 1990; Goodman et al., 1990; Mallya et al., 1992; Black et al., 1993; Hoehn-Saric et al., 1993; van Vliet et al., 1994), headache (Bánk, 1994; Manna et al., 1994), and Wernicke-Korsakoff syndrome (Martin et al., 1989; Anonymous, 1990). Most common side-effects include GI ones like nausea and vomiting and diarrhea as well as

CNS side-effects like drowsiness, insomnia, headache and agitation (Wagner et al., 1994; Committee on Safety of Medicines, 1988; Committee on Safety of Medicines, 1992). Further reported side-effects comprise suicide ideation (Pitchot et al., 1992), effects on the skin (Wolkenstein et al., 1993) and sexual function (Dorevitch and Davis, 1994) as well as epileptogenic effects (Harmant et al., 1990). Fluvoxamine should be avoided in patients with a history of epilepsy; and care should be taken in patients with renal or hepatic impairments. Fluvoxamine is readily absorbed from the GI tract after ingestion and extensively metabolized in the liver. It has about 77% plasma protein binding and elimination half-lifes around 15 hours (de Bree et al., 1983; Overmars, et al., 1983; van Harten et al., 1993; Perucca et al., 1994). Metabolism of fluvoxamine was studied in humans and experimental animals after oral administration. At least sixteen metabolites were identified in urine and feces of the species studied. The main metabolic pathway proceeds via degradation of the aliphatic methoxy group, resulting in fluvoxamine acid (M6) as major metabolite in dogs, rats, hamsters and humans. In mice, both the acid M6 and fluvoxamine alcohol (M9a and its glucuronidated form M5) are important metabolites, the latter being almost absent in either dogs or humans. Other metabolic pathways include acetylation and oxidative removal of the primary amino group of the amino or the removal of the ethanolamino group. Removal of the ethanol group, in combination with oxidation on the methoxy group produces Metabolite-G, Metabolite-C1 and Metabolite-C3 in humans, while the glucuronide of Met-G (M2a) was detected in animals. *N*-acetylated fluvoxamine acid was detected in both animals and humans, while its precursors M8 and M9c were observed in animals. Oxidative elimination at both the amino group and the methoxyl group results in Metabolite-F1, Metabolite-E, M3a, Metabolite-C2 and M2b in humans or animals. A metabolite, derived from the complete cleavage of the side chain followed by glycine conjugation, was reported as M1 in animals. Metabolic clearance studies in healthy volunteers indicated that human CYP2D6 was mainly responsible for the oxidative degradation of the methoxy group, whereas CYP1A2 mainly catalyzed degradations at the amino/ethanolamino group (Hatori et al., 1995; Ruijten et al., 1984; Overmars et al., 1983).

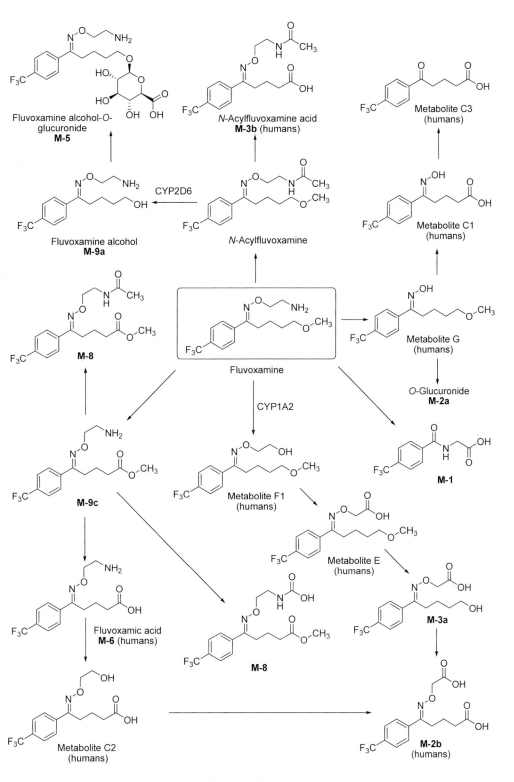

Scheme 26: Metabolism of fluvoxamine

Paroxetine

Paroxetine

[*61869-08-7*], (-)-(3*S*,4*R*)-3-(1,3-Benzodioxol-5-yloxymethyl)-4-(4-fluorophenyl)piperidine, $C_{19}H_{20}FNO_3$, M_r 329.37; [*78246-49-8*] (hydrochloride salt), $C_{19}H_{21}ClFNO_3$, M_r 365.83

- The Danish company Ferrosan A/S discovered paroxetine hydrochloride and its antidepressant qualities in the 1970s (Christensen and Squires, 1973) and licensed its patent rights to Glaxo SmithKline Beecham. First clinical development was performed with the metastable anhydrate polymorph of paroxetine hydrochloride

- In 1985, GSK filed a patent for the use of the more stable hemihydrate polymorphic form in the treatment of depression, thus prolonging the patent protection for paroxetine (Barnes et al., 1987)

- Launched in 1991 for the treatment of depression and obsessive-compulsive disorder

- Launched in 2001 for the treatment of post-traumatic stress and generalized anxiety

- Launched in 2002 for the treatment of major depression and panic disorder

- Launched in 2003 for the treatment social phobia and premenstrual syndrome

Synthesis (Lucas, 2000; Christensen and Squires, 1973; Barnes et al., 1985): From the reaction of 4-(4-fluorophenyl)-1-methyl-1,2,3,6-tetrahydropyridine with formaldehyde and sulfuric acid, racemic 4-(4-fluorophenyl)-3-(hydroxymethyl)-1-methyl-1,2,3,6-tetrahydropyridine is submitted to optical resolution (diastereomeric salt formation) with (−)-dibenzoyltartaric acid, yielding the 3-(*S*)-enantiomer. Hydrogenation over Pd/C affords *trans*-(3*S*,4*R*)-4-(4-fluorophenyl)-3-(hydroxylmethyl)-1-methylpiperidine, which is treated with thyonyl chloride to provide the corresponding chloromethyl derivative. Condensation with 1,3-benzodioxol-5-ol yields trans-(3*S*,4*R*)-3-(1,3-benzodioxol-5-yloxymethyl)-4-(4-fluorophenyl)-1-methylpiperidine, which can be demethylated to the target molecule by one of the following three methods: (*a*) condensation with phenyl chloroformate and removal of the phenoxycarbonyl group by treatment with KOH in refluxing toluene, (*b*) alternatively, with ethyl chloroformate in toluene and treatment with KOH in refluxing ethanol/water in order to eliminate the ethoxycarbonyl group, or (*c*) via treatment with vinyl chloroformate, followed by treatment with dry HCl gas in refluxing dichloromethane/methanol.

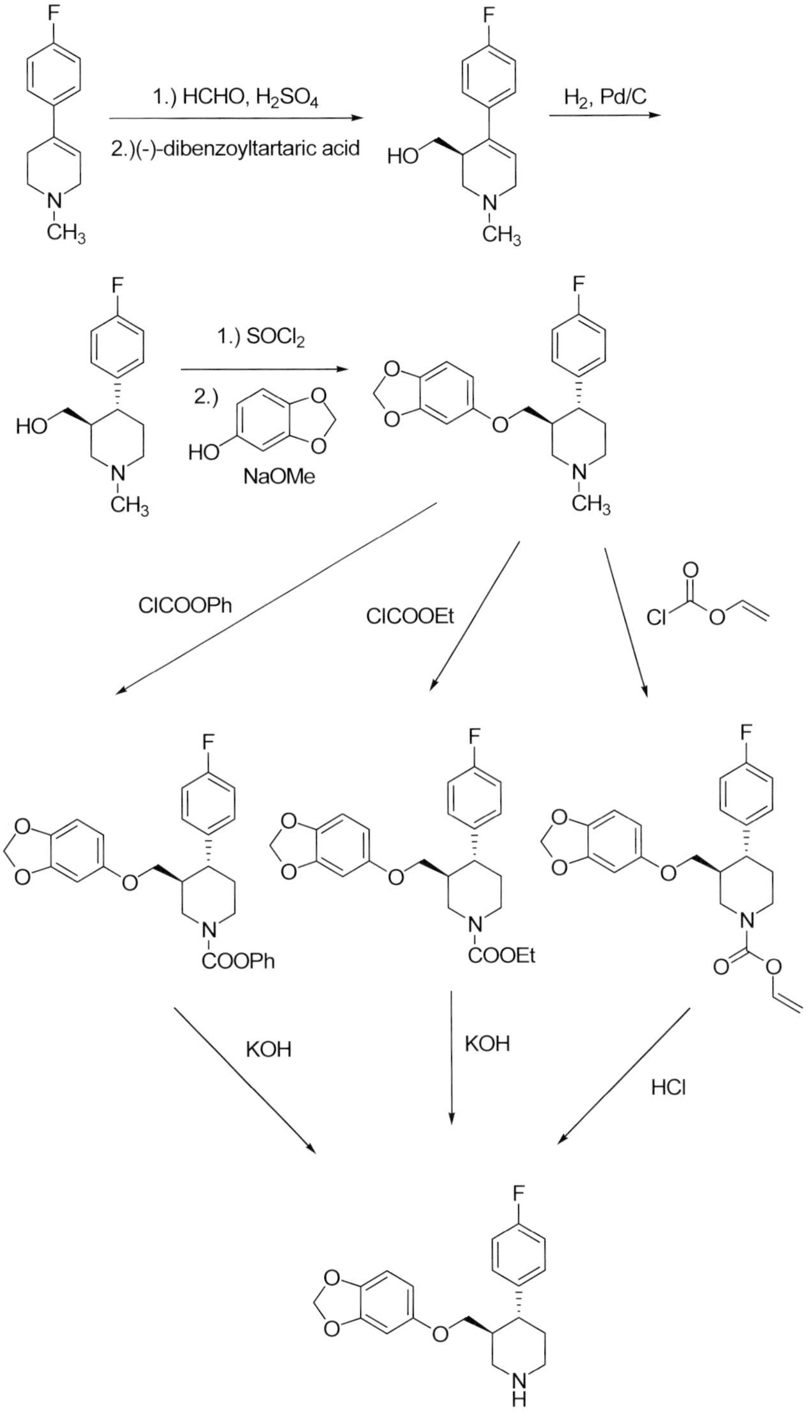

Scheme 27: Synthesis of paroxetine

Several alternative processes have been described in the literature (Yu et al., 2000; Amat et al., 2000; Borrett et al., 2001; Johnson et al., 2001; Cossy et al., 2001), including two for the [14C]-labeled compound (Lawrie and Rustidge, 1993; Willcocks et al., 1993).

Clinical Use, Pharmacokinetics and Metabolism: Paroxetine is a selective serotonin reuptake inhibitor (SSRI) with a clinical profile similar to fluoxetine. It offers, as the other SSRI members an efficient alternative treatment of depression to the TCAs (Dechant and Clissoid, 1991; Grimsley and Jann, 1992; Anonymous, 1993; Caley and Weber, 1993; de Wilde et al., 1993; Arminen et al., 1994; Ansseau et al., 1994; Montgomery et al., 1995). In addition, paroxetine has demonstrated its use in treatment of pain (Sindrup et al., 1990), premenstrual syndrome (Eriksson et al., 1995) and sexual dysfunction (Waldinger et al., 1994). Adverse effects are similar to the ones observed with e.g. fluoxetine. In addition, effects on the endocrine system, eyes (Barrett, 1994), extrapyramidal effects (Committee on Safety of Medicines/Medicines Control Agency, 1993; Choo, 1993) and withdrawal syndromes (Committee on Safety of Medicines/Medicines Control Agency, 1993; Bloch et al., 1995) have been observed. Paroxetine is absorbed from the GI tract leading to plasma peak concentrations 2–8 hours after ingestion. Paroxetine is to a degree of 95% bound to plasma proteins and is distributed widely throughout the body. Elimination half-life is about 21 hours (Dalhoff et al., 1991). Paroxetine is mainly metabolized by hepatic oxidation to inactive intermediary products (BRL-36610, BRL-36583, BRL-35961), via the opening of the 5-membered ring or the cleavage of the benzodioxolan moiety. All these three metabolites undergo subsequent conjugation to form correspondent glucuronide or sulfate. BRL-36610 glucuronide and sulfate were the major urinary metabolites observed in extensive metabolizer of sparteine, whereas BRL-36583 glucuronide was the major urinary metabolites in poor metabolizers, which suggests that the formation of BRL-36610 is mainly mediated by CYP2D6.

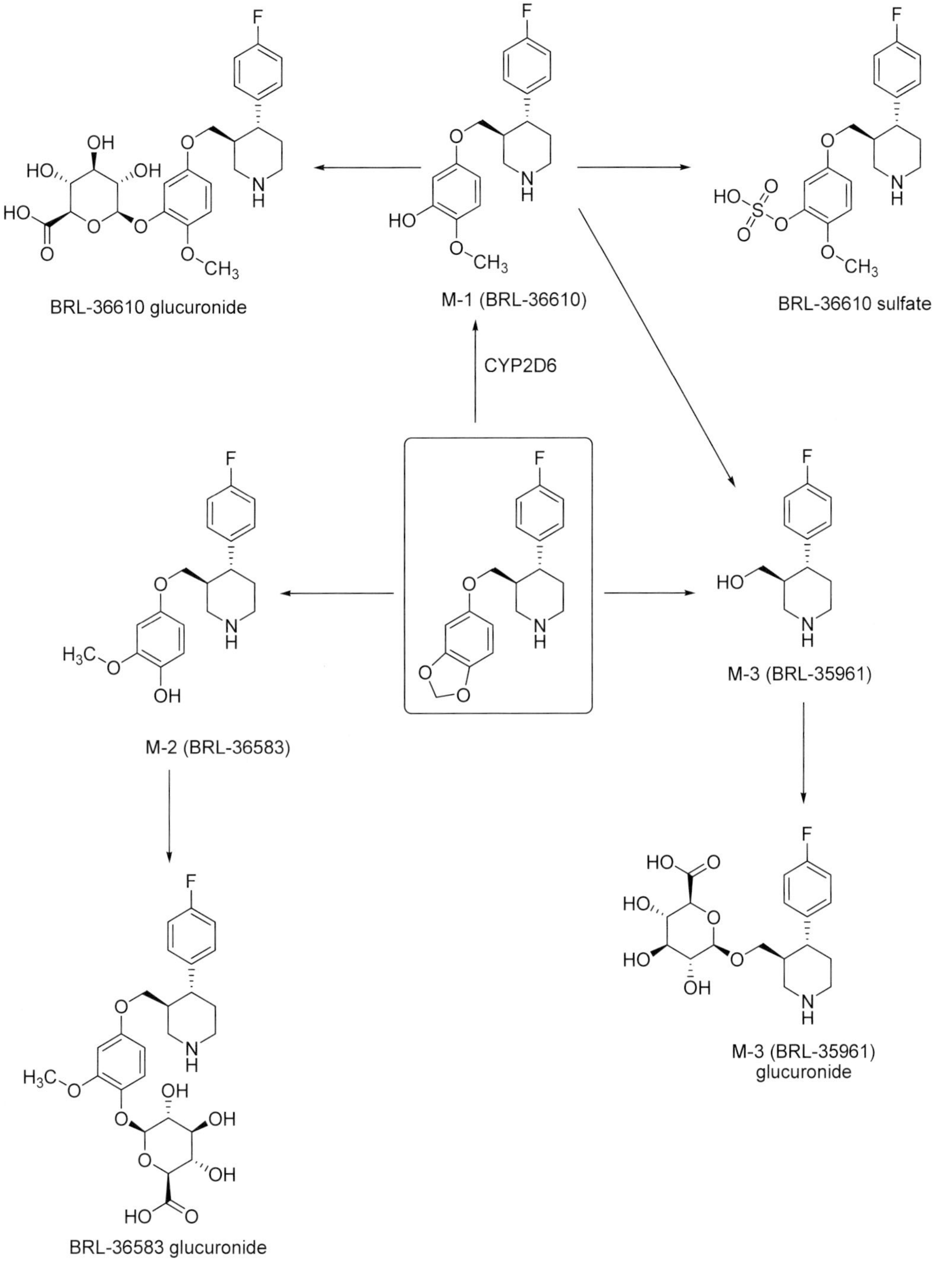

Scheme 28: Metabolism of paroxetine

Sertraline

[*79617-96-2*], (1*S*,4*S*)-4-(3,4-Dichlorophenyl)-*N*-methyl-1,2,3,4-tetrahydro-1-naphthaleneamine, $C_{17}H_{17}Cl_2N$, M_r 306.23; [*79559-97-0*] (hydrochloride salt), $C_{17}H_{18}Cl_3N$, M_r 342.69

Sertraline

- For this compound, the first patent was filed in 1979 by Pfizer (Welch et al., 1981)

- From the four possible stereoisomers of 1-methylamino-4-(3,4-dichlorophenyl) tetraline, the *cis*-1*S*,4*S* isomer is the most selective and potent competitive inhibitor of synaptosomal serotonin uptake (Coe et al., 1983)

- First launched as hydrochloride in 1992 for the treatment of depression

- In 1999 launched for panic disorder (with or without agoraphobia) and obsessive-compulsive disorder (OCD)

- In 2000 launched for the treatment of post-traumatic stress disorder (PTSD)

- In 2003 launched for premenstrual dysphoric mood disorder (PMDD) and approved by the FDA for acute and long-term treatment of social anxiety disorder

Synthesis (Welch et al., 1981; Williams and Quallich, 1990): Sertraline was originally prepared by resolution of racemic 1-methylamino-4-(3,4-dichlorophenyl) tetraline with mandelic acid (diastereomeric salt formation). The synthetic strategy implies first a Friedel-Crafts reaction followed by Stobbe condensation with diethyl succinate, decarboxylation with hydrobromide in acetic acid, double bond reduction (hydrogenation) and again a Friedel-Crafts cyclation to afford the racemic tetralone. Treatment with methylamine followed by reductive hydrogenation affords a 3:1 *cis/trans* mixture of the corresponding amines. The *cis* isomer is separated by hydrochloride salt formation, followed by recrystallisation. Finally, a mandelate resolution gives the target compound.

In the course of the clinical development, this process has been optimized significantly (Williams and Quallich, 1990). Thus, e.g. the synthesis to the intermediate tetralone has been shortened (Quallich and Williams, 1989), improvements of the stereochemical control of the synthesis (e.g., via the enantiomerically pure tetralone) have been reported as well (Quallich and Woodall, 1992; Corey and Gant, 1994). An efficient alternative route to sertraline via anionic imine ring closure has been described by Merck in 1999 (Chen and Reamer, 1999).

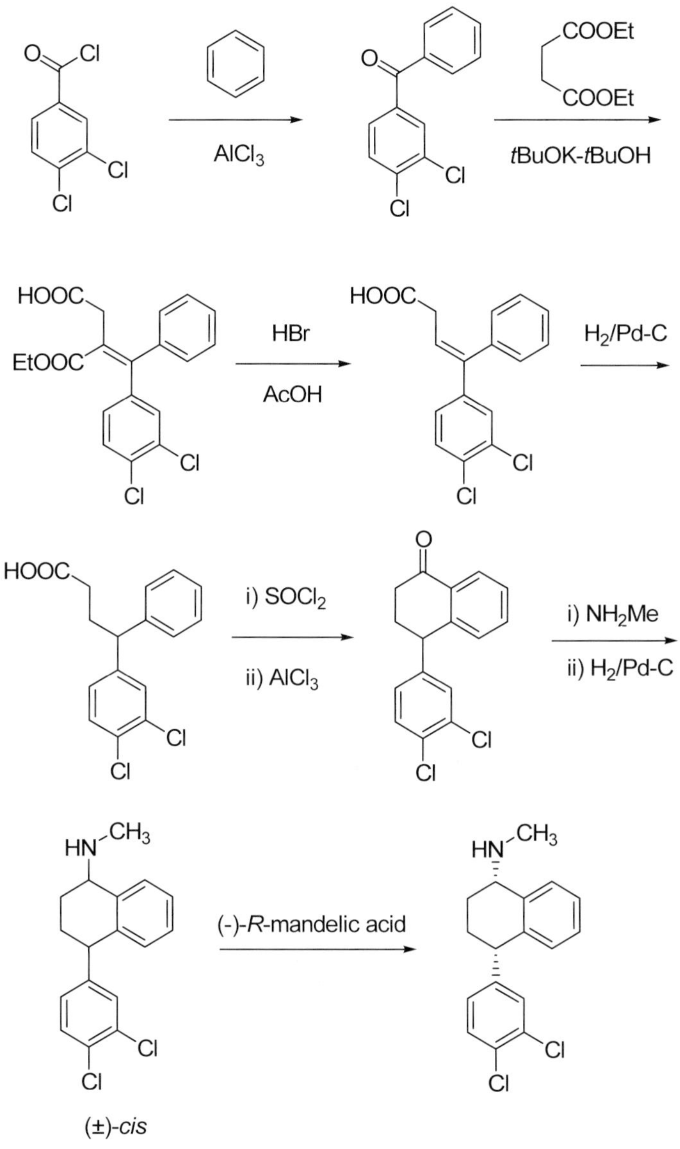

Scheme 29: Synthesis of sertraline

Clinical Use, Pharmacokinetics and Metabolism: Sertraline is a selective serotonin reuptake inhibitor (SSRI) with a clinical profile similar to fluoxetine. It offers, as the other SSRI members an efficient alternative treatment of depression to the TCAs (Murdoch and McTavish, 1992; Grimsley and Jann, 1992). Most common side-effects comprise nausea, diarrhea, dyspepsia, dry mouth, insomnia, somnolence, sweating, tremor, or delayed ejaculation. It is less sedative than the TCAs, e.g.

amitripyline. It should not be used in patients with coronary heart diseases (Iruela, 1994; Berti and Doogan, 1994). Sertraline is slowly absorbed in the GI tract after ingestion and reaches peak plasma concentrations about 4.5–8.5 hours thereafter. It is widely distributed throughout the body and nearly completely (98%) bount to tissue or plasma protein. Elimination half-life is around 25 hours. Sertraline has a linear pharmacokinetic profile and a half-life of about 22–36 hours, and once daily administration is therapeutically effective. Sertraline mildly inhibits the CYP2D6 isoform of cytochrome P450 system but has little effect on CYP1A2, CYP3A3, CYP3A4, CYP2C9, or CYP2C19. Sertraline is highly protein bound and may alter blood levels of other highly protein bound agents. Sertraline is slowly absorbed following oral administration and undergoes extensive first-pass oxidation to N-desmethylsertraline, that accumulates to a greater concentration in plasma than the parent drug at steady state. Sertraline reaches the maximum concentration in plasma at 4–8 hours after oral administration. Absolute bioavailability was estimated to be > 48% (MacQueen et al., 2001). The major metabolite N-desmethylsertraline has a plasma half-life of about 62–104 hours in humans. The metabolite appears to lack a relevant serotonin reuptake inhibiting effect in *in vitro* and *in vivo* studies (DeVane et al., 2002). Sertraline and N-desmethylsertraline undergo oxidative deamination to the corresponding ketone, which is subsequently hydroxylated at the α-carbon, forming a diasteromeric metabolite pair. Another oxidation product of sertraline is N-hydroxysertraline. Sertraline carbamic acid, N-hydroxysertraline, and the α-hydroxy ketone di-astereomers are eliminated by glucuronidation. These glucuronides comprise 45–82 % of the total drug excreted in urine and bile. The steps in the formation of sertraline carbamoyl-O-glucuronide involve an initial, nonenzymatic, and reversible association of sertraline and carbon dioxide to sertraline carbamic acid, followed by the enzymatic conjugation of the carbamic acid by UDP-glucuronyltransferase (Tremaine et al., 1989a; 1989b).

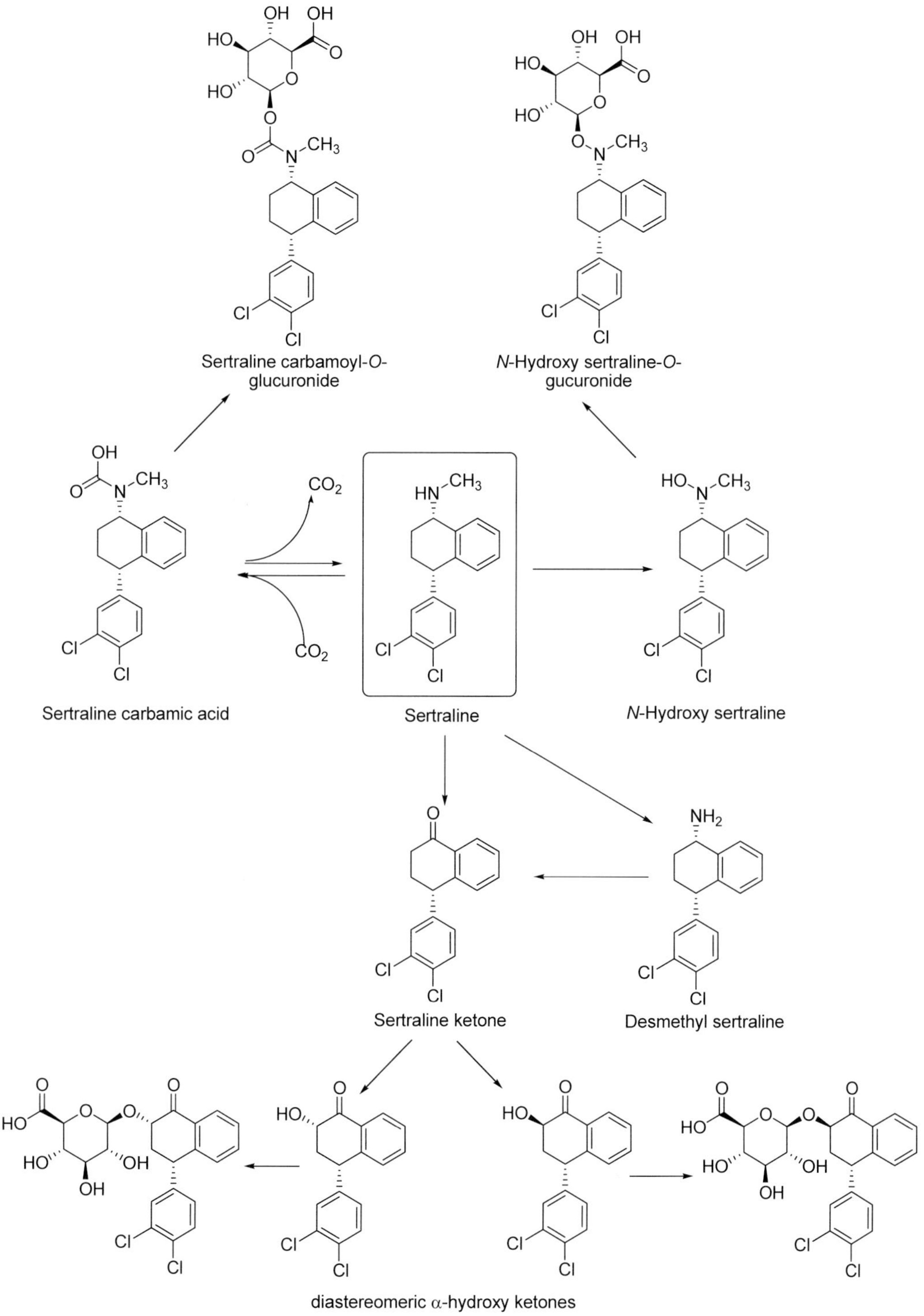

Scheme 30: Metabolism of sertraline

1.5.3.2 Compounds with other serotonergic activity

This group comprises the three structurally related compounds trazodone, etoperidone and nefazodone, all of them bearing a triazolone and a *m*-chlorophenyl-piperazine motif. However, their detailed mode of action is not completely understood. It has been found that they act via central inhibition of serotonine uptake and decrease of peripheral serotonine uptake along with an increase of brain dopamine turnover. Prominent side-effects comprise sedation. In the following, these three compounds are presented summarizing chemical properties, synthesis, important patents, clinical use as well as pharmacokinetics and metabolism.

Trazodone

Trazodone

[19794-93-5], 2-[3-[4-(3-Chlorophenyl)-1-piperazinyl]propyl]-1,2,4-triazolo[4,3-*a*]pyridin-3(2*H*)-one, $C_{19}H_{22}ClN_5O$, M_r 371.86;
[25332-39-2] (hydrochloride salt), $C_{19}H_{23}Cl_2N_5O$, M_r 408.32

- Patent filed in 1965 by Angelini (ACRAF SpA) (Palazzo and Silvestrini, 1968)
- Compound launched in 1972 for depression
- Possesses antidepressant and also some anxiolytic and hypnotic activity
- In 2004 Labopharm Inc has signed an agreement with Angelini, whereby Labopharm will develop a once-daily formulation of trazodone

Synthesis (Chen et al., 2002): Recently, an efficient short preparation method derived from the one patented by Angelini (Palazzo and Silvestrini, 1968) was described for trazodone hydrochloride with an overall yield of 39%. Starting from 2-chloropyridine via substitution with hydrazine hydrate, cyclocondensation with urea, condensation with 1-(3-chlorophenyl)-4-(3-chloropropyl)-piperazine in the presence of alkali, and final treatment with hydrochloric acid, the target compound is obtained.

Clinical Use, Pharmacokinetics and Metabolism: Trazodone is triazolopyridine antidepressant, bearing, like nefazodone and etoperidone, members of the same class, the phenylpiperazine motif. It is an antidepressant with little antimuscarinic, but marked sedative properties. Unlike amitriptyline, it does not block peripheral uptake of NA. Its mode of action comprises central serotonin uptake inhibition and decrease of peripheral serotonin uptake. As a result, it also seems to increase dopamine brain

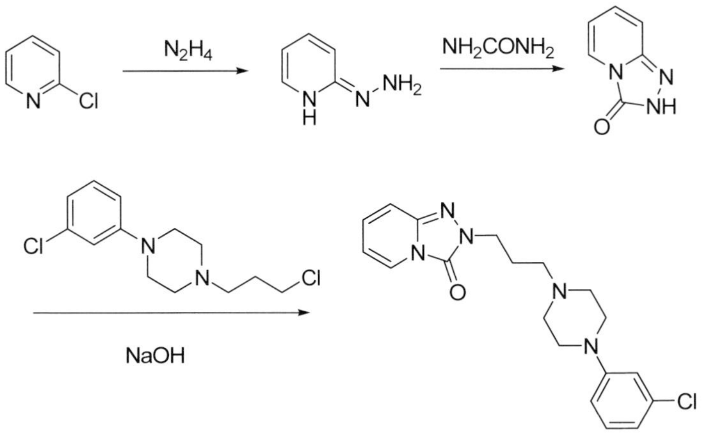

Scheme 31: Synthesis of trazodone

turnover. The complete mode of action of trazodone is not fully understood. Trazodone is often used as an alternate therapy of depression to the conventional TCAs, especially if antimuscarinic side-effects are contraindicated (Weisler et al., 1994). It has demonstrated its usefulness in the treatment of disturbed behavior (Pasion and Kirby, 1993; Lebert et al., 1994) and withdrawal symptoms (Small and Purcell, 1985; Liebowitz and El-Mallakh, 1989; Ansseau and De Roeck, 1993). Adverse effects include mainly sedation and drowsiness, which improve while chronic treatment due to tolerance development. Other side-effects include dizziness, headache, nausea and vomiting, effects on the cardiovascular system (Rausch et al., 1984; Lippmann et al., 1983; Janowsky et al., 1983; Vlay and Friedling, 1983; Johnson, 1985; White and Wong, 1985), effects on the eyes (Cooper and Dening, 1986), liver (Chu et al., 1983; Sheikh and Kies, 1983; Beck et al., 1993; Hull et al., 1994), mental function (Damlouji and Ferguson, 1984; Warren and Bick, 1984; Arana and Kaplan, 1985; Kraft, 1983; Patterson and Srisopark, 1989), sexual function (Committee on Safety of Medicines, 1984; Anonymous, 1984; Jones, 1984; Gartrell, 1986) and skin (Mann et al., 1984; Ford and Jenike, 1985; Barth and Baker, 1986), as well as epileptogenic effects (Bowdan, 1983; Lefkowitz et al., 1985). Trazodone should be used with care in patients with cardiovascular disorders, such as ischaemic heart disease, as well as in patients with a history of epilepsy and renal or hepatic insufficiency. It should not be combined with TCAs in depression therapy. Trazodone is rapidly and extensively absorbed in the GI tract after

Etoperidone

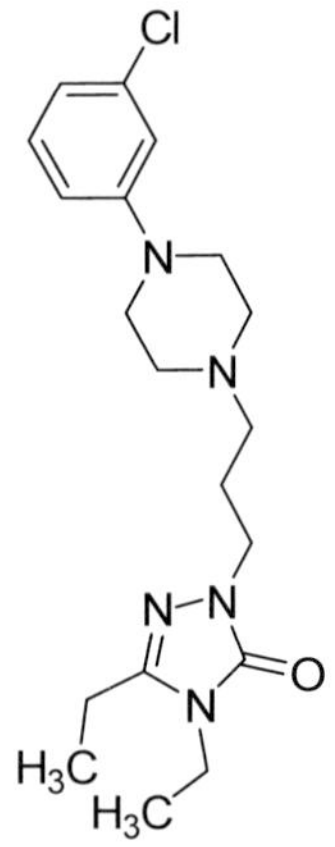

[52942-31-1], 1-[3-[4-(3-Chlorophenyl)-1-piperazinyl]propyl]-3,4-diethyl-Δ^{*2}-1,2,4-triazolin-5-one, $C_{19}H_{28}ClN_5O$, M_r 377.91;
[57775-22-1] (hydrochloride salt), $C_{19}H_{29}Cl_2N_5O$, M_r 414.37

ingestion, experiencing a pronounced food effect on absoption. It is extensively metabolized in the liver, including N-oxidation and hydroxylation steps. One of the active metabolites is m-chlorophenyl-piperazine. Trazodone is excreted in urine, with biphasic elimination half-lifes from plasma around 5–9 hours (Bayer et al., 1983).

Etoperidone

- First patent published in 1974 by Angelini (ACRAF SpA) (Palazzo, 1974). Compound launched for the treatment of depression in 1977

Synthesis (Castaner and de Angelis, 1977): The compound can be prepared in several different ways, all starting from 3,4-diethyl-Δ^2-1,2,4-triazolin-5-one, being the fastest the condensation with N-(3-bromopropyl)-N'-(m-chlorophenyl)piperazine by means of sodium amide, sodium hydride or a sodium alcoholate.

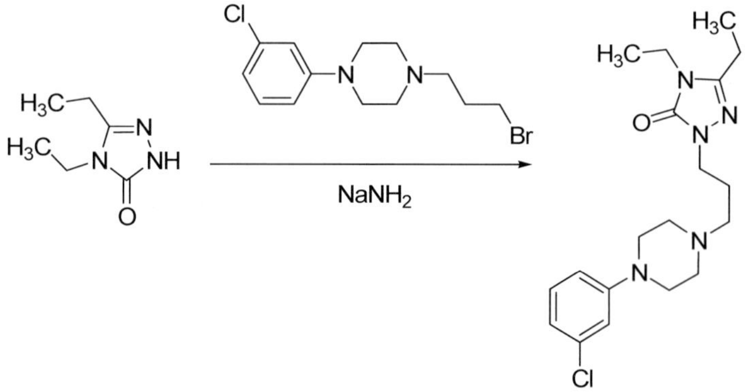

Scheme 32: Synthesis of etoperidone

Clinical Use: Etoperidone is a triazolopyridine serotoninergic antidepressant with a mechanism and clinical spectrum similar to trazodone (Aprile et al., 1983). Etoperidone represents a newer class of antidepressants (as compared e.g. with the TCAs), but as seen with e.g. TCAs it may need some weeks until a clinical benefit in therapy can be seen.

Nefazodone

- Patent applications from Bristol-Myers Squibb (BMS) published first in 1982 for depression (Temple and Lobeck, 1982), then in 1992 for sleep disorder (Gammans, 1992) and in 1998 for migraine prophylaxis (Marcus and Sussman, 1998)
- BMS launched nefazodone in 1994 for depression acting as serotonin reuptake inhibitor and 5-HT$_2$ receptor blocker
- In 1996 nefazodone entered Phase III trials for panic disorder
- In 2003, BMS withdrew the product in Europe due to liver toxicity concerns

Synthesis (Temple and Lobeck, 1982; Nefazodone hydrochloride, 1987): Reaction of 2-ethyloxazoline with phenol affords a propionamide which is treated with phosgene to yield the corresponding imino chloride, which is reacted with methylcarbazate to provide the corresponding amidrazone. Cyclodehydration and alkylation with 1-(3-chlorophenyl)-4-(3-chloropropyl) piperazine affords nefazodone, which can be isolated as its hydrochloride salt.

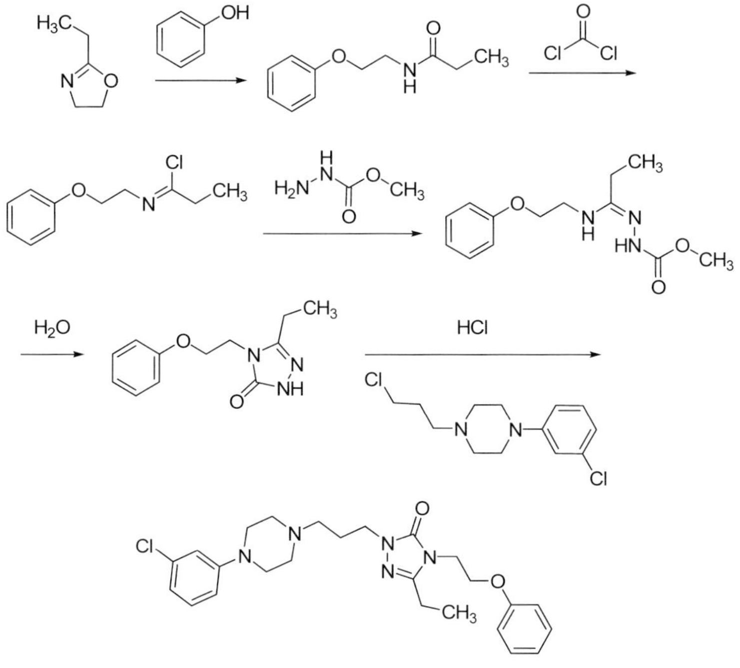

Nefazodone

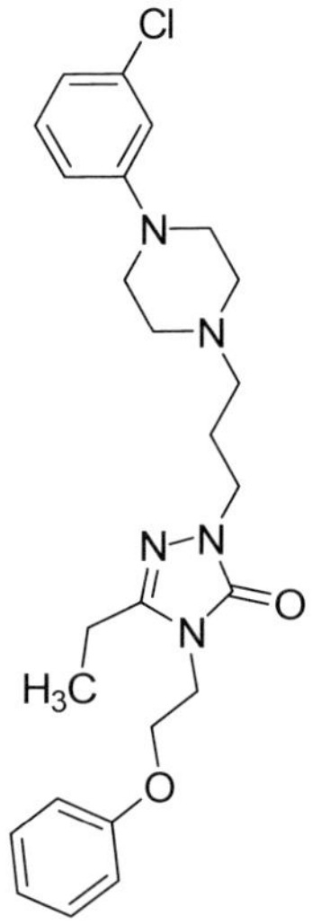

[83366-66-9], 2-[3-[4-(3-Chlorophenyl)piperazin-1-yl]propyl]-5-ethyl-2,4-dihydro-4-(2-phenoxyethyl)-3H-1,2,4-triazol-3-one, C$_{25}$H$_{32}$ClN$_5$O$_2$, M$_r$ 470.01; [82752-99-6] (hydrochloride salt), C$_{25}$H$_{33}$Cl$_2$N$_5$O$_2$, M$_r$ 506.47

Scheme 33: Synthesis of nefazodone

Clinical Use, Pharmacokinetics and Metabolism: Nefazodone is a phelylpiperazine containing antidepressant with a similar clinical spectrum as trazodone. It blocks, like fluoxetine, selectively serotonin reuptake, but postsynaptic 5-HT$_2$ receptors as well (dual mechanism). Nefazodone is used clinically as an antidepressant (also for treatment of premenstrual syndrome (Freeman et al., 1994)) as well as an anxiolytic. In contrast to fluoxetine, it does not show pronounced agitatory side-effects, like agitation or sleep disturbance. Nefazodone represents a newer class of antidepressants, but as seen with e.g. TCAs it may need some weeks until a clinical benefit in therapy can be seen (Rickels et al., 1994; Fontaine et al., 1994). Most common side-effects include asthenia, dry mouth, nausea, somnolence and dizziness. It should not be used in patients with a history of epilepsy, manias, or renal or hepatic impairment. Nefazodone is readily absorbed from the GI tract, reaching peak plasma concentrations 1–3 hours after ingestion. It undergoes excessive first-pass effect and has >99% plasma protein binding. Elimination half-lifes are relatively fast, 2–4 hours (Shea et al., 1988, 1989). Nefazodone undergoes extensive metabolism in all species. *N*-Dealkylation at piperazine ring and aliphatic hydroxylation convert nefazodone to the active metabolites mCPP and hydroxynefazodone, the latter was further biotransformed to another active major plasma metabolite, nefazodone triazoledione, probably via the intermediate ketonefazodone. Other major plasma and urinary metabolites observed are Metabolite D, Metabolite C and Metabolite E, resulting from the oxidative *N*-dealkylation of nefazodone or hydroxynefazodone, of which the alcohol Metabolite E was present exclusively as a conjugate. The *N*-cleavage product mCPP undertook further extensive biotransformation. Para-hydroxylated mCPP and its conjugates were observed in urine. Metabolites para-hydroxynefazodone and ketonefazodone were observed in plasma and /or urine only in trace amounts. In human liver microsomes, CYP3A4 was mainly responsible for the biotransformation of nefazodone and hydroxy-nefazodone to active metabolites, whereas CYP2D6 was responsible for the clearance of mCPP (Staack and Maurer, 2003; Rotzinger and Baker, 2002; Von Moltke et al., 1999; Dodd et al., 1999, Marathe et al., 1995; Mayol et al., 1994; Shukla et al., 1993).

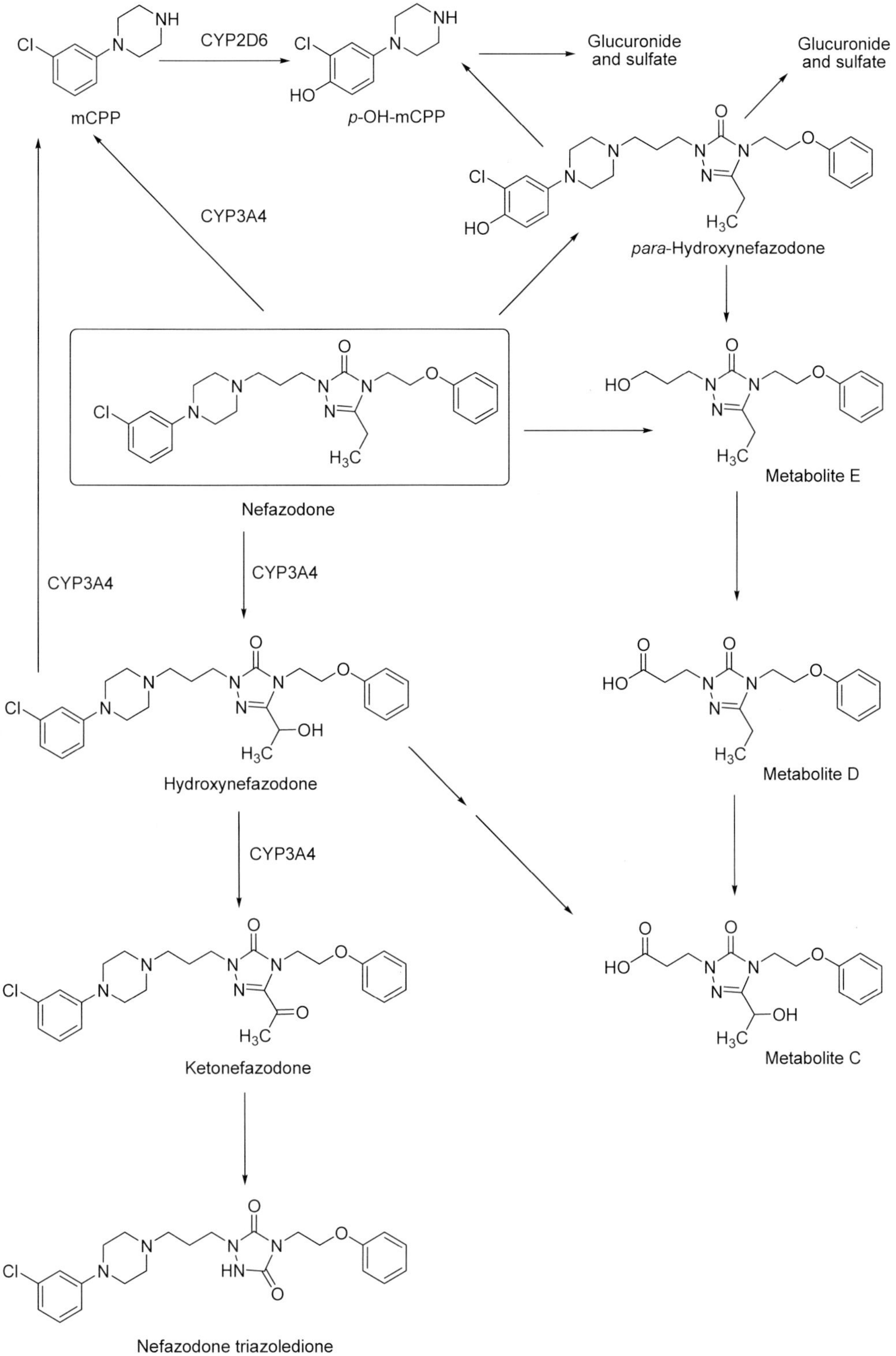

Scheme 34: Metabolism of nefazodone

1.5.3.3 Serotonin and noradrenaline reuptake inhibitors

One of the most recent entries into the family of clinically relevant antidepressants are the SNRIs, represented by milnacipran, duloxetine and venlafaxine. These drugs inhibit selectively serotonin as well as noradrenaline reuptake, and have little to no affinity for muscarinic, histaminergic or α-1-adrenergic receptors, leading to a favorable side-effect profile, as compared e.g. with the TCAs. In the following, these three drugs on the market are presented summarizing chemical properties, synthesis, important patents, clinical use as well as pharmacokinetics and metabolism.

Milnacipran

Milnacipran

[92623-85-3], (1R, 2S)-cis-2-(Aminomethyl)-N,N-diethyl-1-phenylcyclopropane-1-carboxamide, $C_{15}H_{22}N_2O$, M_r 246.35;
[101152-94-7] (hydrochloride salt), $C_{15}H_{23}ClN_2O$, M_r 282.81

- Compound patent filed in 1981 by Pierre Fabre (Mouzin et al., 1984)
- Launched by Pierre Fabre in 1995 for depression
- Cypress and Forest are developing milnacipran for the treatment of fibromyalgia syndrome (FMS) and related chronic pain syndromes
- The *cis* isomer is the active form of the compound, while the *trans* isomer does not inhibit the reuptake of monoamines. The compound is marketed as a racemate

Synthesis (Bonnaud et al., 1986): Milnacipran (sometimes referred to as midalcipran) can e.g. be prepared from benzyl cyanide which is reacted with sodium amide and epichlorhydrin. The resulting lactone can be opened, for example with a 33% hydrobromide solution in acetic acid, to give the corresponding γ-bromo acid which is treated by thionyl chloride and then diethylamine to give the corresponding amide. The γ-bromo function is converted into a primary amine by reaction with potassium phthalimide and subsequent concersion with hydrazine to yield, after salification, the title compound as a racemate. Some minor modifications of this synthetic pathway are also described in the same literature (Bonnaud et al., 1986).

In the course of clinical development, a process for large-scale synthesis of milnacipran was developed (Hascoët and Cousse, 1986).

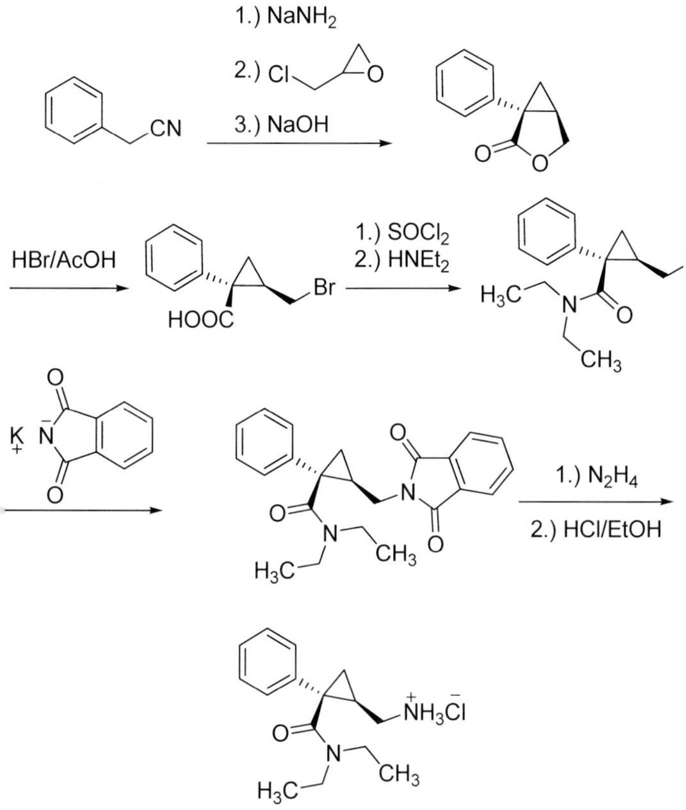

Scheme 35: Synthesis of milnacipran

Clinical Use and Metabolism: Milnacipran is a selective serotonin and noradrenaline reuptake inhibitor, like e.g. venlafaxine, thus providing a modern alternative to classic "tricyclic" treatment of depression (Serre et al., 1986; Ansseau et al., 1989; Palmier et al., 1989). Unchanged milnacipran is the major component in plasma, urine and brain after oral administration in rats. Although only to a limited extent, milnacipran is mainly metabolized via desethylation, oxidative deamination, 4-hydroxylation, and glucuronidation, giving desethylmilnacipran, deaminomilnacipran, 4-hydroxyphenylmilnacipran and milnacipran glucuronide, respectively. Desethylmilnacipran is further converted to didesethylmilnacipran or desethyldeaminomilnacipran. In bile, metabolites like desethylmilnacipran, didesethylmilnacipran and 4-hydroxyphenylmilnacipran are mainly glucuronide-conjugated. Cyclic milnacipran, produced from intramolecular cyclization, was identified as the secondary main metabolite in the brain (Atsushi et al., 1994a, b).

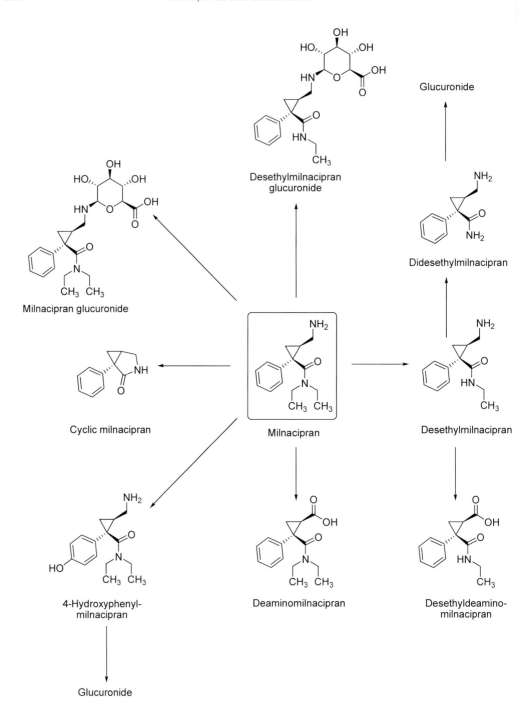

Scheme 36: Metabolism of milnacipran

Venlafaxine

- First patent application filed by American Home Products Corp. (now: Wyeth) in 1982 (Husbands et al., 1984)

- Launched for depression in 1994, in 1999 for anxiety, in 2003 for social phobia and in 2005 registered for panic disorder

Synthesis (Prous and Castaner, 1988; Yardley et al., 1990): Condensation of cyclohexanone with 4-methoxyphenylacetonitrile by means of *n*-butyllithium gives 2-(1-hydroxycyclohexyl)-2-(4-methoxyphenyl) acetonitrile, which is reduced with hydrogen over Rh/Al$_2$O$_3$ yielding 1-[2-amino-1-(4 methoxyphenyl)ethyl]cyclohexanol. Finally, this compound is methylated (Eschweiler-Clarke reductive methylation) with formaldehyde and formic acid in refluxing water.

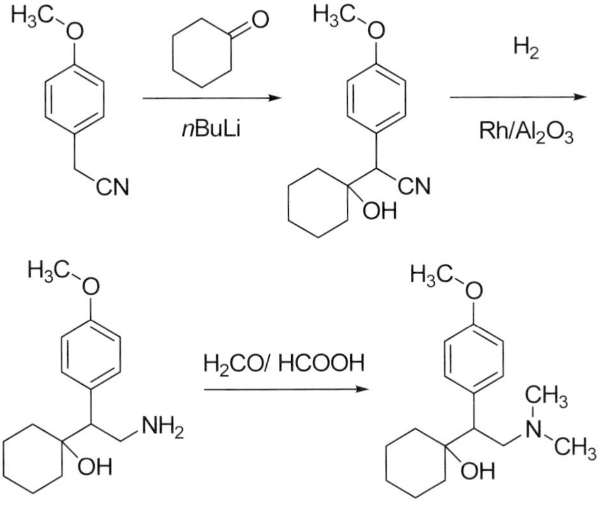

Venlafaxine

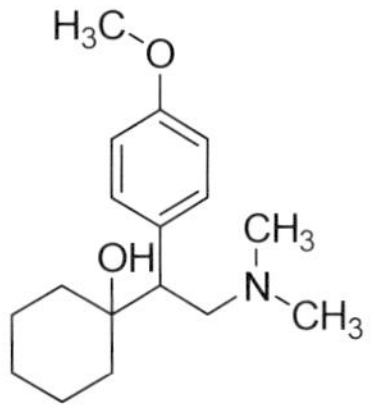

[93413-69-5], 1-[2-(Dimethylamino)-1-(4-methoxyphenyl)ethyl]cyclohexanol, C$_{17}$H$_{27}$NO$_2$, M_r 227.40; [99300-78-4] (hydrochloride salt), C$_{17}$H$_{28}$ClNO$_2$, M_r 313.86

Scheme 37: Synthesis of venlafaxine

In recent years, several new process patents have been filed by competitors, presumably to prepare their entry into the generic prescription market once the original patent protection for venlafaxine expires. Thus, e.g. Medichem S.A. in 1999 (Arnalot i Aquilar et al., 2001) and Sun Pharmaceutical Industries in 2003 (Patel et al., 2005) applied for venlafaxine process patents. Sun Pharmaceutical Industries developed a novel four step process starting from 4-methoxyphenyl acetic acid, employing a

Grignard reagent-mediated (as base) condensation reaction with cyclohexanone and a reduction of the carboxamide function with an aluminium hydride as key steps (Patel et al., 2005).

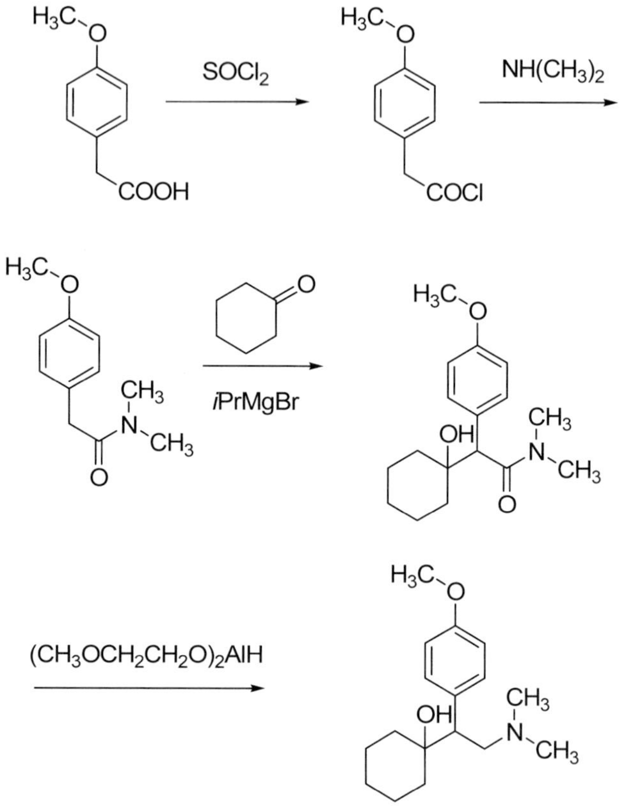

Scheme 38: Synthesis developed by Sun Pharmaceutical Industries

Clinical Use, Pharmacokinetics and Metabolism: Venlafaxine is a phenyethylamine derivative, effecting a selective blockade of serotonin and noradrenaline reuptake. In addition, it weakly inhibits dopamine reuptake and has some affinity to muscarinic, histaminergic and α_1 receptors *in vitro*. It provides, similar to milnacipran and duloxetine, an efficient treatment of depression and an alternative treatment to TCA treatment (Montgomery, 1993; Anonymous, 1994; Ellingrod and Perry, 1994; Holliday and Benfield, 1995; Morton et al., 1995). Most frequent adverse effects include mild ones like nausea, headache, insomnia, somnolescence, dry mouth, dizziness, constipation, asthenia, sweating and nervousness. Venlafaxine should not be used in patients with hepatic or renal impairment, myocardial infection or unstable heart disease. Venlafaxine is well absorbed from the GI tract after oral administration and undergoes extensive first-pass metabolism to its active O-desmethyl metabolite. It has low plasma and tissue protein binding and an elimination half-life of about 5 hours (11 hours for the metabolite). Excretion occurs mainly via kidneys (Troy et al., 1995).

Duloxetine

- The compound patent was filed in 1986 by Eli Lilly (Robertson et al., 1988)

- First pharmacological experiments were performed with the racemate (LY227942; Wong et al., 1988)

- The active (S)-enantiomer was launched in 2004 by Lilly for major depression, urinary incontinence and neurophatic pain. The compound is currently undergoing Phase III clinical trials for the treatment of fibromyalgia and generalized anxiety

Synthesis (Bymaster et al., 2003): After the first synthesis published by Deeter et al. (1990), based on the enantioselective reduction of the intermediate 3-(dimethylamino)-1-(2-thienyl)-1-propanone with (2R,3S)-4-(dimethylamino)-3-methyl-1,2-diphenyl-2-butanol and lithium aluminium hydride, a modified and more efficient strategy has been developed: the key step here represents the enantioselective reduction of 3-chloro-1-(2-thienyl)-1-propanone by means of borane catalyzed by (R)-1-methyl-3,3-diphenyltetrahydropyrrolo[1,2-c][1,3,2]oxazaborole [(R)-2-Methyl-CBS-oxazaborolidine] to give (S)-3-chloro-1-(2-thienyl)propan-1-ol, which is treated with methylamine and sodium iodide to yield (S)-N-methyl-N-[3-(2-thienyl)propyl]amine. This compound is next condensed with 1-fluoronaphthalene by means of sodium hydride in dimethylacetamide to provide duloxetine.

In 2005, a highly enantioselective rhodium-catalyzed hydrogenation of β-secondary ketones was developed and used for the synthesis of (αS)-{[(methyl)amino]ethyl}thiophenemethanol, the direct precursor of (S)-duloxetine in the above depicted reaction scheme (penultimate compound) (Liu et al., 2005).

Duloxetine

[116539-59-4], (+)-(S)-N-Methyl-N-[3-(naphthalen-1-yloxy)-3-(2-thienyl)propyl]amine, $C_{18}H_{19}NOS$, M_r 297.41; [136434-34-9] (hydrochloride salt), $C_{18}H_{20}ClNOS$, M_r 333.88

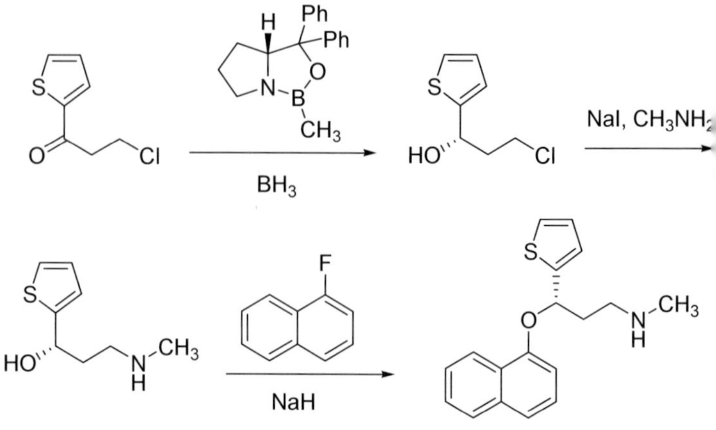

Scheme 39: Synthesis of duloxetine

Clinical Use, Pharmacokinetics and Metabolism: Duloxetine is a potent inhibitor of neuronal serotonin and noradrenaline reuptake and a less potent inhibitor of dopamine reuptake, thus providing effective treatment for depression. MAO is not inhibited by duloxetine. It has extensive metabolism, but major metabolites have no contribution to its activity, the elimination half-life being about 12 hours. Duloxetine undergoes extensive metabolism in humans after oral administration. The major metabolic pathways include oxidation of the naphthyl ring at the 4-, 5-, or 6- positions, followed by further oxidation, methylation and/or conjugation. Mono- or dihydroxylation at the 4-, 5- and 6- positions results in M12, M13, M14, 5,6-dihydroxyduloxetine and 4,6-dihydroxyduloxetine, all of which undergo further glucuronide or sulfate conjugation and/or methylation, forming numerous phase II metabolites (M4, M8, M6, M11, M5, M9, M15, M16, M10, M3 and M7) (Lantz et al., 2003).

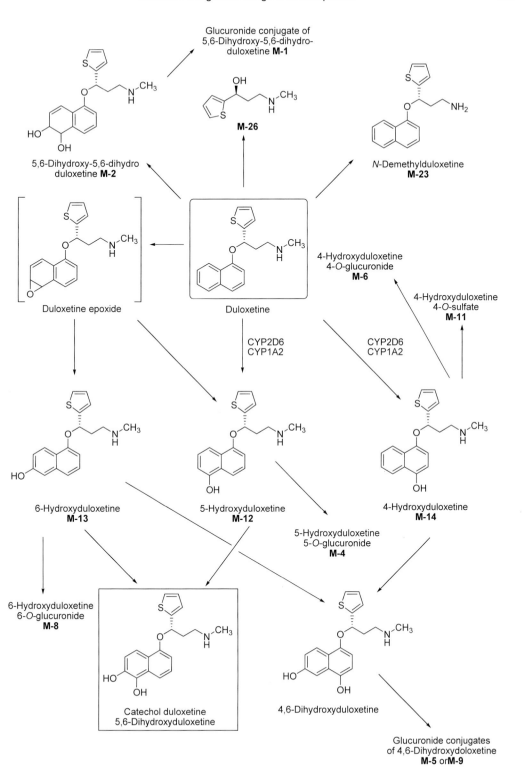

Scheme 40: Metabolism of duloxetine

Duloxetine also undergoes epoxidation at the 5- and 6- positions forming 5,6-dihydroxy-5,6-dihydroduloxetine (M2) and its glucuronide M1.

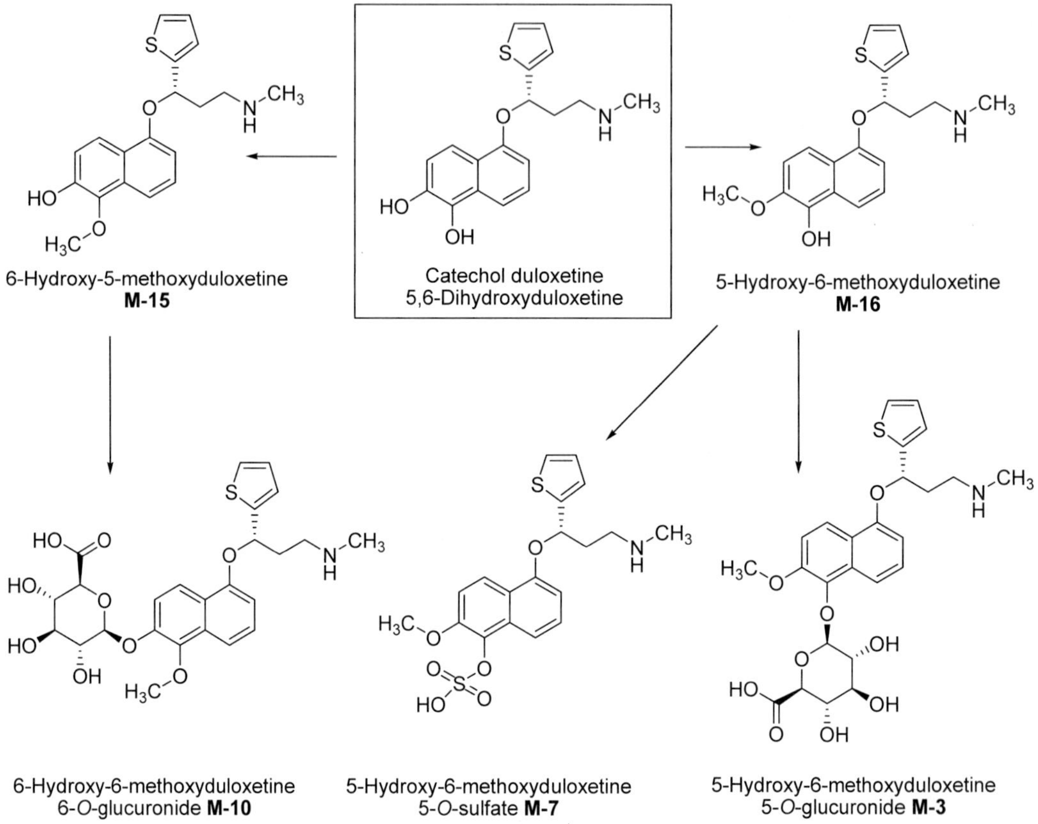

Scheme 41: Biotransformation pathways of 5,6-dihydroxy-5,6-dihydroduloxetine

The major metabolites in plasma and urine are M6, M7, M9, M10. Many additional metabolites were found in urine. Unchanged duloxetine, M14 and an unknown polar metabolite are observed in feces. Other minor metabolites are M26 and M23, formed by the cleavage of the naphthyl ring and *N*-demethylation, respectively. More than 70% of the given duloxetine is excreted in urine in the form of its metabolites. Primary studies showed that human liver CYD2D6 and CYP1A2 were mainly involved with hydroxylation on the naphthyl ring (Lantz et al., 2003).

1.5.3.4 Noradrenaline and specific serotonergic antidepressants

Mirtazapine is the only representative of the class of noradrenaline and specific serotonergic antidepressants (NaSSAs) compounds, acting presumably via blockade of presynaptic α_2 adrenergic receptors as well as post-synaptic 5-HT$_2$ and 5-HT$_3$ receptors, an action which is thought to enhance serotonergic neurotransmission while causing a low incidence of side-effects. Mirtazapine is presented in the following, summarizing chemical properties, synthesis, important patents, clinical use as well as pharmacokinetics and metabolism.

Mirtazapine (Azamianserin)

- Compound patent filed in 1975 by Organon/Akzo Nobel N.V. (van der Burg, 1977)
- Launched in 1994 for the treatment of depression by Organon
- Mirtazapine has a tetracyclic structure similar to mianserin and shows structural resemblance to the class of tricyclic antidepressants
- Complex pharmacology: it blocks noradrenaline α_2 autoreceptors, α_2 heteroreceptors, postsynaptic 5-HT$_2$ and 5-HT$_3$ receptors but not 5-HT$_1$ (reason for the description as *specific* serotonergic antidepressant). The net result is an increase of noradrenaline and 5-HT$_1$ transmission. It also blocks histamine H$_1$ receptors, causing sedation
- The predecessor mianserin, launched in 1975 (also by Organon) as 5-HT$_{2A}$ antagonist, was withdrawn from UK market after introduction of mirtazapine. Mianserin produced weight gain and showed sedation amongst other side-effects

Mirtazapine

[61337-67-5], 2-methyl-1,2,3,4,10,14b-Hexahydropyrazino[2,1-a]pyrido[2,3-c][2]benzazepine, $C_{17}H_{19}N_3$, M_r 265.35

Mianserin

- S-(+)-Mirtazapine is currently under active development for the treatment of sleep disorders (phase II, Organon), however one of the licensees (Cypress

Bioscience) has dropped the project recently (July, 2006)

Synthesis (van der Burg, 1977): Reaction of 2-chloronicotinonitrile with 1-methyl-3-phenylpiperazine gives 2-(4-methyl-2-phenyl-1-piperazinyl)-3-pyridinecarbonitrile, which is next hydrolyzed to the corresponding carboxylic acid. Reduction to the hydroxymethyl derivative and cyclization in concentrated sulfuric acid gives azamianserin after treatment with ammonia.

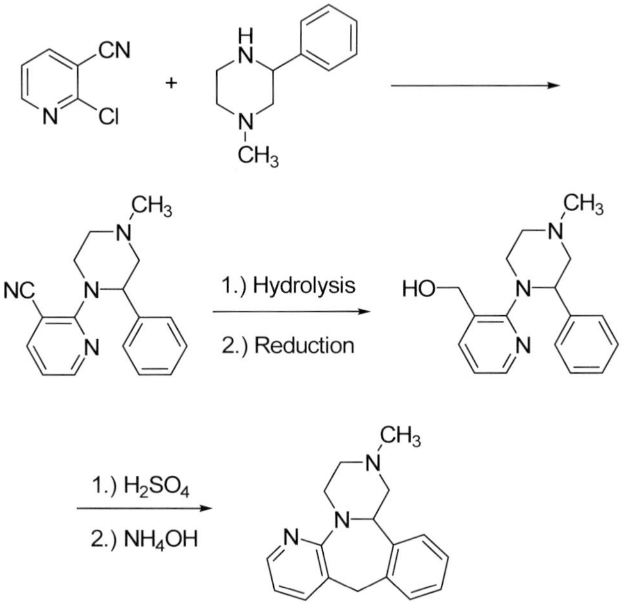

Scheme 42: Synthesis of mirtazapine (azamianserin)

Clinical Use: Mirtazepine has a complex mode of action, resulting in a blockade of noradrenergic and selective serotonin receptors, leading to its classification as NaSSAs. It is thought to work by blocking presynaptic α_2 adrenergic receptors that normally inhibit the release of the neurotransmitters norepinephrine and serotonin, thereby increasing active levels in the synapse. Mirtazapine also blocks post-synaptic 5-HT$_2$ and 5-HT$_3$ receptors, an action which is thought to enhance serotonergic neurotransmission while causing a low incidence of side-effects (Mattila et al., 1989; Sitsen and Moors, 1994; van Moffaert et al., 1995). Being the only representative of this class, it shows structural similarity to mianserin and the class of TCAs and provides effective treatment of depression.

1.5.4 Noradrenaline reuptake inhibitors

Unlike most antidepressants on the market, reboxetine is the only representative of a selective noradrenaline reuptake inhibitor (NARI). It does not inhibit the reuptake of serotonin, though it can be safely combined with an SSRI. In the following, reboxetine is presented summarizing chemical properties, synthesis, important patents, clinical use as well as pharmacokinetics and metabolism.

Reboxetine

- Compound patent filed in 1978 by Pharmacia Corp. (now Pfizer, formerly Pharmacia & Upjohn; Melloni et al., 1980)
- Launched in 1997 as a mesilate salt by Pharmacia for the treatment of depression
- First in its class to be marketed. Another noradrenaline reuptake inhibitor (NARI), tomoxetine hydrochloride, an analogue of fluoxetine (a SSRI) was discontinued in 1990 for the treatment of depression and was finally launched by Lilly for attention deficit/hyperactivity disorder in 2003

Reboxetine

[71620-89-8], (±)-{2-[alpha-(2-Ethoxyphenoxy)benzyl}morpholine, $C_{19}H_{23}NO_3$, M_r 313.39; [98769-84-7] (mesilate salt), $C_{20}H_{27}NO_6S$, M_r 409.50

Tomoxetine

- Reboxetine is prepared by regio- and stereospecific synthesis, the racemic mixture contains only one diastereomer (the R,R-(−) and S,S-(+) stereoisomers)
- The (S,S)-enantiomer is more potent as antidepressant than the (R,R)-enantiomer

Synthesis (Melloni et al., 1980; Riva, 1985): Epoxidation of trans-cinnamyl alcohol and reaction with sodium 2-ethoxyphenate give a diol, whose primary alcoholic group is protected with p-nitrobenzoyl chloride, and the secondary one with methanesulfonyl chloride. By treatment with sodium hydroxide, an epoxide is obtained under inversion of the absolute configuration at the C-2 carbon atom. Treatment with aqueous methanolic ammonia, followed by acylation with chloroacetyl chloride and cyclization with potassium tert-butoxide give the corresponding morpholone. The reduction to reboxetine is performed with [sodium bis(2-methoxyethoxy)aluminum hydride] (RED-AL).

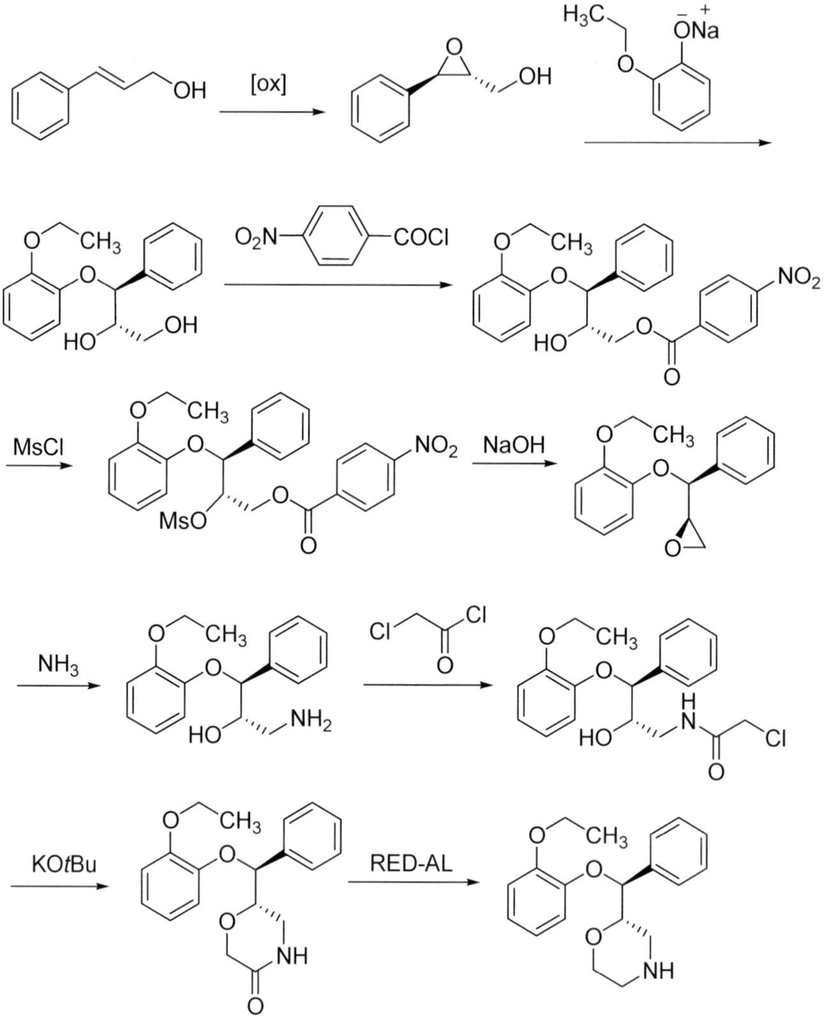

Scheme 43: Synthesis of reboxetine

Clinical Use, Pharmacokinetics and Metabolism: Reboxetine is a selective and potent inhibitor of the reuptake of noradrenaline and it also has a weak effect on serotonin reuptake. It has a slightly different biochemical profile from both the tricyclics and the SSRIs, but it appears to have few side-effects in comparison with the older tryciclics (Montgomery, 1998; Holm and Spencer, 1999; Scates and Doraiswamy, 2000; Andreoli et al., 2002; Montgomery et al., 2003). Reboxetine is well absorbed from the gastrointestinal tract with peak plasma levels after 2 hours being metabolized *in vivo* by the cytochrome CYP3A4. Therefore, fluvoxamine, azole antifungals as ketoconazole or macrolide antibiotics as erythromycin should not be given together with reboxetine (Herman et al., 1999). Dealkylation, hydroxylation and oxidation followed by glucuronide or sulfate conjugation metabolic pathways have been identified being plasma elimination half-life of 13 hours. Animal studies indicate that reboxetine crosses the placenta (Dostert, et al., 1997; Fleishaker, 2000; Coulomb et al., 2000; Poggesi et al., 2000). Reboxetine should not be combined with MAOI either. When is given with other drugs that lower blood pressure a postural hypotension can occur.

1.5.5 Monoamine oxidase inhibitors

Monoamine oxidase inhibitors (MAOIs) represent one of the "traditional" classes of antidepressant drugs. Due to potentially serious dietary and drug interactions they are used less frequently than other classes of antidepressant drugs (e.g., SSRIs and TCAs). However, they have demonstrated their use in cases where patients are unresponsive to other treatment. They are particularly effective in treating atypical depression, and have shown efficacy in smoking cessation. MAOIs act by inhibiting the activity of monoamine oxidase preventing the breakdown of monoamine neurotransmitters, thus increasing the available stores. From monoamine oxidase, two isoforms are known, MAO-A and MAO-B. MAO-A preferentially deaminates serotonin, melatonin, adrenaline and noradrenaline. MAO-B preferentially deaminates phenylethylamine and trace amines. Dopamine is equally deaminated by both types.

1.5.5.1 Non-selective monoamine oxidase inhibitors

Historically, the first MAOIs are inhibiting both isoforms equally and in irreversible manner. When they react with monoamine oxidase, they permanently deactivate the enzyme for up to 2 weeks. The following presents the non-selective MAOIs on the market, summarizing chemical

Phenelzine

Phenelzine

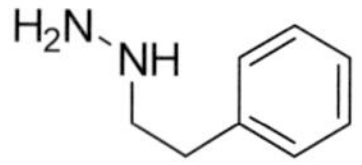

H₂N−NH−CH₂−CH₂−C₆H₅

[51-71-8], 2-(Phenylethyl)hydrazine, $C_8H_{12}N_2$, M_r 136.19;
[156-51-4], (sulfate salt) $C_8H_{14}N_2O_4S$, M_r 234.27

- Non-selective MAO inhibitor firstly described by Lakeside Laboratories (Biel, 1961) and developed by Parke-Davis (Warner-Lambert Pharmaceuticals Co., 1961), division of Warner-Lambert, acquired in 2000 by Pfizer

- Launched in 1957 for the treatment of depression

Synthesis (Biel, 1961): 2-Phenylethylbromide is treated with hydrazine to afford phenelzine.

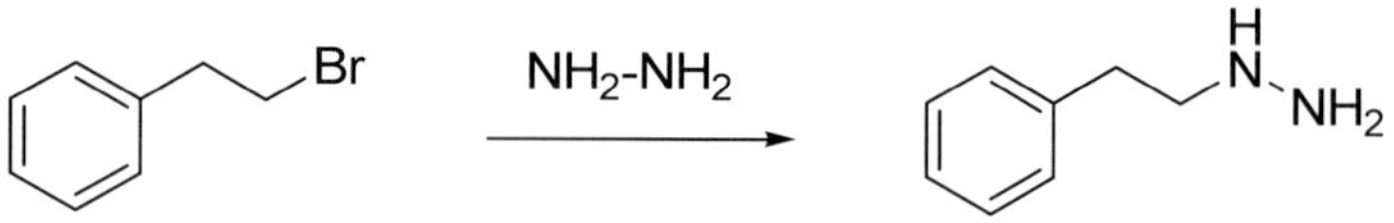

PhCH₂CH₂−Br + NH₂−NH₂ → PhCH₂CH₂−NH−NH₂

Scheme 44: Synthesis of phenelzine

Clinical Use, Pharmacokinetics and Metabolism: Phenelzine is one of the prototypes of the class of classic, non-selective and irreversible MAO-A and MAO-B inhibitors. It has properties which, in general, typify this class of antidepressants and is used in atypical depression and – in some cases – in melancholic depression. As with the TCAs, phenelzine may evoke orthostatic hypotension and has significant interactions with food and other drugs (dietary restrictions necessary). Even though the exact mechanism of action is not completely understood, it is believed that antidepressant activity seems to reside with inhibition of MAO-A. Consequently, side-effects and dietary restrictions are more pronounced with the non-selective inhibitors like phenelzine as for the selective MAO-A inhibitors like e.g. chlorgylin. Because of considerable clinical risk, irreversible MAO inhibitors are in many cases not regarded first choice therapy (Lecrubier, 1994). Phenelzine has been used in combination therapy, e.g. with TCAs, in the treatment of refractory or drug-resistant depression (Katona and Barnes, 1985), even though increased adverse effects have been reported (Beaumont, 1973; Graham et al., 1982). Adverse effects of phenelzine comprise hypotension (Kronig et al., 1983), hepatotoxic reactions (Holdsworth et al., 1961, Wilkinson et al., 1974, Lefebure et al., 1984), hyper-prolactinaemia (Slater et al.,

1977; Segal and Heys, 1969; Baylis, 1986), effects on the nervous system like peripheral neuropathy (Heller and Friedman. 1983; Stewart et al., 1984; Lane and Routledge, 1983), effects on sexual function (Simpson et al., 1965; Wyatt, et al., 1971; Yeragani and Gershon, 1987; Shen and Sata, 1983; Goldwyn and Sevlie, 1993) and epileptogenic (Bhugra and Kaye, 1986) and – in one case – extrapyramidal effects (Gillman and Sandyk, 1986). Interactions of phenelzine with food rich in pressor amines leading to hypertensive crises are possible due to the inhibition of amine metabolism. Interactions of phenelzine with other drugs are very likely and include hypertensive reactions due to enhancement of pressor activity after co-administration of symphatomimetic agents, e.g. amphetamines. Phenelzine also inhibits other drug-metabolizing enzymes, thus enhancing e.g. effects of barbiturates or alcohol. In terms of its pharmacokinetics, phenelzine is readily absorbed and extensively metabolized, reaching peak plasma concentrations 2–4 hours after ingestion.

Isocarboxazid

- Non-selective MAO inhibitor developed by Roche (Gardner et al., 1959) and launched in 1960 as antidepressant

Synthesis (Gardner et al., 1959): From commercially available benzylhydrazin and ethyl 5-methylisoxazol-3-carboxylate.

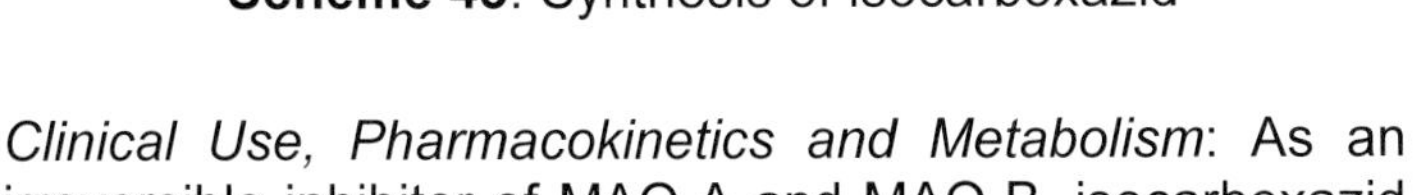

Scheme 45: Synthesis of isocarboxazid

Isocarboxazid

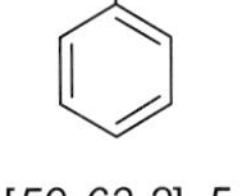

[59-63-2], 5-Methylisoxazole-3-carboxylic acid 2-benzylhydrazide, $C_{12}H_{13}N_3O_2$, M_r 231.25

Clinical Use, Pharmacokinetics and Metabolism: As an irreversible inhibitor of MAO-A and MAO-B, isocarboxazid has a very similar clinical profile as phenelzine. It is used in the treatment of depression, even though the risks associated with this class of *irreversible* inhibitors (Lecrubier, 1994) usually prevent it from being the first choice therapy. Isocarboxazid has very similar adverse effects and food and drug interactions like phenelzine.

Isocarboxazid is readily absorbed in the GI tract and – after extensive metabolism in the liver – is eliminated in the urine in form of its metabolites.

Tranylcypromine

Tranylcypromine

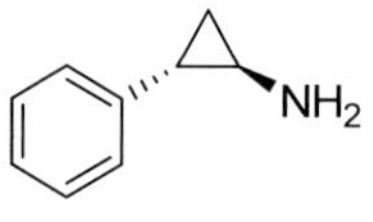

[*155-09-9*], (±)-*trans*-2-Phenylcyclopropanamine, $C_9H_{11}N$, M_r 133.19; [*13492-01-8*], *(sulfate salt)* $C_9H_{13}NO_4S$, M_r 231.27

- Non-selective MAO inhibitor developed by Smith Kline & French (Glaxo) (Tedeschi, 1961)
- Launched in 1961 for the treatment of depression

Synthesis (Tedeschi, 1961): The reaction of styrene with ethyl diazoacetate yields ethyl 2-phenylcyclopropanecarboxylate which is treated with sodium hydroxide to afford the carboxylic acid as a mixture of *trans* and *cis* isomers (3–4 parts of *trans* to 1 part of *cis*). After recrystallization from hot water to afford the pure *trans* isomer, the acid is refluxed with thionyl chloride and the resulting acid chloride is transformed into the final amine by a Curtius rearrangement with sodium azide and acidic treatment to hydrolyze the isocyanate.

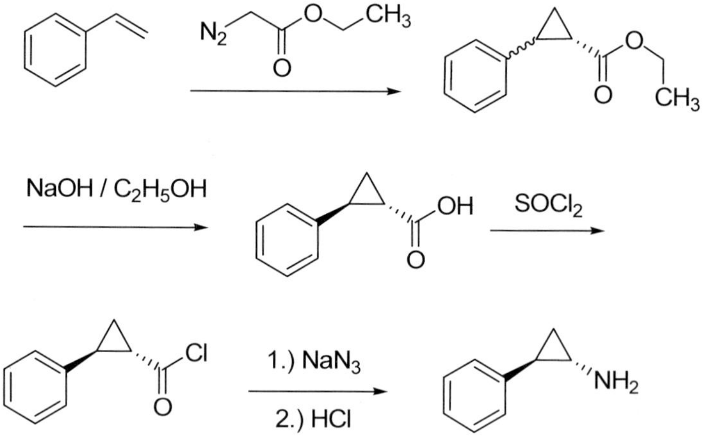

Scheme 46: Synthesis of tranylcypromine

Clinical Use and Pharmacokinetics: Tranylcypromine, even though not being a hydrazine, has a similar clinical profile as phenelzine, even though it produces a less pronounced enzyme inhibition. It is used as an antidepressant, even though given to the risks associated with the class of non-selective MAO inhibitors (Lecrubier, 1994); and in many cases it is not considered as the first choice. Most important adverse effects include stimulant effects like insomnia and hypertensive reactions.

Tranylcypromine is readily absorbed in the gastrointestinal tract, extensively metabolized and peak plasma concentrations are reached 1–3 hours after ingestion. Plasma elimination half-life has been reported to be 2.5 hours (Mallinger et al., 1986).

1.5.5.2 Monoamine oxidase type A inhibitors

A few newer MAOIs, especially moclobemide, have been designed to achieve selectivity for the monoamine oxidase type A subtype and reversibility (sometimes referred to as reversible inhibitors of monoamine oxidase A; RIMAs), meaning that they can inhibit the enzyme for a time, but eventually detach, allowing the enzyme to function once more. The following presents the RIMAs on the market, summarizing chemical properties, synthesis, important patents, clinical use as well as pharmacokinetics and metabolism.

Moclobemide

- MAO-A inhibitor first published in 1977 and launched in 1990 by Roche for the treatment of depression

Moclobemide

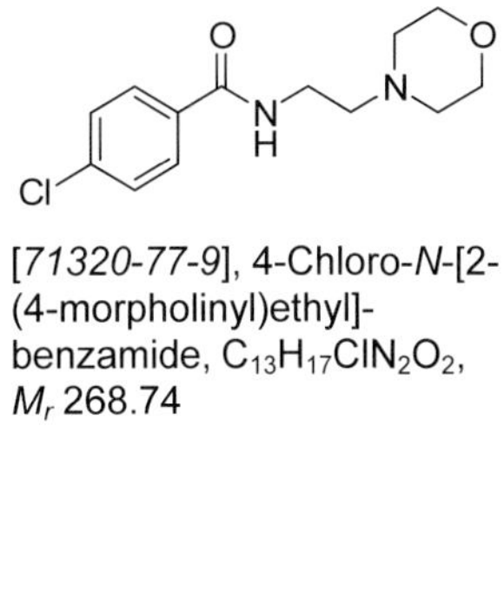

[71320-77-9], 4-Chloro-N-[2-(4-morpholinyl)ethyl]-benzamide, $C_{13}H_{17}ClN_2O_2$, M_r 268.74

Synthesis (Burkard and Wyss, 1977): This compound can be prepared in a huge variety of different ways, for instance by reaction of *p*-chlorobenzoyl chloride with *N*-(2-aminoethyl)-morpholine in pyridine.

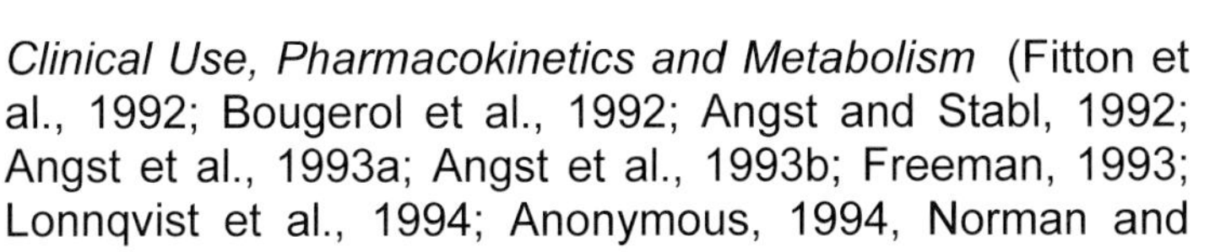

Scheme 47: Synthesis of moclobemide

Clinical Use, Pharmacokinetics and Metabolism (Fitton et al., 1992; Bougerol et al., 1992; Angst and Stabl, 1992; Angst et al., 1993a; Angst et al., 1993b; Freeman, 1993; Lonnqvist et al., 1994; Anonymous, 1994, Norman and Burrows, 1995): As a reversible inhibitor of monoamine oxidase type A, moclobemide is predominantly used as alternate therapy to tricyclic antidepressants for major depressions. Adverse effects comprise sleep disturbances, dizziness, nausea, and headache. Moclobemide is readily absorbed but faces a pronounced hepatic first-

Toloxatone

Toloxatone

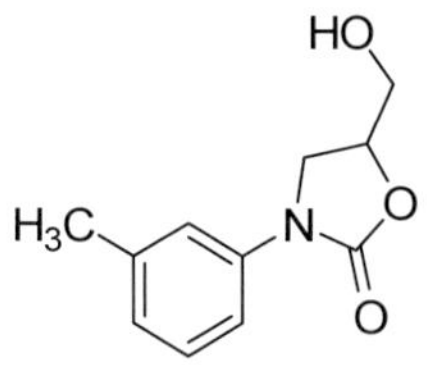

[29218-27-7], (±)-5-(Hydroxymethyl)-3-(3-methylphenyl)-oxazolidin-2-one, $C_{11}H_{13}NO_3$, M_r 207.22

pass effect well as extensive metabolism. Plasma elimination half-lifes detected are between 1 and 2 hours.

Toloxatone

- First patent filed in 1969 (Fauran et al., 1970)
- Reversible and selective MAO-A inhibitor launched in 1984 by Delalande S.A. for the treatment of depression

Synthesis (Castaner and Sungurbey, 1976): first synthesis of toloxatone is afforded by condensation of *m*-toluidine and glycidol to give 3-(*m*-toluidino)-1,2-propanediol, which is then cyclized by condensation with ethyl carbonate and sodium methoxide.

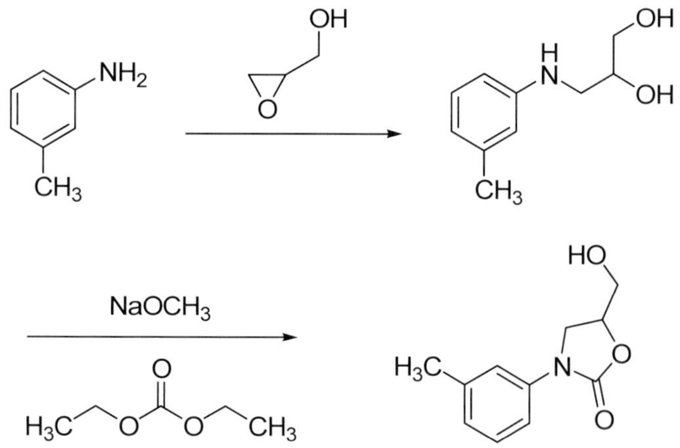

Scheme 48: Synthesis of toloxatone

In 2003, a Buchwald CuI-coupling protocol was used to synthesize toloxatone (Mallesham et al., 2003).

Clinical Use, Pharmacokinetics and Metabolism (Benedetti, 1982): As a member of the class of reversible MAO-A inhibitors, Toloxatone has its use in major depression therapy as an alternative to the classic TCA treatment. Its clinical action profile and use is very similar to Moclobemide. Toloxatone is extensively eliminated in urine as metabolites bieng the plasma half-life of 0.96–1.81 hours. The reduced availability of the drug appears to be related mainly to first-pass metabolism.

1.5.6 Miscellaneous agents

This chapter lists agents which could not be attributed structurally or by their mode of action to one of the other groups. They comprise lithium therapy, carbamazepin, bupropion, divalproex sodium, viloxazine and lamotrigine.

Lithium therapy

- Lithium and lithium salts (especially carbonate or citrate) were the focus of research in depression topics in the 1930s
- Lithium carbonate was launched in 1939 by Solvay for the treatment of depression and schizophrenia
- In 2002, Glaxo launched lithium carbonate for the treatment of mania

Lithium carbonate

[554-13-2], Carbonic acid dilithium salt, CLi_2O_3, M_r 73.89

Clinical Use, Pharmacokinetics and Metabolism: Lithium is usually used in form of its carbonate salt in prophylaxis and treatment of bipolar disorders (manic depression), in mania and in the maintenance treatment of unipolar disorders (recurrent depression) as an alternate treatment to TCAs or newer antidepressants (Schou, 1987). Due to the relatively low toxic lithium blood serum concentration, the therapeutic window is quite narrow, resulting in the need to monitor serum concentrations during therapy (medical supervision). Recommended therapeutic serum concentrations are usually in the range 0.4–1.4 mmol per liter, slightly increased concentrations are needed for the treatment of acute mania. Lithium therapy has been tried in a variety of disorders, e.g. depression (s.a.), anxiety, bipolar disorders (Johnson, 1987; Schou, 1989; Aronson and Reynolds, 1992; Birch et al., 1993; Peet and Pratt, 1993; Price and Heninger, 1994), disturbed behavior (Schou, 1987), headache or schizophrenia. Side-effects with lithium therapy are manifold, the most important ones being GI upsets such as nausea and vomiting, but also more severe ones like effects on the CNS, like tremor, ataxias, or polyuria. Further side-effects described comprise effects on the blood (Collins, 1992), the cardiovascular system (Montalescot et al., 1984; Palileo et al., 1983, Tangedahl and Gau, 1972; Martin and Piascik, 1985; Demers and Henninger, 1971), the endocrine system (Myers and West, 1987; Nordenström et al., 1992; Taylor and Bell, 1993; Pandit et al., 1993), eyes (Fraunfelder et al., 1992), kidneys (Walker and Kincaid-Smith, 1987; George, 1989; Schou, 1988a; Schou,

1988b), musculoskeletal system (Baandrup et al., 1987), nervous and neuromuscular systems (Sansone and Ziegler, 1987; Johns and Harris, 1984; Himmelhoch and Hanin, 1974; Solomon and Vickers, 1975; Worral and Gilham, 1983; McGovern, 1983), respiration (Weiner et al., 1983), sexual function and fertility (Beeley, 1984; Raoof et al., 1989; Salas et al., 1989) and skin and hair (Lambert and Dalac, 1987; Fraunfelder et al., 1992), as well as epileptogenic effects (Demers et al., 1970; Massey and Folger, 1984) and lupus (Johnstone and Whaley, 1975; Presley et al., 1976; Shukla and Borison, 1982). There are several references available for the treatment of side-effects related to lithium therapy, especially when overdosing (Gomolin, 1987; Amdisen, 1988; Okusa et al., 1994; Swartz and Jones, 1994; Wells, 1994). In general, when applying lithium therapy, a lot of precautions need to be considered, especially concerning drug interactions and side-effects. Lithium is readily absorbed from the GI tract, peak serum concentrations being reached between 0.5–3.0 hours after ingestion (for conventional tablets or capsules). Within 6–10 hours, lithium is distributed throughout the whole body. Lithium is excreted in unchanged form mainly in urine, elimination half-lifes ranging from 10 to 50 (!) hours (Johnson, 1987; Ward et al., 1994; Reiss et al., 1994).

Carbamazepin

Carbamazepin

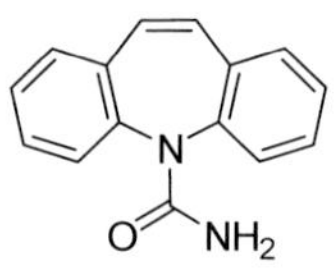

[298-46-4], 5H-Dibenzo[d,f]azepine-5-carboxamide, $C_{15}H_{12}N_2O$, M_r 236.27

- Tricyclic compound first patented in 1960 (Schindler, 1960) and launched in 1964 for the treatment of epilepsy, neuralgia and psychosis

- Patented (Rudnic et al., 1999) and launched for the treatment of bipolar disorder in 2005 by Shire Pharmaceuticals

Synthesis (Schindler, 1960): From 5H-dibenzo[b,f]-azepin and phosgene, and final treatment with ammonia in ethanol.

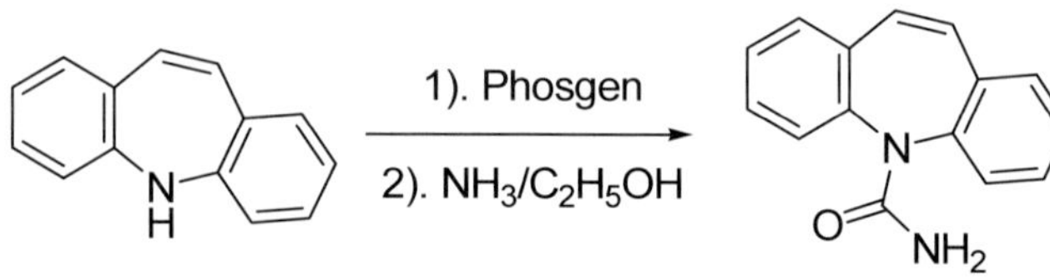

Scheme 49: Synthesis of carbamazepin

Clinical Use, Pharmacokinetics and Metabolism: Structurally related to the TCA class, carbamazepine was originally developped as an anti-epilectic agent (McKee et al., 1991; Anonymous, 1989; Ryan et al., 1990; McKee et al., 1993; Mattson et al., 1985; Gilman, 1991; Mattson et al., 1992; Richens et al., 1994; Heller et al., 1995; Verity et al., 1995). It is also used widely for the treatment of manic depression (prophylaxis of bipolar disorders not responding to lithium therapy; Tobiansky and Shah, 1989; Ballenger, 1988; Frankenburger et al., 1988; Murphy et al.; 1989; Lusznat et al., 1988; Kramlinger and Post, 1989; Di Costanzo and Schifano, 1991; Mattes, 1990; Gleason and Schneider, 1990; Schipperheijn and Dunne, 1991; Wolf et al., 1988; Sramek et al., 1988; Dose et al., 1987), neuralgias (Skelton, 1988), movement disorders (Roig et al., 1988; Roulet and Deonna, 1989; Anonymous, 1988), diabetes insipidus (Seckl and Dunger, 1989), multiple sclerosis (Kuroiwa and Shibasaki, 1967, Espir and Walker, 1967; Albert, 1969; McFarling and Susac, 1974; Ostermann, 1976; Jacome, 1985; Sechi et al., 1987; Sechi et al., 1989), restless leg syndrome (Telstad et al., 1984), tinnitus and withdrawal syndromes (Schweitzer et al., 1991; Halikas et al., 1991; Malcolm et al., 1989; Guthrie, 1989). Common dose-related adverse effects comprise dizziness, drowsiness and visual disturbances (with ataxia, nystagmus and diplopia). In addition, effects on the blood (Sobotka et al., 1990), bones, the endocrine system (Henry et al., 1977; Stephens et al., 1977; Ashton et al., 1977; Smith et al., 1977; Ballardie and Mucklow, 1984; Sørensen and Hammer, 1984; Mucklow, 1991; Stephens et al., 1978), eyes (Anonymous, 1976; Nielsen and Syversen, 1986), heart (Kasarkis et al., 1992; Stone and Lange, 1987; Ambrosi et al., 1993), liver (Hadžić et al., 1990; Zucker et al., 1977; Smith et al., 1988), mental function (Berger, 1971; Mathew, 1988; McKee et al., 1989; Reiss and O'Donnel, 1984; Gardner and Cowdry, 1986), nervous system (Hilton and Stroh, 1989; Simon et al., 1990, Crosley and Swender, 1979; Jacome, 1979; Joyce and Gunderson, 1980; Neglia et al., 1984) and skin (Busch, 1989; Keating and Blahunka, 1995) have been described, as well as hypersensitivity (Anonymous, 1991; Hosoda et al., 1991; Merino et al., 1994) and systemic lupus erythematosus (Jain, 1991). Carbamazepine should be avoided in patients with atrioventricular conduction abnormalities or in patients with a history of bone marrow depression, and given with care to patients with cardiac, renal or hepatic diseases. Many interactions with other drugs are discribed in the literature. Carbamazepine is slowly and irregularly absorbed from the GI tract, extensively metabolized in the liver. One of the main metabolites is the 10,11-epoxide, which is active as well. Excretion happens nearly exclusively in urine in the form of metabolites. Carbamazepine is an inductor of its own metabolism, so that half-lifes after repeated administrations may be considerably lower (10–20 hours) as compared with those observed during the first days of treatment. Carbamazepine shows extense plasma protein binding (Schmidt and Haenel, 1984; Bertilsson and Tomson; 1986; Kodama et al., 1993; Bernus et al., 1994).

Bupropion

Bupropion

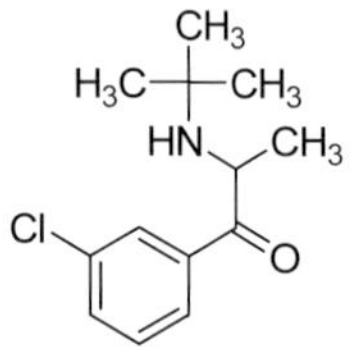

[*34911-55-2*], (±)-1-(3-Chlorophenyl)-2-(tert-butylamino)propan-1-one, $C_{13}H_{18}ClNO$, M_r 239.74; [*31677-93-7*] (hydrochloride salt), $C_{13}H_{19}Cl_2NO$, M_r 276.20

- Compound patent filed in 1969 (Mehta, 1974) and launched in 1989 for depression, and in 2003 for major depression by Glaxo

- The molecular mechanism of bupropion is unknown: it does not inhibit MAO and, compared with classic tricyclic antidepressants, is a weak blocker of the neuronal uptake of serotonin and norepinephrine. Buproprion also weakly inhibits the neuronal reuptake of dopamine

Synthesis (Mehta, 1974; Castaner and Hopkins, 1978): the Grignard condensation of *m*-chlorobenzonitrile with ethyl magnesium bromide gives *m*-chloropropiophenone, which is then treated with sulfuryl chloride or brominated to afford an alpha-halo-*m*-chloropropiophenone which is finally reacted with *tert*-butylamine.

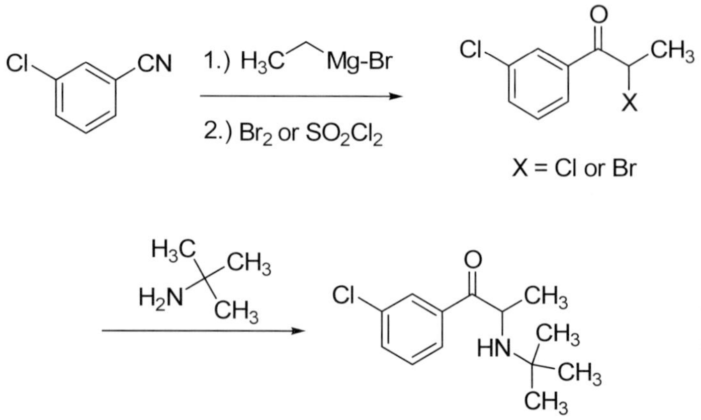

Scheme 50: Synthesis of bupropion

Clinical Use, Pharmacokinetics and Metabolism: The mechanism of action of bupropion is not well understood. It is only a weak inhibitor of neuronal serotonin and noradrenaline uptake and dopamine reuptake. Bupropion is used as alternative antidepressant treatment to the TCAs and usually needs an induction of up to 4 weeks until a clinical effect becomes observable (Anonymous, 1989; Weisler et al., 1994). When treating bipolar disorders, limited evidence suggests that bupropion possess the same risks as other antidepressants to evoke manic episodes (Fogelson et al., 1992). Bupropion has demonstrated use in nicotine withdrawal (smoking cessation; Ferry et al., 1992). As common side-effects, i.a., insomnia, agitation, headache or migraine have been reported. It is contraindicated in patients with epilepsy due to its potency to induce seizures. Bupropion is well absorbed from the GI tract, but may undergo extensive first-pass metabolism. Terminal plasma half-life is relatively long, about 14 h, and some metabolites are pharmacologically active and possess even longer half-lifes. In animals, bupropion has shown to induce its own metabolism.

Divalproex Sodium

- Patented (Fischer, 1983) and launched by Abbott for the treatment of epilepsy (1983), bipolar disorder (1995) and migraine (1996)

Synthesis (Chignac et al., 1978, 1979): 2-propylpentanoic acid synthesis was developed and patented by French laboratories Labaz. The preparation of the sodium salt was finally patented by Abbott (Fischer, 1983). Intermediate for the synthesis of divalproex sodium is 2-propylpentanenitrile that can be obtained by different ways, for instance from reaction of cyanacetic acid ethyl ester with propyl bromide by means of sodium propoxide in propanol to give a dipropylcyanacetic acid ethyl ester, which is decarboxylated at 140–190°C to afford 2-propylpentanenitrile. This is hydrolyzed with sodium hydroxide or sulfuric acid to valproic acid, which is reacted with sodium hydroxide in water-acetonitrile to afford divalproex sodium.

Divalproex sodium

[76584-70-8], 2-Propylpentanoic acid sodium salt (2:1), $C_{16}H_{31}NaO_4$, M_r 310.40

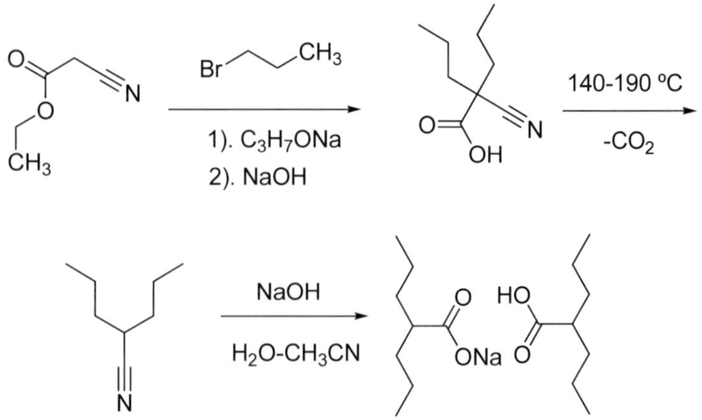

Scheme 51: Synthesis of divalproex sodium

Clinical Use, Pharmacokinetics and Metabolism: Valproic acid and its salts and esters are used in the treatment of various types of epilepsy, in particular primary generalized seizures and absence and myoclonic seizures (Mattson et al., 1992; Davis et al., 1994; Richens et al., 1994; Heller et al., 1995; Verity et al., 1995; Newton, 1988; Bauer and Elger, 1994; Berkovic et al., 1989; Giroud et al., 1993) and for the treatment of bipolar disorders as well as for the control of symptoms like rage or agitation and other psychiatric disorders (Pope et al., 1991; Keck et al., 1993; McElroy and Keck, 1993; Joffe, 1993; Schaff et al., 1993; Narnberg et al., 1994; Anonymous, 1994; Bowden et al., 1994; Stoll et al., 1994; Geracioti, 1994; Primeau et al., 1990; Keck et al., 1993; Woodman and Noyes, 1994; Fesler, 1991). It also has demonstrated use in migraine (Sørensen, 1988; Hering and Kuritzky, 1992; Coria et al., 1994; Mathew et al., 1995; Mathew and Ali, 1991; Elkind and Indelicato, 1995), hiccups (Jacobson et al., 1981) and muscle spasm (Zachariah et al., 1994). The mode of action of valproic acid is not fully understood. Most common adverse effects include considerable liver toxicity, especially in children (Powell-Jackson et al., 1984; Eadie et al., 1988; Driefuss et al., 1989), GI disturbances and effects on blood (Jaecken et al., 1979; Barr et al., 1982; Coulter et al., 1982; Symon and Russell, 1983; MacDoughall, 1982; Watts et al., 1990; Kreuz et al., 1990), bones (Sheth et al., 1995), endocrine system, nervous system (Lautin et al.; Dick and Saunders, 1980; Marescaux et al., 1982), pancreas (Wyllie et al., 1984; Committee on Safety of Medicines/Medicines Control Agency, 1994) and skin (Jeavons et al., 1977; Gupta, 1988; Gillman and Sandyk, 1984; Lewis-Jones et al.,

1985; Kamper et al., 1991) as well as enuresis (Panayiotopoulos, 1985; Choonara, 1985). Concerning precautions while treatment and drug interactions, manyfold references have been published. Valproic acid as well as its salts are rapidly and completely absorbed from the GI tract and extensively metabolized in the liver, partly by glucuronidation, partly by complex metabolic pathways. The metabolism of valproic acid may be enhanced by other drugs, which induce hepatic enzymes. It is excreted nearly exclusively in form of metabolites in urine (Zaccara et al., 1988; Bialer, 1991; Cloyd et al., 1993; Yoshiyama et al., 1989).

Viloxazine

- Patented (Mallion et al., 1973) and launched in 1977 by Imperial Chemical Industries (ICI) in the UK, which demerged its chemical business to a new company, Zeneca Group, in 1993. Zeneca merged with Astra in 1999

Viloxazine

[46817-91-8], (±)-2-(2-Ethoxyphenoxymethyl)morpholine, $C_{13}H_{19}NO_3$, M_r 237.29; [35604-67-2] (hydrochloride salt), $C_{13}H_{20}ClNO_3$, M_r 273.75

Synthesis (Mallion et al., 1973): Reaction of 2-ethoxyphenol with epichlorhydrin in presence of sodium hydroxide gives 1-(2-ethoxyphenoxy)-2,3-epoxypropan which is treated with benzylamine and then with chloroacetyl chloride to give a morpholinone which is reducted with lithium aluminum hydride and then with hydrogen on palladium charcoal to the final viloxazine.

Clinical Use, Pharmacokinetics and Metabolism: Being a bicyclic antidepressant, Viloxazine is devoid of pronounced antimuscarinic or sedative properties. Its antidepressant mechanism of action includes reuptake effects of biogenic amines in the CNS, but is not fully understood. Viloxazine is used as an alternative therapy to classic TCA depression treatment. Typical side-effects comprise nausea, vomiting and headache, and in rare cases epilepsy (Edwards and Glen-Bott, 1984) or migraine (Nightingale, 1985), but significantly fewer typical antimuscarinic side-effects (e.g., dry mouth, tachycardia, constipation, etc.) as e.g. with the TCAs are displayed. Viloxazine is readily absorbed (GI tract) and extensively metabolized by hydroxylation and phase II metabolisation. Unlike the classic TCAs, viloxazine has a short half-life (commonly 2–5 h).

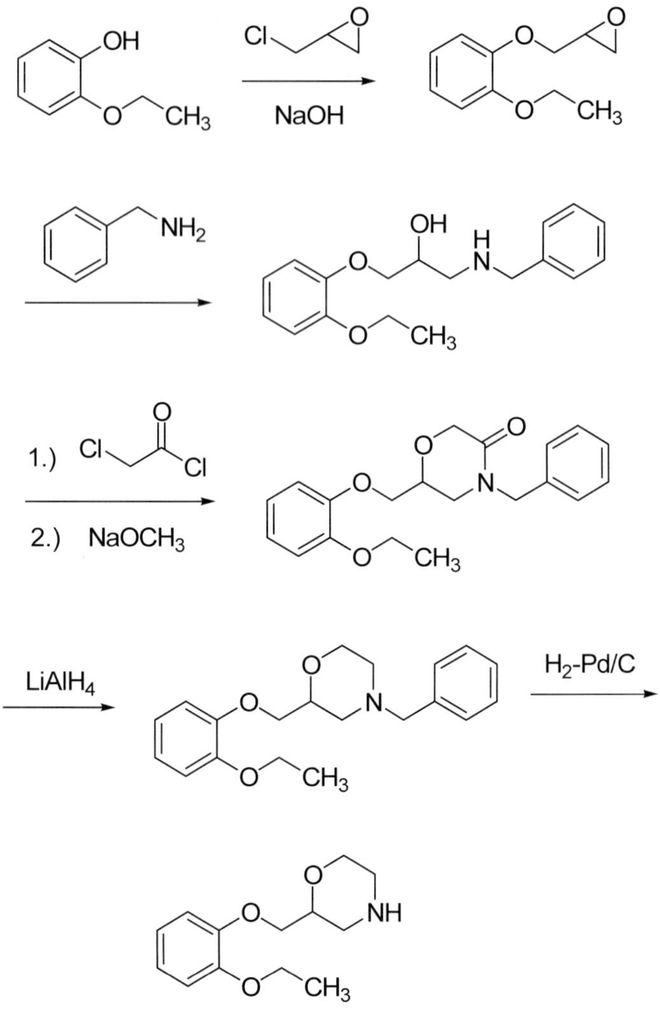

Scheme 52: Synthesis of viloxazine

Lamotrigine

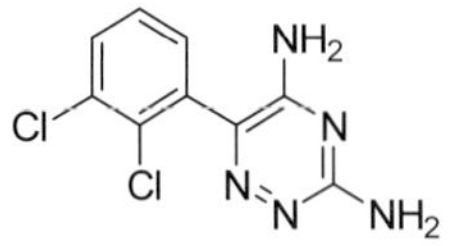

Lamotrigine

[84057-84-1], 6-(2,3-Dichlorophenyl)-1,2,4-triazine-3,5-diamine, $C_9H_7Cl_2N_5$, M_r 256.09

- Compound patent filed in 1979 by GlaxoSmithKline (Baxter et al., 1981)
- Initially launched by GlaxoSmithKline in 1990 for the treatment of epilepsy. In 2003 it was launched as maintenance treatment of bipolar disorder. The product is currently in phase III clinical trials as adjuvant therapy for the once-daily treatment of schizophrenia and neuropathic pain

Synthesis (Baxter et al., 1981): The Grignard reaction of 2,3-dichloroiodobenzene with carbon dioxide gives 2,3-dichlorobenzoic acid, which is converted to 2,3-dichlorobenzoyl chloride by reaction with thionyl chloride. Reaction with cuprous cyanide in the presence of potassium iodide yields 2,3-dichlorobenzoyl cyanide. This is cyclized with aminoguanidine to afford lamotrigine.

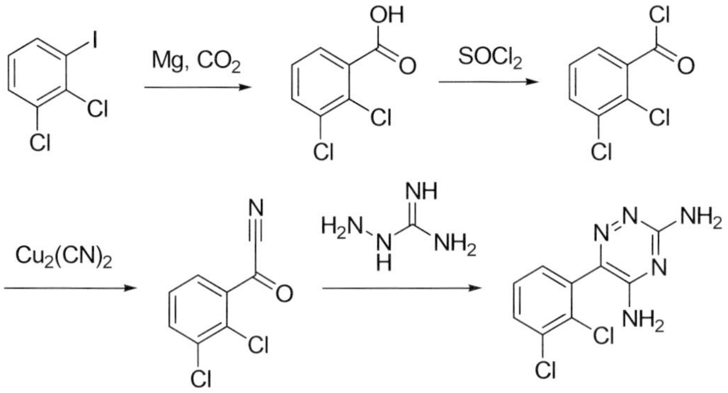

Scheme 53: Synthesis of lamotrigine

Clinical Use, Pharmacokinetics and Metabolism: Lamotrigine is – like valproic acid (divalproex) – mainly used as antiepileptic, but has some use in depression as well. It is a phenyltriazine compound used in the treatment of partial seizures with or without secondary generalization, in monotherapy or as an adjunctive, e.g. together with valproic acid (Reutens et al., 1993; Richens and Yuen, 1991; Brodie, 1992; Goa et al., 1993; Brodie, 1993; Gilman, 1995; Anonymous, 1995; Brodie et al., 1995; Ferrie et al., 1995; Veggiotti et al., 1994). Adverse effects include skin rashes and other skin reactions, GI disturbances, drowsiness and dizziness (Betts et al., 1991; Schaub et al., 1994; Nicholson et al., 1995; Makin et al., 1995). Symptoms of overdosage are relatively mild (Buckley et al., 1993). Contraindications are in patients with hepatic impairment and drug interactions have been described e.g. with paracetamol (Depot et al., 1990). Lamotrigine is well absorbed from the GI tract after ingestion, peak plasma concentrations being reported around 2.5 hours thereafter. It undergoes intense metabolism in the liver and is widely distributed throughout the whole body and is bound to plasma protein in about 55%. Half-life has been reported to be about 24 hours (Rambeck and Wolf, 1993).

1.5.7 Compounds Launched in Single Countries

Some compounds, especially representatives from the tricyclic antidepressants group like dibenzepine (Noveril®, Novartis), dimetacrine (Istonyl®, Nippon Chemiphar), melitracen (Thymeol®, Takeda), metapramine (Timaxel®, Specia), mianserin (Bolvidon®, Organon), quinupramine (Kinupril®, Bellon) and setiptiline (Tecipul®, Mochida) have not been launched worldwide or have been withdrawn in some countries.

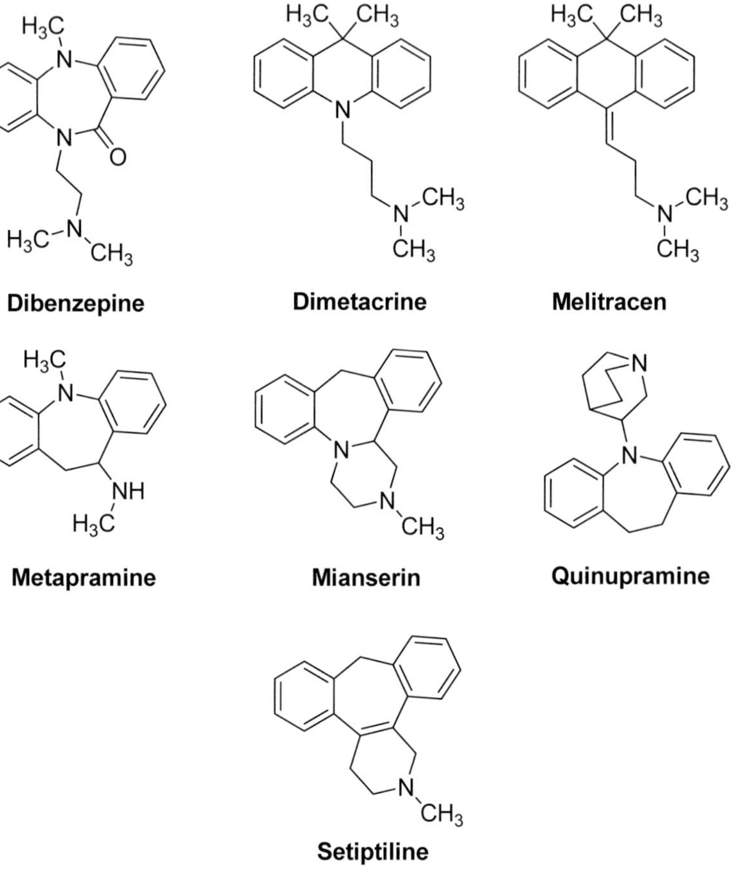

Dibenzepine **Dimetacrine** **Melitracen**

Metapramine **Mianserin** **Quinupramine**

Setiptiline

Scheme 54: Other TCAs

Structurally related to TCAs but with a distinct pharmacological mechanism (5-HT uptake stimulator), tianeptine (Stablon®) has been marketed in France since 1988 and was launched in South Korea in 1998.

Tianeptine (Servier)

1.5.8 New opportunities for marketed drugs

Marketed antipsychotics and anxiolytics used against depression or bipolar disorder

Some compounds initially launched for the treatment of schizophrenia like olanzapine (discovered and developed by Lilly), risperidone (originally developed by Janssen), quetiapine (Astra Zeneca), flupentixol (Lundbeck), 5-HT$_{2A}$ and D$_2$ antagonists, have also been marketed for the treatment of moderate to severe manic episodes associated with bipolar disorder or depression. Tandospirone (Dainippon Sumitomo Pharma/Pfizer), a 5-HT$_{1A}$ agonist has been marketed as antipsychotic, anxiolytic and antidepressant. More information on these compounds can be found in the corresponding sections.

1.5.9 Summary of antidepressants in development

A large number of compounds are currently in different phases under development as new antidepressants. Table 1 details the code, structures (when known) and proposed mechanisms described in different drug databases (Integrity, IDBB3). Some mechanisms are already known from marketed compounds, as in the case of serotonin reuptake inhibitors, but others are novel as for instance corticotrophin-releasing factor-1 (CRF1) antagonists or tachykinin modulators.

New serotonin reuptake inhibitors, selective or non-selective have been described: Ifoxetine (phase II, Novartis) is clearly identified as selective SSRIs, Lu-AA-21004 (phase I, Lundbeck) and Viqualine (phase III) are described only as SRI, LAX-201 (phase II, Amarin/Laxdale) as SNRI, SLV-314 (phase I, Solvay/Wyeth) is a SRI/Dopamine D$_2$ antagonist developed for bipolar disorder by Solvay and Wyeth, DOV-102677 and DOV-21947 (both in phase I, enantiomers of DOV-216303,

DOV/Merck & Co.), SEP-225289 (phase I, Sepracor) and GSK-372475 (phase II, Glaxo/Neurosearch) are serotonin, noradrenaline and Dopamine reuptake inhibitors (SNDRIs), a new group in this field.

Other agents with activity in GPCR serotonin receptors in clinical development trials include 5-HT$_{1A}$ agonists as alnespirone (Servier) in phase II, PRX-00023 (Predix) and gepirone (Fabre-Kramer) in phase III; 5-HT$_{1A}$ antagonists as SB-163090 (Glaxo) in phase I, DU-125530 (Solvay) in phase II; 5-HT$_{1B/1D}$ antagonists as elzasonan (Pfizer) in phase II; 5-HT$_2$ antagonists as asenapine (Pfizer) in phase III and 5-HT$_{2C}$ antagonist as SB-247853 (Glaxo), in phase I. Also combinations with other targets are described, as for example VPI-013 (phase II, Vela), a SRI/5-HT$_{1A}$ agonist/sigma agonist, and bifeprunox (phase III, Lundbeck/Solvay/Wyeth), a 5-HT$_{1A}$ agonist/D$_2$ antagonist. Vilazodone (Merck KGaA/Genaissance) has been described as a SRI and partial 5-HT$_{1A}$ agonist in phase III trials.

Oxaprotiline is the unique representative in clinical trials as a member of the noradrenaline reuptake inhibitor (NARI) family.

Radafaxine, a potent metabolite of bupropion, is a dopamine and noradrenaline reuptake inhibitor that is currently in phase II trials for depression, and in phase I trials for obesity, fibromyalgia and neuropatic pain at Glaxo. Phase III was expected to begin in 2006.

Dopaminergic agents as RGH-188 (Gedeon Richter and Forest), a dopamine D$_{2/3}$ antagonist, was in phase I trials in the United Kingdom for the treatment of schizophrenia and bipolar mania and phase III trials were planned to begin before 2006. Pramipexole (Boehringer Ingelheim), a D$_3$ agonist, is in phase II trials.

Neboglamine and ACPC are two NMDA glycine-site modulators in phase I trials. Neboglamine is in clinical development for the treatment of schizophrenia but preclinical studies are also under way at Rottapharm for the treatment of depression.

At least five drugs have been described as potential antidepressants in phase I trials as CRF1 antagonists: JNJ-19567470 (Taisho and Janssen), ONO-2333MS (Ono), SSR-125543 (Sanofi-Aventis), GW-876008 (Neurocrine/Glaxo), DMP-904 (Bristol-Myers Squibb). In phase II, two agents are under development: CP-316311 (Pfizer) and BMS-562086 (Bristol-Myers Squibb).

Two tachykinin NK1 modulators have been described in phase I development for depression: E-6006/E-6039 from Laboratorios Dr. Esteve, and GW-823296 from Glaxo and three in phase II: Vesipitant and Casopitant (Glaxo), and CP-122721 (Pfizer). Vestipitant has entered phase II trials for the treatment of depression, anxiety, as well as chemotherapy-induced nausea and vomiting. Saredutant, from Sanofi-Aventis described as tachykinin NK2 modulator, is currently in phase III trials.

Ampakine (AMPA) is a novel target also described in some compounds that are currently in phase II trials. Modulators of this target are CX-516 and farampator, developed by Cortex.

A miscellaneous group of agents have been described with alternative mechanisms as peripheral benzodiazepine ligands (emapunil/AC-5216, phase II, Dainippon/Novartis), phosphodiesterase II inhibitors (ND-7001, phase I, Neuro3d), phosphodiesterase IV inhibitors (rolipram, phase III), NMDA antagonists (delucemine, phase I, NPS), vasopressin V1b antagonist (SSR-149415, phase II, Sanofi-Aventis), GABA modulators (DP-VPA, phase II, D-Pharm; adinazolam, phase III, Pfizer), MAO reuptake inhibitor (medifoxamine, phase I, Lipha), MAO-A inhibitor (brofaromine; phase III, Novartis), acetylcholine release enhancer (SA-4503, phase II, M's Science), melatonin agonist (VEC-162, phase II, Vanda/Bristol-Myers Squibb), sodium channel blocker (GW-273293, phase II, Glaxo; licarbazepine and eslicarbazepine, both in phase III, Novartis), L-type calcium channel blocker (MEM-1003, phase II, Memory Pharmaceuticals), glucocorticoid antagonist (ORG-34517, phase II, Organon) and β_3-adrenoceptor agonists (SR-58611, phase III, Sanofi-Aventis).

Finally, other compounds without a clearly defined mechanism in phase I trials are NPS-1776 (NPS), Uridine (Repligen), YKP-581 (Johnson & Johnson and SK Biopharmaceuticals), and Lu-AA-24530 (Lundbeck). In phase II: YKP-10A (Johnson & Johnson and SK Biopharmaceuticals), piberaline (Egis), GPE (Neuren Pharmaceuticals), miraxion (Amarin). In phase II/III: nemifitide (Tetragenex); and in phase III: teniloxazine (Mitsubishi Pharma), PW-4112 (Penwest) and clovoxamine (Duphar). Also St. John's Wort extract from the plant *Hypericum perforatum* has been described as a potential antidepressant in phase III studies.

Drugs near to the market at the moment of writing are the pre-registered SNRI desvenlafaxine, the metabolite of venlafaxine (Effexor®, Wyeth), quetiapine (Astra Zeneca) an antipsychotic, 5-HT$_{2A}$/D$_2$ antagonist launched in 1974 and currently in phase III trials for bipolar disorder, agomelatine, a 5-HT$_{2C/2B}$ antagonist developed by Servier, and selegiline, an approved MAO-B inhibitor of Somerset/Bristol-Myers Squibb. Finally, Virucort (Procaine) from Cortisol, was recently approved for the treatment of depression in an HIV population of children aged 5-14.

Table 1. Compounds currently in different development phases

Code or name	Mechanism	Structure
Phase I		
Lu-AA-21004	SRI	Not known
SEP-225289	SNDRI	Not known
DOV-102677	SNDRI	
DOV-21947	SNDRI	
SLV-314	SRI / D$_2$ antagonist	
SB-163090	5-HT$_{1A}$ antagonist	Not known
SB-247853	5-HT$_{2C}$ antagonist	
RGH-188	D$_2$ / D$_3$ antagonist	Not known

Neboglamine	NMDA glycine-site modulator	(structure)
ACPC	NMDA glycine-site partial agonist	(structure)
JNJ-19567470	CRF1 antagonist	(structure)
ONO-2333MS	CRF1 antagonist	Not known
SSR-125543	CRF1 antagonist	(structure)
GW-876008	CRF1 antagonist	Not known
DMP-904	CRF1 antagonist	(structure)
E-6006/E-6039	Tachykinin modulator	(structure)
GW-823296	Tachykinin NK1 antagonist	Not known
ND-7001	Phosphodiesterase II inhibitor	Not known

Name	Mechanism	Structure
Medifoxamine	Monoamine reuptake inhibitor	(structure)
Delucemine	NMDA antagonist	(structure)
Isovaleramide NPS-1776	Not known	(structure)
Uridine	Not known	(structure)
YKP-581	Not known	Not known
Lu-AA-24530	Not known	Not known
Phase II		
Ifoxetine	SSRI	(structure) · H$_2$SO$_4$
OPC-14523	SRI, 5-HT$_{1A}$ agonist, sigma agonist	(structure)
DU-125530	5-HT$_{1A}$ antagonist	(structure)

Name	Target	Structure
Alnespirone	5-HT$_{1A}$ agonist	
Elzasonan	5-HT$_{1B/1D}$ antagonist	
VPI-013	SRI /5-HT$_{1A}$ agonist/sigma agonist	
GSK-372475 // NS-2359	SNDRI	Not known
DOV-216303	SNDRI	
LAX-201	SNRI	Not known
Radafaxine	Dopamine reuptake inhibitor/NARI	
Pramipexole	D$_3$ agonist	
BMS-562086	CRF1-antagonist	Not known
CP-316311	CRF1-antagonist	Not known

Vestipitant	Tachykinin NK1 antagonist	
Casopitant	Tachykinin NK1 antagonist	
CP-122721	Tachykinin NK1 antagonist	
SSR-149415	Vasopressin V1b antagonist	
DP-VPA	GABA modulator	
SA-4503	Acetylcholine release enhancer/sigma 1 agonist	

Marketed Drugs and Drugs in Development

Name	Target	Structure
VEC-162	Melatonin agonist	Not known
GW-273293	Sodium Channel blocker	(structure: trichlorophenyl-diaminopyrazine)
MEM-1003	L-type calcium channel	Not known
ORG-34517	Glucocorticoid antagonist	(steroid structure with methylenedioxyphenyl and propynyl-hydroxyl groups)
CX-516	AMPA Modulator	(quinoxaline-piperidine carbonyl structure)
Farampator	AMPA Modulator	(benzofurazan-piperidine carbonyl structure)
Emapunil	Peripheral benzodiazepine ligand	(phenyl-purinone with N-benzyl-N-ethyl acetamide)
R-228060 // YKP-10A	Phenylalanine derivative	Not known
LAX-101/Miraxion	Purified Omega-3 (EPA)	(ethyl ester of eicosapentaenoic acid)
Piberaline	Not known	(benzyl-piperazine-pyridine carbonyl structure)

GPE	Not known	

Phase II/III

Nemifitide	Analog of melanocyte-inhibiting factor (MF-1)	

Phase III

Cianopramine	TCA-SSRI	
Viqualine	SRI	
Vilazodone	SRI / 5-HT$_{1A}$ agonist	
PRX-00023	5-HT$_{1A}$ agonist	

Name	Activity	Structure
Gepirone	5-HT$_{1A}$ agonist	
Bifeprunox	5-HT$_{1A}$ agonist/ D$_2$ antagonist	
Asenapine	5-HT$_2$ antagonist	
SLV-308	5-HT$_{1A}$ agonist D$_2$ partial agonist α$_1$-adrenoceptor agonist α$_2$ adrenoceptor antagonist	
Oxaprotiline	NARI	
Clorgyline	MAO-Ai	
Brofaromine	MAO-Ai	
Licarbazepine	Sodium channel blocker	

Name	Target	Structure
Eslicarbazepine	Sodium channel blocker	(structure)
SR-58611	β3-adrenoceptor agonist	(structure)
Saredutant	Tachykinin NK2 antagonist	(structure)
Adinazolam	GABA(A) BZ Site Receptor Agonist	(structure)
Teniloxazine	Not known	(structure)
PW-4112	Not known	(structure)
St. John's Wort extract	Acetylcholinsterase inhibitor / weak SSRI / sigma 1	Not known
Clovoxamine DU-13811	Not known	(structure)

Pre-registered		
Desvenlafaxine	SNRI	
Quetiapine	5-HT$_{2A}$ / D$_2$ antagonist	
Agomelatine S-20098	5-HT$_{2C}$ / 5-HT$_{2B}$ antagonist	
Selegiline	MAO-B inhibitor	
Registered		
Virucort /Procaine	Not known	

References

Abernethy DR, Greenblatt DJ, Shader RI. *Trimipramine kinetics and absolute bioavailability: use of gas-liquid chromatography with nitrogen-phosphorus detection.* Clin Pharmacol Ther. **1984**, 35:348-353.

Ahmad S. *Fluoxetine and glaucoma.* DICP. **1991**, 25:436.

Albert ML. *Treatment of pain in multiple sclerosis--preliminary report.* N Engl J Med. **1969**, 280:1395.

Alderman CP, Moritz CK, Ben Tovim DI. *Abnormal platelet aggregation associated with fluoxetine therapy.* Ann Pharmacother. **1992**, 26:1517-1519.

Alkalay D, Wagner WE, Jr., Carlsen S, Khemani L, Volk J, Bartlett MF, LeSher A. *Bioavailability and kinetics of maprotiline.* Clin Pharmacol Ther. **1980**, 27:697-703.

Altamura AC, Moro AR, Percudani M. *Clinical pharmacokinetics of fluoxetine*. Clin Pharmacokinet. **1994**, 26:201-214.

Amat M, Bosch J, Hidalgo J, Canto M, Perez M, Llor N, Molins E, Miravitlles C, Orozco M, Luque J. *Synthesis of enantiopure trans-3,4-disubstituted piperidines. An enantiodivergent synthesis of (+)- and (-)-paroxetine.* J Org Chem. **2000**, 65:3074-3084.

Ambrosi P, Faugere G, Poggi L, Luccioni R. *Carbamazepine and pacing threshold*. Lancet. **1993**, 342:365.

Amdisen A. *Clinical features and management of lithium poisoning*. Med Toxicol Adverse Drug Exp. **1988**, 3:18-32.

Andersen G, Vestergaard K, Riis JO. *Citalopram for post-stroke pathological crying*. Lancet. **1993**, 342:837-839.

Andersen G, Vestergaard K, Lauritzen L. *Effective treatment of poststroke depression with the selective serotonin reuptake inhibitor citalopram*. Stroke. **1994**, 25:1099-1104.

Anderson IM, Tomenson BM. *Treatment discontinuation with selective serotonin reuptake inhibitors compared with tricyclic antidepressants: a meta-analysis*. BMJ. **1995**, 310:1433-1438.

Anderson RP, Morris BA. *Acrocyanosis due to imipramine*. Arch Dis Child. **1988**, 63:204-205.

Andreoli V, Caillard V, Deo RS, Rybakowski JK, Versiani M. *Reboxetine, a new noradrenaline selective antidepressant, is at least as effective as fluoxetine in the treatment of depression*. J Clin Psychopharmacol. **2002**, 22:393-399.

Angst J, Stabl M. *Efficacy of moclobemide in different patient groups: a meta-analysis of studies*. Psychopharmacology (Berl). **1992**, 106 Suppl:S109-S113.

Angst J, Gachoud JP, Gasser UE, Kohler M. *Antidepressant therapy with moclobemide in primary care practice*. Hum Psychopharmacol Clin Exp. **1993a**, 8:319-325.

Angst J, Scheidegger P, Stabl M. *Efficacy of moclobemide in different patient groups. Results of new subscales of the Hamilton Depression Rating Scale*. Clin Neuropharmacol. **1993b**, 16 Suppl 2:S55-S62.

Anonymous. *Adverse ocular effects of systemic drugs*. Med Lett Drugs Ther. **1976**, 18:63-64.

Anonymous. *Priapism with trazodone (Desyrel)*. Med Lett Drugs Ther. **1984**, 26:35.

Anonymous. *Nefazodone hydrochloride*. Drugs of the future. **1987**, 12:758.

Anonymous. *Dystonia: underdiagnosed and undertreated?* Drug Ther Bull. **1988**, 26:33-36.

Anonymous. *Bupropion for depression*. Med Lett Drugs Ther. **1989a**, 31:97-98.

Anonymous. *Carbamazepine update*. Lancet. **1989b**, 2:595-597.

Anonymous. *Korsakoff's syndrome*. Lancet. **1990**, 336:912-913.

Anonymous. *Anticonvulsants and lymphadenopathy*. WHO Drug Inf. **1991**, 5:11.

Anonymous. *Fluoxetine, suicide and aggression*. Drug Ther Bull. **1992**, 30:5-6.

Anonymous. *Paroxetine for treatment of depression*. Med Lett Drugs Ther. **1993a**, 35:24-25.

Anonymous. *Venlafaxine: a new dimension in antidepressant pharmacotherapy*. J Clin Psychiatry. **1993b**, 54:119-126.

Anonymous. *Fluoxetine (Prozac) and other drugs for treatment of obesity*. Med Lett Drugs Ther. **1994a**, 36:107-108.

Anonymous. *Moclobemide for depression*. Drug Ther Bull. **1994b**, 32:6-8.

Anonymous. *Valproate for bipolar disorder*. Med Lett Drugs Ther. **1994c**, 36:74-75.

Anonymous. *Venlafaxine--a new antidepressant*. Med Lett Drugs Ther. **1994d**, 36:49-50.

Anonymous. *Lamotrigine for epilepsy*. Med Lett Drugs Ther. **1995**, 37:21-23.

Ansseau M, von Frenckell R, Mertens C, de Wilde J, Botte L, Devoitille JM, Evrard JL, De Nayer A, Darimont P, Dejaiffe G, . *Controlled comparison of two doses of milnacipran (F 2207) and amitriptyline in major depressive inpatients*. Psychopharmacology (Berl). **1989**, 98:163-168.

Ansseau M, De Roeck J. *Trazodone in benzodiazepine dependence*. J Clin Psychiatry. **1993**, 54:189-191.

Ansseau M, Gabriels A, Loyens J, Bartholomé F, Evrard JL, De Nayer A, Linhart R, Wirtz J, Bruynooghe F, Surinx K, Clarysse H, Marganne R, Papart P. *Controlled comparison of paroxetine and fluvoxamine in major depression*. Hum Psychopharmacol Clin Exp. **1994**, 9:329-336.

Appelbaum PS, Kapoor W. *Imipramine-induced vasospasm: a case report*. Am J Psychiatry. **1983**, 140:913-915.

Aprile F, et al. *Etoperidone, maprotiline and trazodone for the therapy of severe depressive conditions requiring hospital admision: a standard controlled study*. Acta Ther. **1983**, 9:353-366.

Arana GW, Kaplan GB. *Trazodone-induced mania following desipramine-induced mania in major depressive disorders*. Am J Psychiatry. **1985**, 142:386.

Arminen SL, Ikonen U, Pulkkinen P, Leinonen E, Mahlanen A, Koponen H, Kourula K, Ryyppo J, Korpela V, Lehtonen ML. *A 12-week double-blind multi-centre study of paroxetine and imipramine in hospitalized depressed patients*. Acta Psychiatr Scand. **1994**, 89:382-389.

Arnalot i Aquilar C, Bosch i Llado J, Camps Garcia P, Onrubia Miguel MM, Soldevila MN. *Venlafaxine Production Process*, WO200107397 (**2001**); priority: 1999

Aronson JK, Reynolds DJ. *ABC of monitoring drug therapy. Lithium*. BMJ. **1992**, 305:1273-1274.

Ashton MG, Ball SG, Thomas TH, Lee MR. *Water intoxication associated with carbamazepine treatment*. Br Med J. **1977**, 1:1134-1135.

Atsushi S, Noriko M, Taisuke S, Yoshio E, Katsumi U. *Metabolic fate of the new antidepressant, milnacipran, in rats (1st report) - Absorption, distribution, metabolism and excretion of 14C-milnacipran after a single oral administration in rats*. Clin Rep. **1994a**, 28:7

Atsushi S, Noriko M, Taisuke S, Yoshio E, Katsumi U. *Metabolic fate of the new antidepressant, milnacipran, in rats (2nd report)- Absoprtion, distribution and excretion of 14C-milnacipran after repeated oral administration in rats*. Clin Rep. **1994b**, 28:29.

Baandrup U. Muscle. In: Johnson FN, ed. *Depression & mania: modern lithium therapy.* IRL Press, Oxford, **1987**, pp 236-238.

Balant-Gorgia AE, Gex-Fabry M, Balant LP. *Clinical pharmacokinetics of clomipramine.* Clin Pharmacokinet. **1991**, 20:447-462.

Ballardie FW, Mucklow JC. *Partial reversal of carbamazepine-induced water intolerance by demeclocycline.* Br J Clin Pharmacol. **1984**, 17:763-765.

Ballenger JC. *The clinical use of carbamazepine in affective disorders.* J Clin Psychiatry. **1988**, 49 Suppl:13-21.

Balon R. *Fluoxetine and sexual dysfunction.* JAMA. **1995**, 273:1489.

Bank J. *A comparative study of amitriptyline and fluvoxamine in migraine prophylaxis.* Headache. **1994**, 34:476-478.

Barnes RD, Wood-Kaezmar MW, Richardson JE, Lynch IR, Buxton PC, Curzons AD. *Piperidine derivatives, their preparation and their use as medicaments,* EP-00223403 (**1987**); priority: 1985

Barr RD, Copeland SA, Stockwell ML, Morris N, Kelton JC. *Valproic acid and immune thrombocytopenia.* Arch Dis Child. **1982**, 57:681-684.

Barrett J. *Anisocoria associated with selective serotonin reuptake inhibitors.* BMJ. **1994**, 309:1620.

Barth JH, Baker H. *Generalized pustular psoriasis precipitated by trazodone in the treatment of depression.* Br J Dermatol. **1986**, 115:629-630.

Bauer J, Elger CE. *Management of status epilepticus in adults.* CNS Drugs. **1994**, 1:26-44.

Baxter MG, Elphick AR, Miller AA, Sawyer DA: *1,2,4-Triazine Derivatives, Process For Preparing Such Compounds, Pharmaceutical Compositions And Intermediates Utilized For This Process.* EP-00021121 (**1981**); priority: 1979

Bayer AJ, Pathy MS, Ankier SI. *Pharmacokinetic and pharmacodynamic characteristics of trazodone in the elderly.* Br J Clin Pharmacol. **1983**, 16:371-376.

Baylis CM, et al. *Drug-induced endocrine disorders.* Adverse Drug React Bull. **1986**, 116:432-435.

Beasley CM, Jr., Dornseif BE, Bosomworth JC, Sayler ME, Rampey AH, Jr., Heiligenstein JH, Thompson VL, Murphy DJ, Masica DN. *Fluoxetine and suicide: a meta-analysis of controlled trials of treatment for depression.* BMJ. **1991**, 303:685-692.

Beaumont G. *Drug interactions with clomipramine (Anafranil).* J Int Med Res. **1973**, 1:480-484.

Beck PL, Bridges RJ, Demetrick DJ, Kelly JK, Lee SS. *Chronic active hepatitis associated with trazodone therapy.* Ann Intern Med. **1993**, 118:791-792.

Beeley L. *Drug-induced sexual dysfunction and infertility.* Adverse Drug React Acute Poisoning Rev. **1984**, 3:23-42.

Benedetti MS, Rovei V, Dencker SJ, Nagy A, Johansson R. *Pharmacokinetics of toloxatone in man following intravenous and oral administrations.* Arzneimittelforschung. **1982**, 32:276-280.

Berger H. *An unusual manifestation of Tegretol (carbamazepine) toxicity.* Ann Intern Med. **1971**, 74:449-450.

Berkovic SF, Andermann F, Guberman A, Hipola D, Bladin PF. *Valproate prevents the recurrence of absence status.* Neurology. **1989**, 39:1294-1297.

Bernus I, Dickinson RG, Hooper WD, Eadie MJ. *Early stage autoinduction of carbamazepine metabolism in humans.* Eur J Clin Pharmacol. **1994**, 47:355-360.

Berti CA, Doogan DP. *Sudden chest pain with sertraline.* Lancet. **1994**, 343:1510-1511.

Bertilsson L, et al. *Metabolism of various drugs in subjects with different debrisoquine and sparteine oxidation phenotypes.* Br J Clin Pharmacol. **1982**, 14:602P.

Bertilsson L, Tomson T. *Clinical pharmacokinetics and pharmacological effects of carbamazepine and carbamazepine-10,11-epoxide. An update.* Clin Pharmacokinet. **1986**, 11:177-198.

Betts T, Goodwin G, Withers RM, Yuen AW. *Human safety of lamotrigine.* Epilepsia. **1991**, 32 Suppl 2:S17-S21.

Beumont PJ, Russell JD, Touyz SW. *Treatment of anorexia nervosa.* Lancet. **1993**, 341:1635-1640.

Bhugra DK, Kaye N. *Phenelzine induced grand mal seizure.* Br J Clin Pract. **1986**, 40:173-174.

Bialer M. *Clinical pharmacology of valpromide.* Clin Pharmacokinet. **1991**, 20:114-122.

Biel JH. *Phenylalkylhydrazines and use as psychotherapeutics.* US3000903 (**1961**); priority: 1959

Birch NJ, Grof P, Hullin RP, Kehoe RF, Schou M, Srinivasan DP. *Lithium prophylaxis: proposed guidelines for good clinical practice.* Lithium. **1993**, 4:225-230.

Black DW, Wesner R, Bowers W, Gabel J. *A comparison of fluvoxamine, cognitive therapy, and placebo in the treatment of panic disorder.* Arch Gen Psychiatry. **1993**, 50:44-50.

Bloch M, Stager SV, Braun AR, Rubinow DR. *Severe psychiatric symptoms associated with paroxetine withdrawal.* Lancet. **1995**, 346:57.

Bloom BM, Tretter JR. *Novel dibenzoxepines,* US3420851 (**1969**); priority: 1962

Boegesoe KP, Perregaard J. *New enantiomers and their isolation.* EP347066 (**1989**); priority: 1988

Bogeso KP, Toft AS. *Anti-depressive substituted 1-dimethylaminopropyl-1-phenyl phthalans.* US04136193 (**1979**); priority: 1977

Bolte MA, Avery D. *Case of fluoxetine-induced remission of Raynaud's phenomenon--a case report.* Angiology. **1993**, 44:161-163.

Bonnaud B, Mouzin G, Cousse H, Patoiseau J-F. *Process for the preparation of (Z)-1-phenyl-1-diethyl amino carbonyl 2-amino methyl cyclopropane hydrochloride.* EP0200638 (**1986**); priority: 1985

Borrett GT, Crowe D, Ward N, Wells AS. *Process for the preparation of paroxetine.* WO2001029031 (**2001**); priority: 1999

Bougerol T, Uchida C, Gachoud JP, Kohler M, Mikkelsen H. *Efficacy and tolerability of moclobemide compared with fluvoxamine in depressive disorder (DSM III). A French/Swiss double-blind trial.* Psychopharmacology (Berl). **1992**, 106 Suppl: S102-S108.

Bourne M, Szabadi E, Bradshaw CM. *A comparison of the effects of single doses of amoxapine and amitriptyline on autonomic functions in healthy volunteers.* Eur J Clin Pharmacol. **1993**, 44:57-62.

Bowdan ND. *Seizure possibly caused by trazodone HCl.* Am J Psychiatry. **1983**, 140:642.

Bowden CL, Brugger AM, Swann AC, Calabrese JR, Janicak PG, Petty F, Dilsaver SC, Davis JM, Rush AJ, Small JG. *Efficacy of divalproex vs lithium and placebo in the treatment of mania. The Depakote Mania Study Group.* JAMA. **1994**, 271:918-924.

Böhm H-J, Flohr A, Stahl M. *Scaffold hopping.* Drug Disc Today: Technologies. **2004**, 1:217-224.

Bracher F, Litz T. *An efficient chemoenzymatic route to the antidepressants (R)-fluoxetine and (R)-tomoxetine.* Bioorg Med Chem. **1996**, 4:877-880.

Brannan SK, Talley BJ, Bowden CL. *Sertraline and isocarboxazid cause a serotonin syndrome.* J Clin Psychopharmacol. **1994**, 14:144-145.

Bray GA. *Use and abuse of appetite-suppressant drugs in the treatment of obesity.* Ann Intern Med. **1993**, 119:707-713.

Brodie MJ. *Lamotrigine.* Lancet. **1992**, 339:1397-1400.

Brodie MJ. *Drugs in focus: 10. Lamotrigine.* Prescriber's J. **1993**, 33:212-216.

Brodie MJ, Richens A, Yuen AW. *Double-blind comparison of lamotrigine and carbamazepine in newly diagnosed epilepsy. UK Lamotrigine/Carbamazepine Monotherapy Trial Group.* Lancet. **1995**, 345:476-479.

Brosen K, Gram LF. *Clinical significance of the sparteine/debrisoquine oxidation polymorphism.* Eur J Clin Pharmacol. **1989**, 36:537-547.

Bross R, Hoffer LJ. *Fluoxetine increases resting energy expenditure and basal body temperature in humans.* Am J Clin Nutr. **1995**, 61:1020-1025.

Buckley NA, Whyte IM, Dawson AH. *Self-poisoning with lamotrigine.* Lancet. **1993**, 342:1552-1553.

Burkard W, Wyss P-C. *Benzamide.* DE2706179 (**1977**); priority: 1976

Busch RL. *Generic carbamazepine and erythema multiforme: generic-drug nonequivalency.* N Engl J Med. **1989**, 321:692-693.

Bussing R, Levin GM. *Methamphetamine and fluoxetine treatment of a child with attention deficit hyperactivity disorder and obsessive-compulsive disorder.* J Child- Adolesc Psychopharmacol. **1993**, 3:53-58.

Bymaster FP, Beedle EE, Findlay J, Gallagher PT, Krushinski JH, Mitchell S, Robertson DW, Thompson DC, Wallace L, Wong DT. *Duloxetine (Cymbalta), a dual inhibitor of serotonin and norepinephrine reuptake.* Bioorg Med Chem Lett. **2003**, 13:4477-4480.

Caccia S, Garattini S. *Formation of active metabolites of psychotropic drugs. An updated review of their significance.* Clin Pharmacokinet. **1990**, 18:434-459.

Caley CF, Weber SS. *Paroxetine: a selective serotonin reuptake inhibiting antidepressant.* Ann Pharmacother. **1993**, 27:1212-1222.

Cannon RO, III, Quyyumi AA, Mincemoyer R, Stine AM, Gracely RH, Smith WB, Geraci MF, Black BC, Uhde TW, Waclawiw MA. *Imipramine in patients with chest pain despite normal coronary angiograms.* N Engl J Med. **1994**, 330:1411-1417.

Cantini F, Bellandi F, Niccoli L, Di Munno O. *Fluoxetina associata a ciclobenzaprina nel trattamento della fibromyalgia.* Minerva Med. **1994**, 85:97-100.

Caruso I, Sarzi Puttini PC, Boccassini L, Santandrea S, Locati M, Volpato R, Montrone F, Benvenuti C, Beretta A. *Double-blind study of dothiepin versus placebo in the treatment of primary fibromyalgia syndrome.* J Int Med Res. **1987**, 15:154-159.

Castaner J, Arrigoni-Martelli E. *Lofepramine.* Drugs of the future. **1976**, 1:129.

Castaner J, Playle AC. *Amoxapine.* Drugs of the future. **1976**, 1:511.

Castaner J, Sungurbey K. *Toloxatone.* Drugs of the future. **1976**, 1:569.

Castaner J, de Angelis L. *Etoperidone* Drugs of the future. **1977**, 2:164.

Castaner J, Paton DM. *Fluoxetine.* Drugs of the future. **1977**, 2:27.

Castaner J, Hopkins SJ. *Bupropion.* Drugs of the future. **1978**, 3:124.

Castaner J, Thorpe PJ. *Fluvoxamine.* Drugs of the future. **1978**, 3:288.

Castaner J, Roberts PJ. *Citalopram.* Drugs of the future. **1979**, 4:407.

Castaner J, Grau M. *Moclobemide.* Drugs of the future. **1983**, 8:124.

Chen CY, Reamer RA. *Efficient enantioselective synthesis of sertraline, a potent antidepressant, via a novel intramolecular nucleophilic addition to imine.* Org Lett. **1999**, 1:293-294.

Chen G, Tan D, Xu L, Dong J. *Synthesis of trazodone hydrochloride.* Zhongguo Yaowu Huaxue Zazhi. **2002**, 12:92-93.

Chenevert R, Fortier G, Bel Rhild R. *Asymmetric synthesis of both enantiomers of fluoxetine via microbiological reduction of ethyl benzoylacetate.* Tetrahedron. **2006**, 48:6769-6776.

Chignac M, Grain C, Pigerol C. *Process for preparing acetic acid derivatives.* GB1529786 (**1978**); priority: 1977

Chignac M, Grain C, Pigerol C. *Process for the preparation of an acetonitrile derivative.* US4155929 (**1979**); priority: 1978

Choo V. *Paroxetine and extrapyramidal reactions* . Lancet. **1993**, 341:624.

Choonara IA. *Sodium valproate and enuresis.* Lancet. **1985**, 1:1276.

Christensen JA, Squires RF. *4-Phenylpiperidine compounds.* US3912743 (**1975**); US4007196 (**1977**); priority: 1973

Chu AG, Gunsolly BL, Summers RW, Alexander B, McChesney C, Tanna VL. *Trazodone and liver toxicity.* Ann Intern Med. **1983**, 99:128-129.

Clark DM, Salkovskis PM, Hackmann A, Middleton H, Anastasiades P, Gelder M. *A comparison of cognitive therapy, applied relaxation and imipramine in the treatment of panic disorder.* Br J Psychiatry. **1994**, 164:759-769.

Clark RJ. *The treatment of chemical dependence.* JAMA. **1989**, 261:3239.

Cloyd JC, Fischer JH, Kriel RL, Kraus DM. *Valproic acid pharmacokinetics in children. IV. Effects of age and antiepileptic drugs on protein binding and intrinsic clearance.* Clin Pharmacol Ther. **1993**, 53:22-29.

Coccaro EF, Astill JL, Herbert JL, Schut AG. *Fluoxetine treatment of impulsive aggression in DSM-III-R personality disorder patients.* J Clin Psychopharmacol. **1990**, 10:373-375.

Coe BK, Weissman A, Welch WM, Browne RG. *Sertraline, 1S,4S-N-methyl-4-(3,4-dichlorophenyl)-1,2,3,4-tetrahydro-1- naphthylamine, a new uptake inhibitor with selectivity for serotonin.* J Pharmacol Exp Ther. **1983**, 226:686-700.

Collins S. *Thrombocytopenia associated with lithium carbonate.* BMJ. **1992**, 305:159.

Committe on Safety of Medicines (CSM). *Priapism and trazodone (Molipaxin).* Current problems in pharmacovigilance. **1984**, 13.

Committe on Safety of Medicines (CSM). *Dangers of newer antidepressants.* Current problems in pharmacovigilance. **1985**, 11:1.

Committe on Safety of Medicines (CSM). *Fluvoxamine (Faverin): adverse reaction profile.* Current problems in pharmacovigilance. **1988**, 22:1-2.

Committe on Safety of Medicines (CSM). *Fluvoxamine and Fluoxetine-interaction with monoamine oxidase inhibitors, lithium and Tryptophan.* Current problems in pharmacovigilance. **1989**, 26.

Committe on Safety of Medicines (CSM). *Safety of Fluoxetine (Prozac): comparison with fluvoxamine (Faverin).* Current problems in pharmacovigilance. **1992**, 34.

Committe on Safety of Medicines (CSM). *Dystonia and withdrawal symptoms with paroxetine (Seroxat).* Current problems in pharmacovigilance. **1993**, 19:1.

Committe on Safety of Medicines (CSM). *Drug-induced pancreatitis.* Current problems in pharmacovigilance. **1994**, 20:2-3.

Cooper MA, Dening TR. *Excessive blinking associated with combined antidepressants.* Br Med J. **1986**, 293:1243.

Coquoz D, Porchet HC, Dayer P. *Central analgesic effects of desipramine, fluvoxamine, and moclobemide after single oral dosing: a study in healthy volunteers.* Clin Pharmacol Ther. **1993**, 54:339-344.

Corey EJ. *Enantioselective and practical syntheses of R- and S-fluoxetines.* Tetrahedron Letters. **1989**, 30:5207-5210.

Corey EJ, Gant TG. *A catalytic enantioselective synthetic route to the important antidepressant sertraline.* Tetrahedron Letters. **1994**, 35:5373-5376.

Coria F, Sempere AP, Duarte J, Claveria LE. *Low-dose sodium valproate in the prophylaxis of migraine.* Clin Neuropharmacol. **1994**, 17:569-573.

Cornelius JR, Soloff PH, Perel JM, Ulrich RF. *Fluoxetine trial in borderline personality disorder.* Psychopharmacol Bull. **1990**, 26:151-154.

Cossy J, Mirguet O, Gomez Pardo D, Desmurs J-R. *A short formal synthesis of paroxetine. Diastereoselective cuprate addition to a chiral racemic olefinic amido ester.* Tetrahedron Letters. **2001**, 42:7805-7807.

Coulomb F, Ducret F, Laneury JP, Fiorentini F, Poggesi I, Jannuzzo MG, Fleishaker JC, Houin G, Duchene P. *Pharmacokinetics of single-dose reboxetine in volunteers with renal insufficiency.* J Clin Pharmacol. **2000**, 40:482-487.

Coulter DL, Wu H, Allen RJ. *Valproic acid therapy in childhood epilepsy.* JAMA. **1980**, 244:785-788.

Coutts RT, Su P, Baker GB, Daneshtalab M. *Metabolism of imipramine in vitro by isozyme CYP2D6 expressed in a human cell line, and observations on metabolite stability.* J Chromatogr. **1993**, 615:265-272.

Crome P, Ali C. *Clinical features and management of self-poisoning with newer antidepressants.* Med Toxicol. **1986**, 1:411-420.

Crosley CJ, Swender PT. *Dystonia associated with carbamazepine administration: experience in brain-damaged children.* Pediatrics. **1979**, 63:612-615.

Dalhoff K, Almdal TP, Bjerrum K, Keiding S, Mengel H, Lund J. *Pharmacokinetics of paroxetine in patients with cirrhosis.* Eur J Clin Pharmacol. **1991**, 41:351-354.

Dall'asta L, Casazza U, Petersen H. *Method for the preparation of citalopram.* WO2000023431 (**2000**); priority: 1999

Damlouji NF, Ferguson JM. *Trazodone-induced delirium in bulimic patients.* Am J Psychiatry. **1984**, 141:434-435.

Davis MA, Winthrop SO. *Dibenzo-cycloheptadiene derivatives (wherein R is H or CH3) and their hydrochlorides are new cpds. having pharmacological properties.* BE613750 (**1962**); priority: 1961

Davis MA. *Cyclodehydration process for preparing dibenzocycloheptenes.* CA900471 (**1972**); priority: 1969

Davis R, Peters DH, McTavish D. *Valproic acid. A reappraisal of its pharmacological properties and clinical efficacy in epilepsy.* Drugs. **1994**, 47:332-372.

De Bree H, Van der Schoot JB, Post CL. *Fluvoxamine maleate: disposition in man.* Eur J Drug Metab Pharmacokinet. **1983**, 8:175-179.

De Wilde J, Spiers R, Mertens C, Bartholome F, Schotte G, Leyman S. *A double-blind, comparative, multicentre study comparing paroxetine with fluoxetine in depressed patients.* Acta Psychiatr Scand. **1993**, 87:141-145.

Dechant KL, Clissold SP. *Paroxetine. A review of its pharmacodynamic and pharmacokinetic properties, and therapeutic potential in depressive illness.* Drugs. **1991**, 41:225-253.

Deeter J, Frazier J, Staten G, Staszak M, Weigel L. *Assymmetric synthesis and absolute stereochemistry of LY248686.* Tetrahedron Letters. **1990**, 31:7101-7104.

Demers R, Lukesh R, Prichard J. *Convulsion during lithium therapy.* Lancet. **1970**, 2:315-316.

Demers RG, Heninger GR. *Electrocardiographic T-wave changes during lithium carbonate treatment.* JAMA. **1971**, 218:381-386.

Depot M, Powell JR, Messenheimer JA, Jr., Cloutier G, Dalton MJ. *Kinetic effects of multiple oral doses of acetaminophen on a single oral dose of lamotrigine.* Clin Pharmacol Ther. **1990**, 48:346-355.

DeVane CL, Liston HL, Markowitz JS. *Clinical pharmacokinetics of sertraline.* Clin Pharmacokinet. **2002**, 41:1247-1266.

Devarajan S. *Safety of amoxapine.* Lancet. **1989**, 2:1455.

Di Costanzo E, Schifano F. *Lithium alone or in combination with carbamazepine for the treatment of rapid-cycling bipolar affective disorder.* Acta Psychiatr Scand. **1991**, 83:456-459.

Dick DJ, Saunders M. *Extrapyramidal syndrome with sodium valproate.* Br Med J. **1980**, 280:189.

Dodd S, Buist A, Burrows GD, Maguire KP, Norman TR. *Determination of nefazodone and its pharmacologically active metabolites in human blood plasma and breast milk by high-performance liquid chromatography.* J Chromatogr B Biomed Sci Appl. **1999**, 730:249-255.

Donovan S, Vlottes PW, Min JM. *Dothiepin versus amitriptyline for depression: an analysis of comparative studies.* Drug Invest. **1991**, 3:178-182.

Dorevitch A, Davis H. *Fluvoxamine-associated sexual dysfunction.* Ann Pharmacother. **1994**, 28:872-874.

Dorman T. *Toxicity of tricyclic antidepressants: are there important differences?* J Int Med Res. **1985**, 13:77-83.

Dose M, Apelt S, Emrich HM. *Carbamazepine as an adjunct of antipsychotic therapy.* Psychiatry Res. **1987**, 22:303-310.

Dostert P, Benedetti MS, Poggesi I. *Review of the pharmacokinetics and metabolism of reboxetine, a selective noradrenaline reuptake inhibitor.* Eur Neuropsychopharmacol. **1997**, 7 Suppl 1:S23-S35.

Dreifuss FE, Langer DH, Moline KA, Maxwell JE. *Valproic acid hepatic fatalities. II. US experience since 1984.* Neurology. **1989**, 39:201-207.

Dursun SM, Mathew VM, Reveley MA. *Toxic serotonin syndrome after fluoxetine plus carbamazepine.* Lancet. **1993**, 342:442-443.

Eadie MJ, Hooper WD, Dickinson RG. *Valproate-associated hepatotoxicity and its biochemical mechanisms.* Med Toxicol Adverse Drug Exp. **1988**, 3:85-106.

Edwards JG, Glen-Bott M. *Does viloxazine have epileptogenic properties?* J Neurol Neurosurg Psychiatry. **1984**, 47:960-964.

Edwards JG. *Selective serotonin reuptake inhibitors in the treatment of depression.* Prescriber's J. **1994**, 34:197-204.

Eisenhauer G, Jermain DM. *Fluoxetine and tics in an adolescent.* Ann Pharmacother. **1993**, 27:725-726.

Elkind AH, Indelicato JA. *A retrospective study with divalproex sodium for refractory headache prophylaxis.* Clin Pharmacol Ther. **1995**, 57:201.

Ellingrod VL, Perry PJ. *Venlafaxine: a heterocyclic antidepressant.* Am J Hosp Pharm. **1994**, 51:3033-3046.

Engelhardt EL, Christy ME. *5-Aminoalkylidene- and 5-aminoalkyl-5H-dibenzo[a,d]cycloheptenes.* BE617967 (**1962**); priority: 1961

Eriksoo E, Fex HJ, Högberg KB, Kneip PHO, Mollberg HR, Rohte OA. *Dibenzazepine Derivatives.* US3637660 (**1972**); priority: 1967

Eriksson E, Hedberg MA, Andersch B, Sundblad C. *The serotonin reuptake inhibitor paroxetin is superior to the noradrenaline reuptake inhibitor maprotiline in the treatment of premenstrual syndrome.* Neuropsychopharmacology. **1995**, 12:167-176.

Espir MLE, Walker ME. *Carbamazepine in multiple sclerosis.* Lancet. **1967**, 289:280.

Fairburn CG, Peveler RC. *Bulimia nervosa and a stepped care approach to management.* Gut. **1990**, 31:1220-1222.

Fallon BA, Liebowitz MR, Salman E, Schneier FR, Jusino C, Hollander E, Klein DF. *Fluoxetine for hypochondriacal patients without major depression.* J Clin Psychopharmacol. **1993**, 13:438-441.

Fauran C, Raynaud G, Douzon G, Oliver R. *5-Hydroxymethyl-Oxazolidin-2-On-Derivate.* DE2011333 (**1970**); priority: 1969

Fava M, Rosenbaum JF, McGrath PJ, Stewart JW, Amsterdam JD, Quitkin FM. *Lithium and tricyclic augmentation of fluoxetine treatment for resistant major depression: a double-blind, controlled study.* Am J Psychiatry. **1994**, 151:1372-1374.

Fava M, Rappe SM, Pava JA, Nierenberg AA, Alpert JE, Rosenbaum JF. *Relapse in patients on long-term fluoxetine treatment: response to increased fluoxetine dose.* J Clin Psychiatry. **1995**, 56:52-55.

Feinmann C, Harris M, Cawley R. *Psychogenic facial pain: presentation and treatment.* Br Med J (Clin Res Ed). **1984**, 288:436-438.

Ferrie CD, Robinson RO, Knott C, Panayiotopoulos CP. *Lamotrigine as an add-on drug in typical absence seizures.* Acta Neurol Scand. **1995**, 91:200-202.

Ferry LH, Robbins AS, Abbey DE, Scariati PD, Masterson A, Burchette RJ. *Enhancement of smoking cessation using the antidepressant, bupropion hydrochloride [abstract].* Circulation. **1992**, 86: I-671.

Fesler FA. *Valproate in combat-related posttraumatic stress disorder.* J Clin Psychiatry. **1991**, 52:361-364.

Fichtner CG, Jobe TH, Braun BG. *Does fluoxetine have a therapeutic window?* Lancet. **1991**, 338:520-521.

Fichtner CG, Braun BG. *Hyperphagia and weight loss during fluoxetine treatment.* Ann Pharmacother. **1994**, 28:1350-1352.

Finley PR. *Selective serotonin reuptake inhibitors: pharmacologic profiles and potential therapeutic distinctions.* Ann Pharmacother. **1994**, 28:1359-1369.

Firkusny L, Gleiter CH. *Maprotiline metabolism appears to co-segregate with the genetically-determined CYP2D6 polymorphic hydroxylation of debrisoquine.* Br J Clin Pharmacol. **1994**, 37:383-388.

Fischer FE. *Process for making sodium hydrogen divalproate.* CA1144558 (**1983**); priority: 1979

Fischman MW, Foltin RW, Nestadt G, Pearlson GD. *Effects of desipramine maintenance on cocaine self-administration by humans*. J Pharmacol Exp Ther. **1990**, 253:760-770.

Fitton A, Faulds D, Goa KL. *Moclobemide. A review of its pharmacological properties and therapeutic use in depressive illness*. Drugs. **1992**, 43:561-596.

Flament MF, Rapoport JL, Berg CJ, Sceery W, Kilts C, Mellstrom B, Linnoila M. *Clomipramine treatment of childhood obsessive-compulsive disorder. A double-blind controlled study*. Arch Gen Psychiatry. **1985**, 42:977-983.

Fleishaker JC. *Clinical pharmacokinetics of reboxetine, a selective norepinephrine reuptake inhibitor for the treatment of patients with depression*. Clin Pharmacokinet. **2000**, 39:413-427.

Fogelson DL, Bystritsky A, Pasnau R. *Bupropion in the treatment of bipolar disorders: the same old story?* J Clin Psychiatry. **1992**, 53:443-446.

Fontaine R, Chouinard G. *Fluoxetine in the long-term maintenace treatment of obsessive-compulsive disorder*. Psychiatr Ann. **1989**, 19:88-91.

Fontaine R, Ontiveros A, Elie R, Kensler TT, Roberts DL, Kaplita S, Ecker JA, Faludi G. *A double-blind comparison of nefazodone, imipramine, and placebo in major depression*. J Clin Psychiatry. **1994**, 55:234-241.

Ford HE, Jenike MA. *Erythema multiforme associated with trazodone therapy: case report*. J Clin Psychiatry. **1985**, 46:294-295.

Franchini L, Gasperini M, Smeraldi E. *A 24-month follow-up study of unipolar subjects: a comparison between lithium and fluvoxamine*. J Affect Disord. **1994**, 32:225-231.

Frankenburg FR, Tohen M, Cohen BM, Lipinski JF, Jr. *Long-term response to carbamazepine: a retrospective study*. J Clin Psychopharmacol. **1988**, 8:130-132.

Fraunfelder FT, Fraunfelder FW, Jefferson JW. *The effects of lithium on the human visual system*. J Toxicol Cutan Ocul Toxicol. **1992**, 11:97-169.

Freeman EW, Rickels K, Sondheimer SJ, Denis A, Pfeifer S, Weil S. *Nefazodone in the treatment of premenstrual syndrome: a preliminary study*. J Clin Psychopharmacol. **1994**, 14:180-186.

Freeman H. *Moclobemide*. Lancet. **1993**, 342:1528-1532.

Gammans R. *Treatment of sleep disorders*. US5116852 (**1992**); priority: 1990

Gammon GD, Brown TE. *Fluoxetine and methylphenidate in combination for treatment of attention deficit disorder and comorbid depressive disorder*. J Child- Adolesc Psychopharmacol. **1993**, 3:1-10.

Gao Y, Sharpless KB. *Asymmetric synthesis of both enantiomers of tomoxetine and fluoxetine. Selective reduction of 2,3-epoxycinnamyl alcohol with Red-Al*. J Org Chem. **1988**, 53:4081-4084.

Gardner DL, Cowdry RW. *Development of melancholia during carbamazepine treatment in borderline personality disorder*. J Clin Psychopharmacol. **1986**, 6:236-239.

Gardner TS, Lee J, Wenis E. *5-Methyl-3-isoxazolecarboxylic acid hydrazides*. US2908688 (**1959**); priority: 1958

Gardner TS, Wenis E, Lee J. *Monoamine oxidase inhibitors. I. 1-Alkyl and 1-aralkyl-2-(picolinoyl and 5-methyl-3-isoxazolyl-carbonyl) hydrazines*. J Med Pharm Chem. **1960**, 2:133-145.

Gartrell N. *Increased libido in women receiving trazodone*. Am J Psychiatry. **1986**, 143:781-782.

Gawin FH, Kleber HD, Byck R, Rounsaville BJ, Kosten TR, Jatlow PI, Morgan C. *Desipramine facilitation of initial cocaine abstinence*. Arch Gen Psychiatry. **1989**, 46:117-121.

Geigy JR. *5-dibenzo [b,f] azepines and their production*. GB862297 (**1961**); priority: 1957

Geigy JR. *Substituted 5h-dibenz[b,f]azepines and their 10,11-dihydroderivatives*. GB908788 (**1962**); priority: 1959

Geller SA. *Treatment of fibrositis with fluoxetine hydrochloride (Prozac)*. Am J Med. **1989**, 87:594-595.

George CR. *Renal aspects of lithium toxicity*. Med J Aust. **1989**, 150:291-292.

Geracioti TD, Jr. *Valproic acid treatment of episodic explosiveness related to brain injury*. J Clin Psychiatry. **1994**, 55:416-417.

Gex-Fabry M, Balant-Gorgia AE, Balant LP, Garrone G. *Clomipramine metabolism. Model-based analysis of variability factors from drug monitoring data*. Clin Pharmacokinet. **1990**, 19:241-255.

Ghose K, Huston GJ, Kirby MJ, Witts DJ, Turner P. *Some clinical pharmacological studies with butriptyline, an antidepressive drug*. Br J Clin Pharmacol. **1977**, 4:91-93.

Giannini AJ, Malone DA, Giannini MC, Price WA, Loiselle RH. *Treatment of depression in chronic cocaine and phencyclidine abuse with desipramine*. J Clin Pharmacol. **1986**, 26:211-214.

Gillman MA, Sandyk R. *Nicotinic acid deficiency induced by sodium valproate*. S Afr Med J. **1984**, 65:986.

Gillman MA, Sandyk R. *Parkinsonism induced by a monoamine oxidase inhibitor*. Postgrad Med J. **1986**, 62:235-236.

Gilman JT. *Carbamazepine dosing for pediatric seizure disorders: the highs and lows*. DICP. **1991**, 25:1109-1112.

Gilman JT. *Lamotrigine: an antiepileptic agent for the treatment of partial seizures*. Ann Pharmacother. **1995**, 29:144-151.

Giroud M, Gras D, Escouse A, Dumas R, Venaud G. *Use of injectable valproic acid in status epilepticus: a pilot study*. Drug Invest. **1993**, 5:154-159.

Gleason RP, Schneider LS. *Carbamazepine treatment of agitation in Alzheimer's outpatients refractory to neuroleptics*. J Clin Psychiatry. **1990**, 51:115-118.

Goa KL, Ross SR, Chrisp P. *Lamotrigine. A review of its pharmacological properties and clinical efficacy in epilepsy*. Drugs. **1993**, 46:152-176.

Goldbloom DS, Olmsted MP. *Pharmacotherapy of bulimia nervosa with fluoxetine: assessment of clinically significant attitudinal change*. Am J Psychiatry. **1993**, 150:770-774.

Goldstein BJ, Claghorn JL. *An overview of seventeen years of experience with dothiepin in the treatment of depression in Europe*. J Clin Psychiatry. **1980**, 41:64-70.

Goldstein DJ, Rampey AH, Jr., Potvin JH, Masica DN, Beasley CM, Jr. *Analyses of suicidality in double-blind, placebo-controlled trials of pharmacotherapy for weight reduction*. J Clin Psychiatry. **1993**, 54:309-316.

Goldstein DJ, Wilson MG, Thompson VL, Potvin JH, Rampey AH, Jr. *Long-term fluoxetine treatment of bulimia nervosa. Fluoxetine Bulimia Nervosa Research Group*. Br J Psychiatry. **1995**, 166:660-666.

Golwyn DH, Sevlie CP. *Adventitious change in homosexual behavior during treatment of social phobia with phenelzine*. J Clin Psychiatry. **1993**, 54:39-40.

Gomolin IH. *Coping with excessive doses*. In: Johnson FN, ed. *Depression & mania: modern lithium therapy*. IRL Press, Oxford, **1987**, pp 154-157.

Goodman WK, Price LH, Delgado PL, Palumbo J, Krystal JH, Nagy LM, Rasmussen SA, Heninger GR, Charney DS. *Specificity of serotonin reuptake inhibitors in the treatment of obsessive-compulsive disorder. Comparison of fluvoxamine and desipramine*. Arch Gen Psychiatry. **1990**, 47:577-585.

Goodnick PJ, Henry JH, Buki VM. *Treatment of depression in patients with diabetes mellitus*. J Clin Psychiatry. **1995**, 56:128-136.

Gordon NC, Heller PH, Gear RW, Levine JD. *Temporal factors in the enhancement of morphine analgesia by desipramine*. Pain. **1993**, 53:273-276.

Gorelick DA. *Serotonin uptake blockers and the treatment of alcoholism*. Recent Dev Alcohol. **1989**, 7:267-281.

Graber MA, Hoehns TB, Perry PJ. *Sertraline-phenelzine drug interaction: a serotonin syndrome reaction*. Ann Pharmacother. **1994**, 28:732-735.

Graham PM, Potter JM, Paterson J. *Combination monoamine oxidase inhibitor/tricyclic antidepressants interaction*. Lancet. **1982**, 2:440.

Gram L. *Fluoxetine*. N Engl J Med. **1994**, 331:1354-1361.

Gram LF, Christiansen J. *First-pass metabolism of imipramine in man*. Clin Pharmacol Ther. **1975**, 17:555-563.

Gram LF, Bjerre M, Kragh-Sorensen P, Kvinesdal B, Molin J, Pedersen OL, Reisby N. *Imipramine metabolites in blood of patients during therapy and after overdose*. Clin Pharmacol Ther. **1983**, 33:335-342.

Grimsley SR, Jann MW. *Paroxetine, sertraline, and fluvoxamine: new selective serotonin reuptake inhibitors*. Clin Pharm. **1992**, 11:930-957.

Grubb BP, Samoil D, Kosinski D, Wolfe D, Lorton M, Madu E. *Fluoxetine hydrochloride for the treatment of severe refractory orthostatic hypotension*. Am J Med. **1994**, 97:366-368.

Gupta AK. *'Perming' effects associated with chronic valproate therapy*. Br J Clin Pract. **1988**, 42:75-77.

Guthrie SK. *The treatment of alcohol withdrawal* . Pharmacotherapy. **1989**, 9:131-143.

Haddock RE, Johnson AM, Langley PF, Nelson DR, Pope JA, Thomas DR, Woods FR. *Metabolic pathway of paroxetine in animals and man and the comparative pharmacological properties of its metabolites*. Acta Psychiatr Scand Suppl. **1989**, 350:24-26.

Hadzic N, Portmann B, Davies ET, Mowat AP, Mieli-Vergani G. *Acute liver failure induced by carbamazepine*. Arch Dis Child. **1990**, 65:315-317.

Haefliger F, Schindler W. *Tertiary-aminoalkyliminodibenzyls*. US2554736 (**1951**); priority: 1949

Halikas JA, Crosby RD, Carlson GA, Crea F, Graves NM, Bowers LD. *Cocaine reduction in unmotivated crack users using carbamazepine versus placebo in a short-term, double-blind crossover design*. Clin Pharmacol Ther. **1991**, 50:81-95.

Hanger HC. *Emotionalism after stroke*. Lancet. **1993**, 342:1235-1236.

Harmant J, Rijckevorsel-Harmant K, de Barsy T, Hendrickx B. *Fluvoxamine: an antidepressant with low (or no) epileptogenic effect*. Lancet. **1990**, 336:386.

Harris B, Szulecka TK, Anstee JA. *Fluvoxamine versus amitriptyline in depressed hospital outpatients: a multicentre double-blind comparative trial*. Br J Clin Res. **1991**, 2:89-99.

Hascöet P, Cousse H. *Procede Industriel D'Obtention Du Midalcipran*. FR2581060 (**1986**); priority: 1985

Hatori Y, Hatori A, Aihara M, Mukai S, Shigematsu A, Kawai Y, Edanami K, Sato N, Aizawa K. *Studies on the pharmacokinetics of fluvoxamine maleate (1) - Absoprtion, distribution, metabolism and excretion after single oral administration in rats*. Jpn Pharmacol Ther. **1995**, 23:59.

Healy D. *The fluoxetine and suicide controversy: a review of the evidence*. CNS Drugs. **1994**, 1:223-231.

Heath A. *Suicidal overdoses of antidepressants, with special reference to lofepramine*. Int Med. **1984**, Suppl. 10:27-30.

Heller AJ, Chesterman P, Elwes RD, Crawford P, Chadwick D, Johnson AL, Reynolds EH. *Phenobarbitone, phenytoin, carbamazepine, or sodium valproate for newly diagnosed adult epilepsy: a randomised comparative monotherapy trial*. J Neurol Neurosurg Psychiatry. **1995**, 58:44-50.

Heller CA, Friedman PA. *Pyridoxine deficiency and peripheral neuropathy associated with long-term phenelzine therapy*. Am J Med. **1983**, 75:887-888.

Henry DA, Lawson DH, Reavey P, Renfrew S. *Hyponatraemia during carbamazepine treatment*. Br Med J. **1977**, 1:83-84.

Hering R, Kuritzky A. *Sodium valproate in the prophylactic treatment of migraine: a double-blind study versus placebo*. Cephalalgia. **1992**, 12:81-84.

Herman BD, Fleishaker JC, Brown MT. *Ketoconazole inhibits the clearance of the enantiomers of the antidepressant reboxetine in humans*. Clin Pharmacol Ther. **1999**, 66:374-379.

Hilborn JW, Jurgens AR. *Fluoxetine process from benzoylacetonitrile*. WO2000007976 (**2000**); priority: 1998.

Hilborn JW, Lu Z-H, Jurgens AR, Fang QK, Byers P, Wald SA, Senanayake CH. *A practical asymmetric synthesis of (R)-fluoxetine and its major metabolite (R)-norfluoxetine*. Tetrahedron Letters. **2001**, 42:8919-8921.

Hilton E, Stroh EM. *Aseptic meningitis associated with administration of carbamazepine*. J Infect Dis. **1989**, 159:363-364.

Himmelhoch JM, Hanin I. *Letter: Side effects of lithium carbonate*. Br Med J. **1974**, 4:233.

Hoehn-Saric R, McLeod DR, Hipsley PA. *Effect of fluvoxamine on panic disorder*. J Clin Psychopharmacol. **1993**, 13:321-326.

Hoffman-La Roche F. *Tricyclic amines*. BE613362 (**1962**); priority: 1961

Hoffsommer RD, Taub D, Wendler NL. *The Homoallylic Rearrangement in the Synthesis of Amitriptyline and Related Systems*. J Org Chem. **1962**, 27:4134-4137.

Holdsworth CD, Atkinson M, Goldie W. *Hepatitis caused by the newer amine-oxidase-inhibiting drugs*. Lancet. **1961**, 2:621-623.

Hollander JB. *Fluoxetine and sexual dysfunction*. JAMA. **1994**, 272:242.

Hollander JB. *Fluoxetine and sexual dysfunction*. JAMA. **1995**, 273:1490.

Holliday SM, Benfield P. *Venlafaxine. A review of its pharmacology and therapeutic potential in depression*. Drugs. **1995**, 49:280-294.

Holm KJ, Spencer CM. *Reboxetine: A Review of its Use in Depression*. CNS Drugs. **1999**, 12:65-83.

Hopkins HS, Gelenberg AJ. *Fluoxetine and sexual dysfunction*. JAMA. **1995**, 273:1489-1490.

Hosoda N, Sunaoshi W, Shirai H, Bando Y, Miura H, Igarashi M. *Anticarbamazepine antibody induced by carbamazepine in a patient with severe serum sickness*. Arch Dis Child. **1991**, 66:722-723.

Howell CF, Hardy RA, Quinones NQ. *Process for 11 – Aminodibenz [b,f][1,4]oxazepines and analogous thiazepines*. US3444169 (**1969**); priority: 1966.

Hull M, Jones R, Bendall M. *Fatal hepatic necrosis associated with trazodone and neuroleptic drugs*. BMJ. **1994**, 309:378.

Hummel T, Hummel C, Friedel I, Pauli E, Kobal G. *A comparison of the antinociceptive effects of imipramine, tramadol and anpirtoline*. Br J Clin Pharmacol. **1994**, 37:325-333.

Hunt-Fugate AK, Zander J, Lesar TS. *Adverse reactions due to dopamine blockade by amoxapine. A case report and review of the literature*. Pharmacotherapy. **1984**, 4:35-39.

Husbands GEM, Yardley JP, Muth EA. *Phenethylamine derivatives and intermediates therefor.*, EP0112669 (**1984**); US4535186 (**1985**); priority: 1962

Ilett KF, Lebedevs TH, Wojnar-Horton RE, Yapp P, Roberts MJ, Dusci LJ, Hackett LP. *The excretion of dothiepin and its primary metabolites in breast milk*. Br J Clin Pharmacol. **1992**, 33:635-639.

Insel TR, Murphy DL, Cohen RM, Alterman I, Kilts C, Linnoila M. *Obsessive-compulsive disorder. A double-blind trial of clomipramine and clorgyline*. Arch Gen Psychiatry. **1983**, 40:605-612.

Iruela LM. *Sudden chest pain with sertraline*. Lancet. **1994**, 343:1106.

Jabbari B, Bryan GE, Marsh EE, Gunderson CH. *Incidence of seizures with tricyclic and tetracyclic antidepressants*. Arch Neurol. **1985**, 42:480-481.

Jacobson PL, Messenheimer JA, Farmer TW. *Treatment of intractable hiccups with valproic acid*. Neurology. **1981**, 31:1458-1460.

Jacome D. *Carbamazepine-induced dystonia*. JAMA. **1979**, 241:2263.

Jacome DE. *La toux diabolique: neurogenic tussive crisis*. Postgrad Med J. **1985**, 61:515-516.

Jaeken J, van Goethem C, Casaer P, Devlieger H, Eggermont E, Pilet M. *Neutropenia during sodium valproate treatment*. Arch Dis Child. **1979**, 54:986-987.

Jaffe IA. *Serotonin reuptake inhibitors in Raynaud's phenomenon*. Lancet. **1995**, 345:1378.

Jain KK. *Systemic lupus erythematosus (SLE)-like syndromes associated with carbamazepine therapy*. Drug Saf. **1991**, 6:350-360.

Janowsky D, Curtis G, Zisook S, Kuhn K, Resovsky K, Le Winter M. *Ventricular arrhythmias possibly aggravated by trazodone*. Am J Psychiatry. **1983**, 140:796-797.

Jeavons PM, Clark JE, Harding GF. *Valproate and curly hair*. Lancet. **1977**, 1:359.

Jenike MA, Baer L, Summergrad P, Weilburg JB, Holland A, Seymour R. *Obsessive-compulsive disorder: a double-blind, placebo-controlled trial of clomipramine in 27 patients*. Am J Psychiatry. **1989**, 146:1328-1330.

Jenike MA, Hyman S, Baer L, Holland A, Minichiello WE, Buttolph L, Summergrad P, Seymour R, Ricciardi J. *A controlled trial of fluvoxamine in obsessive-compulsive disorder: implications for a serotonergic theory*. Am J Psychiatry. **1990**, 147:1209-1215.

Jerling M, Alvan G. *Nonlinear kinetics of nortriptyline in relation to nortriptyline clearance as observed during therapeutic drug monitoring*. Eur J Clin Pharmacol. **1994**, 46:67-70.

Joffe RT. *Valproate in bipolar disorder: the Canadian perspective*. Can J Psychiatry. **1993**, 38:S46-S50.

Johns S, Harris B. *Tremor*. Br Med J (Clin Res Ed). **1984**, 288:1309.

Johnson BA. *Trazodone toxicity*. Br J Hosp Med. **1985**, 33:298.

Johnson FN ed. *Depression & mania: modern lithium therapy*. IRL Press; Oxford, **1987**.

Johnson TA, Curtis MD, Beak P. *Highly diastereoselective and enantioselective carbon-carbon bond formations in conjugate additions of lithiated N-Boc allylamines to nitroalkenes: enantioselective synthesis of 3,4- and 3,4,5-substituted piperidines including (-)-paroxetine*. J Am Chem Soc. **2001**, 123:1004-1005.

Johnstone EC, Whaley K. *Antinuclear antibodies in psychiatric illness: their relationship to diagnosis and drug treatment*. Br Med J. **1975**, 2:724-725.

Jones SD. *Ejaculatory inhibition with trazodone*. J Clin Psychopharmacol. **1984**, 4:279-281.

Joyce RP, Gunderson CH. *Carbamazepine-induced orofacial dyskinesia*. Neurology. **1980**, 30:1333-1334.

Jue SG, Dawson GW, Brogden RN. *Amoxapine: a review of its pharmacology and efficacy in depressed states*. Drugs. **1982**, 24:1-23.

Kamal A, Khanna GBR, Ramu R. *Chemoenzymatic synthesis2 of both enantiomers of fluoxetine, tomoxetine and nisoxetine: lipase-catalyzed resolution of 3-aryl-3-hydroxypropanenitriles.* Tetrahedron: Asymm. **2002**, 13:2039-2051.

Kamper AM, Valentijn RM, Strickler BH, Purcell PM. *Cutaneous vasculitis induced by sodium valproate.* Lancet. **1991**, 337:497-498.

Kasarskis EJ, Kuo CS, Berger R, Nelson KR. *Carbamazepine-induced cardiac dysfunction. Characterization of two distinct clinical syndromes.* Arch Intern Med. **1992**, 152:186-191.

Katona CL, Barnes TR. *Pharmacological strategies in depression.* Br J Hosp Med. **1985**, 34:168-171.

Keating A, Blahunka P. *Carbamazepine-induced Stevens-Johnson syndrome in a child.* Ann Pharmacother. **1995**, 29:538-539.

Keck PE, Jr., McElroy SL, Tugrul KC, Bennett JA. *Valproate oral loading in the treatment of acute mania.* J Clin Psychiatry. **1993a**, 54:305-308.

Keck PE, Jr., Taylor VE, Tugrul KC, McElroy SL, Bennett JA. *Valproate treatment of panic disorder and lactate-induced panic attacks.* Biol Psychiatry. **1993b**, 33:542-546.

Kelly MW, Myers CW. *Clomipramine: a tricyclic antidepressant effective in obsessive compulsive disorder.* DICP. **1990**, 24:739-744.

Kishore-Kumar R, Max MB, Schafer SC, Gaughan AM, Smoller B, Gracely RH, Dubner R. *Desipramine relieves postherpetic neuralgia.* Clin Pharmacol Ther. **1990**, 47:305-312.

Kline SS, Mauro LS, Scala-Barnett DM, Zick D. *Serotonin syndrome versus neuroleptic malignant syndrome as a cause of death.* Clin Pharm. **1989**, 8:510-514.

Kodama Y, Tsutsumi K, Kuranari M, Kodama H, Fujii I, Takeyama M. *In vivo binding characteristics of carbamazepine and carbamazepine-10,11-epoxide to serum proteins in paediatric patients with epilepsy.* Eur J Clin Pharmacol. **1993**, 44:291-293.

Kraft TB. *Psychoses following trazodone administration.* Am J Psychiatry. **1983**, 140:1383-1384.

Kramlinger KG, Post RM. *Adding lithium carbonate to carbamazepine: antimanic efficacy in treatment-resistant mania.* Acta Psychiatr Scand. **1989**, 79:378-385.

Kreuz W, Linde R, Funk M, Meyer-Schrod R, Foll E, Nowak-Gottl U, Jacobi G, Vigh Z, Scharrer I. *Induction of von Willebrand disease type I by valproic acid.* Lancet. **1990**, 335:1350-1351.

Kronig MH, Roose SP, Walsh BT, Woodring S, Glassman AH. *Blood pressure effects of phenelzine.* J Clin Psychopharmacol. **1983**, 3:307-310.

Kuhn R. *[Treatment of depressive states with an iminodibenzyl derivative (G 22355).].* Schweiz Med Wochenschr. **1957**, 87:1135-1140.

Kumar A, Ner DH, Dike SY. *A new chemoenzymatic enantioselective synthesis of R-(-)-tomoxetine, (R)- and (S)-fluoxetine.* Tetrahedron Letters. **1991**, 32:1901-1904.

Kuroiwa Y, Shibasaki H. *Carbamazepine for tonic seizure in multiple sclerosis.* Lancet. **1967**, 289:116.

Kvinesdal B, Molin J, Froland A, Gram LF. *Imipramine treatment of painful diabetic neuropathy.* JAMA. **1984**, 251:1727-1730.

Lambert D, Dalac S. Skin, hair and nails. In: Johnson FN, ed. *Depression & mania: modern lithium therapy*. IRL Press, Oxford, **1987**, pp 232-234.

Lancaster SG, Gonzalez JP. *Dothiepin. A review of its pharmacodynamic and pharmacokinetic properties, and therapeutic efficacy in depressive illness*. Drugs. **1989a**, 38:123-147.

Lancaster SG, Gonzalez JP. *Lofepramine. A review of its pharmacodynamic and pharmacokinetic properties, and therapeutic efficacy in depressive illness*. Drugs. **1989b**, 37:123-140.

Lane RJ, Routledge PA. *Drug-induced neurological disorders*. Drugs. **1983**, 26:124-147.

Langdon N, Shindler J, Parkes JD, Bandak S. *Fluoxetine in the treatment of cataplexy*. Sleep. **1986**, 9:371-373.

Lantz RJ, Gillespie TA, Rash TJ, Kuo F, Skinner M, Kuan HY, Knadler MP. *Metabolism, excretion, and pharmacokinetics of duloxetine in healthy human subjects*. Drug Metab Dispos. **2003**, 31:1142-1150.

Lautin A, Stanley M, Angrist B, Gershon S. *Extrapyramidal syndrome with sodium valproate*. Br Med J. **1979**, 2:1035-1036.

Lawrie KW, Rustidge DC. *The synthesis of [methylene-14C]paroxetine BRL 29060A*. J Label Compd Radiopharm. **1993**, 33:777-781.

Lebert F, Pasquier F, Petit H. *Behavioral effects of trazodone in Alzheimer's disease*. J Clin Psychiatry. **1994**, 55:536-538.

Lecrubier Y. *Risk-benefit assessment of newer versus older monoamine oxidase (MAO) inhibitors*. Drug Saf. **1994**, 10:292-300.

Lefebure B, Castot A, Danan G, Elmalem J, Jean-Pastor MJ, Efthymiou ML. *[Hepatitis from antidepressants. Evaluation of cases from the French Association of Drug Surveillance Centers and the Technical Committee]*. Therapie. **1984**, 39:509-516.

Lefkowitz D, Kilgo G, Lee S. *Seizures and trazodone therapy*. Arch Gen Psychiatry. **1985**, 42:523.

Leonard H, Swedo S, Rapoport JL, Coffey M, Cheslow D. *Treatment of childhood obsessive compulsive disorder with clomipramine and desmethylimipramine: a double-blind crossover comparison*. Psychopharmacol Bull. **1988**, 24:93-95.

Lepola UM, Rimon RH, Riekkinen PJ. *Three-year follow-up of patients with panic disorder after short-term treatment with alprazolam and imipramine*. Int Clin Psychopharmacol. **1993**, 8:115-118.

Levine LR, Enas GG, Thompson WL, Byyny RL, Dauer AD, Kirby RW, Kreindler TG, Levy B, Lucas CP, McIlwain HH. *Use of fluoxetine, a selective serotonin-uptake inhibitor, in the treatment of obesity: a dose-response study (with a commentary by Michael Weintraub)*. Int J Obes. **1989**, 13:635-645.

Lewis-Jones MS, Evans S, Culshaw MA. *Cutaneous manifestations of zinc deficiency during treatment with anticonvulsants*. Br Med J (Clin Res Ed). **1985**, 290:603-604.

Li Wan Po, A. *Fluoxetine and suicide. meta-analysis and Monte-Carlo simulations*. Pharmacoepidemiol Drug Safety. **1993**, 2:79-84.

Liebowitz NR, el Mallakh RS. *Trazodone for the treatment of anxiety symptoms in substance abusers*. J Clin Psychopharmacol. **1989**, 9:449-451.

Lipinski JF, Jr., Mallya G, Zimmerman P, Pope HG, Jr. *Fluoxetine-induced akathisia: clinical and theoretical implications*. J Clin Psychiatry. **1989**, 50:339-342.

Lippmann S, Bedford P, Manshadi M, Mather S. *Trazodone cardiotoxicity*. Am J Psychiatry. **1983**, 140:1383.

Liu D, Gao W, Wang C, Zhang X. *Practical synthesis of enantiopure gamma-amino alcohols by rhodium-catalyzed asymmetric hydrogenation of beta-secondary-amino ketones*. Angew Chem Int Ed Engl. **2005**, 44:1687-1689.

LLerena A, Herraiz AG, Cobaleda J, Johansson I, Dahl ML. *Debrisoquin and mephenytoin hydroxylation phenotypes and CYP2D6 genotype in patients treated with neuroleptic and antidepressant agents*. Clin Pharmacol Ther. **1993**, 54:606-611.

Lonnqvist J, Sihvo S, Syvalahti E, Kiviruusu O. *Moclobemide and fluoxetine in atypical depression: a double-blind trial*. J Affect Disord. **1994**, 32:169-177.

Lucas E. *Process for the preparation of paroxetine and structurally related compounds*, WO2000078753 (**2000**); priority: 1999

Lusznat RM, Murphy DP, Nunn CM. *Carbamazepine vs lithium in the treatment and prophylaxis of mania*. Br J Psychiatry. **1988**, 153:198-204.

MacDougall LG. *Pure red cell aplasia associated with sodium valproate therapy*. JAMA. **1982**, 247:53-54.

Macfarlane JG, Jalali S, Grace EM. *Trimipramine in rheumatoid arthritis: a randomized double-blind trial in relieving pain and joint tenderness*. Curr Med Res Opin. **1986**, 10:89-93.

MacQueen G, Born L, Steiner M. *The selective serotonin reuptake inhibitor sertraline: its profile and use in psychiatric disorders*. CNS Drug Rev. **2001**, 7:1-24.

Maguire KP, Norman TR, McIntyre I, Burrows GD. *Clinical pharmacokinetics of dothiepin. Single-dose kinetics in patients and prediction of steady-state concentrations*. Clin Pharmacokinet. **1983**, 8:179-185.

Makin AJ, Fitt S, Williams R, Duncan JS. *Fulminant hepatic failure induced by lamotrigine*. BMJ. **1995**, 311:292.

Malcolm R, Ballenger JC, Sturgis ET, Anton R. *Double-blind controlled trial comparing carbamazepine to oxazepam treatment of alcohol withdrawal*. Am J Psychiatry. **1989**, 146:617-621.

Malen C, Danree B, Poignant J-C. *Tricyclic compounds*. US3758528 (**1973**); priority: 1970

Mallesham B, Rajesh BM, Reddy PR, Srinivas D, Trehan S. *Highly efficient CuI-catalyzed coupling of aryl bromides with oxazolidinones using Buchwald's protocol: a short route to linezolid and toloxatone*. Org Lett. **2003**, 5:963-965.

Mallinger AG, Edwards DJ, Himmelhoch JM, Knopf S, Ehler J. *Pharmacokinetics of tranylcypromine in patients who are depressed: relationship to cardiovascular effects*. Clin Pharmacol Ther. **1986**, 40:444-450.

Mallion KB, Turner RW, Todd AH. *Morpholine derivatives*. US3714161 (**1973**); priority: 1966

Mallya GK, White K, Waternaux D, Quay S. *Short- and long-term treatment of obsessive-compulsive disorder with fluvoxamine*. Ann Clin Psychiatry. **1992**, 4:77-80.

Mandrioli R, Cantelli Forti G, Raggi MA. *Fluoxetine metabolism and pharmacological interactions: the role of cytochrome P450*. Current Drug Metabolism **2006**, 7:127-133.

Mann SC, Walker MM, Messenger GG, Greenstein RA. *Leukocytoclastic vasculitis secondary to trazodone treatment*. J Am Acad Dermatol. **1984**, 10:669-670.

Manna V, Bolino F, Di Cicco L. *Chronic tension-type headache, mood depression and serotonin: therapeutic effects of fluvoxamine and mianserine*. Headache. **1994**, 34:44-49.

Marathe PH, Salazar DE, Greene DS, Brennan J, Shukla UA, Barbhaiya RH. *Absorption and presystemic metabolism of nefazodone administered at different regions in the gastrointestinal tract of humans*. Pharm Res. **1995**, 12:1716-1721.

Marcus RN, Sussman NM: *Nefazodone: use in migraine prophylaxis*. WO9804261 (**1998**); priority: 1996

Marescaux C, Warter JM, Micheletti G, Rumbach L, Coquillat G, Kurtz D. *Stuporous episodes during treatment with sodium valproate: report of seven cases*. Epilepsia. **1982**, 23:297-305.

Marks IM, Lelliott P, Basoglu M, Noshirvani H, Monteiro W, Cohen D, Kasvikis Y. *Clomipramine, self-exposure and therapist-aided exposure for obsessive-compulsive rituals*. Br J Psychiatry. **1988**, 152:522-534.

Martin CA, Piascik MT. *First degree A-V block in patients on lithium carbonate*. Can J Psychiatry. **1985**, 30:114-116.

Martin PR, Adinoff B, Eckardt MJ, Stapleton JM, Bone GA, Rubinow DR, Lane EA, Linnoila M. *Effective pharmacotherapy of alcoholic amnestic disorder with fluvoxamine. Preliminary findings*. Arch Gen Psychiatry. **1989**, 46:617-621.

Massey EW, Folger WN. *Seizures activated by therapeutic levels of lithium carbonate*. South Med J. **1984**, 77:1173-1175.

Mathew G. *Psychiatric symptoms associated with carbamazepine*. Br Med J (Clin Res Ed). **1988**, 296:1071.

Mathew NT, Ali S. *Valproate in the treatment of persistent chronic daily headache. An open label study*. Headache. **1991**, 31:71-74.

Mathew NT, Saper JR, Silberstein SD, Rankin L, Markley HG, Solomon S, Rapoport AM, Silber CJ, Deaton RL. *Migraine prophylaxis with divalproex*. Arch Neurol. **1995**, 52:281-286.

Mattes JA. *Comparative effectiveness of carbamazepine and propranolol for rage outbursts*. J Neuropsychiatry Clin Neurosci. **1990**, 2:159-164.

Mattila M, Mattila MJ, Vrijmoed-de Vries M, Kuitunen T. *Actions and interactions of psychotropic drugs on human performance and mood: single doses of ORG 3770, amitriptyline, and diazepam*. Pharmacol Toxicol. **1989**, 65:81-88.

Mattson RH, Cramer JA, Collins JF, Smith DB, Delgado-Escueta AV, Browne TR, Williamson PD, Treiman DM, McNamara JO, McCutchen CB, . *Comparison of carbamazepine, phenobarbital, phenytoin, and primidone in partial and secondarily generalized tonic-clonic seizures*. N Engl J Med. **1985**, 313:145-151.

Mattson RH, Cramer JA, Collins JF. *A comparison of valproate with carbamazepine for the treatment of complex partial seizures and secondarily generalized tonic-clonic seizures in adults. *The Department of Veterans Affairs Epilepsy Cooperative Study No. 264 Group*. N Engl J Med. **1992**, 327:765-771.

Maurer H. *Metabolism of trimipramine in man*. Arzneimittelforschung. **1989**, 39:101-103.

Max MB, Lynch SA, Muir J, Shoaf SE, Smoller B, Dubner R. *Effects of desipramine, amitriptyline, and fluoxetine on pain in diabetic neuropathy*. N Engl J Med. **1992**, 326:1250-1256.

Mayol RF, Cole CA, Luke GM, Colson KL, Kerns EH. *Characterization of the metabolites of the antidepressant drug nefazodone in human urine and plasma*. Drug Metab Dispos. **1994**, 22:304-311.

McElroy SL, Keck PE, Jr. *Treatment guidelines for valproate in bipolar and schizoaffective disorders*. Can J Psychiatry. **1993**, 38:S62-S66.

McFarling DA, Susac JO. *Letter: Carbamazepine for hiccoughs*. JAMA. **1974**, 230:962.

McGovern GP. *Lithium induced constructional dyspraxia*. Br Med J (Clin Res Ed). **1983**, 286:646.

McKee PJ, Blacklaw J, Butler E, Gillham RA, Brodie MJ. *Monotherapy with conventional and controlled-release carbamazepine: a double-blind, double-dummy comparison in epileptic patients*. Br J Clin Pharmacol. **1991**, 32:99-104.

McKee PJ, Blacklaw J, Carswell A, Gillham RA, Brodie MJ. *Double dummy comparison between once and twice daily dosing with modified-release carbamazepine in epileptic patients*. Br J Clin Pharmacol. **1993**, 36:257-261.

McKee RJ, Larkin JG, Brodie MJ. *Acute psychosis with carbamazepine and sodium valproate*. Lancet. **1989**, 1:167.

McTavish D, Benfield P. *Clomipramine. An overview of its pharmacological properties and a review of its therapeutic use in obsessive compulsive disorder and panic disorder*. Drugs. **1990**, 39:136-153.

Mehta NB. *Meta chloro substituted-alpha-butylamino-propiophenones*. US3819706 (**1974**); priority: 1969

Melloni P, Torre AD, Carniel GC, Rossi A. *Substituted morpholine derivatives and compositions*. US4229449 (**1980**); priority: 1978

Mendlewicz J. *Efficacy of fluvoxamine in severe depression*. Drugs. **1992**, 43 Suppl 2:32-37.

Menkes DB, Taghavi E, Mason PA, Spears GF, Howard RC. *Fluoxetine treatment of severe premenstrual syndrome*. BMJ. **1992**, 305:346-347.

Merino N, Duran JA, Jimenez MC, Ravella R. *Multisystem hypersensitivity reaction to carbamazepine*. Ann Pharmacother. **1994**, 28:402-403.

Milne RJ, Goa KL. *Citalopram. A review of its pharmacodynamic and pharmacokinetic properties, and therapeutic potential in depressive illness*. Drugs. **1991**, 41:450-477.

Molloy BB, Schmiegel KK. *Arloxyphenylpropylamines*. US4314081 (**1982**); priority: 1974

Montalescot G, Levy Y, Hatt PY. *Serious sinus node dysfunction caused by therapeutic doses of lithium*. Int J Cardiol. **1984**, 5:94-96.

Montgomery S, Ferguson JM, Schwartz GE. *The antidepressant efficacy of reboxetine in patients with severe depression.* J Clin Psychopharmacol. **2003**, 23:45-50.

Montgomery SA, McIntyre A, Osterheider M, Sarteschi P, Zitterl W, Zohar J, Birkett M, Wood AJ. *A double-blind, placebo-controlled study of fluoxetine in patients with DSM-III-R obsessive-compulsive disorder. The Lilly European OCD Study Group.* Eur Neuropsychopharmacol. **1993**, 3:143-152.

Montgomery SA, Henry J, McDonald G, Dinan T, Lader M, Hindmarch I, Clare A, Nutt D. *Selective serotonin reuptake inhibitors: meta-analysis of discontinuation rates.* Int Clin Psychopharmacol. **1994**, 9:47-53.

Montgomery SA, Dunner DL, Dunbar GC. *Reduction of suicidal thoughts with paroxetine in comparison with reference antidepressants and placebo.* Eur Neuropsychopharmacol. **1995**, 5:5-13.

Montgomery SA. *Chairman's overview. The place of reboxetine in antidepressant therapy.* J Clin Psychiatry. **1998**, 59 Suppl 14:26-29.

Mortola JF. *A risk-benefit appraisal of drugs used in the management of premenstrual syndrome.* Drug Saf. **1994**, 10:160-169.

Morton WA, Sonne SC, Verga MA. *Venlafaxine: a structurally unique and novel antidepressant.* Ann Pharmacother. **1995**, 29:387-395.

Mouzin G, Cousse H, Bonnaud B, Morre M, Stenger A. *1-Aryl 2-Aminomethyl Cyclopropane Carboxyamide (Z) Derivatives And Their Use As Useful Drugs In The Treatment Of Disturbances Of The Central Nervous System.* US4478836 (**1984**); priority: 1981

Mucklow J. *Selected side-effects 2: carbamazepine and hyponatraemia.* Prescriber's J. **1991**, 31:61-64.

Murdoch D, McTavish D. *Sertraline. A review of its pharmacodynamic and pharmacokinetic properties, and therapeutic potential in depression and obsessive-compulsive disorder.* Drugs. **1992**, 44:604-624.

Murphy DJ, Gannon MA, McGennis A. *Carbamazepine in bipolar affective disorder.* Lancet. **1989**, 2:1151-1152.

Musa MN. *Nonlinear kinetics of trimipramine in depressed patients.* J Clin Pharmacol. **1989**, 29:746-747.

Myers DH, West TET. Hormone systems. In: Johnson FN, ed. *Depression & mania: modern lithium therapy.* Oxford: IRL Press, Oxford, **1987**, pp 220-226.

Nakanishi S. *Cis-11-[(3-Dimethylamino)propylidiene]-6,11-dihydrodibenz[b,e]oxepin.* DE215318 (**1972**); priority: 1970

Nakielny J. *Fluoxetine and suicide.* Lancet. **1994**, 343:1359.

Naranjo CA, Sellers EM, Sullivan JT, Woodley DV, Kadlec K, Sykora K. *The serotonin uptake inhibitor citalopram attenuates ethanol intake.* Clin Pharmacol Ther. **1987**, 41:266-274.

Naranjo CA, Kadlec KE, Sanhueza P, Woodley-Remus D, Sellers EM. *Fluoxetine differentially alters alcohol intake and other consummatory behaviors in problem drinkers.* Clin Pharmacol Ther. **1990**, 47:490-498.

Naranjo CA, Poulos CX, Bremner KE, Lanctot KL. *Citalopram decreases desirability, liking, and consumption of alcohol in alcohol-dependent drinkers*. Clin Pharmacol Ther. **1992**, 51:729-739.

Neglia JP, Glaze DG, Zion TE. *Tics and vocalizations in children treated with carbamazepine*. Pediatrics. **1984**, 73:841-844.

Neuvonen PJ, Pohjola-Sintonen S, Tacke U, Vuori E. *Five fatal cases of serotonin syndrome after moclobemide-citalopram or moclobemide-clomipramine overdoses*. Lancet. **1993**, 342:1419.

Newton RW. *Randomised controlled trials of phenobarbitone and valproate in febrile convulsions*. Arch Dis Child. **1988**, 63:1189-1191.

Nicholson RJ, Kelly KP, Grant IS. *Leucopenia associated with lamotrigine*. BMJ. **1995**, 310:504.

Nielsen KK, Brosen K, Hansen MG, Gram LF. *Single-dose kinetics of clomipramine: relationship to the sparteine and S-mephenytoin oxidation polymorphisms*. Clin Pharmacol Ther. **1994**, 55:518-527.

Nielsen KK, Flinois JP, Beaune P, Brosen K. *The biotransformation of clomipramine in vitro, identification of the cytochrome P450s responsible for the separate metabolic pathways*. J Pharmacol Exp Ther. **1996**, 277:1659-1664.

Nielsen NV, Syversen K. *Possible retinotoxic effect of carbamazepine*. Acta Ophthalmol (Copenh). **1986**, 64:287-290.

Nightingale S. *Management of migraine*. Prescriber's J. **1985**, 25:129-134.

Norden MJ. *Buspirone treatment of sexual dysfunction associated with selective serotonin re-uptake inhibitors*. Depression. **1994**, 2:109-112.

Nordenström J, Strigard K, Perbeck L, Willems J, Bagedahl-Strindlund M, Linder J. *Hyperparathyroidism associated with treatment of manic-depressive disorders by lithium*. Eur J Surg. **1992**, 158:207-211.

Nordin C, Siwers B, Benitez J, Bertilsson L. *Plasma concentrations of nortriptyline and its 10-hydroxy metabolite in depressed patients--relationship to the debrisoquine hydroxylation metabolic ratio*. Br J Clin Pharmacol. **1985**, 19:832-835.

Nordin C, Bertilsson L. *Active hydroxymetabolites of antidepressants. Emphasis on E-10-hydroxy-nortriptyline*. Clin Pharmacokinet. **1995**, 28:26-40.

Norman TR, Burrows GD. *A risk-benefit assessment of moclobemide in the treatment of depressive disorders*. Drug Saf. **1995**, 12:46-54.

Nurnberg HG, Martin GA, Karajgi BM, Roskin JK, Longshore CT. *Response to anticonvulsant substitution among refractory bipolar manic patients*. J Clin Psychopharmacol. **1994**, 14:207-209.

Oakley AM, Hodge L. *Cutaneous vasculitis from maprotiline*. Aust N Z J Med. **1985**, 15:256-257.

Ogilvie AD. *Hair loss during fluoxetine treatment*. Lancet. **1993**, 342:1423.

Okusa MD, Crystal LJ. *Clinical manifestations and management of acute lithium intoxication*. Am J Med. **1994**, 97:383-389.

Osterman PO. *Paroxysmal itching in multiple sclerosis*. Br J Dermatol. **1976**, 95:555-558.

Ottevanger EA. *The efficacy of fluvoxamine in patients with severe depression*. Br J Clin Res. **1991**, 2:125-132.

Overmars H, Scherpenisse PM, Post LC. *Fluvoxamine maleate: metabolism in man*. Eur J Drug Metab Pharmacokinet. **1983**, 8:269-280.

Palazzo G, Silvestrini B. *Triazole-(4,3-a)-pyridines*. US3381009 (**1968**); priority: 1965

Palazzo G. *1-(3-(4-Metrachlorophenyl-1-piperazinyl)-propyl)-3,4-diethyl-delta-1,2,4-triazolin-5-one*. US3857845 (**1974**); priority: 1972

Palileo EV, Coelho A, Westveer D, Dhingra R, Rosen KM. *Persistent sinus node dysfunction secondary to lithium therapy*. Am Heart J. **1983**, 106:1443-1444.

Palmer KJ, Benfield P. *Fluvoxamine: an overview of its pharmacological properties and review of its therapeutic potential in non-depressive disorders*. CNS Drugs. **1994**, 1:57-87.

Palmier C, Puozzo C, Lenehan T, Briley M. *Monoamine uptake inhibition by plasma from healthy volunteers after single oral doses of the antidepressant milnacipran*. Eur J Clin Pharmacol. **1989**, 37:235-238.

Panayiotopoulos CP. *Nocturnal enuresis associated with sodium valproate*. Lancet. **1985**, 1:980-981.

Pandit MK, Burke J, Gustafson AB, Minocha A, Peiris AN. *Drug-induced disorders of glucose tolerance*. Ann Intern Med. **1993**, 118:529-539.

Panunzio M, Rossi K, Tamanini E, Campana E, Martelli G. *Synthesis of enantiomerically pure (S)- and (R)-fluoxetine (Prozac®) via a hetero Diels-Alder strategy*. Tetrahedron: Asymm. **2004**, 15:3489-3493.

Park BK, Kitteringham NR. *Adverse reactions and drug metabolism*. Adverse Drug React Bull. **1987**, 122:456-459.

Pasion RC, Kirby SG. *Trazodone for screaming*. Lancet. **1993**, 341:970.

Pataki CS, Carlson GA, Kelly KL, Rapport MD, Biancaniello TM. *Side effects of methylphenidate and desipramine alone and in combination in children*. J Am Acad Child Adolesc Psychiatry. **1993**, 32:1065-1072.

Patel VM, Kansara RR, Patel NV, Rehani RB, Thennati R. *Process for the preparation of antidepressant compound*. WO2005049560 (**2005**); priority: 2003

Patterson BD, Srisopark MM. *Severe anorexia and possible psychosis or hypomania after trazodone-tryptophan treatment of aggression*. Lancet. **1989**, 1:1017.

Patterson JF. *Amoxapine-induced chorea*. South Med J. **1983**, 76:1077.

Pearlstein TB, Stone AB. *Long-term fluoxetine treatment of late luteal phase dysphoric disorder*. J Clin Psychiatry. **1994**, 55:332-335.

Peet M, Pratt JP. *Lithium. Current status in psychiatric disorders*. Drugs. **1993**, 46:7-17.

Pento JT. *WY-45030*. Drugs of the future. **1988**, 13:839-840.

Perucca E, Gatti G, Spina E. *Clinical pharmacokinetics of fluvoxamine*. Clin Pharmacokinet. **1994**, 27:175-190.

Peters LR, Hennion GF. *Nortriptylene*. BE628904 (**1963**); priority: 1962

Petersen H, Bregnedal P, Boegesoe KP. *Method for the preparation of citalopram*. WO1998019512 (**1998**); priority: 1997.

Petersen H, Dancer R. *Method for the preparation of citalopram*. WO2001085712 (**2001**); priority: 2000.

Petersen H. *Method for the preparation of Citalopram*. WO2001068630 (**2001**); priority: 2000.

Pigott TA, Pato MT, Bernstein SE, Grover GN, Hill JL, Tolliver TJ, Murphy DL. *Controlled comparisons of clomipramine and fluoxetine in the treatment of obsessive-compulsive disorder. Behavioral and biological results*. Arch Gen Psychiatry. **1990**, 47:926-932.

Pitchot W, Gonzalez-Moreno A, Ansseau M. *Therapeutic window for 5-HT reuptake inhibitors*. Lancet. **1992**, 339:689.

Poggesi I, Pellizzoni C, Fleishaker JC. *Pharmacokinetics of reboxetine in elderly patients with depressive disorders*. Int J Clin Pharmacol Ther. **2000**, 38:254-259.

Pollock BG, Perel JM, Shostak M. *Rapid achievement of antidepressant effect with intravenous chlorimipramine*. N Engl J Med. **1985**, 312:1130.

Pope HG, Jr., McElroy SL, Keck PE, Jr., Hudson JI. *Valproate in the treatment of acute mania. A placebo-controlled study*. Arch Gen Psychiatry. **1991**, 48:62-68.

Powell-Jackson PR, Tredger JM, Williams R. *Hepatotoxicity to sodium valproate: a review*. Gut. **1984**, 25:673-681.

Power AC, Cowen PJ. *Fluoxetine and suicidal behaviour. Some clinical and theoretical aspects of a controversy*. Br J Psychiatry. **1992**, 161:735-741.

Presley AP, Kahn A, Williamson N. *Antinuclear antibodies in patients on lithium carbonate*. Br Med J. **1976**, 2:280-281.

Price LH, Heninger GR. *Lithium in the treatment of mood disorders*. N Engl J Med. **1994**, 331:591-598.

Primeau F, Fontaine R, Beauclair L. *Valproic acid and panic disorder*. Can J Psychiatry. **1990**, 35:248-250.

Qian MR, Zeng S. *Biosynthesis of imipramine glucuronide and characterization of imipramine glucuronidation catalyzed by recombinant UGT1A4*. Acta Pharmacol Sin. **2006**, 27:623-628.

Quallich GJ, Williams MT. *Process For Preparing a 4,4-Diphenylbutanoic Acid Derivative*. US4777288 (**1988**); priority: 1987.

Quallich GJ, Woodall TM. *Synthesis of 4(S)-(3,4-dichlorophenyl)-3,4-dihydro-1(2H)-naphthalenone by SN2 cuprate displacement of an activated chiral benzylic alcohol*. Tetrahedron. **1992**, 48:10239-10248.

Rahman MK, Akhtar MJ, Savla NC, Sharma RR, Kellett JM, Ashford JJ. *A double-blind, randomised comparison of fluvoxamine with dothiepin in the treatment of depression in elderly patients.* Br J Clin Pract. **1991**, 45:255-258.

Rambeck B, Wolf P. *Lamotrigine clinical pharmacokinetics.* Clin Pharmacokinet. **1993**, 25:433-443.

Raoof NT, Pearson RM, Turner P. *Lithium inhibits human sperm motility in vitro.* Br J Clin Pharmacol. **1989**, 28:715-717.

Rapport MD, Carlson GA, Kelly KL, Pataki C. *Methylphenidate and desipramine in hospitalized children: I. Separate and combined effects on cognitive function.* J Am Acad Child Adolesc Psychiatry. **1993**, 32:333-342.

Rausch JL, Pavlinac DM, Newman PE. *Complete heart block following a single dose of trazodone.* Am J Psychiatry. **1984**, 141:1472-1473.

Reiss AL, O'Donnell DJ. *Carbamazepine-induced mania in two children: case report.* J Clin Psychiatry. **1984**, 45:272-274.

Reiss RA, Haas CE, Karki SD, Gumbiner B, Welle SL, Carson SW. *Lithium pharmacokinetics in the obese.* Clin Pharmacol Ther. **1994**, 55:392-398.

Reutens DC, Duncan JS, Patsalos PN. *Disabling tremor after lamotrigine with sodium valproate.* Lancet. **1993**, 342:185-186.

Rey-Bellet G, Spiegelberg G. *Verfahren zur Herstellung von Dibenzo-cycloheptaenen.* CH356759 (**1961**); priority: 1958.

Reynolds JE. *MARTINDALE: The Extra Pharmacopoeia.* 31st ed. The Royal Pharmaceutical Society; London, **1996**.

Rhone-Poulenc. *Nouveau procédé de préparation d'amines tertiaires aminoalcoylées.* FR1172014 (**1959**); priority: 1955

Rhone-Poulenc. *5-(3-dimethylamino-2-methylpropyl)iminodibenzyl.* FR1380404 (**1964**); priority: 1959

Richens A, Yuen AW. *Overview of the clinical efficacy of lamotrigine.* Epilepsia. **1991**, 32 Suppl 2:S13-S16.

Richens A, Davidson DL, Cartlidge NE, Easter DJ. *A multicentre comparative trial of sodium valproate and carbamazepine in adult onset epilepsy. Adult EPITEG Collaborative Group.* J Neurol Neurosurg Psychiatry. **1994**, 57:682-687.

Richer M, Crismon ML. *Pharmacotherapy of sexual offenders.* Ann Pharmacother. **1993**, 27:316-320.

Rickels K, Schweizer E, Clary C, Fox I, Weise C. *Nefazodone and imipramine in major depression: a placebo-controlled trial.* Br J Psychiatry. **1994**, 164:802-805.

Riva F. *FCE-20124.* Drugs of the future. **1985**, 10:905-906.

Robertson DW, Krushinski JH, Fuller RW, Leander JD. *Absolute configurations and pharmacological activities of the optical isomers of fluoxetine, a selective serotonin-uptake inhibitor.* J Med Chem. **1988**, 31:1412-1417.

Robertson DW, Krushinski JH, Fuller RW, Leander JD. *3-Aryloxy-3-substituted propanamines.* EP273658 (**1988**); priority: 1986

Rock MH, Petersen H, and Ellegaard P: Method for the preparation of citalopram, WO2000012044 (**2000**); priority: 1999

Roig M, Montserrat L, Gallart A. *Carbamazepine: an alternative drug for the treatment of nonhereditary chorea.* Pediatrics. **1988**, 82:492-495.

Roose SP, Glassman AH, Attia E, Woodring S. *Comparative efficacy of selective serotonin reuptake inhibitors and tricyclics in the treatment of melancholia.* Am J Psychiatry. **1994**, 151:1735-1739.

Rothschild AJ, Locke CA. *Reexposure to fluoxetine after serious suicide attempts by three patients: the role of akathisia.* J Clin Psychiatry. **1991**, 52:491-493.

Rotzinger S, Baker GB. *Human CYP3A4 and the metabolism of nefazodone and hydroxynefazodone by human liver microsomes and heterologously expressed enzymes.* Eur Neuropsychopharmacol. **2002**, 12:91-100.

Roulet E, Deonna T. *Successful treatment of hereditary dominant chorea with carbamazepine.* Pediatrics. **1989**, 83:1077.

Rudnic EM, Belendiuk GW, McCarty J, Wassink S, Couch RA. *Advanced drug delivery system and method of treating psychiatric, neurological and other disorders with carbamazepine.* US5912013 (**1999**); priority: 1995.

Rudolph R, Entsuah R, Derivan A. *Early clinical response in depression to venlafaxine hydrochloride.* Biol Psychiatry. **1991**, 29:630S.

Ruijten HM, De Bree H, Borst AJ, de Lange N, Scherpenisse PM, Vincent WR, Post LC. *Fluvoxamine: metabolic fate in animals.* Drug Metab Dispos. **1984**, 12:82-92.

Ryan SW, Forsythe I, Hartley R, Haworth M, Bowmer CJ. *Slow release carbamazepine in treatment of poorly controlled seizures.* Arch Dis Child. **1990**, 65:930-935.

Salas IG, Pearson RM, Lawson M. *Lithium carbonate concentration in cervicovaginal mucus and serum after repeated oral dose administration.* Br J Clin Pharmacol. **1989**, 28:751P.

Sallee FR, Pollock BG. *Clinical pharmacokinetics of imipramine and desipramine.* Clin Pharmacokinet. **1990**, 18:346-364.

Sansone ME, Ziegler DK. Brain and nervous system. In: Johnson FN, ed. *Depression & mania: modern lithium therapy.* IRL Press, Oxford, **1987**, pp 240-245.

Saper JR, Silberstein SD, Lake AE, III, Winters ME. *Double-blind trial of fluoxetine: chronic daily headache and migraine.* Headache. **1994**, 34:497-502.

Scates AC, Doraiswamy PM. *Reboxetine: a selective norepinephrine reuptake inhibitor for the treatment of depression.* Ann Pharmacother. **2000**, 34:1302-1312.

Schaff MR, Fawcett J, Zajecka JM. *Divalproex sodium in the treatment of refractory affective disorders.* J Clin Psychiatry. **1993**, 54:380-384.

Schaub JE, Williamson PJ, Barnes EW, Trewby PN. *Multisystem adverse reaction to lamotrigine.* Lancet. **1994**, 344:481.

Schindler W. *New N-heterocyclic compounds*. US2948718 (**1960**); priority: 1957

Schindler W. *Verfahren zur Herstellung von 5-{ -(4'-Hydroxyaethylpiperazino)-propyl}-dibenzo[b,f] azepin*. DE1133729 (**1962**); priority: 1957

Schindler W. *Verfahren zur Herstellung von Aminderivaten des 5H-Dibenzo[a,d]-cycloheptens*. DE1288599 (**1962**); priority: 1961

Schindler W, Dietrich H. *Verfahren zur Herstellung von neuen N-heterocyclischen Verbindungen*. CH371799 (**1963**); priority: 1958

Schipperheijn JA, Dunne FJ. *Managing violence in psychiatric hospitals*. BMJ. **1991**, 303:71-72.

Schmidt D, Haenel F. *Therapeutic plasma levels of phenytoin, phenobarbital, and carbamazepine: individual variation in relation to seizure frequency and type*. Neurology. **1984**, 34:1252-1255.

Schmidt HS, Clark RW, Hyman PR. *Protriptyline: an effective agent in the treatment of the narcolepsy-cataplexy syndrome and hypersomnia*. Am J Psychiatry. **1977**, 134:183-185.

Schmolka SJ, Zimmer H. *N-Dimethylaminopropylation in a solid-liquid two phase system: synthesis of chlorpromazine, its analogs, and related compounds*. Synthesis. **1984**, 1:29-31.

Schneier FR, Liebowitz MR, Davies SO, Fairbanks J, Hollander E, Campeas R, Klein DF. *Fluoxetine in panic disorder*. J Clin Psychopharmacol. **1990**, 10:119-121.

Schou M. *Use in non-psychiatric conditions*. In: Johnson FN, ed. Depression & mania: modern lithium therapy. IRL Press, Oxford, **1987**, pp 46-50.

Schou M. *Lithium treatment of manic-depressive illness. Past, present, and perspectives*. JAMA. **1988a**, 259:1834-1836.

Schou M. *Serum lithium monitoring of prophylactic treatment. Critical review and updated recommendations*. Clin Pharmacokinet. **1988b**, 15:283-286.

Schou M. *Lithium treatment of manic-depressive illness: a practical guide*. 4th rev.ed. Karger AG; Basel, **1989**.

Schulz P, Dick P, Blaschke TF, Hollister L. *Discrepancies between pharmacokinetic studies of amitriptyline*. Clin Pharmacokinet. **1985**, 10:257-268.

Schweizer E, Rickels K, Case WG, Greenblatt DJ. *Carbamazepine treatment in patients discontinuing long-term benzodiazepine therapy. Effects on withdrawal severity and outcome*. Arch Gen Psychiatry. **1991**, 48:448-452.

Sechi GP, Piras MR, Demurtas A, Tanca S, Rosati G. *Dexamethasone-induced schizoaffective-like state in multiple sclerosis: prophylaxis and treatment with carbamazepine*. Clin Neuropharmacol. **1987**, 10:453-457.

Sechi GP, Zuddas M, Piredda M, Agnetti V, Sau G, Piras ML, Tanca S, Rosati G. *Treatment of cerebellar tremors with carbamazepine: a controlled trial with long-term follow-up*. Neurology. **1989**, 39:1113-1115.

Seckl J, Dunger D. *Postoperative diabetes insipidus*. Br Med J. **1989**, 298:2-3.

Segal M, Heys RF. *Inappropriate lactation*. Br Med J. **1969**, 4:236.

Seliger GM, Hornstein A, Flax J, Herbert J, Schroeder K. *Fluoxetine improves emotional incontinence.* Brain Inj. **1992**, 6:267-270.

Serre C, Clerc G, Escande M, Feline A, Ginestet D, Tignol J, Van Ameringen P. *An early clinical trial of midalcipran, 1-phenyl-1-diethyl aminocarbonyl 2-aminomethylcyclopropane (Z) hydrochloride, a potential fourth generation antidepressant.* Curr Ther Res. **1986**, 39:156-164.

Shea JP, Shukla UA, Pittman KA. *Single dose pharmacokinetics of nefazodone in elderly subjects, renally impaired patients, and patients with hepatic cirrhosis in comparison to healthy volunteers.* Clin Pharmacol Ther. **1988**, 43:146.

Shea JP, et al. *Single dose pharmacokinetics of nefazodone in poor and extensive metabolizers of dextromethorphan.* J Clin Pharmacol. **1989**, 29:841.

Sheikh KH, Nies AS. *Trazodone and intrahepatic cholestasis.* Ann Intern Med. **1983**, 99:572.

Shen WW, Sata LS. *Inhibited female orgasm resulting from psychotropic drugs. A clinical review.* J Reprod Med. **1983**, 28:497-499.

Sheth RD, Wesolowski CA, Jacob JC, Penney S, Hobbs GR, Riggs JE, Bodensteiner JB. *Effect of carbamazepine and valproate on bone mineral density.* J Pediatr. **1995**, 127:256-262.

Shukla UA, Marathe PH, Pittman KA, Barbhaiya RH. *Pharmacokinetics, absolute bioavailability, and disposition of [14C]nefazodone in the dog.* Drug Metab Dispos. **1993**, 21:502-507.

Shukla VR, Borison RL. *Lithium and lupuslike syndrome.* JAMA. **1982**, 248:921-922.

Simon LT, Hsu B, Adornato BT. *Carbamazepine-induced aseptic meningitis.* Ann Intern Med. **1990**, 112:627-628.

Simpson GM, Blair JH, Amuso D. *Effects of anti-depressants on genito-urinary function.* Dis Nerv Syst. **1965**, 26:787-789.

Sindrup SH, Gram LF, Brosen K, Eshoj O, Mogensen EF. *The selective serotonin reuptake inhibitor paroxetine is effective in the treatment of diabetic neuropathy symptoms.* Pain. **1990**, 42:135-144.

Sindrup SH, Bjerre U, Dejgaard A, Brosen K, Aaes-Jorgensen T, Gram LF. *The selective serotonin reuptake inhibitor citalopram relieves the symptoms of diabetic neuropathy.* Clin Pharmacol Ther. **1992a**, 52:547-552.

Sindrup SH, Brosen K, Gram LF, Hallas J, Skjelbo E, Allen A, Allen GD, Cooper SM, Mellows G, Tasker TC. *The relationship between paroxetine and the sparteine oxidation polymorphism.* Clin Pharmacol Ther. **1992b**, 51:278-287.

Singer HS, Brown J, Quaskey S, Rosenberg LA, Mellits ED, Denckla MB. *The treatment of attention-deficit hyperactivity disorder in Tourette's syndrome: a double-blind placebo-controlled study with clonidine and desipramine.* Pediatrics. **1995**, 95:74-81.

Sitsen MJA, Moors J. *Mirtazapine: a novel antidepressant in the treatment of anxiety symptoms: results from a placebo-controlled trial.* Drug Invest. **1994**, 8:339-344.

Skelton WP, III. *Neuropathic beriberi and carbamazepine.* Ann Intern Med. **1988**, 109:598-599.

Slater SL, Lipper S, Shiling DJ, Murphy DL. *Elevation of plasma-prolactin by monoamine-oxidase inhibitors.* Lancet. **1977**, 2:275-276.

Slywka S, Hart LL. *Fluoxetine in alcoholism.* Ann Pharmacother. **1993**, 27:1066-1067.

Small GW, Purcell JJ. *Trazodone and cocaine abuse.* Arch Gen Psychiatry. **1985**, 42:524.

Smith DW, Cullity GJ, Silberstein EP. *Fatal hepatic necrosis associated with multiple anticonvulsant therapy.* Aust N Z J Med. **1988**, 18:575-581.

Smith NJ, Espir ML, Baylis PH. *Raised plasma arginine vasopressin concentration in carbamazepine-induced water intoxication.* Br Med J. **1977**, 2:804.

Sobotka JL, Alexander B, Cook BL. *A review of carbamazepine's hematologic reactions and monitoring recommendations.* DICP. **1990**, 24:1214-1219.

Soelberg SP, Hammer M. *Effects of long-term carbamazepine treatment on water metabolism and plasma vasopressin concentration.* Eur J Clin Pharmacol. **1984**, 26:719-722.

Solomon K, Vickers R. *Dysarthria resulting from lithium carbonate. A case report.* JAMA. **1975**, 231:280.

Song F, Freemantle N, Sheldon TA, House A, Watson P, Long A, Mason J. *Selective serotonin reuptake inhibitors: meta-analysis of efficacy and acceptability.* BMJ. **1993**, 306:683-687.

Sorbera LA, Revel L, Martin L, Castaner J. *Escitalopram oxalate. Antidepressant 5-HT reuptake inhibitor.* Drugs of the future. **2001**, 26:115-120.

Sorensen KV. *Valproate: a new drug in migraine prophylaxis.* Acta Neurol Scand. **1988**, 78:346-348.

Sosin D. *Clinical efficacy of fluoxetine vs. sertraline in a headache clincia population.* Headache. **1993**, 33:284.

Spofa. *6,11- Dihydrodibenzo[b,e]thiepins.* BE618591 (**1962**); priority: 1961

Sramek J, Herrera J, Costa J, Heh C, Tran-Johnson T, Simpson G. *A carbamazepine trial in chronic, treatment-refractory schizophrenia.* Am J Psychiatry. **1988**, 145:748-750.

Staack RF, Maurer HH. *Piperazine-derived designer drug 1-(3-chlorophenyl) piperazine (mCPP): GC-MS studies on its metabolism and its toxicological detection in rat urine including analytical differentiation from its precursor drugs trazodone and nefazodone.* J Anal Toxicol. **2003**, 27:560-568.

Steiner M, Steinberg S, Stewart D, Carter D, Berger C, Reid R, Grover D, Streiner D. *Fluoxetine in the treatment of premenstrual dysphoria. Canadian Fluoxetine/Premenstrual Dysphoria Collaborative Study Group.* N Engl J Med. **1995**, 332:1529-1534.

Stephens WP, Espir ML, Tattersall RB, Quinn NP, Gladwell SR, Galbraith AW, Reynolds EH. *Water intoxication due to carbamazepine .* Br Med J. **1977**, 1:754-755.

Stephens WP, Coe JY, Baylis PH. *Plasma arginine vasopressin concentrations and antidiuretic action of carbamazepine.* Br Med J. **1978**, 1:1445-1447.

Stewart JW, Harrison W, Quitkin F, Liebowitz MR. *Phenelzine-induced pyridoxine deficiency.* J Clin Psychopharmacol. **1984**, 4:225-226.

Stoll AL, Banov M, Kolbrener M, Mayer PV, Tohen M, Strakowski SM, Castillo J, Suppes T, Cohen BM. *Neurologic factors predict a favorable valproate response in bipolar and schizoaffective disorders.* J Clin Psychopharmacol. **1994**, 14:311-313.

Stone AB, Pearlstein TB, Brown WA. *Fluoxetine in the treatment of late luteal phase dysphoric disorder.* J Clin Psychiatry. **1991**, 52:290-293.

Stone S, Lange LS. *Syncope and sudden unexpected death attributed to carbamazepine in a 20-year-old epileptic.* J Neurol Neurosurg Psychiatry. **1986**, 49:1460-1461.

Sullivan M, Katon W, Russo J, Dobie R, Sakai C. *A randomized trial of nortriptyline for severe chronic tinnitus. Effects on depression, disability, and tinnitus symptoms.* Arch Intern Med. **1993**, 153:2251-2259.

Swartz CM, Jones P. *Hyperlithemia correction and persistent delirium.* J Clin Pharmacol. **1994**, 34:865-870.

Swedo SE, Leonard HL, Rapoport JL, Lenane MC, Goldberger EL, Cheslow DL. *A double-blind comparison of clomipramine and desipramine in the treatment of trichotillomania (hair pulling).* N Engl J Med. **1989**, 321:497-501.

Swedo SE, Lenane MC, Leonard HL. *Long-term treatment of trichotillomania (hair pulling).* N Engl J Med. **1993**, 329:141-142.

Sweetman SC. *MARTINDALE: The Complete drug reference.* 34th ed. Pharmaceutical Press; London, **2004**.

Symon DN, Russell G. *Sodium valproate and neutropenia.* Arch Dis Child. **1983**, 58:235.

Tangedahl TN, Gau GT. *Myocardial irritability associated with lithium carbonate therapy.* N Engl J Med. **1972**, 287:867-869.

Tao GK, Harada DT, Kootsikas ME, Gordon MN, Brinkman JH. *Amoxapine-induced tardive dyskinesia.* Drug Intell Clin Pharm. **1985**, 19:548-549.

Task Force on the Use of Laboratory Tests in Psychiatry. *Tricyclic antidepressants--blood level measurements and clinical outcome: an APA Task Force report.* Am J Psychiatry. **1985**, 142:155-162.

Taylor JW, Bell AJ. *Lithium-induced parathyroid dysfunction: a case report and review of the literature.* Ann Pharmacother. **1993**, 27:1040-1043.

Tedeschi RE. *Monoamine oxidase inhibition.* US2997422 (**1961**); priority: 1959

Telstad W, Sorensen O, Larsen S, Lillevold PE, Stensrud P, Nyberg-Hansen R. *Treatment of the restless legs syndrome with carbamazepine: a double blind study.* Br Med J (Clin Res Ed). **1984**, 288:444-446.

Temple DL, Lobeck WG. *Phenoxyethyl-1,2,4,-triazol-3-one antidepressants.* US4338317 (**1982**); priority: 1981

Tishler M, Chemerda JM, and Kollonitsch JM. *5h-dibenzo [a, d] cycloheptenes.* US3244748 (**1966**); priority: 1962

Tobiansky RI, Shah AK, Benzer A, Hasibeder M. *Carbamazepine update.* Lancet. **1989**, 334:867.

Tremaine LM, Stroh JG, Ronfeld RA. *Characterization of a carbamic acid ester glucuronide of the secondary amine sertraline.* Drug Metab Dispos. **1989a**, 17:58-63.

Tremaine LM, Welch WM, Ronfeld RA. *Metabolism and disposition of the 5-hydroxytryptamine uptake blocker sertraline in the rat and dog.* Drug Metab Dispos. **1989b**, 17:542-550.

Tretter JR. *Aminoalkylphosphonium compounds and process for the use thereof.* US3354155 (**1967**); priority: 1963

Troy SM, Parker VD, Fruncillo RJ, Chiang ST. *The pharmacokinetics of venlafaxine when given in a twice-daily regimen.* J Clin Pharmacol. **1995**, 35:404-409.

Turner SM, Jacob RG, Beidel DC, Himmelhoch J. *Fluoxetine treatment of obsessive-compulsive disorder.* J Clin Psychopharmacol. **1985**, 5:207-212.

Van Ameringen M, Mancini C, Streiner DL. *Fluoxetine efficacy in social phobia.* J Clin Psychiatry. **1993**, 54:27-32.

van der Burg W. *Tetracyclic compounds.* US4062848 (**1977**); priority: 1975

van der Kolk BA, Dreyfuss D, Michaels M, Shera D, Berkowitz R, Fisler R, Saxe G. *Fluoxetine in posttraumatic stress disorder.* J Clin Psychiatry. **1994**, 55:517-522.

van Harten J, Duchier J, Devissaguet JP, van Bemmel P, de Vries MH, Raghoebar M. *Pharmacokinetics of fluvoxamine maleate in patients with liver cirrhosis after single-dose oral administration.* Clin Pharmacokinet. **1993**, 24:177-182.

van Moffaert M, de Wilde J, Vereecken A, Dierick M, Evrard JL, Wilmotte J, Mendlewicz J. *Mirtazapine is more effective than trazodone: a double-blind controlled study in hospitalized patients with major depression.* Int Clin Psychopharmacol. **1995**, 10:3-9.

van Vliet IM, den Boer JA, Westenberg HG. *Psychopharmacological treatment of social phobia; a double blind placebo controlled study with fluvoxamine.* Psychopharmacology (Berl). **1994**, 115:128-134.

Veggiotti P, Cieuta C, Rex E, Dulac O. *Lamotrigine in infantile spasms.* Lancet. **1994**, 344:1375-1376.

Verity CM, Hosking G, Easter DJ. *A multicentre comparative trial of sodium valproate and carbamazepine in paediatric epilepsy. The Paediatric EPITEG Collaborative Group.* Dev Med Child Neurol. **1995**, 37:97-108.

Vexiau P, Gourmel B, Julien R, Husson C, Fiet J, Puissant A, Dreux C, Cathelineau G. *Severe acne-like lesions caused by amineptine overdose.* Lancet. **1988**, 1:585.

Vlay SC, Friedling S. *Trazodone exacerbation of VT.* Am Heart J. **1983**, 106:604.

von Moltke LL, Greenblatt DJ, Granda BW, Grassi JM, Schmider J, Harmatz JS, Shader RI. *Nefazodone, meta-chlorophenylpiperazine, and their metabolites in vitro: cytochromes mediating transformation, and P450-3A4 inhibitory actions.* Psychopharmacology (Berl). **1999**, 145:113-122.

Wagner W, Zaborny BA, Gray TE. *Fluvoxamine. A review of its safety profile in world-wide studies.* Int Clin Psychopharmacol. **1994**, 9:223-227.

Waldinger MD, Hengeveld MW, Zwinderman AH. *Paroxetine treatment of premature ejaculation: a double-blind, randomized, placebo-controlled study.* Am J Psychiatry. **1994**, 151:1377-1379.

Walker RG, Kincaid-Smith P. Kidneys and the fluid regulatory system. In: Johnson FN ed. *Depression & mania: modern lithium therapy.* IRL Press, Oxford, **1987**, pp 206-213.

Walsh TD. *Controlled study of imipramine and morphine in chronic pain due to advanced cancer.* Proc Am Soc Clin Oncol. **1986**, 5:237.

Wang G, Liu X, Zhao G. *Polymer-supported chiral sulfonamide catalyzed one-pot reduction of ß-keto nitriles: a practical synthesis of (R)-fluoxetine and (R)-duloxetine.* Tetrahedron: Asymm. **2005**, 16:1873-1879.

Ward ME, Musa MN, Bailey L. *Clinical pharmacokinetics of lithium.* J Clin Pharmacol. **1994**, 34:280-285.

Ware MR, Stewart RB. *Seizures associated with fluoxetine therapy.* DICP. **1989**, 23:428.

Warner-Lambert Pharmaceutical Co. *Therapeutic antidepressant compositions comprising hydrazine derivatives.* GB877464 (**1961**); priority: 1957.

Warren M, Bick PA. *Two case reports of trazodone-induced mania.* Am J Psychiatry. **1984**, 141:1103-1104.

Watts RG, Emanuel PD, Zuckerman KS, Howard TH. *Valproic acid-induced cytopenias: evidence for a dose-related suppression of hematopoiesis.* J Pediatr. **1990**, 117:495-499.

Weber JJ. *Seizure activity associated with fluoxetine therapy.* Clin Pharm. **1989**, 8:296-298.

Weigmann H, Hartter S, Bagli M, Hiemke C. *Steady state concentrations of clomipramine and its major metabolite desmethylclomipramine in rat brain and serum after oral administration of clomipramine.* Eur Neuropsychopharmacol. **2000**, 10:401-405.

Weiner M, Chausow A, Wolpert E, Addington W, Szidon P. *Effect of lithium on the responses to added respiratory resistances.* N Engl J Med. **1983**, 308:319-322.

Weisler RH, Johnston JA, Lineberry CG, Samara B, Branconnier RJ, Billow AA. *Comparison of bupropion and trazodone for the treatment of major depression.* J Clin Psychopharmacol. **1994**, 14:170-179.

Weiss RD. *Relapse to cocaine abuse after initiating desipramine treatment.* JAMA. **1988**, 260:2545-2546.

Welch WM, Harbert CA, Koe BK, Kraska AR. *Antidepressant derivatives of cis-4-phenyl-1,2,3,4-tetrahydro-1-naphthalenamine and pharmaceutical compositions thereof.* EP0030081 (**1981**); US4536518 (**1985**); priority: 1979.

Welle HBA, Claassen V. *Oxime ethers having anti-depressive activity.* US4085225 (**1978**); priority: 1975.

Wells BG. *Amiloride in lithium-induced polyuria.* Ann Pharmacother. **1994**, 28:888-889.

White WB, Wong SH. *Rapid atrial fibrillation associated with trazodone hydrochloride.* Arch Gen Psychiatry. **1985**, 42:424.

Wilde MI, Plosker GL, Benfield P. *Fluvoxamine. An updated review of its pharmacology, and therapeutic use in depressive illness.* Drugs. **1993**, 46:895-924.

Wilhelm M, Schmidt P. *Verfahren zur Herstellung eines neuen Amins.* CH467747 (**1969**); priority: 1964

Wilkinson SP, Blendis LM, Williams R. *Frequency and type of renal and electrolyte disorders in fulminant hepatic failure.* Br Med J. **1974**, 1:186-189.

Willcocks K, Barnes RD, Rustidge DC, Tidy DJD. *The synthesis of [^{14}C]-3S,4R-4-(4-flurophenyl)-3-(3,4-methylenedioxyphenoxymethyl)piperidine hydrochloride (BRL 29060A), and mechanistic studies using carbon-13 labelling.* J Label Compd Radiopharm. **1993**, 33:783-794.

Williams MT, Quallich.G.J. *Sertraline: development of a chiral inhibitor of serotonin uptake.* Chemistry and Industry. **1990**, 10:315-319.

Winthrop SO, Davis MA, Myers GS, Gavin JG, Thomas R, Barber R. *New Psychotropic Agents. Derivatives of Dibenzo[a,d]-1,4-cycloheptadiene.* J Org Chem. **1962**, 27:230-240.

Wolf ME, Alavi A, Mosnaim AD. *Posttraumatic stress disorder in Vietnam veterans clinical and EEG findings; possible therapeutic effects of carbamazepine.* Biol Psychiatry. **1988**, 23:642-644.

Wolkenstein P, Revuz J, Diehl JL, Langeron O, Roupie E, Machet L. *Toxic epidermal necrolysis after fluvoxamine.* Lancet. **1993**, 342:304-305.

Wong DT, Horng JS, Bymaster FP, Hauser KL, Molloy BB. *A selective inhibitor of serotonin uptake: Lilly 110140, 3-(p-trifluoromethylphenoxy)-N-methyl-3-phenylpropylamine.* Life Sci. **1974**, 15:471-479.

Wong DT, Robertson DW, Bymaster FP, Krushinski JH, Reid LR. *LY227942, an inhibitor of serotonin and norepinephrine uptake: biochemical pharmacology of a potential antidepressant drug.* Life Sci. **1988**, 43:2049-2057.

Wong DT, Perry KW, Bymaster FP. *Case history: the discovery of fluoxetine hydrochloride (Prozac).* Nat Rev Drug Discov. **2005**, 4:764-774.

Wood A, Tollefson GD, Birkett M. *Pharmacotherapy of obsessive compulsive disorder--experience with fluoxetine.* Int Clin Psychopharmacol. **1993**, 8:301-306.

Wood AJ, Zhou HH. *Ethnic differences in drug disposition and responsiveness.* Clin Pharmacokinet. **1991**, 20:350-373.

Wood SH, Mortola JF, Chan YF, Moossazadeh F, Yen SS. *Treatment of premenstrual syndrome with fluoxetine: a double-blind, placebo-controlled, crossover study.* Obstet Gynecol. **1992**, 80:339-344.

Woodman CL, Noyes R, Jr. *Panic disorder: treatment with valproate.* J Clin Psychiatry. **1994**, 55:134-136.

Worrall EP, Gillham RA. *Lithium-induced constructional dyspraxia.* Br Med J (Clin Res Ed). **1983**, 286:189.

Wyatt RJ, Fram DH, Buchbinder R, Snyder F. *Treatment of intractable narcolepsy with a monoamine oxidase inhibitor.* N Engl J Med. **1971**, 285:987-991.

Wyllie E, Wyllie R, Cruse RP, Erenberg G, Rothner AD. *Pancreatitis associated with valproic acid therapy.* Am J Dis Child. **1984**, 138:912-914.

Yanchick JK, Barton TL, Kelly MW. *Efficacy of fluoxetine in trichotillomania.* Ann Pharmacother. **1994**, 28:1245-1246.

Yardley JP, Husbands GE, Stack G, Butch J, Bicksler J, Moyer JA, Muth EA, Andree T, Fletcher H, III, James MN. *2-Phenyl-2-(1-hydroxycycloalkyl)ethylamine derivatives: synthesis and antidepressant activity.* J Med Chem. **1990**, 33:2899-2905.

Yeragani VK, Gershon S. *Priapism related to phenelzine therapy*. N Engl J Med. **1987**, 317:117-118.

Yoshiyama Y, Nakano S, Ogawa N. *Chronopharmacokinetic study of valproic acid in man: comparison of oral and rectal administration*. J Clin Pharmacol. **1989**, 29:1048-1052.

Yu DK, Dimmitt DC, Lanman RC, Giesing DH. *Pharmacokinetics of dothiepin in humans: a single dose dose-proportionality study*. J Pharm Sci. **1986**, 75:582-585.

Yu MS, Lantos I, Peng Z-Q, Yu J, Cacchio T. *Asymmetric synthesis of (-)-paroxetine using PLE hydrolysis*. Tetrahedron Letters. **2000**, 41:5647-5651.

Zaccara G, Messori A, Moroni F. *Clinical pharmacokinetics of valproic acid--1988*. Clin Pharmacokinet. **1988**, 15:367-389.

Zachariah SB, Borges EF, Varghese R, Cruz AR, Ross GS. *Positive response to oral divalproex sodium (Depakote) in patients with spasticity and pain*. Am J Med Sci. **1994**, 308:38-40.

Zucker P, Daum F, Cohen MI. *Fatal carbamazepine hepatitis*. J Pediatr. **1977**, 91:667-668.

2 Schizophrenia and Other Psychoses

2.1 Introductory and Basic Aspects

Current Status and Challenges in Schizophrenia Research

Francesc Artigas

Dept. of Neurochemistry and Neuropharmacology, Institut d' Investigacions Biomèdiques de Barcelona, CSIC (IDIBAPS), Roselló, 161. 08036 Barcelona, Spain

2.1.1 Introduction

The first descriptions of schizophrenia symptoms can be already found in Greek and Roman books, yet the illness was not recognized as a single pathological condition. At the end of the 19th century, the German psychiatrist Emil Kraepelin was the first to identify what we know today as schizophrenia as a brain disorder different from other psychiatric conditions, particularly from manic depression. Together with his colleague, Alois Alzheimer, he was firmly convinced that this disorder had a neuropathological basis and called it *dementia praecox* (early or premature dementia) to distinguish it from other forms of dementia which occur in later stages of life, such as Alzheimer's disease. In 1908, this disorder was termed "schizophrenia" (from the Greek *schizo* = split, and *phrenos* = mind) by the Swiss psychiatrist, Eugen Bleuler who felt that the term *dementia paecox* was misleading. During the early decades of the 20th century, schizophrenia was considered a hereditary disease and schizophrenic patients were sterilized in various European countries and in United States, notably in the Nazi Germany where many patients were also murdered during the course of the T-4 Euthanasia program.

Schizophrenia is characterized by a profound disruption in cognition and emotion, affecting the most fundamental human attributes: language, thought, perception, affect, and sense of self. The socioeconomic impact of schizophrenia is high in developed countries, due to its chronic character, the huge disability of the patients and the disturbances produced in patients' relatives. Health costs are also high due to the chronic nature of treatments as well as the social and medical care, including periodic hospitalizations. Schizophrenia and related psychotic disorders have and annual cost of 35.000 million euros in the 25 countries of the European Community (Andlin-Sobocki et al., 2005).

Typically, the first clinical manifestations of the illness appear during the second and third decades of life and involve a constellation of symptoms, which have been classified as positive, negative and cognitive. Positive symptoms include hallucinations or false perceptions (typically auditory: patients hear internal "voices" or experience other sensations not connected to an obvious source) as well as delusions (patients often assign an unusual significance or meaning to normal events or hold fixed false personal beliefs). Negative symptoms involve apathy, loss of motivation, social withdrawal, etc., whereas cognitive symptoms involve disturbances in basic cognitive functions: attention, executive functions (ability to plan, initiate and regulate goal-directed behaviors), problem solving and specific forms of memory, particularly working (short-term) memory. Cognitive symptoms are currently viewed as key features in the behavioral disturbances and functional disability of schizophrenic patients. The lack of improvement of these symptoms by the classic and modern antipsychotic drugs poses a major problem in the treatment of schizophrenia. Moreover, many patients also exhibit "affective" symptoms (i.e., anxiety, depression) which markedly contribute to the high lifetime incidence (10%) of suicide in schizophrenic patients.

2.1.2 Clinical diagnosis and assessment of schizophrenia

Schizophrenia is diagnosed clinically and symptoms of hallucinations and thought disorder play a major role (Andreasen, 1995). The most commonly used criteria for diagnosing schizophrenia are from the American Psychiatric Association's "Diagnostic and Statistical Manual of Mental Disorders" (DSM) and the World Health Organization's "International Statistical Classification of Diseases and Related Health Problems" (ICD). The most recent versions are ICD-10 (http://www.who.int/classifications/icd/en/) and DSM-IVR (http://www.psych.org/research/dor/dsm/index.cfm). These classification systems identify several psychotic disorders and forms of schizophrenia. Box 1 lists the various forms of psychoses in the ICD-10.

In general, schizophrenic disorders are characterized by fundamental and characteristic distortions of thinking and perception, and affects that are inappropriate or blunted. Clear consciousness and intellectual capacity are usually maintained although cognitive deficits may evolve with time. The most important symptoms are shown in Box 2.

> **Box 1. Classification of psychotic disorders according to the World Health Organization (ICD-10)**
>
> F06.0 Organic hallucinosis
> F06.1 Organic catatonic disorder
> F06.2 Organic delusional [schizophrenia-like] disorder
> F20 Schizophrenia
> F20.0 Paranoid schizophrenia
> F20.1 Hebephrenic schizophrenia
> F20.2 Catatonic schizophrenia
> F20.3 Undifferentiated schizophrenia
> F20.5 Residual schizophrenia
> F20.6 Simple schizophrenia
> F20.8 Other schizophrenia
> F20.9 Schizophrenia, unspecified
> F20-F29 Schizophrenia, schizotypal and delusional disorders
> F21 Schizotypal disorder
> F22.0 Delusional disorder
> F22.8 Other persistent delusional disorders
> F23.0 Acute polymorphic psychotic disorder without symptoms of schizophrenia
> F23.1 Acute polymorphic psychotic disorder with symptoms of schizophrenia
> F23.2 Acute schizophrenia-like psychotic disorder
> F23.3 Other acute predominantly delusional psychotic disorders
> F25 Schizoaffective disorders
> F25.0 Schizoaffective disorder, manic type
> F25.1 Schizoaffective disorder, depressive type
> F25.2 Schizoaffective disorder, mixed type
> F28 Other non organic psychotic disorders
> F44.3 Trance and possession disorders
> F45.4 Persistent somatoform pain disorder
> F60.0 Paranoid personality disorder
> F60.1 Schizoid personality disorder
> F62.1 Enduring personality change after psychiatric illness
> F63.1 Pathological fire-setting [pyromania]
> F90 Hyperkinetic disorders
> F91 Conduct disorders
> F94.0 Elective mutism

The criteria of the DSM-IVR (American Psychiatric Association, 2000), used for most research studies, require symptoms to have been initially present for 1 month and persist for at least 6 months; there must also be impaired personal functioning, and the symptoms must not be secondary to another disorder (e.g., depression, substance abuse (Box 3). The DSM now contains five sub-classifications of schizophrenia whereas the ICD-10 identifies seven (Box 4).

Box 2. Main symptoms of schizophrenia

Positive Symptoms

Delusions are firmly held erroneous beliefs due to distortions or exaggerations of reasoning and/or misinterpretations of perceptions or experiences. Delusions of being followed or watched are common, as are beliefs that comments, radio or TV programs, etc., are directing special messages directly to him/her.

Hallucinations are distortions or exaggerations of perception in any of the senses, although auditory hallucinations ("hearing voices" within, distinct from one's own thoughts) are the most common, followed by visual hallucinations.

Disorganized speech/thinking, also described as "thought disorder" or "loosening of associations," is a key aspect of schizophrenia. Disorganized thinking is usually assessed primarily based on the person's speech. Therefore, tangential, loosely associated, or incoherent speech severe enough to substantially impair effective communication is used as an indicator of thought disorder by the DSM-IV.

Grossly disorganized behavior includes difficulty in goal-directed behavior (leading to difficulties in activities in daily living), unpredictable agitation or silliness, social disinheriting, or behaviors that are bizarre to onlookers. Their purposelessness distinguishes them from unusual behavior prompted by delusional beliefs.

Catatonic behaviors are characterized by a marked decrease in reaction to the immediate surrounding environment, sometimes taking the form of motionless and apparent unawareness, rigid or bizarre postures, or aimless excess motor activity.

Other symptoms sometimes present in schizophrenia but not often enough to be definitional alone include affect inappropriate to the situation or stimuli, unusual motor behavior (pacing, rocking), depersonalization, derealisation, and somatic preoccupations.

Negative Symptoms

Affective flattening is the reduction in the range and intensity of emotional expression, including facial expression, voice tone, eye contact, and body language.

Alogia, or poverty of speech, is the lessening of speech fluency and productivity, thought to reflect slowing or blocked thoughts, and often manifested as laconic, empty replies to questions.

Avolition is the reduction, difficulty, or inability to initiate and persist in goal-directed behavior; it is often mistaken for apparent disinterest.

Box 3. DSM-IV criteria for schizophrenia

A. *Characteristic symptoms*: Two (or more) of the following, each present for a significant portion of time during a 1-month period (or less if successfully treated):

    1. Delusions
    2. Hallucinations
    3. Disorganized speech (e.g., frequent derailment or incoherence)
    4. Grossly disorganized or catatonic behaviors
    5. Negative symptoms, i.e., affective flattening, alogia, or avolition

    Note: Only one Criterion A symptom is required if delusions are bizarre or hallucinations consist of a voice keeping up a running commentary on the person's behaviors or thoughts, or two or more voices conversing with each other.

B. *Social/occupational dysfunction*: For a significant portion of the time since the onset of the disturbance, one or more major areas of functioning such as work, interpersonal relations, or self-care are markedly below the level achieved prior to the onset (or when the onset is in childhood or adolescence, failure to achieve expected level of interpersonal, academic, or occupational achievement).

C. *Duration*: Continuous signs of the disturbance persist for at least 6 months. This 6-month period must include at least 1 month of symptoms (or less if successfully treated) that meet Criterion A (i.e., active-phase symptoms) and may include periods of prodromal or residual symptoms. During these prodromal or residual periods, the signs of the disturbance may be manifested by only negative symptoms or two or more symptoms listed in Criterion A present in an attenuated form (e.g., odd beliefs, unusual perceptual experiences).

D. *Schizoaffective and mood disorder exclusion*: Schizoaffective disorder and mood disorder with psychotic features have been ruled out because either: (1) no major depressive, manic, or mixed episodes have occurred concurrently with the active-phase symptoms; or (2) if mood episodes have occurred during active-phase symptoms, their total duration has been brief relative to the duration of the active and residual periods.

E. *Substance/general medical condition exclusion*: The disturbance is not due to the direct physiological effects of a substance (e.g., a drug of abuse, a medication) or a general medical condition.

F. *Relationship to a pervasive developmental disorder*: If there is a history of autistic disorder or another pervasive developmental disorder, the additional diagnosis of schizophrenia is made only if prominent delusions or hallucinations are also present for at least a month (or less if successfully treated).

Box 4. Subtypes of schizophrenia and their main characteristics

Classification system	Subtype
DSM-IV (295.2)/ICD-10 (F20.2)	**Catatonic type**: marked absences or peculiarities of movement are present
DSM-IV (295.1)/ICD-10 (F20.1)	**Disorganized**: where thought disorder and flat affect are present together
DSM-IV (295.3)/ICD-10 (F20.0)	**Paranoid type**: where delusions and hallucinations are present but thought disorder, disorganized behaviors, and affective flattening is absent
DSM-IV (295.6)/ICD-10 (F20.5)	**Residual type**: where positive symptoms are present at a low intensity only
DSM-IV (295.9)/ICD-10 (F20.3)	**Undifferentiated type**: psychotic symptoms are present but the criteria for paranoid, disorganized, or catatonic types has not been met

Usually, the diagnosis of schizophrenia is reliable, but it is sometimes complicated by the prior treatment of schizophrenia's positive symptoms. Antipsychotic medications, particularly the classic antipsychotics that block the actions of dopamine, often produce side-effects that closely resemble the negative symptoms of affective flattening and avolition. In addition, other negative symptoms are sometimes present in schizophrenia but not often enough to satisfy diagnostic criteria: loss of usual interests or pleasures (anhedonia), disturbance of sleep and eating, dysphoric mood (depressed, anxious, irritable, or angry mood), and difficulty concentrating or focusing attention. Moreover, antipsychotic mediations may induce negative symptoms (Artaloytia et al., 2006), which adds further complications to the diagnosis. However, diagnosis of schizophrenia should not be made in the presence of extensive depressive or manic symptoms nor in the presence of overt brain disease or during states of drug abuse or withdrawal.

The severity of schizophrenia is assessed usually by two different scales. The BPRS (Brief Psychiatric Rating Scale) consists of 24 items ranked from 1 to 7 according to the most used version (Ventura et al., 1993). The PANSS (Positive and Negative Symptoms in Schizophrenia; Kay et al., 1987) is made up of three different subscales which assess, respectively, the positive symptoms (7 items), the negative symptoms (7 items) and the global psychopathology (16 items) with a score from 0 (absent) to 6 (extreme) for each of the items.

2.1.2.1 Cognitive and functional impairment in schizophrenia

Cognitive difficulties in schizophrenic patients have become recently the focus of interest because of their relevance in the functional impairment experienced by schizophrenic patients and because of the poor efficacy of the current medications on these deficits, including second generation antipsychotic drugs (Harvey and Keefe, 2001; Remillard et al., 2005). Indeed, some authors place the dysfunction of fundamental cognitive processes at the *centre* of schizophrenia, rather than as one more type of symptoms (Andreasen, 1997a, 1997b; Elevåg and Goldberg, 2000).

Cognitive problems include information processing, abstract categorization, planning and regulating goal-directed behavior (the so-called "executive functions"), cognitive flexibility, attention and working memory. Cognitive problems vary from person to person and can change over time. In some situations it is unclear whether such deficits are due to the illness or to the side-effects of certain neuroleptic medications (King, 1994; Zalewski et al., 1998).

The relevance of cognitive deficits, in particular that in working memory, and the presence of some neuro-anatomical, histological and neurochemical alterations (see below) have led to postulate that schizophrenia is a disorder of the prefrontal cortex (PFC; Goldman-Rakic and Selemon, 1997). Problems in such a fundamental area may explain the cognitive disturbances but also, due to the "top-down" control of other cortical and subcortical areas by the PFC (Miller and Cohen, 2001), alterations in PFC function may translate into an abnormal function of all these areas which may account for the wide array of schizophrenic symptoms.

The diagnostic of schizophrenia includes functional impairment in addition to the constellation of symptoms outlined above. Schizophrenic patients experience a significant dysfunction in one or more major areas of life activities such as their interpersonal relationships, work or education, family life, communication, or self-care. They have serious economic, social, and psychological consequences: unemployment, disrupted education, limited social relationships, isolation, family stress, substance abuse, etc. The distress produced by all these aspects of the illness leads to an increased risk of suicide among schizophrenic patients.

2.1.3 Epidemiology

The accepted prevalence rate for schizophrenia is typically 1% worldwide, although a recent review on incidence and prevalence studies published between 1980 and 2000 reduced this value to 0.55% (Goldner et al., 2002). This study also found that prevalence may vary from country to country, despite the widely held view that schizophrenia occurs at the same rate in different geographical locations and cultures. Yet these differences may partly reflect differences in the diagnostic of schizophrenia in different countries (e.g., Europe vs. USA). Schizophrenia incidence varies between 7.5 and 16.3 cases per year per 100,000 population.

The incidence of schizophrenia increases markedly in relatives of schizophrenic patients (Fig. 1). Family and twin studies have clearly documented that this increased incidence has a clear genetic basis (see below).

Figure 1. The risk of schizophrenia

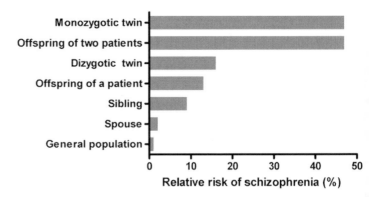

The risk of schizophrenia increases with the genetic proximity to an schizophrenic patient, ranging from a nearly 1% in the general population to 16–17% in dizygotic twins or offsprings of a patient to nearly 50% in the case of monozygotic twins or offsprings of two patients. Adoption studies with monozygotic twins have permitted to establish a genetic basis for the risk of the illness. Thus, the risk of schizophrenia is related to the occurrence of the illness in biological parents and not in the adoptive parents. Redrawn from Lewis and Lieberman (2000).

The incidence of schizophrenia is approximately equal in both genders yet the onset tends to be earlier in men, who also have a less favorable course of the illness and outcome. Cases of childhood schizophrenia and late-onset schizophrenia also occur, although in a very low proportion.

2.1.4 Course of schizophrenia

Schizophrenia is typically diagnosed in late adolescence or early adulthood, typically early or mid-20s for men, late 20s for women, and follows a time-course characterized by different phases. It can be either continuous, or episodic with progressive or stable deficit, or there can be one or more episodes with complete or incomplete remission. Overall, research indicates that schizophrenia's course over time varies considerably from person to person. The variability may arise from the underlying complexity and heterogeneity of the disease process itself, as well as from biological and genetic vulnerability, cognitive impairment, sociocultural stressors, and personal and social factors that confer protection against stress and vulnerability (Nuechterlein et al., 1994).

Despite individual variability, four main periods can be distinguished (Lewis and Lieberman, 2000). First, a premorbid period which extends during childhood. Research conducted on high-risk cohorts has identified mild deficits in social, motor, and cognitive functions during childhood and adolescence that may represent a premorbid phase of the illness. This premorbid phase is followed by a prodromal phase during adolescence, characterized by exacerbation of bizarre behaviors and the emergence of some symptoms that will later be fully present. In this stage, symptoms are less pronounced but may include the whole array of schizophrenia manifestations: psychotic (e.g., illusions, magical thoughts, superstitious etc.), mood (e.g., anxiety, dysphasia, irritability) and cognitive symptoms (e.g., distractibility, concentration difficulties), as well as social withdrawal, or obsessive behaviors (McGlashan, 1996; Yung et al., 1996).

These prodromal symptoms appear also in high-risk individuals who later do not develop the illness and therefore cannot be considered as diagnostic. However, in the vast majority of cases, this prodromal phase is followed by a first psychotic episode which signals the onset of the third phase (progression) of the illness, i.e., that in which overt manifestations are present in the form of a full-blown symptomatology. Typically, the first psychotic episode occurs during the second and third decades of life and results from the action of environmental stressors on a high susceptibility neurobiological background, determined by genetic and/or developmental factors. Environmental stressors include a variety of factors of adult life, such as military duties, difficulties of adaptation to work or high-level education, as well as drug abuse. These factors may act as external triggers of the first overt manifestations of the illness due to the inability of certain brain circuits to cope with these stressful stimuli (Lieberman et al., 1997).

This progressive phase of the illness is characterized by the presence of one or typically more psychotic episodes. Most individuals experience periods of symptom exacerbation and remission, while others maintain a steady level of symptoms and disability which can range from moderate to severe. Typically, negative symptoms remain between psychotic episodes (Gupta et al., 1997). Early response to antipsychotic medications in the first psychotic episode has been found to predict better long-term outcomes, compared to people who have been treated in more advanced stages of the illness (Lieberman et al., 1996; Wyatt and Henter, 1998). Finally, this progressive phase is followed by a stabilization phase during which the clinical deterioration produced by the illness in the early stages (i.e., 5–10 years after onset) tends to be milder.

2.1.5 Brain pathology in schizophrenia

For several decades, neuropathologists failed to successfully understand the neuropathology of schizophrenia. This led to the unproven assumption that schizophrenia was a "functional" psychosis with no brain damage or alteration. However, in the last two decades, evidence has accumulated indicating the occurrence of macroscopic and microscopic changes in the brain of schizophrenic patients. Yet this is a still controversial field due to several reasons. On the one hand, the variables examined and the histological techniques used differ from investigator to investigator, as the brain regions examined do. On the other hand, post-mortem studies have often been conducted on a limited number of tissue samples from individuals who had different clinical features, different course of the illness and were exposed to different medications.

Moreover, the illness is sometimes accompanied by neurological diseases that produce brain pathology. Overall, these confounding factors have hampered to reach consensus on what are the neuropathological manifestations of schizophrenia. Despite these difficulties, it seems clear that schizophrenia is associated with brain alterations at macroscopic and microscopic level. These changes seem to be more the result of an aberrant neuroanatomy (altered neurodevelopment) than neuropathology *per se* (Harrison, 1999).

2.1.5.1 Macroscopic studies: neuroimaging and neuropathology

A turning point in the understanding of the brain pathology in schizophrenia was the observation that chronic schizophrenic patients examined with computerized tomography scan (CT) had larger cerebral ventricles than control individuals (Johnstone et al., 1976). This is probably the most replicated finding in schizophrenia and many studies have reported an enlargement of the third and lateral ventricles in schizophrenic patients. Typically, the change amounts 20–75% of the ventricle/brain volume ratio (VBR). A median 40% increase was reported using MRI (magnetic resonance imaging; for reviews see Harrison, 1999; McCarley et al., 1999). Interestingly, VBR values follow a normal distribution, which indicates that this is common to all classes, subtypes and clinical courses of schizophrenia and cannot be ascribed to a particular subtype of the illness. The distribution of VBR values in schizophrenic patients overlaps with that in healthy controls, which indicates that enlargement of the lateral and third ventricles cannot be used for diagnostic purposes.

The increased VBR is accompanied by a reduction of the cortical volume, a determined by volumetric MRI studies and confirmed in brain autopsies. In general these studies have found reductions in brain weight and in the volume of the cerebral hemispheres. This reduction is more moderate (typically 3–8%) than the enlargement of the ventricles. Despite both observations are likely to be related, no consistent correlation is found between reduction of the cortical size and ventricle enlargement, possibly because the large differences in size between both and the difficulties to measure small changes in ventricle volume.

Neuroimaging studies have documented that the cortical reduction is not homogenous but restricted to certain areas in the temporal and frontal lobes: the superior temporal gyrus, the prefrontal cortex (PFC; mostly the dorsolateral part, DLPFC, Broadman area 46), and limbic areas such as the hippocampus and the cingulate cortex (Nelson et al., 1998; Harrison, 1999; McCarley et al., 1999; Gur et al., 2000; Shenton et al., 2001). Interestingly, studies in monozygotic twins discordant for schizophrenia, the affected twin shows the larger ventricles and smaller cortical and hippocampal size. These studies have been instrumental in determining that the illness is associated with a structural brain change. They also show that the main factor for these changes is the expression of the schizophrenia phenotype and not the underlying genotype, common to both twins. Nevertheless, studies in relatives who are obligate carriers but who do not develop the illness show that these individuals have large ventricle volumes than relatives with a lower genetic load, indicating that – to some extent – the structural pathology of schizophrenia is determined by the genetic liability (for a review, see Harrison,1999).

Contrary to what happens with the changes in temporal and – to a lesser extent – frontal lobes, the change in subcortical structures seems less clear. Some magnetic resonance imaging (MRI) studies have reported changes in the basal ganglia and the thalamus (McCarley et al., 1999). Of these areas, it seems that the medial thalamus is particularly affected (Andreasen et al., 1994; Buchsbaum et al., 1996), a finding that has been confirmed by neuropathological studies showing a reduced neuronal number in this area (Harrison, 1999). Likewise, positron emission tomography (PET) scan studies have found a lower metabolic rate in the right thalamus of schizophrenic patients, which is consistent with functional studies showing a poor filtering of sensory information (Buchsbaum et al., 1996). More recent MRI studies have documented a reduction of the mediodorsal and pulvinar thalamic nuclei in schizophrenic patients (Byne et al., 2001).

Hence, pathology appears to occur in key brain areas involved in the processing of information along the cognitive and limbic circuits of the basal ganglia (Fig. 2). In particular, the PFC, although not exhibiting the larger degree of macroscopic pathology, displays a large number of cellular abnormalities, and schizophrenic patients have a poor performance in cognitive tasks involving the PFC, which suggests that this cortical area plays a crucial role in the pathophysiology of schizophrenia (Goldman-Rakic, 1999; Selemon and Goldman-Rakic, 1999; Weinberger et al., 2001).

Figure 2. Scheme of the basal ganglia circuits involved in the control of cognitive and affective processes.

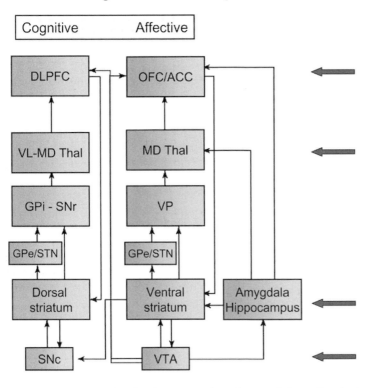

A similar circuitry exists for the processing of motor information (not shown). In the cognitive circuit, pyramidal neurons in the dorsolateral part of the prefrontal cortex (DLPFC) excite medium-size spiny GABAergic neurons the caudate nucleus (dorsal striatum). These, in turn, project to the internal part of the globus pallidus (GPi) and substantia nigra *pars reticulata* (SNr) either monosynaptically (direct pathway) or through the external part of the globus pallidus (GPe) and subthalamic nucleus (STN) (indirect pathway). The GPi/SNr or output nuclei project to the ventrolateral and mediodorsal (VL, MD) nuclei of the thalamus, which send excitatory afferents to the DLPFC, thus closing the circuit. Similarly, in the "affective" circuit, the orbitofrontal cortex (OFC) and the anterior cingulate (ACC) project to the ventral part of the striatum (nucleus accumbens and adjacent areas), which in turn projects directly or indirectly to the ventral pallidum (VP). The latter nucleus projects to the MD nucleus of the thalamus, which in turn projects to OFC and ACC. The activity of some areas of the circuit, such as the ventral striatum, the mediodorsal thalamus and cortical areas, is modulated by external inputs from temporal structures, such as the hippocampus and the amygdala, which in turn are innervated by the PFC. Dopaminergic neurons of the *substantia nigra pars compacta* (SNc) and the ventral tegmental area (VTA) provide ascending modulatory inputs to these circuits through connections with the dorsal and ventral striatum (mesolimbic pathway). Also, dopaminergic neurons from the VTA project to DLPFC and OFC/ACC (mesocortical pathway). Arrows show the different levels of the circuit at which macroscopic and microscopic abnormalities have been found in schizophrenic patients, in particular (from top downwards): (a) reduced cortical volume and reduced thickness of gray matter, reduced neuropil, (b) reduced neuronal number in some thalamic nuclei (e.g., mediodorsal, MD), (c) reduced size and connectivity of temporal lobe, including the hippocampus, and (d) increased activity of dopaminergic neurons of the VTA.

Interestingly, the PFC is reciprocally connected with the hippocampal formation and receives excitatory afferents from the mediodorsal nucleus of the thalamus, two of the areas showing overt pathological changes. Moreover untreated schizophrenic patients show a reduced energy metabolism in the PFC (Andreasen et al., 1997) which has been related with negative symptoms (Potkin et al., 2002). Given the top-down control exerted by the PFC on other cortical and subcortical areas (Miller and Cohen, 2001), a pathological change in the activity of PFC may result in downstream changes in these other areas, eventually altering the output of the circuit. Alternatively, pathological changes may simultaneously occur in several brain areas which may contribute independently to the emergence of schizophrenia symptoms.

In addition to the PFC, the mediodorsal thalamus and the temporal lobe, it has been suggested that the cerebellum may also play a role in the cognitive deficits of schizophrenic patients, based on the engagement of a prefrontal–thalamic–cerebellar circuit during certain cognitive tasks (Andreasen et al., 1996).

There are two important questions related to the brain pathology found in schizophrenic patients: (a) when do the changes appear? and (b) what causes them? Several studies have documented the presence of these macroscopic changes in first-episode patients, which suggests that they are not due to the course of the illness or the treatments used. Moreover, adolescents and young adults with a high genetic load also show enlarged ventricles and smaller temporal lobes (Sharma et al., 1998; Lawrie et al., 1999) indicating that brain pathology is present before the emergence of clinical symptoms.

However, the question whether brain pathology in schizophrenia is progressive and what causes it is still matter of debate. In general, there is consensus on the lack of a neurodegenerative process. Some studies have reported the existence of small brain infarcts and white matter changes that could be interpreted as signs of neurodegenerative and vascular processes. However, these findings seem to be restricted to patients with concurrent neurological diseases and have not been replicated in other samples. In contrast, some early studies reported the presence of reactive astrocytosis (gliosis) yet this finding has not been replicated in more recent studies, which suggests that, if present, gliosis is a sign of a concurrent pathological process but not an aethiopathogenic sign of schizophrenia. Overall, the lack of clear signs of neurodegeneration in schizophrenia gives further support to the notion that brain pathology results from an altered neurodevelopment of the schizophrenic's brain.

2.1.5.2 Microscopic changes: cellular abnormalities in schizophrenia

Given the lack of a clear neurodegenerative process in schizophrenia, research has focused on how microscopic cellular changes can result in those observed at a macroscopic level. Hence many studies have examined the cortical cytoarchitecture in post-mortem samples of schizophrenic patients. Most studies have focused in the hippocampus and the dorsolateral PFC (DLPFC), which are two of the brain structures showing a larger degree of anatomical or functional abnormalities. Hence, neuronal number, size and density have been studied, as well as size and shape of synapses, dendrites and axons, which are the major components of the neuropil.

The reduced hippocampal volume in schizophrenic patients (Harrison, 1999; McCarley et al., 1999) is accompanied by a decrease in hippocampal *N*-acetyl aspartate, a putative marker of neuronal integrity (Bertolino et al., 1998). In addition, PET scan studies have provided evidence of hippocampal dysfunction during episodic memory retrieval in subjects with schizophrenia (Heckers et al., 1998). These anatomical and functional changes are not due to a reduced neuronal number, but probably to reductions in the size of neuronal cell bodies (Harrison, 1999; Lewis and Lieberman, 2000). Furthermore, there seems to be a reduction in neuropil and synaptic connectivity, as assessed by various presynaptic and dendritic markers such as microtubule-associated protein, in certain subdivisions of the hippocampus (for a review, see Weinberger, 1999)

Despite the magnitude of the gray matter changes being lower in PFC than in the hippocampus, a large number of studies have focused on the cytoarchitecture of this cortical area. The interest in the PFC stems from the knowledge on its role in higher brain functions which are altered in schizophrenic patients, such as cognition, action planning and behavioral inhibition (Fuster, 1997; Miller and Cohen, 2001). Hence, schizophrenic patients show a poor performance in cognitive tests assessing working memory, which is dependent on changes in the neuronal activity in the DLPFC (Fuster, 1997; Weinberger and Gallhofer, 1997; Elvevag and Goldberg, 2000). Moreover, the long-term prognosis for schizophrenic patients appears to be best predicted by the degree of cognitive impairment than by the severity of psychotic symptoms, (Green, 1996; Weinberger and Gallhofer, 1997), which adds a further element of interest in the understanding of the pathological changes in this area.

As observed in the hippocampus, the reduction in cortical gray matter is not accompanied by a decrease in cell number but by an increased cell density and a reduction in the neuropil (Harrison, 1999; Selemon and Goldman-Rakic, 1999). Hence, a reduction in the size of cell bodies of certain neuronal populations, particularly of pyramidal neurons of layer III and – to a lesser extent – layers V–VI has been documented (Harrison, 1999; Lewis and Lieberman, 2000). Likewise, there is a reduced neuropil volume which reflects abnormalities of the axonal and dendritic neuronal compartments. Hence, there seems to be a reduced corticocortical connectivity and lesser axon collaterals of intrinsic neuronal populations. Consistent with this reduced connectivity, the levels of synaptophysin, a presynaptic terminal protein, have been found to be decreased in the PFC of subjects with schizophrenia (for a review, see Lewis and Lieberman, 2000) and reductions in the density of dendritic spines of pyramidal neurons, where synaptic contacts take place, is lower in schizophrenic patients than in controls (Glantz and Lewis, 2000). Furthermore, the levels of *N*-acetyl aspartate, a putative marker of axonal and/or neuronal integrity, are reduced in the PFC of schizophrenic patients, as also observed in the hippocampus (Callicott et al., 2000).

In addition to a putative loss of intracortical connectivity, a reduced density of thalamic fibres might also account for the decreased neuropil in PFC. Hence, the mediodorsal nucleus or the thalamus, which is the main non-cortical input to the PFC appears to have a decreased size and metabolic activity in individuals with schizophrenia (Andreasen et al., 1994; Buchsbaum et al., 1996; Byne et al., 2001; Hazlett et al., 2004). Moreover, unlike in PFC and hippocampus, the size reduction of the mediodorsal nucleus appears to be accompanied by a decrease in cell number (Harrison, 1999, Selemon and Goldman-Rakic, 1999). Finally, the size of the thalamus has been found to correlate with white matter and with ventricle enlargement schizophrenic patients (Portas et al., 1998) and the density of parvalbumin-positive axon terminals (a putative marker from thalamic inputs) is reduced in the PFC of schizophrenic patients (Lewis and Lieberman, 2000). These abnormalities in the thalamic–cortical circuitry may account for the deficits in sensory gating exhibited by schizophrenic patients, which show an abnormal PPI (prepulse inhibition) of startle response (Geyer et al., 2001).

In addition to the afore mentioned changes in the density and size of principal (pyramidal) cortical neurons and reduced neuropil, other abnormalities have been identified in the PFC of subjects with schizophrenia, such as a reduced expression of the mRNA encoding glutamic acid decarboxylase (GAD67), the synthesizing enzyme for GABA, in a subset of PFC GABA neurons and a decreased density of GABA transporter (GAT-1)-immunoreactive axon cartridges (Lewis and Lieberman, 2000). The latter are vertically oriented axon terminals of GABAergic chandelier neurons, which synapse exclusively on the axon hillock of pyramidal neurons and control the generation of nerve impulses through the activation of $GABA_A$ receptors (De Felipe et al., 1989, 2001). More recent studies have identified a selective reduction in the mRNA of parvalbumin (a marker of certain subtypes of GABAergic interneurons, including chandelier cells) in layers III-IV of the DLPFC of schizophrenic patients (Lewis et al., 2005). This change is not accompanied by a concurrent decrease in the mRNA of calretinin, a marker of other types of GABAergic neurons, which is predominantly expressed in layer II (Lewis et al., 2005). Likewise, a concurrent reduction in brain-derived neurotrophic factor (BDNF) and its receptor TrkB has also been found in the same specimens, which has led to postulate a deficit in the BDNF-mediated, TrkB-dependent signaling as the main pathogenetic factor in schizophrenia (Lewis et al., 2005). The BDNF/TrkB reduction in developmental stages might result in a loss of $GABA_A$ receptor-mediated inhibition exerted by chandelier neurons, with the subsequent disruption of thalamocortical circuitry (Lewis and Lieberman, 2000; Lewis et al., 2005). Finally, a reduced density of dopaminergic nerve terminals in layer VI of the DLPC has also been documented which might further contribute to the dysregulation of the thalamocortical connectivity (Akil et al., 1999).

2.1.6 Pathogenesis and pathophysiology of schizophrenia

It is beyond doubt that genetic factors play an important role in influencing susceptibility to schizophrenia. The results of numerous family, twin, and adoption studies have consistently shown that the risk of the illness is increased among the relatives of affected individuals and that this is the result of the inherited genes rather than of a shared environment (Fig. 1). The attempts to find a dominant "schizophrenia gene" were unsuccessful, and most studies neither achieved stringent "genome-wide" levels of significance nor replicated previous findings. These disappointing findings are probably attributable to a combination of limited genetic effects, small sample sizes and the use of marker maps of insufficient density to fully extract the genetic information (Owen et al., 2005). However, recently, a large number of genetic studies have identified susceptibility genes in schizophrenia and several well established linkages have been documented. Hence, regions on several chromosomes (e.g., 1, 6, 8, 10, 13, and 22) have been implicated as sites of potential vulnerability genes (Pulver, 2000). Several studies of schizophrenia pedigrees have found evidence for linkage at genome-wide significant level. Strongly supported regions are 6p24–22, 1q21–22, and 13q32–34, while other promising regions include 8p21–22, 6q16–25, 22q11–12, 5q21–q33, 10p15–p11, and 1q42 (for reviews, see Craddock et al., 2005; Owen et al., 2005). Moreover, specific genes or loci have been implicated. Current evidence supports a number of genes such as *NRG1* (neuregulin 1), *DTNBP1* (dysbindin), *DISC1* (disrupted in schizophrenia 1), *DAOA* (formerly *G72*) (D-amino-acid oxidase activator), *DAO* (D-Amino-acid oxidase), *RGS4* (regulator of G-protein signaling 4) and *COMT* (catechol-O-methyltransferase) as schizophrenia susceptibility loci (Harrison and Owen, 2003; Craddock et al., 2005; Owen et al., 2005). Moreover, some genes related to neurotransmitter function, such as the dopamine receptor genes *DRD2* and *DRD3* and the serotonin receptor gene *5-HT2AR* have also been identified as susceptibility genes yet with a very low effect size.

However, one of the most difficult challenges in schizophrenia research is to explain the interaction between genetic and environmental influences to produce the typical course of the illness (e.g., emergence of the first psychotic episode in the second and third decades of life followed by a progressive deterioration). The neurodevelopmental nature of schizophrenia (for a review, see Lewis and Levitt, 2002) has often been called upon to explain such a temporal course. Thus, certain genes may confer a susceptibility background so that exposure to pre- and perinatal life events (e.g., maternal infections, hypoxia, obstetric complications, etc.) may act to produce subtle changes in brain circuitry, such as a reduced connectivity (e.g., the "reduced neuropil" hypothesis; Selemon and Goldman-Rakic, 1999) which would result in the abnormal function of one or more brain circuits. Stressful life events in early adulthood may then trigger the appearance of clinical symptoms due to the inability of the schizophrenic brain circuits to cope with such stressful situations. To some extent, this can be exemplified by a chemical buffer (circuit), whose acid, base and salt concentrations (activity of the neurotransmitters in the circuit) permit to maintain a given pH value (optimal circuit output leading to "normal" behavior) during the addition of H_3O^+ or OH^- ions (stress). However, a poorly buffered solution, with an incorrect acid/base concentration or balance (disturbed circuit) will not tolerate the same amount of ions (stress) and pH will change dramatically (aberrant circuit output: clinical symptoms).

Over the years, several hypotheses have been postulated to explain the clinical symptoms of schizophrenia which are based in alterations of certain neurotransmitters. Three main hypotheses have dominated research over the years, which have focused on dopamine (DA), serotonin (5-HT) and glutamate. More recently, GABA has been incorporated as a new player. These hypotheses are by no means exclusive to each other, since alterations in one or more cerebral areas and/or neurotransmitter systems may subsequently give rise to adaptive mechanisms in other areas or cell groups, resulting in alterations of the function of other neurotransmitter systems (Carlsson et al., 2001).

2.1.6.1 Dopamine

The dopamine hypothesis of schizophrenia was put forward by Arvid Carlsson (Carlsson and Lindqvist., 1963; Carlsson, 1988) and proposes that the symptoms of schizophrenia are due to a hyperactivity of dopaminergic neurons of the ventral tegmental area (VTA). This system gives rise to two ascending systems, the mesolimbic pathway, which innervates the ventral striatum (nucleus accumbens and adjacent structures) and the limbic system (hippocampus, amygdala, etc.) and the mesocortical system which has a more circumscribed distribution and innervates the frontal lobe, in particular the PFC (see Section 2.3.1.1. *Dopaminergic Pathways and Dopamine Receptor Physiology* for a more detailed description of the anatomy and neurochemistry of the dopaminergic system).

Two early fundamental observations supported the dopaminergic theory of schizophrenia. On the one hand, abuse of amphetamine (a DA releaser) and other drugs enhancing massively dopaminergic activity could produce psychotic symptoms. On the other hand, the classic antipsychotic drugs (chlorpromazine, haloperidol, etc.) are potent DA receptor blockers and therapeutic daily doses correlate with the affinity for DA D_2 receptors (Seeman and Lee, 1975, Creese et al., 1976; Peroutka and Synder, 1980). This suggested that an overactivity of the ascending DA neurons could induce psychotic symptoms and that this effect could be reversed by blocking the excessive activation of DA D_2 receptors. This view was further strengthened by the observation that the antipsychotic activity of flupentixol resided in its α-isomer (DA receptor antagonist) and not in its β-isomer (non-DA antagonist; Johnstone et al., 1978). More recent neuroimaging studies in previously untreated patients confirm that DA D_2 receptor occupancy is a crucial step in the achievement of the therapeutic activity of antipsychotic drugs (Kapur et al., 2000).

A refinement of the classic DA theory states that the subcortical DA system is overactive, giving raise to an exacerbated mesolimbic activity, responsible for the positive symptoms whereas there is a hypoactivity of the mesocortical system, which accounts for the negative and cognitive symptoms (Carter and Pycock, 1980; Pycock et al., 1980; Weinberger, 1987). Evidence in support of this differential behavior of cortical and subcortical DA systems has been produced by neuroimaging studies, yet some are controversial. Hence, the administration of a challenge dose of amphetamine to controls and schizophrenic patients produced a greater than normal displacement of the binding of postsynaptic DA D_2 receptor ligands to striatal receptors, as assessed by SPECT (Laruelle et al., 1996) and PET techniques (Breier et al., 1997). The displacement of he radioactive ligand from DA D_2 receptors by endogenous DA, produces a reduction of the blinding potential proportional to the amount of DA released by the amphetamine challenge. Further studies revealed a basal occupancy of striatal DA D_2 receptors higher than normal in untreated schizophrenic patients (Abi-Dargham et al., 2000) which may be consistent with an increased DA release in non-stimulated conditions. Overall, these results support the notion of an hyperactive subcortical DA system yet they have several important limitations. On the one hand, there is a marked overlap between the data from controls and schizophrenic patients, which raises questions about the predictive value of such measures. On the other hand, since the ligands used do not discriminate between dorsal and ventral parts of the striatum the observed changes are not specific for the mesolimbic pathway (e.g., they may also reflect similar changes in the nigrostriatal pathway).

Several changes in dopaminergic markers have been reported in the brains of schizophrenic patients, either using post-mortem tissue or imaging techniques such as SPECT or PET scan. A highly replicated observation is the greater density of DA D_2 receptors in post-mortem samples compared with normal controls (for a review, see Harrison, 1999). Yet, it is likely that this change results from the long-term treatment effects on such receptors since first-episode schizophrenic patients do not show such differences in PET scan studies (Nordstrom et al., 1995). Indeed, treatment effects are one of the confounding factors in the understanding the role of DA in schizophrenia since all antipsychotic drugs used, either conventional DA D_2 blockers or the second generation atypical antipsychotics produce an antagonist occupancy of DA D_2 receptors that may lead to receptor up-regulation and/or supersensitivity.

Regarding DA D_1 receptor density, a decreased striatal binding was reported in neuroleptic-naïve schizophrenics using PET scan (Okubo et al., 1997). These changes were not replicated in another study, which however reported a specific increase in density of DA D_1 receptors in the DLPFC of schizophrenic patients which negatively correlated with their score in a working memory task (Abi-

Dargham et al., 2002). The increased DA D_1 receptor density was interpreted as resulting from a decreased dopaminergic innervation of the PFC, in apparent support of a reduced dopaminergic cortical function.

2.1.6.2 Serotonin

The cell bodies of serotonergic neurons are located in the dorsal and median raphe nuclei of the midbrain and project densely to the whole forebrain. To date, 14 different serotonergic receptors have been identified and are grouped in seven subfamilies, as follows: $5\text{-}HT_1$ ($5\text{-}HT_{1A}$, $5\text{-}HT_{1B}$, $5\text{-}HT_{1D}$, $5\text{-}HT_{1E}$ and $5\text{-}HT_{1F}$), $5\text{-}HT_2$ ($5\text{-}HT_{2A}$, $5\text{-}HT_{2B}$ and $5\text{-}HT_{2C}$), $5\text{-}HT_3$, $5\text{-}HT_4$, $5\text{-}HT_5$ ($5\text{-}HT_{5A}$ and $5\text{-}HT_{5B}$), $5\text{-}HT_6$ and $5\text{-}HT_7$. Some of them have splice variants, which adds a further element of the already complicated neurochemistry of 5-HT (for reviews of the anatomical and functional characteristics of the serotonergic system, see Jacobs and Azmitia, 1992; Barnes and Sharp, 1999; Adell et al., 2002).

The involvement of serotonin in the pathophysiology of schizophrenia is supported mainly by two main – yet indirect – lines of evidence. On the one hand, certain non-selective 5-HT receptor agonists, such as lysergic acid diethyilamide (LSD) or $5\text{-}HT_2$ receptor agonists display hallucinogenic properties (for a review, see Nichols, 2004). On the other hand, the so-called atypical antipsychotics display more affinity for the $5\text{-}HT_2$ receptor family ($5\text{-}HT_{2A}$ and $5\text{-}HT_{2C}$ receptors) than for DA D_2 receptors (Meltzer, 1999). Some members of the atypical antipsychotics, and in particular, clozapine, display also some affinity or behave as indirect agonists of other 5-HT receptors, such as $5\text{-}HT_7$ or $5\text{-}HT_{1A}$ receptors, among others. Moreover, polymorphism of the $5\text{-}HT_{2A}$ receptor gene has been reported as a risk factor in schizophrenia (Williams et al., 1997) and in the response to the atypical antipsychotic drug clozapine (Arranz et al., 1998).

Several studies have reported a decreased expression of the $5\text{-}HT_{2A}$ receptor in the frontal cortex of schizophrenic patients (Harrison, 1999), and there is a blunted neuroendocrine response to $5\text{-}HT_2$ agonists (Abi-Dargham et al., 1997). Likewise, an increased expression of cortical $5\text{-}HT_{1A}$ receptors has also been reported (Burnet et al., 1996). However, PET scan analysis of $5\text{-}HT_2$ receptors have failed to show such differences in neuroleptic-naïve patients (Lewis et al., 1999; Okubo et al., 2000) whereas previously treated patients exhibited a trend towards a decrease in 5-HT$_2$ receptor binding potential (Okubo et al., 2000), which suggests that the above changes may be due to previous treatments. Likewise, no differences in 5-HT$_{1A}$ receptor binding potential in pre- or postsynaptic regions (e.g., midbrain raphe or cortex and limbic system) have been found in a group of schizophrenic patients treated with clozapine or with neuroleptics lacking 5-HT$_{1A}$ receptor affinity (Bantick et al., 2004). However, another study found a decreased 5-HT$_{1A}$ receptor binding potential in the amygdala of schizophrenic patients, including untreated and antipsychotic-treated patients (Yasuno et al., 2004).

Overall, these results do not support the existence of marked alterations of 5-HT receptors in schizophrenic patients. However, it is unclear whether a decreased 5-HT neuron activity or 5-HT release could be associated with negative/affective symptoms in schizophrenia. One of the difficulties to directly test this hypothesis is the inability of current PET scan techniques to measure 5-HT release *in vivo*, in a manner similar to that of DA (e.g., measuring the difference in binding potential of a postsynaptic PET ligand in basal conditions and after a drug-evoked release of 5-HT). The reasons for such a difference between the DA and 5-HT systems remain obscure but may be related to the difficulty of endogenous 5-HT to displace the binding of antagonist ligands (e.g., ^{11}C-WAY-100635, ^{11}C-M100907) from their corresponding receptors. Alternatively, it could be that the 5-HT system is operating at a lower occupancy level than the DA system so that the amount of 5-HT released by a drug challenge may not be sufficient to produce a measurable PET scan signal.

Interestingly, drugs acting at 5-HT$_{1A}$ and 5-HT$_{2A}$ receptors, including the atypical antipsychotics, can modulate DA neuron activity and DA release in PFC (Arborelius et al., 1993 a, b; Ichikawa et al., 2001; Gessa et al., 2000). This effect appears to take place predominantly through the activation of such receptors in PFC (Bortolozzi et al., 2005; Díaz-Mataix et al., 2005), which appears to be a key area for the action of atypical antipsychotics.

2.1.6.3 Glutamate

Glutamate is the main excitatory neurotransmitter in the brain and is used by the vast majority of projection neurons, with the exception of the basal ganglia (e.g., striatum, pallidum, *substantia nigra pars reticulata*), where projection neurons use GABA as the main neurotransmitter and the brainstem modulatory systems, whose projection neurons use monoamines (dopamine, noradrenaline, serotonin, acetylcholine, histamine) as neurotransmitters. But even in the latter cases, there is evidence in the literature that serotonergic and dopaminergic neurons can store and release glutamate as a co-transmitter (Johnson, 1994; Trudeau, 2004,). Overall, it has been estimated that glutamate is used in 40% of all synapses in brain. Hence, the cerebral cortex and other areas potentially involved in the pathophysiology and treatment of schizophrenia talk to each other using glutamate as a language. Therefore, it is likely that alterations in glutamatergic neurotransmission have an impact on the communication between these areas, which may underlie schizophrenia symptoms.

Glutamate uses a large number of receptors for this communication. The first glutamate receptors discovered were the ionotropic receptors, which are ion channels that permeate cations such as Na^+ and Ca^{2+}. These have been classified in three different subgroups: as AMPA, kainate (KA) and N-methyl-D-aspartate (NMDA) receptors. More recently, a large family of G-protein coupled (metabotropic) receptors with eight members has been identified (Cartmell and Schoepp, 2000; Schoepp, 2001). These are classified in three subfamilies, mGluRI, mGluRII and mGluRIII according to their homology and signal transduction mechanisms (Fig. 3). Figure 4 shows a scheme of a glutamatergic synapse with the putative main localization of the different receptor subtypes.

Figure 3. Classification of glutamatergic receptors

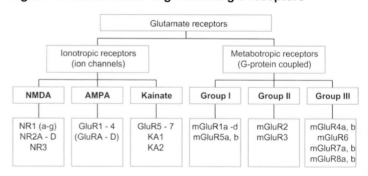

Ionotropic receptors comprise three main subtypes (AMPA, kainite and NMDA) which are composed by different subunits. The characteristics of the ion channel depend on the number and ratio of these subunits. Metabotropic receptors (mGluR1 to mGluR8) are classified in three different subfamilies and are coupled to different signal transduction mechanisms: mGlu I receptors are positively coupled to phospholipase C, mGlu II are negatively coupled to adenylate cyclase and mGlu III receptors are positively coupled to adenylate cyclase.

The involvement of glutamate in the pathophysiology of schizophrenia is mainly supported by the NMDA receptor hypofunction (Javitt and Zukin, 1991; Tsai and Coyle, 2002). Hence, NMDA receptor antagonists such as phencyclidine (PCP) and ketamine can evoke positive and negative symptoms as well as the cognitive deficits that resemble clinical symptoms of schizophrenia (Javitt and Zukin, 1991; Krystal et al., 1994). In addition, these drugs usually aggravate clinical symptoms in schizophrenic patients (for a review, see Krystal et al., 2003).

Figure 4. Scheme of a glutamatergic synapse

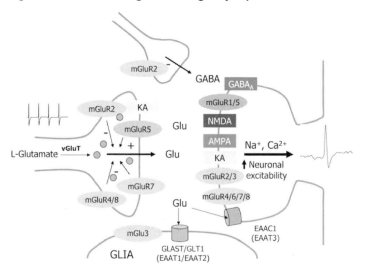

The putative localization of the various glutamate receptors is shown. Once synthesized, glutamate (Glu) is stored in synaptic vesicles through the vesicular transporters (vGluT1 to vGluT3). The arrival of action potentials to the axon terminal opens voltage-gated Na^+ channels, which leads to the fusion of synaptic vesicles with the plasma membrane and the exocytotic release of glutamate. Various presynaptic metabotropic receptors modulate positively (mGluR5) or negatively (mGluR2, 4, 6, 8) the release process. Ionotropic receptors (AMPA, KA, NMDA) are mainly localized in postsynaptic dendritic spines, although there is also evidence for a presynaptic localization. The activation of ionotropic receptors leads to the input of Na^+ and Ca^{2+} ions, which enhances neuronal excitability and results in synaptic plasticity processes, such as long-term potentiation (LTP) .$GABA_A$ receptor-mediated inputs, which enhance intracellular [Cl⁻] oppose to glutamate-mediated ionotropic effects. Ionotropic receptors mediate rapid (in the millisecond-second scale) changes in neuronal excitability whereas metabotropic receptors exert slower, long-lasting modulatory effects which involve changes in various protein kinases and gene expression. Metabotropic glutamate receptors can also modulate the release of other neurotransmitters such as GABA, catecholamines, serotonin and acetylcholine. Some receptors, such as mGluR3 or mGluR5 display a glial localization. The excess glutamate is removed from the synaptic cleft and the presynaptic space by various transporters (EAAT1–EAAT3 in human brain) through a Na^+- and energy-dependent reuptake process. Redrawn and modified from Schoepp (2001).

These drugs have been shown to act as non-competitive antagonists of NMDA glutamate receptors. In rodents, PCP and the more potent NMDA channel blocker dizocilpine (MK-801) produce a behavioral syndrome characterized by increased locomotor activity, head weaving, ataxia, body rolling, and stereotypic motor patterns. This behavioral syndrome has been used as an animal model of psychosis, and both typical and atypical antipsychotics showed to be effective in reducing NMDA receptor antagonist-induced hyperactivity (Krystal et al., 2003).

NMDA receptor antagonists induce also a deficit in the pre-pulse inhibition of startle response (PPI) in rodents, primates and humans. Normally, a low-intensity auditory tone or a warning light emitted before a high-intensity tone attenuates or prevents the startle response produced by the latter sound. However, schizophrenic patients present a deficit in this behavioral paradigm and display an exaggerated response, i.e.., as if the low-intensity tone were ineffective (Braff et al., 2001). This behavioral deficit can be mimicked in experimental animals by the administration of NMDA receptor antagonists and the effect can be prevented by antipsychotic drugs, particularly the atypical drugs (clozapine, olanzapine, risperidone, etc.).

The administration of NMDA receptor antagonists paradoxically increases glutamate release in PFC (Moghaddam and Adams, 1998) and the activity of putative pyramidal neurons in the mPFC (Suzuki et al., 2002). Interestingly, this effect is not produced when the NDMA receptor antagonist is locally applied in the PFC, which suggests that the effects of systemically administered NMDA receptor antagonists on PFC activity are mediated by NMDA receptors outside the PFC, possibly n afferent areas (e.g., hippocampus, thalamus). Hence, the application of MK-801 in the CA1/subiculum, which projects to mPFC, enhances the firing activity of PFC neurons (Jodo et al., 2005). The widespread distribution of

NMDA receptors and the different unit composition of the receptors expressed in glutamatergic and GABAergic neurons has led to suggest that non-competitive NMDA receptor antagonists (MK-801, ketamine, phencyclidine) may preferentially block NMDA inputs on GABAergic interneurons (Tsai and Coyle, 2002).

It is unclear whether schizophrenic patients have a reduced activation of certain NMDA receptor subpopulations in brain. Postmortem studies of glutamate receptor binding suggest that there is an increase in KA receptors in the PFC and decrease in AMPA and KA receptors in the hippocampus, but no changes in NMDA receptor density. Immunohistochemical studies have also found a decrease in MAPA receptors in the medial temporal lobe, but have failed to replicate the decrease in hippocampus. Some attention has been also paid to the glycine modulatory site within the NMDA receptor complex, with conflicting results (for a review, see Tsai and Coyle 2002). Finally, a decrease of the mRNA encoding KA and AMPA receptor subunits have been found in the hippocampus and parahippocampus of schizophrenics patients whereas the expression of the NMDA NR-2D subunit was found to be increased in the PFC and the NMDA NR-1 subunit was found to be decreased in the hippocampus of schizophrenic patients. Unfortunately, the lack of suitable PET scan or SPECT ligands to directly measure NMDA receptors *in vivo* has hampered the direct testing of the hypothesis of NMDA receptor hypofunction in humans. Even with the development of appropriate ligands, it is possible that only a minority of receptors can be affected (e.g., those in certain GABAergic neurons), which would not produce a sufficient PET scan signal to be detected.

2.1.6.4 GABA

Although there is no formal "GABAergic hypothesis" of schizophrenia, there is a growing interest in possible changes of GBAergic transmission in schizophrenia. On the one hand, GABAergic interneurons make up 15–20 % of the total neuronal neuronal number in the neocortex and exert a profound and widespread on the activity of principal (pyramidal) neurons through the activation of $GABA_A$ and $GABA_B$ receptors in dendritic and somatic fields ($GABA_A$) and axon terminals ($GABA_B$). Numerous types of GABAergic interneurons have been described and classified according to their morphology, the Ca^{2+}-binding protein or peptide expressed as well as the type of synaptic contacts made with pyramidal neurons (De Felipe, 2002; Kawaguchi and Kondo, 2002). Furthermore, GABAergic neurons may establish local networks through electrical synapses (Hestrin and Galarreta, 2005), which enables their coordinated activity in cortical microcircuits and enhances their capacity of control much above of that suggested by their relatively low proportion with respect to pyramidal neurons.

In recent years, a substantial number of alterations in the GABAergic system of schizophrenic patients have been reported (for reviews, see Lewis and Lieberman, 2000; Lewis et al., 2005). These include: (a) reduction in the mRNA of parvalbumin (a marker of certain subtypes of GABAergic interneurons, including chandelier cells) in layers III-IV of the DLPFC, (b) a reduced expression and immunoreactivity of GAT-1 (GABA transporter 1) in the axon cartridges of chandelier neurons, and (c) an up-regulation of the α_2 subunit of $GABA_A$ receptor immunoreactivity in the axon hillock (initial segment) of pyramidal neurons. This is a crucial synapse in cortical microcircuits since chandelier neurons make synaptic contacts with $GABA_A$ receptors in the initial segment of the axon of pyramidal neurons, where action potentials are generated. In this manner, they can modulate cortical activity in a more powerful manner than other GABAergic interneurons that synapse on distal dendrites or the cell body.

The concurrent reduction in the BDNF-TrkB signaling pathway has led to postulate that loss of BDNF activity may act as a pathogenetic factor in schizophrenia during development, resulting in a reduced activity of parvalbumin-containing chandelier neurons and the subsequent disruption of neocortical activity (Lewis et al., 2005).

2.1.7 Concluding remarks

Schizophrenia is a highly complex mental disease whose underlying genetic, anatomical, cellular and molecular basis is just beginning to be understood. It is manifested by a plethora of symptoms with devastating effects in the individuals who suffer from this illness. Although positive symptoms are undoubtedly the most evident behavioral manifestation of schizophrenia, cognitive deficits appear to play a central role during the course of the illness. Yet, despite several decades of research, the pathogenesis of schizophrenia remains obscure and current treatments are far from optimal, particularly with regards to cognitive disturbances. More research is needed to study the genetics of schizophrenia, its anatomical and cellular abnormalities and the neurotransmitter systems involved. It is hoped that new information coming from these human studies, as well from new data from experimental models in animals can provide new insights that will help in the prevention and treatment of schizophrenia, overcoming the limitations of the currently available antipsychotic drugs.

Acknowledgements

This work was supported by the Ministry of Science and Education (Grant SAF2004-05525). Support from the Generalitat de Catalunya (SGR00758) is also acknowledged.

References

Abi-Dargham A, Laruelle M, Aghajanian GK, Charney D, Krystal J. *The role of serotonin in the pathophysiology and treatment of schizophrenia*. J Neuropsychiatry Clin Neurosci. **1997**, 9:1-17.

Abi-Dargham A, Rodenhiser J, Printz D, Zea-Ponce Y, Gil R, Kegeles LS, Weiss R, Cooper TB, Mann JJ, Van Heertum RL, Gorman JM, Laruelle M. *Increased baseline occupancy of D2 receptors by dopamine in schizophrenia*. Proc Natl Acad Sci U S A. **2000**, 97:8104-8109.

Abi-Dargham A, Mawlawi O, Lombardo I, Gil R, Martinez D, Huang Y, Hwang DR, Keilp J, Kochan L, Van Heertum R, Gorman JM, Laruelle M. *Prefrontal dopamine D1 receptors and working memory in schizophrenia*. J Neurosci. **2002**, 22:3708-3719.

Adell A, Celada P, Abellan MT, Artigas F. *Origin and functional role of the extracellular serotonin in the midbrain raphe nuclei*. Brain Res Brain Res Rev. **2002**, 39:154-180.

Akil M, Pierri JN, Whitehead RE, Edgar CL, Mohila C, Sampson AR, Lewis DA. *Lamina-specific alterations in the dopamine innervation of the prefrontal cortex in schizophrenic subjects*. Am J Psychiatry. **1999**, 156:1580-1589.

American Psychiatric Association. *Diagnostic Statistical Manual of Mental Disorders*. 4th ed. American Psychiatric Press; Washington DC, **2000**.

Andlin-Sobocki P, Jonsson B, Wittchen HU, Olesen J. *Cost of disorders of the brain in Europe*. Eur J Neurol. **2005**, 12 Suppl 1:1-27.

Andreasen NC, Arndt S, Swayze V, Cizadlo T, Flaum M, O'Leary D, Ehrhardt JC, Yuh WT. *Thalamic abnormalities in schizophrenia visualized through magnetic resonance image averaging*. Science. **1994**, 266:294-298.

Andreasen NC. *Symptoms, signs, and diagnosis of schizophrenia*. Lancet. **1995**, 346:477-481.

Andreasen NC, O'Leary DS, Cizadlo T, Arndt S, Rezai K, Ponto LL, Watkins GL, Hichwa RD. *Schizophrenia and cognitive dysmetria: a positron-emission tomography study of dysfunctional prefrontal-thalamic-cerebellar circuitry*. Proc Natl Acad Sci U S A. **1996**, 93:9985-9990.

Andreasen NC. *Linking mind and brain in the study of mental illnesses: a project for a scientific psychopathology*. Science. **1997**, 275:1586-1593.

Andreasen NC, O'Leary DS, Flaum M, Nopoulos P, Watkins GL, Boles Ponto LL, Hichwa RD. *Hypofrontality in schizophrenia: distributed dysfunctional circuits in neuroleptic-naive patients*. Lancet. **1997**, 349:1730-1734.

Arborelius L, Chergui K, Murase S, Nomikos GG, Hook BB, Chouvet G, Hacksell U, Svensson TH. *The 5-HT1A receptor selective ligands, (R)-8-OH-DPAT and (S)-UH-301, differentially affect the activity of midbrain dopamine neurons*. Naunyn Schmiedebergs Arch Pharmacol. **1993a**, 347:353-362.

Arborelius L, Nomikos GG, Hacksell U, Svensson TH. *(R)-8-OH-DPAT preferentially increases dopamine release in rat medial prefrontal cortex*. Acta Physiol Scand. **1993b**, 148:465-466.

Arranz MJ, Munro J, Sham P, Kirov G, Murray RM, Collier DA, Kerwin RW. *Meta-analysis of studies on genetic variation in 5-HT2A receptors and clozapine response*. Schizophr Res. **1998**, 32:93-99.

Artaloytia JF, Arango C, Lahti A, Sanz J, Pascual A, Cubero P, Prieto D, Palomo T. *Negative signs and symptoms secondary to antipsychotics: a double-blind, randomized trial of a single dose of placebo, haloperidol, and risperidone in healthy volunteers*. Am J Psychiatry. **2006**, 163:488-493.

Bantick RA, Montgomery AJ, Bench CJ, Choudhry T, Malek N, McKenna PJ, Quested DJ, Deakin JF, Grasby PM. *A positron emission tomography study of the 5-HT1A receptor in schizophrenia and during clozapine treatment*. J Psychopharmacol. **2004**, 18:346-354.

Barnes NM, Sharp T. *A review of central 5-HT receptors and their function*. Neuropharmacology. **1999**, 38:1083-1152.

Bertolino A, Callicott JH, Elman I, Mattay VS, Tedeschi G, Frank JA, Breier A, Weinberger DR. *Regionally specific neuronal pathology in untreated patients with schizophrenia: a proton magnetic resonance spectroscopic imaging study*. Biol Psychiatry. **1998**, 43:641-648.

Bortolozzi A, Diaz-Mataix L, Scorza MC, Celada P, Artigas F. *The activation of 5-HT receptors in prefrontal cortex enhances dopaminergic activity*. J Neurochem. **2005**, 95:1597-1607.

Braff DL, Geyer MA, Swerdlow NR. *Human studies of prepulse inhibition of startle: normal subjects, patient groups, and pharmacological studies*. Psychopharmacology (Berl). **2001**, 156:234-258.

Breier A, Su TP, Saunders R, Carson RE, Kolachana BS, de Bartolomeis A, Weinberger DR, Weisenfeld N, Malhotra AK, Eckelman WC, Pickar D. *Schizophrenia is associated with elevated amphetamine-induced synaptic dopamine concentrations: evidence from a novel positron emission tomography method*. Proc Natl Acad Sci U S A. **1997**, 94:2569-2574.

Buchsbaum MS, Someya T, Teng CY, Abel L, Chin S, Najafi A, Haier RJ, Wu J, Bunney WE, Jr. *PET and MRI of the thalamus in never-medicated patients with schizophrenia*. Am J Psychiatry. **1996**, 153:191-199.

Burnet PW, Eastwood SL, Lacey K, Harrison PJ. *The distribution of 5-HT1A and 5-HT2A receptor mRNA in human brain*. Brain Res. **1995**, 676:157-168.

Burnet PW, Eastwood SL, Harrison PJ. *5-HT1A and 5-HT2A receptor mRNAs and binding site densities are differentially altered in schizophrenia*. Neuropsychopharmacology. **1996**, 15:442-455.

Byne W, Buchsbaum MS, Kemether E, Hazlett EA, Shinwari A, Mitropoulou V, Siever LJ. *Magnetic resonance imaging of the thalamic mediodorsal nucleus and pulvinar in schizophrenia and schizotypal personality disorder.* Arch Gen Psychiatry. **2001**, 58:133-140.

Callicott JH, Bertolino A, Egan MF, Mattay VS, Langheim FJ, Weinberger DR. *Selective relationship between prefrontal N-acetylaspartate measures and negative symptoms in schizophrenia.* Am J Psychiatry. **2000**, 157:1646-1651.

Carlsson A, Lindqvist M. *Effect of chlorpromazine or haloperidol on formation of 3-methoxytyramine and normetanephrine in mouse brain.* Acta Pharmacol Toxicol (Copenh). **1963**, 20:140-144.

Carlsson A. *The current status of the dopamine hypothesis of schizophrenia.* Neuropsychopharmacology. **1988**, 1:179-186.

Carlsson A, Waters N, Holm-Waters S, Tedroff J, Nilsson M, Carlsson ML. *Interactions between monoamines, glutamate, and GABA in schizophrenia: new evidence.* Annu Rev Pharmacol Toxicol. **2001**, 41:237-260.

Carter CJ, Pycock CJ. *Behavioural and biochemical effects of dopamine and noradrenaline depletion within the medial prefrontal cortex of the rat.* Brain Res. **1980**, 192:163-176.

Cartmell J, Schoepp DD. *Regulation of neurotransmitter release by metabotropic glutamate receptors.* J Neurochem. **2000**, 75:889-907.

Craddock N, O'Donovan MC, Owen MJ. *The genetics of schizophrenia and bipolar disorder: dissecting psychosis.* J Med Genet. **2005**, 42:193-204.

Creese I, Burt DR, Snyder SH. *Dopamine receptor binding predicts clinical and pharmacological potencies of antischizophrenic drugs.* Science. **1976**, 192:481-483.

DeFelipe J, Hendry SH, Jones EG. *Visualization of chandelier cell axons by parvalbumin immunoreactivity in monkey cerebral cortex.* Proc Natl Acad Sci U S A. **1989**, 86:2093-2097.

DeFelipe J, Arellano JI, Gomez A, Azmitia EC, Munoz A. *Pyramidal cell axons show a local specialization for GABA and 5-HT inputs in monkey and human cerebral cortex.* J Comp Neurol. **2001**, 433:148-155.

DeFelipe J. *Cortical interneurons: from Cajal to 2001.* Prog Brain Res. **2002**, 136:215-238.

Diaz-Mataix L, Scorza MC, Bortolozzi A, Toth M, Celada P, Artigas F. *Involvement of 5-HT1A receptors in prefrontal cortex in the modulation of dopaminergic activity: role in atypical antipsychotic action.* J Neurosci. **2005**, 25:10831-10843.

Elevag B, Goldberg TE. *Cognitive impairment in schizophrenia is the core of the disorder.* Crit Rev Neurobiol. **2000**, 14:1-21.

Fuster JM. *The Prefrontal Cortex: Anatomy, Physiology, and Neuropsychology of the Frontal Lobe.* 3rd ed. Lippincott Williams & Wilkins; Philadelphia-New York, **1997.**

Gessa GL, Devoto P, Diana M, Flore G, Melis M, Pistis M. *Dissociation of haloperidol, clozapine, and olanzapine effects on electrical activity of mesocortical dopamine neurons and dopamine release in the prefrontal cortex.* Neuropsychopharmacology. **2000**, 22:642-649.

Geyer MA, Krebs-Thomson K, Braff DL, Swerdlow NR. *Pharmacological studies of prepulse inhibition models of sensorimotor gating deficits in schizophrenia: a decade in review*. Psychopharmacology (Berl). **2001**, 156:117-154.

Glantz LA, Lewis DA. *Decreased dendritic spine density on prefrontal cortical pyramidal neurons in schizophrenia*. Arch Gen Psychiatry. **2000**, 57:65-73.

Goldman-Rakic PS, Selemon LD. *Functional and anatomical aspects of prefrontal pathology in schizophrenia*. Schizophr Bull. **1997**, 23:437-458.

Goldman-Rakic PS. *The physiological approach: functional architecture of working memory and disordered cognition in schizophrenia*. Biol Psychiatry. **1999**, 46:650-661.

Goldner EM, Hsu L, Waraich P, Somers JM. *Prevalence and incidence studies of schizophrenic disorders: a systematic review of the literature*. Can J Psychiatry. **2002**, 47:833-843.

Green MF. *What are the functional consequences of neurocognitive deficits in schizophrenia?* Am J Psychiatry. **1996**, 153:321-330.

Gupta S, Andreasen NC, Arndt S, Flaum M, Hubbard WC, Ziebell S. *The Iowa Longitudinal Study of Recent Onset Psychosis: one-year follow-up of first episode patients*. Schizophr Res. **1997**, 23:1-13.

Gur RE, Turetsky BI, Cowell PE, Finkelman C, Maany V, Grossman RI, Arnold SE, Bilker WB, Gur RC. *Temporolimbic volume reductions in schizophrenia*. Arch Gen Psychiatry. **2000**, 57:769-775.

Harrison PJ. *The neuropathology of schizophrenia. A critical review of the data and their interpretation*. Brain. **1999**, 122 (Pt 4):593-624.

Harrison PJ, Owen MJ. *Genes for schizophrenia? Recent findings and their pathophysiological implications*. Lancet. **2003**, 361:417-419.

Harvey PD, Keefe RS. *Studies of cognitive change in patients with schizophrenia following novel antipsychotic treatment*. Am J Psychiatry. **2001**, 158:176-184.

Hazlett EA, Buchsbaum MS, Kemether E, Bloom R, Platholi J, Brickman AM, Shihabuddin L, Tang C, Byne W. *Abnormal glucose metabolism in the mediodorsal nucleus of the thalamus in schizophrenia*. Am J Psychiatry. **2004**, 161:305-314.

Heckers S, Rauch SL, Goff D, Savage CR, Schacter DL, Fischman AJ, Alpert NM. *Impaired recruitment of the hippocampus during conscious recollection in schizophrenia*. Nat Neurosci. **1998**, 1:318-323.

Hestrin S, Galarreta M. *Electrical synapses define networks of neocortical GABAergic neurons*. Trends Neurosci. **2005**, 28:304-309.

Ichikawa J, Ishii H, Bonaccorso S, Fowler WL, O'Laughlin IA, Meltzer HY. *5-HT(2A) and D(2) receptor blockade increases cortical DA release via 5-HT(1A) receptor activation: a possible mechanism of atypical antipsychotic-induced cortical dopamine release*. J Neurochem. **2001**, 76:1521-1531.

Jacobs BL, Azmitia EC. *Structure and function of the brain serotonin system*. Physiol Rev. **1992**, 72:165-229.

Javitt DC, Zukin SR. *Recent advances in the phencyclidine model of schizophrenia*. Am J Psychiatry. **1991**, 148:1301-1308.

Jodo E, Suzuki Y, Katayama T, Hoshino KY, Takeuchi S, Niwa S, Kayama Y. *Activation of medial prefrontal cortex by phencyclidine is mediated via a hippocampo-prefrontal pathway.* Cereb Cortex. **2005**, 15:663-669.

Johnson MD. *Synaptic glutamate release by postnatal rat serotonergic neurons in microculture.* Neuron. **1994**, 12:433-442.

Johnstone EC, Crow TJ, Frith CD, Husband J, Kreel L. *Cerebral ventricular size and cognitive impairment in chronic schizophrenia.* Lancet. **1976**, 2:924-926.

Johnstone EC, Crow TJ, Frith CD, Carney MW, Price JS. *Mechanism of the antipsychotic effect in the treatment of acute schizophrenia.* Lancet. **1978**, 1:848-851.

Kapur S, Zipursky R, Jones C, Remington G, Houle S. *Relationship between dopamine D(2) occupancy, clinical response, and side effects: a double-blind PET study of first-episode schizophrenia.* Am J Psychiatry. **2000**, 157:514-520.

Kawaguchi Y, Kondo S. *Parvalbumin, somatostatin and cholecystokinin as chemical markers for specific GABAergic interneuron types in the rat frontal cortex.* J Neurocytol. **2002**, 31:277-287.

Kay SR, Fiszbein A, Opler LA. *The positive and negative syndrome scale (PANSS) for schizophrenia.* Schizophr Bull. **1987**, 13:261-276.

King DJ. *Psychomotor impairment and cognitive disturbances induced by neuroleptics.* Acta Psychiatr Scand Suppl. **1994**, 380:53-58.

Krystal JH, Karper LP, Seibyl JP, Freeman GK, Delaney R, Bremner JD, Heninger GR, Bowers MB, Jr., Charney DS. *Subanesthetic effects of the noncompetitive NMDA antagonist, ketamine, in humans. Psychotomimetic, perceptual, cognitive, and neuroendocrine responses.* Arch Gen Psychiatry. **1994**, 51:199-214.

Krystal JH, D'Souza DC, Mathalon D, Perry E, Belger A, Hoffman R. *NMDA receptor antagonist effects, cortical glutamatergic function, and schizophrenia: toward a paradigm shift in medication development.* Psychopharmacology (Berl). **2003**, 169:215-233.

Laruelle M, Abi-Dargham A, van Dyck CH, Gil R, D'Souza CD, Erdos J, McCance E, Rosenblatt W, Fingado C, Zoghbi SS, Baldwin RM, Seibyl JP, Krystal JH, Charney DS, Innis RB. *Single photon emission computerized tomography imaging of amphetamine-induced dopamine release in drug-free schizophrenic subjects.* Proc Natl Acad Sci U S A. **1996**, 93:9235-9240.

Lawrie SM, Whalley H, Kestelman JN, Abukmeil SS, Byrne M, Hodges A, Rimmington JE, Best JJ, Owens DG, Johnstone EC. *Magnetic resonance imaging of brain in people at high risk of developing schizophrenia.* Lancet. **1999**, 353:30-33.

Lewis DA, Lieberman JA. *Catching up on schizophrenia: natural history and neurobiology.* Neuron. **2000**, 28:325-334.

Lewis DA, Levitt P. *Schizophrenia as a disorder of neurodevelopment.* Annu Rev Neurosci. **2002**, 25:409-432.

Lewis DA, Hashimoto T, Volk DW. *Cortical inhibitory neurons and schizophrenia.* Nat Rev Neurosci. **2005**, 6:312-324.

Lewis R, Kapur S, Jones C, DaSilva J, Brown GM, Wilson AA, Houle S, Zipursky RB. *Serotonin 5-HT2 receptors in schizophrenia: a PET study using [18F]setoperone in neuroleptic-naive patients and normal subjects*. Am J Psychiatry. **1999**, 156:72-78.

Lieberman JA, Alvir JM, Koreen A, Geisler S, Chakos M, Sheitman B, Woerner M. *Psychobiologic correlates of treatment response in schizophrenia*. Neuropsychopharmacology. **1996**, 14:13S-21S.

Lieberman JA, Sheitman BB, Kinon BJ. *Neurochemical sensitization in the pathophysiology of schizophrenia: deficits and dysfunction in neuronal regulation and plasticity*. Neuropsychopharmacology. **1997**, 17:205-229.

McCarley RW, Wible CG, Frumin M, Hirayasu Y, Levitt JJ, Fischer IA, Shenton ME. *MRI anatomy of schizophrenia*. Biol Psychiatry. **1999**, 45:1099-1119.

McGlashan TH. *Early detection and intervention in schizophrenia: research*. Schizophr Bull. **1996**, 22:327-345.

Meltzer HY. *The role of serotonin in antipsychotic drug action*. Neuropsychopharmacology. **1999**, 21:106S-115S.

Miller EK, Cohen JD. *An integrative theory of prefrontal cortex function*. Annu Rev Neurosci. **2001**, 24:167-202.

Moghaddam B, Adams BW. *Reversal of phencyclidine effects by a group II metabotropic glutamate receptor agonist in rats*. Science. **1998**, 281:1349-1352.

Nelson MD, Saykin AJ, Flashman LA, Riordan HJ. *Hippocampal volume reduction in schizophrenia as assessed by magnetic resonance imaging: a meta-analytic study*. Arch Gen Psychiatry. **1998**, 55:433-440.

Nichols DE. *Hallucinogens*. Pharmacol Ther. **2004**, 101:131-181.

Nordstrom AL, Farde L, Eriksson L, Halldin C. *No elevated D2 dopamine receptors in neuroleptic-naive schizophrenic patients revealed by positron emission tomography and [11C]N-methylspiperone*. Psychiatry Res. **1995**, 61:67-83.

Nuechterlein KH, Dawson ME, Ventura J, Gitlin M, Subotnik KL, Snyder KS, Mintz J, Bartzokis G. *The vulnerability/stress model of schizophrenic relapse: a longitudinal study*. Acta Psychiatr Scand Suppl. **1994**, 382:58-64.

Okubo Y, Suhara T, Suzuki K, Kobayashi K, Inoue O, Terasaki O, Someya Y, Sassa T, Sudo Y, Matsushima E, Iyo M, Tateno Y, Toru M. *Decreased prefrontal dopamine D1 receptors in schizophrenia revealed by PET*. Nature. **1997**, 385:634-636.

Okubo Y, Suhara T, Suzuki K, Kobayashi K, Inoue O, Terasaki O, Someya Y, Sassa T, Sudo Y, Matsushima E, Iyo M, Tateno Y, Toru M. *Serotonin 5-HT2 receptors in schizophrenic patients studied by positron emission tomography*. Life Sci. **2000**, 66:2455-2464.

Owen MJ, Craddock N, O'Donovan MC. *Schizophrenia: genes at last?* Trends Genet. **2005**, 21:518-525.

Peroutka SJ, Synder SH. *Relationship of neuroleptic drug effects at brain dopamine, serotonin, alpha-adrenergic, and histamine receptors to clinical potency*. Am J Psychiatry. **1980**, 137:1518-1522.

Portas CM, Goldstein JM, Shenton ME, Hokama HH, Wible CG, Fischer I, Kikinis R, Donnino R, Jolesz FA, McCarley RW. *Volumetric evaluation of the thalamus in schizophrenic male patients using magnetic resonance imaging.* Biol Psychiatry. **1998**, 43:649-659.

Potkin SG, Alva G, Fleming K, Anand R, Keator D, Carreon D, Doo M, Jin Y, Wu JC, Fallon JH. *A PET study of the pathophysiology of negative symptoms in schizophrenia. Positron emission tomography.* Am J Psychiatry. **2002**, 159:227-237.

Pulver AE. *Search for schizophrenia susceptibility genes.* Biol Psychiatry. **2000**, 47:221-230.

Pycock CJ, Carter CJ, Kerwin RW. *Effect of 6-hydroxydopamine lesions of the medial prefrontal cortex on neurotransmitter systems in subcortical sites in the rat.* J Neurochem. **1980**, 34:91-99.

Remillard S, Pourcher E, Cohen H. *The effect of neuroleptic treatments on executive function and symptomatology in schizophrenia: a 1-year follow up study.* Schizophr Res. **2005**, 80:99-106.

Schoepp DD. *Unveiling the functions of presynaptic metabotropic glutamate receptors in the central nervous system.* J Pharmacol Exp Ther. **2001**, 299:12-20.

Seeman P, Lee T. *Antipsychotic drugs: direct correlation between clinical potency and presynaptic action on dopamine neurons.* Science. **1975**, 188:1217-1219.

Selemon LD, Goldman-Rakic PS. *The reduced neuropil hypothesis: a circuit based model of schizophrenia.* Biol Psychiatry. **1999**, 45:17-25.

Selemon LD. *Regionally diverse cortical pathology in schizophrenia: clues to the etiology of the disease.* Schizophr Bull. **2001**, 27:349-377.

Sharma T, Lancaster E, Lee D, Lewis S, Sigmundsson T, Takei N, Gurling H, Barta P, Pearlson G, Murray R. *Brain changes in schizophrenia. Volumetric MRI study of families multiply affected with schizophrenia--the Maudsley Family Study 5.* Br J Psychiatry. **1998**, 173:132-138.

Shenton ME, Dickey CC, Frumin M, McCarley RW. *A review of MRI findings in schizophrenia.* Schizophr Res. **2001**, 49:1-52.

Suzuki Y, Jodo E, Takeuchi S, Niwa S, Kayama Y. *Acute administration of phencyclidine induces tonic activation of medial prefrontal cortex neurons in freely moving rats.* Neuroscience. **2002**, 114:769-779.

Trudeau LE. *Glutamate co-transmission as an emerging concept in monoamine neuron function.* J Psychiatry Neurosci. **2004**, 29:296-310.

Tsai G, Coyle JT. *Glutamatergic mechanisms in schizophrenia.* Annu Rev Pharmacol Toxicol. **2002**, 42:165-179.

Ventura J, Green MF, Shaner A, Liberman RP. *Training and Quality Assurance with the brief psychiatric rating scale - the drift busters.* Int J Methods Psychiatr. **1993**, 3:221-244.

Weinberger DR. *Implications of normal brain development for the pathogenesis of schizophrenia.* Arch Gen Psychiatry. **1987**, 44:660-669.

Weinberger DR, Gallhofer B. *Cognitive function in schizophrenia.* Int Clin Psychopharmacol. **1997**, 12 Suppl 4:S29-S36.

Weinberger DR. *Cell biology of the hippocampal formation in schizophrenia*. Biol Psychiatry. **1999**, 45:395-402.

Weinberger DR, Egan MF, Bertolino A, Callicott JH, Mattay VS, Lipska BK, Berman KF, Goldberg TE. *Prefrontal neurons and the genetics of schizophrenia*. Biol Psychiatry. **2001**, 50:825-844.

Williams J, McGuffin P, Nothen M, Owen MJ. *Meta-analysis of association between the 5-HT2a receptor T102C polymorphism and schizophrenia. EMASS Collaborative Group. European Multicentre Association Study of Schizophrenia*. Lancet. **1997**, 349:1221.

Wyatt RJ, Henter ID. *The effects of early and sustained intervention on the long-term morbidity of schizophrenia*. J Psychiatr Res. **1998**, 32:169-177.

Yasuno F, Suhara T, Ichimiya T, Takano A, Ando T, Okubo Y. *Decreased 5-HT1A receptor binding in amygdala of schizophrenia*. Biol Psychiatry. **2004**, 55:439-444.

Yung AR, McGorry PD, McFarlane CA, Jackson HJ, Patton GC, Rakkar A. *Monitoring and care of young people at incipient risk of psychosis*. Schizophr Bull. **1996**, 22:283-303.

Zalewski C, Johnson-Selfridge MT, Ohriner S, Zarrella K, Seltzer JC. *A review of neuropsychological differences between paranoid and nonparanoid schizophrenia patients*. Schizophr Bull. **1998**, 24:127-145.

2.2 Clinics

Schizophrenia: A Clinical Review

Salvador Ros[1] and Francisco Javier Arranz[2]

[1]Dept. of Psychiatry, Autonomous University of Barcelona, Hospital del Mar, Pg. Marítim, 25. 08003 Barcelona, Spain.

[2]Medical Dept., CNS Area, Laboratorios Esteve, Av. Mare de Déu de Montserrat, 221. 08041 Barcelona, Spain.

2.2.1 Introduction

Schizophrenia is an illness or group of illnesses characterized by the sudden appearance, or an appearance after a prodromic period, of a deterioration predominantly in thought, language, emotion and social skills, which in the majority of cases appears within the context of a psychotic syndrome characterized by hallucinations, deliriums and extravagant behavior. The illness appears in adolescence or youth, and progresses with psychotic outbreaks within the context of steadier deterioration in emotional and social terms over time. Awareness and intellectual skills are generally preserved, although some deficits of a cognitive nature appear as the illness evolves.

Schizophrenia is a serious mental disorder that significantly compromises the quality of life of those suffering from it, as well as their relatives. It generally starts in adolescence or youth, preventing sufferers from achieving their educational and occupational goals. It is present in all countries and cultures of the world and is one of the ten leading causes of incapacity in the population aged between 18 and 44 (Murray and López, 1996). Its clinical characteristics, the nature of the illness, the particular seriousness of some of its symptoms and the frequent tendency of it becoming a chronic illness mean that these disorders are difficult to handle in healthcare terms and sufferers are at great risk of aggression to themselves and others, with frequent hospitalization and high costs. It is also frequent for some kind of dependency to appear during its evolution, especially alcohol (10%), although many of the sufferers, between 40% and 50%, meet dependence criteria for other drugs, worsening the course of the illness as well as the possibilities of following the therapy and aggravating the prognosis.

Schizophrenia involves a significant degree of stigma and discrimination that increases the burden for patients and their relatives. Those diagnosed with schizophrenia are often faced with great social isolation, certain discrimination with regard to access to housing or job opportunities and other kinds of prejudices. Stigmatization extends also to the rest of the family and to those health professionals involved in caring for patients with this diagnosis. The treatments they receive can also contribute to magnifying the problem, particularly if there are significant side-effects, such as those associated with taking certain antipsychotic drugs and specifically those known as typical or conventional antipsychotics.

2.2.2 Background

Although it was Kahlbaum who started the nosological trend of psychosis with the publication in 1863 of his classic "Monograph", Kraepelin was the true founder of modern nosology. He describes the clinical history, classifies and constructs the supports for the natural history and prognosis of schizophrenia. Kraepelin even takes contributions from his most intense critics, such as Hoche; from Morel he takes the Latinate term "dementia praecox", having surpassed the concept of degeneration; from Magnan he takes the idea of predisposition and from Griesinger, founder of true scientific psychiatry, the belief of the existence of an underlying cerebral foundation to the illness; from Kahlbaum himself, in addition to the idea of a symptomatic complex and of how people become schizophrenic, he takes catatonia; and through his disciple, Hecker, hebephrenic; from Moebius he takes the concept of endogenous and exogenous and, finally, from Wundt, the experimental method and its associating psychology. Later Bleuler added the simple form of schizophrenia and broadened the concept of *dementia praecox* by including relatively slight and non-psychotic forms (latent schizophrenia, which opens the doors to the complicated dialectic between core schizophrenia and latent cases), a concept that was abused, partly due to the expansion of psychoanalytical literature.

Kleist and Leonhard, following the postulates of Wernicke, propose a complex classificatory system that attempts to reflect the diversity of the different clinical forms. Both these authors and Kraepelin and Bleuler based their work on the idea that different sub-types of schizophrenia would be a reflection of injuries in different cerebral areas.

Later, attempts were made to differentiate other entities, such as the pseudo-neurotic form of Hoche and Polantin, the schizoaffective form of Kasanin, Knight's borderline states and Langfeldt's schizophreniform psychoses that, according to the author, correlate with schizoaffective forms, oneirophrenia, brief reactive psychosis, atypical schizophrenia and others.

The thinking of Bleuler was a powerful influence in the United States. For Bleuler, the disassociation of thought, ambivalence and affective rigidity are primary symptoms of the fundamental disorder; deliriums, hallucinations and catatonic symptoms would be adaptive or secondary consequences of the fundamental disorder. Only autism would remain at an intermediate level between primary and secondary symptoms. With Bleuler's criteria, the group of schizophrenia, in addition to changing their name, was over-extended, in European psychiatry even reaching the schizoid personalities of Kretschmer. The concept of schizophrenic reaction was abused in American psychiatry. In this respect, the work of Kurt Schneider, who had more influence in Europe and who demanded the existence of first range symptoms to justify diagnosis of schizophrenia, guided clinical practice and avoided, in European psychiatry, many of the diagnostic abuses committed in American psychiatry.

The diagnostic abuses observed in American psychiatry also required, quite a few years later, a similar rectification, which however were the consequence of other interests. In the presentation of Schneider's first range symptoms, the main interest was in helping the clinic carry out its diagnosis. As sometimes has been criticized, it is possible that European clinical practice at the time was content with diagnosing and then remaining immersed in a certain therapeutic nihilism. What does appear evident is that there were two clear priorities in the interest in the American rectification: the first was to specify the objective of scientific research as clearly as possible, so that the second and more decisive priority could be achieved, namely that of having therapeutically effective resources, similar to those used in all medical specialities. The key work consisted in researching the different diagnoses existing on both sides of the Atlantic.

By the middle of the 20th century, schizophrenia had been defined in highly diverse ways which, as we have already mentioned, led to diverging conceptions. It is therefore not surprising that a series of studies carried out in the 1950s, 1960s and 1970s showed that clinicians were often not in agreement when diagnosing schizophrenia and diagnostic reliability was extremely low. A significant example of this was the study by Cooper, United States–United Kingdom, in which the comparative schizophrenia and mood disorder rates were examined in hospitals in New York and London. It was found that schizophrenia was much more frequent in New York than mood disorders, while in London the frequency of both pathologies was more or less similar. If there were really a geographic variation in this illness, there might have been significant etiological implications (e.g., genetic, environmental or cultural factors). However, when research psychiatrists applied standardized operational criteria to all the cases (Transversal National Project United States–United Kingdom, 1974), the relative rates of schizophrenia and affective disorders were identical in both countries. In the 1960s, it is evident that the concept of schizophrenia in America was broader than in Europe, to the extent that people joked that all you had to do was to cross the sea in order to be cured of the illness, where one supposed the person would be diagnosed with something else, presumably a manic depressive psychosis.

The findings of other studies (Andreasen and Flaum, 1991) have indicated that the prevalence of schizophrenia is relatively constant throughout the world, although the conceptualization (and therefore the diagnosis) differs notoriously across geographical borders.

Other studies have shown that the conception of schizophrenia has also changed over time. For example, in the hospitals in New York, the rate of schizophrenics in a period of 10 years in the 1950s almost doubled the rates in the same hospital in the previous decade. Such significant changes in ten years, if true, would suggest etiologies such as those of viral syndromes (e.g., post-influenza epidemics). However, once again, no differences were observed in the rates when researchers applied operational diagnostic criteria to the clinical records.

Without doubt, schizophrenia is a markedly heterogeneous entity with variables ranging from the symptoms per se and how it presents itself to pre-morbid adjustment and genetic vulnerability, up to the response to treatment and prognostic.

Studies during the 1960s and 1970s underlined the need to establish a common language in psychiatry. The aim of optimizing diagnostic reliability has been important not only to facilitate research but also for its direct application in the clinical and administrative field. One method to improve diagnostic reliability, used by the research community for a time, adopted the application of operational diagnostic criteria based on the theory that diagnosis deriving from the application of specific algorithms of signs and symptoms was probably more stable in individuals and over time than those assigned based on theoretical constructs or personal beliefs.

It was in this context when, in the 1960s and 1970s, diagnostic systems based on operational criteria started to proliferate, such as: Langfeldt's criteria (1960), Present State Examination (Wing, 1966), the criteria of Feighner et al. (St. Louis criteria or Washington University criteria; 1972), the New Haven Schizophrenia Rate (Astrachan et al., 1972), Flexible System (Carpenter et al., 1973), the criteria of Taylor and Abrams (1978), Research Diagnostic Criteria: RDC (Spitzer et al., 1978), CATEGO System (Wing et al., 1977), Diagnostic Criteria in the Tenth Review of the International Classification of Diseases ICD-10 (WHO, 1992) and Diagnostic Criteria of the DSM-IV-R for Schizophrenia.

As we can see, there is heterogeneity and marked conceptual differences depending on the author and the historical and cultural context of the time. How many sub-types of schizophrenia are still valid and can be justified nowadays? What is the name now for the hebephrenic forms, the pseudo-neurotic, latent, simple, disorganized, catatonic, undifferentiated, paranoid, residual, deficient or defective or schizoaffective forms? Can a productive versus non-productive dualism be established, or paranoid versus non-paranoid, positive versus negative?

In those medical disorders whose mechanism has been classified (e.g., pneumococcal pneumonia), making a correct diagnosis accounts for the most important part of patient care. Once this has been done, the right treatment can be prescribed and followed. In psychiatry, establishing the diagnosis is only the beginning. Diagnoses group individuals into categories that are still relatively broad and certainly heterogeneous with regard to the underlying etiology and mechanisms. The researcher must consider this heterogeneity when designing and interpreting studies and must avoid the tendency to restrict these designs to comparisons between schizophrenic patients and control groups, or comparisons with regard to a few given variables. Clinicians must also take this into account, namely the fact that the illness can be heterogeneous in its response to treatment and evolution. The process of establishing a correct diagnosis is therefore an essential beginning but in no way an end to psychiatric practice.

New questions arise, of difficult consensus after the appearance of antipsychotic drug complexes, resulting from the significant advances made in neuroscience research and forcing us to draw up new guides for action including therapeutic procedures (early treatment of schizophrenia, intervention in the acute episode, rapid neuroleptization, handling of side-effects associated with antipsychotic treatment, dose equivalencies for different antipsychotics, the conversion from oral to depot medication, parenteral medication, use of polypharmacy, megadoses, prevention of relapses, resistance to treatment and dual pathology, among others). Increasingly more attention is being paid to the incessant growth in treatment costs for serious mental disorders and the need to contain these costs. The new treatment guides and algorithms are of growing interest to those in charge of managing health policy. Based on a combination of empirical studies and expert consensus, the more general guides and more specific algorithms attempt to define a rational focus of treatment that includes both clinical and economic issues. Examples of this new kind of guide are the Practice Guidelines for the Treatment of Patients With Schizophrenia from the APA and the Texas Medication Algorithm Project (TMAP). APA's guidelines focus on the most rigorously established clinical findings and therefore tend to be general in their recommendations for treatment. For example, in the therapy for schizophrenia they only specify that a combination of pharmacological and psychosocial measures should be employed. However, the TMAP freely incorporates expert opinion but its aim is for doctors to maintain a high degree of flexibility (e.g., for treating schizophrenia it recommends mono-therapy with atypical antipsychotic drugs for at least three weeks, up to the recommended maximum dose in the product specifications; for patients who do not respond to the treatment, it encourages the use of tests with at least three first line atypical antipsychotic drugs and a conventional antipsychotic, followed by a clozapin test). Some additional algorithms are also provided to control the side-effects and use of complementary drugs. The TMAP has been designed not only as an aid in treatment for doctors but also to study the effectiveness of treatment algorithms in optimizing health and management costs.

Last, the clinical handling of schizophrenia is becoming increasingly more complex and must integrate new concepts from different areas of knowledge, of unquestionable interest but which entail considerable effort in terms of systemisation. Structured interviews, contributions from neuroimaging, new psychometric tools for cognitive assessment, assessment of disability and quality of life, rehabilitation models, new health care resources and handling special clinical situations (pregnancy and sexuality, suicide and violence), among others, all entail a constant effort of consensus between psychiatry professionals.

2.2.3 Epidemiology

The rate of schizophrenia varies from 10 to 60 cases per 100,000 inhabitants/years (Jablensky, 1986), without varying substantially in different geographical regions, races or cultures, although it is currently being debated whether this rate should be lowered (Eaton et al.,1991). With regard to prevalence, different studies show annual figures ranging from 0.6% to 17% with higher rates in developed countries compared to non-developed. It is supposed that this divergence is explained by a different numerical estimation of the cases, a different mortality rate and different course.

The starting age differs between men (15–25 years) and women (25–35 years). Sex has also been seen to influence the evolution, more favorable and with less deterioration in the case of women, in spite of having the same prevalence. The start of schizophrenia before age 10 or after 50 is exceptional.

The time of year of birth has been seen to increase prevalence. Aspects of bad nutrition, a cold climate, greater perinatal problems and viral infections are adduced that would disturb how the pregnancy develops (Mednick et al., 1987). The finding of greater incidence in low socio-economic groups (Eaton, 1985) seems to be due more to the consequences of the illness than to its cause.

The higher mortality of these patients, particularly via violent death (suicide, homicide or accident) is an empirically proven fact. Globally, the risk of death doubles that of the normal population. References have been made to long periods spent in hospital, iatrogenicity and sedentariness, among others, as factors leading to a greater number of deaths by infectious disease.

2.2.4 General semiology

Nowadays, we could consider that schizophrenia is characterized at least by three main kinds of symptoms that we can group into three psycho-pathological dimensions: positive, negative and cognitive. Positive symptoms are characterized by the loss of contact with reality, false delusional beliefs and alternations in senses and perception or hallucinations. These generally occur episodically, although they can occasionally become chronic over time. Negative symptoms are deficient states where the emotional and behavioral processes are reduced or absent. Frequent in this group are: flat affect, anhedonia, abulia and mutism. In general, these are usually more persistent than positive symptoms and are associated with a functional deterioration of the illness.

Historically, cognitive symptoms were considered as epiphenomena of the positive and negative symptoms but, given their growing importance in the prognosis of the illness, they are now considered as main symptoms of schizophrenia, a psycho-pathological neo-Kraepelinian movement occurring that includes the concept of *dementia praecox*, significantly moving away from excessive emphasis on the appearance of psycho-pharmacology impregnated in the model of Kurt Schneider.

Cognitive deficits are characterized by problems with attention and concentration, in learning, memory and executive functions and have been correlated with 61–78% of schizophrenic patients, a higher prevalence than that estimated in neurological illnesses associated with cognitive deterioration such as epilepsy, Parkinson's disease or multiple sclerosis. Cognitive decline, as well as the negative symptoms, have been associated with the functional deterioration of schizophrenia.

2.2.5 Positive symptoms in schizophrenia

The DSM-IV-TR system defines positive symptoms as an excess or distortion of normal functions, manifested in the form of deliriums, hallucinations, disorganized speech and seriously disorganized or catatonic behavior.

For the DSM, deliriums are erroneous beliefs that generally entail a bad interpretation of perceptions or experiences. The difference between a delirium and an overvalued idea is, however, difficult to appreciate in clinical practice and depends on the degree of conviction with which the idea is held in the face of evidence to the contrary. The presence of strange or bizarre delirious ideas (implausible for any person in their cultural milieu) supposes a greater probability of suffering from schizophrenia, so that its mere presence would fulfil criterion A, without the need for a second symptom. The concept of rigidity attributed to the delirious belief, which would make it untreatable with psychological methods and would allow us to differentiate between delirious ideas and other overvalued ideas, is today being revised, as well as some studies allowing the possibility for treatment using cognitive-behavioral techniques (Walkup, 1995; Sensky et al., 2000).

Hallucinations are perceptive alterations that can occur in any sensory mode. Auditory hallucinations are the most frequent form and the most characteristic of schizophrenia. Certain types of auditory hallucinations, such as two or more voices talking together or voices providing a continuous commentary on what the person is doing, saying or thinking, are considered as characteristic of schizophrenia.

Disorganized speech is characterized by a tendency to present weak associations, to separate or disrupt language, which would translate that of thought, and also by the tangential nature of the language, i.e., the tendency to present associations based on marginal aspects.

Seriously disorganized behavior can manifest itself via a large variety of clinical forms, ranging from agitation to infantile or catatonic behavior. This criterion must nonetheless be applied restrictively, as the possible causes of alterations in behavior are many and these behaviors are similar, although some are specific to schizophrenia, such as shouting for no reason, while others are due to hallucinatory or delirious phenomena, in which case they should not be considered as a primary alteration in behavior.

Classically, it has been pointed out that positive symptoms do not allow us to predict the prognosis of Schizophrenia in the medium or long term (Hafner and Heiden, 2003). However, there are some studies that find the seriousness of the positive symptoms do predict situations of social disability in the adult stage (Curson et al., 1988). It is an accepted figure that between 25% and 50% of patients with hallucinations are resistant to treatment with antipsychotics (Kane et al., 1988; Meltzer, 1992).

With regard to the persistence of voices, some authors have referred to a persistence of the hallucinatory symptoms of up to 50% (Miller, 1996). Sanjuán and Aguilar (2005), in a sample of 260 patients with a clinical history of having suffered from auditory hallucinations, all under treatment for at least one year with antipsychotics, found that 124 (47.7%) had hallucinations persistently.

The persistence of hallucinations is not a trivial figure. In a classic work on this area, Falloon and Talbot (1981) warned that persistent hallucinations were associated with a greater frequency of suicidal thoughts and, more recently, they have also been related to violent behavior (Cheung et al., 1997).

In a cluster analysis of the dimensions of hallucinations (González et al., 2003), it can be seen that the emotional factor is the first that divides patients into two broad groups. Patients with a high degree of anxiety and with a negative content in hallucinations and patients in whom hallucinations are less emotionally intense, where the content is neutral or positive. This differentiation had connotations for the prognosis. In a previous study, Hustig and Hafner (1990), in a sample of 12 patients with auditory hallucinations, found significant associations between a high degree of anxiety and depression and the persistence of hallucinations.

A recent analysis of the data (González et al., 2005) pointed to various factors as possible predictors of the chronic nature of auditory hallucinations. The emotional variable is once again stressed. The most interesting data in this publication support the idea that auditory hallucinations experienced in a pleasurable way influence the chronic nature of the hallucinations. Does this mean that hearing pleasant voices has a worse prognosis? The answer is no. Earlier studies (Sanjuán et al., 2004) suggested that experiencing this kind of voices offsets, on the one hand, the negative effect of disagreeable hallucinations and that the subject feeds them back, generating a tolerable emotional atmosphere. However, patients whose emotional repercussions are intense and where this is experienced negatively evolve differently. For example, persistence united with negative emotional intensity has been related with a high frequency of violent behavior and therefore a worse prognosis (Cheung et al., 1997).

The duration and degree of control over the voices can also be variables that predict whether the patient will have persistent hallucinations. Benjamin (1989) claimed that the phenomenological evolution of hallucinations, whether they become more complex, detailed and long-lasting, increases the possibility of the patient suffering from chronic hallucinations. However, this author did not make it clear whether greater complexity or duration results in a worse prognosis in the functional sense of the illness.

Evolution over time means that individuals with hallucinations generate greater control over them. Voices that are powerful and threatening for the patient become less stressful due to a greater degree of control. A greater control over the voices is undoubtedly related to a better prognosis. In fact, learning how to control the voices is one of the strategies that has been widely investigated by therapists of the cognitive-behavioral treatment of hallucinations (Romme and Escher, 1989; Meltzer, 1992; Garety et al., 2000). When the variables of hallucination persistence, pleasurable hallucinations and the degree of control are associated with each other, the prognosis of the psychotic hallucinator improves.

2.2.6 Negative symptoms in schizophrenia

A negative symptom can be defined from a psychopathological point of view as a state of not possessing a behavior that people who are not ill possess, a deficit or absence of an original function or signal. There is some agreement in differentiating between what we would call primary negative symptoms that appear from the onset of the illness and are associated with a structural, irreversible abnormality at a cerebral level from secondary negative symptoms associated with positive symptoms, environmental causes or the effect of the treatment, which would be partially or totally reversible. However, this classification is not always easy, as sometimes patients start with negative, depressive symptoms.

In spite of the difficulties involved in this kind of definition, this is not the only problem we are faced with as, in addition, if we do not have a list of symptoms that are considered as positive or negative, if there are no precise, more operational definitions for each of these symptoms, if their boundaries with other symptoms (e.g., depression) or with behaviors that are difficult to catalogue as pathological (e.g., differentiating apathy or abulia from boredom) are not clear, in general we will have limits that are not very sharp and measurement scales will therefore vary significantly from one to the other. This explains the difference in studies using more sophisticated technologies such as neuroimaging or genetics (Box 1).

There is therefore no consensus on what we must include within the group of negative symptoms, although the commonly accepted ones are: flat affect, blunted affect, mutism, anhedonia, apathy and abulia. Simplistically, anhedonia has been described as a lack of pleasure, apathy as a lack of interest and abulia as a lack of motivation; in this way these symptoms have been conceptualised superficially as a reduction or abolition of a function that does not appear to be a single unit.

Box 1: Brain structures associated with the presence of negative symptoms (brain areas possibly involved according to various studies)

MR:

- White substance of the left front lobe
- Both temporal horns
- Ventricular dilatation
- Dilatation in both caudate nuclei
- Higher frontal/temporal ratio in left dorsolateral prefrontal lobe
- Higher ventricle/brain and fissure of Silvius/brain ratio

PET:

- Reduction in perfusion in prefrontal cortex and associative parietal cortex

---

However, it has never been specified whether the mechanisms, involved in the formation of these symptoms, are due to a deficit in brain signals (in the case of anhedonia, the pleasure centres would be altered), in the parameters of a feeling (the centres would be fine but the patient would not know how to interpret them at a conscious level) or in some quality inherent in this (the patient would abnormally recognise part of the brain signals).

In summary, negative symptoms might be difficult to evaluate because they occur in close contact with normality, are not specific and could be secondary to other factors (positive symptoms, extrapyramidal symptoms associated with the therapy, a mood disorder or environmental hypostimulation).

In a long term follow-up study, Moller et al. (2000) analyzed the importance of negative symptoms in a sample of 85 schizophrenic subjects, 69 schizoaffective disorders and 48 subjects with affective psychosis. The schizophrenic subjects have the worst prognosis of the three kinds of psychosis and the presence of negative symptoms in the group of schizophrenics, assessed at the time of the first hospitalization, leads to expectations of the worst prognosis for this group.

In the analysis of the stability of negative symptoms over a period of 5 years, Hafner and Heiden (2003) found hardly any variation in the prevalence of the same symptoms assessed using the SANS scale. The question is whether this scale reflects the whole clinical and functional situation of the patient. A recent study by Milev et al. (2005), with 99 patients followed over 7 years, found that negative symptoms can predict greater difficulties in interpersonal relations.

2.2.7 Cognitive alterations in schizophrenia

The first psychometric studies referred to a loss in IC of around 33 points before and after the illness (Webb, 1950). However, this and other works with similar findings can be criticized for not being very reliable in their retrospective measurement of IC prior to the psychotic episode. Russell et al. (1997) suggested that any cognitive deficit in schizophrenia is actually due to a prior deterioration and not to the psychotic episode per se. Two fundamental questions arise: When does the deterioration begin? And how does it evolve in the patient's lifetime?

Based on pre-schizophrenic children, there are signs that cognitive deficits are present in early infancy, existing a long time before the start of the illness (Erlenmayer-Kimling et al., 1984). Longitudinal studies in children at a high risk of suffering schizophrenia suggest that these children have a lower IC than normal (Nelson et al., 1990; David et al., 1997). Studies in soldiers (Mason, 1956) showed that subjects who later developed schizophrenia could be differentiated from normal soldiers in a capacity test prior to recruitment. These data are coherent with the hypothesis of early neurodevelopment, suggesting an alteration probably in the neural connections from birth, either due to genetic causes or pre-natal injury. However, other studies have found sub-groups of patients with a neuropsychological performance that is indistinguishable from normal subjects (Goldstein and Shemansky, 1995).

In their first episode, schizophrenic patients show significant deficits in the measurements of language, executive function, memory, attention, concentration and motor speed, indicating that significant cognitive defects are present from the start of the illness.

Longitudinal studies in adult patients have also produced contradictory findings. With regard to the evolution of the cognitive function in schizophrenia (Table 1), most of the data available at present indicate an absence of deterioration as the illness evolves, except for small subgroups of patients. In fact, evidence supports findings of no change or improvement throughout the illness.

Table 1: Most relevant longitudinal studies on cognitive deterioration in schizophrenia (taken from Sanjuán and Balanzá 2002, modified).

Study	Period (years)	Sample	Tasks	Results
Klonoff et al., 1970	8	66 sch.	WAIS_R	Significant improvement
Bilder et al., 1991	1.5	28 sch. (FE).	Wide battery, 4 domains	High stability of deficits, with tendency to overall improvement. Decline: attention
Hoff et al., 1992	2	17 sch. (FE)	Wide battery, 7 domains	Stability: VF, verbal and spatial memory. Improvement: problem solving, attention, concentration and motor speed.
Rund et al., 1997	1-2	15 sch. 14 controls	Mini-battery of 4 tests	High stability: backward masking and LTM. Improvement in STM
Gold et al, 1999	5	54 sch. (FE)	Wide battery	Stability: verbal IC, VF, TM-B, Rey figure. Improvement: Total IC and manipulative, WCST, verbal memory. Decline: finger tapping.
Harvey et al., 1999	2.5	57 sch. elderly	Mini-mental	Deterioration in subgroup of institutionalized patients, with poor diagnosis.
Heatin et al., 2001	3	142 sch. 206 controls	Wide battery, 7 domains	General stability
Hill et al., 2004	2	45 sch. 33 controls	Wide battery	No general change in deficits. Improvement in verbal memory

FE: first episodes, VF: verbal fluency, LTM: long-term memory; STM: short-term memory; IC: intellectual coefficient, WCST: Wisconsin Card Sorting Test.

According to a transversal study (Hoff et al., 1992), schizophrenics in their first episode were as affected cognitively as chronic schizophrenics. In some longitudinal findings, patients with a first episode showed improvements in the measurements of executive function, concentration and motor speed after 2 years of the illness.

Preliminary evidence indicates that cognitive improvement could be related to a reduction in symptoms. Some indications point to the improvement coming about long before recovery after an acute psychotic episode and that this improvement is not related to a reduction in the symptoms (Sweeney et al., 1991). Cognitive improvement might be related to a reduction in medication or with a re-establishment of the brain physiology due to treatment with psychoactive drugs.

Goldstein and Zubin (1990) found little evidence of cognitive deterioration in schizophrenia. Only one group of schizophrenics with documented neurological dysfunction (cranial trauma, epilepsy, vascular disorders) showed signs of deterioration, suggesting a possible interaction between neurological and schizophrenic factors and the ageing process in producing this deterioration.

Currently some researchers have found a subgroup of patients (around 25%) who do show progressive deterioration (Knoll et al., 1998), coherent with some findings in MRI where this patient subgroup show evolution in the ventricles and atrophy in the hippocampus that differs from the classic concept of "static encephalopathy" studied to date. The explanation for how this deterioration evolves is not clear, and speculations could be made ranging from genetic differences, environmental factors and the effect of the medication administered.

A relevant piece of data could be the relation between the course of the cognitive dysfunction and gender differences in schizophrenia. In general there are few neuropsychological studies on this. Haas et al. (1991) observed that men evolved worse than women when verbal memory was measured. Perlick et al. (1992) found women to be more affected in conceptualization and attention. Hoff et al. (1991) found greater affect on the executive function in women with chronic schizophrenia but in the first episodes there were no significant gender differences.

2.2.8 Characteristics of cognitive deterioration in schizophrenia

Overall, schizophrenic patients have always shown a type of neuropsychological deficit halfway between neurotic patients and patients with acknowledged brain damage. The areas in which there is consensus concerning the poorer performance of these patients are the capacity for abstraction and executive function, attention, memory and language. The main objections that can be made to this renaissance in the importance of cognitive deficit is that, as occurs with negative symptoms, this deficit is not specific. However, one of the reasons why the cognitive deficit has gained in importance is because it is one of the best predictors of the degree of social function.

The first neuropsychological studies on schizophrenia (Heaton et al., 1978) focused principally on the level of functioning in order to differentiate schizophrenic patients from others with brain damage, based on a multitude of neuropsychological batteries. In general, it was observed that acute schizophrenic patients and other psychotic patients had higher performance rates than those with organic brain dysfunction, while chronic schizophrenic patients showed similar performance rates to those with organic brain dysfunction. Heaton et al. (1978) stated that these two groups cannot be differentiated from each other except at a random level. Some of these studies were limited by little diagnostic reliability; there were also distortions because schizophrenic patients were compared with others whose organic brain dysfunction had various etiologies and evolutions.

The work of Flor-Henry in 1975 on epilepsy in the temporal lobe was decisive in stimulating neuropsychological research to focus on whether psychiatric illnesses had characteristic patterns in the tests, thereby reflecting a disorder in functional lateralization or damage in one of the brain hemispheres. In general it was postulated that schizophrenia was a disorder involving a dysfunction in the left hemisphere, while bipolar affective disorder and other affective disorders were more related with problems in the right hemisphere. More recently, Taylor and Abrams (1984) displaced attention from the influence on the left or right side of the brain towards a hypothesis concerning cognitive dysfunction in schizophrenia more related to the role of the frontal regions of the brain compared with the rear, as well as the medial and lower structures such as the temporal cortex and the hippocampus region.

Currently, in general it is accepted that, in schizophrenia, there is a generalized affect on the cognitive functions with greater repercussions on three areas: attention, memory and executive functions. The scores obtained in neuropsychological assessments for these functions in schizophrenic patients are, on average, two standard deviations below the normal population. Although some factors, such as psychotic symptoms and the lack of motivation, may play a part in the assessments, it has been demonstrated that cognitive deficits observed in schizophrenia are not epiphenomena of these.

Last, there are four aspects for which there is consensus regarding the follow-up studies on cognitive deterioration in schizophrenia:

- This deterioration may exist prior to the first outbreak or identification of the illness.

- After diagnosis, the deficits are relatively stable and do not undergo significant variation, unlike with positive symptoms.

- Available data support the view that cognitive deterioration is not progressive, unlike degenerative diseases, but there is debate as to whether there could be a degenerative course with age in a subgroup of patients.

- It has been confirmed, although not in all studies, that cognitive deficits appearing in the first episode can predict the degree of social functioning and adaptation in the long term.

2.2.8.1 Alterations in attention

A stable deficit in attention and information processing has been observed in schizophrenic patients, consisting of defects in their capacity to apprehend, in shadowing, reaction times and their susceptibility to interference (Lieh-Mak and Lee 1997).

The first studies by Shakow (1962) using a simple "reaction time paradigm", and by Kornetsky and Mirsky (1966) using the "continuous performance test", underlined the proposal that schizophrenic patients had a key deficit in continuous attention. Later studies using Backward Masking or the "apprehension lapse test" showed that schizophrenic subjects processed elementary information anomalously, involving sensory recording or storage in the short term that was deficient compared to control subjects (Atkinson and Shiffrin, 1968).

This attention and information processing deficit has been observed in schizophrenics in remission (Asarnow and McCrimmon, 1982), first-degree relatives of schizophrenic patients (Asarnow et al., 1977) and schizophrenic and bipolar patients (Saccuzzo and Braff, 1986). It could be interpreted as a vulnerability factor in serious psychiatric illnesses and psychosis in particular. The finding of these deficits in first degree relatives suggests a probable phenotypic marker of schizophrenia, in addition to a predictor of prognosis (Nuerchterlein and Dawson 1984).

Some researchers point to the fact that patients with negative symptoms process information relatively poorly compared with those with positive symptoms (Weiner et al., 1990).

Attention and information processing deficits have been demonstrated in schizophrenia using neuropsychological tests of immediate attention, sustained attention, reaction time and visual search. Frequently used tests are Span Digital, Double Stimulation, the Stroop Test and Reaction Time to a Visual or Sound Stimulus. The findings indicated that consistently less than 50% of patients diagnosed with schizophrenia have an attention deficit.

2.2.8.2 Alterations in memory

It would seem reasonable to suppose that schizophrenic patients have memory problems. In post-mortem studies, schizophrenic subjects show changes in the temporal areas and the hippocampus (Bogerts et al., 1985), and studies with MRI have also documented reductions in the size of the hippocampus, amygdala and temporal lobes (Dauphinais et al., 1990; Shenton et al., 1992).

Kraepelin stated that the memory was one of the cognitive functions conserved in early memory; however, current research points in another direction. Aleman (1999) in a meta-analysis of 70 studies, found consistent data on memory deficit in schizophrenia. The most significant deficit is related to working memory, both verbal and spatial. Working memory can be defined as that which allows you to retain useful information over a short period of time; this information cannot be transferred immediately to the long-term memory store.

Some researchers (McKenna et al., 1990) state that the memory dysfunction of schizophrenia is no different from the classic amnesic syndrome with immediate preserved memory and deficient memory recall.

Koh (1978) suggested that schizophrenic patients perform worse in recall tasks than in recognition tasks. Within recall tasks, they make more mistakes in those of deferred recall compared with those of immediate recall and greater deterioration is observed in the tasks requiring information processing and organization (voluntary and conscious) than in those that only require automatic or unconscious processing, such as implicit learning (Schmand et al., 1992a) and procedural learning (Schmand et al., 1992b), both in visual and verbal memory (Calev, 1990; Rund and Landro, 1995; McKay et al., 1996). Calev et al. (1983) suggested that memory difficulties in schizophrenic patients are attributable to organizational defects present at the time of coding; they also point out that chronic patients may suffer deficit both in recognition and recall tasks. However, Gold et al. (1992) found defects in both voluntary and automatic memory processes.

Heinrichs and Zakzanis (1998), in a meta-analysis of 204 studies, stated that the deterioration in memory can be of a different nature than the rest of the deteriorated neurocognitive functions in schizophrenia; in addition to being one of the fields where the deficit is most significant in quantitative terms.

Follow-up studies demonstrate the long-term stability of memory deterioration (Censits et al., 1997). One study with magnetic resonance found a weak correlation between neuropsychological performance and clinical improvement, but a strong relationship between the volumetric variability of the cerebral parenchyma (Gur et al., 1998). Alterations in neuropsychological tests could be related to stable neuroanatomical alterations. Frontal and temporal mechanisms might be involved in the memory dysfunction of schizophrenia. Goldman-Rakic (1990) and Park and Holzman (1992) suggested that pre-frontal cortical dysfunction contributes to working memory deficits, conceptualized as a bank of information connected and continuously updated that directs the response of the patient in resolving problems.

The basic and elementary neuropsychological battery to assess working memory would currently be Verbal Recall (similar to the Mini-Mental memory test), Spatial Recall (Complex Rey Figure) and the Weschler Memory Scale (Test I and II and the Figure Memory Test).

2.2.8.3 Executive function

The behavior of schizophrenic patients, in particular those with negative symptoms, show great similarity to that of patients with frontal lobe disorders (Levin, 1984). The capacity to plan or resolve problems is one of the typical functions of the pre-frontal lobe. We could define the executive functions as the cognoscitive processes that allow us to respond appropriately to the demands of the environment, i.e. all the cognoscitive skills that allow anticipation and goal-setting, the designing of plans and programmes, start of activities and mental operations, self-regulation and task monitoring, the precise selection of behaviors and conduct, flexibility in cognoscitive work and its organization in time and space (Zandio-Zorrilla et al., 2005).

Weinberger et al. (1986) observed that schizophrenic patients showed a pattern of hypofrontality in the PET while taking the Wisconsin Card Test, a test to assess abstract thought and the flexibility or capacity to change lines of thought. These findings were interpreted as proof of the affect on the dorsolateral prefrontal cortex in schizophrenic patients.

Other neuropsychological batteries: the Stroop Test (response inhibition and selective attention), Category Test (abstraction, problem solving, flexibility), Hanoi Tower (planning), COWAT (verbal search strategy) and the Trail Making Test (flexibility and working memory) and other, less used tests, such as: the Word or Drawing Fluidity Test (Lezak, 1983), the Olfactory Identification Test (Kopala and Clark, 1990), Delayed Recall (Freedman et al., 1986) and Time Order Measurements and Frequency Estimation (Smith and Milner 1988). These have identified the slowing up and greater error rate in problem solving when complex information processing is required (Cuesta et al., 1996).

However, other researchers have argued against the supposed incapacity of schizophrenics to carry out the Wisconsin Card Test and have pointed out that the instructions (Goldman et al., 1992) and financial reward (Summerfelt et al., 1991), or a combination of both (Bellack et al., 1990) significantly improve performance in patients, suggesting that other variables associated with the frontal lobe, such as apathy, lack of motivation, flat affect or attention can interfere with executive function.

Other studies question the location of the schizophrenic deficit in the frontal lobe. Hoff et al. (1992), in a group of

56 schizophrenic patients, observed that the functions of the frontal lobe were significant damaged, but not relatively more affected than the functions of language, verbal memory, spatial memory, concentration or motor performance. Nor did Saykin et al. (1991) find evidence of selective alterations in frontal functioning in a subgroup of non-medicated patients who were possibly less seriously ill. Heinrichs (1990) found more evident alterations between hospitalized, seriously ill schizophrenic patients as well as a lower intellectual performance, and Morrison-Stewart et al. (1992) related this to the seriousness of the positive symptoms.

The scientific literature undoubtedly provides contradictory data and it has not be clarified whether the differences in results are a reflection of certain types of schizophrenia or the degree of neuropathology in the illness.

At the same time, doubts have also been raised concerning the specific nature of the frontal lobe deficit in schizophrenia. Morice (1990) suggested that bipolar patients obtain as poor results as schizophrenics in tests assessing executive function, supporting Crow's hypothesis of a continuum within psychosis with frequent overlaps between schizophrenia and bipolar disorder.

Given the relatively large size of the cerebral cortex, it is not surprising that these measurements have no significant correlation.

2.2.8.4 Language disorders

Morice et al. (1983) observe that, when the tables of spontaneous speech for schizophrenic patients are compared with those of a non-psychiatric control subject, the group of patients show a reduction in syntactic complexity, evident in simpler and shorter sentences, fewer interrelated dependent clauses, less depth of interrelations and less verbal fluency. King et al. (1990) provided evidence of deterioration in the syntax of schizophrenic individuals in a period of 2 or 3 years.

Controversy surrounds the idea that schizophrenic language might be like that of aphasia, where the cerebral pathology underlying language mechanisms is a key factor (Chaika, 1974), or whether language dysfunction in schizophrenia is more a secondary epiphenomenon to a primary disorder in thinking or related to a general deficit in information processing (Asarnow, 1982). Schizophrenic and aphasic language may have phenomenological similarities, but a review of the literature indicates that only a subgroup of schizophrenic patients resembles aphasic patients, both in language reception and production (Landre et al., 1992). Some research (Andreasen et al., 1979; Halpern et al., 1984) showed that, unlike aphasic subjects, schizophrenics do not have any difficulty with simple measurements of language regarding syntax, fluidity or denomination.

Landre et al. (1992) suggested that schizophrenic language disorder forms part of a more generalized cognitive deficit. Frith and Allen (1988) stated that language deficits in schizophrenics are caused more by errors in execution than by linguistic incompetence. Morice (1986) correlated changes in the complex structure of language with a cortical–prefrontal dysfunction.

Most non-experimental psychometric tests used in the study of language in schizophrenic patients are based on the battery used in aphasic patients (assessment of spontaneous language via open questions, perceptive tests with tonal perception thresholds associated with distraction manoeuvres or verbal fluency tests with the animals test or FAS).

2.2.9 Methods to evaluate cognitive deterioration in schizophrenia

The large majority of psychometric instruments used in the cognitive assessment of schizophrenia have been designed initially for something as non-specific as intelligence or for something as specific as patients with focal brain damage. These scales are no doubt interesting but they are not specific to schizophrenia and are complex, extensive and difficult to handle. In the last few years new cognitive assessment tools have been developed with a good psychometric profile, efficient and short and easy to apply in everyday clinical practice.

There are a priori significant methodological problems in the neuropsychological exploration of schizophrenic patients, some related to the measurement instrument per se and its application and others associated with the frequency with which they are used as longitudinal studies.

The most frequent traits in cognitive assessment are:

- The importance of motivation. The attitude of the patient towards the test, demotivation, suspicion and apathy.

- The influence of cultural level or prior learning. No cognitive assessment is free from the influence of skill or prior training. That is why it is fundamental to attempt to specify, with the utmost care, the usual capacity of the subject. Family information is essential, although it must be considered that, on some occasions, the relative might over-value prior cognitive capacity.

- Contamination of the functions. If, for example, there is an attention deficit, this deficit will impregnate and affect, negatively, any other performance in any other test.

- Other variables to be considered. In any neuropsychological exploration it will be important to bear in mind the age and handedness. Performance differences according to gender may also be important.

In addition to these problems, there can also be other limitations due to the follow-up studies per se. A lot of research has been carried out with small samples and the follow-up periods are relatively short, also losing a significant percentage of the patients. In many of these studies no control group is used, nor are other confusing variables taken into account, such as possible training between tests. Neither is there consensus on what kind of cognitive test is the most appropriate in order to observe sensitivity to change. Finally, part of the variability of the results is due to the heterogeneous nature of schizophrenia per se and to the way the sample is selected. For example, it is not the same if the subject in the study is a chronic institutionalized patient or an out patient.

Gold et al. (1999) drew up the first series of short neuropsychological tests specifically designed for the cognitive evaluation of schizophrenic patients, the RBANS (Repeatable Battery for the Assessment of Neuropsychological Status), and a little later Keefe et al. (2004) developed the BACS (Brief Assessment of Cognition in Schizophrenia).

Aimed at Spanish-speaking communities, a series of short neuropsychological tests has been developed called NEUROPSI, which includes a broad spectrum of cognitive domains, such as orientation, attention, memory, language, visuo-spatial skills and executive functions (Otrosky-Solis et al., 1999).

However, a reliable, valid and agreed method of cognitive assessment is still necessary for research. In the first conference in April 2003, as part of the MATRICS initiative

(Measurement and Treatment Research to Improve Cognition in Schizophrenia) of the NIMH, the basis was decided for a series of agreed cognitive tests. It was defined which cognitive areas had to be represented in the agreed series of neuropsychological tests (working memory, attention or vigilance, learning and verbal memory, learning and visual memory, reasoning and problem solving, processing speed and social cognition), and the key criteria were established to select the tests in this series (good test-retest reliability, repetition measures, functional prognosis, potential response to pharmacological treatments or tolerability) (Zandio-Zorrilla et al., 2005).

An interesting proposal would be the consideration of scales containing an overall subjective rating regarding cognitive deterioration by of the patient him or herself and by the relatives. In fact, the GEOPTE group has drawn up a self-applied scale (Sanjuán et al., 2003) that contains the subjective opinion of the patient and relatives concerning cognitive deterioration, adding questions on social cognition. Preliminary results with this scale find that the information from the relative or carer correlate highly with the impression of the clinician, the response of patients being less reliable in general, especially in the items of cognition and social functioning.

2.2.10 Affective symptoms in schizophrenia

Alterations in affective quality or intensity are an integral part of the clinical picture of schizophrenia. Bleuler already considered affective anomalies as one of the four primary symptoms of the illness, accompanying association disorders, autism and ambivalence. All schizophrenia instruments and assessment scales include areas reflecting different aspects of affective alterations in the subject specifically.

Data from ECA studies on the comorbidity of schizophrenia and depression (Judd, 1998) indicate an increase of 14 times in the risk of unipolar affective disorder in patients who, sometime in their lives, have been diagnosed as schizophrenic. However, a depressive clinical picture can appear in any phase of the illness' evolution, both in early episodes as well as in chronic patients, and both in acute episodes and in stable periods of the illness.

Bottlender et al. (2000) found depressive mood in 40% of a sample of almost 1,000 schizophrenic patients in their first time in a German hospital. Hafner et al. (1999) in a retrospective study of 232 first episodes, found a prevalence of depressed mood in 81% of the cases. In a prospective follow-up of 70 patients, Sands et al. (1999) found between 30% and 40% complete depressive syndromes at each monitoring point.

Affective symptoms complicate the clinical symptoms of schizophrenic patients and have significant influence on different areas of the patient's life, in their response to treatment and in their evolution, in addition to having a complex relation with suicide attempts and suicides carried out. Hoffman et al. (2000) related the presence of negative self-concepts, low expectations and external control locus with the appearance of despair, and they stress the influence of this on the patient's rehabilitation. Olfson et al. (1999) pointed out the presence of greater depression as one of the best predictors of early rehospitalization (in under 3 months) of patients when they leave hospital. Depressive symptoms often overlap with symptoms related to negative symptoms; anhedonia, difficulties in relating to others and social isolation and low energy deepen the patient's alienation and make it easier for despair to appear and for a lack of involvement in the treatment.

The causes of depressive symptoms in schizophrenic patients are far from being defined. In part this is due to the difficulty in discriminating between depressive symptoms per se and negative symptoms, the adverse affects of the treatment or positive symptoms per se. In an attempt to resolve this problem, Muller et al. (1999) propose a three-dimensional model (retardation, core and accessory depressive symptoms) of depressive symptoms in schizophrenic patients. Peralta and Cuesta (1999), in a study examining the relationship between negative, Parkinson's, catatonic and depressive symptoms, concluded that, while the first three types are highly related, depressive symptoms seem to constitute a relatively independent dimension from the psychopathology of schizophrenia. Siris (2000) proposed a model of vulnerability-stress as a way to integrate depression in the schizophrenic process, considering depression as a possible precipitating factor of a psychotic episode in vulnerable patients. Joseph (1999) pointed out the relationship between anomalies in the frontal lobe and certain negative symptoms that could overlap with depressive symptoms. Kamali et al. (2000) correlated substance abuse and the appearance of depressive symptoms, and Dollfus et al. (2000) suggested a relationship between extrapyramidal signs, negative symptoms and depressive symptoms. Smith et al. (2000) showed the relationship between a greater awareness of the illness on the part of the schizophrenic and a greater level of depressive symptoms.

2.2.11 Schizophrenia and suicide

Traditionally, self-destructive behavior has been considered as belonging to perturbed mental states, attributing 60-80% of suicides to the last psychopathological manifestation of various psychiatric illnesses (depressive disorders, drug dependence and serious personal disorders and psychosis).

Among all these depression has always been considered as the psychopathological basis that is most favorable for suicidal behavior (Sainsbury 1986); but the latest epidemiological studies have provided data that we clinicians already knew intuitively: schizophrenia, due to its bad prognosis, chronic nature, frequent comorbidity with cognitive and depressive disorders and progressive wear on the patient's resources and of his/her environment, constitutes an entity whose risk of suicide is comparable to larger affective disorders (Sartorius, 1987).

2.2.11.1 Epidemiology

In spite of the fact that schizophrenic disorders involve mainly alterations in cognition and thought, more than in mood, suicide is a frequent behavior in this pathology. The risk of suicide in schizophrenia is between 10% and 15%, with a yearly incidence of 350–600 per 100,000 compared with 10–15 per 100,000 in the general population (Caldwell and Gottesman, 1992). Some 20% of hospitalized schizophrenic patients committed suicidal acts and, of these, half did so successfully (Caldwell and Gottesmank, 1990; Miles 1977).

Meltzer (1999) recently found a suicide rate of between 9% and 13% of schizophrenic patients, with an annual rate of between 0.4% and 0.8%. In the study by Osby et al. (2000) carried out in Stockholm between 1973 and 1995, it was concluded that suicide is the main cause of premature death in schizophrenic men and the second in women.

With regard to attempted suicide, Harkaky-Friedman et al. (1999) showed a rate of 38% for their sample of schizophrenic patients; and in Spain Gutiérrez Rodríguez et al. (2000) obtained similar results with 34.7% of their sample.

2.2.11.2 Risk factors

Our knowledge of suicide risk factors associated with schizophrenia are limited, in spite of them being very frequently present. Some of them are no different from those to be expected in the population at large, such as: male, white, single, unemployed, living alone or socially isolated and depressed. However, some specific points can be determined. Schizophrenic suicides are young, with an average age of 33, while in the general population the greater risk is around 65. In Denmark, Mortensen and Juel (1993) placed the age of greatest risk in schizophrenia between 20 and 29. In the USA the ratio of male:female suicide in the general population is 3:1, while in schizophrenia it comes close to 2:1 and for some authors it is comparable (Allebeck et al., 1986).

Male schizophrenics who commit suicide share similar personality traits to those who do so in the general population: impulsive, aggressive and a low tolerance to frustration. Men also use more violent methods than women.

It is possible that the higher risk in schizophrenia does not occur in the acute phase, during the hallucinatory phase or active delirium, but when the psychosis is under control, in the remission phase. Sarró and De la Cruz (1991) stated that the periods of greatest risk are in the first week of hospitalization and the first month after leaving hospital. Drake and Cotton (1984) observed that schizophrenics who commit suicide often suffer from a chronic form of the illness, with numerous relapses and remissions, but they pointed out that, at the time of committing suicide, none presented acute psychotic activity. Schizophrenics who killed themselves had a higher social-occupational adjustment and a high premorbid self-esteem. There was a higher percentage of university graduates with a good awareness of the seriousness of their pathology and with a fear of deterioration. The associated depressive symptoms were characterized mainly by despair, cognitive decline and deterioration, more than by symptoms that could be diagnosed as Great Depression. In the study by Cheng et al. (1990), at least 36% committed suicide in a period of remission. In the work by Breier and Astracham (1984) only three out of the 20 patients who committed suicide were in an acute exacerbation of their psychosis at the time of the suicide.

Bousoño et al. (1997) stated the chronic nature, high psychological suffering in acute episodes, high degree of deterioration in functioning, destructuring of the patient's life, depressive comorbidity, high levels of aggression and impulsiveness were the elements of risk for suicide in schizophrenic patients, as well as specific factors associated to clinical subtypes of schizophrenia (motor disinhibition in catatonic forms, lack of control in paranoid forms or automatic obedience in clinical pictures with a strong hallucinatory component).

For Rodríguez (1999), schizophrenics have two kinds of risk factors with regard to suicidal conduct, some which they share with the population at large and others that are specific to them:

1. Common risks:

- Gender. Greater incidence among men. This could be explained by the worse prognosis of the illness in men, linked to an earlier onset of the process, greater deterioration in personality, higher rate of inactivity and loneliness, use of more violent methods, explainable by greater impulsiveness and worse tolerance to frustration.
- Race. Predominantly white.
- Civil status. Higher incidence in those who are single, separated or widowed (90%).
- Employment. Greater risk among the unemployed (70%).
- Deteriorated physical health.
- Consumption of alcohol or other toxic substances.
- Social isolation.
- History of other suicide attempts.
- Early loss of one of the parents.
- Recent affective, economic or ideological loss.
- Instability or stressful atmosphere in the family.
- Feeling of despair.
- Depressed state.

2. Specific risks:

- Youth. Average age lower than other pathologies around 50.

- Chronic nature and number of relapses. The average amount of time hospitalized of schizophrenic suicides is 470 days, far from the next group, bipolar disorder with 370 days (Roy 1982).

- Marked awareness of the illness. A schizophrenic suicide normally commits suicide in periods of stability, with a clear awareness of clarity and evidence of his or her dark expectations for the future (Amador and Friedman, 1996).

- Fear of deterioration: both mental disintegration and loss of independence, feelings and image with a clear awareness of the disadvantages suffered in terms of social competence (finding a partner, studies, work, finance; Drake et al., 1985).

- Excessive dependence on the treatment or therapy. Suicide coinciding with a change in therapist or loss of trust in a drug or a person on the therapeutic team is not infrequent.

- Existence of positive symptoms. In spite of what is supposed to be an anecdotal appearance (Breier and Astrachan, 1984; Drake et al., 1984 Wilkinson and Bacon, 1984).

- Existence of negative symptoms. Not fear of deterioration but confirmation of their cognitive deterioration and the repercussions on their lives in mental, social and financial terms (Wilkinson, 1982; Roy et al., 1984; Roy 1986).

- Last hospitalization for symptoms other than schizophrenia or simultaneous with those of their primary illness (frequently depressive symptoms) (Wolfersdorf and Hole, 1991).

2.2.11.3 Time of suicide

The risk of suicide usually falls, although it does not disappear, as from the third decade of life. In the first decade of the illness the rate is 40%, 22% in the second and 22% in the third (Pokorny and Kaplan, 1976). The risk is greater during the first week of hospitalization and after 6 months after leaving hospital (30% after the first month and 50% after three months).

Approximately 40% commit suicide during hospitalization, half during the first few days after being admitted and rarely a year after being admitted. Suicide outside hospital predominates in patients who are released from hospital without their consent or with an inadequate integration programme. (Nynan and Jonsson, 1986).

Early suicides generally coincide with a clear awareness of the incapacitating illness and after remission of acute symptoms. Suicides that appear late coincide with severe deterioration, particularly social. Suicides are generally premeditated, although they can be triggered by an event experienced under stress (real or imagined loss of a relative, of home).

2.2.11.4 Method of suicide

Highly lethal methods are usually used (hanging, jumping out of windows), particularly among men, with little doubt as to the desire for death and little possibility of rescue, unlike methods used by the population at large (Drake, 1986; Modestin and Kamm, 1990; Asnis and Friedman, 1993).

2.2.11.5 Place of suicide

Schizophrenics commit suicide both inside and outside hospital. Crammer (1984) found that 28% commit suicide during their stay in hospital, 19% at home, 31% during their trip home from hospital and in 14% of the cases the location is unknown.

Some 50% of suicides occurring in general hospitals are by schizophrenics. In specialist hospitals the percentage of suicides in depressed patients is one or two times higher than the rate among schizophrenics (Wolfersdorf et al., 1988).

Schizophrenics who commit suicide in hospital have particular characteristics that set them apart from patients committing suicide outside (Modestin et al., 1992) .

Radomsky et al. (1999), in a sample of 1048 hospitalized psychotic patients, found a prevalence throughout life of 30.2% of suicide attempts, and 7.2% in the preceding month. In the follow-up study of 9,000 patients by Rossau and Montersen (1997), the risk of suicide in schizophrenia seems to be maximum at the time immediately following admission into hospital, during times of temporary permits and one month after leaving hospital.

2.2.12 Onset and states

Classically, the criterion of time is used to define how schizophrenia starts and, in this context, the onset is divided into acute and insidious, understanding that there is a continuum between the two poles and the use of other criteria may provide clinical and evolutionary data that should be taken into account.

An acute onset appears suddenly, in a few hours or, more usually, over a few days. Intrusion, impression and delirious certainty, leading to inappropriate or extravagant behavior, are explained almost immediately by the clinical picture that appears. The whole mental experience disassociates itself and a new "existing in the world" emerges. When a correct anamnesis is carried out, sometimes small behavioral disorders can be seen, as well as alterations in work performance or in the affective state, which were not appreciated in the past (probably due to their little importance and fleeting nature), which make sense in the light of the current set of symptoms, presupposing a prodromic phase.

In the two types of onset, there is the question as to whether this previously observed set of symptoms corresponds to the premorbid personality, to early signs of the process per se or to a factor of predisposition. In 1982, Parnas et al. published a longitudinal study over 20 years following high risk children of schizophrenic mothers. At age 10, 40% of these could be included within the so-called schizophrenic spectrum. The most important point of this study, however, are perhaps the disorders presented by these children, in the opinion of the parents, before the illness started, 35% were considered passive, 33% showed little tolerance to criticism, 32% showed little attention and enthusiasm in play and 29% had inappropriate behavior and 26% showed isolation.

In the insidious onset, the profile described by Parnas is more evident, correlating with the description of a schizoid personality and the classic concept of a leptomsomic biotype. A profile appears that is characterized by being midway between an overlapping onset and schizoid disorder with a lack of friendships, low sexual interest, little affective involvement and avoidance of groups.

The long evolution of the clinical picture provides different syndromic aspects, among which of note are the pseudoneurotic, pseudodepressive and pseudopsychopathic types. As their name indicates, they acquire the appearance of other psychiatric disorders but there are other latent characteristics in all of these. There is also a torpid course that leads to isolation and social deterioration, lack of concordance with the rest of the symptoms of each entity and null response to symptomatic treatment. In the pseudoneurotic form, somatic symptoms frequently appear, as well as a feeling of depersonalisation and detachment from reality, as well as experiences of change. In pseudodepressive forms inhibition is added, as well as difficulties in interpersonal communication. Alterations in behavior appear in pseudopsychopathic forms, without apparent logic or cause, as well as a blunted affect in a personality previously healthy.

Having reached the state phase of schizophrenia, four classic forms have been described: catatonic, hebephrenic, paranoid and simple, which have recently been extended with other clinical forms described as undifferentiated and residual:

- *Catatonic schizophrenia.* A subtype of schizophrenia characterized by strong motor inhibition and evidenced by catalepsy, which can lead to total immobility and alternate with episodes of agitation. Stuporous catatonia occurs together with negativism, mutism, automatic obedience, mannerisms, echopraxia or echolalia. Today it is rather rare in developed countries and usually occurs earlier than other subgroups and follows a more chronic course. Some subtypes have been described, such as Stauder's mortal catatonia, of organic origin and currently almost disappeared, as well as periodic catatonia, related to the metabolising of nitrogen. In any case, to diagnose schizophrenic catatonia, organic etiologies must be ruled out (toxic, metabolic, hormonal infections, tumours, etc.), as well as iatrogenic etiologies. Sodium amobarbital has been used intravenously for differential diagnosis, as this makes functional catatonia improve temporarily, unlike secondary catatonia with organic problems, whose response is insignificant.

- *Hebephrenic schizophrenia (disorganized).* This starts insidiously in puberty. It is a highly aggressive, deteriorating and chronic form of the illness, characterized by disorganized language and behavior, flat or inappropriate affect (typical empty laughter or smiles, dull and inappropriate) and, in general, a regressive conduct (infantile aspect). Delusions and hallucinations, if present, are fragmentary and not often manifested in the body.

- *Paranoid schizophrenia.* Appears later with better premorbid activity and evolution. The positive symptoms, for example deliriums and hallucinations, occupy the core of the clinical symptoms. Unlike other clinical forms, here disorganized language, catatonic behavior and flat affect, although possible, are not relevant clinical aspects. Deterioration, also possible, is not as marked as in other clinical forms.

- *Simple schizophrenia.* Extremely serious form of schizophrenia that tends to disappear from current classifications, given that the diagnosis is established due to the absence of positive symptoms (deliriums, hallucinations) and the presence of negative symptoms such as little emotional response, reduction in impulse, interest and initiative and tendency towards isolation. It starts insidiously and progresses with significant deterioration and autism. Response to treatment is quite bad and sometimes it is mistakenly diagnosed as a schizoid personality disorder.

- *Residual or undifferentiated schizophrenia.* Variety of schizophrenia whose symptoms do not fit with the previous forms, except partially with simple schizophrenia. The diagnosis of residual schizophrenia is attributed to patients who have had a prior episode of symptoms that comply with the criteria for schizophrenia but who currently do not have positive psychotic symptoms, such as deliriums, hallucinations or disorganized language but do have negative symptoms and other types, such as flat affect, poor language, apathy, mutism, etc., together with strange experiences and extravagant behavior.

2.2.13 Etiopathogeny

Schizophrenia constitutes a heterogeneous disorder of a neurobiological nature for which no common causal factor has been found to date. Although biological theories of schizophrenia are old, in recent years those of neurodevelopment have acquired relevance, formulated under two decades ago by Murray and Lewis. This theory combines genetic causality with the existence of physical environmental factors that probably act in pre- and perinatal stages of the development of the human being. The theoretical model proposes a structural alteration at some time in the anatomical-functional development of the brain of the schizophrenic subject and claims that this alteration is the determining factor in the later appearance of the psychotic clinical picture. This model explains the possible existence of morbid alterations that can precede delirious-hallucinatory clinical symptoms, which would allow a formal diagnosis of schizophrenia. This theory means that the most deficient clinical forms can be explained fundamentally, evolving towards the so-called negative symptoms. However, there is no empirical clinical evidence to support this single theory in all forms of schizophrenia, particularly those subgroups with more favorable evolution, episodic subgroups and those with less deterioration.

2.2.13.1 Genetic hypotheses

In a meta-analysis of family studies, Gottesman and Shields (1982) found an increased risk of suffering from schizophrenia in relatives of patients diagnosed with this illness. The average morbidity factor was calculated at 12%, whichever parent was schizophrenic, indicating that maternal factors are not critical. The risk factor when both parents are schizophrenic is approximately 4 times higher than the expected figure in the population at large.

In studies of twins, Gottesman and Shields (1966), in a sample of patients at Mausdley Hospital obtained concordance in the prevalence of schizophrenia, 62% in homozygote twins and 15% in heterozygotes.

In order to analyze the possible environmental influence on the incidence of schizophrenia, two paradigms are usually used: "adoptee study method" and "adoptee family study method". In the first method, the parents who adopt are not affected and the adoptee has someone affected in his/her biological family. In this way, the incidence of schizophrenia is measured in adoptees, establishing the biological parents and healthy adoptees as control groups. The second paradigm studies whether there is a different incidence among affected and unaffected adoptees compared with their biological and adopted parents. The results in both kinds of studies once again stress the importance of genetics and not the environment in the aetiopathogeny of schizophrenia.

The evidence of a genetic factor in schizophrenia has led to several studies on the contribution of chromosomes and the identification of one or more loci of susceptibility to the illness. The genetic map of schizophrenics and their relatives has been studied, as well as studies of loci that might be linked to the illness and studies of candidate loci.

The first linkage studies were carried out on dopamine receptors D_1, D_2, D_3 and D_4. Coon et al. (1993) studied the GABA receptor gene without any results. Any study of the molecular genetics of schizophrenia is complicated, at present unavoidably so, as we do not know through which or how many phenotypic forms what we understand as schizophrenic illness is manifested.

Wright et al. (1996) describe that the DRB1*04 allele has a very partial influence on schizophrenia. Pulver et al. (1995) found clear evidence of a locus of vulnerability to schizophrenia and schizoaffective disorder associated with changes in the 6P chromosome. With regard to sexual chromosomes, D'Amato et al. (1992) did not find any kind of relation between sexual chromosomes and schizophrenia. With regard to how it is transmitted, Cloninger defended an oligogenic model with some genetic interactions, not additive, such as environmental factors; while other authors propose a polygenic hypothesis. Beyond transmission models, Basset (1994) observed that the prognosis worsens progressively in relatives affected by schizophrenia; it is considered that this could be due to the expansion of repeated sequences of unstable tetranucleotides.

2.2.13.2 Biochemical hypotheses

The more classic theories suggest that hyperactivity of the D_2 dopamine system underlines schizophrenia, corrected with antipsychotic drugs. However, other neurotransmitters are also involved, particularly acetylcholine and serotonin, and other neuromodulators (e.g., glutamate hypofunction) which, in turn, modulate the dopamine system and other brain paths and circuits. Without doubt alterations in the neurotransmission in schizophrenia are varied and complex.

- Dopamine (DA):

  The dopamine hypothesis suggests that functional hyperactivity of the brain's dopamine system is responsible, at least partially, for the development of schizophrenic symptoms. It is based on the findings that drugs stimulating DA availability, such as amphetamine, can lead to psychotic symptoms, and on the high affinity of antipsychotic drugs in blocking D_2 receptors and on the fact that their efficacy is proportional to their degree of affinity with these receptors. Similarly, animal models of dopamine agonism show alterations reminiscent of psychosis. However, there is little direct evidence of DA abnormalities in the neuronal activity of schizophrenics and the results of post-mortem studies and with PET *in vivo* have provided conflicting results. Current interest in DA alterations in schizophrenia are due to the recent discovery of new receptors whose blocking seems to depend partially on the action of the so-called atypical antipsychotic drugs.

- Serotonin:

Some hallucinogenics, such as LSD, are 5-HT_{2A} and 5-HT_{2C} agonists. Similarly, atypical antipsychotics such as clozapine or risperidone are powerful agonists of 5-HT_2 receptors and have a greater affinity for these than for D2 dopamine receptors. 5-HT_2 receptors are widely distributed in the frontal cortex and modulate the dopamine neurons. Many studies suggest that the action of some atypical antipsychotics could be due to the combination of a strong anti-5-HT_2 effect and a weaker anti-D_2 effect. There is growing interest in the study of response patterns to antipsychotics according to the variability in 5-HT_2 receptors polymorphism.

- Dopamine/glutamate:

The hypofunction of the neurotransmission system by glutamate was proposed as a mechanism causing schizophrenia by Kim et al. (1980), based on the finding of low glutamate concentrations in schizophrenics' CSF. Other authors did not confirm this reduction, so this etiological proposal lost interest. Lodge and Anis (1982) report the psychotomimetic effect of phencyclidine, more popularly known as PCP (Peace Pill) or "angel dust", produced by the non-competitive blocking of the ionic channel of the *N*-methyl-*D*-aspartate (NMDA) receptor, a subtype of glutamate receptors. Later, phencyclidine was introduced as an anaesthetic in human clinical practice, being withdrawn shortly afterwards due to the high rate of psychotic reactions in healthy people. An increase in acute episodes was also observed in chronic schizophrenic patients with a stable clinical picture.

Olney et al. (1989) show that antagonists, whether they are competitive or not and in addition to psychotic reactions in humans, also produce neurodegenerative changes in the cortex–limbic region in rat brains. When antagonists are given over a series of days, the structural changes extend to the cingulate cortex, hippocampus, hippocampal gyrus and entorhinal cortex which, together with limbic alteration, are the areas mostly involved in the neuropathology of schizophrenia. When phencyclidine is used as an anaesthetic, the result depends on the age, so that it has no psychotogenic effect in children but it does in adolescents and adults. This latency could explain the typical age when schizophrenia appears in adolescence, although the neurostructural changes may occur pre-, peri- and postnatal, but will be clinically silent.

2.2.13.3 Neuroimaging

Classic theories concerning cerebral anomalies in schizophrenics have been confirmed in the last two decades. Studies with structural techniques such as CT and MR have confirmed the existence of ventricular dilation, sulcus enlargement and cerebral atrophy, as well as a reduction in the size of the temporal and frontal lobes and in the hippocampus, and an increase in the size of the basal ganglia. There is sufficient evidence to prove a relationship between ventricular dilation and mediocre premorbid functioning, negative symptoms, bad response to treatment and deterioration. Functional neuroimaging studies such as Single Photon Emission Computerized Tomography (SPECT), Regional Cerebral Blood Flow (RCBF) and Positron Emission Tomography (PET) have focused on the hypothesis of hypofrontality, associated with the predominance of negative symptoms. The involvement of the frontal lobe has been extended to the thalamus and cerebellum, confirming that schizophrenia is an illness affecting several brain circuits.

2.2.13.4 Neurophysiology

It has been seen that the brain processes of coding, processing and transferring information are damaged in schizophrenia, as anomalies are observed in the potentials evoked, specifically a reduction in the P300 wave, alterations in the P50 and slow ocular movements, suggesting a slowing-up in reaction times and deterioration in neurosensory transmission, as well as an overall alteration in sensory regulation, located in the cortex–pallidum–thalamus circuit.

2.2.13.5 Psychosocial factors

Social theory, which attributes a causal dimension in schizophrenia to negative social conditions, has inverted to such an extent that such social conditions are considered a consequence and not a cause of the illness. However, psychosocial factors are involved in a model of diathesis stress, according to which a subject with schizophrenia has a high specific biological vulnerability that is set off by stress and leads to schizophrenic symptoms. The type of psycho-environmental stress that is most likely to lead to psychotic decompensation is not precisely known. Similarly, the type of family is linked to the prognosis of the illness, so that more frequent relapses occur in subjects with families expressing a high degree of hostility and lack of understanding towards the illness.

2.2.13.6 Other etiopathogenic theories

The autoimmune hypothesis proposes the relationship of certain forms of schizophrenia with a group of autoimmune illnesses that have in common immunological deregulation (such as asthma, myasthenia gravis and rheumatoid arthritis). In spite of the negative results when repeating Heath's work on taraxein in 1967, he persisted in claiming that taraxein was impure and unstable but it was a specific fraction of immunoglobulins in schizophrenic patients that would produce an antigen in the septal areas of the brain.

Crow et al. (1994) suggest the possibility of a viral etiology that would affect groups of genetically predisposed subjects. In some it would cause neurochemical disorders (type I) and in others encephalitis, which would result in structural changes and neuronal degeneration (type II). The hypothesis of a latent virus put forward by Waltrip et al. (1990) is interesting, insofar as these viruses could be asymptomatic throughout life. As a theoretical model, this concept would be closely linked to the recent findings concerning stress and immunosuppression and vulnerability to stress.

Some studies have found endocrine alterations in subgroups of schizophrenic patients. A depressed hypothalamic-pituitary function has been described, showing a prepuberal pattern (increase in gonadotrophins on being stimulated with GNRH, FSH and low levels of testosterone). Given that catecholamines stimulate the release of FSH, LH and GH and inhibit prolactin and insulin, these findings suggest the existence of a deficit in the catecholaminergic function at a hypothalamic level.

Neuropathology suggests a relationship between alterations in psychomotricity in schizophrenia and its possible relation with alterations in basal ganglia; even more so when, later, the interconnection was observed between the basal ganglia and the frontal lobes, which in neuroimaging studies seem to be altered in schizophrenia.

The most convincing results have been established in the limbic system, where a reduction is found in the size of this region, comprising the amygdala, hippocampus and hippocampal gyrus.

2.2.14 Prognosis

Once a psychotic disorder has been diagnosed, one of the first questions asked by the clinician and the family concerns the patient's prognosis. Will the patient be able to continue with his or her work? Will he or she continue relating to others as before? Will he or she lead a normal life? There have been many studies looking at ways to establish the prognosis in patients suffering from schizophrenia. We must remember that one of the main criteria that first led E. Kraepelin to differentiate schizophrenia as a clinical entity was that the illness had a defective course. However, in his later work Kraepelin himself recognized that it actually evolved in a wide variety of ways over time (Kraepelin, 1920).

The most important longitudinal studies (Bleuler et al., 1976; Ciompi, 1980; Hwu et al., 1988) showed a wide variation of prognoses depending particularly on how the illness is defined and the time it is monitored. In general terms, the longer the monitoring the worse the prognosis. It is also evident that the prognosis will get worse if we include a specific time of evolution in the diagnostic criteria (as in the DSM-IV TR). Another aspect that will influence the results is the system used to define this prognosis (clinical scales, hospitalization, functionality, etc.). However, some data can be picked out as common findings from all these studies. There is certain consensus concerning the three overall kinds of evolution: 30% of patients evolve favorably with an almost complete remission of symptoms and good social adjustment; 50% have episodic evolution with periods of instability and their social integration irregularly affected but with clear functioning; and the remaining 20% evolve badly with clear deterioration and their social integration severely affected.

Both classic longitudinal studies and recent studies have established up to eight different types of course for the illness that are actually various combinations of the episodic or chronic forms (Ciompi, 1980; Jablensky et al., 1992; Harrison et al., 2001).

Over the past few years the variables have been refined that seem to be associated with a worse prognosis of the illness: starting at an early age, male, cognitive deterioration, prior personality alterations, lack of social-family support, slow, progressive start, absence of triggers, absence of affective factors and substance abuse, among others. We should note that, in the international study on schizophrenia (Jablensky et al., 1992), the prognosis of the illness was clearly more favorable in developing countries compared with developed countries. The main argument used to explain this difference is the greater social-family support patients in poor countries may have (Murray et al., 2003).

But perhaps the most important study in the past 20 years on the prognosis of schizophrenia has been the work carried out by Hegarty et al. (1994). These authors revised all the longitudinal research published between 1894 and 1994. They found 320 studies that met with a minimum of methodological requirements. The work focuses on comparing over time the rates of patients with good evolution. They find an average recovery rate of 35% in the period of 1895–1955, increasing to 49% in the following period of 1956–1985. This increase could be explained by the appearance of antipsychotics at the end of the 1950s. The discovery of neuroleptics helped the deinstitutionalisation process and was the most decisive factor in being able to apply programs aimed at social reintegration (Mancevski et al., 2005).

However, surprisingly the percentage of patients recovering fell again in the last period analyzed, 1986–1992. For the authors this is due to the appearance of criteria (DSM-III) that are more limiting as they include a minimum evolution time (6 months) to carry out the diagnosis. The kind of criteria used could therefore result in the percentage of patients recovering varying between 27% and 47%. In any case, the overall panorama of prognosis for schizophrenia, according to this article, is that it has not changed substantially in the past 100 years, at least in terms of the percentage of patients with the worst evolution, cognitive deterioration and social adjustment.

Ten years have passed since the article by Hegarty and his collaborators. The question now is: has the prognosis for schizophrenia changed over the past 10 years? Although this is undoubtedly a very short period to analyze longitudinal studies and to reach definitive conclusions, it is nonetheless true that this period has seen significant advances in the methodology of longitudinal studies. Patients are diagnosed operationally and symptoms are gathered in a standardized way, on some occasions adding neuroimaging or cognitive variables. However, how these patients are handled therapeutically has also changed. The rediscovery of clozapin has led to the arrival of new antipsychotics, the so-called atypical or second generation antipsychotics. It has been suggested that these new drugs could be helping social reintegration and therefore changing the prognosis of the illness (Litrell, 2001). In addition to these new molecules, in recent years new specific psychotherapeutic strategies have also been consolidated (fundamentally based on cognitive therapy) to tackle psychotic symptoms (Jenner, 2002; Garety and Freeman, 1999).

2.2.15 Schizophrenia therapy

As a neurobiologically based illness, the fundamental treatment is biological (antipsychotic drugs, electroconvulsive therapy), although psychological treatment and rehabilitation act as an excellent complement; they have no affect, however, on the core cause.

2.2.15.1 Pharmacological treatment

Antipsychotic drugs form the basis of the treatment of schizophrenia. Since chlorpromazine was introduced in 1950, the therapeutic possibilities for this serious illness changed radically and the number of hospitalizations in mental hospitals fell drastically. Some 10–30% of patients have been shown to respond badly to these drugs, but they have allowed present-day community psychiatry to be implemented, in which the schizophrenic lives as part of society and not isolated in closed institutions. The undeniable benefits of classic antipsychotic drugs (Table 2) have been superseded by the new antipsychotics (Table 3), with similar effectiveness but much better tolerance. They have few side-effects, especially of the extrapyramidal type, only a small increase in prolactin levels and therefore a low rate of galactorrhea and menstrual disorders.

Table 3. Dose and side-effects of classic antipsychotics

	Dose (mg/day)	Sedation	Autonomic	Extrapyramidal symptoms
Pimozide	2–20	(+)	+	++
Flupentixol	2–20	+	+	++
Haloperidol	2–20	+	+	+++
Fluphenazine	2–20	+	+	+++
Zuclopentixol	4–50	++	++	++
Perphenazine	8–64	++	+(+)	++
Chlorpromazine	100–600	+++	++(+)	+
Levomepromazine	100–600	+++	+++	+

Relapses in these patients are frequent (50% in the first year after being released from hospital and 85% in the first 5 years), especially if the medication is abandoned. For this reason, it is advisable to carry out maintenance treatment for 2 or 3 years after the first acute episode and up to 5 years after a second episode. However, most patients require indefinite treatment throughout their lives. For this reason, pharmacological treatment now tends to be complemented with another kind of psycho-educational and rehabilitating treatment, making the patient and his or her family aware of the need for continuous treatment. Occasionally, in patients unwilling to take medication, prolonged released antipsychotics (depot) could be very useful and effective, with intramuscular administration every 2 or 4 weeks (fluphenazine, zuclopentixol, pipotiazine, risperidone). Every 4 weeks the antipsychotic effect can be evaluated, so that if the response is bad the treatment must be changed and, if it is partial, it can be maintained another 4 weeks. In any event, it is useful to assess thoroughly the withdrawal of the drug after a few years and, in any case, this withdrawal must be progressive and very slow.

It is debatable whether classic antipsychotics were only effective in positive symptoms and whether the new antipsychotics also act, albeit moderately, on the negative symptoms (particularly clozapin). However, although the new antipsychotics have meant a significant advance, given their good tolerance and their possible action on negative symptoms, in many cases the classic drugs still have a respectable role, since they can be as useful as the new ones and can be used in cases resistant to the latter.

Table 4. Range of dose and side-effects of new antipsychotics

	Dose (mg/day)	Sedation	Autonomic	Extrapyramidal symptoms
Amisulpride	200–600	+	+	++
Aripiprazole	10–30	+	+	+
Clozapine	200–600	+++	+++	(+)
Olanzapine	10–30	++	+(+)	+
Quetiapine	300–900	++	++	+
Risperidone	4–12	+	+	+(+)
Sertindole	8–24	+	+	+(+)
Ziprasidone	80–160	++	++	++

Another important point to be tackled in treating schizophrenia is the ECT, highly effective in acute forms and most particularly in catatonic schizophrenia and in those patients with associated affective symptoms. Although generally the action of ECT is not higher than that of antipsychotic drugs, it is recommended in cases resistant to pharmacological treatment and specifically in catatonic schizophrenia.

2.2.15.2 Psycho-social therapies

Psychological support is essential in these patients, either individually or in groups, either as a guidance or support therapy or, in a more structured way, of a cognitive-behavioral type. The aims of this therapy are fundamentally to acquire a good awareness of the illness, helping the treatment to be followed successfully, although on occasion this is aimed at improving functional capacity or social interaction.

One of the most successful programmes implemented is that aimed at treating psycho-educational aspects of the families, as a reduction in emotional expression and hostility, as well as an understanding of the illness, leading to a better prognosis.

In general, psycho-social rehabilitation including three phases (evaluation, planning and action) must depend on the individual characteristics of the patient and is an excellent complement to biological therapies. In order to be able to carry out some of these treatments, special resources or centres are required (day or outpatient centres, sheltered housing, etc.).

References

Aleman A, Hijman R, de Haan EH, Kahn RS. *Memory impairment in schizophrenia: a meta-analysis.* Am J Psychiatry. **1999**, 156:1358-1366.

Allebeck P, Wistedt B. *Mortality in schizophrenia. A ten-year follow-up based on the Stockholm County inpatient register.* Arch Gen Psychiatry. **1986**, 43:650-653.

Amador XF, Friedman JH, Kasapis C, Yale SA, Flaum M, Gorman JM. *Suicidal behavior in schizophrenia and its relationship to awareness of illness.* Am J Psychiatry. **1996**, 153:1185-1188.

Andreasen NC, Grove W. The relationship between schizophrenics language, manic language, and aphasia. In: Gruzelier J, Flor-Henry P, eds. *Hemisphere asymmetries of function in psychopathology.* ElSevier, Amsterdam, **1979**, pp 377-390.

Andreasen NC, Flaum M. *Schizophrenia: the characteristic symptoms.* Schizophr Bull. **1991**, 17:27-49.

Asarnow RF, Steffy RA, MacCrimmon DJ, Cleghorn JM. *An attentional assessment of foster children at risk for schizophrenia.* J Abnorm Psychol. **1977**, 86:267-275.

Asarnow RF, Watkins J. *Schizophrenic thought disorder: Linguistic incompetence or information-processing impairment?* The Behav Brain Sci. **1982**, 5:589-590.

Asarnow RF, MacCrimmon DJ. *Attention/information processing, neuropsychological functioning, and thought disorder during the acute and partial recovery phases of schizophrenia: a longitudinal study.* Psychiatry Res. **1982**, 7:309-319.

Asnis GM, Friedman TA, Sanderson WC, Kaplan ML, van Praag HM, Harkavy-Friedman JM. *Suicidal behaviors in adult psychiatric outpatients, I: Description and prevalence.* Am J Psychiatry. **1993**, 150:108-112.

Astrachan BM, Harrow M, Adler D, Brauer L, Schwartz A, Schwartz C, Tucker G. *A checklist for the diagnosis of schizophrenia.* Br J Psychiatry. **1972**, 121:529-539.

Atkinson R, Shiffrin R. Human memory: A proposed system and its control processes. In: Spence KW, Spence JT, eds. *Advances in the psychology of learning and motivation.* Academic Press, New York., **1968**, pp .

Bassett AS, Honer WG. *Evidence for anticipation in schizophrenia.* Am J Hum Genet. **1994**, 54:864-870.

Bellack AS, Mueser KT, Morrison RL, Tierney A, Podell K. *Remediation of cognitive deficits in schizophrenia.* Am J Psychiatry. **1990**, 147:1650-1655.

Benjamin LS. *Is chronicity a function of the relationship between the person and the auditory hallucination?* Schizophr Bull. **1989**, 15:291-310.

Bilder RM, Lipschutz-Broch L, Reiter G, Geisler S, Mayerhoff D, Lieberman JA. *Neuropsychological deficits in the early course of first episode schizophrenia.* Schizophr Res. **1991**, 5:198-199.

Bleuler M, Huber G, Gross G, Schuttler R. *[Long-term course of schizophrenic psychoses. Joint results of two studies].* Nervenarzt. **1976**, 47:477-481.

Bogerts B, Meertz E, Schonfeldt-Bausch R. *Basal ganglia and limbic system pathology in schizophrenia. A morphometric study of brain volume and shrinkage.* Arch Gen Psychiatry. **1985**, 42:784-791.

Bottlender R, Strauss A, Moller HJ. *Prevalence and background factors of depression in first admitted schizophrenic patients.* Acta Psychiatr Scand. **2000**, 101:153-160.

Bousoño M, Bobes J, Gonzalez Quiros P. Tratamiento psicofarmacológico. In: Bobes J, González Seijo JC, Saiz Martínez PA, eds. *Prevención de las conductas suicidas y parasuicidas.* Masson, Barcelona, **1997**, pp .

Breier A, Astrachan BM. *Characterization of schizophrenic patients who commit suicide.* Am J Psychiatry. **1984**, 141:206-209.

Caldwell CB, Gottesman II. *Schizophrenics kill themselves too: a review of risk factors for suicide.* Schizophr Bull. **1990**, 16:571-589.

Caldwell CB, Gottesman II. *Schizophrenia--a high-risk factor for suicide: clues to risk reduction.* Suicide Life Threat Behav. **1992**, 22:479-493.

Calev A, Venables PH, Monk AF. *Evidence for distinct verbal memory pathologies in severely and mildly disturbed schizophrenics.* Schizophr Bull. **1983**, 9:247-264.

Calev A. Memory in schizophrenia. In: Weller M, ed. *International perspective in schizophrenia: Biological, social and epidemiological findings.* John Libbey & Co, London, **1990**, pp 29-41.

Carpenter WT, Jr., Strauss JS, Bartko JJ. *Flexible system for the diagnosis of schizophrenia: report from the WHO International Pilot Study of Schizophrenia.* Science. **1973**, 182:1275-1278.

Censits DM, Ragland JD, Gur RC, Gur RE. *Neuropsychological evidence supporting a neurodevelopmental model of schizophrenia: a longitudinal study.* Schizophr Res. **1997**, 24:289-298.

Chaika E. *A linguist looks at "schizophrenic" language.* Brain Lang. **1974**, 1:257-276.

Cheng KK, Leung CM, Lo WH, Lam TH. *Risk factors of suicide among schizophrenics.* Acta Psychiatr Scand. **1990**, 81:220-224.

Cheung P, Schweitzer I, Crowley K, Tuckwell V. *Violence in schizophrenia: role of hallucinations and delusions.* Schizophr Res. **1997**, 26:181-190.

Ciompi L. *Catamnestic long-term study on the course of life and aging of schizophrenics.* Schizophr Bull. **1980**, 6:606-618.

Coon H, Byerley W, Holik J, Hoff M, Myles-Worsley M, Lannfelt L, Sokoloff P, Schwartz JC, Waldo M, Freedman R, . *Linkage analysis of schizophrenia with five dopamine receptor genes in nine pedigrees.* Am J Hum Genet. **1993**, 52:327-334.

Crammer JL. *The special characteristics of suicide in hospital in-patients.* Br J Psychiatry. **1984**, 145:460-463.

Crow TJ, Delisi LE, Lofthouse R, Poulter M, Lehner T, Bass N, Shah T, Walsh C, Boccio-Smith A, Shields G, . *An examination of linkage of schizophrenia and schizoaffective disorder to the pseudoautosomal region (Xp22.3).* Br J Psychiatry. **1994**, 164:159-164.

Cuesta MJ, Peralta V, de Leon J. *Neurological frontal signs and neuropsychological deficits in schizophrenic patients.* Schizophr Res. **1996**, 20:15-20.

Curson DA, Patel M, Liddle PF, Barnes TR. *Psychiatric morbidity of a long stay hospital population with chronic schizophrenia and implications for future community care*. BMJ. **1988**, 297:819-822.

d'Amato T, Campion D, Gorwood P, Jay M, Sabate O, Petit C, Abbar M, Malafosse A, Leboyer M, Hillaire D, . *Evidence for a pseudoautosomal locus for schizophrenia. II: Replication of a non-random segregation of alleles at the DXYS14 locus*. Br J Psychiatry. **1992**, 161:59-62.

Dauphinais ID, Delisi LE, Crow TJ, Alexandropoulos K, Colter N, Tuma I, Gershon ES. *Reduction in temporal lobe size in siblings with schizophrenia: a magnetic resonance imaging study*. Psychiatry Res. **1990**, 35:137-147.

David AS, Malmberg A, Brandt L, Allebeck P, Lewis G. *IQ and risk for schizophrenia: a population-based cohort study*. Psychol Med. **1997**, 27:1311-1323.

Dollfus S, Ribeyre JM, Petit M. *Objective and subjective extrapyramidal side effects in schizophrenia: their relationships with negative and depressive symptoms*. Psychopathology. **2000**, 33:125-130.

Drake RE, Gates C, Cotton PG, Whitaker A. *Suicide among schizophrenics. Who is at risk?* J Nerv Ment Dis. **1984**, 172:613-617.

Drake RE, Gates C, Whitaker A, Cotton PG. *Suicide among schizophrenics: a review*. Compr Psychiatry. **1985**, 26:90-100.

Drake RE, Cotton PG. *Depression, hopelessness and suicide in chronic schizophrenia*. Br J Psychiatry. **1986**, 148:554-559.

Drake RE, Gates C, Cotton PG. *Suicide among schizophrenics: a comparison of attempters and completed suicides*. Br J Psychiatry. **1986**, 149:784-787.

Eaton WW. *Epidemiology of schizophrenia*. Epidemiol Rev. **1985**, 7:105-126.

Eaton WW, Romanoski A, Anthony JC, Nestadt G. *Screening for psychosis in the general population with a self-report interview*. J Nerv Ment Dis. **1991**, 179:689-693.

Erlenmayer-Kimling L, Kestenbaum C, Bird H, Hildoff U. Assessment of the New York high-risk project subjects in sample A who are now clinical deviants. In: Watt NF, Anthony EJ, Wyne LC, Rolf JE, eds. *Children at risk in schizophrenia*. Cambridge University Press, New York, **1984**, pp 227-240.

Falloon IR, Talbot RE. *Persistent auditory hallucinations: coping mechanisms and implications for management*. Psychol Med. **1981**, 11:329-339.

Feighner JP, Robins E, Guze SB, Woodruff RA, Jr., Winokur G, Munoz R. *Diagnostic criteria for use in psychiatric research*. Arch Gen Psychiatry. **1972**, 26:57-63.

Flor-Henry P, Yeudall L, Stefanyk W, Howarth B. *The neuropsychological correlates of functional psychoses*. IRCS Medical Sci : neurosugey, psychiatry and clinical psychology. **1975**, 3:34.

Freedman M, Oscar-Berman M. *Bilateral frontal lobe disease and selective delayed response deficits in humans*. Behav Neurosci. **1986**, 100:337-342.

Frith CD, Allen HA. Language disorders in schizophrenia and their implications for neuropsychology. In: Bebbington P, McGuffin P, eds. *Schizophrenia. The major issues*. Butterworth-Heinemann, Oxford, **1988**, pp 172-186.

Garety PA, Freeman D. *Cognitive approaches to delusions: a critical review of theories and evidence*. Br J Clin Psychol. **1999**, 38 (Pt 2):113-154.

Garety PA, Fowler D, Kuipers E. *Cognitive-behavioral therapy for medication-resistant symptoms*. Schizophr Bull. **2000**, 26:73-86.

Gold J, Randolph C, Carpenter C, Goldberg TE, Weinberger DR. *The performance of patients with schizophrenia on the Wechsler Memory Scale-Revised*. Clin Neuropsychol. **1992**, 6:367-373.

Gold JM, Queern C, Iannone VN, Buchanan RW. *Repeatable battery for the assessment of neuropsychological status as a screening test in schizophrenia I: sensitivity, reliability, and validity*. Am J Psychiatry. **1999**, 156:1944-1950.

Goldman-Rakic PS. Prefrontal cortical dysfunction in schizophrenia: the relevance of working memory. In: Carroll BJ, Barrett JE, eds. *Psychopathology and the brain*. Raven Press, New York, **1991**, pp 1-23.

Goldman RS, Axelrod BN, Tompkins LM. *Effect of instructional cues on schizophrenic patients' performance on the Wisconsin Card Sorting Test*. Am J Psychiatry. **1992**, 149:1718-1722.

Goldstein G, Zubin J. *Neuropsychological differences between young and old schizophrenics with and without associated neurological dysfunction*. Schizophr Res. **1990**, 3:117-126.

Goldstein G, Shemansky WJ. *Influences on cognitive heterogeneity in schizophrenia*. Schizophr Res. **1995**, 18:59-69.

Gonzalez JC, Aguilar EJ, Berenguer V, Leal C, Sanjuan J. *Persistent auditory hallucinations*. Psychopathology. **2006**, 39:120-125.

González JC, Sanjuán J, Aguilar EJ, Berenguer V, Leal C. *Dimensiones clínicas de las alucinaciones auditivas*. Archivos de Psiquiatría. **2003**, 66:231-246.

Gottesman II, Shields J. *Schizophrenia in twins: 16 years' consecutive admissions to a psychiatric clinic*. Br J Psychiatry. **1966**, 112:809-818.

Gottesman IL, Shields J, Hanson DR. *Schizophrenia: The epigenetic puzzle*. Cambridge University Press; New York, **1982**.

Gur RE, Cowell P, Turetsky BI, Gallacher F, Cannon T, Bilker W, Gur RC. *A follow-up magnetic resonance imaging study of schizophrenia. Relationship of neuroanatomical changes to clinical and neurobehavioral measures*. Arch Gen Psychiatry. **1998**, 55:145-152.

Gutiérrez Rodríguez M, García Cabeza I, Sánchez Díaz EI. *Experiencias depresivas en el curso de la esquizofrenia*. Archivos de Psiquiatría. **2000**, 63:81-92.

Haas GL, Sweeney JA, Hien DA, Goldman D, Deck MDF. *Sex differences in the course of schizophrenia*. Clin Neuropsychol. **1991**, 5:281-282.

Hafner H, Loffler W, Maurer K, Hambrecht M, an der HW. *Depression, negative symptoms, social stagnation and social decline in the early course of schizophrenia*. Acta Psychiatr Scand. **1999**, 100:105-118.

Hafner H, Heiden W. *Schizophrenia*. 2nd ed. Blackwell Science; Oxford, **2003**.

Halpern H, McCartin-Clark M. *Differential language characteristics in adult aphasic and schizophrenic subjects*. J Commun Disord. **1984**, 17:289-307.

Harkavy-Friedman JM, Restifo K, Malaspina D, Kaufmann CA, Amador XF, Yale SA, Gorman JM. *Suicidal behavior in schizophrenia: characteristics of individuals who had and had not attempted suicide.* Am J Psychiatry. **1999**, 156:1276-1278.

Harrison G, Hopper K, Craig T, Laska E, Siegel C, Wanderling J, Dube KC, Ganev K, Giel R, an der HW, Holmberg SK, Janca A, Lee PW, Leon CA, Malhotra S, Marsella AJ, Nakane Y, Sartorius N, Shen Y, Skoda C, Thara R, Tsirkin SJ, Varma VK, Walsh D, Wiersma D. *Recovery from psychotic illness: a 15- and 25-year international follow-up study.* Br J Psychiatry. **2001**, 178:506-517.

Harvey PD, Parrella M, White L, Mohs RC, Davidson M, Davis KL. *Convergence of cognitive and adaptive decline in late-life schizophrenia.* Schizophr Res. **1999**, 35:77-84.

Heath RG. *Schizophrenia: pathogenetic theories.* Int J Psychiatry. **1967**, 3:407-410.

Heaton RK, Baade LE, Johnson KL. *Neuropsychological test results associated with psychiatric disorders in adults.* Psychol Bull. **1978**, 85:141-162.

Hegarty JD, Baldessarini RJ, Tohen M, Waternaux C, Oepen G. *One hundred years of schizophrenia: a meta-analysis of the outcome literature.* Am J Psychiatry. **1994**, 151:1409-1416.

Heinrichs RW. *Variables associated with Wisconsin Card Sorting Test performance in neuropsychiatric patients referred for assessment.* Neuropsychiatry Neuropsychol Behav Neurol. **1990**, 3:107-112.

Heinrichs RW, Zakzanis KK. *Neurocognitive deficit in schizophrenia: a quantitative review of the evidence.* Neuropsychology. **1998**, 12:426-445.

Hill SK, Beers SR, Kmiec JA, Keshavan MS, Sweeney JA. *Impairment of verbal memory and learning in antipsychotic-naive patients with first-episode schizophrenia.* Schizophr Res. **2004**, 68:127-136.

Hoff AL, Riordan H, DeLisi L. *Influence of gender on neuropsychological testing and MRI measures of schizophrenic in-patients.* Clin Neuropsychol. **1991**, 5:281.

Hoff AL, Riordan H, O'Donnell DW, Morris L, Delisi LE. *Neuropsychological functioning of first-episode schizophreniform patients.* Am J Psychiatry. **1992**, 149:898-903.

Hoffmann H, Kupper Z, Kunz B. *Hopelessness and its impact on rehabilitation outcome in schizophrenia -an exploratory study.* Schizophr Res. **2000**, 43:147-158.

Hustig HH, Hafner RJ. *Persistent auditory hallucinations and their relationship to delusions and mood.* J Nerv Ment Dis. **1990**, 178:264-267.

Hwu HG, Chen CC, Strauss JS, Tan KL, Tsuang MT, Tseng WS. *A comparative study on schizophrenia diagnosed by ICD-9 and DSM-III: course, family history and stability of diagnosis.* Acta Psychiatr Scand. **1988**, 77:87-97.

Jablensky A. *Epidemiology of schizophrenia: a European perspective.* Schizophr Bull. **1986**, 12:52-73.

Jablensky A, Sartorius N, Ernberg G, Anker M, Korten A, Cooper JE, Day R, Bertelsen A. *Schizophrenia: manifestations, incidence and course in different cultures. A World Health Organization ten-country study.* Psychol Med Monogr Suppl. **1992**, 20:1-97.

Jenner JA. *An integrative treatment for patients with persistent auditory hallucinations.* Psychiatr Serv. **2002**, 53:897-898.

Joseph R. *Frontal lobe psychopathology: mania, depression, confabulation, catatonia, perseveration, obsessive compulsions, and schizophrenia.* Psychiatry. **1999**, 62:138-172.

Judd LL. *Mood disorders in schizophrenia: epidemiology and cormobidity.* J Clin Pschiatr Monogr Ser. **1998**, 16:2-4.

Kamali M, Kelly L, Gervin M, Browne S, Larkin C, O'Callaghan E. *The prevalence of comorbid substance misuse and its influence on suicidal ideation among in-patients with schizophrenia.* Acta Psychiatr Scand. **2000**, 101:452-456.

Kane J, Honigfeld G, Singer J, Meltzer H. *Clozapine for the treatment-resistant schizophrenic. A double-blind comparison with chlorpromazine.* Arch Gen Psychiatry. **1988**, 45:789-796.

Keefe RS, Goldberg TE, Harvey PD, Gold JM, Poe MP, Coughenour L. *The Brief Assessment of Cognition in Schizophrenia: reliability, sensitivity, and comparison with a standard neurocognitive battery.* Schizophr Res. **2004**, 68:283-297.

Kim JS, Kornhuber HH, Schmid-Burgk W, Holzmuller B. *Low cerebrospinal fluid glutamate in schizophrenic patients and a new hypothesis on schizophrenia.* Neurosci Lett. **1980**, 20:379-382.

King K, Fraser WI, Thomas P, Kendell RE. *Re-examination of the language of psychotic subjects.* Br J Psychiatry. **1990**, 156:211-215.

Klonoff H, Fibiger CH, Hutton GH. *Neuropsychological patterns in chronic schizophrenia.* J Nerv Ment Dis. **1970**, 150:291-300.

Knoll JL, Garver DL, Ramberg JE, Kingsbury SJ, Croissant D, McDermott B. *Heterogeneity of the psychoses: is there a neurodegenerative psychosis?* Schizophr Bull. **1998**, 24:365-379.

Koh S. *Remembering of verbal materials by schizophrenia adults.* In: Schwartz S, Hillsdale NJ, eds. *Language and cognition in schizophrenia.* Erlbaum Associates, New York, **1978**, pp 59-69.

Kopala L, Clark C. *Implications of olfactory agnosia for understanding sex differences in schizophrenia.* Schizophr Bull. **1990**, 16:255-261.

Kornetsky C, Mirsky AF. *On certain psychopharmacological and physiological differences between schizophrenic and normal persons.* Psychopharmacologia. **1966**, 8:309-318.

Kraepelin E. *Die Erscheinnungsfromen des irreseins.* Neurologie und Psychiatrie. **1920**, 62:1-29.

Landre N, Taylor M, Kearns K. *Language functioning in schizophrenic and aphasic patients.* Neuropsychiatry Neuropsychol Behav Neurol. **1992**, 5:7-14.

Langfeldt G. *Diagnosis and prognosis of schizophrenia.* Proc R Soc Med. **1960**, 53:1047-1052.

Levin S. *Frontal lobe dysfunctions in schizophrenia--II. Impairments of psychological and brain functions.* J Psychiatr Res. **1984**, 18:57-72.

Lezak MD. *Neuropsychological assessment (Oxford Medicine Publications).* 2nd ed. Oxford University Press; New York, **1983**.

Lieh-Mak F, Lee PW. *Cognitive deficit measures in schizophrenia: factor structure and clinical correlates*. Am J Psychiatry. **1997**, 154:39-46.

Litrell KH. Psychopharmacology and social reintegration. In: Breier A, Tran P, Bymaster FP, Lewis M, eds. *Current issues in the psychopharmacology of schizophrenia*. Lippincott Williams & Wilkins, Philadelphia, **2001**, pp 556-574.

Lodge D, Anis NA. *Effects of phencyclidine on excitatory amino acid activation of spinal interneurones in the cat*. Eur J Pharmacol. **1982**, 77:203-204.

Mancevski B, Rosoklija G, Kurzon M, Serafimova T, Ortakov V, Trencevska I, Keilp J, Dwork AJ. *Effects of introduction of neuropleptics on symtomatology in chronic schizophrenia inpatients*. Schizophr Bull. **2005**, 31:495-496.

MASON CF. *Pre-illness intelligence of mental hospital patients*. J Consult Psychol. **1956**, 20:297-300.

McKay AP, McKenna PJ, Bentham P, Mortimer AM, Holbery A, Hodges JR. *Semantic memory is impaired in schizophrenia*. Biol Psychiatry. **1996**, 39:929-937.

McKenna PJ, Tamlyn D, Lund CE, Mortimer AM, Hammond S, Baddeley AD. *Amnesic syndrome in schizophrenia*. Psychol Med. **1990**, 20:967-972.

Mednick SA, Parnas J, Schulsinger F. *The Copenhagen High-Risk Project, 1962-86*. Schizophr Bull. **1987**, 13:485-495.

Meltzer HY. *Treatment of the neuroleptic-nonresponsive schizophrenic patient*. Schizophr Bull. **1992**, 18:515-542.

Meltzer HY. *Suicide and schizophrenia: clozapine and the InterSePT study. International Clozaril/Leponex Suicide Prevention Trial*. J Clin Psychiatry. **1999**, 60 Suppl 12:47-50.

Miles CP. *Conditions predisposing to suicide: a review*. J Nerv Ment Dis. **1977**, 164:231-246.

Milev P, Ho BC, Arndt S, Andreasen NC. *Predictive values of neurocognition and negative symptoms on functional outcome in schizophrenia: a longitudinal first-episode study with 7-year follow-up*. Am J Psychiatry. **2005**, 162:495-506.

Miller LJ. *Qualitative changes in hallucinations*. Am J Psychiatry. **1996**, 153:265-267.

Modestin J, Kamm A. *Parasuicide in psychiatric inpatients: results of a controlled investigation*. Acta Psychiatr Scand. **1990**, 81:225-230.

Modestin J, Zarro I, Waldvogel D. *A study of suicide in schizophrenic in-patients*. Br J Psychiatry. **1992**, 160:398-401.

Moller HJ, Bottlender R, Wegner U, Wittmann J, Strauss A. *Long-term course of schizophrenic, affective and schizoaffective psychosis: focus on negative symptoms and their impact on global indicators of outcome*. Acta Psychiatr Scand Suppl. **2000**, 54-57.

Morice R, McNicol D. *Language changes in schizophrenia: a limited replication*. Schizophr Bull. **1986**, 12:239-251.

Morice R. *Cognitive inflexibility and pre-frontal dysfunction in schizophrenia and mania*. Br J Psychiatry. **1990**, 157:50-54.

Morice RD, Igram JC. *Language complexity and age of onset of schizophrenia*. Psychiatry Res. **1983**, 9:233-242.

Morrison-Stewart SL, Williamson PC, Corning WC, Kutcher SP, Snow WG, Merskey H. *Frontal and non-frontal lobe neuropsychological test performance and clinical symptomatology in schizophrenia*. Psychol Med. **1992**, 22:353-359.

Mortensen PB, Juel K. *Mortality and causes of death in first admitted schizophrenic patients*. Br J Psychiatry. **1993**, 163:183-189.

Muller MJ, Wetzel H, Szegedi A, Benkert O. *Three dimensions of depression in patients with acute psychotic disorders: a replication study*. Compr Psychiatry. **1999**, 40:449-457.

Murray CJL, López AD. *The Global burden of disease*. Harvard University Press; Cambridge, **1996**.

Murray RM, Jones PB, Susser E, van Os J, Cannon M. *The Epidemiology of Schizophrenia*. Cambridge University Press; Cambridge, **2003**.

Nelson HE, Pantelis C, Carruthers K, Speller J, Baxendale S, Barnes TR. *Cognitive functioning and symptomatology in chronic schizophrenia*. Psychol Med. **1990**, 20:357-365.

Nuechterlein KH, Dawson ME. *Information processing and attentional functioning in the developmental course of schizophrenic disorders*. Schizophr Bull. **1984**, 10:160-203.

Nyman AK, Jonsson H. *Patterns of self-destructive behaviour in schizophrenia*. Acta Psychiatr Scand. **1986**, 73:252-262.

Olfson M, Mechanic D, Boyer CA, Hansell S, Walkup J, Weiden PJ. *Assessing clinical predictions of early rehospitalization in schizophrenia*. J Nerv Ment Dis. **1999**, 187:721-729.

Olney JW, Labruyere J, Price MT. *Pathological changes induced in cerebrocortical neurons by phencyclidine and related drugs*. Science. **1989**, 244:1360-1362.

Osby U, Correia N, Brandt L, Ekbom A, Sparen P. *Mortality and causes of death in schizophrenia in Stockholm county, Sweden*. Schizophr Res. **2000**, 45:21-28.

Ostrosky-Solis F, Ardila A, Rosselli M. *NEUROPSI: a brief neuropsychological test battery in Spanish with norms by age and educational level*. J Int Neuropsychol Soc. **1999**, 5:413-433.

Park S, Holzman PS. *Schizophrenics show spatial working memory deficits*. Arch Gen Psychiatry. **1992**, 49:975-982.

Parnas J, Schulsinger F, Schulsinger H, Mednick SA, Teasdale TW. *Behavioral precursors of schizophrenia spectrum. A prospective study*. Arch Gen Psychiatry. **1982**, 39:658-664.

Peralta V, Cuesta MJ. *Negative parkinsonian, depressive and catatonic symptoms in schizophrenia: a conflict of paradigms revisited*. Schizophr Res. **1999**, 40:245-253.

Perlick D, Mattis S, Stastny P, Teresi J. *Gender differences in cognition in schizophrenia*. Schizophr Res. **1992**, 8:69-73.

Pokorny AD, Kaplan HB. *Suicide following psychiatric hospitalization*. J Nerv Ment Dis. **1976**, 162:119-125.

Pulver AE, Lasseter VK, Kasch L, Wolyniec P, Nestadt G, Blouin JL, Kimberland M, Babb R, Vourlis S, Chen H, . *Schizophrenia: a genome scan targets chromosomes 3p and 8p as potential sites of susceptibility genes.* Am J Med Genet. **1995**, 60:252-260.

Radomsky ED, Haas GL, Mann JJ, Sweeney JA. *Suicidal behavior in patients with schizophrenia and other psychotic disorders.* Am J Psychiatry. **1999**, 156:1590-1595.

Rodríguez A. El suicidio en la esquizofrenia. In: Ros SM, ed. *La conducta suicida.* Arán, Madrid, **1998**, pp .

Romme MA, Escher AD. *Hearing voices.* Schizophr Bull. **1989**, 15:209-216.

Rossau CD, Mortensen PB. *Risk factors for suicide in patients with schizophrenia: nested case-control study.* Br J Psychiatry. **1997**, 171:355-359.

Roy A. *Suicide in chronic schizophrenia.* Br J Psychiatry. **1982**, 141:171-177.

Roy A, Mazonson A, Pickar D. *Attempted suicide in chronic schizophrenia.* Br J Psychiatry. **1984**, 144:303-306.

Roy A. *Depression, attempted suicide, and suicide in patients with chronic schizophrenia.* Psychiatr Clin North Am. **1986**, 9:193-206.

Rund BR, Landro NI. *Memory in schizophrenia and affective disorders.* Scand J Psychol. **1995**, 36:37-46.

Rund BR, Landro NI, Orbeck AL. *Stability in cognitive dysfunctions in schizophrenic patients.* Psychiatry Res. **1997**, 69:131-141.

Russell AJ, Munro JC, Jones PB, Hemsley DR, Murray RM. *Schizophrenia and the myth of intellectual decline.* Am J Psychiatry. **1997**, 154:635-639.

Saccuzzo DP, Braff DL. *Information-processing abnormalities: trait- and state-dependent components.* Schizophr Bull. **1986**, 12:447-459.

Sainsbury P. Depression, suicide and suicide prevention. In: Roy A, ed. *Suicide.* Williams and Wilkins, Baltimore, **1986**, pp 73-88.

Sands JR, Harrow M. *Depression during the longitudinal course of schizophrenia.* Schizophr Bull. **1999**, 25:157-171.

Sanjuán J, Prieto L, Olivares JM, Ros S, Montejo A, Ferrer F, Mayoral F, González-Torres MA, Bousono M. *Escala GEOPTE para la cognición social en la psicosis.* Actas Esp Psiquiatr. **2003**, 31:120-128.

Sanjuán J, González JC, Aguilar EJ, Leal C, van Os J. *Pleasurable auditory hallucinations.* Acta Psychiatr Scand. **2004**, 110:273-278.

Sanjuán J, Balanzá V. Deterioro cognitivo en la esquizofrenia. Asociación Gallega de Psiquiatría. In: Investigación en Psiquiatría (V) Vigo, ed. **2002**, pp 51-72.

Sanjuán J, Aguilar EJ. Tratamiento de la esquizofrenia resistente. In: Sánchez Planell L, Vallejo J, Menchón JM, Díez C, eds. *Patologías resistentes en psiquiatría.* Ars Medica, Barcelona, **2005**, pp .

Sarró B, De la Cruz C. *Los suicidios.* Martínez-Roca; Barcelona, **1991**.

Sartorius N, Jablensky A, Ernberg G, Leff J, Korten A, Gulbinat WH. Course of schizophrenia in different countries: Some results of a WHO international comparative 5 years follow up study. In: Hafner H, Gattaz WF, Janzarik W, eds. *Search for the causes of schizophrenia.* Springer Verlag, Berlin, **1987**, pp 107-113.

Saykin AJ, Gur RC, Gur RE, Mozley PD, Mozley LH, Resnick SM, Kester DB, Stafiniak P. *Neuropsychological function in schizophrenia. Selective impairment in memory and learning.* Arch Gen Psychiatry. **1991**, 48:618-624.

Schmand B, Kop WJ, Kuipers T, Bosveld J. *Implicit learning in psychotic patients.* Schizophr Res. **1992**, 7:55-64.

Schmand B, Brand N, Kuipers T. *Procedural learning of cognitive and motor skills in psychotic patients.* Schizophr Res. **1992**, 8:157-170.

Sensky T, Turkington D, Kingdon D, Scott JL, Scott J, Siddle R, O'Carroll M, Barnes TR. *A randomized controlled trial of cognitive-behavioral therapy for persistent symptoms in schizophrenia resistant to medication.* Arch Gen Psychiatry. **2000**, 57:165-172.

Shakow D. *Segmental set.* Arch Gen Psychiatry. **1962**, 6:1-17.

Shenton ME, Kikinis R, Jolesz FA, Pollak SD, LeMay M, Wible CG, Hokama H, Martin J, Metcalf D, Coleman M, . *Abnormalities of the left temporal lobe and thought disorder in schizophrenia. A quantitative magnetic resonance imaging study.* N Engl J Med. **1992**, 327:604-612.

Siris SG. *Depression in schizophrenia: perspective in the era of "Atypical" antipsychotic agents.* Am J Psychiatry. **2000**, 157:1379-1389.

Smith ML, Milner B. *Estimation of frequency of occurrence of abstract designs after frontal or temporal lobectomy.* Neuropsychologia. **1988**, 26:297-306.

Smith TE, Hull JW, Israel LM, Willson DF. *Insight, symptoms, and neurocognition in schizophrenia and schizoaffective disorder.* Schizophr Bull. **2000**, 26:193-200.

Spitzer RL, Endicott J, Robins E. *Research diagnostic criteria: rationale and reliability.* Arch Gen Psychiatry. **1978**, 35:773-782.

Summerfelt AT, Alphs LD, Funderburk FR, Strauss ME, Wagman AM. *Impaired Wisconsin Card Sort performance in schizophrenia may reflect motivational deficits.* Arch Gen Psychiatry. **1991**, 48:282-283.

Sweeney JA, Keilp JG, Haas GL, Hill J, Weiden PJ. *Relationships between medication treatments and neuropsychological test performance in schizophrenia.* Psychiatry Res. **1991**, 37:297-308.

Taylor MA, Abrams R. *The prevalence of schizophrenia: a reassessment using modern diagnostic criteria.* Am J Psychiatry. **1978**, 135:945-948.

Taylor MA, Abrams R. *Cognitive impairment in schizophrenia.* Am J Psychiatry. **1984**, 141:196-201.

Walkup J. *A clinically based rule of thumb for classifying delusions.* Schizophr Bull. **1995**, 21:323-331.

Waltrip RW, Carrigan DR, Carpenter WT, Jr. *Immunopathology and viral reactivation. A general theory of schizophrenia.* J Nerv Ment Dis. **1990**, 178:729-738.

Webb WB. *An attempt to study intellectual deterioration by premorbid and psychotic testing.* J Consult Psychol. **1950**, 14:95-98.

Weinberger DR, Berman KF, Zec RF. *Physiologic dysfunction of dorsolateral prefrontal cortex in schizophrenia. I. Regional cerebral blood flow evidence.* Arch Gen Psychiatry. **1986**, 43:114-124.

Weiner RU, Opler LA, Kay SR, Merriam AE, Papouchis N. *Visual information processing in positive, mixed, and negative schizophrenic syndromes.* J Nerv Ment Dis. **1990**, 178:616-626.

Wilkinson DG. *The suicide rate in schizophrenia .* Br J Psychiatry. **1982**, 140:138-141.

Wilkinson G, Bacon NA. *A clinical and epidemiological survey of parasuicide and suicide in Edinburgh schizophrenics.* Psychol Med. **1984**, 14:899-912.

Wing J, Nixon J, von Cranach M, Strauss A. *Further developments of the 'present state examination' and CATEGO system.* Arch Psychiatr Nervenkr. **1977**, 224:151-160.

Wing JK. *The measurement of psychiatric diagnosis.* Proc R Soc Med. **1966**, 59:1030-1032.

Wolfersdorf M, Vogel R, Hole G. Suicide in psychiatric hospitals. Selected results of a study on suicides committed during treatment in fi ve psychiatric hospitals in southern Germany with special regard to therapy success and persuicidal symptoms. In: Moller HJ, Schmidtke A, Weltz R, eds. *Current issues of suicidology.* Springer Verlag, Berlin, **1988**, pp 83-100.

Wolfersdorf M, Hole G. Hospital suicides in psychiatric institution. In: Seva A, ed. *The European Handbook of Psychiatry and Mental Health.* Anthropos, Barcelona, **1991**, pp 1804-1829.

World Health Organization. *ICD-10.Draft of chapter V. Mental, Behavioural and Developmental disorders. Clinical descriptions and diagnostic guidelines.* Word Health Organization; Geneva, **1992**.

Wright P, Donaldson PT, Underhill JA, Choudhuri K, Doherty DG, Murray RM. *Genetic association of the HLA DRB1 gene locus on chromosome 6p21.3 with schizophrenia.* Am J Psychiatry. **1996**, 153:1530-1533.

Zandio-Zorrilla E, García de Jalón MS, Campos-Burgui M. *Tratamiento del déficit cognitivo en la esquizofrenia.* Med Clín. **2005**, 5:17-26.

2.3 Pharmacology

Pharmacotherapy of Schizophrenia

Analía Bortolozzi, Llorenç Díaz-Mataix and Francesc Artigas

Dept. of Neurochemistry and Neuropharmacology, Institut d' Investigacions Biomèdiques de Barcelona, CSIC (IDIBAPS), Roselló, 161. 08036 Barcelona, Spain.

2.3.1 Antipsychotic drugs: introduction

The term antipsychotic is applied to a group of drugs used to treat a variety of disorders such as schizophrenia, mania and delusional disorder, although antipsychotics can also be used to treat psychotic symptoms present in a wide range of other diagnostic categories (e.g., schizoaffective disorders, personality disorders, dementia, etc.). There are currently two main types of antipsychotics, the classic or typical antipsychotics and the atypical antipsychotics. Antipsychotics (particularly the atypical ones) are also used in the treatment of certain mood disorders such the bipolar disorder and also display some augmenting properties in treatment-resistant depression.

All antipsychotic drugs block to a different extent the dopaminergic D_2 receptor family (D_2, D_3 and D_4 subtypes) in the various brain dopaminergic pathways. The blockade of these receptors in the mesolimbic dopaminergic pathway (see below) is thought to account for the therapeutic properties of classic antipsychotics. However, the blockade of the same receptors in other brain dopaminergic pathways, such as the nigrostriatal system (involved in motor behavior) or the tuberoinfundibular pathway, restricted to the hypothalamus and involved in hormonal secretion, results in important side-effect that limit the effectiveness of these drugs, reducing compliance and making some patients to abandon the treatment once psychotic symptoms have remitted. Moreover, the blockade of dopamine (DA) actions in prefrontal cortex (PFC) is detrimental for cognitive function, which does not help to improve the cognitive deficits in schizophrenic patients. Similarly, blockade of DA actions in cortical and limbic areas may have a negative impact on negative and affective symptoms.

Unlike the typical (or classic) antipsychotics, the so-called "atypical" antipsychotics, represented by clozapine and similar compounds, inhibit the actions of DA at D_2 receptors to a much lower extent and therefore, they do not produce the motor and hormonal side-effects of classic compounds. Moreover, clozapine was found to be superior that classic antipsychotics in chronic schizophrenics (Kane et al., 1988) which raised a large expectancy on th potential therapeutic action of the second generation of antipsychotics.

The action of atypical antipsychotics takes place predominantly through the blockade of serotonergic receptors (5-HT$_{2A}$ and 5-HT$_{2C}$) for which they display a higher *in vitro* affinity (see below). Some compounds are also direct or indirect agonists of other serotonin (5-HT) receptors, the 5-HT$_{1A}$ class. *In vivo* neuroimaging studies using PET (positron emission tomography) scan or SPECT (single photon emission computed tomography) have confirmed that therapeutic doses of atypical antipsychotics produce a much greater occupancy of 5-HT$_2$ receptors than of DA D$_2$ receptors. Moreover, except for certain drugs (e.g., risperidone), the D$_2$ receptor occupancy never reaches the threshold to elicit extrapyramidal side-effects (typically 70–75%), which explains the favorable side-effect profile of atypical compounds. However, it remains to be determined how atypical antipsychotics (which were discovered by serendipity) can exert their therapeutic action with a subthreshold occupancy of DA D$_2$ receptors and what is the role of the blockade of 5-HT$_{2A}$ receptors in their antipsychotic clinical action. Although the 5-HT$_{2A}$/D$_2$ affinity ratio is considered the main criterion to define the "atypicality" of antipsychotics (Meltzer, 1989), there are discrepant views that focus on the way that these drugs interact with DA D$_2$ receptors (Kapur and Seeman, 2001). Moreover, a recently marketed drug (aripiprazole) shows high affinity for D$_2$ receptors and displays supra-threshold occupancy but does not produce extrapyramidal side-effects, due to its partial agonist character at D$_2$ receptors (contrary to the classic antipsychotics, which are antagonists at this receptor). Hence, a variety of pharmacological mechanisms seem to account for both the therapeutic and undesirable effects of antipsychotic drugs.

Yet, the atypical family is not devoid of problems, since clozapine – the gold standard in terms of overall efficacy – must be used under strict medical control since it can induce agranulocitosis in some patients. Other compounds produce metabolic side-effects, as well as weight gain. Moreover, the potential benefits of atypical antipsychotics in the improvement of negative, affective and cognitive symptoms are less than initially expected from clozapine's data. Hence a National Institute of Mental Health (NIMH)-sponsored naturalistic study (CATIE) conducted in a large sample of USA schizophrenic patients found a superior efficacy of the atypical drug olanzapine but similar efficacy of other atypical drugs (quetiapine, risperidone, ziprasidone) when compared with a classic antipsychotic such as pherphenazine (Lieberman et al., 2005). Interestingly, patients in the CATIE study who did not respond to either of the atypical drugs used, responded to clozapine, showing that the latter drug has an unsurpassed efficacy in chronic schizophrenia (McEvoy et al., 2006).

Indeed, the limited effectiveness of the classic and atypical drugs on negative/affective symptoms of schizophrenia and on the core cognitive dysfunction of the illness indicates that there is ample room for improving the therapeutic action of antipsychotics. The present chapter will review the pharmacology and mechanism of action of these drugs, with a particular emphasis on the neurobiological mechanisms involved in their therapeutic action.

2.3.1.1 Dopaminergic pathways and dopamine receptor physiology

The anatomy, neurochemistry, physiology and pharmacology of the dopaminergic systems are extensively described in a large number of reviews and textbooks (see for instance Hoffman, 2001; Kuhar et al., 2005; Grace, 2002). Here we give a brief summary of these aspects to serve as a background from a better comprehension of the pharmacology of antipsychotic drugs.

The ascending dopaminergic pathways in the central nervous system (CNS) of mammals include two main systems: the nigrostriatal system that originates in the *substantia nigra pars compacta* (A9) and innervate predominantly the dorsal striatum (caudate and putamen), and the mesocorticolimbic system that originates in the ventral tegmental area (VTA; A10) (Fig. 1). The nigrostriatal system innervates the dorsal striatum (caudate + putamen) and is mainly – though not exclusively – involved in motor functions. Most DA-containing neurons of the *substantia nigra pars compacta* degenerates in patients with Parkinson's disease, causing the motor symptoms of this neurological disease (akinesia, tremor, rigidity, etc.). In contrast, the dopaminergic neurons in the VTA project to cortical and limbic structures and are involved in a large number of relevant brain functions and derangements of the VTA system are suspected in various psychiatric conditions, such as attention deficit hyperactivity disorder (ADHD) and schizophrenia.

The ascending VTA system is actually subdivided in two subsystems, one that projects to the prefrontal cortex (mesocortical) and another projecting to a variety of subcortical structures, in particular the ventral striatum (nucleus accumbens, olfactory tubercle, islands of Calleja), the hippocampus and the amygdala (Albanese and Minciacchi, 1983; Björklund and Lindvall, 1984; Sesack et al., 1995; Krimer et al., 1997; Adell and Artigas, 2004). The mesolimbic system is involved in reward and motivational processes (Le Moal and Simon, 1991; Schultz, 1998; Spanagel and Weiss, 1999) whereas the mesocortical system is intimately involved in cognitive processes such as working memory (Williams and Goldman-Rakic, 1998). Different physiological or pathological conditions like stress, drug addiction or neuropsychiatric disorders have been associated with changes in the activity of the mesocorticolimbic DA system (Goldstein and Deutch, 1992; Carlsson et al., 2001; Tzschentke, 2001; Volkow et al., 2002). In addition to DA-containing neurons, the VTA contains projection GABAergic neurons though in a lower proportion (~20% vs 80%; Swanson, 1982) which share projection areas with dopaminergic neurons (Carr and Sesack, 2000).

The VTA is one of the several monoaminergic modulatory systems of the brainstem, together with the serotonergic and noradrenergic systems, which originate in the midbrain raphe nuclei and the locus coeruleus and adjacent nuclei, respectively (Fig. 1). Forebrain neurons in cortical and limbic structures express the receptors to DA, noradrenaline (NA) and 5-HT which are sensitive to antipsychotic drugs, in particular DA D_2 receptors, α_1 and α_2-adrenoceptors and serotonergic receptors of various subtypes (5-HT_{2A}, 5-HT_{2C}, 5-HT_{1A}, etc.).

A third DA-containing system originates in the arcuate and periventricular nuclei of the hypothalamus which innervates the intermediate lobe of the pituitary and the median eminence. This system is critically involved in regulating the release of pituitary hormones, especially prolactin. In addition to these major pathways, dopaminergic neurons have been found in the olfactory bulb and in the retina.

Figure 1. Scheme of the projections of the brainstem monoaminergic nuclei in the rat brain

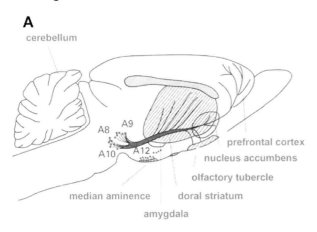

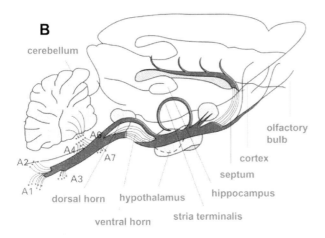

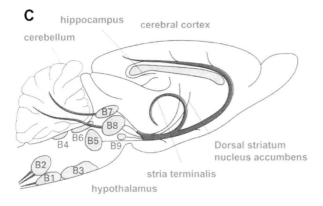

Projections of the brainstem monoaminergic nuclei in the rat brain. (A) Dopamine-containing neurons of the *substantia nigra pars compacta* project to the dorsal striatum (caudate + putamen) whereas those in the adjacent ventral tegmental area project to the ventral striatum (nucleus accumbens and anatomically related structures), limbic structures (hippocampus, amygdala, etc.) and prefrontal cortex. (B) and (C) show the projections of noradrenaline- and serotonin-containing neurons, whose cell bodies are located in the *locus coeruleus* and raphe nuclei, respectively. Note the similar ascending pathways and projection areas of noradrenergic and serotonergic neurons, whose axons exhibit a more widespread distribution than those of dopaminergic neurons.

Five different subtypes of DA receptors have been identified and cloned, which belong to the superfamily of G-protein coupled receptors. According to their homology and signaling mechanisms, they have been classified in two different families, the D_1 (or D_1-like) family, with two members (D_1 and D_5) and the D_2 (or D_2-like) family, which includes the subtypes D_2, D_3 and D_4. From a biochemical point of view, they are coupled to adenilyl cyclase; the D_1 family has a stimulatory action whereas the D_2 family has an inhibitory action. However, the physiological effects of the activation of D_1 and D_2 receptor families are, by far, much more complex and there are numerous and conflicting reports, which are possibly due to the existence of alternative signaling mechanisms and/or the presence of both receptor subtypes in different neuronal subtypes (e.g., glutamatergic and GABAergic neurons). The activation of the autoreceptors D_2 induces an increase in the conductance of the potassium and the subsequent hyperpolarization of the dopaminergic neuron (Grace, 2002) producing an inhibition of the frequency of discharge. Besides the role of the D_2 receptors as autoreceptors they can also be working like heteroreceptors.

D_1-like receptors are mainly located post-sinaptically and DA-containing neurons do not express the mRNA encoding D_1 receptors (Mansour et al., 1992). On the contrary, some members of the D_2 family, like the D_2 and D_3 receptors, are highly expressed in the VTA (Wamsley et al., 1989; Chen et al., 1991; Sesack et al., 1994; Diaz et al., 2000). These two receptors, together with the D_1 receptor are expressed in a large density in various brain areas, particularly the dorsal striatum (caudate + putamen), the ventral striatum (nucleus accumbens, islands of Calleja) and the olfactory bulb, although there is a predominance for the expression of D_3 receptors in ventral striatum. In contrast, the D_4 and D_5 receptors exhibit mostly an extra-striatal location. Figure 2 shows a scheme of a brain dopaminergic synapse.

Figure 2. Scheme of a dopaminergic synapse

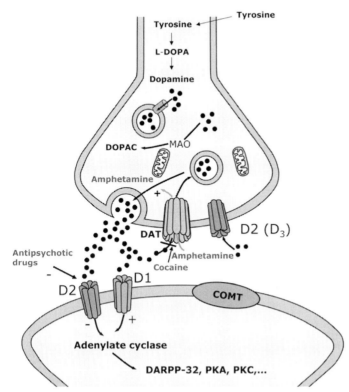

Once synthesized from L-tyrosine and L-DOPA, dopamine (DA) is stored in synaptic vesicles through a VMAT (*vesicular monoamine transporter*) which internalizes the monoamine and protects it from degradation. DA can be metabolized to DOPAC by monoamine oxidase (MAO). Upon arrival of a nerve impulse to the axon terminal, DA is released by a Ca^{2+}-dependent mechanism. DA can act on pre- and postsynaptic receptors of two different families (D_1 and D_2; from D_1 to D_5) whose activation results in changes of cAMP and further activation/inhibition of several protein kinases (PK, PKC, DARPP-32). The extracellular concentration of DA is regulated by a membrane transporter (DAT). This transporter is the molecular target of several drugs of abuse such as cocaine, which inhibits its activity and enhances extracellular DA concentration. Amphetamine has a dual effect on DA terminals: it modestly inhibits DAT but produces a nerve impulse-independent release of DA from non-vesicular pools. Once release, DA can also be metabolized by COMT (catechol-*O*-methyl transferase).

Lesion studies have shown that the great majority of D_2 receptors of the VTA are located on dopaminergic neurons in rodent brain (Chen et al., 1991) more precisely in dendrites (Pickel et al., 2002). However, while nigrostriatal dopaminergic neurons express DA D_2 receptors in rodent, primate and human brain, this does not seem to be the case for mesolimbic and mesocortical neurons in primate and human brain, since the VTA shows a conspicuous lack of D_2 receptor mRNA in monkey and human brain, suggesting that DA neurons are self-regulated via different mechanisms in rodents and primates (Meador-Woodruff et al., 1994). In the rat, DA neurons also express the DA D_3 receptor, suggesting a multiplicity of autoreceptors, as it also occurs in noradrenergic and serotonergic neurons, where at least two different receptor subtypes play the role of autoreceptors.

2.3.1.2 Classic antipsychotic drugs

The first antipsychotic drug discovered was chlorpromazine, which was first marketed as an antiemetic. Its sedating properties were discovered in 1951 by Henri Laborit, who was searching for new agents to induce sedation for use during surgery and observed that chlorpromazine had strong sedating properties. Soon, the drug was tested by Jean Delay and Pierrer Deniker, at the Hôpital Sainte-Anne, Paris, in schizophrenic patients where its antipsychotic activity became evident. In 1954, the drug was approved for psychiatric treatment. Chlorpromazine and new antipsychotics such as haloperidol (marketed in 1957) opened a new era in Psychiatry, replacing less effective and sometimes unethical treatments such as insulin shocks, electroconvulsive therapy or frontal lobotomy and enabling to discharge internalized patients, who experienced a considerable improvement in their quality of life.

Typical antipsychotic drugs are also referred to as conventional, standard, classic, traditional or first-generation antipsychotic drugs, as well as neuroleptics. They can be divided in various groups attending to their chemical structure (Martindale, 1993):

1. *Phenothiazines*:

- Chlorpromazine was the first antipsychotic drug. It acts as antagonist of different receptors such as DA D2, 5-HT_2, histamine H_1, α_1-adrenergic and acetylcholine muscarinic receptors. Due to its high activity against the actions of histamine, chlorpromazine was first used as an antihistaminergic drug.
- Fluphenazine possess anticholinergic and α_1-adrenergic blocking activity as well as the dopaminergic antagonism.

- Perphenazine is an old phenothiazine antipsychotic with a potency similar to haloperidol. It has been used for many years and is popular in the northern European countries and Japan for the treatment of psychotic patients and for patients presenting with manic phases of bipolar disorder.
- Prochlorperazine is relatively seldom used as antipsychotic, in contrast is more frequently used for the treatment of nausea and vertigo caused by cancer therapy due to its pharmacological profile.
- Thioridazine has a low incidence of extrapyramidal symptoms and has a surprising antimicrobial activity (Amaral et al., 2004).
- Trifluoperazine is little-used due to the highly frequent extrapyramidal symptoms and tardive dyskinesia. Some studies have suggested that this compound can be useful to reverse addiction to opioids (Tang et al., 2006).

2. *Butyrophenones:*

- Haloperidol is the classic antipsychotic drug more used worldwide. Due to its strong central antidopaminergic action it is approximately 50 times more potent than chlorpromazine. It has high α_1-adrenergic receptor affinity and minor antihistaminic and anticholinergic activity.
- Pimozide is used for patients failing to respond to other medications. It has also neurologic indications for Gilles de la Tourette syndrome and resistant tics.
- Droperidole is used as antiemetic and antipsychotic (Scuderi, 2003).

3. *Thioxanthenes*

Thioxanthenes drugs are poorly used today due to their important and severe adverse effects but are being studied for other uses such as substance dependence (Soyka and De Vry, 2000).

- Chlorprothixene
- Flupentixol
- Thiothixene

It is important to note that all classic antipsychotics are a very complex pharmacological profile, with moderate-high *in vitro* affinity for many monoaminergic receptors. The therapeutic action – as well as the motor side-effects – of classic antipsychotics derives from their blockade of the actions of DA at D2 receptors whereas the other pharmacological activities are responsible of several of the side-effects exhibited by these agents (sedation, constipation, cardiovascular problems, etc.).

2.3.1.3 Dopamine receptors and antipsychotic drug action

Without exception, effective antipsychotic drugs act as antagonists of the dopamine D_2 receptors to some extent (Jones and Pilowsky, 2002), thus altering both the pre- and postsynaptic components of dopaminergic transmission. The development of ligand binding techniques in the 1970s led to the well known relationship between DA D_2 receptor affinity and average daily doses of treatment (Seeman and Lee, 1975, Creese et al., 1976; Peroutka and Snyder, 1980). These observations strongly confirmed the views that their antipsychotic action derived from the blockade of DA D_2 receptors and provided a direct confirmation of the dopaminergic theory of schizophrenia as postulated earlier by Arvid Carlsson (Carlsson and Lindqvist, 1963). This view was also supported by the fact that the antipsychotic activity of flupentixol resided in its α-isomer (DA receptor antagonist) and not in its β-isomer (non-DA antagonist; Johnstone et al., 1978).

Hence, the blockade of dopaminergic transmission in the mesolimbic and in the mesocortical pathways could be the responsible of the therapeutic action of classic antipsychotic drugs. However, the precise way by which blockade of the actions of DA at D_2 receptors translates into an antipsychotic action remains unclear. New views on this matter suggest that DA plays an important role in assigning the importance or "salience" of external stimuli and internal representations. An excessive dopaminergic transmission can result in an abnormal or aberrant assignment of the salience of certain stimuli. Hallucinations may be a direct experience of these aberrant experiences, whereas delusions can result from the cognitive effort of the patient to make sense of these experiences. Antipsychotic drugs may attenuate psychotic symptoms by reducing dopaminergic transmission and restoring a "normal" level of motivational salience (Kapur,

2003). The neurobiological background of this behavioral effect is still poorly understood, but most accepted views suggest that schizophrenic patients have deficient information filtering, as suggested by prepulse inhibition deficits. This possibly reflects abnormal information processing in thalamocortical circuits, resulting eventually in an altered cortical descending input on limbic structures (e.g., ventral striatum). Hence, electrophysiological observations indicate that antipsychotic drugs may act by normalizing excessive limbic inputs onto the nucleus accumbens through the blockade of D_2 receptors (Grace, 2000).

A highly debated question in this field is whether antipsychotic drugs show a therapeutic delay, i.e., whether repeated treatment is required to achieve their antipsychotic action. This question has a remarkable interest to explain the action of antipsychotic drugs just in terms of DA receptor blockade. Hence, several rat studies have shown that repeated antipsychotic drug treatment results in a delayed inactivation of dopamine neuron firing in the midbrain due to depolarization block (Grace and Bunney, 1986). That effect can be useful to modulate the dysregulated dopaminergic system that possibly occurs in the schizophrenic brain (Grace et al., 1997). However, despite clinical trials show a slow onset of the action of antipsychotics, most of their clinical action takes place in the first two weeks of treatment, which clearly argues in favor of the view that DA receptor blockade is sufficient to evoke an antipsychotic effect (Agid et al., 2006).

Notwithstanding the above observations, it is still unclear whether the therapeutic action of classic antipsychotics is only accounted for by DA D_2 receptor blockade or other pharmacological activities may also participate. Hence, most antipsychotic drugs, either classic or atypical, display high affinity for α_1-adrenoceptors, and some classic antipsychotics, like chlorpromazine share with most atypical drugs a high affinity for 5-HT$_{2A}$ receptors (Meltzer et al., 1989; Trichard et al., 1998). The advent of neuroimaging techniques and in particular PET scan has enabled to directly examine the relationship between DA D_2 receptor occupancy and clinical response in the living brain. Hence, numerous studies, starting with the initial work by Farde and colleagues (Farde et al., 1989) have shown that conventional antipsychotic drugs produced a dose-related, substantial occupancy (65–85%) of DA D_2 receptors in schizophrenic patients. Further studies have permitted to establish a relationship between DA D_2

receptor occupancy and clinical improvement, with a lower limit (threshold) of receptor occupancy for the establishment of therapeutic effects (Nordstrom et al., 1993; Kapur et al., 2000). However, deviations to this rule have been reported, either of patients with high receptor occupancy and poor response (Wolkin et al., 1989) or viceversa (Pilowsky et al., 1993).

In contrast, PET scan studies with clozapine, olanzapine and other atypical antipsychotic drugs have revealed that these agents produce their therapeutic effects with DA D_2 receptor occupancy substantially lower than that caused by conventional antipsychotic drugs (Farde et al., 1989; Kapur et al., 1999, 2000). In particular, the occupancy of DA D_2 receptors by very high doses of clozapine never reaches more than 70%, below the threshold to elicit motor side-effects (Kapur et al., 2000).

Overall, these observations suggest that other pharmacological activities in addition to DA D_2 receptor antagonism can also contribute to the antipsychotic effect although no clinically effective antipsychotic drug, either classic or atypical, is free from DA D_2 receptor affinity. Hence, the selective 5-HT_{2A} receptor antagonist M100907 failed to show antipsychotic activity, although it had a remarkable effect on negative symptoms in some patients (Potkin et al., 2001; H.Y. Meltzer, personal communication). Hence, DA D_2 receptor blockade appears to be the main mechanism by which conventional antipsychotics exert their therapeutic action, yet this mechanism may not be the only one able to produce a clinically meaningful antipsychotic action.

2.3.1.4 Side effects

The therapeutic actions of classic antipsychotics are due to the blockade of D_2-mediated dopamine transmission in mesocorticolimbic pathway. However, DA D_2 receptors are also expressed by dorsal striatal neurons which project to the pallidum and to the *substantia nigra pars reticulata* and are involved in motor control. Obviously, conventional antipsychotics do not discriminate between DA D_2 receptors in dorsal (motor) and ventral (limbic) striatum. Thus, their therapeutic action is necessarily associated with motor side-effects derived from DA D_2 receptor blockade in other brain areas, and particularly in motor structures. Likewise, the antagonism of DA actions in the hypothalamus results in a rise in prolactin since DA tonically suppresses the secretion of this hormone.

Several PET scan studies have examined the relationship between DA D_2 receptor occupancy and emergence of motor side-effects (e.g., Kapur et al., 2000). As a result, it is currently accepted that extrapyramidal side-effects occur at a level of occupancy close to or above 80%.

The extrapiramidal reactions and the tardive dyskinesia are the two major motor side-effects. Extrapyramidal symptoms include including acute dystonias, akathisia, and parkinsonism (rigidity and tremor). However, one of the more serious side-effects is tardive dyskinesia, characterized by repetitive, involuntary, purposeless movements often of the lips, face, legs or torso. Another serious side-effect is neuroleptic malignant syndrome, characterized by a loss of temperature regulation, which requires urgent medical care.

2.3.2 Atypical antipsychotics: introduction

As described above, neuroleptic action was first discovered and defined in 1952 with the clinical use of chlorpromazine, known before as an antihistaminergic and antiemetic agent (Delay et al., 1956). Following the introduction of chlorpromazine, a large number of neuroleptic (or typical antipsychotic) drugs were developed for the treatment of schizophrenia. The identification of dopamine (DA) as a neurotransmitter in the brain and neurochemical and pharmacological studies revealed that all typical neuroleptic are potent D_2 DA receptor antagonists (Carlsson and Lindqvist, 1963; Lee and Seeman, 1975; Creese et al., 1976; Seeman et al., 1976; Peroutka and Snyder, 1980).

Although typical antipsychotic drugs are effective at reducing the positive symptoms of schizophrenia (e.g., delusions, hallucinations, conceptual disorganization) in many patients, they are not without serious side-effects (see above). For many years following the discovery of chlorpromazine, it was assumed that the motor side-effects of typical antipsychotic drugs were intrinsically connected with their therapeutic potential. However, the observation that clozapine, a drug discovered in 1958 (for a historical perspective, see Hippius, 1999), had a full antipsychotic clinical action devoid of extrapyramidal side-effects (EPS) provided the first evidence that the therapeutic action could be clearly differentiated from the emergence of side-effects. The presence of an unequivocal antipsychotic efficacy in absence of EPS led to the concept of "atypical" antipsychotic. The observation that clozapine is a potent serotonin $5\text{-}HT_{2A}$ receptor antagonist, which down-regulates $5\text{-}HT_{2A}$ receptors after chronic treatment (Reynolds et al., 1983) brought the attention of the scientific community to 5-HT as a potential mediator of the antipsychotic action and indicated that there was life outside DA in schizophrenia research.

There are eight atypical antipsychotics commercially available in the USA: clozapine, risperidone, olanzapine, quetiapine, ziprasidone, aripiprazole, and two more in other countries (zotepine and amisulpride). A ninth atypical drug, sertindole, had to be withdrawn from the market due to a potential cardiovascular side-effect (prolongation of the Q–T interval), but after additional data from Lundbeck, it was again approved in the EU in early 2006. With the exception of quetiapine and amisulpride, all currently approved atypical antipsychotic drugs display a high affinity for $5\text{-}HT_{2A}$ receptors (e.g., K_i values between 0.1 nM and 30 nM; yet quetiapine, 96–696 nM; amisulpride >2000 nM; see htpp://kidb.bioc.cwru.edu/pdsp.php).

Compared with the conventional antipsychotic drugs, the atypical drugs have a lower propensity to induce EPS and elevation of serum prolactin (except for risperidone) and have a broader spectrum of efficacy (Remington and Kapur, 2000). Additionally, some atypical antipsychotic drugs, in particular clozapine, are also effective in "treatment-resistant" schizophrenia (Kane et al., 1988), significantly reduce the incidence of suicide (Meltzer, 1998), modestly improve cognition (Stip, 2000) and diminish the risk of depression (Farah, 2005). However, the label "atypical antipsychotics" designates a group of chemically and pharmacologically heterogeneous compounds with varying clinical efficacy and side-effect profile, of which clozapine remains the gold standard in terms of effectiveness.

Clozapine and other atypical antipsychotic drugs are not without serious *drug-specific side-effects* including weight gain (e.g., clozapine, olanzapine, risperidone, quetiapine) and associated metabolic disorders (e.g., hypercholesterolemia, diabetes). Others important effects include orthostasic or postural hypotension (clozapine, olanzapine, ziprasidone, risperidone, quetiapine), prolongation of the Q–T interval (ziprasidone), constipation and significant lowering of seizure threshold. Orthostatic hypotension likely results from α_1-adrenergic receptor blockade, while constipation results from the antimuscarinic effects of atypical drugs. A particularly severe side-effect is agranulocitosis, only produced by clozapine, which unfortunately has led to the restricted use of this antipsychotic.

Because atypical antipsychotic drugs interact with a large number of G-protein coupled receptors and transporters (Box 1, Table 1), it has proven difficult to assign precise pharmacologic mechanisms for the various therapeutic and toxic side-effects of atypical drugs. In the following sections we summarize our current understanding of the pharmacological mechanism of "atypicality". Additionally, we integrate this information in an attempt to identify the neuronal circuits involved in the therapeutic action of atypical antipsychotic drugs.

Box 1. Classification of antipsychotic drugs (marketed and in development)

Family	Compound	Mode of action*
Typical antipsychotics	Haloperidol Fluphenazine Raclopride Pherphenazine Eticlopride	Predominant D_2 antagonism
Atypical antipsychotics	Clozapine Risperidone Olanzapine Ziprasidone** Quetiapine Sertindole Amisulpride Remoxipride Sulpiride Aripiprazole	Multiple actions: predominant D_2 and 5-HT_{2A} antagonism
New generation putative antipsychotics	SLV314 SLV313 Sarizotan Bifeprunox SSR181507	Mixed D_2 partial agonism or antagonism and 5-HT_{1A} agonism

*Based on an abbreviated binding profile; **Predominant D_2 and 5-HT_{1A} partial agonism. Adapted from Bardin et al., 2006.

2.3.2.1 The 5-HT_{2A}/D_2 hypothesis of atypicality

There is currently no consensus on a true definition of atypicality for antipsychotic drugs. Originally, the term was used to describe effective antipsychotic agent associated with a minimal risk of causing EPS. However, a broader definition of atypicality is used today. One of the first systematic explorations related to predicting the pharmacologic criteria for atypicality was put forward by Meltzer et al. (1989). In this study, the affinities of a large number of "typical" and "atypical" antipsychotic drugs were examined at D_1- and D_2-dopamine and 5-HT_{2A}-serotonin receptors. Function analysis revealed that atypical antipsychotic drugs could be distinguished by two criteria: (1) lower affinities for D_2 DA receptors than conventional antipsychotics, and (2) relatively higher affinities for 5-HT_{2A} receptors than typical drugs. Using a combined criterion of the 5-HT_{2A}/D_2 affinity ratios, Meltzer et al. (1989) were able to correctly classify antipsychotic drugs as typical or atypical.

Table 1. *In vitro* aminergic receptor binding profile (K_i value, nM) for some antipsychotic drugs

Receptor	Haloperidol	Clozapine	Olanzapine	Risperidone	Aripiprazole	Ziprasidone	Quetiapine
D_1	210	85	31	430	265*	130*	4240
D_2	0.7	126	11	4	0.45*	3.1*	310
D_3	2	473	49	10	0.8*	7.2*	650
D_4	3	35	27	9	44*	32*	1600
5-HT_{1A}	1100	875	>10,000	210	4.4*	2.5*	230*
5-HT_{2A}	45	16	4	0.5	3.4*	0.39*	120
5-HT_{2C}	>10,000	16	23	25	15*	0.72*	3820
5-HT_6	6000*	11*	10*	2000*	160#	76*	4100*
5-HT_7	1100*	66*	250*	3*	15#	9.3*	1800*
α_1	6	7	19	0.7	47	13*	58
α_2	1200*	50*	470*	23*	n.a	310*	1000*
H_1	440	6	7	20	61*	47*	8.7*
M_1	>1,500	1.9	1.9	>10,000	>10,000	5100*	1020

Data taken from Bymaster et al., 1996; Arnt and Skarsfeldt, 1998; Miyamoto et al., 2003.
* *In vitro* receptor binding profile in cloned human receptors and # HEK cells

Since then, several drugs characterized by favorable 5-HT_{2A}/D_2 affinity ratios have been demonstrated to be effective in the treatment of schizophrenia. However, there are some interesting exceptions in the relationship between the 5-HT_{2A}/D_2 affinity ratio and "atypicality". Thus, aripiprazole is a novel atypical antipsychotic (Inoue et al., 1996) with a complex pharmacology (Lawler et al., 1999; Burris et al., 2002; Jordan et al., 2002), which has a high affinity for D_2-dopamine receptors and a relatively lower affinity for 5-HT_{2A} 5-HT receptors. Thus, aripiprazole would be classified as a typical antipsychotic drug using 5-HT$_{2A}$/D$_2$ affinity ratio criterion. However, since aripiprazole is a partial D$_2$ DA receptor agonist, it produces a weak DA D$_2$ receptor blockade at a high occupancy level (see below).

One intriguing and still unresolved question is how the blockade of 5-HT$_{2A}$ receptors translates into an antipsychotic clinical action. Indeed, 5-HT$_{2A}$ receptors can be seen as potential targets for antipsychotic drug actions because 5-HT$_{2A}$ agonists, such as lysergic acid and related compounds induce hallucinations. However, the hallucinogenic experience is qualitatively different from that seen in schizophrenia (Hollister, 1962). Moreover, there is no evidence that schizophrenic patients may have an endogenous serotonergic tone higher than that in normal individuals, as one would predict from the action of an antagonist. A large number of anatomical, neurochemical and electrophysiological studies have been conducted in an attempt to identify the molecular and cellular basis of the atypical antipsychotic action, as well as the brain circuits potentially involved.

5-HT$_{2A}$ receptors are densely expressed by pyramidal neurons in layers III and V in the neocortex of rodents (Willins et al., 1997; Amargós-Bosch et al., 2004; Santana et al., 2004) and primates (Jakab and Goldman-Rakic, 1998). 5-HT$_{2A}$ receptors are also present in GABA interneurons (Santana et al., 2004). Pyramidal cells in layer V are involved in integrating cognitive and perceptual information from many diverse cortical and subcortical regions and thus, are exquisitely positioned to mediate the effects of antipsychotic drugs. Interestingly, psychotomimetic NMDA antagonists, such as MK-801, PCP and ketamine increase cortical glutamate, DA and 5-HT release (Adams and Moghaddan, 2001; Amargós-Bosch et al., 2006), and some behavioral effects of PCP and MK-801 such as hyperlocomotion and stereotypy are attenuated by the selective 5-HT$_{2A}$ receptor M100907 antagonist (Martin et al., 1997; Adams and Moghaddan, 2001). Finally, activation of 5-HT$_{2A}$ receptors depolarizes and increases spontaneous synaptic activity of pyramidal neurons in the medial prefrontal cortex (mPFC; Araneda and Andrade, 1991; Aghajanian and Marek, 1997) which results in an increased firing activity of these cells (Puig et al., 2003; Amargós-Bosch et al., 2004). Hence, one possible cellular mechanism for the blockade of 5-HT$_{2A}$ receptors exerted by atypical antipsychotic drugs is to dampen an excessive (e.g., glutamate-induced) activity of pyramidal neurons which could take place during psychotic states. Likewise, it has been suggested that these compounds might antagonize an excessive serotonergic tone on 5-HT$_{2A}$ receptors (Martin et al., 1998).

5-HT$_{2A}$ receptors are also associated with dopaminergic pathways, either directly or indirectly. Thus, Bubser et al. (2001) showed that striatal afferents including corticostriatal glutamatergic neurons and striatopallidal GABAergic neurons are enriched in 5-HT$_{2A}$ receptors. Thus, it is possible that the modulation of striatal neurotransmission via 5-HT$_{2A}$ receptors located on the corticostriatal and striatopallidal neurons are important for the reduced EPS seen with atypical antipsychotic drugs. Additionally, Nocjar et al. (2002) have mapped the distribution of 5-HT$_{2A}$ receptors on midbrain DA neurons in the A10 nuclei. This study reported the existence of 5-HT$_{2A}$ receptors on dopaminergic neurons that project to forebrain regions involved in schizophrenia (e.g., prefrontal cortex, cingulated cortex, nucleus accumbens). Moreover, recent studies have demonstrated that 5-HT$_{2A}$ receptors are also found on putative presynaptic dopaminergic terminals in the rat mPFC, a key position to modulate DA release (Miner et al., 2003). However, the vast majority of excitatory 5-HT$_{2A}$ receptors in the neocortex are located on pyramidal neurons. In particular, they are enriched in prefrontal cortex, an area which projects to the monoaminergic midbrain nuclei and which is involved in the regulation of the activity of monoamine-containing neurons (Thierry et al., 1983, Sesack et al., 1989: Hajós et al., 1998; Peyron et al., 1998; Celada et al., 2001; Martín-Ruiz et al., 2001).

Interestingly, 5-HT$_{2A}$ receptor activation increases the activity of dopaminergic neurons in the VTA as well as the DA release in mPFC and VTA (Pehek et al., 2001; Bortolozzi et al., 2005) and the selective 5-HT$_{2A}$ antagonist M100907 reduced the firing of DA neurons (Bortolozzi et al., 2005). These results suggest that 5-HT$_{2A}$ antagonism by atypical antipsychotic drugs decreases the stimulated dopaminergic activity, a mechanism by which these drugs could exert their therapeutic effect.

Additionally, midbrain serotonergic neurons appear to be under an excitatory distal control of 5-HT$_{2A}$ receptors. In particular, the stimulation of prefrontal 5-HT$_{2A}$ receptors by the 5-HT$_{2A/2C}$ agonist DOI increased the firing rate of serotonergic neurons in the dorsal raphe nucleus (DR) and 5-HT release in the mPFC (Martín-Ruiz et al., 2001)

This effect was dependent on the 5-HT$_{2A}$ receptor-mediated activation of pyramidal neurons in mPFC and involved AMPA-mediated glutamatergic inputs (Martín-Ruiz et al., 2001; Puig et al., 2003). Recently, Bortolozzi et al. (2003) confirmed and extended these previous observations in rat brain to the mouse brain, showing that the local application of the partial 5-HT$_{2A/2C}$ agonist DOI in mPFC enhanced the local 5-HT release. This effect was reversed by antipsychotic drugs, either atypical (clozapine and olanzapine) or classic (chlorpromazine, haloperidol) (Fig. 3A and Fig. 3B, respectively). These results suggest that antipsychotic drugs may contribute to modulate cortical 5-HT release, an effect putatively involved in their therapeutic action.

Figure 3

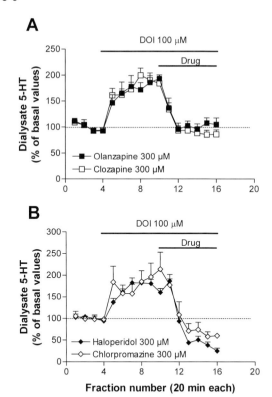

(A) The local application of the atypical antipsychotics clozapine ($n=11$ mice) and olanzapine ($n=6$ mice; 300 µM each) completely reversed the 5-HT elevation induced by DOI (partial 5-HT$_{2A/2C}$ receptor agonist). (B) Likewise, the classic antipsychotic drugs chlorpromazine ($n=5$ mice) and haloperidol ($n=4$ mice; 300 µM each) fully reversed the effect of DOI. The period of drug application is shown by horizontal bars. $P<0.05$ vs DOI alone. Adapted from data in Bortolozzi et al. (2003).

Further support for the involvement of 5-HT$_{2A}$ receptors in the action of atypical antipsychotic agents comes from studies showing down-regulation of 5-HT$_{2A}$ receptors after chronic treatment (Gray and Roth, 2001). Additionally, the chronic treatment with clozapine and olanzapine, but not with the conventional antipsychotic drug haloperidol, was reported to induce internalization of 5-HT$_{2A}$ receptors in individual cortical neurons (Willins et al., 1999). Taken together, these results suggest that one of the possible ways by which chronic 5-HT$_{2A}$ receptor blockade could translate into an antipsychotic action is via the forced internalization and down regulation of 5-HT$_{2A}$ receptors from active synaptic zones to intracellular sites. Yet, several recent studies show that the onset of the antipsychotic effect is within the first day of treatment and that most of the clinical action takes place during the first two weeks of treatment (for a review, see Agid et al., 2006). This effect is distinguishable from behavioral sedation, is specific to antipsychotic drugs, is seen with oral and parenteral preparations, and is seen with typical and atypical antipsychotic drugs. Thus, according to these new views, the relevance of the effects of chronically administered antipsychotics is questionable.

Because all clinically approved atypical antipsychotic drugs have potent 5-HT$_{2A}$ antagonist actions, agents directly targeting the 5-HT$_{2A}$ receptor have been suggested as potential new antipsychotics. At least three drugs with varying 5-HT$_{2A}$ receptor selectivity have been tested in schizophrenia including ritanserin, M100907 and SR46349B (eplivanserin). Ritanserin is a potent 5-HT$_{2A/2B/2C}$ antagonist with moderate affinity for a number of other G protein coupled receptors including 5-HT$_6$, 5-HT$_7$ and D$_1$, D$_2$, D$_3$ and D$_4$ DA receptors (see on-line database htpp://kidb.bioc.cwru.edu/pdsp.php). Ritanserin induced a moderate reduction in core symptoms of schizophrenia (Duinkerke et al., 1993; Wiesel et al., 1994). In contrast, the development of the highly selective 5-HT$_{2A}$ receptor antagonist M100907 was discontinued after two phase III studies in the USA. These studies demonstrated the superiority of M100907 over placebo but its efficacy was below that of haloperidol. Moreover, a European phase III study in schizophrenic patients with predominant negative symptoms failed to observe separation of M100907 from placebo. Furthermore, a phase II study of SR46349B, another 5-HT$_2$ receptor antagonist with approximately 20-fold selectivity for 5-HT$_{2A}$ over 5-HT$_{2C}$ receptors, also showed antipsychotic efficacy intermediate between that of placebo and haloperidol (Marek and Merchant, 2005).

Taken together, these clinical findings demonstrated that 5-HT$_{2A}$ blockade alone results in some antipsychotic activity, yet for optimal efficacy in schizophrenia, some degree of D$_2$ blockade may be also necessary (Fig. 4).

Figure 4. Schematic representation of the relationship between DA D$_2$ receptor occupancy, antipsychotic effects and extrapyramidal side-effects (EPS) produced by conventional and atypical antipsychotics.

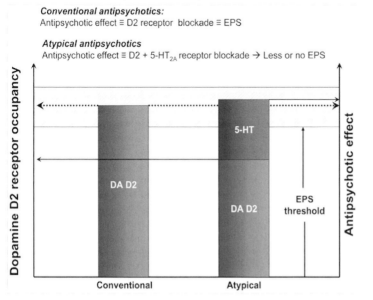

Most of the pharmacological activity of the conventional antipsychotics derives from blockade of DA D$_2$ receptors. This effect is responsible for the therapeutic efficacy (due to blockade of limbic DA D$_2$ receptors) and the side-effects (due to the concurrent blockade of DA D$_2$ receptors in dorsal striatum). DA D$_2$ receptor occupancy for conventional drugs is often above the threshold to produce EPS. In contrast, atypical drugs produce a lower receptor occupancy and did not reach the threshold to evoke EPS (yet some atypical drugs like risperidone can reach this threshold). The pharmacological actions at 5-HT receptors, in particular the 5-HT$_{2A}$ receptor, account for the additional therapeutic effect through still poorly known mechanisms.

2.3.2.2 The role of the 5-HT$_{1A}$ receptor in antipsychotic drug action

Among the various monoaminergic receptors, there is growing interest in 5-HT$_{1A}$ receptors as potential targets for antipsychotic drug action (Wadenberg and Ahlenius, 1991; Millan 2000). These receptors seem to contribute to the ability of atypical (but not classic) antipsychotic to increase cortical DA release, an effect potentially involved in the improvement of negative symptoms and cognitive dysfunctions in schizophrenia (Rollema et al., 1997, 2000; Kuroki et al., 1999). Of the various atypical drugs only two, aripiprazole and ziprasidone, are partial 5-HT$_{1A}$ receptor agonists (Lawler et al., 1999; Burris at al., 2002; Newman-Tancredi et al., 2005). However, clozapine occupies *in vivo* 5-HT$_{1A}$ receptors in primate brain at clinically representative plasma levels despite it showing a negligible *in vitro* affinity for these receptors (Chou et al., 2003). A similar study in human brain has given negative results (Bantick et al., 2004) and hence it is still controversial whether it behaves as a 5-HT$_{1A}$ agonist *in vivo*.

Interestingly, several atypical antipsychotics, including risperidone, olanzapine, clozapine, ziprasidone and aripiprazole, with markedly different *in vitro* affinities for 5-HT$_{1A}$ receptors increase DA release in the mPFC of rats or mice by a 5-HT$_{1A}$ receptor-dependent mechanism. This has been assessed either using the selective antagonist WAY-100635 or mice lacking 5-HT$_{1A}$ receptors (Ichikawa et al., 2001; Díaz-Mataix et al., 2005; Bortolozzi et al., unpublished observations). This effect appears to be shared by atypical drugs but not by haloperidol, since its systemic or local (in mPFC) administration did not elevate DA release in this cortical area. These results suggest that agonism at 5-HT$_{1A}$ receptors is an important component of the action of all members of the 5-HT$_{2A}$/D$_2$ antagonist family as well as aripiprazole. Thus, both 5-HT$_{2A}$ antagonism and 5-HT$_{1A}$ agonism may be the most important of the 5-HT receptors for atypical antipsychotic action.

The 5-HT$_{1A}$ receptor is perhaps the best characterized 5-HT receptor subtype from a functional point of view, and plays an important role in controlling the activity of monoaminergic neurons (Barnes and Sharp, 1999). 5-HT$_{1A}$ receptors can be considered as functionally antagonistic to the 5-HT$_{2A}$ receptor, as they mediate cellular hyperpolarization through a G protein-coupled potassium channel and suppression of cell firing, whereas 5-HT$_{2A}$ receptors mediate neuronal depolarization and increased firing activity (Araneda and Andrade, 1991; Puig et al., 2005). The 5-HT$_{1A}$ receptor is localized on midbrain raphe 5-HT neurons where it acts as an autoreceptor and it is also localized to a larger extent in cortical and limbic areas, postsynaptically to 5-HT axons. In the mPFC, the transcripts 5-HT$_{1A}$ and 5-HT$_{2A}$ receptors are expressed by a very large proportion of neurons (mostly pyramidal) and show a massive co-localization (around 80%) (Amargós-

Bosch et al., 2004; Santana et al., 2004) (Fig. 5). 5-HT$_{1A}$ receptors are also localized in GABAergic interneurons in various brain areas, including the PFC (Aznar et al., 2003; Santana et al., 2004). Interestingly, DA cell firing and DA release have been shown to be modulated by 5-HT$_{1A}$ receptors agonists (Arborelius et al., 1993a, b; Prisco et al., 1994; Ichikawa et al., 1995, Lejeune and Millan, 1998).

Figure 5. Localization of 5-HT$_{1A}$ and 5-HT$_{2A}$ receptor mRNAs in the medial prefrontal cortex of the rat using double in situ hybridization histochemistry

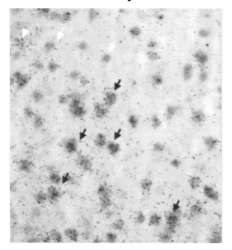

Coronal sections of prelimbic cortex show a large number of cells expressing both 5-HT$_{1A}$ receptors (Dig-labeled oligonucleotide) and 5-HT$_{2A}$ receptors (^{33}P-labeled oligonucleotide) (black arrows). Occasional cells profiles containing only 5-HT$_{1A}$ (white arrows) or 5-HT$_{2A}$ receptor mRNAs (white arrowhead). Adapted from Amargós Bosch et al. (2004).

Local administration of WAY100635 into the rat prefrontal cortex blocked the effect of subcutaneous MKC-242, a potent selective 5-HT$_{1A}$ agonist, to increase cortical DA release in the cortex (Sakaue et al., 2000). These authors also found that buspirone, a 5-HT$_{1A}$ partial agonist, increased DA release in the PFC. However, the mechanism(s) involved and the localization of the 5-HT$_{1A}$ receptors responsible for this effect have not been fully elucidated. In a recent study, we reported that postsynaptic 5-HT$_{1A}$ receptors in PFC are involved in the modulation of DA release in the mesocortical pathway. Hence, BAY x 3702, a selective 5-HT$_{1A}$ receptor agonist increased the activity of VTA dopaminergic neurons and the release of DA in mPFC and VTA. However, these effects disappeared in rats subjected to cortical transaction (Fig. 6)

(Díaz-Mataix et al., 2005). The same study showed that the local application of various atypical antipsychotic drugs in the mPFC enhanced the local DA release in control mice but not in mice lacking 5-HT$_{1A}$ receptors. Altogether, these data suggest that the activation of 5-HT$_{1A}$ receptors in PFC is critically involved in the regulation of DA neuron activity and DA release, one of the crucial targets to modulate cognitive function in schizophrenic patients.

Figure 6

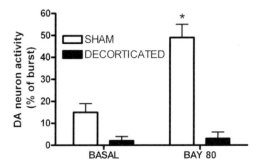

Bar diagram showing the effect of the i.v. administration of the 5-HT$_{1A}$ receptor agonist BAY x 3702 on the burst activity of dopaminergic neurons in the rat VTA. BAY x 3702 (0.08 mg/kg i.v.) markedly enhanced the percentage of spikes fired in bursts in control (sham-operated) rats. The surgical stereotaxic transection of afferents from mPFC to VTA reduced the basal burst activity and completely prevented the action of BAY x 3702 on dopaminergic neurons. Redrawn from data in Díaz-Mataix et al. (2005).

The mechanism(s) by which atypical antipsychotic drugs with low or no *in vitro* affinity for 5-HT$_{1A}$ receptors can behave as agonists of this receptor *in vivo* are unclear. Given the high co-expression of 5-HT$_{1A}$ and 5-HT$_{2A}$ receptors in the mPFC (approximately 80%; Amargós-Bosch et al., 2004), and their opposite role on neuronal function, one might think that blockade of 5-HT$_{2A}$ receptors could shift the physiological balance of 5-HT activation toward 5-HT$_{1A}$ receptors. However, this possibility seems unlikely because the selective 5-HT$_{2A}$ receptor antagonist M100907 did not increase the cortical DA release (Pehek et al., 2001; Bortolozzi et al., 2005). Thus, the exact way in which clozapine and related compounds interact with 5-HT$_{1A}$-mediated neurotransmission remains to be determined. However, the effect on cortical DA release of these drugs was cancelled by the cortical GABA$_A$ antagonist bicuculline application, with suggests that 5-HT$_{1A}$ receptors in GABA interneurons may be affected.

This action might eventually result in an increased excitatory cortical output to the VTA to enhance DA neuron activity, as observed previously with atypical antipsychotic drugs (Gessa et al., 2000). Taken together, these results suggest that atypical antipsychotic exert their effects on dopaminergic neurotransmission, at least in part, via activation of 5-HT$_{1A}$ receptors (Millan, 2000). Previous views suggest that it could be presumably due to concurrent 5-HT$_{2A}$ and relatively weak D$_2$ receptor antagonism (Ichikawa and Meltzer, 1999). However, the results obtained in 5-HT$_{1A}$ receptor *knockout* mice indicate that 5-HT$_{1A}$ receptors play a permissive role in the increase of DA release produced by atypical antipsychotics, which does not seem to depend on blockade of DA D$_2$ receptors, as this property is not shared by haloperidol (Díaz-Mataix et al., 2005) (Fig. 7).

Converging preclinical and clinical evidence has increasingly drawn attention on the role of 5-HT$_{1A}$ receptors, suggesting that the combined antagonist activity at DA D$_2$ receptors with an agonist activity at 5-HT$_{1A}$ receptors offers a promising strategy for the design of novel antipsychotic with a wide spectrum of action (Bantick et al., 2001; Meltzer et al., 2003). In addition to the above preclinical results, a key observation in support of this strategy is that the 5-HT$_{1A}$ receptor partial agonist tandospirone augmented the effect of a low haloperidol dose in two small pilot studies, (Sumiyoshi et al., 2001a, b). Both studies showed a beneficial effect of tandospirone augmentation on cognitive function over placebo when added to a conventional antipsychotic treatment alone. Moreover, a new generation of antipsychotic drugs is in development, which selectively targets DA D2 and 5-HT$_{1A}$ receptors without significant interactions with other pharmacological sites. These drugs, now undergoing clinical development, include bifeprunox (Wolf, 2003; Assié et al., 2005), SSR181507 (Claustre et al., 2003; Cosi et al., 2006), and SLV313 (Glennon et al., 2002) as well as the antidyskinetic agent, sarizotan (Bartoszyk et al., 2004). However, only sparse information is available concerning the pharmacological properties of these agents, and the relative influence of serotonergic and dopaminergic transmission remains still poorly known (Assié et al., 2005).

Figure 7

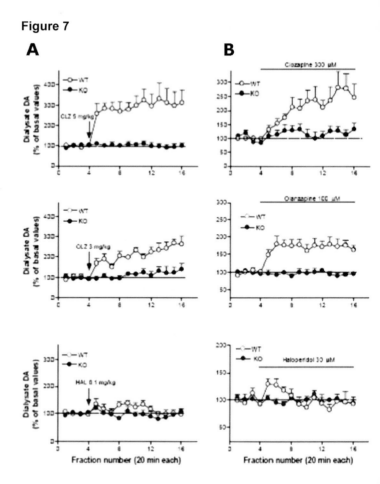

Effects of the **(A)** intraperitoneal administration or **(B)** local application of antipsychotic drugs on extracellular DA levels in the mPFC of control (wild-type, WT) mice (o) or 5-HT_{1A} receptor knockout (KO) (•) mice. In WT mice, the systemic and local administration of clozapine (CLZ) and olanzapine (OLZ) increased the DA release in the mPFC. However none of these effects was observed when the antipsychotic drugs were applied in 5-HT_{1A} KO mice. Haloperidol (HAL) induced a moderate and transient increase of cortical DA levels in WT mice after local (but not systemic) administration. Data are expressed as means ± SEM, n = 6-9 mice for genotype and treatment. Redrawn from Díaz-Mataix et al. (2005).

2.3.2.3 Other serotonin receptors: 5-HT_{2C}, 5-HT_6, 5-HT_7 involved in antipsychotic drug action

2.3.2.3.1 5-HT_{2C} receptors

Many drugs that bind to 5-HT_{2A} receptors with high affinity also bind to some extent to the structurally related HT_{2C} receptor. This receptor has received relatively little attention in psychopharmacology. This may be partly due to the inadequacy of specific techniques (e.g., radioligand binding and immunocytochemistry) to determine the regional and cellular expression of 5-HT$_{2C}$ receptors in brain tissue. However evidence from a variety of sources has implicated this receptor in several important physiological and psychological processes including motor function, anxiety, feeding behavior and appetite. In addition, some interesting pharmacogenetic observations have focused on the potential importance of this receptor in several aspects of antipsychotic drug actions (Meltzer et al., 2003; Reynolds, 2004, Reynolds et al., 2005; Giorgetti and Tecott, 2004).

The 5-HT$_{2C}$ receptor is found in the prefrontal cortex, limbic structures, striatum and in VTA and substantia nigra, where they have a key role in regulating the tonic activity of DA neurons (Pompeiano et al., 1994; Abramowski et al., 1995, Mengod et al., 1996; Di Matteo et al., 2001). Like 5-HT$_{2A}$, these receptors are excitatory and positively coupled receptors via Gq to phospholipase A$_2$ and C. Thus, 5-HT$_{2C}$ receptor stimulation leads to an accumulation of inositol phosphates and Ca^{2+} mobilization in cells expressing this receptor. Interestingly, the 5-HT$_{2C}$ receptor was the first monoaminergic receptor for which an agonist-directed trafficking of intracellular signaling was discovered (Berg et al., 1998) and recent studies have revealed a constitutive activity of 5-HT$_{2C}$ receptors that modulate DA release (De Deurwaerdere et al., 2004).

Striatal 5-HT$_{2C}$ receptors exert both tonic and phasic facilitatory control on basal DA release (Lucas and Spampinato, 2000). 5-HT$_{2C}$ receptors also appear to mediate the tonic inhibitory serotonergic tone on DA neurons in the VTA (Prisco et al., 1994; Millan et al., 1998; Di Matteo et al., 1999; Gobert et al., 2000a). Consequently, the systemic administration of 5-HT$_{2C}$ receptor antagonists can directly increase DA release in the nucleus accumbens and in the PFC (Di Matteo et al., 1998) and 5-HT$_{2C}$ agonists such as Ro 60-0175 markedly suppress dialysate levels of DA and noradrenaline in the frontal cortex of rats (Millan et al., 1998). Stimulation of DA release in these areas has been demonstrated with SB242084 (Millan et al., 1998) and SB206553 (Di Matteo et al., 1998; Gobert et al., 2000a). Most atypical antipsychotic drugs show high affinity for 5-HT$_{2C}$ receptors (see Table 1). Specifically, clozapine reverses the inhibition of accumbal DA release induced by the 5-HT$_{2C}$ agonist Ro 60-0175 (Di Matteo et al., 2002). It is noteworthy that clozapine, like several antipsychotics drugs, behaves as a 5-HT$_{2C}$ inverse agonist in heterologus expression *in vitro* (Herrick-Davis et al., 2000). A recent study suggests that clozapine alters the constitutive activity of 5-HT$_{2C}$ receptors and may be behaving as a 5-HT$_{2C}$ inverse agonist *in vivo* (Navailles et al., 2006).

Thus, the combination of 5-HT$_{2A}$ and 5-HT$_{2C}$ receptor blockade may be a more efficient way of augmenting antipsychotic action than blocking either receptor alone. Despite this evidence, at present, there are no selective 5-HT$_{2C}$ receptor antagonists or mixed 5-HT$_{2A/2C}$ receptor antagonists in development for the treatment of schizophrenia, possibly because the extended feeling that 5-HT$_{2C}$ receptors mediate some of the undesirable effects of atypical antipsychotics such as weight gain. This view is supported, among other observations by the association between antipsychotic-induced weight gain and the 759C/T promoter region polymorphism of the 5-HT$_{2C}$ receptor gene (Reynolds et al., 2005).

2.3.2.3.2 5-HT$_6$ and 5-HT$_7$ receptors

5-HT$_6$ and 5-HT$_7$ receptors have been suggested to be involved in schizophrenia (East et al., 2002). Both are putative targets of atypical antipsychotics (Roth et al., 1994; Branchek and Blackburn, 2000) and a 5-HT$_6$ receptor polymorphism (267C/T) has a minor effect on clozapine response (Yu et al., 1999). The same polymorphism has also been associated with schizophrenia (Tsai et al., 1999) although this finding has not been replicated (Masellis et al., 2001). A role for 5-HT$_6$ and 5-HT$_7$ receptors in schizophrenia is also suggested by their predominantly cortical and limbic distribution, although 5-HT$_6$ receptors are also densely expressed in dorsal striatum (Gerard et al., 1996, 1997). The few existing human data show mRNA distributions for both receptors similar to the rodent (Bard et al., 1993; Kohen et al., 1996).

SB271046, a selective 5-HT$_6$ receptor antagonist, can improve cognition (Rogers and Hagan, 2001) possibly via facilitation of cortical and hippocampal glutamatergic activity and acetylcholine (ACh) release (Dawson et al., 2001). Several studies suggest that 5-HT6 receptors may play a role in the control of cholinergic transmission. For example, 5-HT$_6$ receptors are not located on 5-HT or DA neurons, but possibly on cholinergic and GABAergic neurons (Branchek and Blackburn, 2000). Scopolamine, a nonselective muscarinic receptor antagonist, reversed "stretching" behavior in rats produced by 5-HT$_6$ receptor antagonists (Bentley et al., 1999) or its gene deficiency in mice (for a review, see Branchek and Blackburn, 2000).

Clozapine and 8-OH-DPAT, both of which increase ACh release in the mPFC (Ichikawa et al., 2002a, b), have appreciable affinity (K$_i$ of 61 nM and 52 nM, respectively) for 5-HT$_7$ receptors (Ruat et al., 1993). However, clozapine is a 5-HT$_7$ receptor antagonist and R(+)-8-OH-DPAT appears to be a 5-HT$_7$ receptor agonist (Meltzer et al., 2003) as well as a 5-HT$_{1A}$ receptor agonist. There is no clear evidence to date that 5-HT$_7$ receptor stimulation enhances ACh release (Nakai et al., 1998). Furthermore, the 5-HT$_{2A/2C}$ receptor antagonist ritanserin, which has also high affinity for 5-HT$_7$ receptors compared with that of clozapine or 8-OH-DPAT (Ruat et al., 1993), did not affect ACh release in the mPFC (Consolo et al., 1996).

2.3.2.4 The α-adrenoceptor modulation hypothesis of antipsychotic atypicality

2.3.2.4.1 Significance of α$_1$-adrenoceptor antagonistic activity for the antipsychotic effect

Although many typical and atypical antipsychotic drugs (see Table 1) possess α$_1$-adrenoceptor blocking properties, the putative significance of this effect for their clinical efficacy in schizophrenia has remained unclear (Peroutka and Snyder, 1980; Cohen and Lipinski, 1986). Noradrenergic (NA) neurotransmission, via actions in the PFC and other corticolimbic loci, play an important role in the control of mood and cognition (Arnsten, 1997, Coull et al., 1997). Furthermore, a perturbation of noradrenergic transmission has been related to psychotic states, intensification of negative symptoms, and the risk of relapse after treatment discontinuation (Maas et al., 1993). A previous clinical, double-blind, placebo-controlled trial of the α$_1$-adrenoceptor antagonist prazosin alone in schizophrenia showed no effect, although its limited penetration of the blood–brain barrier in man makes definitive conclusions from this study difficult (Hommer et al., 1984).

Several sets of experimental data demonstrate the existence of deep and complex interactions between brain noradrenergic and dopaminergic neurotransmission. To start with, one should not forget that DA is the precursor of NA in noradrenergic neurons. Both DA and NA are stored in synaptic vesicles and can be co-released by noradrenergic axons (Devoto et al., 2004). Further, reciprocal receptor-mediated interactions have been described. Thus, the systemic administration of prazosin has been found to exert an inhibitory effect on DA neuron activity (Grenhoff and Svensson, 1993). Moreover, electrical stimulation of the noradrenergic nucleus locus coeruleus (LC) has been shown to exert a α_1-adrenoceptor mediated stimulatory effect on VTA DA neurons (Grenhoff et al., 1993). Interestingly, prazosin can inhibit the MK-801-induced hyperlocomotion and DA release in the nucleus accumbens (Mathe et al., 1996). Thus, the blockade of α_1-adrenoceptor preferentially suppresses mesolimbic vs. nigrostriatal DA transmission (Svensson et al., 1995) and facilitates thalamic gating of sensory input to the cortex, a process compromised in psychotic patients (McCormick and Pape, 1990; Bakshi and Geyer, 1997). In contrast, α_1-adrenoreceptor blockade by prazosin potentiates the antipsychotic effects of the DA D_2 antagonist raclopride in rats (Wadenberg et al., 2000).

Additionally, the neocortex is enriched in various α_1-adrenoceptor subtypes (α_{1A}, α_{1B}, α_{1D}) (McCune et al., 1993; Pieribone et al., 1994; Day et al., 1997). The stimulation of α_1-adrenoceptors activates phospolipase C, which results in IP3 production and mobilization of Ca^{2+} stores (Molinoff, 1984; Claro et al., 1993; Bartrup and Newberry, 1994; Berg et al., 1998). α_1-Adrenoceptors mediate the excitatory actions of NA on pyramidal neurons of mPFC (Araneda and Andrade, 1991; Marek and Aghajanian, 1999). The axons of prefrontal pyramidal neurons project to various subcortical brain areas, including the brainstem aminergic nuclei, and controls the activity of these neuronal groups (Aghajanian and Wang, 1977; Thierry et al., 1983; Sesack et al., 1989; Takagishi and Chiba, 1991; Sesack and Pickel, 1992; Murase et al., 1993; Hajós et al., 1998;; Jodo et al., 1998; Peyron et al., 1998). In particular, the mPFC controls the activity of brainstem serotonergic neurons (Hajós et al., 1998; Celada et al., 2001; Martín-Ruiz et al., 2001). In a recent study, Amargós-Bosch et al. (2003) reported that: (a) the activation of α_1-adrenoceptor in mPFC increases the local release of serotonin (5-HT) by an impulse-dependent mechanism, and (b) typical (haloperidol and chlorpromazine) and atypical (clozapine and olanzapine) antipsychotic drugs reduce the basal 5-HT release and reverse the effect of α_1-adrenoceptor activation, an observation possibly related to their therapeutic actions (Fig. 8).

Figure 8

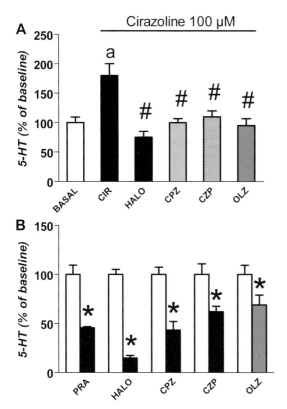

(A) Reversal of the increase in prefrontal 5-HT release produced by the local perfusion of cirazoline (CIR) 100 μM in mPFC by the co-perfusion of 300 μM of the classic antipsychotics haloperidol (HALO, n=5), chlorpromazine (CPZ, n=6) and atypical clozapine (CZP, n=6) and olanzapine (OLZ, n=4). Shown is also the effect of cirazoline alone (n=8). (B) Effects of the perfusion of prazosin (PRA 100 μM, n=5), haloperidol (HAL 300 μM, n=4), chlorpromazine (CPZ 300 μM, n=4), clozzpine (CLZ 300 μM, n=4) and olanzapine (OLZ 300 μM, n=4) on non-stimulated dialysate 5-HT in mPFC of rat. Data are averaged 5-HT values over the last four samples (once the effect was stabilized) and expressed as the percentage change from the corresponding basal (predrug). a $P< 0.05$ vs basal, # $P< 0.05$ vs cirazoline alone, * $P<0.05$ vs. basal. Redrawn from Amargós-Bosch et al. (2003).

From the above considerations, the potential interest of antipsychotics sharing the adrenergic receptor profile appears evident. Thus, α_1-adrenoceptor blockade may contribute to the therapeutical effects of antipsychotic drugs, i.e. reducing an increased excitability of prefrontal pyramidal neurons and inhibits dorsal raphe-derived serotonergic pathways (Millan et al., 2000; Amargós-Bosch et al., 2003). However, the lack of selective α_1-

adrenoreceptor agents devoid of cardiovascular actions limits the applicability of this approach.

2.3.2.4.2 Significance of α_2-adrenoceptor antagonistic activity for the antipsychotic effect

In contrast to most antipsychotic drugs, atypical or classic, clozapine possesses a high affinity for α_2 adrenoceptor (Ashby and Wang, 1996) and clinical trials demonstrate that adjuvant treatment with α_2 adrenoceptor antagonist may augment the clinical efficacy of classic D_2-antagonists (Litman et al., 1996). Accordingly, the α_2 adrenoceptor blocking effect of clozapine and also, to some extent, risperidone, has been hypothesized to be important for their clinical profiles. The mechanisms are not fully understood, but it has been previously shown that the combination of α_2 adrenoceptor antagonist idazoxan with the $D_{2/3}$ receptor antagonist raclopride, similarly to clozapine, increased DA efflux in the PFC and also enhanced suppression of the conditioned avoidance response (CAR, an animal model of antipsychotic activity; Hertel et al., 1997, 1999a).

In contrast, the systemic administration of the non-competitive NMDA receptor antagonist MK-801, markedly disturbs the activity of DA neurons in the VTA, generating an impaired prefrontal DA function concurrently with a hyperactive mesolimbic DA projection. This occurs in parallel with a significant regularization of the firing pattern of the DA neurons (Svensson, 2003). Interestingly, idazoxan, a α_2 adrenoceptor antagonist selectively modulates the firing variability of VTA DA neurons and causes a major increase in DA output in the mPFC but not in the NAc (Hertel et al., 1999b). Neurochemical studies had shown that clozapine and other atypical antipsychotic drugs, in contrast to conventional neuroleptics, cause a marked increase in DA release in rodent mPFC (Moghaddam and Bunney, 1990, Nomikos et al., 1994; Rollema et al., 1997, Ichikawa et al., 2001; Díaz-Mataix et al., 2005), an area of significant importance for cognitive functions, which are fundamental for social outcome in schizophrenia (Arnsten et al., 1994; Sawaguchi and Goldman-Rakic, 1994). Both the augmenting effect of idazoxan on variability of firing of the DA neurons and its selective activation of prefrontal DA output suggests that the α_2 adrenoceptor antagonistic effect of clozapine may substantially contribute to its clinical effects in schizophrenia (Devoto et al., 2003), in addition to the involvement of 5-HT$_{1A}$ receptors (see above).

Since idazoxan does not block DA receptors these results provide a substantial challenge to the original DA hypothesis of schizophrenia (see above), as an enhanced antipsychotic effect thus may be obtained by an actual increase of dopaminergic neurotransmission in the PFC, in spite of an only modest degree of D$_2$ receptor blockage. Recently, it has been reported that the adjunctive treatment with idazoxan to low doses of a typical (haloperidol) or an atypical (olanzapine) antipsychotic drugs, both lacking noticeable α$_2$ adrenoceptor affinity, enhanced suppression of CAR, increased DA output in the prefrontal cortex, and reversed haloperidol-induced catalepsy (Wadenberg et al., 2006). Additionally, the adjunct treatment with the selective NA reuptake inhibitor reboxetine was found to increase the antipsychotic-like effect of raclopride and preferentially increased DA release in the mPFC (Linnér et al., 2002). The increase in prefrontal DA output, also seen with other NA reuptake inhibitors (Carboni et al., 1990; Pozzi et al., 1994; Westerink et al., 2001), may be related to the fact that a significant proportion of DA in the PFC appears to be removed from the extracellular space by the NA transporter (Pozzi et al., 1994; Gresch et al., 1995; Linnér et al., 2002). Indeed, DA is coreleased with NA from noradrenergic axon terminals containing α$_2$ adrenoceptors in certain brain areas lacking a prominent dopaminergic innervation (e.g., parietal cortex; Devoto et al., 2004), although in PFC, extracellular DA may arise from both dopaminergic and noradrenergic axon terminals.

Clinically, initial attempts to use reboxetine as add-on therapy to haloperidol in the treatment of schizophrenia failed to show any augmentation (Schutz and Berk, 2001). However, this was effective in patients with prominent negative symptoms who showed significant improvement on all clinical ratings (Raedler et al., 2004). The novel compound S-18327 displays marked antagonist properties at α$_1$- and α$_2$ adrenoceptors and displays a broad-based pattern of potential antipsychotic activity at doses appreciably lower than those eliciting EPSs (Millan et al., 2000). Antagonism by S-18327 of α$_2$ adrenoceptor enhances noradrenergic transmission and reinforces front cortical DA pathways, whereas blockade of α$_1$-adrenoceptor inhibits dorsal rape serotonergic pathways, although the putative therapeutic significance of the α$_1$-

and α_2 adrenoceptors antagonistic activity remains poorly known.

2.3.2.5 Other hypothesis of atypicality

2.3.2.5.1 The "rapid dissociation" hypothesis of atypical antipsychotic drug actions

Clinically effective doses of antipsychotic drugs occupy 60–80% of brain striatal DA D_2 receptors in treated patients, as measured by PET or SPECT scan (Martinot et al., 1996; Dresel et al., 1999; Meisenzahl et al., 2000; Schmitt et al., 2002; Tauscher et al., 2002). Clozapine and quetiapine, however, produce a substantially lower occupancy. For instance, patients taking therapeutic doses of clozapine or quetiapine, show an occupancy of 40–60% of striatal dopamine D_2 receptors. This might indicate that only a part of their antipsychotic efficacy is explained by D_2 receptor blockade, and interactions with the 5-HT system or other transmitter systems may account for the rest, as shown in Fig. 4.

However, another interesting hypothesis of atypicality has been proposed by S. Kapur and colleagues. This hypothesis is mainly based on the fact that amisulpride, an atypical antipsychotic used in Europe is a selective DA D2/D3 receptor antagonist – with no known affinity for other monoaminergic receptors – which does not show EPS or other motor side-effects. Hence, Kapur's view is that compounds with a fast dissociation from D_2 receptors may display antipsychotic affinity without concurrently showing EPS, prolactin elevation or secondary negative symptoms, all them produced by a strong blockade of DA D_2 receptors (Kapur and Seeman, 2001). This may be the case for amisulpride and remoxipride. Hence, [^{3}H]clozapine, [^{3}H]quetiapine, [^{3}H]remoxipride and [^{3}H]amisulpride dissociate from human cloned DA D_2 receptors at least 100 times faster than [^{3}H]haloperidol, with olanzapine and sertindole having intermediate rates (Seeman and Tallerico 1999; Kapur and Seeman, 2001).

Hence, the fast dissociation or "fast-off" theory of atypicality is based on the assumption that atypical drugs having low affinity for the dopamine D_2 receptors are loosely bound and would rapidly dissociate from these receptors (Kapur and Seeman 2001; Möller 2005). In contrast, the 5-HT$_{2A}$ occupancy may not be a necessary condition for atipycality or even antipsychotic action. Indeed, two reports show that fanaserin (Truffinet et al., 1999) and M100907 (Potkin et al., 2001), both with high 5-

HT_{2A} occupancy but devoid of D_2 occupancy, showed a very moderate, if any, antipsychotic activity. However, If DA D_2 receptor occupancy is excessive, atypicality is lost even in the presence of high 5-HT_2 occupancy (Kapur et al., 1998, 1999) as it occurs with moderate-high doses of risperidone. Thus, according to this hypothesis the multireceptor profile of atypical antipsychotic drugs is not required for low incidence of EPS and other dopaminergic side-effects. This does not imply that other receptors and neurotransmitter pathways can also be involved (Kapur and Mamo, 2003).

2.3.2.5.2 The D_2/D_3 only hypothesis

All antipsychotic agents, irrespective of their overall receptor binding profiles, interact with dopaminergic mechanisms that are presumably perturbed in schizophrenic patients. DA exerts its actions via five different receptors, offering a broad palette of targets for the conception of novel antipsychotic agents. Based on the atypical features of several drugs with predominant actions at DA D_2 and DA D_3 receptors including amisulpride, remoxipride and sulpiride, a D_2/D_3-only hypothesis had been put forward for atypical antipsychotic drug actions (Roth et al., 2003; Millan, 2005). Experimental studies suggest that, as compared to other drugs, antipsychotic agents which preferentially block presynaptic D_2/D_3 receptors lead to an enhanced release of DA from dopaminergic terminals, an effect that may improve negative symptoms and cognitive deficits whereas the concurrent blockade of the same postsynaptic receptors in limbic structures (e.g., nucleus accumbens, islands of Calleja, etc.) may improve cognitive symptoms.

There is good clinical evidence supporting the view that amisulpride is effective in treating schizophrenia with few EPS, having been classified as an atypical antipsychotic drug (Leucht et al., 2002). Amisulpride has high affinity for D_2 and D_3 receptors. Sulpiride and remoxipride are other atypical antipsychotic drugs with relative selectivity for DA D_2 and D_3 receptors and are virtually devoid of actions at tested 5-HT receptors including 5-HT_{2A}. Additionally, these drugs might rapidly dissociate from DA D_2 receptors, as it occurs with some compounds of this kind (e.g., amisulpride, remoxipride, etc.; Roth et al., 2003).

2.3.2.5.3 Partial agonism at DA D_2 receptors

The idea of using partial DA D_2 receptor agonists was initially put forward by Arvid Carlsson in the early 1980s but the compounds developed never reached the clinical setting (for a review, see Carlsson et al., 2004). Aripiprazole (OPC-14597) is distinct from all other known antipsychotic drugs by virtue of its partial agonist effect at a number of G protein-coupled receptors including DA D_2, D_3 and D_4 as well as 5-HT_{1A} receptors. In contrast, it has an antagonist action at 5-HT_{2A} and 5-HT_{2C} serotonin receptors (Jordan et al., 2002; Burris et al., 2002). Aripiprazole is a partial agonist with relative high efficacy (~40–80% of DA) at DA D3 and DA D4 and 5-HT_{1A} receptors and moderate efficacy (~20–30% of DA) at DA D_2 receptors. Additionally, it has a weak efficacy at 5-HT_{2A} receptors (<5% of 5-HT), and lacks noticeable affinity for D_1 receptors (Kikuchi et al., 1995; Semba et al., 1995; Lawler et al., 1999; Roth et al., 2003). Concerning the dopaminergic system, aripiprazole decreased striatal DA release (Semba et al., 1995) and inhibited the DA neurons when applied locally to the rat VTA (Momiyana et al., 1996) or did not evoke consistent changes when applied systemically (Bortolozzi et al., unpublished observations).

Animal behavioral studies showed that aripiprazole exhibited a weak cataleptogenic effect compared to haloperidol and chlorpromazine (typical antipsychotic drugs) despite the fact that it has almost identical D_2 receptor affinity (Kikuchi et al., 1995). Aripiprazole has been marketed in most countries, including the USA, Europe and Japan, after completing clinical trials for registration. Early clinical studies already showed efficacy in treating both positive and negative symptoms of schizophrenia.

Clinically effective doses of conventional antipsychotic drugs correlate well with their affinity for binding to and blocking post-synaptic DA D_2 receptors (Seeman et al., 1976). Yet, as described above, this effect is associated with the incidence of EPS and prolactin elevation. These side-effects are not seen with aripiprazole despite it produces antipsychotic effects that would be equivalent to 70–80% occupancy of D_2 receptors by an antagonist (Petrie et al., 1998). Actually, PET scan data shows that DA D_2 receptor occupancy by aripiprazole can reach up to 95% in treated patients without showing EPS (Grunder et al., 2003). Based on available data, it appears that aripiprazole is the first compound with partial D_2 agonist properties to be clinically effective antipsychotic effect

(Tamminga, 2005). Similarly, S-33592 is a benzopyrano-pyrrole partial agonist at D_2/D_3 receptors, has shown a promising antipsychotic profile in the preclinical study (Gobert et al., 2000b).

2.3.3 Other major investigational approaches

2.3.3.1 Modulation of glutamatergic neurotransmission

The postulated hypofunction of NMDA receptors in schizophrenia has given rise to the use of potential therapeutic agents that enhance NMDA receptor function (for a review, see Coyle and Tsai, 2004). These interventions have focused on agents that would activate the glycine modulatory site on the NMDA receptor (glycine B receptor), thereby avoiding the potential excitotoxic effects of direct agonists at the glutamate recognition site. Preclinical studies have shown that such drugs can reverse the behavioral effects of MK-801 and have cognitive enhancing effects. One of the first agents examined, D-cycloserine, is partial agonist at the glycine site with approximately 50% efficacy. Placebo-controlled trials with drugs that directly or indirectly activate the glycine modulatory site on the NMDA receptor have shown a reduction in negative symptoms, improvement in cognition and in some cases, reduction in positive symptoms, in schizophrenic patients receiving concurrent antipsychotic medications (Marek and Merchant, 2005; Coyle, 2006). Trials with glycine, with doses ranging from 30 g to 60 g per day, also revealed significant reductions in negative symptoms and improvement in cognitive functions without effect on positive symptoms (Heresco-Levy et al., 1999). A trial with the full agonist, D-serine at 2 g per day demonstrated improvement not only in negative symptoms and cognition but also in positive symptoms (Tsai et al., 1998). This impact on positive symptoms in contrast to D-cycloserine and glycine may reflect better brain access and greater efficacy of D-serine at the glycine B receptor. Finally, sarcosine, an endogenous antagonist at the glycine transporter 1 (GlyT1), was also reported to reduce negative symptoms, improve cognitive and reduce positive symptoms in patients suffering from chronic schizophrenia receiving neuroleptic treatment (Tsai et al., 2004). It should be also noted that a relatively large multicenter, double blind, placebo-controlled study of D-cycloserine and glycine failed to show symptomatic improvement for negative or cognitive symptoms (Carpenter et al., 2004).

A number of studies have indicated that administration of relatively low doses of NMDA antagonists induces behavioral and brain metabolic activation in experimental animals and human. Consistent with these data, non-competitive NMDA antagonists increase glutamate release in the PFC of rats (Moghaddam et al., 1997). Increased glutamate levels could mediate some of the behavioral actions of the drugs by activation of non-NMDA receptors, including α-amino-3-hydroxy-5-methyl-isoxazole-4-propionic acid (AMPA) and kainite receptors (Moghaddam et al., 1997). In support of this, the AMPA/kainite receptor antagonists (e.g., GYKI 52466) reduced NMDA antagonist-induced hyperlocomotion (Willins et al., 1993) and neurodegeneration (Sharp et al., 1995). These data suggest that AMPA/kainite receptor antagonists may have utility for treatment of cognitive deficits in which NMDA receptor hypofunction is suspected (Moghaddam et al., 1997).

In apparent contrast to the postulated utility of AMPA/kainite receptor antagonist as antipsychotics, ampakines, a class of compounds that allosterically enhances AMPA receptor function, have also been suggested to represent potential adjunctive treatments of schizophrenia. Preliminary results suggest that chronic administration of an ampakine (CX-516) can improve negative and cognitive symptoms in schizophrenia patients who also received clozapine (Goff et al., 1999). Such findings appear to be paradoxical with the observations that AMPA antagonists can reduce the effects induced by NMDA receptor hypofunction in some preclinical models. Further clinical experience on NMDA and AMPA receptor modulators is necessary to clarify the therapeutic potential of these targets in the treatment of schizophrenia.

Finally, metabotropic glutamate receptors have been also considered as potential targets in the treatment of schizophrenia (Moghaddam, 2003) and recent data are suggestive that one such compound (Ly-354740, a mGluR II agonist) partly reverses the cognitive deficits produced by NMDA receptor blockade in humans (Krystall et al., 2005).

2.3.3.2 Dopamine D$_4$ receptor antagonists

This receptor, which belong to the D2 (or D2-like) receptor family, is also a target for various antipsychotic drugs (Oak et al., 2000). Soon after the D$_4$ receptor was identified and cloned, it was realized that clozapine had a high affinity for this receptor (Van Tol et al., 1991). It was also observed that the D$_4$ receptor was more prominently expressed in limbic regions like frontal cortex and hippocampus than in motor brain regions (e.g., dorsal striatum). The high affinity of clozapine and the anatomical distribution of the D$_4$ receptor were features that made it a potential target in the development of novel antipsychotic medications that would not induce the debilitating motor side-effects of classic neuroleptics. Yet to date this receptor has not fulfilled this promise (Kramer et al., 1997; Truffinet et al., 1999) and currently it seems unlikely that drugs targeting the D$_4$ receptor alone have significant antipsychotic properties (Wong and Van Tol, 2003). A relatively small phase II study reported a slight worsening of patients relative to the placebo group following treatment with the selective DA D$_4$ receptor antagonist L-745,870 (Kramer et al., 1997) These results were confirmed in a recent, large multicenter, placebo- and active comparator (olanzapine)-controlled, study testing a 40-fold dose range of sonepiprazole, a Pfizer (USA) DA D$_4$ receptor antagonist. Given these results, it is not surprising that a DA D$_{4/5}$-HT$_{2A}$/α_1 adrenergic receptor antagonist also did not exhibit antipsychotic efficacy in a relatively small (97 patients) phase II study in the treatment of schizophrenia patients (Marek and Merchant, 2005).

2.3.3.3 Neurokinin 3 receptor antagonism

Another mechanism of action that may be associated with a relatively modest degree of efficacy is the blockade of neurokinin 3 (NK$_3$) receptors. Recent data from clinical trials of selective neurokinin 3 (NK$_3$) receptor antagonists in schizophrenia have shown a significant improvement in positive symptoms, with no reported major side-effects. This might represent a new approach for the treatment of schizophrenia and possibly other neuropsychiatric disorders.

The clinical efficacy of two NK$_3$ receptor antagonists — osanetant and talnetant — has been evaluated in double-blind, placebo-controlled clinical trials in schizophrenic patients. Interestingly, these two agents represent distinct chemical classes and their pharmacological properties are also markedly different, albeit they share an antagonist character at NK_3 receptors and clinical efficacy in schizophrenia (Marek and merchant, 2005; Spooren et al., 2005).

2.3.3.4 Neurotensin NTS1 antagonist

The placebo-controlled meta-trial using haloperidol as a positive comparator that demonstrated moderate efficacy for the Sanolfi-Aventis NK_3 receptor antagonist and 5-HT_2 receptor antagonist failed to see any efficacy for a neurotensin NTS1 antagonist (SR48692) (Meltzer et al., 2004). Therefore, it is unlikely that this receptor may become a new therapeutic target in schizophrenia.

2.3.3.5 Cannabinoid CB_1 receptor antagonism

There appears to be a relationship between abuse to cannabis and risk of schizophrenia (Degenhardt and Hall, 2002). Moreover, schizophrenic patients abusing cannabis appeared to have a higher degree of psychopathology (Meltzer et al., 2004). This has led to test the cannabinoid-1 (CB_1) receptor antagonist, SR141716 (rimonabant), in a clinical trial. Rimonabant was used at a dose (20 mg/day) which is efficacious at reducing weight, yet it did not show any antipsychotic effect (Marek and Merchant, 2005).

2.3.3.6 Muscarinic, nicotinic and other miscellaneous targets

Muscarinic receptors are anatomically positioned in cortical and subcortical areas and modulate dopaminergic and glutamatergic neurotransmission. Neurochemical studies have shown that DA and muscarinic receptors reciprocally modulate each another. Hence, the muscarinic receptor agonist xanomeline increases extracellular DA levels and Fos expression in cortical areas more than in subcortical areas. Indeed, N-desmethylclozapine, a major metabolite of clozapine, is a M1 receptor agonist and can contribute to the release of cortical acetylcholine and DA release (Li et al., 2005). In electrophysiological studies, acute and chronic administration of xanomeline decreased the activity of the mesocorticolimbic dopamine A10 tract, but not the nigrostriatal dopamine A9 tract. Behavioral data also showed that muscarinic agonists inhibit conditioned-avoidance response (CAR) and DA-agonist-induced behaviors including hyperactivity, climbing behavior and disruption of prepulse inhibition, which are models for positive symptoms of schizophrenia. Furthermore, muscarinic agonists are active in animal models of cognitive dysfunction and affective disorders, symptoms that are prominent in schizophrenic patients. Preliminary clinical investigation indicates that muscarinic agonists like xanomeline may also be effective in the pharmacotherapy of schizophrenia and perhaps other neuropsychiatric disorders (Bymaster et al., 2002).

Further, a number of additional investigational therapies are currently in progress. Thus, a small phase II study is underway for the α_7 nicotinic agonist 3-2,4 dimethoxy benylidene to improve cognitive dysfunction in schizophrenic patients employing the MATRICS battery (www.ClinicalTrials.gov identifier NCT00100165).

The National Institute of Mental Health (NIMH) is sponsoring a USA study examining whether the pharmacogenomic status may be informative for the effects of the COMT inhibitor tolcapone. Other investigational putative antipsychotic agents currently in phase III include bifeprunox (Wyeth & Solvay Pharmaceuticals, USA; DU-127090, a partial agonist at dopamine D_2 and 5-HT_{1A} receptors), paliperidone extended release (Johnson & Johnson; an active risperidone metabolite), asenapine (Organon, Netherlands, USA; Pfizer; ORG 5222, a 5-HT_2 antagonist/dopamine D_2 partial agonist). Other phase II investigational compounds include ORG 24448 (Organon, Cortex, USA; an AMPA potentiator), ACP-103 (Acadia, USA; a 5-HT_{2A} inverse agonist; www.ClinicalTrials.gov identifier NCT00087542), lonasen (blonanserin, Dainippon Pharmaceuticals, Japan; a D_2/5-HT_{2A} receptor antagonist), talnetant (the GlaxoSmithKline NK_3 receptor antagonist discussed above, SB-223412), secretin (Repligen RG1068 endogenous pancreatic hormone for psychosis and autism) (Marek and Merchant, 2005).

2.3.4 Concluding remarks: challenges in drug discovery

The first family of antipsychotic drugs appeared during the 1950s and some members are still used 50 years later due to their robust effectiveness for the acute and chronic treatment of schizophrenia and related psychotic disease. These agents are successful in improving the symptoms of schizophrenia in 60–70% of patients. However, the toll to pay with these agents is a the incidence of severe side-effects (EPS, hyperporlactinemia, tardive dyskinesia, etc.) in up to 40 % of the patients. The so-called "atypical" drugs (whatever mechanism is involved in "atypicality") show an equal or slightly better efficacy than older drugs and present with a different profile of side-effects, in which weight gain and metabolic problems rank first, and some drugs, like risperidone, show EPS at relatively moderate doses. Nevertheless, they have an overall better tolerance than conventional drugs and they have reached a great commercial success. Hence, to some extent, the situation is similar to that in the antidepressant field: selective serotonin reuptake inhibitors (SSRIs), which make up ~80% of the market worldwide are better tolerated drugs than first-generation tricyclic antidepressants due to the absence of severe side-effects but they are not more effective than the older drugs. With one exception: clozapine. This drug, which is used in a restricted manner due to occasional agranulocitosis, remains the gold standard in the treatment of schizophrenia. Olanzapine also shows some advantage over classic DA D_2 blockers, according to the CATIE study (Lieberman et al., 2005) but overall, both classic and atypical antipsychotics are far from what should be demanded to a true antischizophrenic treatment: full efficacy in all kind of symptoms, including the cognitive deficits which are central to the psychopathology of schizophrenia. Therefore, there is ample room for developing new agents that can overcome the limitations of existing antipsychotics, either conventional or atypical drugs.

The complex and intriguing pharmacological profile of clozapine has been the drive for the development of multitarget agents and it is possible that it continues to be so for a number of years. Thus, ongoing drug development is to a large extent based on "classic" targets, which include DA and other monoaminergic receptors. In this regard, a new generation of DA D_2 antagonists (or partial agonists)/ partial $5-HT_{1A}$ agonists is merging that could substitute some of the current atypical drugs in case these new agents demonstrate: (a) absence of motor and metabolic side-effects and/or (b) higher efficacy than conventional and second generation antipsychotics. Interestingly, new $5-HT_{1A}$ agonists are being developed that can modulate DA release in PFC with an extremely high potency and selectivity. These compounds could exhibit an antipsychotic profile *per se* or be used in add-on strategies. Other approaches, based on selectively targeting DA receptors in limbic areas may also be successful if these drugs can prove regionally selective *in vivo*.

However, future drug discovery approaches need to take risk leaving aside the "classic" targets and taking into account the pathogenesis of the disease in more detail. For instance, genetic research in the last decade has produced important information on the factors that may determine response variability to antipsychotic treatment. Pharmacogenetic research has succeeded in identifying genes with a major contribution into response variability and the combination of genetic information can help in the prediction of response to antipsychotic treatment. However, the wealth of information produced by pharmacogenomic research on genetic factors, both at a sequence and expression level, that influences a treatment response may take longer before it is translated into successful clinical applications (Arranz and Kerwin, 2003; Müller et al., 2003; Wilffert et al., 2005). Nevertheless, genetic information does not cure the illness by itself, so that a great effort will be needed in terms of identifying new therapeutic targets and developing the appropriate compounds to be used in humans. Likewise, considering the neurodevelopmental character of schizophrenia, those drugs should be ideally used to stop the progression of the disease in early stages, before the emergence of psychotic episodes and the subsequent cognitive decline of the patients.

Acknowledgements

This work was supported by the Spanish Ministry of Science and Education (grant SAF-CICYT 2004-05525). Support from the Generalitat de Catalunya is also acknowledged (2005 SGR00758). A. B. is recipient of a Ramón y Cajal contract from the Ministry of Science and Education. L. D.-M. is recipient of a predoctoral fellowship from IDIBAPS.

References

Abramowski D, Rigo M, Duc D, Hoyer D, Staufenbiel M. *Localization of the 5-hydroxytryptamine2C receptor protein in human and rat brain using specific antisera.* Neuropharmacology. **1995**, 34:1635-1645.

Adams BW, Moghaddam B. *Effect of clozapine, haloperidol, or M100907 on phencyclidine-activated glutamate efflux in the prefrontal cortex.* Biol Psychiatry. **2001**, 50:750-757.

Adell A, Artigas F. *The somatodendritic release of dopamine in the ventral tegmental area and its regulation by afferent transmitter systems.* Neurosci Biobehav Rev. **2004**, 28:415-431.

Aghajanian GK, Wang RY. *Habenular and other midbrain raphe afferents demonstrated by a modified retrograde tracing technique.* Brain Res. **1977**, 122:229-242.

Aghajanian GK, Marek GJ. *Serotonin induces excitatory postsynaptic potentials in apical dendrites of neocortical pyramidal cells.* Neuropharmacology. **1997**, 36:589-599.

Agid O, Seeman P, Kapur S. *The "delayed onset" of antipsychotic action--an idea whose time has come and gone.* J Psychiatry Neurosci. **2006**, 31:93-100.

Albanese A, Minciacchi D. *Organization of the ascending projections from the ventral tegmental area: a multiple fluorescent retrograde tracer study in the rat.* J Comp Neurol. **1983**, 216:406-420.

Amaral L, Viveiros M, Molnar J. *Antimicrobial activity of phenothiazines.* In Vivo. **2004**, 18:725-731.

Amargós-Bosch M, Adell A, Bortolozzi A, Artigas F. *Stimulation of alpha1-adrenoceptors in the rat medial prefrontal cortex increases the local in vivo 5-hydroxytryptamine release: reversal by antipsychotic drugs.* J Neurochem. **2003**, 87:831-842.

Amargós-Bosch M, Bortolozzi A, Puig MV, Serrats J, Adell A, Celada P, Toth M, Mengod G, Artigas F. *Co-expression and in vivo interaction of serotonin1A and serotonin2A receptors in pyramidal neurons of prefrontal cortex.* Cereb Cortex. **2004**, 14:281-299.

Amargós-Bosch M, Lopez-Gil X, Artigas F, Adell A. *Clozapine and olanzapine, but not haloperidol, suppress serotonin efflux in the medial prefrontal cortex elicited by phencyclidine and ketamine.* Int J Neuropsychopharmacol. **2006**, 9:565-573.

Araneda R, Andrade R. *5-Hydroxytryptamine2 and 5-hydroxytryptamine 1A receptors mediate opposing responses on membrane excitability in rat association cortex.* Neuroscience. **1991**, 40:399-412.

Arborelius L, Chergui K, Murase S, Nomikos GG, Hook BB, Chouvet G, Hacksell U, Svensson TH. *The 5-HT1A receptor selective ligands, (R)-8-OH-DPAT and (S)-UH-301, differentially affect the activity of midbrain dopamine neurons.* Naunyn Schmiedebergs Arch Pharmacol. **1993a**, 347:353-362.

Arborelius L, Nomikos GG, Hacksell U, Svensson TH. *(R)-8-OH-DPAT preferentially increases dopamine release in rat medial prefrontal cortex.* Acta Physiol Scand. **1993b**, 148:465-466.

Arnsten AF, Cai JX, Murphy BL, Goldman-Rakic PS. *Dopamine D1 receptor mechanisms in the cognitive performance of young adult and aged monkeys.* Psychopharmacology (Berl). **1994**, 116:143-151.

Arnsten AF. *Catecholamine regulation of the prefrontal cortex.* J Psychopharmacol. **1997**, 11:151-162.

Arnt J, Skarsfeldt T. *Do novel antipsychotics have similar pharmacological characteristics? A review of the evidence.* Neuropsychopharmacology. **1998**, 18:63-101.

Arranz MJ, Kerwin RW. *Pharmacogenetic and pharmacogenomic research for the prediction of response to antipsychotics in schizophrenia.* Drug Dev Res. **2003**, 60:104-110.

Ashby CR, Jr., Wang RY. *Pharmacological actions of the atypical antipsychotic drug clozapine: a review.* Synapse. **1996**, 24:349-394.

Assie MB, Ravailhe V, Faucillon V, Newman-Tancredi A. *Contrasting contribution of 5-hydroxytryptamine 1A receptor activation to neurochemical profile of novel antipsychotics: frontocortical dopamine and hippocampal serotonin release in rat brain.* J Pharmacol Exp Ther. **2005**, 315:265-272.

Aznar S, Qian Z, Shah R, Rahbek B, Knudsen GM. *The 5-HT1A serotonin receptor is located on calbindin- and parvalbumin-containing neurons in the rat brain.* Brain Res. **2003**, 959:58-67.

Bakshi VP, Geyer MA. *Phencyclidine-induced deficits in prepulse inhibition of startle are blocked by prazosin, an alpha-1 noradrenergic antagonist.* J Pharmacol Exp Ther. **1997**, 283:666-674.

Bantick RA, Deakin JF, Grasby PM. *The 5-HT1A receptor in schizophrenia: a promising target for novel atypical neuroleptics?* J Psychopharmacol. **2001**, 15:37-46.

Bantick RA, Montgomery AJ, Bench CJ, Choudhry T, Malek N, McKenna PJ, Quested DJ, Deakin JF, Grasby PM. *A positron emission tomography study of the 5-HT1A receptor in schizophrenia and during clozapine treatment.* J Psychopharmacol. **2004**, 18:346-354.

Bard JA, Zgombick J, Adham N, Vaysse P, Branchek TA, Weinshank RL. *Cloning of a novel human serotonin receptor (5-HT7) positively linked to adenylate cyclase.* J Biol Chem. **1993**, 268:23422-23426.

Bardin L, Kleven MS, Barret-Grevoz C, Depoortere R, Newman-Tancredi A. *Antipsychotic-like vs cataleptogenic actions in mice of novel antipsychotics having D2 antagonist and 5-HT1A agonist properties.* Neuropsychopharmacology. **2006**, 31:1869-1879.

Barnes NM, Sharp T. *A review of central 5-HT receptors and their function.* Neuropharmacology. **1999**, 38:1083-1152.

Bartoszyk GD, Van Amsterdam C, Greiner HE, Rautenberg W, Russ H, Seyfried CA. *Sarizotan, a serotonin 5-HT1A receptor agonist and dopamine receptor ligand. 1. Neurochemical profile.* J Neural Transm. **2004**, 111:113-126.

Bartrup JT, Newberry NR. *5-HT2A receptor-mediated outward current in C6 glioma cells is mimicked by intracellular IP3 release.* Neuroreport. **1994**, 5:1245-1248.

Bentley JC, Bourson A, Boess FG, Fone KC, Marsden CA, Petit N, Sleight AJ. *Investigation of stretching behaviour induced by the selective 5-HT6 receptor antagonist, Ro 04-6790, in rats.* Br J Pharmacol. **1999**, 126:1537-1542.

Berg KA, Maayani S, Goldfarb J, Scaramellini C, Leff P, Clarke WP. *Effector pathway-dependent relative efficacy at serotonin type 2A and 2C receptors: evidence for agonist-directed trafficking of receptor stimulus.* Mol Pharmacol. **1998**, 54:94-104.

Björklund A, Lindvall O. Dopamine-containing systems in the CNS. In: Björklund A, Hökfelt T, eds. *Handbook of chemical neuroanatomy. Vol.2: Classical transmitters in the CNS.* ElSevier, Amsterdam, **1984**, pp 55-122.

Bortolozzi A, Amargos-Bosch M, Adell A, Diaz-Mataix L, Serrats J, Pons S, Artigas F. *In vivo modulation of 5-hydroxytryptamine release in mouse prefrontal cortex by local 5-HT(2A) receptors: effect of antipsychotic drugs*. Eur J Neurosci. **2003**, 18:1235-1246.

Bortolozzi A, Diaz-Mataix L, Scorza MC, Celada P, Artigas F. *The activation of 5-HT receptors in prefrontal cortex enhances dopaminergic activity*. J Neurochem. **2005**, 95:1597-1607.

Branchek TA, Blackburn TP. *5-ht6 receptors as emerging targets for drug discovery*. Annu Rev Pharmacol Toxicol. **2000**, 40:319-334.

Bubser M, Backstrom JR, Sanders-Bush E, Roth BL, Deutch AY. *Distribution of serotonin 5-HT(2A) receptors in afferents of the rat striatum*. Synapse. **2001**, 39:297-304.

Burris KD, Molski TF, Xu C, Ryan E, Tottori K, Kikuchi T, Yocca FD, Molinoff PB. *Aripiprazole, a novel antipsychotic, is a high-affinity partial agonist at human dopamine D2 receptors*. J Pharmacol Exp Ther. **2002**, 302:381-389.

Bymaster FP, Calligaro DO, Falcone JF, Marsh RD, Moore NA, Tye NC, Seeman P, Wong DT. *Radioreceptor binding profile of the atypical antipsychotic olanzapine*. Neuropsychopharmacology. **1996**, 14:87-96.

Bymaster FP, Felder C, Ahmed S, McKinzie D. *Muscarinic receptors as a target for drugs treating schizophrenia*. Curr Drug Targets CNS Neurol Disord. **2002**, 1:163-181.

Carboni E, Tanda GL, Frau R, Di Chiara G. *Blockade of the noradrenaline carrier increases extracellular dopamine concentrations in the prefrontal cortex: evidence that dopamine is taken up in vivo by noradrenergic terminals*. J Neurochem. **1990**, 55:1067-1070.

Carlsson A, Lindqvist M. *Effect of chlorpromazine or haloperidol on formation of 3-methoxytyramine and normetanephrine in mouse brain*. Acta Pharmacol Toxicol (Copenh). **1963**, 20:140-144.

Carlsson A, Waters N, Holm-Waters S, Tedroff J, Nilsson M, Carlsson ML. *Interactions between monoamines, glutamate, and GABA in schizophrenia: new evidence*. Annu Rev Pharmacol Toxicol. **2001**, 41:237-260.

Carlsson ML, Carlsson A, Nilsson M. *Schizophrenia: from dopamine to glutamate and back*. Curr Med Chem. **2004**, 11:267-277.

Carpenter WT, Buchanan RW, Javitt DC, Marder SR, Schooler NR, Heresco-Levy V, Gold JM. *Is glutamatergic therapy efficacious in schizophrenia?* Neuropsychopharmacology. **2004**, 29:S110.

Carr DB, Sesack SR. *Projections from the rat prefrontal cortex to the ventral tegmental area: target specificity in the synaptic associations with mesoaccumbens and mesocortical neurons*. J Neurosci. **2000**, 20:3864-3873.

Celada P, Puig MV, Casanovas JM, Guillazo G, Artigas F. *Control of dorsal raphe serotonergic neurons by the medial prefrontal cortex: Involvement of serotonin-1A, GABA(A), and glutamate receptors*. J Neurosci. **2001**, 21:9917-9929.

Chen JF, Qin ZH, Szele F, Bai G, Weiss B. *Neuronal localization and modulation of the D2 dopamine receptor mRNA in brain of normal mice and mice lesioned with 6-hydroxydopamine*. Neuropharmacology. **1991**, 30:927-941.

Chou YH, Halldin C, Farde L. *Occupancy of 5-HT1A receptors by clozapine in the primate brain: a PET study*. Psychopharmacology (Berl). **2003**, 166:234-240.

Claro E, Sarri E, Picatoste F. *Endogenous phosphoinositide precursors of inositol phosphates in rat brain cortical membranes*. Biochem Biophys Res Commun. **1993**, 193:1061-1067.

Claustre Y, Peretti DD, Brun P, Gueudet C, Allouard N, Alonso R, Lourdelet J, Oblin A, Damoiseau G, Francon D, Suaud-Chagny MF, Steinberg R, Sevrin M, Schoemaker H, George P, Soubrie P, Scatton B. *SSR181507, a dopamine D(2) receptor antagonist and 5-HT(1A) receptor agonist. I: Neurochemical and electrophysiological profile*. Neuropsychopharmacology. **2003**, 28:2064-2076.

Cohen BM, Lipinski JF. *In vivo potencies of antipsychotic drugs in blocking alpha 1 noradrenergic and dopamine D2 receptors: implications for drug mechanisms of action*. Life Sci. **1986**, 39:2571-2580.

Consolo S, Baronio P, Guidi G, Di Chiara G. *Role of the parafascicular thalamic nucleus and N-methyl-D-aspartate transmission in the D1-dependent control of in vivo acetylcholine release in rat striatum*. Neuroscience. **1996**, 71:157-165.

Cosi C, Carilla-Durand E, Assie MB, Ormiere AM, Maraval M, Leduc N, Newman-Tancredi A. *Partial agonist properties of the antipsychotics SSR181507, aripiprazole and bifeprunox at dopamine D2 receptors: G protein activation and prolactin release*. Eur J Pharmacol. **2006**, 535:135-144.

Coull JT, Frith CD, Dolan RJ, Frackowiak RS, Grasby PM. *The neural correlates of the noradrenergic modulation of human attention, arousal and learning*. Eur J Neurosci. **1997**, 9:589-598.

Coyle JT, Tsai G. *The NMDA receptor glycine modulatory site: a therapeutic target for improving cognition and reducing negative symptoms in schizophrenia*. Psychopharmacology (Berl). **2004**, 174:32-38.

Coyle JT. *Glutamate and Schizophrenia: Beyond the Dopamine Hypothesis*. Cell Mol Neurobiol. **2006**.

Creese I, Burt DR, Snyder SH. *Dopamine receptor binding predicts clinical and pharmacological potencies of antischizophrenic drugs*. Science. **1976**, 192:481-483.

Dawson LA, Nguyen HQ, Li P. *The 5-HT(6) receptor antagonist SB-271046 selectively enhances excitatory neurotransmission in the rat frontal cortex and hippocampus*. Neuropsychopharmacology. **2001**, 25:662-668.

Day HE, Campeau S, Watson SJ, Jr., Akil H. *Distribution of alpha 1a-, alpha 1b- and alpha 1d-adrenergic receptor mRNA in the rat brain and spinal cord*. J Chem Neuroanat. **1997**, 13:115-139.

De Deurwaerdere P, Navailles S, Berg KA, Clarke WP, Spampinato U. *Constitutive activity of the serotonin2C receptor inhibits in vivo dopamine release in the rat striatum and nucleus accumbens*. J Neurosci. **2004**, 24:3235-3241.

Degenhardt L, Hall W. *Cannabis and psychosis*. Curr Psychiatry Rep. **2002**, 4:191-196.

Delay J, Deniker P, Ropert R. *Study of 300 case histories of psychotic patients treated with chlorpromazine in closed wards since 1952*. Encephale. **1956**, 45:528-535.

Devoto P, Flore G, Vacca G, Pira L, Arca A, Casu MA, Pani L, Gessa GL. *Co-release of noradrenaline and dopamine from noradrenergic neurons in the cerebral cortex induced by clozapine, the prototype atypical antipsychotic*. Psychopharmacology (Berl). **2003**, 167:79-84.

Devoto P, Flore G, Pira L, Longu G, Gessa GL. *Alpha2-adrenoceptor mediated co-release of dopamine and noradrenaline from noradrenergic neurons in the cerebral cortex.* J Neurochem. **2004**, 88:1003-1009.

Di Matteo, V, Di Giovanni G, Di Mascio M, Esposito E. *Selective blockade of serotonin2C/2B receptors enhances dopamine release in the rat nucleus accumbens.* Neuropharmacology. **1998**, 37:265-272.

Di Matteo, V, Di Giovanni G, Di Mascio M, Esposito E. *SB 242084, a selective serotonin2C receptor antagonist, increases dopaminergic transmission in the mesolimbic system.* Neuropharmacology. **1999**, 38:1195-1205.

Di Matteo, V, De Blasi A, Di Giulio C, Esposito E. *Role of 5-HT(2C) receptors in the control of central dopamine function.* Trends Pharmacol Sci. **2001**, 22:229-232.

Di Matteo, V, Cacchio M, Di Giulio C, Di Giovanni G, Esposito E. *Biochemical evidence that the atypical antipsychotic drugs clozapine and risperidone block 5-HT(2C) receptors in vivo.* Pharmacol Biochem Behav. **2002**, 71:607-613.

Díaz-Mataix L, Scorza MC, Bortolozzi A, Toth M, Celada P, Artigas F. *Involvement of 5-HT1A receptors in prefrontal cortex in the modulation of dopaminergic activity: role in atypical antipsychotic action.* J Neurosci. **2005**, 25:10831-10843.

Diaz J, Pilon C, Le Foll B, Gros C, Triller A, Schwartz JC, Sokoloff P. *Dopamine D3 receptors expressed by all mesencephalic dopamine neurons.* J Neurosci. **2000**, 20:8677-8684.

Dresel S, Mager T, Rossmuller B, Meisenzahl E, Hahn K, Moller HJ, Tatsch K. *In vivo effects of olanzapine on striatal dopamine D(2)/D(3) receptor binding in schizophrenic patients: an iodine-123 iodobenzamide single-photon emission tomography study.* Eur J Nucl Med. **1999**, 26:862-868.

Duinkerke SJ, Botter PA, Jansen AA, van Dongen PA, van Haaften AJ, Boom AJ, van Laarhoven JH, Busard HL. *Ritanserin, a selective 5-HT2/1C antagonist, and negative symptoms in schizophrenia. A placebo-controlled double-blind trial.* Br J Psychiatry. **1993**, 163:451-455.

East SZ, Burnet PW, Kerwin RW, Harrison PJ. *An RT-PCR study of 5-HT(6) and 5-HT(7) receptor mRNAs in the hippocampal formation and prefrontal cortex in schizophrenia.* Schizophr Res. **2002**, 57:15-26.

Farah A. *Atypicality of atypical antipsychotics.* Prim Care Companion J Clin Psychiatry. **2005**, 7:268-274.

Farde L, Wiesel FA, Nordstrom AL, Sedvall G. *D1- and D2-dopamine receptor occupancy during treatment with conventional and atypical neuroleptics.* Psychopharmacology (Berl). **1989**, 99 Suppl:S28-S31.

Gérard C, el Mestikawy S, Lebrand C, Adrien J, Ruat M, Traiffort E, Hamon M, Martres MP. *Quantitative RT-PCR distribution of serotonin 5-HT6 receptor mRNA in the central nervous system of control or 5,7-dihydroxytryptamine-treated rats.* Synapse. **1996**, 23:164-173.

Gérard C, Martres MP, Lefevre K, Miquel MC, Verge D, Lanfumey L, Doucet E, Hamon M, el Mestikawy S. *Immuno-localization of serotonin 5-HT6 receptor-like material in the rat central nervous system.* Brain Res. **1997**, 746:207-219.

Gessa GL, Devoto P, Diana M, Flore G, Melis M, Pistis M. *Dissociation of haloperidol, clozapine, and olanzapine effects on electrical activity of mesocortical dopamine neurons and dopamine release in the prefrontal cortex.* Neuropsychopharmacology. **2000**, 22:642-649.

Giorgetti M, Tecott LH. *Contributions of 5-HT(2C) receptors to multiple actions of central serotonin systems.* Eur J Pharmacol. **2004**, 488:1-9.

Glennon J, McCreary AC, Ronken E, Siarey R, Hesselink MB, Feenstra R, Van Vliet B, Long SK, Kruse CG. *SLV313 is a dopamine D2 receptor antagonist and serotonin 5-HT1A receptor agonist: In vitro and in vivo neuropharmacology.* Eur Neuropsychopharmacol. **2002**, 12:277.

Gobert A, Rivet JM, Lejeune F, Newman-Tancredi A, Adhumeau-Auclair A, Nicolas JP, Cistarelli L, Melon C, Millan MJ. *Serotonin(2C) receptors tonically suppress the activity of mesocortical dopaminergic and adrenergic, but not serotonergic, pathways: a combined dialysis and electrophysiological analysis in the rat.* Synapse. **2000a**, 36:205-221.

Gobert A, Rivet JM, Cussac D, Newman-Tancredi A, Lejeune F, Bosc C, Dubuffet T, Lavielle G, Millan MJ. *S33592, a benzopyranopyrrole partial agonist dopamine D2/D3 receptors and potential antipsychotic agent. I.Modulation of dopaminergic transmission in comparison to aripiprazole, preclamol and raclopride.* Society of neuroscience, **2000b**, Abstract 26.

Goff D, Berman I, Posever T, Herz L, Leahy L, Lynch G. *A preliminary dose-escalation trial of CX 516 (Ampakine) added to clozapine in schizophrenia.* Schizophrenia Res. **1999**, 36:280.

Goldstein M, Deutch AY. *Dopaminergic mechanisms in the pathogenesis of schizophrenia.* Faseb J. **1992**, 6:2413-2421.

Grace AA, Bunney BS. *Induction of depolarization block in midbrain dopamine neurons by repeated administration of haloperidol: analysis using in vivo intracellular recording.* J Pharmacol Exp Ther. **1986**, 238:1092-1100.

Grace AA, Bunney BS, Moore H, Todd CL. *Dopamine-cell depolarization block as a model for the therapeutic actions of antipsychotic drugs.* Trends Neurosci. **1997**, 20:31-37.

Grace AA. *Gating of information flow within the limbic system and the pathophysiology of schizophrenia.* Brain Res Brain Res Rev. **2000**, 31:330-341.

Grace AA. *Dopamine.* In: Davis KL, Charney D, Coyle JT, eds. *Neuropsychopharmacology: The fifth generation of progress.* Lippincott Williams & Wilkins, Philadelphia, **2002**, pp 119-132.

Gray JA, Roth BL. *Paradoxical trafficking and regulation of 5-HT(2A) receptors by agonists and antagonists.* Brain Res Bull. **2001**, 56:441-451.

Grenhoff J, Svensson TH. *Prazosin modulates the firing pattern of dopamine neurons in rat ventral tegmental area.* Eur J Pharmacol. **1993**, 233:79-84.

Grenhoff J, Nisell M, Ferre S, Aston-Jones G, Svensson TH. *Noradrenergic modulation of midbrain dopamine cell firing elicited by stimulation of the locus coeruleus in the rat.* J Neural Transm Gen Sect. **1993**, 93:11-25.

Gresch PJ, Sved AF, Zigmond MJ, Finlay JM. *Local influence of endogenous norepinephrine on extracellular dopamine in rat medial prefrontal cortex.* J Neurochem. **1995**, 65:111-116.

Grunder G, Carlsson A, Wong DF. *Mechanism of new antipsychotic medications: occupancy is not just antagonism.* Arch Gen Psychiatry. **2003**, 60:974-977.

Hajos M, Richards CD, Szekely AD, Sharp T. *An electrophysiological and neuroanatomical study of the medial prefrontal cortical projection to the midbrain raphe nuclei in the rat.* Neuroscience. **1998**, 87:95-108.

Heresco-Levy U, Javitt DC, Ermilov M, Mordel C, Silipo G, Lichtenstein M. *Efficacy of high-dose glycine in the treatment of enduring negative symptoms of schizophrenia.* Arch Gen Psychiatry. **1999**, 56:29-36.

Herrick-Davis K, Grinde E, Teitler M. *Inverse agonist activity of atypical antipsychotic drugs at human 5-hydroxytryptamine2C receptors.* J Pharmacol Exp Ther. **2000**, 295:226-232.

Hertel P, Nomikos GG, Schilstrom B, Arborelius L, Svensson TH. *Risperidone dose-dependently increases extracellular concentrations of serotonin in the rat frontal cortex: role of alpha 2-adrenoceptor antagonism.* Neuropsychopharmacology. **1997**, 17:44-55.

Hertel P, Fagerquist MV, Svensson TH. *Enhanced cortical dopamine output and antipsychotic-like effects of raclopride by alpha2 adrenoceptor blockade.* Science. **1999a**, 286:105-107.

Hertel P, Nomikos GG, Svensson TH. *Idazoxan preferentially increases dopamine output in the rat medial prefrontal cortex at the nerve terminal level.* Eur J Pharmacol. **1999b**, 371:153-158.

Hippius H. *A historical perspective of clozapine.* J Clin Psychiatry. **1999**, 60 Suppl 12:22-23.

Hoffman BB. Catecholamines, sympathomimetic drugs and adrenergic receptor antagonist. In: Hardman JG, Limbird LE, Gilman AG, eds. *Goodman & Gilman's The Pharmacological Basis of Therapeutics.* 10th ed. McGraw-Hill Professional, **2001**, pp 215-268.

Hollister LE. *Drug-induced psychoses and schizophrenic reactions: a critical comparison.* Ann N Y Acad Sci. **1962**, 96:80-92.

Hommer DW, Zahn TP, Pickar D, van Kammen DP. *Prazosin, a specific alpha 1-noradrenergic receptor antagonist, has no effect on symptoms but increases autonomic arousal in schizophrenic patients.* Psychiatry Res. **1984**, 11:193-204.

Ichikawa J, Kuroki T, Kitchen MT, Meltzer HY. *R(+)-8-OH-DPAT, a 5-HT1A receptor agonist, inhibits amphetamine-induced dopamine release in rat striatum and nucleus accumbens.* Eur J Pharmacol. **1995**, 287:179-184.

Ichikawa J, Meltzer HY. *Relationship between dopaminergic and serotonergic neuronal activity in the frontal cortex and the action of typical and atypical antipsychotic drugs.* Eur Arch Psychiatry Clin Neurosci. **1999**, 249 Suppl 4:90-98.

Ichikawa J, Ishii H, Bonaccorso S, Fowler WL, O'Laughlin IA, Meltzer HY. *5-HT(2A) and D(2) receptor blockade increases cortical DA release via 5-HT(1A) receptor activation: a possible mechanism of atypical antipsychotic-induced cortical dopamine release.* J Neurochem. **2001**, 76:1521-1531.

Ichikawa J, Dai J, O'Laughlin IA, Fowler WL, Meltzer HY. *Atypical, but not typical, antipsychotic drugs increase cortical acetylcholine release without an effect in the nucleus accumbens or striatum.* Neuropsychopharmacology. **2002a**, 26:325-339.

Ichikawa J, Dai J, Meltzer HY. *5-HT(1A) and 5-HT(2A) receptors minimally contribute to clozapine-induced acetylcholine release in rat medial prefrontal cortex.* Brain Res. **2002b**, 939:34-42.

Inoue T, Domae M, Yamada K, Furukawa T. *Effects of the novel antipsychotic agent 7-(4-[4-(2,3-dichlorophenyl)-1-piperazinyl]butyloxy)-3,4-dihydro -2(1H)-quinolinone (OPC-14597) on prolactin release from the rat anterior pituitary gland.* J Pharmacol Exp Ther. **1996**, 277:137-143.

Jakab RL, Goldman-Rakic PS. *5-Hydroxytryptamine2A serotonin receptors in the primate cerebral cortex: possible site of action of hallucinogenic and antipsychotic drugs in pyramidal cell apical dendrites.* Proc Natl Acad Sci U S A. **1998**, 95:735-740.

Jodo E, Chiang C, Aston-Jones G. *Potent excitatory influence of prefrontal cortex activity on noradrenergic locus coeruleus neurons.* Neuroscience. **1998**, 83:63-79.

Johnstone EC, Crow TJ, Frith CD, Carney MW, Price JS. *Mechanism of the antipsychotic effect in the treatment of acute schizophrenia.* Lancet. **1978**, 1:848-851.

Jones HM, Pilowsky LS. *Dopamine and antipsychotic drug action revisited.* Br J Psychiatry. **2002**, 181:271-275.

Jordan S, Koprivica V, Chen R, Tottori K, Kikuchi T, Altar CA. *The antipsychotic aripiprazole is a potent, partial agonist at the human 5-HT1A receptor.* Eur J Pharmacol. **2002**, 441:137-140.

Kane J, Honigfeld G, Singer J, Meltzer H. *Clozapine for the treatment-resistant schizophrenic. A double-blind comparison with chlorpromazine.* Arch Gen Psychiatry. **1988**, 45:789-796.

Kapur S, Zipursky RB, Remington G, Jones C, DaSilva J, Wilson AA, Houle S. *5-HT2 and D2 receptor occupancy of olanzapine in schizophrenia: a PET investigation.* Am J Psychiatry. **1998**, 155:921-928.

Kapur S, Zipursky RB, Remington G. *Clinical and theoretical implications of 5-HT2 and D2 receptor occupancy of clozapine, risperidone, and olanzapine in schizophrenia.* Am J Psychiatry. **1999**, 156:286-293.

Kapur S, Zipursky R, Jones C, Remington G, Houle S. *Relationship between dopamine D(2) occupancy, clinical response, and side effects: a double-blind PET study of first-episode schizophrenia.* Am J Psychiatry. **2000**, 157:514-520.

Kapur S, Seeman P. *Does fast dissociation from the dopamine d(2) receptor explain the action of atypical antipsychotics?: A new hypothesis.* Am J Psychiatry. **2001**, 158:360-369.

Kapur S, Mamo D. *Half a century of antipsychotics and still a central role for dopamine D2 receptors.* Prog Neuropsychopharmacol Biol Psychiatry. **2003**, 27:1081-1090.

Kapur S. *Psychosis as a state of aberrant salience: a framework linking biology, phenomenology, and pharmacology in schizophrenia.* Am J Psychiatry. **2003**, 160:13-23.

Kikuchi T, Tottori K, Uwahodo Y, Hirose T, Miwa T, Oshiro Y, Morita S. *7-(4-[4-(2,3-Dichlorophenyl)-1-piperazinyl]butyloxy)-3,4-dihydro-2(1H)-qui nolinone (OPC-14597), a new putative antipsychotic drug with both presynaptic dopamine autoreceptor agonistic activity and postsynaptic D2 receptor antagonistic activity.* J Pharmacol Exp Ther. **1995**, 274:329-336.

Kohen R, Metcalf MA, Khan N, Druck T, Huebner K, Lachowicz JE, Meltzer HY, Sibley DR, Roth BL, Hamblin MW. *Cloning, characterization, and chromosomal localization of a human 5-HT6 serotonin receptor.* J Neurochem. **1996**, 66:47-56.

Kramer MS, Last B, Getson A, Reines SA. *The effects of a selective D4 dopamine receptor antagonist (L-745,870) in acutely psychotic inpatients with schizophrenia. D4 Dopamine Antagonist Group.* Arch Gen Psychiatry. **1997**, 54:567-572.

Krimer LS, Jakab RL, Goldman-Rakic PS. *Quantitative three-dimensional analysis of the catecholaminergic innervation of identified neurons in the macaque prefrontal cortex.* J Neurosci. **1997**, 17:7450-7461.

Krystal JH, Abi-Saab W, Perry E, D'Souza DC, Liu N, Gueorguieva R, McDougall L, Hunsberger T, Belger A, Levine L, Breier A. *Preliminary evidence of attenuation of the disruptive effects of the NMDA glutamate receptor antagonist, ketamine, on working memory by pretreatment with the group II metabotropic glutamate receptor agonist, LY354740, in healthy human subjects .* Psychopharmacology (Berl). **2005**, 179:303-309.

Kuhar MJ, Minneman K, Muly CE. Catecholamines. In: Siegel G, Wayne-Albers R, Brady S, Price D, American Society Neurochemistry., eds. *Basic Neurochemistry, Seventh edition: Molecular, cellular and medical aspects.* ElSevier, Canada, **2005**, pp 211-226.

Kuroki T, Meltzer HY, Ichikawa J. *Effects of antipsychotic drugs on extracellular dopamine levels in rat medial prefrontal cortex and nucleus accumbens.* J Pharmacol Exp Ther. **1999**, 288:774-781.

Lawler CP, Prioleau C, Lewis MM, Mak C, Jiang D, Schetz JA, Gonzalez AM, Sibley DR, Mailman RB. *Interactions of the novel antipsychotic aripiprazole (OPC-14597) with dopamine and serotonin receptor subtypes.* Neuropsychopharmacology. **1999**, 20:612-627.

Le Moal M, Simon H. *Mesocorticolimbic dopaminergic network: functional and regulatory roles.* Physiol Rev. **1991**, 71:155-234.

Lejeune F, Millan MJ. *Induction of burst firing in ventral tegmental area dopaminergic neurons by activation of serotonin (5-HT)1A receptors: WAY 100,635-reversible actions of the highly selective ligands, flesinoxan and S 15535.* Synapse. **1998**, 30:172-180.

Leucht S, Pitschel-Walz G, Engel RR, Kissling W. *Amisulpride, an unusual "atypical" antipsychotic: a meta-analysis of randomized controlled trials.* Am J Psychiatry. **2002**, 159:180-190.

Li Z, Huang M, Ichikawa J, Dai J, Meltzer HY. *N-desmethylclozapine, a major metabolite of clozapine, increases cortical acetylcholine and dopamine release in vivo via stimulation of M1 muscarinic receptors.* Neuropsychopharmacology. **2005**, 30:1986-1995.

Lieberman JA, Stroup TS, McEvoy JP, Swartz MS, Rosenheck RA, Perkins DO, Keefe RS, Davis SM, Davis CE, Lebowitz BD, Severe J, Hsiao JK. *Effectiveness of antipsychotic drugs in patients with chronic schizophrenia.* N Engl J Med. **2005**, 353:1209-1223.

Linner L, Wiker C, Wadenberg ML, Schalling M, Svensson TH. *Noradrenaline reuptake inhibition enhances the antipsychotic-like effect of raclopride and potentiates D2-blockage-induced dopamine release in the medial prefrontal cortex of the rat.* Neuropsychopharmacology. **2002**, 27:691-698.

Litman RE, Su TP, Potter WZ, Hong WW, Pickar D. *Idazoxan and response to typical neuroleptics in treatment-resistant schizophrenia. Comparison with the atypical neuroleptic, clozapine.* Br J Psychiatry. **1996**, 168:571-579.

Lucas G, Spampinato U. *Role of striatal serotonin2A and serotonin2C receptor subtypes in the control of in vivo dopamine outflow in the rat striatum.* J Neurochem. **2000**, 74:693-701.

Maas JW, Contreras SA, Miller AL, Berman N, Bowden CL, Javors MA, Seleshi E, Weintraub S. *Studies of catecholamine metabolism in schizophrenia/psychosis--I.* Neuropsychopharmacology. **1993**, 8:97-109.

Mansour A, Meador-Woodruff JH, Zhou Q, Civelli O, Akil H, Watson SJ. *A comparison of D1 receptor binding and mRNA in rat brain using receptor autoradiographic and in situ hybridization techniques.* Neuroscience. **1992**, 46:959-971.

Marek G, Merchant K. *Developing therapeutics for schizophrenia and other psychotic disorders.* NeuroRx. **2005**, 2:579-589.

Marek GJ, Aghajanian GK. *5-HT2A receptor or alpha1-adrenoceptor activation induces excitatory postsynaptic currents in layer V pyramidal cells of the medial prefrontal cortex.* Eur J Pharmacol. **1999**, 367:197-206.

Martín-Ruiz R, Puig MV, Celada P, Shapiro DA, Roth BL, Mengod G, Artigas F. *Control of serotonergic function in medial prefrontal cortex by serotonin-2A receptors through a glutamate-dependent mechanism.* J Neurosci. **2001**, 21:9856-9866.

Martin P, Waters N, Waters S, Carlsson A, Carlsson ML. *MK-801-induced hyperlocomotion: differential effects of M100907, SDZ PSD 958 and raclopride.* Eur J Pharmacol. **1997**, 335:107-116.

Martin P, Waters N, Schmidt CJ, Carlsson A, Carlsson ML. *Rodent data and general hypothesis: antipsychotic action exerted through 5-Ht2A receptor antagonism is dependent on increased serotonergic tone.* J Neural Transm. **1998**, 105:365-396.

Martindale W. *Martindale: The Extra Pharmacopoeia.* Pharmaceutical Press; London, **1993**.

Martinot JL, Paillere-Martinot ML, Poirier MF, Dao-Castellana MH, Loc'h C, Maziere B. *In vivo characteristics of dopamine D2 receptor occupancy by amisulpride in schizophrenia.* Psychopharmacology (Berl). **1996**, 124:154-158.

Masellis M, Basile VS, Meltzer HY, Lieberman JA, Sevy S, Goldman DA, Hamblin MW, Macciardi FM, Kennedy JL. *Lack of association between the T-->C 267 serotonin 5-HT6 receptor gene (HTR6) polymorphism and prediction of response to clozapine in schizophrenia.* Schizophr Res. **2001**, 47:49-58.

Mathe JM, Nomikos GG, Hildebrand BE, Hertel P, Svensson TH. *Prazosin inhibits MK-801-induced hyperlocomotion and dopamine release in the nucleus accumbens.* Eur J Pharmacol. **1996**, 309:1-11.

McCormick DA, Pape HC. *Noradrenergic and serotonergic modulation of a hyperpolarization-activated cation current in thalamic relay neurones.* J Physiol. **1990**, 431:319-342.

McCune SK, Voigt MM, Hill JM. *Expression of multiple alpha adrenergic receptor subtype messenger RNAs in the adult rat brain.* Neuroscience. **1993**, 57:143-151.

McEvoy JP, Lieberman JA, Stroup TS, Davis SM, Meltzer HY, Rosenheck RA, Swartz MS, Perkins DO, Keefe RS, Davis CE, Severe J, Hsiao JK. *Effectiveness of clozapine versus olanzapine, quetiapine, and risperidone in patients with chronic schizophrenia who did not respond to prior atypical antipsychotic treatment.* Am J Psychiatry. **2006**, 163:600-610.

Meador-Woodruff JH, Damask SP, Watson SJ, Jr. *Differential expression of autoreceptors in the ascending dopamine systems of the human brain.* Proc Natl Acad Sci U S A. **1994**, 91:8297-8301.

Meisenzahl EM, Dresel S, Frodl T, Schmitt GJ, Preuss UW, Rossmuller B, Tatsch K, Mager T, Hahn K, Moller HJ. *D2 receptor occupancy under recommended and high doses of olanzapine: an iodine-123-iodobenzamide SPECT study.* J Psychopharmacol. **2000**, 14:364-370.

Meltzer HY, Matsubara S, Lee JC. *Classification of typical and atypical antipsychotic drugs on the basis of dopamine D-1, D-2 and serotonin2 pKi values.* J Pharmacol Exp Ther. **1989**, 251:238-246.

Meltzer HY. *Suicide in schizophrenia: risk factors and clozapine treatment.* J Clin Psychiatry. **1998**, 59 Suppl 3:15-20.

Meltzer HY, Li Z, Kaneda Y, Ichikawa J. *Serotonin receptors: their key role in drugs to treat schizophrenia.* Prog Neuropsychopharmacol Biol Psychiatry. **2003**, 27:1159-1172.

Meltzer HY, Arvanitis L, Bauer D, Rein W. *Placebo-controlled evaluation of four novel compounds for the treatment of schizophrenia and schizoaffective disorder.* Am J Psychiatry. **2004**, 161:975-984.

Mengod G, Vilaro MT, Raurich A, Lopez-Gimenez JF, Cortes R, Palacios JM. *5-HT receptors in mammalian brain: receptor autoradiography and in situ hybridization studies of new ligands and newly identified receptors.* Histochem J. **1996**, 28:747-758.

Millan MJ, Dekeyne A, Gobert A. *Serotonin (5-HT)2C receptors tonically inhibit dopamine (DA) and noradrenaline (NA), but not 5-HT, release in the frontal cortex in vivo.* Neuropharmacology. **1998**, 37:953-955.

Millan MJ, Gobert A, Newman-Tancredi A, Lejeune F, Cussac D, Rivet JM, Audinot V, Adhumeau A, Brocco M, Nicolas JP, Boutin JA, Despaux N, Peglion JL. *S18327 (1-[2-[4-(6-fluoro-1, 2-benzisoxazol-3-yl)piperid-1-yl]ethyl]3-phenyl imidazolin-2-one), a novel, potential antipsychotic displaying marked antagonist properties at alpha(1)- and alpha(2)-adrenergic receptors: I. Receptorial, neurochemical, and electrophysiological profile.* J Pharmacol Exp Ther. **2000**, 292:38-53.

Millan MJ. *Improving the treatment of schizophrenia: focus on serotonin (5-HT)(1A) receptors.* J Pharmacol Exp Ther. **2000**, 295:853-861.

Millan MJ. *Dopamine D3 receptors as a novel target for improving the treatment of schizophrenia.* Med Sci (Paris). **2005**, 21:434-442.

Miner LA, Backstrom JR, Sanders-Bush E, Sesack SR. *Ultrastructural localization of serotonin2A receptors in the middle layers of the rat prelimbic prefrontal cortex.* Neuroscience. **2003**, 116:107-117.

Miyamoto S, Stroup TS, Duncan GE, Aoba A, Lieberman JA. Acute pharmacological treatment of schizophrenia. In: Hirsch SR, ed. *Schizophrenia.* 2nd ed. Blackwell Science, Massachussetts, USA, **2003**, pp 442-473.

Moghaddam B, Bunney BS. *Acute effects of typical and atypical antipsychotic drugs on the release of dopamine from prefrontal cortex, nucleus accumbens, and striatum of the rat: an in vivo microdialysis study .* J Neurochem. **1990**, 54:1755-1760.

Moghaddam B, Adams B, Verma A, Daly D. *Activation of glutamatergic neurotransmission by ketamine: a novel step in the pathway from NMDA receptor blockade to dopaminergic and cognitive disruptions associated with the prefrontal cortex.* J Neurosci. **1997**, 17:2921-2927.

Moghaddam B. *Bringing order to the glutamate chaos in schizophrenia.* Neuron. **2003**, 40:881-884.

Molinoff PB. *Alpha- and beta-adrenergic receptor subtypes properties, distribution and regulation.* Drugs. **1984**, 28 Suppl 2:1-15.

Möller HJ. *Antipsychotic and antidepressive effects of second generation antipsychotics: two different pharmacological mechanisms?* Eur Arch Psychiatry Clin Neurosci. **2005**, 255:190-201.

Momiyama T, Amano T, Todo N, Sasa M. *Inhibition by a putative antipsychotic quinolinone derivative (OPC-14597) of dopaminergic neurons in the ventral tegmental area.* Eur J Pharmacol. **1996**, 310:1-8.

Murase S, Grenhoff J, Chouvet G, Gonon FG, Svensson TH. *Prefrontal cortex regulates burst firing and transmitter release in rat mesolimbic dopamine neurons studied in vivo.* Neurosci Lett. **1993**, 157:53-56.

Müller D, Vincenzo DL, James L. *Overview: Towards individualized treatment in schizophrenia.* Drug Dev Res. **2003**, 60:75-94.

Nakai K, Fujii T, Fujimoto K, Suzuki T, Kawashima K. *Effect of WAY-100135 on the hippocampal acetylcholine release potentiated by 8-OH-DPAT, a serotonin1A receptor agonist, in normal and p-chlorophenylalanine-treated rats as measured by in vivo microdialysis.* Neurosci Res. **1998**, 31:23-29.

Navailles S, De Deurwaerdere P, Spampinato U. *Clozapine and haloperidol differentially alter the constitutive activity of central serotonin2C receptors in vivo.* Biol Psychiatry. **2006**, 59:568-575.

Newman-Tancredi A, Assie MB, Leduc N, Ormiere AM, Danty N, Cosi C. *Novel antipsychotics activate recombinant human and native rat serotonin 5-HT1A receptors: affinity, efficacy and potential implications for treatment of schizophrenia.* Int J Neuropsychopharmacol. **2005**, 8:341-356.

Nocjar C, Roth BL, Pehek EA. *Localization of 5-HT(2A) receptors on dopamine cells in subnuclei of the midbrain A10 cell group.* Neuroscience. **2002**, 111:163-176.

Nomikos GG, Iurlo M, Andersson JL, Kimura K, Svensson TH. *Systemic administration of amperozide, a new atypical antipsychotic drug, preferentially increases dopamine release in the rat medial prefrontal cortex.* Psychopharmacology (Berl). **1994**, 115:147-156.

Nordstrom AL, Farde L, Wiesel FA, Forslund K, Pauli S, Halldin C, Uppfeldt G. *Central D2-dopamine receptor occupancy in relation to antipsychotic drug effects: a double-blind PET study of schizophrenic patients.* Biol Psychiatry. **1993**, 33:227-235.

Oak JN, Oldenhof J, Van Tol HH. *The dopamine D(4) receptor: one decade of research.* Eur J Pharmacol. **2000**, 405:303-327.

Pehek EA, McFarlane HG, Maguschak K, Price B, Pluto CP. *M100,907, a selective 5-HT(2A) antagonist, attenuates dopamine release in the rat medial prefrontal cortex.* Brain Res. **2001**, 888:51-59.

Peroutka SJ, Synder SH. *Relationship of neuroleptic drug effects at brain dopamine, serotonin, alpha-adrenergic, and histamine receptors to clinical potency.* Am J Psychiatry. **1980**, 137:1518-1522.

Petrie JL, Saha AR, McEvoy JP. *Acute and long-term efficacy and safety of aripiprazole: A new atypical antipsychotic.* Schizophr Res. **1998**, 29:155.

Peyron C, Petit JM, Rampon C, Jouvet M, Luppi PH. *Forebrain afferents to the rat dorsal raphe nucleus demonstrated by retrograde and anterograde tracing methods.* Neuroscience. **1998**, 82:443-468.

Pickel VM, Chan J, Nirenberg MJ. *Region-specific targeting of dopamine D2-receptors and somatodendritic vesicular monoamine transporter 2 (VMAT2) within ventral tegmental area subdivisions.* Synapse. **2002**, 45:113-124.

Pieribone VA, Nicholas AP, Dagerlind A, Hokfelt T. *Distribution of alpha 1 adrenoceptors in rat brain revealed by in situ hybridization experiments utilizing subtype-specific probes.* J Neurosci. **1994**, 14:4252-4268.

Pilowsky LS, Costa DC, Ell PJ, Murray RM, Verhoeff NP, Kerwin RW. *Antipsychotic medication, D2 dopamine receptor blockade and clinical response: a 123I IBZM SPET (single photon emission tomography) study.* Psychol Med. **1993**, 23:791-797.

Pompeiano M, Palacios JM, Mengod G. *Distribution of the serotonin 5-HT2 receptor family mRNAs: comparison between 5-HT2A and 5-HT2C receptors.* Brain Res Mol Brain Res. **1994**, 23:163-178.

Potkin SG, Shipley J, Bera RB, Carreon D, Fallon J, Alva G, Keator D. *Clinical and PET effects of M100905; a selective 5HT-2A receptor antagonist.* Schizophr Res. **2001**, 49:242.

Pozzi L, Invernizzi R, Cervo L, Vallebuona F, Samanin R. *Evidence that extracellular concentrations of dopamine are regulated by noradrenergic neurons in the frontal cortex of rats.* J Neurochem. **1994**, 63:195-200.

Prisco S, Pagannone S, Esposito E. *Serotonin-dopamine interaction in the rat ventral tegmental area: an electrophysiological study in vivo.* J Pharmacol Exp Ther. **1994**, 271:83-90.

Puig MV, Celada P, Diaz-Mataix L, Artigas F. *In vivo modulation of the activity of pyramidal neurons in the rat medial prefrontal cortex by 5-HT2A receptors: relationship to thalamocortical afferents.* Cereb Cortex. **2003**, 13:870-882.

Puig MV, Artigas F, Celada P. *Modulation of the activity of pyramidal neurons in rat prefrontal cortex by raphe stimulation in vivo: involvement of serotonin and GABA.* Cereb Cortex. **2005**, 15:1-14.

Raedler TJ, Jahn H, Arlt J, Kiefer F, Schick M, Naber D, Wiedemann K. *Adjunctive use of reboxetine in schizophrenia.* Eur Psychiatry. **2004**, 19:366-369.

Remington G, Kapur S. *Atypical antipsychotics: are some more atypical than others?* Psychopharmacology (Berl). **2000**, 148:3-15.

Report, S. *Management decisions on priority pipeline products: MDL 100907 (Company Report).* Frankfurt, Germany, Hoechst Marion Russel, **1999**, pp 2-3.

Reynolds GP, Garrett NJ, Rupniak N, Jenner P, Marsden CD. *Chronic clozapine treatment of rats down-regulates cortical 5-HT2 receptors.* Eur J Pharmacol. **1983**, 89:325-326.

Reynolds GP. *Receptor mechanisms in the treatment of schizophrenia.* J Psychopharmacol. **2004**, 18:340-345.

Reynolds GP, Templeman LA, Zhang ZJ. *The role of 5-HT2C receptor polymorphisms in the pharmacogenetics of antipsychotic drug treatment.* Prog Neuropsychopharmacol Biol Psychiatry. **2005**, 29:1021-1028.

Rogers DC, Hagan JJ. *5-HT6 receptor antagonists enhance retention of a water maze task in the rat.* Psychopharmacology (Berl). **2001**, 158:114-119.

Rollema H, Lu Y, Schmidt AW, Zorn SH. *Clozapine increases dopamine release in prefrontal cortex by 5-HT1A receptor activation*. Eur J Pharmacol. **1997**, 338:R3-R5.

Rollema H, Lu Y, Schmidt AW, Sprouse JS, Zorn SH. *5-HT(1A) receptor activation contributes to ziprasidone-induced dopamine release in the rat prefrontal cortex*. Biol Psychiatry. **2000**, 48:229-237.

Roth BL, Craigo SC, Choudhary MS, Uluer A, Monsma FJ, Jr., Shen Y, Meltzer HY, Sibley DR. *Binding of typical and atypical antipsychotic agents to 5-hydroxytryptamine-6 and 5-hydroxytryptamine-7 receptors*. J Pharmacol Exp Ther. **1994**, 268:1403-1410.

Roth BL, Sheffler D, Potkin SG. *Atypical antipsychotic drug actions: unitary or multiple mechanisms for 'atypicality'?* Clin NeuroSci Res. **2003**, 3:108-117.

Ruat M, Traiffort E, Leurs R, Tardivel-Lacombe J, Diaz J, Arrang JM, Schwartz JC. *Molecular cloning, characterization, and localization of a high-affinity serotonin receptor (5-HT7) activating cAMP formation*. Proc Natl Acad Sci U S A. **1993**, 90:8547-8551.

Sakaue M, Somboonthum P, Nishihara B, Koyama Y, Hashimoto H, Baba A, Matsuda T. *Postsynaptic 5-hydroxytryptamine(1A) receptor activation increases in vivo dopamine release in rat prefrontal cortex*. Br J Pharmacol. **2000**, 129:1028-1034.

Santana N, Bortolozzi A, Serrats J, Mengod G, Artigas F. *Expression of serotonin1A and serotonin2A receptors in pyramidal and GABAergic neurons of the rat prefrontal cortex*. Cereb Cortex. **2004**, 14:1100-1109.

Sawaguchi T, Goldman-Rakic PS. *The role of D1-dopamine receptor in working memory: local injections of dopamine antagonists into the prefrontal cortex of rhesus monkeys performing an oculomotor delayed-response task*. J Neurophysiol. **1994**, 71:515-528.

Schmitt GJ, Meisenzahl EM, Dresel S, Tatsch K, Rossmuller B, Frodl T, Preuss UW, Hahn K, Moller HJ. *Striatal dopamine D2 receptor binding of risperidone in schizophrenic patients as assessed by 123I-iodobenzamide SPECT: a comparative study with olanzapine*. J Psychopharmacol. **2002**, 16:200-206.

Schultz W, Tremblay L, Hollerman JR. *Reward prediction in primate basal ganglia and frontal cortex*. Neuropharmacology. **1998**, 37:421-429.

Schutz G, Berk M. *Reboxetine add on therapy to haloperidol in the treatment of schizophrenia: a preliminary double-blind randomized placebo-controlled study*. Int Clin Psychopharmacol. **2001**, 16:275-278.

Scuderi PE. *Droperidol: many questions, few answers*. Anesthesiology. **2003**, 98:289-290.

Seeman P, Lee T. *Antipsychotic drugs: direct correlation between clinical potency and presynaptic action on dopamine neurons*. Science. **1975**, 188:1217-1219.

Seeman P, Lee T, Chau-Wong M, Wong K. *Antipsychotic drug doses and neuroleptic/dopamine receptors*. Nature. **1976**, 261:717-719.

Seeman P, Tallerico T. *Rapid release of antipsychotic drugs from dopamine D2 receptors: an explanation for low receptor occupancy and early clinical relapse upon withdrawal of clozapine or quetiapine*. Am J Psychiatry. **1999**, 156:876-884.

Semba J, Watanabe A, Kito S, Toru M. *Behavioural and neurochemical effects of OPC-14597, a novel antipsychotic drug, on dopaminergic mechanisms in rat brain*. Neuropharmacology. **1995**, 34:785-791.

Sesack SR, Deutch AY, Roth RH, Bunney BS. *Topographical organization of the efferent projections of the medial prefrontal cortex in the rat: an anterograde tract-tracing study with Phaseolus vulgaris leucoagglutinin.* J Comp Neurol. **1989**, 290:213-242.

Sesack SR, Pickel VM. *Dual ultrastructural localization of enkephalin and tyrosine hydroxylase immunoreactivity in the rat ventral tegmental area: multiple substrates for opiate-dopamine interactions.* J Neurosci. **1992**, 12:1335-1350.

Sesack SR, Aoki C, Pickel VM. *Ultrastructural localization of D2 receptor-like immunoreactivity in midbrain dopamine neurons and their striatal targets.* J Neurosci. **1994**, 14:88-106.

Sesack SR, Snyder CL, Lewis DA. *Axon terminals immunolabeled for dopamine or tyrosine hydroxylase synapse on GABA-immunoreactive dendrites in rat and monkey cortex.* J Comp Neurol. **1995**, 363:264-280.

Sharp JW, Petersen DL, Langford MT. *DNQX inhibits phencyclidine (PCP) and ketamine induction of the hsp70 heat shock gene in the rat cingulate and retrosplenial cortex.* Brain Res. **1995**, 687:114-124.

Soyka M, De Vry J. *Flupenthixol as a potential pharmacotreatment of alcohol and cocaine abuse/dependence.* Eur Neuropsychopharmacol. **2000**, 10:325-332.

Spanagel R, Weiss F. *The dopamine hypothesis of reward: past and current status.* Trends Neurosci. **1999**, 22:521-527.

Spooren W, Riemer C, Meltzer H. *Opinion: NK3 receptor antagonists: the next generation of antipsychotics?* Nat Rev Drug Discov. **2005**, 4:967-975.

Stip E. *Novel antipsychotics: issues and controversies. Typicality of atypical antipsychotics.* J Psychiatry Neurosci. **2000**, 25:137-153.

Sumiyoshi T, Matsui M, Nohara S, Yamashita I, Kurachi M, Sumiyoshi C, Jayathilake K, Meltzer HY. *Enhancement of cognitive performance in schizophrenia by addition of tandospirone to neuroleptic treatment.* Am J Psychiatry. **2001a**, 158:1722-1725.

Sumiyoshi T, Matsui M, Yamashita I, Nohara S, Kurachi M, Uehara T, Sumiyoshi S, Sumiyoshi C, Meltzer HY. *The effect of tandospirone, a serotonin(1A) agonist, on memory function in schizophrenia.* Biol Psychiatry. **2001b**, 49:861-868.

Svensson TH, Mathe JM, Andersson JL, Nomikos GG, Hildebrand BE, Marcus M. *Mode of action of atypical neuroleptics in relation to the phencyclidine model of schizophrenia: role of 5-HT2 receptor and alpha 1-adrenoceptor antagonism [corrected].* J Clin Psychopharmacol. **1995**, 15:11S-18S.

Svensson TH. *Preclinical effects of conventional and atypical antipsychotic drugs: defining the mechanisms of action.* Clin NeuroSci Res. **2003**, 3:34-46.

Swanson LW. *The projections of the ventral tegmental area and adjacent regions: a combined fluorescent retrograde tracer and immunofluorescence study in the rat.* Brain Res Bull. **1982**, 9:321-353.

Takagishi M, Chiba T. *Efferent projections of the infralimbic (area 25) region of the medial prefrontal cortex in the rat: an anterograde tracer PHA-L study.* Brain Res. **1991**, 566:26-39.

Tamminga C. *Partial Dopamine Agonists and the Treatment of Psychosis.* Curr Neuropharm. **2005**, 3:3-8.

Tang L, Shukla PK, Wang ZJ. *Trifluoperazine, an orally available clinically used drug, disrupts opioid antinociceptive tolerance.* Neurosci Lett. **2006**, 397:1-4.

Tauscher J, Kufferle B, Asenbaum S, Tauscher-Wisniewski S, Kasper S. *Striatal dopamine-2 receptor occupancy as measured with [123I]iodobenzamide and SPECT predicted the occurrence of EPS in patients treated with atypical antipsychotics and haloperidol.* Psychopharmacology (Berl). **2002**, 162:42-49.

Thierry AM, Deniau JM, Chevalier G, Ferron A, Glowinski J. *An electrophysiological analysis of some afferent and efferent pathways of the rat prefrontal cortex.* Prog Brain Res. **1983**, 58:257-261.

Trichard C, Paillere-Martinot ML, Attar-Levy D, Recassens C, Monnet F, Martinot JL. *Binding of antipsychotic drugs to cortical 5-HT2A receptors: a PET study of chlorpromazine, clozapine, and amisulpride in schizophrenic patients.* Am J Psychiatry. **1998**, 155:505-508.

Truffinet P, Tamminga CA, Fabre LF, Meltzer HY, Riviere ME, Papillon-Downey C. *Placebo-controlled study of the D4/5-HT2A antagonist fananserin in the treatment of schizophrenia.* Am J Psychiatry. **1999**, 156:419-425.

Tsai G, Yang P, Chung LC, Lange N, Coyle JT. *D-serine added to antipsychotics for the treatment of schizophrenia.* Biol Psychiatry. **1998**, 44:1081-1089.

Tsai G, Lane HY, Yang P, Chong MY, Lange N. *Glycine transporter I inhibitor, N-methylglycine (sarcosine), added to antipsychotics for the treatment of schizophrenia.* Biol Psychiatry. **2004**, 55:452-456.

Tsai SJ, Chiu HJ, Wang YC, Hong CJ. *Association study of serotonin-6 receptor variant (C267T) with schizophrenia and aggressive behavior.* Neurosci Lett. **1999**, 271:135-137.

Tzschentke TM. *Pharmacology and behavioral pharmacology of the mesocortical dopamine system.* Prog Neurobiol. **2001**, 63:241-320.

Van Tol HH, Bunzow JR, Guan HC, Sunahara RK, Seeman P, Niznik HB, Civelli O. *Cloning of the gene for a human dopamine D4 receptor with high affinity for the antipsychotic clozapine.* Nature. **1991**, 350:610-614.

Volkow ND, Fowler JS, Wang GJ, Goldstein RZ. *Role of dopamine, the frontal cortex and memory circuits in drug addiction: insight from imaging studies.* Neurobiol Learn Mem. **2002**, 78:610-624.

Wadenberg ML, Ahlenius S. *Antipsychotic-like profile of combined treatment with raclopride and 8-OH-DPAT in the rat: enhancement of antipsychotic-like effects without catalepsy.* J Neural Transm Gen Sect. **1991**, 83:43-53.

Wadenberg ML, Hertel P, Fernholm R, Hygge BK, Ahlenius S, Svensson TH. *Enhancement of antipsychotic-like effects by combined treatment with the alpha1-adrenoceptor antagonist prazosin and the dopamine D2 receptor antagonist raclopride in rats.* J Neural Transm. **2000**, 107:1229-1238.

Wadenberg ML, Wiker C, Svensson TH. *Enhanced efficacy of both typical and atypical antipsychotic drugs by adjunctive alpha 2 adrenoceptor blockade: experimental evidence.* Int J Neuropsychopharmacol. **2006**, 1-12.

Wamsley JK, Gehlert DR, Filloux FM, Dawson TM. *Comparison of the distribution of D-1 and D-2 dopamine receptors in the rat brain.* J Chem Neuroanat. **1989**, 2:119-137.

Westerink BH, Kawahara Y, De Boer P, Geels C, De Vries JB, Wikstrom HV, Van Kalkeren A, Van Vliet B, Kruse CG, Long SK. *Antipsychotic drugs classified by their effects on the release of dopamine and noradrenaline in the prefrontal cortex and striatum.* Eur J Pharmacol. **2001**, 412:127-138.

Wiesel FA, Nordstrom AL, Farde L, Eriksson B. *An open clinical and biochemical study of ritanserin in acute patients with schizophrenia.* Psychopharmacology (Berl). **1994**, 114:31-38.

Wilffert B, Zaal R, Brouwers JR. *Pharmacogenetics as a tool in the therapy of schizophrenia.* Pharm World Sci. **2005**, 27:20-30.

Williams SM, Goldman-Rakic PS. *Widespread origin of the primate mesofrontal dopamine system.* Cereb Cortex. **1998**, 8:321-345.

Willins DL, Narayanan S, Wallace LJ, Uretsky NJ. *The role of dopamine and AMPA/kainate receptors in the nucleus accumbens in the hypermotility response to MK801.* Pharmacol Biochem Behav. **1993**, 46:881-887.

Willins DL, Deutch AY, Roth BL. *Serotonin 5-HT2A receptors are expressed on pyramidal cells and interneurons in the rat cortex.* Synapse. **1997**, 27:79-82.

Willins DL, Berry SA, Alsayegh L, Backstrom JR, Sanders-Bush E, Friedman L, Roth BL. *Clozapine and other 5-hydroxytryptamine-2A receptor antagonists alter the subcellular distribution of 5-hydroxytryptamine-2A receptors in vitro and in vivo.* Neuroscience. **1999**, 91:599-606.

Wolf W. *DU-127090 Solvay/H Lundbeck.* Curr Opin Investig Drugs. **2003**, 4:72-76.

Wolkin A, Barouche F, Wolf AP, Rotrosen J, Fowler JS, Shiue CY, Cooper TB, Brodie JD. *Dopamine blockade and clinical response: evidence for two biological subgroups of schizophrenia.* Am J Psychiatry. **1989**, 146:905-908.

Wong AH, Van Tol HH. *The dopamine D4 receptors and mechanisms of antipsychotic atypicality.* Prog Neuropsychopharmacol Biol Psychiatry. **2003**, 27:1091-1099.

Yu YW, Tsai SJ, Lin CH, Hsu CP, Yang KH, Hong CJ. *Serotonin-6 receptor variant (C267T) and clinical response to clozapine.* Neuroreport. **1999**, 10:1231-1233.

2.4 Experimental Research

Modeling Schizophrenia in Experimental Animals

Pau Celada, Anna Castañé, Albert Adell and Francesc Artigas

Dept. of Neurochemistry and Neuropharmacology, Institut d' Investigacions Biomèdiques de Barcelona, CSIC (IDIBAPS), Roselló, 161. 08036 Barcelona, Spain.

2.4.1 Introduction

Schizophrenia is a very complex psychiatric disorder characterized by a cluster of positive (psychotic), negative, affective and cognitive symptoms. Psychotic episodes are indeed one of the most characteristic features of the illness and have been attributed to an overactivity of the subcortical (mesolimbic) dopaminergic pathways. As a result of early theories on the etiology of schizophrenia based on dopaminergic overactivity, the first experimental models of the illness relied upon drug-induced increases of dopaminergic function. However, contrary to positive symptoms, negative symptoms (e.g., social withdrawal, anhedonia) and cognitive disturbances are thought to be due in part to a hypofunction of the mesocortical dopaminergic pathway (Carlsson, 1988; Weinberger et al., 1994, Laruelle et al., 1996; Abi-Dargham et al., 2000). However, numerous findings support the contention that schizophrenia involves macro- and microanatomical changes in several key brain areas such as prefrontal cortex (PFC), medial temporal lobe (hippocampus) and mediodorsal thalamus (Harrison, 1999; Selemon et al., 2002; Selemon et al., 2003; Selemon and Goldman-Rakic, 1999; Selemon 2004). Overall, these findings support that schizophrenia is something more complex than up or down changes in dopaminergic transmission. Indeed, schizophrenia is characterized by a profound disruption of information processing along brain circuits that control cognition and emotion, involving changes at cellular and molecular levels that eventually modify the activity of different neurotransmitter systems. Thus, it becomes apparent that schizophrenia is the most difficult psychiatric disorders to be modelled in experimental animals.

Several authors have discussed the validity of animal models of psychiatric disorders such as schizophrenia (Geyer and Markou, 1995, 2002). Indeed, there is no valid genotype, cellular phenotype or a biological marker that can aid in the diagnostic of schizophrenia and no animal model can fully reproduce the perceptual, cognitive and emotional features of this human illness. Ideally, a model of the disease should mimic the neuropathological, neurophysiological, neurochemical and behavioral changes observed in schizophrenia. In absence of such a general model, multiple animal models (discrete models) are necessary to study the neurobiological basis of different classes of symptoms and to evaluate new therapies.

A large number of different behavioral paradigms have been used in schizophrenia research (Box 1) and selected symptoms of the disease can be mimicked in experimental animals with predictive validity and reliability (van den Buuse et al., 2005) (Box 2). *Predictive validity* is defined as the ability of a test to reliably discriminate a clinically effective drug from a non-effective one. Usually, in antipsychotic drug screening, a positive result in a single preclinical test possessing predictive validity is not sufficient to establish a drug's potential antipsychotic efficacy. Then, *reliability* refers to the accuracy with which both the experimental and clinical observations are made. Another criterion to characterize an animal model is *face validity*, which refers to the phenomenological similarity between the behavior exhibited by the animal and selected symptoms of the human condition. In the case of schizophrenia, it seems very difficult, if not impossible, to reproduce in rodents the whole constellation of symptoms (positive, negative, affective, cognitive) and therefore it appears more relevant to search for similarities in the underlying mechanisms involved in certain behaviors, which is referred to as *construct validity* (Geyer and Markou, 1995, 2002).

2.4.2 Neurophysiology of schizophrenia

One of the very few methodologies that can be concurrently used in living humans and experimental animals in the study of schizophrenia is electrophysiology. Indeed, neuronal activity can be studied at various complexity levels, ranging from single channel or single units to non-invasive global assessments using electroencephalography (EEG). EEG methodology can be applied to patients and experimental animals in order to link the results of clinical studies to findings in basic neuroscience. It offers also the possibility to concurrently study drug effects in humans and experimental animals.

Schizophrenic patients frequently show EEG abnormalities (Matsura et al., 1994, Winterer et al., 2004). A close inspection of these changes can provide some clues about

Box 1. Behavioral paradigms used in schizophrenia research

A. Locomotion testing

Suppression of hyperactivity is probably one of the most widely used preclinical behavioral measures of antipsychotic drug action. Most, if not all, clinically effective antipsychotic drugs – regardless of their typical/atypical nature – reduce hyperactivity in animals. However, not all drugs that suppress activity have antipsychotic properties in the clinic. Therefore, the major limitation of hyperactivity models is the presence of false positive results.

B. Prepulse inhibition

Prepulse inhibition (PPI) of the startle response has also been used to examine deficits in sensorimotor gating. Schizophrenic patients display PPI deficits, which can also be produced in laboratory animals using pharmacological, neurodevelopmental and genetic approaches. PPI models have face, predictive, and construct validity. Briefly, PPI experiments are conducted as follows. Animals or humans are subjected to a discrete high intensity stimulus (pulse) that evokes a startle response. Typically, the startle response is reduced by the presentation of a low intensity stimulus (normally a tone) 100-200 ms before the pulse. Drugs of several classes, such as dopaminergic D2 receptor agonists, 5-HT$_2$ receptor agonists or NMDA receptor agonists produce a deficit in PPI that resembles that occurring in schizophrenic patients (see Geyer et al., 2001 for review). Compared to locomotion testing, the PPI test shows less false positives.

C. Conditioned avoidance response test

Conditioned avoidance response (CAR) test is used to predict drug's antipsychotic potential. In this test, animals are placed in a box with two compartments and a grid floor, and are trained to move into the adjacent compartment (response) within few seconds upon presentation of the conditioned stimuli (tone and light) in order to avoid the appearance of the unconditioned stimulus (mild food-shock). Although CAR does not emulate any aspect of the clinical syndrome of schizophrenia, it has predictive validity. Most clinically effective antipsychotics reduce performance in this test.

D. Animal models of negative symptoms

In schizophrenia, negative symptoms encompass, among others, anhedonia, flat affect, abolition and social withdrawal. Two models of **social withdrawal** have been developed in animals: amphetamine-induced social isolation (Ellenbroek, 1991) and social withdrawal in animals treated with phencyclidine (PCP) (Sams-Dodd, 1996). However, there is an urgent need for animal models focusing on these symptoms.

E. Animal models of cognitive symptoms

One of the main objectives of the NIMH-MATRICS program is to facilitate the development of effective therapeutic approaches to treat the cognitive deficits in schizophrenia, as the clinically efficacy of current typical and atypical antipsychotics is modest (see Hagan and Jones, 2005 for review). There are several primary cognitive domains that are affected in schizophrenic patients and that could potentially be modelled in animals: attention/vigilance, speed of processing, working memory, visual learning and memory, reasoning and problem solving, and social cognition (Nuechterlein et al., 2004). The Morris water maze has been used to determine long-term spatial memory, a behavior particularly dependent on hippocampal function. In this procedure, rats or mice are trained to detect and remember the presence of an invisible platform in a water tank. The T-maze test has been used to assess working memory in rodents. In this test, rats or mice are trained to run down the stem of the T-maze and choose an arm of the T configuration to obtain a reward. After a delay, the animal is returned to the beginning of the stem of the T-maze and it is again rewarded if it chooses the previously unvisited arm. Other tests like the radial arm maze or the 5-choice serial reaction time task (5-CSRTT) have also been used to examine working memory performance. In general, antipsychotics do not improve cognitive performance in the majority of preclinical cognitive tests (for a review, see Hagan and Jones, 2005).

Box 2. Clinical aspects of schizophrenia and relevant behavioral changes in animals	
Schizophrenia: clinical phenomena	Animal models: behavioral changes
1. Psychotic symptoms	Behaviors related to increased dopaminergic transmission: -Hyperlocomotion induced by DA agonists -Reduced haloperidol-induced catalepsy
2. Stereotypic behaviors	DA agonists-induced stereotypies (e.g., grooming, sniffing, etc.)
3. Psychotic symptoms induced by NMDA antagonists	NMDA antagonists-induced hyperlocomotion
4. Vulnerability to stress	Stress-induced hyperlocomotion
5. Information processing deficits	Sensorimotor gating (PPI, P50) deficits
6. Attention deficits	Deficits in latent inhibition
7. Cognitive deficits	Impaired performance in delayed alternation and spatial memory tests
8. Social withdrawal	Reduced contacts with unfamiliar partners

PPI, pre-pulse inhibition. Modified after Lipska and Weinberger (2003)

the pathogenesis of schizophrenia. For instance, neuronal oscillations within the EEG β and γ bands (15-80 Hz) are associated with intense mental activity and cognitive function (Whittington et al., 2000) and several studies suggest that the γ band oscillations are disrupted in schizophrenic patients (Waberski et al., 2004; Winterer et al., 2004). Fast oscillations are thought to depend on microcircuits involving local GABAergic interneurons. Hence, alterations in the γ band can be related to recent observations on neuropathology of schizophrenia involving a decrease in the function of chandelier GABAergic neurons (Lewis et al., 2005).

In addition to high frequency oscillations, low frequency oscillations (α, θ and δ) are also related to cognitive processing. Hence, the α wave range (8–12 Hz) is related to working memory (Basar et al., 1997) whereas the θ range (5–8 Hz) is involved in attention and signal detection (Basar-Eroglu et al., 1992) reflecting cortico-hippocampal interactions (Miller, 1991; Klimesch et al., 1994; Basar et al., 1999). Likewise, δ waves are involved in signal matching and decision making (Basar-Eroglu et al., 1992).

The capacity of antipsychotics to synchronize the EEG may be a critical factor for their therapeutic efficacy. The EEG profiles of a range of different antipsychotic agents such as haloperidol, chlorpromazine, levomepromazine and risperidone have been evaluated in conscious rats. In general, these agents synchronize the EEG, increasing the power spectra between 2 Hz and 30 Hz although there were marked differences among the individual profiles of EEG effects for each drug. Also, other drugs such as raclopride and sulpiride evoke a desynchronization of the EEG in the PFC, showing a complex drug interaction with cortical activity (Sebban at al., 1999a, b).

2.4.3 Pharmacological models of schizophrenia

2.4.3.1 NMDA/glutamate receptor antagonists: PCP, MK-801 and ketamine

The hypothesis of the *N*-methyl-D-aspartate (NMDA) receptor hypofunction in schizophrenia (Javitt and Zukin, 1991; Tamminga, 1998) stems from the observation that NMDA receptor antagonists such as phencyclidine (PCP) and ketamine can evoke positive and negative symptoms as well as the characteristic cognitive deficits that closely resemble those of the illness (Javitt and Zukin, 1991; Krystal et al., 1994; Malhotra et al., 1996; Newcomer et al., 1999). In addition, these drugs usually aggravate this array of symptoms in schizophrenic patients (for a review, see Krystal et al., 2003).

Dissociative anaesthetic drugs, such as ketamine and phencyclidine (PCP) can produce hallucinations and delusions in humans. These drugs have been shown to act as non-competitive antagonists of NMDA-type glutamate receptors. In rodents, PCP and the more potent NMDA channel blocker dizocilpine (MK-801) produce a behavioral syndrome that is characterized by increased locomotor activity, head weaving, ataxia, body rolling, and stereotyped motor patterns. This behavioral syndrome has been used as an animal model of psychosis, and both typical and atypical antipsychotics showed to be effective in reducing NMDA receptor antagonist-induced hyperactivity (Ogren, 1996). Although both PCP and MK-801 increase activity in rats and mice, it should be pointed out that the behavioral pharmacology of the two drugs is not identical and similarities between them should be considered carefully.

Blockade of NMDA receptors by antagonists induces PPI deficits in rodents and primates (Box 1, Fig. 1). Mansbach and Geyer (1989) first demonstrated that the non competitive NMDA antagonists, PCP and dizocilpine (MK-801), disrupt PPI strongly in rats. Other authors have replicated these findings and have demonstrated that competitive NMDA antagonists also produce PPI deficits. NMDA antagonists-induced PPI deficits appear to be more sensitive to clozapine-like atypical antipsychotics than to conventional antipsychotics (for a review, see Swerdlow and Geyer, 1998; see Geyer et al., 2001).

Figure 1. Effects of clozapine and haloperidol on PCP-induced deficits in PPI in monkeys.

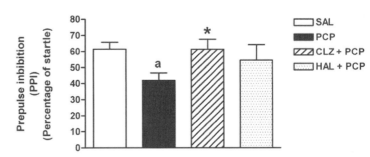

Results are expressed as MEAN ± SEM of the percentage of prepulse inhibition (PPI) in saline (SAL), phencyclidine (PCP) (0.12 mg/kg, i.m.), clozapine (CLZ) (2.5 mg/kg, i.m.) + PCP (0.12 mg/kg, i.m.) and haloperidol (HAL) (0.035 mg/kg, i.m.) + PCP (0.12 mg/kg, i.m.) treated monkeys (n=8). [a]differs from saline, P=0.001; *differs from PCP, P=0.009. Redrawn from Linn et al. (2003).

Acute NMDA receptor antagonism has been reported to increase the release of glutamate (Moghaddam et al., 1997; Adams and Moghaddam, 2001; Lorrain et al., 2003a), dopamine (Moghaddam and Adams, 1998; Mathé et al., 1999; Schmidt and Fadayel, 1996), serotonin (5-HT) (Martin et al., 1998; Millan et al., 1999; Adams and Moghaddam, 2001; Amargós-Bosch et al., 2006) and acetylcholine (Schiffer et al., 2001; Nelson et al., 2002) in the medial prefrontal cortex (mPFC) of rats (Fig. 2). Altogether these findings indicate that a hyperactivity of different transmitter systems in the mPFC is a general response to NMDA receptor hypofunction and can account for the behavioral effects elicited by NMDA receptor antagonists. An interesting finding is that this effect does not seem to result from a direct action of these compounds on NMDA receptors located in the mPFC because the local application of PCP or ketamine did not produce any change in the concentration of 5-HT or glutamate (Lorrain et al., 2003a; Amargós-Bosch et al., 2006). This suggests that the NMDA receptors responsible for such actions are located outside the mPFC, possibly in GABA neurons that would tonically inhibit glutamatergic efferents in areas that project densely to mPFC such as hippocampus, thalamus or amygdala. According to the hypothesis of the disinhibition of glutamatergic input to the mPFC, PCP and ketamine would increase glutamate release onto non-NMDA receptors in the mPFC (Moghaddam et al., 1997; Krystal et al., 2003; Lorrain et al., 2003a).

Figure 2. Effect of the systemic administration of phencyclidine (PCP, 5 mg/kg i.p.) on the 5-HT concentration in dialysates from medial prefrontal cortex (mPFC) in freely moving rats.

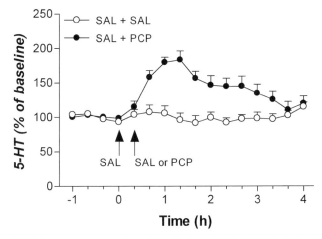

The 5-HT increase was not produced when PCP (10, 30, 100 or 300 µM) was locally applied in mPFC, suggesting that the main action of PCP to produced this effect takes place outside the PFC, possibly in afferent areas such as the hippocampus and/or certain thalamic nuclei, which project to mPFC. However, the increase in dialysate 5-HT was cancelled by the prior treatment with 1 mg/kg of the atypical antipsychotic clozapine. Redrawn from Amargós-Bosch et al. (2006).

In addition, neurochemical changes have been observed in other brain areas. For instance, noncompetitive NMDA receptor antagonists also enhance the efflux of glutamate in the nucleus accumbens (Razoux et al., 2006), acetylcholine in the retrosplenial cortex and hippocampus (Kim et al., 1999; Hutson and Hogg, 1996), 5-HT in the nucleus accumbens (Millan et al., 1999), noradrenaline in nucleus accumbens and hippocampus (Yan et al., 1997; Lorrain et al., 2003b; Swanson and Schoepp, 2003), and dopamine in limbic areas such as nucleus accumbens, hippocampus and ventral pallidum (Whitton et al., 1994; Millan et al., 1999; Mathe et al., 1999; Kretschmer, 2000; Schiffer et al., 2001; Greenslade and Mitchell, 2004), albeit to a much lower extent than that in the mPFC. This latter finding is in line with data in humans showing that ketamine does not increase the availability of dopamine at D_2 receptors in the striatum (Kegeles et al., 2002). The increase of dopamine release in the nucleus accumbens produced by systemic MK-801 is suppressed by the local application of the AMPA/kainate receptor antagonist CNQX in the ventral tegmental area (Mathe et al., 1998), which suggests that this effect of MK-801 is mediated through an increased glutamatergic stimulation of the mesolimbic dopaminergic system.

The effects of antipsychotic drugs on the neurochemical changes elicited by NMDA antagonists are much less known than those of behavioral consequences. In a recent work, we showed that the increased efflux of 5-HT produced by the systemic administration of PCP and ketamine was reversed by the atypical antipsychotics clozapine and olanzapine, but not by the classic antipsychotic haloperidol (Amargós-Bosch et al., 2006). These effects are shared by the 5-HT$_{2A/2C}$ antagonist, ritanserin, and the selective α_1-adrenoceptor antagonist, prazosin (Amargós-Bosch et al., 2006). Prazosin also blocked the MK-801-induced increase in dopamine release in the nucleus accumbens (Mathe et al., 1996). Thus, the blockade of α_1-adrenoceptors may play a role in reducing the enhanced dopaminergic transmission seen in schizophrenia. It is also likely that 5-HT$_{2A/2C}$ receptors also play an important role in the therapeutic effects of atypical antipsychotic drugs.

However, the effects of antipsychotic drugs on glutamate release are controversial. Thus, neither clozapine nor haloperidol seem to affect the increase in cortical glutamate evoked by PCP (Adams and Moghaddam, 2001) although it has been shown that the novel compound NRA0045, which potently blocks D_4, 5-HT$_{2A}$ and α_1-adrenergic receptors, inhibited PCP-induced glutamate efflux in the medial prefrontal cortex (Abekawa et al., 2003). In contrast, the congener compound NRA0160, which selectively blocks D_4 receptors, was unable to produce the same effect (Abekawa et al., 2003).

Early *in vivo* electrophysiological studies of the effects of PCP on mPFC neuronal activity were conducted using microiontophoretical drug application and showed a differential effect of PCP in superficial and deep layers (Gratton et al., 1987). More recently, Suzuki et al. (2002) showed that systemic administration of PCP dramatically increased the spontaneous firing rate of mPFC neurons (see also Fig. 3) with a timing similar to that of increased behavioral activity, which suggests that the exacerbation of psychotic symptoms produced by PCP in schizophrenic patients may be related to an uncontrolled increase of the activity of PFC neurons. As observed above in neurochemical studies, the local application of PCP was unable to elicit such an effect, which further supports the notion that the site of action of PCP was outside the PFC, likely in afferent excitatory areas to this cortical region. Among these areas, the hippocampus may play an important role, since the local application of PCP or MK-801 into the ventral hippocampus increased spontaneous firing of mPFC neurons together with augmentation of locomotor activity (Jodo et al., 2005). However, it remains to be established whether the systemic effects of PCP and MK-801 on cortical activity are only accounted for by an effect in the hippocampus or other areas are also involved.

Despite the similarities in behavioral effects between PCP and methamphetamine in animals, the latter drug did not modify the firing activity of mPFC neurons (Jodo et al., 2003), in concordance with clinical studies indicating that PCP reproduces the symptom profile of schizophrenia more accurately that amphetamine or methamphetamine (Javitt and Zukin, 1991).

Administration of NMDA receptor antagonists may model two pathophysiologic aspects of schizophrenia: the disturbance of cortical connectivity (deficient or aberrant connectivity) and the disinhibition of glutamatergic networks (by producing a transient neural network dysfunction) (Krystal et al., 2003). In PFC, these effects are reflected by an increased and disorganized activity of pyramidal neurons (Jackson et al., 2004).

Studies in freely moving rats using multiple unit recordings have found that NMDA receptor blockade by MK-801, at doses that impair working memory, increase the firing rate of most PFC neurons recorded, with a concurrent reduction of the organized bursting activity. This suggests that NMDA receptor deficiency may disrupt frontal lobe function by two distinct mechanisms: an increase in disorganized spike activity, which may enhance cortical noise and information overflow; and a decrease in burst activity, which may reduce transmission efficiency in cortical neurons (Jackson et al., 2004).

Additional evidence supporting that NMDA receptor antagonists produce their psychotomimetic effects, at least in part, by affecting PFC function comes from whole cell recordings of PFC neurons in rat brain slices where ketamine and subanaesthetic concentrations of PCP decrease dendritic glutamate-induced bursting. Changes in bursting would alter information processing in the PFC thereby contributing to PCP's effects (Shi and Zhang, 2003).

Figure 3. Example of the increase in the activity of an identified pyramidal neuron in the rat mPFC produced by the i.v. administration of phencyclidine (PCP; 0.25 mg/kg).

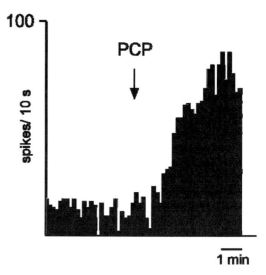

This unit exhibited a spontaneous firing rate of 1.2 spikes/s and was maximally increased by PCP near to 8 spikes/s (almost 7-fold increase) (Kargieman et al., unpublished observations).

In accordance with clinical evidence showing EEG abnormalities in schizophrenic patients (Matsuura et al., 1994), PCP was shown to decrease the power spectrum in the range 9–30 Hz in the rat. Moreover, in close analogy with neurochemical observations (see above), this effect was antagonized by clozapine, prazosin and M100907 (selective 5-HT$_{2A}$ receptor antagonist) (Sebban et al., 2002) suggesting that 5-HT$_{2A}$ receptor and α_1-adrenoreceptor blockade in PFC by antipsychotics may contribute to their therapeutic effect.

PCP and MK-801 also alter the activity of the dopaminergic neurons in the ventral tegmental area. The systemic administration of these NMDA antagonists decreases the burst activity of DA neurons mostly projecting to PFC and increases the burst activity of DA neurons projecting to nucleus accumbens (Murase et al., 1993). However, these observations have not been replicated and have to be taken with some caution given the complex projection pattern of dopaminergic neurons (Mathe et al., 1999; Carr and Sesack, 2000).

2.4.3.2 Dopaminergic agonists

The involvement of dopaminergic transmission in the pathophysiology of schizophrenia originated from two fundamental observations. On the one hand, drugs of abuse such as amphetamine and cocaine that raise extracellular DA concentration in the brain, can cause psychotic episodes similar to the positive symptoms of schizophrenia (Kornetsky and Markowitz, 1978; Ellenbroek and Cools, 1990). However, all effective antipsychotic drugs (both classic and atypical) are antagonists of DA D_2 receptors (yet atypical antipsychotics are preferential 5-HT_2 receptor antagonists; Meltzer, 1999). Psychotic symptoms can result from an aberrant reward prediction induced by an excessive dopaminergic transmission, an effect that can be reversed by DA receptor blockade (Kapur, 2004). Hence, it is not surprising that dopaminergic drugs have been widely used as pharmacological models of schizophrenia, although their popularity has declined in parallel with the increasing use of NMDA receptor antagonists.

The effect of psychotropic drugs such as amphetamine on locomotor activity in rodents has been used to model positive symptoms of schizophrenia, particularly psychosis. While lacking face validity, this model has served as a good preclinical predictor of antipsychotic activity. It has been suggested that amphetamine-induced hyperactivity may be used to distinguish typical versus atypical antipsychotics (Bardgett, 2004). The induction of psychotic symptoms by amphetamine is most likely due to an enhancement of dopaminergic transmission in all brain areas innervated by dopaminergic terminals. This effect takes place concurrently in cortical and limbic circuits, where the excess DA can evoke psychotic symptoms, and in motor circuits of the basal ganglia, where the increased activation of DA receptors evokes hyperactivity. Indeed, amphetamine markedly increased DA release throughout the brain in a tetrodotoxin-resistant manner (Nishijima et al., 1996). However, little is known on the effects of antipsychotics on the DA release induced by amphetamine. It is likely that these agents can reverse the behavioral effects of amphetamine postsynaptically (see below), i.e., by blocking DA D_2 receptors, without exerting a marked effect on the presynaptic component of dopaminergic transmission. However, in the nucleus accumbens and striatum the amphetamine-evoked DA release is reduced by selective 5-HT$_{2A}$, but not by 5-HT$_{2C}$ receptor antagonists (Porras et al., 2002). Furthermore, local perfusion of amphetamine in the mPFC produced a dose-dependent increase in norepinephrine, acetylcholine and serotonin levels, although the magnitude of this effect depended on the are within the mPFC (Hedou et al., 2000). Also, haloperidol potentiated the amphetamine-induced release of dopamine in the mPFC and striatum in a D$_2$-dependent manner (Pehek, 1999).

Although typical and atypical antipsychotic drugs have been shown to reduce amphetamine-induced hyperactivity (Ellenbroek, 1993), it is possible to distinguish these two groups of compounds by comparing the dose ratio required to produce such effect. Thus, while haloperidol was shown to be equipotent in reducing the hyperactivity produced by moderate (2 mg/kg) and low (0.5 mg/kg) doses of amphetamine, the dose of atypical drugs such as clozapine, risperidone, olanzapine, ziprasidone and quetiapine required to inhibit the hyperactivity produced by a moderate dose of amphetamine was around ten times higher than the dose required to inhibit the locomotor responses to a lower dose of amphetamine (Arnt, 1995). This difference can be related to the different pharmacological profile of haloperidol and atypical antipsychotics and the fact that amphetamine induces its locomotor effects by activating postsynaptic DA receptors. Hence, whereas haloperidol shows high affinity for DA D$_2$ receptors and occupies a large proportion of such receptors at standard doses, atypical antipsychotics need to be administered at higher doses to elicit a moderate-high occupancy of DA D$_2$ receptors (Schotte et al., 1993).

Similar to the indirect DA agonist amphetamine, direct DA agonists also produce stimulant locomotor effects that are reversed by antipsychotic drugs. Hence, the mixed D$_1$/D$_2$ receptor agonist apomorphine induces hyperactivity in rodents, which is reversed by both typical and atypical antipsychotics (Ellenbroek, 1993). Moreover, apomorphine can induce climbing behavior in mice (Protais et al., 1976), an effect also blocked by antipsychotics (Migler et al., 1993).

Dopaminergic agonists also produce disruptions in PPI of startle (Box 1). Mansbach et al., (1988) reported the first examples of disruptions in PPI produced by systemic administration of apomorphine and amphetamine in intact rats. The effects of direct and indirect DA agonists on PPI have been extensively replicated (for a review, see Geyer et al., 2001). These studies have demonstrated that PPI is more sensitive to the effects of direct DA agonists, such as apomorphine than to DA releasers, such as amphetamine. Selective DA $D_{2/3}$ agonists such as quinpirole (Peng et al., 1990), but not D_1 agonists, also evoke the disruption of PPI produced by apomorphine and amphetamine. However, it is well documented that both typical and atypical antipsychotics restore PPI in rats treated with DA agonists (for a review, see Geyer et al., 2001).

2.4.3.3 Serotonergic agents: lysergic acid diethylamide and related compounds

D-Lysergic acid diethylamide (LSD) was first observed to have hallucinogenic properties 50 years ago. It was proposed that the hallucinogenic effects of LSD might result from the antagonism of 5-HT actions in the CNS based on its effect on the actions of 5-HT in smooth muscle preparations (Gaddum and Hameed, 1954; Woolley and Shaw, 1954) Comparison of the hallucinogenic properties of LSD and certain psychotic symptoms of schizophrenia led to the serotonergic hypothesis of the pathogenesis of schizophrenia, either suggesting a deficit (van Praag, 1992) or an excess of 5-HT, since it was reported that LSD could also mimic certain effects of 5-HT (Shaw and Woolley, 1956). Consistent with the initial hypothesis of schizophrenia it was found that LSD potently inhibited the firing activity of 5-HT neurons in the dorsal raphe (Aghajanian et al., 1968) through a direct action on 5-HT autoreceptors (later known as 5-HT$_{1A}$ subtype) located in the somatodendritic region of the 5-HT neuron (Aghajanian et al., 1972). Yet, despite the high density of somatodendritic 5-HT$_{1A}$ autoreceptors and the high affinity of LSD for these receptors, further work indicated that LSD's hallucinogenic action depends on the activation of 5-HT$_2$ receptors (Titeler et al., 1988; Glennon et al., 1984; Glennon, 1990). Also, electrophysiological studies showed that indolamine and phenethylamine hallucinogens (LSD and DOI) display a partial agonist action at 5-HT$_{2A}$ receptors (Marek and Aghajanian, 1996).

5-HT$_{2A}$ receptors are located in various brain regions (López-Giménez et al., 1997; Pazos et al., 1985), but particularly in the cerebral cortex, where they show a layered distribution and a rostrocaudal decline, being particularly enriched in the PFC (Pompeiano et al., 1994; Amargós-Bosch et al., 2004). They are expressed in pyramidal neurons and in large and medium-size GABAergic interneurons (Santana et al., 2004). Despite that the physiological role of cortical 5-HT$_{2A}$ receptors remains largely unknown, recent data suggest their involvement in working memory (Williams et al., 2002). Indeed, the hallucinogenic effects of serotonergic drugs are consistent with the high density of 5-HT$_{2A}$ receptors in cortical areas which are involved in cognition and perception.

The action of hallucinogens (LSD and DOI), via 5-HT$_{2A}$ receptors, have been found to enhance glutamatergic transmission in PFC (Aghajanian and Marek, 1999). Consistent with the existence of descending excitatory pathways from PFC to the midbrain aminergic nuclei, the local application of DOI in mPFC enhanced the activity of serotonergic neurons of the dorsal raphe and of dopaminergic neurons of the ventral tegmental area as well as the release of the corresponding neurotransmitter in mPFC (Martín-Ruiz et al., 2001; Bortolozzi et al., 2005). These effects are presumably mediated by an increased excitatory transmission through pyramidal axons projecting to serotonergic and dopaminergic neurons (Aghajanian and Wang, 1977; Thierry et al., 1979; Thierry et al., 1983; Sesack et al., 1989; Hajós et al., 1998; Peyron et al., 1998; Carr and Sesack, 2000; Celada et al., 2001).

Regarding the effect of 5-HT$_{2A}$ receptor activation on cortical pyramidal neurons conflicting results have been reported, thus, the microiontophoretic application of DOI to anesthetized rats predominantly inhibited prefrontal neurons, although low ejection currents potentiated the excitatory effect of glutamate (Ashby et al., 1989, 1990). Using *in vitro* intracellular recordings of pyramidal neurons, depolarizing and hyperpolarizing effects of 5-HT through the activation of 5-HT$_{2A}$ receptors have been reported (Araneda and Andrade, 1991; Tanaka and North, 1993; Aghajanian and Marek, 1997, 1999, 2000; Arvanov et al., 1999; Zhou and Hablitz, 1999). Likewise, the systemic administration of DOI markedly increased the firing rate of nearly 40% of pyramidal neurons in the mPFC, inhibited ~20% and left the rest unaffected (Puig et al., 2003) (Fig. 4). These contrasting effects of DOI are probably due to the presence of 5-HT$_{2A}$ receptors in pyramidal and GABAergic neurons (Santana et al., 2004).

Figure 4

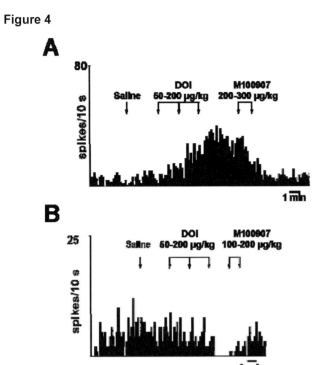

(A) The systemic administration of DOI (50–200 μg i.v.) markedly enhanced the firing activity of a 40% of pyramidal neurons in the mPFC of the chloral hydrate anesthetized rat. (B) A smaller proportion of neurons (ca. 20%) had their firing rate fully suppressed by the same DOI doses. Both effects were antagonized by the selective 5-HT$_{2A}$ receptor antagonist M100907 and the suppressing effect of DOI was also sensitive to the GABA$_A$ receptor antagonist picrotoxin (not shown), suggesting the involvement of 5-HT$_{2A}$ receptors located on GABAergic interneurons in his effect. Redrawn from Puig et al., 2003.

Likewise, *in vivo* extracellular recordings of cortical pyramidal neurons showed that the electrical stimulation of the dorsal raphe inhibits –via 5-HT$_{1A}$ receptors– and excites – via 5-HT$_{2A}$ receptors – pyramidal neurons in the rat mPFC (Hajós et al., 2003; Puig et al., 2003, 2005; Amargós-Bosch et al., 2004).

Hence, the involvement of dysfunctional brain serotonergic systems in schizophrenia is supported by studies on the behavioral effects of serotonergic hallucinogens such as LSD, DOI, mescaline, psilocybin and *N,N*-dimethyl-tryptamine. These compounds can produce important sensory distortions and visual hallucinations, but less frequently evoke severe delusions or auditory hallucinations, characteristic of schizophrenia. From a neurochemical point of view, these compounds (albeit having different chemical structures) are potent $5\text{-}HT_{2A}$ agonists. However, most atypical antipsychotics are $5\text{-}HT_{2A}$ receptor antagonists (Kroeze and Roth, 1998; Meltzer, 1999). In rodents, DOI enhances the release of 5-HT in the mPFC, an effect blocked by classic (chlorpromazine, haloperidol) and atypical (clozapine, olanzapine) antipsychotics, as well as by $5\text{-}HT_{2A}$ and α_1-adrenergic receptor antagonists (Bortolozzi et al., 2003; Amargós-Bosch et al., 2003, 2004).

The involvement of other 5-HT receptors in the symptoms of schizophrenia is supported by PPI studies in rodents. Thus, direct $5\text{-}HT_{1A}$ agonists (8-OH-DPAT), $5\text{-}HT_{1A/1B}$ (RU24969) and $5\text{-}HT_{2A}$ (DOI) are capable of disrupting PPI. Compared to PPI models involving dopaminergic agents, those employing 5-HT agents are less generally sensitive to antipsychotic drugs (for a review, see Geyer et al., 2001).

2.4.4 Neurodevelopmental models

Neurodevelopmental models are based on experimentally induced disruption of brain development that becomes evident in an adult animal in the form of an altered brain neurochemistry and aberrant behavior. Some models rely upon lesions of selective areas which later affect a common neural circuitry comprising cortical regions, particularly the PFC, the hippocampal formation and the nucleus accumbens. The timing of brain damage seems critical for the behavioral consequences since the same type of lesion performed during adulthood does not elicit the same effect than that produced during early development (Bertolino et al., 1997).

Neurodevelopmental models of schizophrenia include: (a) lesions of the ventral hippocampus in neonatal rats (Lipska et al., 1993), (b) impaired neurogenesis (Flagstad et al., 2004), (c) early-life stressful experiences such as isolating animals post-weaning (Geyer et al., 1993), and (d) prenatal exposure to influenza virus (Mednik et al., 1988). These manipulations are associated with: (a) post-pubertal behavioral and cognitive abnormalities that resemble those observed in schizophrenia, and (b) neuroanatomical and cellular alterations in cortical and subcortical systems implicated in the pathophysiology of the disease.

2.4.4.1 Neonatal hippocampal lesion model

Lipska and colleagues (1993) developed an interesting model wherein an excitotoxin (typically ibotenic acid) is injected into the ventral region of the rat hippocampus one week after birth (P7). The purpose of this model is to interfere with the development of the hippocampus, which is strongly linked to human schizophrenia and is extremely sensitive to damage produced by several insults, such as hypoxia or excitotoxins, including excessive glutamatergic transmission. The rationale is that a transient damage produced early in postnatal life in this brain area may later result in persistent behavioral abnormalities in adult life, as observed in schizophrenic patients, who present a higher incidence of obstetric complications than the normal population (Lewis and Murray, 1987) and experience the first psychotic episodes in early adulthood. Although the lesion is restricted to the ventral hippocampus and the subiculum, cellular and neurochemical abnormalities have been identified in PFC of adult rats, where these two hippocampal areas project. This suggests that a transient hippocampal lesion in early postnatal life can profoundly alter the brain circuitry presumably involved in schizophrenia in adulthood.

In lesioned rats, several behavioral abnormalities emerge around adolescence. Hence, juvenile rats subjected to the lesion have less social contacts than controls. During adolescence and adulthood, lesioned animals display a marked motor hyperresponsiveness to stress and stimulants, enhanced stereotypies, enhanced sensitivity to MK-801 and PCP, deficits in PPI and latent inhibition, impaired social behaviors as well as working memory deficits (Lipska and Weinberger, 2000). The delayed emergence of behavioral changes in rats with neonatal damage seems consistent with the delayed onset of schizophrenia in late adolescence and early adulthood. Interestingly, antipsychotic drugs normalize some lesion-induced behaviors. Thus, a three-weeks treatment with haloperidol eliminates the hyperlocomotion induced by neonatal hippocampal lesions in rats (Lipska and Weinberger, 1993). However, lesioned rats are less sensitive than non-lesioned rats (sham) to the catalepsy produced by high doses of haloperidol (Lipska and Weinberger, 1993).

More recently, the same group reported that the application of tetrodotoxin in the ventral hippocampus (in the CA1 and CA2 area) during a critical period in maturation of intracortical connection (P7) also produces a wide range of behavioral and cellular changes relates to certain dopaminergic- and NMDA receptor blockade-induced behaviors (Lipska et al., 2002). Tetrodotoxin does not damage brain cells but temporarily prevents the correct function of neural networks by suppressing nerve impulse propagation. It is therefore interesting to observe that such a mild lesion performed at a critical developmental time can produce delayed disturbances of brain function.

The impact of the neonatal hippocampal lesion on the electrophysiological properties of prefrontal cortical neurons in adult rats has been studied by analyzing their responses to the electrical stimulation of specific afferents. Lesioned animal show an overresponse (excessive firing) to mesocortical (midbrain dopaminergic and GABAergic projections) stimulation during adulthood which is not observed in rats subjected to the same lesion in adulthood (O'Donnell et al., 2002).

Rats with neonatal excitotoxic hippocampal lesion display behaviors potentially linked to an increased mesolimbic and nigrostriatal dopaminergic transmission. Hence, neurochemical, molecular and anatomic alterations in the mesocorticolimbic systems have been reported in these animals (Flores et al., 1996a; Lipska and Weinberger 2000; O'Donnell et al., 2002). The nucleus accumbens receives a dense synaptic input from the ventral hippocampus (subiculum) which is responsible of the spontaneous changes in up-down states in membrane potential, which determine the firing probability (Goto and O'Donnell, 2001). Although these spontaneous changes are not altered in lesioned rats at adulthood, there is dramatic increase in firing in response to the stimulation of the ventral tegmental area (Goto and O'Donnell, 2002), an effect reversed by subchronic antipsychotic treatment and by PFC lesion (Goto and O'Donnell, 2004). These results illustrate that neonatal hippocampal lesions can profoundly alter mesolimbic dopaminergic transmission, possibly by affecting the input from PFC to the ventral tegmental area-nucleus accumbens circuit. The key role of PFC is also supported by the fact that certain abnormal behaviors (hyperlocomotion to novelty and amphetamine) produced by early neonatal damage are abolished by PFC lesions in adulthood (Lipska et al., 1998).

In addition to the above functional changes, neonatal hippocampal lesions also produce neurochemical alterations which are evident in adulthood, such as reduced levels of the mRNA encoding the dopamine transporter (DAT) in the substantia nigra and ventral tegmental area, glutamic acid decarboxylase-67 (GAD_{67}; the synthesizing enzyme for GABA) in the mPFC and the D_2 receptor in the striatum. However, no changes in the expression of tyrosine hydroxylase, neurotensin, proenkephalin or D_1 and D_3, receptors were observed in any brain region examined (Lipska et al., 2003a, b). GAD_{67} mRNA was up-regulated by chronic clozapine and haloperidol in the nucleus accumbens and the striatum and by chronic haloperidol in the prefrontal cortex (Lipska et al., 2003a). Taken together, these findings suggest that the disruption of the hippocampo-prefrontal connectivity plays an important role in the pathophysiology of schizophrenia.

Developmental lesions of other brain corticolimbic structures involved in schizophrenia (e.g., thalamus, PFC) have also been considered as models (Rajakumar et al., 1996; Flores et al., 1996b) yet with much less widespread use than the neonatal hippocampal model.

2.4.4.2 Models based on the disruption of neurogenesis and discrete developmental lesions in the limbic-neocortical circuit

Postmortem studies of schizophrenia have reported alterations in cortical cytoarchitecture (Akbarian et al., 1993; Kirkpatrick et al., 1999; for reviews, see Harrison, 1999 and Lewis et al., 2005) which may result from an abnormal cortical development in early embryonic stages. Consequently, animal models have been devised to mimic such cellular alterations such as that produced by mitotic toxin administration. These manipulations induce morphological changes in cortex (Talamini et al., 1998) and in hippocampus which are accompanied by a variety of behavioral signs such as locomotor hyperactivity, stereotypies, cognitive impairments, disruption of latent inhibition and PPI as well as electrophysiological abnormalities such as an altered responsiveness to DA (Lavin et al., 2005; Moore et al., 2006).

Other models are based on the systemic administration of nitric oxide synthase (NOS) inhibitors (Black et al., 1999). Male rats exposed to the NOS inhibitor, L-nitroarginine at P3-P5, show a significant locomotor hypersensitivity to amphetamine and deficits in PPI when adults. Finally, the exposure to mitotic toxins constitutes another model of neurogenesis disruption and subsequent disarrangement of brain circuits potentially involved in schizophrenia (Johnston et al., 1988).

2.4.4.3 Perinatal exposure to infections or malnutrition

In the search of early developmental factors that may predispose to schizophrenia there have been reports linking this disorder to gestational complications (Woerner et al., 1973; Hultman et al., 1997; De Lisi et al., 1988; Dalman et al., 1999), perinatal exposure to alcohol (Lohr and Bracha, 1989), malnutrition (Susser and Lin, 1992) or prenatal exposure to infections (Adams et al., 1993). Moreover, schizophrenic patients seem to have a lower weight at birth than the corresponding controls (Kunugi et al., 2001). In this context, different studies have been designed to test the role of specific gestational factors in the pathogenesis of the disease and to develop animal models.

Models based in gestational malnutrition result in alterations in brain development by affecting cell migration/ differentiation and formation of neuronal circuits (Lewis et al., 1977; Lewis et al., 1979; Cintra et al., 1997, Granados-Rojas et al., 2002). Several epidemiological studies have shown that prenatal exposure to influenza virus is a predisposing factor in schizophrenia (Mednick et al., 1988). Prenatal exposure of mothers to mouse-adapted and non-adapted influenza virus suggests the presence of a time window (around gestational day 13) for an effect of the virus on the organization of hippocampal pyramidal neurons of mice born from infected mothers, yet the effect size was small (Cotter et al., 1995). In addition, reductions of the thickness of the neocortex and the hippocampus, suggestive of alterations in corticogenesis, have been reported in mice infected perinatally with human influenza virus (Fatemi et al., 1999). Likewise, the peripheral administration of the bacterial endotoxin lipopolysacharide to pregnant rats disrupts sensorimotor gating in adult rats born from treated mothers. This effect is reversed by the typical (haloperidol) and atypical (clozapine) antipsychotic treatments (Borrell et al., 2002).

2.4.5 Genetic models

Taking advantage of technologies involving gene-targeted deletions or gene transfer techniques, new animal models of schizophrenia are emerging. Genetic manipulations offer the advantage, compared with previous models, to selectively target specific proteins and to study the cascade of subsequent pathological changes induced by the gene deletion or overexpression. In contrast, these models face two main problems. On the one hand, schizophrenia is a polygenic disease, which makes difficult to model the constellation of schizophrenia symptoms with changes in a single gene. On the other hand, gene changes are present since conception in some models (e.g., permanent knock-outs or knock-ins) what makes that redundant or adaptive mechanisms are operant to reduce the impact of the gene change. Several reviews have summarized the rationale, neurobiological background and usefulness of these models (Gainetdinov et al., 2001; Geyer et al., 2002; Lipska and Weinberger 2000, 2002; Ellenbroek, 2003). The first models used were based in inactivation of genes of neurotransmitter receptors or transporters classically related to schizophrenia and its treatment (e.g., DA, glutamate/NMDA, etc.) Further, the finding of human genes related to schizophrenia allowed to generate mouse models based on the human genetic susceptibility (Weinberger et al., 2001; Stefansson et al., 2002; Goldberg and Weinberger 2004; Kirov et al., 2005). A particular emphasis has been put on cognitive dysfunction, one of the key symptoms of schizophrenia, given the availability of models to assess working memory in rodents (Dalley et al., 2004).

Several schizophrenia models have been developed based on the total or partial genetic inactivation of neurotransmitter neurotransmitter receptors, such as DA, A_{2A} adenosine, α_2-adrenergic, and NMDA receptors (Sibley, 1999; Ralph-Williams et al., 2002; Wang et al., 2003; Lahdesmaki et al., 2004; Mohn et al., 1999). Genetic inactivation of G proteins associated with dopamine D_2-like receptors has also been studied. Thus, mice deficient in the α-subunit of G(z) show enhanced sensitivity to the disruption of PPI and locomotor hyperactivity caused by dopaminergic stimulation (van den Buuse et al., 2005).

Another promising genetic approach is the identification of predisposing candidate genes by selecting rodent strains that exhibit behavioral deficits linked to schizophrenia. These genes may then be used to identify homologous human genes relevant for the etiology of schizophrenia. Using this strategy, the genetic defect in α_7 nicotinic cholinergic receptors is being tested as a predisposing factor in schizophrenia (Stevens et al., 1998; Freedman et al., 1997). Finally, there are currently in progress new animal models of schizophrenia that combine genetic and neurodevelopmental approaches.

2.4.5.1 Targeting glutamate signaling pathways

Based on the NMDA hypofunction model of schizophrenia (see above), several genetic models have been developed with the general aim to reduce the expression or function of glutamatergic NMDA receptors. Functional NMDA receptors are composed of a common NR1 subunit and one of the various NR2 subunits (NR2A to NR2D) and contain several binding and regulatory sites. In particular, the glycine binding site has received much attention given its ability to modulate the activity of the NMDA channel.

The role of NR1 in schizophrenia is supported by changes in NR1 expression in the brain of schizophrenics (Sokolov, 1998; Meador-Woodruff and Healy, 2000). Therefore, attempts were made to generate mutants with deletions of one of the various subunits of the NMDA receptor. However, knockout mice for the NR1 subunit are not viable because they die within 8–15 h after birth (Forrest et al., 1994; Li et al., 1994). For this reason, a new mutant expressing only 5–10% of the NR1 subunit was generated. This mutant mice (*NR1neo –/–*) reaches adulthood and displays several behavioral abnormalities, including increased motor activity, stereotypies and deficits in social and sexual interactions. These behavioral alterations are similar to those observed in pharmacologically (PCP or MK-801) induced animal models of schizophrenia and can be ameliorated by treatment with haloperidol or clozapine, antipsychotic drugs that antagonize dopaminergic and serotonergic receptors (Mohn et al., 1999). Furthermore, DA release and metabolism are unchanged in striatum of *NR1neo –/–* mice (Mohn et al., 1999). Hence, a low expression of the NMDA receptor subunit 1 (NR1) in mice results in abnormal brain development, hyperlocomotion, stereotypy, altered social behavior and cognitive dysfunction (Mohn et al., 1999; Duncan et al., 2004; Fradley et al., 2005).

Further, NR2A-knockout mice exhibit increased locomotor activity and increased turnover of DA and 5-HT in frontal cortex and striatum (Miyamoto et al., 2001). The hyperlocomotion is dependent upon activation of DA and 5-HT transmission (Miyamoto et al., 2001). Other mutant mouse such as the NR2D-knockout showed reduced DA, NA and 5-HT in the hippocampus (Miyamoto et al., 2002). Altogether, only two mutant mice can be considered as a model of schizophrenia: the *NR1neo −/−* and the NR2A-KO since they show the characteristic hyperlocomotion and stereotypies seen in other animal models of the illness.

The role of glycine on the NMDA receptor function has been studied using heterozygous mice expressing 50% of the gene encoding for the glycine transporter (GlyT1). This protein maintains extracellular glycine concentrations below the threshold to activate its modulatory site on the NMDA receptor complex. Electrophysiological and behavioral studies showed that reduced expression of GlyT1 enhances hippocampal NMDA receptor function and memory retention and protects against the amphetamine-induced disruption of sensory gating, suggesting that drugs which inhibit GlyT1 might have both cognitive enhancing and antipsychotic effects (Tsai et al., 2004).

Another model targeting a key molecule in the glutamate signaling pathway is the calcineurin knockout. The calcineurin A gamma subunit gene (PPP3CC) has been identified as a susceptibility gene for schizophrenia in human genetic association studies (Gerber et al., 2003). Mice lacking calcineurin show deficits in PPI, increased locomotor activity, decreased social interactions and impaired working memory. Also, they show increased locomotor response to a challenge of the non-competitive NMDA receptor antagonist MK-801 (Miyakawa et al., 2003).

2.4.5.2 Targeting DA signaling pathways

DA has been involved in the pathophysiology of schizophrenia for decades. Hence, it is not surprising that various animal models have been developed based on changes in the function of genes regulating dopaminergic transmission. One such model is the DA transporter knockout mouse. These mice show hyperactivity in a novel environment, stereotyped behavior and PPI deficits (Giros et al., 1996). At the neurochemical level it shows a decrease in brain tissue DA levels and a 300 times slower clearance of DA extracellular levels compared to control mice. D_1 knockout mouse (Xu et al., 1994) and D_2 knockout mouse (Baik et al., 1995) have also been developed. Behavioral studies revealed that the D_1 knockout mouse do not appear to be a good candidate to model schizophrenia symptoms since – despite showing locomotor hyperactivity – they show a reduced sensitivity to cocaine and amphetamine and normal PPI response (Xu et al., 2000).

D_2 knockout mice have been proposed as a model to study the mechanism of action of antipsychotic drugs. Interestingly, D_2 knockout mice are not sensitive to amphetamine-induced PPI deficits (Ralph-Williams et al., 2002). An isoform-specific knockout of the D_{2L} (the major isoform of brain D_2 receptors) has also been generated. The cataleptic effects of haloperidol are absent in D_{2L}-deficient mice, suggesting that this drug acts specifically on this isoform of the D_2 receptor. Likewise, the absence of D_{2L} receptors revealed that D_{2S} receptors inhibit D_1 receptor-mediated functions (Usiello et al., 2000).

DARPP-32 is a dopamine and cyclic adenosine monophosphate (cAMP) regulated phosphoprotein that regulates the efficacy of dopaminergic neurotransmission (Fienberg et al., 1998). Postmortem studies from schizophrenic patients revealed a decrease of DARPP-32 in layer II of prefrontal cortex (Albert et al., 2002). Hence a DARPP-32 knockout mouse has been generated (Fienberg and Greengard, 2000) that shows PPI deficits and attenuated stereotypic behavior induced by psychotomimetics (Svenningsson et al., 2003).

Neuregulin-1 has been identified as a potential susceptibility gene in schizophrenia (Stefansson et al., 2002) and there is anatomical evidence of the presence of the functional neuregulin-1 receptor, ErbB4 on DA neurons of the substantia nigra and ventral tegmental area. Neuregulin1-beta, a neuregulin-1 gene isoform that preferentially activates the ErbB4 receptor, elicits an overflow of DA in the striatum when injected close to the ipsilateral substantia nigra. These data are indicative that neuregulins can modulate the activity of mesostriatal dopaminergic neurons (Yurek et al., 2004).

2.4.5.3 Other genetic models

Reelin is a protein secreted by several neuronal populations throughout the brain which binds to transmembrane receptors located on adjacent cells triggering a tyrosine kynase cascade. This allows neurons to complete migration during development and to adopt their ultimate positions in laminar structures in the CNS. Thus, reelin plays a major role in the development of layered structures such as the cerebral cortex, the hippocampus or the cerebellum (Rice and Curran, 2001). The expression of reelin and its encoding mRNA appears to be significantly reduced in brain post-mortem brain samples of schizophrenic patients (Impagnatiello et al., 1998; Guidotti et al., 2000). Therefore, the *Reeler* mouse (a naturally occurring mutant mouse strain showing a mutation of the reelin gene) has been suggested as a neurodevelopmental genetic model of schizophrenia (Costa et al., 2002). This mutant mouse shows a decrease in reelin and in GAD67 expression, a higher density in neuronal packing, reduction in dendritic spine density and deficits in PPI, as observed in schizophrenic patients.

As mentioned above, the finding of human schizophrenia genes allow to develop newer genetic mouse models based on the human genetic susceptibility. Most of these genes have been associated to cognitive dysfunction and have mouse homologues. Manipulation of these genes in mice may alter their cognitive function and cognitive dysfunction may be tested in these animal model a key phenotype. Some examples of these genetics models based in susceptibility genes are the following (for a review, see Chen et al., 2006): (a) the transgenic model that overexpresses *COMT* (catechol-*O*-methyltransferase; Chen et al., 2005); (b) a heterozygous *NRG/1* knockout mouse lacking the transmembrane domain of the gene and showing a decrease in the expression of the receptor (Stefansson et al., 2002), and (c) the *Disc1* knockout still in process (Ishizuka et al., 2006).

2.4.6 Other animal models

Hippocampal damage in adult rats and mice as a consequence of the surgical infusion of neurotoxic compounds, such as the kainic acid, ibotenic acid, or NMDA directly into the hippocampus has been proposed as an animal model of schizophrenia (Bardgett et al., 1995). Rats with kainic acid-induced hippocampal damage show an increase in locomotor activity within the three-week period after surgery (Bardgett et al., 1995), an effect that atypical antipsychotics such as clozapine, risperidone and olanzapine suppresses with more efficacy than the typical drug haloperidol (Bardgett et al., 1998, 2002).

Several animal models have been developed to study the possible role of stress in the pathogenesis of schizophrenia. Hence, stressful life events in rodents, like maternal separation or social isolation produce a variety of hormonal, neurochemical and behavioral changes, including locomotor hyperactivity in a novel environment, learning impairments, anxiety, latent inhibition and PPI deficits. Some of these changes emerge in adulthood and are restored by typical and atypical antipsychotic treatments (Varty and Higgins, 1995; Ellenbroek et al., 1998; Bakshi et al., 1998).

2.4.7 Concluding remarks

The difficulties of modeling a complex psychiatric disease such as schizophrenia in experimental animals are beyond doubt. However, a number of different experimental paradigms based on pharmacological, neurodevelopmental or genetic approaches have been established over the years in order to mimic discrete aspects of the illness. Some of these models have face, predictive and construct validity and have been extensively used in antipsychotic drug research and also to investigate the possible pathological entity or entities causing the illness. However, we still lack a unitary perspective on the neurobiological changes occurring in these models due to the enormous complexity of the task. It is expected that further research will better characterize the impact of lesions, drugs and gene changes on the cellular and molecular elements potentially involved in schizophrenia. Moreover, it is expected that more refined mouse transgenic models based on human findings (e.g., susceptibility gene complexes) can shed more light on the pathogenesis and pathophysiology of schizophrenia, thus helping to develop new preventive and therapeutic approaches.

Acknowledgements

This work was supported by the Spanish Ministry of Science and Education (grant SAF-CICYT 2004-05525). Support from the Generalitat de Catalunya is also acknowledged (2005 SGR00758). A.C: is the recipient of a postdoctoral contract from the Fondo de Investigación Sanitaria (ISCIII-Ministry of Health). P. C. is recipient of a Ramón y Cajal contract from the Ministry of Science and Education.

References

Abekawa T, Honda M, Ito K, Koyama T. *Effects of NRA0045, a novel potent antagonist at dopamine D4, 5-HT2A, and alpha1 adrenaline receptors, and NRA0160, a selective D4 receptor antagonist, on phencyclidine-induced behavior and glutamate release in rats.* Psychopharmacology (Berl). **2003**, 169:247-256.

Abi-Dargham A, Rodenhiser J, Printz D, Zea-Ponce Y, Gil R, Kegeles LS, Weiss R, Cooper TB, Mann JJ, Van Heertum RL, Gorman JM, Laruelle M. *Increased baseline occupancy of D2 receptors by dopamine in schizophrenia.* Proc Natl Acad Sci U S A. **2000**, 97:8104-8109.

Adams BW, Moghaddam B. *Effect of clozapine, haloperidol, or M100907 on phencyclidine-activated glutamate efflux in the prefrontal cortex.* Biol Psychiatry. **2001**, 50:750-757.

Adams W, Kendell RE, Hare EH, Munk-Jorgensen P. *Epidemiological evidence that maternal influenza contributes to the aetiology of schizophrenia. An analysis of Scottish, English, and Danish data.* Br J Psychiatry. **1993**, 163:522-534.

Aghajanian GK, Foote WE, Sheard MH. *Lysergic acid diethylamide: sensitive neuronal units in the midbrain raphe.* Science. **1968**, 161:706-708.

Aghajanian GK, Haigler HJ, Bloom FE. *Lysergic acid diethylamide and serotonin: direct actions on serotonin-containing neurons in rat brain.* Life Sci I. **1972**, 11:615-622.

Aghajanian GK, Wang RY. *Habenular and other midbrain raphe afferents demonstrated by a modified retrograde tracing technique.* Brain Res. **1977**, 122:229-242.

Aghajanian GK, Marek GJ. *Serotonin induces excitatory postsynaptic potentials in apical dendrites of neocortical pyramidal cells.* Neuropharmacology. **1997**, 36:589-599.

Aghajanian GK, Marek GJ. *Serotonin, via 5-HT2A receptors, increases EPSCs in layer V pyramidal cells of prefrontal cortex by an asynchronous mode of glutamate release.* Brain Res. **1999**, 825:161-171.

Aghajanian GK, Marek GJ. *Serotonin model of schizophrenia: emerging role of glutamate mechanisms.* Brain Res Brain Res Rev. **2000**, 31:302-312.

Akbarian S, Bunney WE, Jr., Potkin SG, Wigal SB, Hagman JO, Sandman CA, Jones EG. *Altered distribution of nicotinamide-adenine dinucleotide phosphate-diaphorase cells in frontal lobe of schizophrenics implies disturbances of cortical development.* Arch Gen Psychiatry. **1993**, 50:169-177.

Albert KA, Hemmings HC, Jr., Adamo AI, Potkin SG, Akbarian S, Sandman CA, Cotman CW, Bunney WE, Jr., Greengard P. *Evidence for decreased DARPP-32 in the prefrontal cortex of patients with schizophrenia.* Arch Gen Psychiatry. **2002**, 59:705-712.

Amargós-Bosch M, Adell A, Bortolozzi A, Artigas F. *Stimulation of alpha1-adrenoceptors in the rat medial prefrontal cortex increases the local in vivo 5-hydroxytryptamine release: reversal by antipsychotic drugs.* J Neurochem. **2003**, 87:831-842.

Amargós-Bosch M, Bortolozzi A, Puig MV, Serrats J, Adell A, Celada P, Toth M, Mengod G, Artigas F. *Co-expression and in vivo interaction of serotonin1A and serotonin2A receptors in pyramidal neurons of prefrontal cortex.* Cereb Cortex. **2004**, 14:281-299.

Amargós-Bosch M, López-Gil X, Artigas F, Adell A. *Clozapine and olanzapine, but not haloperidol, suppress serotonin efflux in the medial prefrontal cortex elicited by phencyclidine and ketamine.* Int J Neuropsychopharmacol. **2006**, 9:565-573.

Araneda R, Andrade R. *5-Hydroxytryptamine2 and 5-hydroxytryptamine 1A receptors mediate opposing responses on membrane excitability in rat association cortex.* Neuroscience. **1991**, 40:399-412.

Arnt J. *Differential effects of classical and newer antipsychotics on the hypermotility induced by two dose levels of D-amphetamine.* Eur J Pharmacol. **1995**, 283:55-62.

Arvanov VL, Liang X, Magro P, Roberts R, Wang RY. *A pre- and postsynaptic modulatory action of 5-HT and the 5-HT2A, 2C receptor agonist DOB on NMDA-evoked responses in the rat medial prefrontal cortex.* Eur J Neurosci. **1999**, 11:2917-2934.

Ashby CR, Jr., Edwards E, Harkins K, Wang RY. *Effects of (+/-)-DOI on medial prefrontal cortical cells: a microiontophoretic study.* Brain Res. **1989**, 498:393-396.

Ashby CR, Jr., Jiang LH, Kasser RJ, Wang RY. *Electrophysiological characterization of 5-hydroxytryptamine2 receptors in the rat medial prefrontal cortex.* J Pharmacol Exp Ther. **1990**, 252:171-178.

Baik JH, Picetti R, Saiardi A, Thiriet G, Dierich A, Depaulis A, Le Meur M, Borrelli E. *Parkinsonian-like locomotor impairment in mice lacking dopamine D2 receptors.* Nature. **1995**, 377:424-428.

Bakshi VP, Swerdlow NR, Braff DL, Geyer MA. *Reversal of isolation rearing-induced deficits in prepulse inhibition by Seroquel and olanzapine.* Biol Psychiatry. **1998**, 43:436-445.

Bardgett ME, Jackson JL, Taylor GT, Csernansky JG. *Kainic acid decreases hippocampal neuronal number and increases dopamine receptor binding in the nucleus accumbens: an animal model of schizophrenia.* Behav Brain Res. **1995**, 70:153-164.

Bardgett ME, Jackson JL, Taylor BM, Csernansky JG. *The effects of kainic acid lesions on locomotor responses to haloperidol and clozapine.* Psychopharmacology (Berl). **1998**, 135:270-278.

Bardgett ME, Humphrey WM, Csernansky JG. *The effects of excitotoxic hippocampal lesions in rats on risperidone- and olanzapine-induced locomotor suppression.* Neuropsychopharmacology. **2002**, 27:930-938.

Bardgett ME. Behavioral models of atypical antipsychotic drug action in rodents. In: Csernansky JG, Lauriello J, eds. *Atypical Antipsychotics: From Bench to Bedside (Medical Psychiatry).* Marcel Dekker, New York, **2004**, pp 61-92.

Basar-Eroglu C, Basar E, Demiralp T, Schurmann M. *P300-response: possible psychophysiological correlates in delta and theta frequency channels. A review.* Int J Psychophysiol. **1992**, 13:161-179.

Basar-Eroglu C, Kolev V, Ritter B, Aksu F, Basar E. *EEG, auditory evoked potentials and evoked rhythmicities in three-year-old children.* Int J Neurosci. **1994**, 75:239-255.

Basar E, Schurmann M, Basar-Eroglu C, Karakas S. *Alpha oscillations in brain functioning: an integrative theory.* Int J Psychophysiol. **1997**, 26:5-29.

Basar E, Basar-Eroglu C, Karakas S, Schurmann M. *Are cognitive processes manifested in event-related gamma, alpha, theta and delta oscillations in the EEG?* Neurosci Lett. **1999**, 259:165-168.

Bertolino A, Saunders RC, Mattay VS, Bachevalier J, Frank JA, Weinberger DR. *Altered development of prefrontal neurons in rhesus monkeys with neonatal mesial temporo-limbic* lesions: a proton magnetic resonance spectroscopic imaging study. Cereb Cortex. **1997**, 7:740-748.

Black MD, Selk DE, Hitchcock JM, Wettstein JG, Sorensen SM. *On the effect of neonatal nitric oxide synthase inhibition in rats: a potential neurodevelopmental model of schizophrenia.* Neuropharmacology. **1999**, 38:1299-1306.

Borrell J, Vela JM, Arévalo-Martin A, Molina-Holgado E, Guaza C. *Prenatal immune challenge disrupts sensorimotor gating in adult rats. Implications for the etiopathogenesis of schizophrenia.* Neuropsychopharmacology. **2002**, 26:204-215.

Bortolozzi A, Amargós-Bosch M, Adell A, Díaz-Mataix L, Serrats J, Pons S, Artigas F. *In vivo modulation of 5-hydroxytryptamine release in mouse prefrontal cortex by local 5-HT(2A) receptors: effect of antipsychotic drugs.* Eur J Neurosci. **2003**, 18:1235-1246.

Bortolozzi A, Díaz-Mataix L, Scorza MC, Celada P, Artigas F. *The activation of 5-HT receptors in prefrontal cortex enhances dopaminergic activity.* J Neurochem. **2005**, 95:1597-1607.

Carlsson A. *The current status of the dopamine hypothesis of schizophrenia.* Neuropsychopharmacology. **1988**, 1:179-186.

Carr DB, Sesack SR. *Projections from the rat prefrontal cortex to the ventral tegmental area: target specificity in the synaptic associations with mesoaccumbens and mesocortical neurons.* J Neurosci. **2000**, 20:3864-3873.

Celada P, Puig MV, Casanovas JM, Guillazo G, Artigas F. *Control of dorsal raphe serotonergic neurons by the medial prefrontal cortex: Involvement of serotonin-1A, GABA(A), and glutamate receptors.* J Neurosci. **2001**, 21:9917-9929.

Chen J, Wu J, Song J, et al. *Overexpression of human catechol-O-methyltransferase transgene impairs cognitive function in inducible tissue-specific transgenic mice.* Society of neuroscience, **2005**, Program n° 1021.14.

Chen J, Lipska BK, Weinberger DR. *Genetic mouse models of schizophrenia: from hypothesis-based to susceptibility gene-based models.* Biol Psychiatry. **2006**, 59:1180-1188.

Cintra L, Granados L, Aguilar A, Kemper T, DeBassio W, Galler J, Morgane P, Duran P, Diaz-Cintra S. *Effects of prenatal protein malnutrition on mossy fibers of the hippocampal formation in rats of four age groups.* Hippocampus. **1997**, 7:184-191.

Costa E, Davis J, Pesold C, Tueting P, Guidotti A. *The heterozygote reeler mouse as a model for the development of a new generation of antipsychotics.* Curr Opin Pharmacol. **2002**, 2:56-62.

Cotter D, Takei N, Farrell M, Sham P, Quinn P, Larkin C, Oxford J, Murray RM, O'Callaghan E. *Does prenatal exposure to influenza in mice induce pyramidal cell disarray in the dorsal hippocampus?* Schizophr Res. **1995**, 16:233-241.

Dalley JW, Cardinal RN, Robbins TW. *Prefrontal executive and cognitive functions in rodents: neural and neurochemical substrates.* Neurosci Biobehav Rev. **2004**, 28:771-784.

Dalman C, Allebeck P, Cullberg J, Grunewald C, Koster M. *Obstetric complications and the risk of schizophrenia: a longitudinal study of a national birth cohort.* Arch Gen Psychiatry. **1999**, 56:234-240.

DeLisi LE, Dauphinais ID, Gershon ES. *Perinatal complications and reduced size of brain limbic structures in familial schizophrenia.* Schizophr Bull. **1988**, 14:185-191.

Duncan GE, Moy SS, Perez A, Eddy DM, Zinzow WM, Lieberman JA, Snouwaert JN, Koller BH. *Deficits in sensorimotor gating and tests of social behavior in a genetic model of reduced NMDA receptor function*. Behav Brain Res. **2004**, 153:507-519.

Ellenbroek BA, Cools AR. *Animal models with construct validity for schizophrenia*. Behav Pharmacol. **1990**, 1:469-490.

Ellenbroek BA. The ethological analysis of monkeys in a social setting as an animal model of schizophrenia. In: Olivier B, Mos J, Slangen JL, eds. *Animal models in psychopharmacology*. Birkhäuser, Basel, **1991**, pp 265-284.

Ellenbroek BA. *Treatment of schizophrenia: a clinical and preclinical evaluation of neuroleptic drugs*. Pharmacol Ther. **1993**, 57:1-78.

Ellenbroek BA, van den Kroonenberg PT, Cools AR. *The effects of an early stressful life event on sensorimotor gating in adult rats*. Schizophr Res. **1998**, 30:251-260.

Ellenbroek BA. *Animal models in the genomic era: possibilities and limitations with special emphasis on schizophrenia*. Behav Pharmacol. **2003**, 14:409-417.

Fatemi SH, Emamian ES, Kist D, Sidwell RW, Nakajima K, Akhter P, Shier A, Sheikh S, Bailey K. *Defective corticogenesis and reduction in Reelin immunoreactivity in cortex and hippocampus of prenatally infected neonatal mice*. Mol Psychiatry. **1999**, 4:145-154.

Fienberg AA, Hiroi N, Mermelstein PG, Song W, Snyder GL, Nishi A, Cheramy A, O'Callaghan JP, Miller DB, Cole DG, Corbett R, Haile CN, Cooper DC, Onn SP, Grace AA, Ouimet CC, White FJ, Hyman SE, Surmeier DJ, Girault J, Nestler EJ, Greengard P. *DARPP-32: regulator of the efficacy of dopaminergic neurotransmission*. Science. **1998**, 281:838-842.

Fienberg AA, Greengard P. *The DARPP-32 knockout mouse*. Brain Res Brain Res Rev. **2000**, 31:313-319.

Flagstad P, Mork A, Glenthoj BY, van Beek J, Michael-Titus AT, Didriksen M. *Disruption of neurogenesis on gestational day 17 in the rat causes behavioral changes relevant to positive and negative schizophrenia symptoms and alters amphetamine-induced dopamine release in nucleus accumbens*. Neuropsychopharmacology. **2004**, 29:2052-2064.

Flores G, Barbeau D, Quirion R, Srivastava LK. *Decreased binding of dopamine D3 receptors in limbic subregions after neonatal bilateral lesion of rat hippocampus*. J Neurosci. **1996a**, 16:2020-2026.

Flores G, Wood GK, Liang JJ, Quirion R, Srivastava LK. *Enhanced amphetamine sensitivity and increased expression of dopamine D2 receptors in postpubertal rats after neonatal excitotoxic lesions of the medial prefrontal cortex*. J Neurosci. **1996b**, 16:7366-7375.

Forrest D, Yuzaki M, Soares HD, Ng L, Luk DC, Sheng M, Stewart CL, Morgan JI, Connor JA, Curran T. *Targeted disruption of NMDA receptor 1 gene abolishes NMDA response and results in neonatal death*. Neuron. **1994**, 13:325-338.

Fradley RL, O'Meara GF, Newman RJ, Andrieux A, Job D, Reynolds DS. *STOP knockout and NMDA NR1 hypomorphic mice exhibit deficits in sensorimotor gating*. Behav Brain Res. **2005**, 163:257-264.

Freedman R, Coon H, Myles-Worsley M, Orr-Utrtreger A, Olincy A, Davis A, Polymeropoulos M, Holik J, Hopkins J, Hoff M, Rosenthal J, Waldo MC, Reimherr F, Wender P, Yaw J, Young DA, Breese CR, Adams C, Patterson D, Adler LE, Kruglyak L, Leonard S, Byerley W. *Link-*

age of a neurophysiological deficit in schizophrenia to a chromosome 15 locus. Proc Natl Acad Sci U S A. **1997**, 94:587-592.

Gaddum JH, Hameed KA. *Drugs which antagonize 5-hydroxytryptamine*. Br J Pharmacol Chemother. **1954**, 9:240-248.

Gainetdinov RR, Caron MG. *An animal model of attention deficit hyperactivity disorder*. Mol Med Today. **2000**, 6:43-44.

Gainetdinov RR, Mohn AR, Caron MG. *Genetic animal models: focus on schizophrenia*. Trends Neurosci. **2001**, 24:527-533.

Gerber DJ, Hall D, Miyakawa T, Demars S, Gogos JA, Karayiorgou M, Tonegawa S. *Evidence for association of schizophrenia with genetic variation in the 8p21.3 gene, PPP3CC, encoding the calcineurin gamma subunit*. Proc Natl Acad Sci U S A. **2003**, 100:8993-8998.

Geyer MA, Wilkinson LS, Humby T, Robbins TW. *Isolation rearing of rats produces a deficit in prepulse inhibition of acoustic startle similar to that in schizophrenia*. Biol Psychiatry. **1993**, 34:361-372.

Geyer MA, Markou A. Animal models of psychiatric disorders. In: Bloom FE, Kupfer FE, eds. *Psychopharmacology: the Fourth Generation of Progress*. 4th rev ed. Lippincott Williams & Wilkins, Philadelphia, **1995**, pp 787-798.

Geyer MA, Krebs-Thomson K, Braff DL, Swerdlow NR. *Pharmacological studies of prepulse inhibition models of sensorimotor gating deficits in schizophrenia: a decade in review*. Psychopharmacology (Berl). **2001**, 156:117-154.

Geyer MA, McIlwain KL, Paylor R. *Mouse genetic models for prepulse inhibition: an early review*. Mol Psychiatry. **2002**, 7:1039-1053.

Geyer MA, Markou A. The role of preclinical models in the development of psychotropic drugs. In: Davis KL, Charney D, Coyle JT, Nemeroff C, eds. *Neuropsychopharmacology: The Fifth Generation of Progress*. Lippincott Williams & Wilkins, Philadelphia., **2002**, pp 445-455.

Giros B, Jaber M, Jones SR, Wightman RM, Caron MG. *Hyperlocomotion and indifference to cocaine and amphetamine in mice lacking the dopamine transporter*. Nature. **1996**, 379:606-612.

Glennon RA, Titeler M, McKenney JD. *Evidence for 5-HT2 involvement in the mechanism of action of hallucinogenic agents*. Life Sci. **1984**, 35:2505-2511.

Glennon RA. *Do classical hallucinogens act as 5-HT2 agonists or antagonists?* Neuropsychopharmacology. **1990**, 3:509-517.

Goldberg TE, Weinberger DR. *Genes and the parsing of cognitive processes*. Trends Cogn Sci. **2004**, 8:325-335.

Goto Y, O'Donnell P. *Network synchrony in the nucleus accumbens in vivo*. J Neurosci. **2001**, 21:4498-4504.

Goto Y, O'Donnell P. *Delayed mesolimbic system alteration in a developmental animal model of schizophrenia*. J Neurosci. **2002**, 22:9070-9077.

Goto Y, O'Donnell P. *Prefrontal lesion reverses abnormal mesoaccumbens response in an animal model of schizophrenia*. Biol Psychiatry. **2004**, 55:172-176.

Granados-Rojas L, Larriva-Sahd J, Cintra L, Gutiérrez-Ospina G, Rondan A, Díaz-Cintra S. *Prenatal protein malnutrition decreases mossy fibers-CA3 thorny excrescences asymmetrical synapses in adult rats*. Brain Res. **2002**, 933:164-171.

Gratton A, Hoffer BJ, Freedman R. *Electrophysiological effects of phencyclidine in the medial prefrontal cortex of the rat*. Neuropharmacology. **1987**, 26:1275-1283.

Greenslade RG, Mitchell SN. *Selective action of (-)-2-oxa-4-aminobicyclo[3.1.0]hexane-4,6-dicarboxylate (LY379268), a group II metabotropic glutamate receptor agonist, on basal and phencyclidine-induced dopamine release in the nucleus accumbens shell*. Neuropharmacology. **2004**, 47:1-8.

Guidotti A, Auta J, Davis JM, Giorgi-Gerevini V, Dwivedi Y, Grayson DR, Impagnatiello F, Pandey G, Pesold C, Sharma R, Uzunov D, Costa E. *Decrease in reelin and glutamic acid decarboxylase67 (GAD67) expression in schizophrenia and bipolar disorder: a postmortem brain study*. Arch Gen Psychiatry. **2000**, 57:1061-1069.

Hagan JJ, Jones DN. *Predicting drug efficacy for cognitive deficits in schizophrenia*. Schizophr Bull. **2005**, 31:830-853.

Hajós M, Richards CD, Szekely AD, Sharp T. *An electrophysiological and neuroanatomical study of the medial prefrontal cortical projection to the midbrain raphe nuclei in the rat*. Neuroscience. **1998**, 87:95-108.

Hajós M, Gartside SE, Varga V, Sharp T. *In vivo inhibition of neuronal activity in the rat ventromedial prefrontal cortex by midbrain-raphe nuclei: role of 5-HT1A receptors*. Neuropharmacology. **2003**, 45:72-81.

Harrison PJ. *The neuropathology of schizophrenia. A critical review of the data and their interpretation*. Brain. **1999**, 122 (Pt 4):593-624.

Harrison PJ, Weinberger DR. *Schizophrenia genes, gene expression, and neuropathology: on the matter of their convergence*. Mol Psychiatry. **2005**, 10:40-68.

Hedou G, Homberg J, Martin S, Wirth K, Feldon J, Heidbreder CA. *Effect of amphetamine on extracellular acetylcholine and monoamine levels in subterritories of the rat medial prefrontal cortex*. Eur J Pharmacol. **2000**, 390:127-136.

Hultman CM, Ohman A, Cnattingius S, Wieselgren IM, Lindstrom LH. *Prenatal and neonatal risk factors for schizophrenia*. Br J Psychiatry. **1997**, 170:128-133.

Hutson PH, Hogg JE. *Effects of and interactions between antagonists for different sites on the NMDA receptor complex on hippocampal and striatal acetylcholine efflux in vivo*. Eur J Pharmacol. **1996**, 295:45-52.

Impagnatiello F, Guidotti AR, Pesold C, Dwivedi Y, Caruncho H, Pisu MG, Uzunov DP, Smalheiser NR, Davis JM, Pandey GN, Pappas GD, Tueting P, Sharma RP, Costa E. *A decrease of reelin expression as a putative vulnerability factor in schizophrenia*. Proc Natl Acad Sci U S A. **1998**, 95:15718-15723.

Ishizuka K, Paek M, Kamiya A, Sawa A. *A review of Disrupted-In-Schizophrenia-1 (DISC1): neurodevelopment, cognition, and mental conditions*. Biol Psychiatry. **2006**, 59:1189-1197.

Jackson ME, Homayoun H, Moghaddam B. *NMDA receptor hypofunction produces concomitant firing rate potentiation and burst activity reduction in the prefrontal cortex*. Proc Natl Acad Sci USA. **2004**, 101:8467-8472.

Javitt DC, Zukin SR. *Recent advances in the phencyclidine model of schizophrenia*. Am J Psychiatry. **1991**, 148:1301-1308.

Jentsch JD, Roth RH. *The neuropsychopharmacology of phencyclidine: from NMDA receptor hypofunction to the dopamine hypothesis of schizophrenia*. Neuropsychopharmacology. **1999**, 20:201-225.

Jodo E, Suzuki Y, Takeuchi S, Niwa S, Kayama Y. *Different effects of phencyclidine and methamphetamine on firing activity of medial prefrontal cortex neurons in freely moving rats*. Brain Res. **2003**, 962:226-231.

Jodo E, Suzuki Y, Katayama T, Hoshino KY, Takeuchi S, Niwa S, Kayama Y. *Activation of medial prefrontal cortex by phencyclidine is mediated via a hippocampo-prefrontal pathway*. Cereb Cortex. **2005**, 15:663-669.

Johnston MV, Barks J, Greenamyre T, Silverstein F. *Use of toxins to disrupt neurotransmitter circuitry in the developing brain*. Prog Brain Res. **1988**, 73:425-446.

Kapur S. *How antipsychotics become anti-"psychotic"--from dopamine to salience to psychosis*. Trends Pharmacol Sci. **2004**, 25:402-406.

Kegeles LS, Martinez D, Kochan LD, Hwang DR, Huang Y, Mawlawi O, Suckow RF, Van Heertum RL, Laruelle M. *NMDA antagonist effects on striatal dopamine release: positron emission tomography studies in humans*. Synapse. **2002**, 43:19-29.

Kim SH, Price MT, Olney JW, Farber NB. *Excessive cerebrocortical release of acetylcholine induced by NMDA antagonists is reduced by GABAergic and alpha2-adrenergic agonists*. Mol Psychiatry. **1999**, 4:344-352.

Kirkpatrick B, Conley RC, Kakoyannis A, Reep RL, Roberts RC. *Interstitial cells of the white matter in the inferior parietal cortex in schizophrenia: An unbiased cell-counting study*. Synapse. **1999**, 34:95-102.

Kirov G, O'Donovan MC, Owen MJ. *Finding schizophrenia genes*. J Clin Invest. **2005**, 115:1440-1448.

Klimesch W, Schimke H, Schwaiger J. *Episodic and semantic memory: an analysis in the EEG theta and alpha band*. Electroencephalogr Clin Neurophysiol. **1994**, 91:428-441.

Kornetsky C, Markowitz R. *Animal models of schizophrenia*. In: Lipton MA, DiMascio A, Killam KF, eds. *Psychopharmacology : a generation of progress*. Raven Press, New York, **1978**, pp 583-593.

Kretschmer BD. *NMDA receptor antagonist-induced dopamine release in the ventral pallidum does not correlate with motor activation*. Brain Res. **2000**, 859:147-156.

Kroeze WK, Roth BL. *The molecular biology of serotonin receptors: therapeutic implications for the interface of mood and psychosis*. Biol Psychiatry. **1998**, 44:1128-1142.

Krystal JH, Karper LP, Seibyl JP, Freeman GK, Delaney R, Bremner JD, Heninger GR, Bowers MB, Jr., Charney DS. *Subanesthetic effects of the noncompetitive NMDA antagonist, ketamine, in humans. Psychotomimetic, perceptual, cognitive, and neuroendocrine responses*. Arch Gen Psychiatry. **1994**, 51:199-214.

Krystal JH, D'Souza DC, Mathalon D, Perry E, Belger A, Hoffman R. *NMDA receptor antagonist effects, cortical glutamatergic function, and schizophrenia: toward a paradigm shift in medication development*. Psychopharmacology (Berl). **2003**, 169:215-233.

Kunugi H, Nanko S, Murray RM. *Obstetric complications and schizophrenia: prenatal underdevelopment and subsequent neurodevelopmental impairment.* Br J Psychiatry Suppl. **2001**, 40:s25-s29.

Lahdesmaki J, Sallinen J, MacDonald E, Scheinin M. *Alpha2A-adrenoceptors are important modulators of the effects of D-amphetamine on startle reactivity and brain monoamines.* Neuropsychopharmacology. **2004**, 29:1282-1293.

Laruelle M, Abi-Dargham A, van Dyck CH, Gil R, D'Souza CD, Erdos J, McCance E, Rosenblatt W, Fingado C, Zoghbi SS, Baldwin RM, Seibyl JP, Krystal JH, Charney DS, Innis RB. *Single photon emission computerized tomography imaging of amphetamine-induced dopamine release in drug-free schizophrenic subjects.* Proc Natl Acad Sci U S A. **1996**, 93:9235-9240.

Lavin A, Moore HM, Grace AA. *Prenatal disruption of neocortical development alters prefrontal cortical neuron responses to dopamine in adult rats.* Neuropsychopharmacology. **2005**, 30:1426-1435.

Lewis DA, Hashimoto T, Volk DW. *Cortical inhibitory neurons and schizophrenia.* Nat Rev Neurosci. **2005**, 6:312-324.

Lewis PD, Patel AJ, Balazs R. *Effect of undernutrition on cell generation in the adult rat brain.* Brain Res. **1977**, 138:511-519.

Lewis PD, Patel AJ, Balazs R. *Effect of undernutrition on cell generation in the rat hippocampus.* Brain Res. **1979**, 168:186-189.

Lewis SW, Murray RM. *Obstetric complications, neurodevelopmental deviance, and risk of schizophrenia.* J Psychiatr Res. **1987**, 21:413-421.

Li Y, Erzurumlu RS, Chen C, Jhaveri S, Tonegawa S. *Whisker-related neuronal patterns fail to develop in the trigeminal brainstem nuclei of NMDAR1 knockout mice.* Cell. **1994**, 76:427-437.

Linn GS, Negi SS, Gerum SV, Javitt DC. *Reversal of phencyclidine-induced prepulse inhibition deficits by clozapine in monkeys.* Psychopharmacology (Berl). **2003**, 169:234-239.

Lipska BK, Weinberger DR. *Delayed effects of neonatal hippocampal damage on haloperidol-induced catalepsy and apomorphine-induced stereotypic behaviors in the rat.* Brain Res Dev Brain Res. **1993**, 75:213-222.

Lipska BK, Jaskiw GE, Weinberger DR. *Postpubertal emergence of hyperresponsiveness to stress and to amphetamine after neonatal excitotoxic hippocampal damage: a potential animal model of schizophrenia.* Neuropsychopharmacology. **1993**, 9:67-75.

Lipska BK, al Amin HA, Weinberger DR. *Excitotoxic lesions of the rat medial prefrontal cortex. Effects on abnormal behaviors associated with neonatal hippocampal damage.* Neuropsychopharmacology. **1998**, 19:451-464.

Lipska BK, Weinberger DR. *To model a psychiatric disorder in animals: schizophrenia as a reality test.* Neuropsychopharmacology. **2000**, 23:223-239.

Lipska BK, Halim ND, Segal PN, Weinberger DR. *Effects of reversible inactivation of the neonatal ventral hippocampus on behavior in the adult rat.* J Neurosci. **2002**, 22:2835-2842.

Lipska BK, Weinberger DR. *A neurodevelopmental model of schizophrenia: neonatal disconnection of the hippocampus.* Neurotox Res. **2002**, 4:469-475.

Lipska BK, Lerman DN, Khaing ZZ, Weickert CS, Weinberger DR. *Gene expression in dopamine and GABA systems in an animal model of schizophrenia: effects of antipsychotic drugs*. Eur J Neurosci. **2003a**, 18:391-402.

Lipska BK, Lerman DN, Khaing ZZ, Weinberger DR. *The neonatal ventral hippocampal lesion model of schizophrenia: effects on dopamine and GABA mRNA markers in the rat midbrain*. Eur J Neurosci. **2003b**, 18:3097-3104.

Lipska BK, Weinberger DR. Animal models of schizophrenia. In: Mitchell PR, Hirsch SR, Weinberger DR, eds. *Schizophrenia.* Blackwell Science, Oxford, **2003**, pp 388-402.

Lohr JB, Bracha HS. *Can schizophrenia be related to prenatal exposure to alcohol? Some speculations*. Schizophr Bull. **1989**, 15:595-603.

López-Giménez JF, Mengod G, Palacios JM, Vilaro MT. *Selective visualization of rat brain 5-HT2A receptors by autoradiography with [3H]MDL 100,907*. Naunyn Schmiedebergs Arch Pharmacol. **1997**, 356:446-454.

Lorrain DS, Baccei CS, Bristow LJ, Anderson JJ, Varney MA. *Effects of ketamine and N-methyl-D-aspartate on glutamate and dopamine release in the rat prefrontal cortex: modulation by a group II selective metabotropic glutamate receptor agonist LY379268*. Neuroscience. **2003a**, 117:697-706.

Lorrain DS, Schaffhauser H, Campbell UC, Baccei CS, Correa LD, Rowe B, Rodriguez DE, Anderson JJ, Varney MA, Pinkerton AB, Vernier JM, Bristow LJ. *Group II mGlu receptor activation suppresses norepinephrine release in the ventral hippocampus and locomotor responses to acute ketamine challenge*. Neuropsychopharmacology. **2003b**, 28:1622-1632.

Malhotra AK, Pinals DA, Weingartner H, Sirocco K, Missar CD, Pickar D, Breier A. *NMDA receptor function and human cognition: the effects of ketamine in healthy volunteers*. Neuropsychopharmacology. **1996**, 14:301-307.

Mansbach RS, Geyer MA, Braff DL. *Dopaminergic stimulation disrupts sensorimotor gating in the rat*. Psychopharmacology (Berl). **1988**, 94:507-514.

Mansbach RS, Geyer MA. *Effects of phencyclidine and phencyclidine biologs on sensorimotor gating in the rat*. Neuropsychopharmacology. **1989**, 2:299-308.

Marek GJ, Aghajanian GK. *LSD and the phenethylamine hallucinogen DOI are potent partial agonists at 5-HT2A receptors on interneurons in rat piriform cortex*. J Pharmacol Exp Ther. **1996**, 278:1373-1382.

Martín-Ruiz R, Puig MV, Celada P, Shapiro DA, Roth BL, Mengod G, Artigas F. *Control of serotonergic function in medial prefrontal cortex by serotonin-2A receptors through a glutamate-dependent mechanism*. J Neurosci. **2001**, 21:9856-9866.

Martin P, Carlsson ML, Hjorth S. *Systemic PCP treatment elevates brain extracellular 5-HT: a microdialysis study in awake rats*. Neuroreport. **1998**, 9:2985-2988.

Mathe JM, Nomikos GG, Hildebrand BE, Hertel P, Svensson TH. *Prazosin inhibits MK-801-induced hyperlocomotion and dopamine release in the nucleus accumbens*. Eur J Pharmacol. **1996**, 309:1-11.

Mathe JM, Nomikos GG, Schilstrom B, Svensson TH. *Non-NMDA excitatory amino acid receptors in the ventral tegmental area mediate systemic dizocilpine (MK-801) induced hyperlocomotion and dopamine release in the nucleus accumbens*. J Neurosci Res. **1998**, 51:583-592.

Mathe JM, Nomikos GG, Blakeman KH, Svensson TH. *Differential actions of dizocilpine (MK-801) on the mesolimbic and mesocortical dopamine systems: role of neuronal activity.* Neuropharmacology. **1999**, 38:121-128.

Matsuura M, Yoshino M, Ohta K, Onda H, Nakajima K, Kojima T. *Clinical significance of diffuse delta EEG activity in chronic schizophrenia.* Clin Electroencephalogr. **1994**, 25:115-121.

Meador-Woodruff JH, Healy DJ. *Glutamate receptor expression in schizophrenic brain.* Brain Res Brain Res Rev. **2000**, 31:288-294.

Mednick SA, Machon RA, Huttunen MO, Bonett D. *Adult schizophrenia following prenatal exposure to an influenza epidemic.* Arch Gen Psychiatry. **1988**, 45:189-192.

Meltzer HY. *The role of serotonin in antipsychotic drug action.* Neuropsychopharmacology. **1999**, 21:106S-115S.

Migler BM, Warawa EJ, Malick JB. *Seroquel: behavioral effects in conventional and novel tests for atypical antipsychotic drug.* Psychopharmacology (Berl). **1993**, 112:299-307.

Millan MJ, Brocco M, Gobert A, Joly F, Bervoets K, Rivet J, Newman-Tancredi A, Audinot V, Maurel S. *Contrasting mechanisms of action and sensitivity to antipsychotics of phencyclidine versus amphetamine: importance of nucleus accumbens 5-HT2A sites for PCP-induced locomotion in the rat.* Eur J Neurosci. **1999**, 11:4419-4432.

Miller R. *Cortico-Hippocampal Interplay and the Representation of Contexts in the Brain.* Springer Verlag; Berlin, **1991**.

Miyakawa T, Leiter LM, Gerber DJ, Gainetdinov RR, Sotnikova TD, Zeng H, Caron MG, Tonegawa S. *Conditional calcineurin knockout mice exhibit multiple abnormal behaviors related to schizophrenia.* Proc Natl Acad Sci USA. **2003**, 100:8987-8992.

Miyamoto Y, Yamada K, Noda Y, Mori H, Mishina M, Nabeshima T. *Hyperfunction of dopaminergic and serotonergic neuronal systems in mice lacking the NMDA receptor epsilon1 subunit.* J Neurosci. **2001**, 21:750-757.

Miyamoto Y, Yamada K, Noda Y, Mori H, Mishina M, Nabeshima T. *Lower sensitivity to stress and altered monoaminergic neuronal function in mice lacking the NMDA receptor epsilon 4 subunit.* J Neurosci. **2002**, 22:2335-2342.

Moghaddam B, Adams B, Verma A, Daly D. *Activation of glutamatergic neurotransmission by ketamine: a novel step in the pathway from NMDA receptor blockade to dopaminergic and cognitive disruptions associated with the prefrontal cortex.* J Neurosci. **1997**, 17:2921-2927.

Moghaddam B, Adams BW. *Reversal of phencyclidine effects by a group II metabotropic glutamate receptor agonist in rats.* Science. **1998**, 281:1349-1352.

Mohn AR, Gainetdinov RR, Caron MG, Koller BH. *Mice with reduced NMDA receptor expression display behaviors related to schizophrenia.* Cell. **1999**, 98:427-436.

Moore H, Jentsch JD, Ghajarnia M, Geyer MA, Grace AA. *A neurobehavioral systems analysis of adult rats exposed to methylazoxymethanol acetate on E17: implications for the neuropathology of schizophrenia.* Biol Psychiatry. **2006**, 60:253-264.

Murase S, Mathe JM, Grenhoff J, Svensson TH. *Effects of dizocilpine (MK-801) on rat midbrain dopamine cell activity: differential actions on firing pattern related to anatomical localization.* J Neural Transm Gen Sect. **1993**, 91:13-25.

Nelson CL, Burk JA, Bruno JP, Sarter M. *Effects of acute and repeated systemic administration of ketamine on prefrontal acetylcholine release and sustained attention performance in rats.* Psychopharmacology (Berl). **2002**, 161:168-179.

Newcomer JW, Farber NB, Jevtovic-Todorovic V, Selke G, Melson AK, Hershey T, Craft S, Olney JW. *Ketamine-induced NMDA receptor hypofunction as a model of memory impairment and psychosis.* Neuropsychopharmacology. **1999**, 20:106-118.

Nishijima K, Kashiwa A, Hashimoto A, Iwama H, Umino A, Nishikawa T. *Differential effects of phencyclidine and methamphetamine on dopamine metabolism in rat frontal cortex and striatum as revealed by in vivo dialysis.* Synapse. **1996**, 22:304-312.

Nuechterlein KH, Barch DM, Gold JM, Goldberg TE, Green MF, Heaton RK. *Identification of separable cognitive factors in schizophrenia.* Schizophr Res. **2004**, 72:29-39.

O'Donnell P, Lewis BL, Weinberger DR, Lipska BK. *Neonatal hippocampal damage alters electrophysiological properties of prefrontal cortical neurons in adult rats.* Cereb Cortex. **2002**, 12:975-982.

Ogren SO. The behavioral pharmacology of typical and atypical antipsychotic drugs. In: Csernansky JG, ed. *Antipsychotics.* 1st ed. Springer-Verlag Telos, Berlin, **1996**, pp 225-266.

Pazos A, Cortes R, Palacios JM. *Quantitative autoradiographic mapping of serotonin receptors in the rat brain. II. Serotonin-2 receptors.* Brain Res. **1985**, 346:231-249.

Pehek EA. *Comparison of effects of haloperidol administration on amphetamine-stimulated dopamine release in the rat medial prefrontal cortex and dorsal striatum.* J Pharmacol Exp Ther. **1999**, 289:14-23.

Peng RY, Mansbach RS, Braff DL, Geyer MA. *A D2 dopamine receptor agonist disrupts sensorimotor gating in rats. Implications for dopaminergic abnormalities in schizophrenia.* Neuropsychopharmacology. **1990**, 3:211-218.

Peyron C, Petit JM, Rampon C, Jouvet M, Luppi PH. *Forebrain afferents to the rat dorsal raphe nucleus demonstrated by retrograde and anterograde tracing methods.* Neuroscience. **1998**, 82:443-468.

Pompeiano M, Palacios JM, Mengod G. *Distribution of the serotonin 5-HT2 receptor family mRNAs: comparison between 5-HT2A and 5-HT2C receptors.* Brain Res Mol Brain Res. **1994**, 23:163-178.

Porras G, Di M, V, Fracasso C, Lucas G, De Deurwaerdere P, Caccia S, Esposito E, Spampinato U. *5-HT2A and 5-HT2C/2B receptor subtypes modulate dopamine release induced in vivo by amphetamine and morphine in both the rat nucleus accumbens and striatum.* Neuropsychopharmacology. **2002**, 26:311-324.

Protais P, Costentin J, Schwartz JC. *Climbing behavior induced by apomorphine in mice: a simple test for the study of dopamine receptors in striatum.* Psychopharmacology (Berl). **1976**, 50:1-6.

Puig MV, Celada P, Diaz-Mataix L, Artigas F. *In vivo modulation of the activity of pyramidal neurons in the rat medial prefrontal cortex by 5-HT2A receptors: relationship to thalamocortical afferents.* Cereb Cortex. **2003**, 13:870-882.

Puig MV, Artigas F, Celada P. *Modulation of the activity of pyramidal neurons in rat prefrontal cortex by raphe stimulation in vivo: involvement of serotonin and GABA.* Cereb Cortex. **2005**, 15:1-14.

Rajakumar N, Williamson PC, Stoessl JA. *Neurodevelopmental pathogenesis of schizophrenia.* Soc.Neurosci Abstract **1996**, 22, 1187.

Ralph-Williams RJ, Lehmann-Masten V, Otero-Corchon V, Low MJ, Geyer MA. *Differential effects of direct and indirect dopamine agonists on prepulse inhibition: a study in D1 and D2 receptor knock-out mice.* J Neurosci. **2002**, 22:9604-9611.

Razoux F, Garcia R, Lena I. *Ketamine, at a Dose that Disrupts Motor Behavior and Latent Inhibition, Enhances Prefrontal Cortex Synaptic Efficacy and Glutamate Release in the Nucleus Accumbens.* Neuropsychopharmacology. **2006**.

Rice DS, Curran T. *Role of the reelin signaling pathway in central nervous system development.* Annu Rev Neurosci. **2001**, 24:1005-1039.

Sams-Dodd F. *Phencyclidine-induced stereotyped behaviour and social isolation in rats: a possible animal model of schizophrenia.* Behav Pharmacol. **1996**, 7:3-23.

Santana N, Bortolozzi A, Serrats J, Mengod G, Artigas F. *Expression of serotonin1A and serotonin2A receptors in pyramidal and GABAergic neurons of the rat prefrontal cortex.* Cereb Cortex. **2004**, 14:1100-1109.

Schiffer WK, Gerasimov M, Hofmann L, Marsteller D, Ashby CR, Brodie JD, Alexoff DL, Dewey SL. *Gamma vinyl-GABA differentially modulates NMDA antagonist-induced increases in mesocortical versus mesolimbic DA transmission.* Neuropsychopharmacology. **2001**, 25:704-712.

Schmidt CJ, Fadayel GM. *Regional effects of MK-801 on dopamine release: effects of competitive NMDA or 5-HT2A receptor blockade.* J Pharmacol Exp Ther. **1996**, 277:1541-1549.

Schotte A, Janssen PF, Megens AA, Leysen JE. *Occupancy of central neurotransmitter receptors by risperidone, clozapine and haloperidol, measured ex vivo by quantitative autoradiography.* Brain Res. **1993**, 631:191-202.

Sebban C, Tesolin-Decros B, Millan MJ, Spedding M. *Contrasting EEG profiles elicited by antipsychotic agents in the prefrontal cortex of the conscious rat: antagonism of the effects of clozapine by modafinil.* Br J Pharmacol. **1999a**, 128:1055-1063.

Sebban C, Zhang XQ, Tesolin-Decros B, Millan MJ, Spedding M. *Changes in EEG spectral power in the prefrontal cortex of conscious rats elicited by drugs interacting with dopaminergic and noradrenergic transmission.* Br J Pharmacol. **1999b**, 128:1045-1054.

Sebban C, Tesolin-Decros B, Ciprian-Ollivier J, Perret L, Spedding M. *Effects of phencyclidine (PCP) and MK 801 on the EEGq in the prefrontal cortex of conscious rats; antagonism by clozapine, and antagonists of AMPA-, alpha(1)- and 5-HT(2A)-receptors.* Br J Pharmacol. **2002**, 135:65-78.

Selemon LD, Goldman-Rakic PS. *The reduced neuropil hypothesis: a circuit based model of schizophrenia.* Biol Psychiatry. **1999**, 45:17-25.

Selemon LD, Kleinman JE, Herman MM, Goldman-Rakic PS. *Smaller frontal gray matter volume in postmortem schizophrenic brains.* Am J Psychiatry. **2002**, 159:1983-1991.

Selemon LD, Mrzljak J, Kleinman JE, Herman MM, Goldman-Rakic PS. *Regional specificity in the neuropathologic substrates of schizophrenia: a morphometric analysis of Broca's area 44 and area 9.* Arch Gen Psychiatry. **2003**, 60:69-77.

Selemon LD. *Increased cortical neuronal density in schizophrenia.* Am J Psychiatry. **2004**, 161:1564.

Sesack SR, Deutch AY, Roth RH, Bunney BS. *Topographical organization of the efferent projections of the medial prefrontal cortex in the rat: an anterograde tract-tracing study with Phaseolus vulgaris leucoagglutinin.* J Comp Neurol. **1989**, 290:213-242.

Shaw E, Woolley DW. *Some serotoninlike activities of lysergic acid diethylamide.* Science. **1956**, 124:121-122.

Shi WX, Zhang XX. *Dendritic glutamate-induced bursting in the prefrontal cortex: further characterization and effects of phencyclidine.* J Pharmacol Exp Ther. **2003**, 305:680-687.

Sibley DR. *New insights into dopaminergic receptor function using antisense and genetically altered animals.* Annu Rev Pharmacol Toxicol. **1999**, 39:313-341.

Sokolov BP. *Expression of NMDAR1, GluR1, GluR7, and KA1 glutamate receptor mRNAs is decreased in frontal cortex of "neuroleptic-free" schizophrenics: evidence on reversible up-regulation by typical neuroleptics.* J Neurochem. **1998**, 71:2454-2464.

Stefansson H, Sigurdsson E, Steinthorsdottir V, Bjornsdottir S, Sigmundsson T, Ghosh S, Brynjolfsson J, Gunnarsdottir S, Ivarsson O, Chou TT, Hjaltason O, Birgisdottir B, Jonsson H, Gudnadottir VG, Gudmundsdottir E, Bjornsson A, Ingvarsson B, Ingason A, Sigfusson S, Hardardottir H, Harvey RP, Lai D, Zhou M, Brunner D, Mutel V, Gonzalo A, Lemke G, Sainz J, Johannesson G, Andresson T, Gudbjartsson D, Manolescu A, Frigge ML, Gurney ME, Kong A, Gulcher JR, Petursson H, Stefansson K. *Neuregulin 1 and susceptibility to schizophrenia.* Am J Hum Genet. **2002**, 71:877-892.

Stevens KE, Kem WR, Mahnir VM, Freedman R. *Selective alpha7-nicotinic agonists normalize inhibition of auditory response in DBA mice.* Psychopharmacology (Berl). **1998**, 136:320-327.

Susser ES, Lin SP. *Schizophrenia after prenatal exposure to the Dutch Hunger Winter of 1944-1945.* Arch Gen Psychiatry. **1992**, 49:983-988.

Suzuki Y, Jodo E, Takeuchi S, Niwa S, Kayama Y. *Acute administration of phencyclidine induces tonic activation of medial prefrontal cortex neurons in freely moving rats.* Neuroscience. **2002**, 114:769-779.

Svenningsson P, Tzavara ET, Carruthers R, Rachleff I, Wattler S, Nehls M, McKinzie DL, Fienberg AA, Nomikos GG, Greengard P. *Diverse psychotomimetics act through a common signaling pathway.* Science. **2003**, 302:1412-1415.

Swanson CJ, Schoepp DD. *A role for noradrenergic transmission in the actions of phencyclidine and the antipsychotic and antistress effects of mGlu2/3 receptor agonists.* Ann N Y Acad Sci. **2003**, 1003:309-317.

Swerdlow NR, Geyer MA. *Using an animal model of deficient sensorimotor gating to study the pathophysiology and new treatments of schizophrenia.* Schizophr Bull. **1998**, 24:285-301.

Talamini LM, Koch T, Ter Horst GJ, Korf J. *Methylazoxymethanol acetate-induced abnormalities in the entorhinal cortex of the rat; parallels with morphological findings in schizophrenia.* Brain Res. **1998**, 789:293-306.

Tamminga CA. *Schizophrenia and glutamatergic transmission.* Crit Rev Neurobiol. **1998**, 12:21-36.

Tanaka E, North RA. *Actions of 5-hydroxytryptamine on neurons of the rat cingulate cortex.* J Neurophysiol. **1993**, 69:1749-1757.

Thierry AM, Deniau JM, Feger J. *Effects of stimulation of the frontal cortex on identified output VMT cells in the rat.* Neurosci Lett. **1979**, 15:102-107.

Thierry AM, Chevalier G, Ferron A, Glowinski J. *Diencephalic and mesencephalic efferents of the medial prefrontal cortex in the rat: electrophysiological evidence for the existence of branched axons.* Exp Brain Res. **1983**, 50:275-282.

Titeler M, Lyon RA, Glennon RA. *Radioligand binding evidence implicates the brain 5-HT2 receptor as a site of action for LSD and phenylisopropylamine hallucinogens.* Psychopharmacology (Berl). **1988**, 94:213-216.

Tsai G, Ralph-Williams RJ, Martina M, Bergeron R, Berger-Sweeney J, Dunham KS, Jiang Z, Caine SB, Coyle JT. *Gene knockout of glycine transporter 1: characterization of the behavioral phenotype.* Proc Natl Acad Sci USA. **2004**, 101:8485-8490.

Usiello A, Baik JH, Rouge-Pont F, Picetti R, Dierich A, LeMeur M, Piazza PV, Borrelli E. *Distinct functions of the two isoforms of dopamine D2 receptors.* Nature. **2000**, 408:199-203.

van den Buuse M, Garner B, Gogos A, Kusljic S. *Importance of animal models in schizophrenia research.* Aust N Z J Psychiatry. **2005**, 39:550-557.

van Praag HM. Serotonergic mechanisms in the pathogenesis of schizophrenia. In: Lindenmayer JP, Kay SR, eds. *New Biological Vistas on Schizophrenia (Clinical & Experimental Psychiatry).* Bruner-Mazel, New York, **1992**, pp 182-206.

Varty GB, Higgins GA. *Examination of drug-induced and isolation-induced disruptions of prepulse inhibition as models to screen antipsychotic drugs.* Psychopharmacology (Berl). **1995**, 122:15-26.

Waberski TD, Norra C, Kawohl W, Thyerlei D, Hock D, Klostermann F, Curio G, Buchner H, Hoff P, Gobbele R. *Electrophysiological evidence for altered early cerebral somatosensory signal processing in schizophrenia.* Psychophysiology. **2004**, 41:361-366.

Wang JH, Short J, Ledent C, Lawrence AJ, van den BM. *Reduced startle habituation and prepulse inhibition in mice lacking the adenosine A2A receptor.* Behav Brain Res. **2003**, 143:201-207.

Weinberger DR, Aloia MS, Goldberg TE, Berman KF. *The frontal lobes and schizophrenia.* J Neuropsychiatry Clin Neurosci. **1994**, 6:419-427.

Weinberger DR, Egan MF, Bertolino A, Callicott JH, Mattay VS, Lipska BK, Berman KF, Goldberg TE. *Prefrontal neurons and the genetics of schizophrenia.* Biol Psychiatry. **2001**, 50:825-844.

Whittington MA, Faulkner HJ, Doheny HC, Traub RD. *Neuronal fast oscillations as a target site for psychoactive drugs.* Pharmacol Ther. **2000**, 86:171-190.

Whitton PS, Maione S, Biggs CS, Fowler LJ. *N-methyl-d-aspartate receptors modulate extracellular dopamine concentration and metabolism in rat hippocampus and striatum in vivo.* Brain Res. **1994**, 635:312-316.

Williams GV, Rao SG, Goldman-Rakic PS. *The physiological role of 5-HT2A receptors in working memory*. J Neurosci. **2002**, 22:2843-2854.

Winterer G, Coppola R, Goldberg TE, Egan MF, Jones DW, Sanchez CE, Weinberger DR. *Prefrontal broadband noise, working memory, and genetic risk for schizophrenia*. Am J Psychiatry. **2004**, 161:490-500.

Woerner MG, Pollack M, Klein DF. *Pregnancy and birth complications in psychiatric patients: a comparison of schizophrenic and personality disorder patients with their siblings*. Acta Psychiatr Scand. **1973**, 49:712-721.

Woolley DW, Shaw E. *A Biochemical and pharmacological suggestion about certain mental disorders*. Proc Natl Acad Sci U S A. **1954**, 40:228-231.

Xu M, Moratalla R, Gold LH, Hiroi N, Koob GF, Graybiel AM, Tonegawa S. *Dopamine D1 receptor mutant mice are deficient in striatal expression of dynorphin and in dopamine-mediated behavioral responses*. Cell. **1994**, 79:729-742.

Xu M, Guo Y, Vorhees CV, Zhang J. *Behavioral responses to cocaine and amphetamine administration in mice lacking the dopamine D1 receptor*. Brain Res. **2000**, 852:198-207.

Yan QS, Reith ME, Jobe PC, Dailey JW. *Dizocilpine (MK-801) increases not only dopamine but also serotonin and norepinephrine transmissions in the nucleus accumbens as measured by microdialysis in freely moving rats*. Brain Res. **1997**, 765:149-158.

Yurek DM, Zhang L, Fletcher-Turner A, Seroogy KB. *Supranigral injection of neuregulin1-beta induces striatal dopamine overflow*. Brain Res. **2004**, 1028:116-119.

Zhou FM, Hablitz JJ. *Activation of serotonin receptors modulates synaptic transmission in rat cerebral cortex*. J Neurophysiol. **1999**, 82:2989-2999.

2.5 Chemistry

Marketed Drugs and Drugs in Development

Antonio Párraga, Jörg Holenz and Helmut Buschmann

Laboratorios Esteve, Av. Mare de Déu de Montserrat, 221. 08041 Barcelona, Spain.

2.5.1 Summary of drug classes

Following the Anatomical Therapeutic Chemical (ATC) classification system by WHO Collaborating Centre for Drug Statistics Methodology (http://www.whocc.no/atcddd/), antipsychotics (ATC N05A) are classified as:

- N05AA Phenothiazines with aliphatic side-chain: Chlorpromazine, Levomepromazine, Promazine, Triflupromazine, Cyamemazine, Chlorproethazine

- N05AB Phenothiazines with piperazine structure: Dixyrazine, Fluphenazine, Perphenazine, Prochlorperazine, Thiopropazate, Trifluoperazine, Acetophenazine, Thioproperazine, Butaperazine, Perazine

- N05AC Phenothiazines with piperidine structure: Periciazine, Thioridazine, Mesoridazine, Pipotiazine

- N05AD Butyrophenone derivatives: Haloperidol, Trifluperidol, Melperone, Moperone, Pipamperone, Bromperidol, Benperidol, Droperidol, Fluanisone

- N05AE Indole derivatives: Molindone, Sertindole, Ziprasidone, Perospirone

- N05AF Thioxanthene derivatives: Flupentixol, Clopenthixol, Chlorprothixene, Thiothixene, Zuclopenthixol

- N05AG Diphenylbutylpiperidine derivatives: Fluspirilene, Pimozide, Penfluridol

- N05AH Diazepines, oxazepines and thiazepines: Loxapine, Clozapine, Olanzapine, Quetiapine

- N05AL Benzamides: Sulpiride, Sultopride, Tiapride, Remoxipride, Amisulpride, Veralipride, Levosulpiride, Nemonapride

- N05AN Lithium: Lithium

- N05AX Other antipsychotics: Prothipendyl, Risperidone, Clotiapine, Mosapramine, Zotepine, Aripiprazole, Carpipramine, Tandospirone

The antipsychotic compounds are also classified as typical (also known as classic) or atypical based on the main mechanism of action and the appearance of side-effects associated with the clinical action. The typical (neuroleptic) antipsychotics were developed after chlorpromazine, and they are potent dopamine D_2 receptor antagonists (Jones and Pilowsky, 2002) (Table 1).

Table 1. Typical antipsychotics on the market for the treatment of schizophrenia

Drug name	Company	Year of first launch
Lithium	Solvay	1939
Chlorpromazine	GlaxoSmithKline	1952
Prochlorperazine	GlaxoSmithKline	1956
Promazine	Wyeth	1956
Levomepromazine	Wyeth	1957
Perphenazine	Schering-Plough	1957
Triflupromazine	Bristol-Myers Squibb	1957
Thioridazine	Novartis	1958
Trifluoperazine	GlaxoSmithKline	1958
Fluphenazine	Bristol-Myers Squibb	1959
Haloperidol	Janssen	1959
Zuclopenthixol	Lundbeck	1962
Droperidol	Janssen-Cilag	1963
Periciazine	Sanofi-Aventis	1964
Thiothixene	Pfizer	1965
Benperidol	Janssen	1966
Flupentixol	Lundbeck	1966
Sulpiride	Astellas Pharma	1968
Pimozide	Janssen	1969
Fluspirilene	Janssen	1970
Cyamemazine	Sanofi-Aventis	1972
Pipotiazine	Sanofi-Aventis	1973
Loxapine	Wyeth	1975
Sultopride	Sanofi-Aventis	1976
Carpipramine	Pierre Fabre	1977
Tiapride	Sanofi-Aventis	1977
Veralipride	Sanofi-Aventis	1980
Bromperidol	Janssen	1981
Levosulpiride	Abbott	1987
Remoxipride	AstraZeneca	1990
Mosapramine	Mitsubishi Pharma	1991
Nemonapride	Yamanouchi	1991
Tandospirone	Sumitomo Pharmaceuticals	1996

The typical antipsychotic drugs are effective at reducing the positive symptoms of schizophrenia, but they are associated with motor side-effects known as extrapyramidal side-effects (EPS). The absence of EPS in the presence of antipsychotic efficacy observed in clozapine (a potent $5\text{-}HT_{2A}$ antagonist) led to the concept of atypical antipsychotics (Kerwin, 1994) (Table 2).

The group of atypical antipsychotics interacts with a large number of targets, so it is difficult to assign precise pharmacological mechanisms for the various therapeutic and toxic side-effects.

Table 2. Atypical antipsychotics on the market for the treatment of schizophrenia		
Drug name	Company	Year of first launch
Dopamine D_4 antagonists		
Clozapine	Novartis	1972
Dopamine D_2 antagonists and $5\text{-}HT_{2A}$ antagonists		
Zotepine	Fujisawa	1982
Risperidone	Janssen	1993
Olanzapine	Eli Lilly and Company	1996
Quetiapine	AstraZeneca	1997
Ziprasidone	Pfizer	2000
Perospirone	Sumitomo Pharmaceuticals/ Mitsubishi Pharma	2001
Aripiprazole	Otsuka/Bristol-Myers Squibb	2002
Sertindole	Lundbeck	2006
Dopamine D_2 and D_3 antagonists		
Amisulpride	Synthélabo	1986

In the following sections, the history, structure, synthesis, clinical use, pharmacokinetics and metabolism of marketed compounds are detailed. Drugs under active development at clinical phases are also detailed, including structure and mechanism of action, when known. In order to search for the primary literature, the authors used, i. a., Prous Integrity database (http://integrity.prous.com), "Martindale – The Complete Drug Reference" (Sweetman, 2004) and "Pharmaceutical Substances – Synthesis, Patents, Applications" (Kleeman et al., 2001).

2.5.2 Typical antipsychotics

Lithium

Lithium

[554-13-2], Carbonic acid dilithium salt, CLi_2O_3, M_r 73.89

- Lithium and lithium salts (especially carbonate or citrate) were the focus of research in depression topic in the 1930's. Lithium carbonate was launched in 1939 by Solvay for the treatment of depression and schizophrenia, and in 2002 Glaxo launched it for the treatment of mania. (see Chapter 1.5, Section 1.5.6, for additional details)

- For many years, the effect of lithium carbonate on patients remained a mystery until Dr. P. S. Klein and his colleagues at the University of Pennsylvania discovered in 2006 that lithium, a natural salt, deactivated the GSK-3B enzyme (Yin et al., 2006). When the GSK-3B enzyme is activated, the protein Bmal1 is unable to reset the "master clock" inside the brain which disrupts the body's natural cycle. When the cycle is disrupted, the routine schedule of many functions (metabolism, sleep, body temperature) are disturbed

Clinical Use, Pharmacokinetics and Metabolism: Mainly in the form of its carbonic acid salt (sometimes, citrate salt), lithium is used in prophylaxis and treatment of bipolar disorders (manic depression) and schizophrenia, in mania and in the maintenance treatment of unipolar disorders (recurrent depression) as an alternative to conventional treatments (Schou, 1987). Due to the narrow therapeutic window, blood serum concentrations need to be monitored during therapy (medical supervision). Recommended therapeutic serum concentrations are usually in the range 0.4–1.4 mmol per liter, slightly increased concentrations are needed for the treatment of acute mania. Lithium therapy has been tried in a variety of disorders, e.g., depression, anxiety, bipolar disorders (Johnson, 1987; Schou, 1989; Aronson and Reynolds, 1992; Birch et al., 1993; Peet and Pratt, 1993; Price and Heninger, 1994), disturbed behavior (Schou, 1987), headache and schizophrenia. Side-effects with lithium therapy are manifold, the most important ones being GI upsets such as nausea and vomiting, but also more severe ones such as effects on the CNS, including tremor, ataxias, or polyuria. Further side-effects described comprise effects on the blood (Collins, 1992), the cardiovascular system (Demers and Henninger, 1971; Tangedahl and Gau, 1972; Palileo et al., 1983; Montalescot et al., 1984; Martin and

Piascik, 1985), the endocrine system (Myers and West, 1987; Nordenström et al., 1992; Taylor and Bell, 1993; Pandit et al., 1993), eyes (Fraunfelder et al., 1992), kidneys (Walker and Kincaid-Smith, 1987; Schou, 1988a, b; George, 1989), musculoskeletal system (Baandrup et al., 1987), nervous and neuromuscular systems (Himmelhoch and Hanin, 1974; Solomon and Vickers, 1975; McGovern, 1983; Worral and Gilham, 1983; Johns and Harris, 1984; Sansone and Ziegler, 1987), respiration (Weiner et al., 1983), sexual function and fertility (Beeley, 1984; Raoof et al., 1989; Salas et al., 1989) and skin and hair (Lambert and Dalac, 1987; Fraunfelder et al., 1992), as well as epileptogenic effects (Demers et al., 1970; Massey and Folger, 1984) and lupus (Johnstone and Whaley, 1975; Presley et al., 1976; Shukla and Borison, 1982). There are several references available for the treatment of side-effects related to lithium therapy, especially when overdosing (Gomolin, 1987; Amdisen, 1988; Okusa et al., 1994; Swartz and Jones, 1994; Wells, 1994). In general, when applying lithium therapy, many precautions need to be considered, especially concerning drug interactions and side-effects. Lithium is readily absorbed from the GI tract, peak serum concentrations being reached between 0.5–3.0 hours after ingestion (for conventional tablets or capsules). Within 6–10 hours, lithium is distributed throughout the whole body. Lithium is excreted in unchanged form mainly in urine, elimination half-lifes ranging between 10 hours to 50 hours (Johnson, 1987; Reiss et al., 1994; Ward et al., 1994).

Chlorpromazine

- The drug had been developed by Laboratoires Rhône-Poulenc in 1950 (Charpentier, 1953) but they had sold the rights in 1952 to Smith-Kline & French. The drug was being sold as an antiemetic when its other use was noted. Smith-Kline quickly started clinical trials and in 1954 the drug was approved in the USA for psychiatric treatment

- Chlorpromazine is a phenothiazine and was the first antipsychotic drug, used during the 1950s and 1960s. It has sedative, hypotensive, antiemetic and anxiolytic properties

- Chlorpromazine substituted and eclipsed the old therapies of electro- and insulin shocks and other methods such as psychosurgical means (lobotomy) causing permanent brain injury

Chlorpromazine

[50-53-3], 2-Chloro-N,N-dimethyl-10H-phenothiazine-10-propanamine, $C_{17}H_{19}ClN_2S$, M_r 318.86;
[69-09-0] hydrochloride salt, $C_{17}H_{20}Cl_2N_2S$, M_r 355.32;
Pamoate ester, $C_{40}H_{35}ClN_2O_6S$, M_r 1026.10

- Chlorpromazine is a dopamine D_2 receptor antagonist. In addition, it has α-adrenergic blocking, antimuscarinic, serotonin-blocking and antihistaminic properties

Synthesis (Charpentier, 1953): The compound is obtained by condensation of 2-chlorophenotiazine with 3-dimethylaminopropyl by means of sodamide in refluxing xylene.

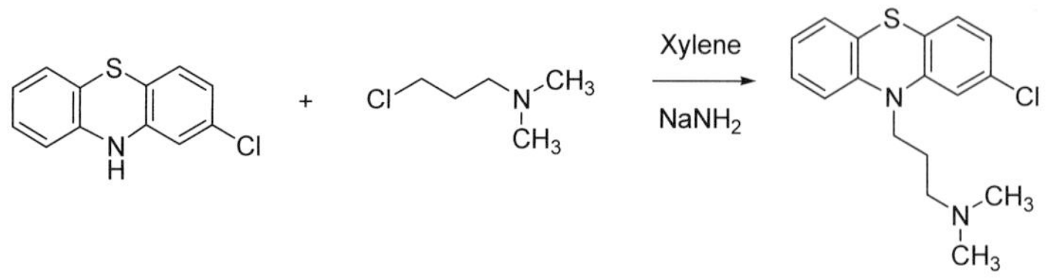

Scheme 1: Synthesis of chlorpromazine

Clinical Use, Pharmacokinetics and Metabolism: As a phenothiazine antipsychotic, chlorpromazine possesses a wide range of CNS activities, based on its receptor profile. Its main antipsychotic mode of action seems to occur via central dopamine (D_2) inhibition in combination with an increase in central dopamine turnover. In addition, α-adrenergic blocking, antimuscarinic, serotonin blocking (e.g., via 5-HT_2 receptors; depressant action), antiemetic and weak antihistaminic and ganglion-blocking properties have been detected. One of the most pronounced side-effects is sedation, even though patients usually develop tolerance during long-term therapy. Chlorpromazine inhibits the heat regulating centre so that patients tend to acquire ambient temperature (poikilothermy). It sometimes has skeletal muscle relaxing properties. Chlorpromazine is widely used in the management of psychotic and other conditions, such as: acute and chronic schizophrenia, acute mania as e.g., in bipolar disorder, control of severely disturbed, agitated or violent behavior, in autistic children, in the short term therapy of acute anxiety, to reduce pre-operative anxiety, and sometimes in the treatment of some forms of hiccup or as adjunct in tetanus therapy, to control acute intermittent porphyria. Chlorpromazine is administered by mouth either in form of its hydrochloride or as its embolate salt. The hydrochloride salt can also be applied intravenously. Dosage varies with the individual and the treatment, typically commencing with three 25 mg doses HCl salt per day as a starting dose for antipsychotic oral treatment. Maintenance doses usually range from 25 mg to 100 mg three times daily p.o. The therapeutic effects of antipsychotics appear to be at least in part mediated by interference with dopamine transmission in the brain. Chlorpromazine has about equal affinities for the D_1 and D_2 receptor, but its metabolites tend to bind more potently to D_2 receptors. The traditional hypothesis for the action of antipsychotics comprises the blockade of D_2 receptors in the limbic and cortical regions in combination with the D_2 blockade in the striatum, a typical motor region of basal ganglia, responsible for extrapyramidal motor side-effects. In addition, the balance between 5-HT_2 and D_2 antagonism or the kinetics of dissociation of the antipsychotic drug from the D_2 receptor have been proposed to be important to classify an antipsychotic drug as "typical" (greater D_2 than 5-HT_2 antagonism) or "atypical" (greater 5-HT_2 than D_2 antagonism). Glutamate or calcium antagonism have been discussed similarly in this respect. The complex mechanism of action of many antipsychotic drugs is not completely understood. According to their pharmacological potency, they can be divided into low- (phenothiazines with an aliphatic or piperidine side chain or thioxanthenes with an aliphatic side chain) and high-potency (butyrophenones, diphenylbutylpiperidines, and phenothiazines or thioxanthenes with a piperazine side-chain) drugs. At equipotent doses, low-potency antipsychotic drugs are more prone to have side-effects, like sedation, antimuscarinic effects, or α-adrenergic mediated effects as compared with high-potency drugs (Baldessarini et al., 1988; Sramek et al., 1988; Snyder, 1989; Anonymous, 1990; Ereshefsky et al., 1990; Thompson, 1994; Remington, 2003). Adverse effects with chlorpromazine treatment comprise mainly sedation (but tolerance development under chronic treatment is seen with most patients) and depression (but not as much as with e.g., barbiturates and benzodiazepines). Chlorpromazine shows antimuscarinic properties like dry mouth, constipation, difficulty with micturation, blurred vision and mydriasis. In addition, ECG changes, tachycardia, and orthostatic hypotension, and in rare cases cardiac arrhythmias are seen (Mehta et al., 1979; Møgelvang et al., 1980; Bett and Holt, 1983; Zee-Cheng et al., 1985; DiGiacomo, 1989; Hui et al., 1990; Henderson et al., 1991; Wilt et al., 1993; O'Brian et al., 1999; Zornberg and Jick, 2000; Reilly et al., 2000, 2002; Ray et al., 2001; Haddad and Anderson, 2002). Other rare adverse effects include delirium, convulsions/EEG abnormalities (Cold et al., 1990; Zacara et al., 1990), agitation, catatonic-like states, insomnia or drowsiness, nightmares, miosis, nasal congestion, minor abnormalities in liver function tests (Sherlock, 1986; Watson et al.; 1988; Fukuzako et al., 1991), neuroleptic malignant syndrome (Wells et al., 1988; Bristow and Kohen, 1993; Kornhuber and Weller, 1994; Velamoor et al., 1995; Ebadi and Srinivasan, 1995; Bristow and Kohen, 1996; Velamoor, 1998; Adnet et al., 2000), effects on sexual function (Kotin et al., 1976; Beeley, 1984; Segraves, 1988 Banlos et al., 1989; Chan et al., 1990; Fabian, 1993; Patel et al., 1996; Salado et al., 2002), effects on the eye (Crombie, 1981; Spiteri and James, 1983; Lam and Remick, 1985; Ngen and Singh, 1988; Marmor, 1990; Power et al., 1991), on the blood (Committee on Safety of Medicines/Medicines Control Agency, 1993a), and on fluid and electrolyte homoeostasis (Witz et al., 1987; Spigset and Hedenmalm, 1995; Rider et al., 1995) and on the skin (Huang and Sands, 1967; Ferguson et al., 1986; Hay, 1995; Harth and Rapoport, 1996). Various haematological disorders and hypersensitivity reactions, as well as extrapyramidal dysfunctions (Moleman et al., 1986; Bateman et al., 1989; Koek and Pi, 1989; Barnes, 1990; Ereshefsky et al., 1990; WHO, 1990; Haag et al., 1992; Committee on Safety of Medicines/Medicines Control Agency, 1994; Ebadi and Srinivasan, 1995; Egan et al., 1997; Holloman and Marder, 1997; Jímenez-Jímenez et al., 1997; Adler et al., 1998; Raja, 1998; Casey, 1999; Kane, 1999; Mamo et al., 1999; Najib, 1999; van Harten et al., 1999; Miller and Fleischhacker, 2000; Soares et al., 2004a; Soares and McGrath, 2004) and alterations of endocrine (Rosenblatt et al., 1978; Brown et al., 1981; Asplund et al., 1982; Lilford et al., 1984; Gunnest and Moore, 1988; Mortensen, 1989, 1994; Wieck and Haddad, 2002; Howes and Smith, 2002; Wang et al., 2002) and metabolic functions (Henkin et al., 1992; Allison et al., 1999; Wetterling, 2001) have been described along with chlorpromazine treatment. The frequency and the patterns of phenothiazine related adverse effects can be classified in three groups: group 1 with pronounced sedative and only moderate antimuscarinic and extrapyramidal side-effects (e.g., chlorpromazine, levomepromazine and promazine), group 2 with moderate sedative, marked antimuscarinic and few extrapyramidal side-effects (e.g., pericyazine, pipotiazine and thioridazine) and group 3 with few sedative and antimuscarinic, but more pronounced extrapyramidal side-effects (e.g., fluphenazine, perphenazine, prochlorperazine and trifluoperazine). Classic antipsychotics of other chemical classes (e.g., butyrophenones like benperidol and haloperidol, diphenylbutylpiperidines like pimozide, thioxanthenes like flupentixol and zuclophenthixol, substituted benzamides like sulpiride, and miscellaneous structures like oxypertine and loxapide) resemble in most cases group 3 in their side-effect patterns. Treatment of side-effects related to chlorpromazine therapy, especially when overdosing, is mainly symptomatic and supportive (intensive care). Abrupt discontinuation of an antipsychotic may be accompanied by withdrawal symptoms (Dilsaver, 1994). Chlorpromazine, like most other phenothiazines, is contraindicated in patients with a history of CNS depression, coma, bone-marrow suppression, phaeochromocytoma or prolactin-dependent tumours, as well as patients with impaired liver, kidney, cardiovascular, cerebrovascular and respiratory function, amongst others. Depressant side-effects may be enhanced by other drugs with CNS-depressant properties like alcohol, general anaesthetics, hypnotics, anxiolytics, and opioids. Chlorpromazine is readily absorbed from the GI tract, reaching peak plasma concentrations about 2–4 hours after ingestion. It suffers considerable and extensive first-pass metabolism in the gut wall and the liver and is excreted in urine and bile in form of numerous active and inactive metabolites. Chlorpromazine shows wide interindividual differences in plasma concentrations, so that the optimum dose needs to be adjusted for each patient. Paths of metabolism include hydroxylation/glucuronidation, N-oxidation, S-oxidation, and dealkylation. Elimination of chlorpromazine (half-life about 30 hours) and its metabolites is slow. Chlorpromazine shows high (95–90%) plasma protein binding and is widely distributed throughout the body and into the brain. Chlorpromazine brain concentrations are usually higher than plasma concentrations (Rivera-Calimlim et al., 1979; Furlanut et al., 1990; Caccia and Garattini, 1990; Yeung et al., 1993).

Prochlorperazine

- Prochlorperazine patent was filed in 1955 by Rhone-Poulenc (Horclois, 1958) and the compound was introduced to the USA market in 1956

- Prochlorperazine is a phenothiazine antipsychotic with general properties similar to those of chlorpromazine

- It is now relatively seldom used for the treatment of psychosis and the manic phase of bipolar disorder, as well as an adjunct in the short-term management of

Prochlorperazine

[58-38-8], 2-Chloro-10-[3-(4-methylpiperazin-1-yl)propyl]-10H-phenothiazine, $C_{20}H_{24}ClN_3S$, M_r 373.95;
[84-02-6], maleate salt, $C_{24}H_{28}ClN_3O_4S$, M_r 490.02;
[5132-55-8], methansulfonate salt, $C_{21}H_{28}ClN_3O_3S_2$, M_r 470.06;
[1257-78-9], edisilate salt, $C_{22}H_{30}ClN_3O_6S_3$, M_r 564.10

severe anxiety. It has a prominent antiemetic/antivertignoic activity and is more often used for the (short-time) treatment of nauseas/emesis and vertigo. Quite recently, in the UK prochlorperazine maleate has been made available as an OTC-treatment for migraine

- Prochlorperazine is a dopamine D_2 receptor antagonist

Synthesis (Horclois, 1958; 1959): The compound is obtained by condensation of 2-chloro-10-(13-chloropropyl)-phenotiazine with 1-methylpiperazine.

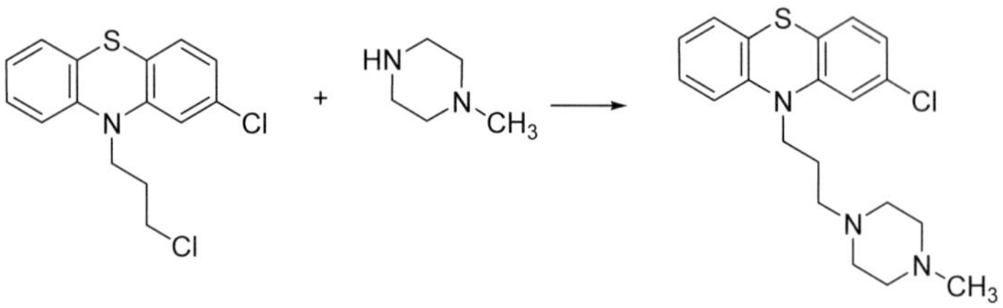

Scheme 2: Synthesis of prochlorperazine

Clinical Use, Pharmacokinetics and Metabolism: As a phenothiazine antipsychotic, prochlorperazine has a similar pharmacological spectrum as chlorpromazine. Prochlorperazine and its salts are used in the treatment of nausea and vomiting (also associated with migraine) and drug-associated emesis, as well as in the short-term symptomatic treatment of vertigo, labyrinthitis, and in the short-term management of schizophrenia, mania and other psychoses. It is used as adjunct in severe anxiety short-term treatment. Usual starting doses for the treatment of psychoses comprise 5.0–2.5 mg doses maleate or mesilate salt b.i.d. According to the response, the dose may be increased gradually up to 100 mg daily. In comparison to chlorpromazine, prochlorperazine causes less sedation and fewer antimuscarinic effects, but frequently pronounced extrapyramidal side-effects (group 3 classification, see chlorpromazine chapter). Severe dystonic reactions have followed the use of prochlorperazine, so that it should be used with care especially in children. Hypertension (Roche et al., 1985) and effects on the mouth like local irritations (Duxburry et al., 1982; Reilly and Wood, 1984) have been described for prochlorperazine. Drug interactions are similar to those described with chlorpromazine. In terms of its pharmacokinetics, prochlorperazine showed significant interindividual differences in a study following intravenous administration. Terminal half life was determined to be around 7 hours and apparent volume of distribution was very high in that study. The study suggests that the liver may not be the only site of metabolism. After oral administration, plasma peak levels were reached at 1.5–5.0 hours, while bioavailability was estimated to range from 0–16%. High plasma clearance indicates a high first-pass metabolism (Taylor and Bateman, 1987). The bioavailability can be increased by buccal administration (Hessell et al., 1989).

Promazine

- The compound patent was filed in 1945 by Rhone-Poulenc (Charpentier, 1950)

- First launched in the USA by Wyeth for the treatment of schizophrenia in 1956

- Promazine is a phenothiazine with general properties similar to those of chlorpromazine. It has relatively weak antypsichotic activity and is mainly used for the short-term management of agitated or disturbed behavior and for the alleviation of nausea and emesis

Promazine

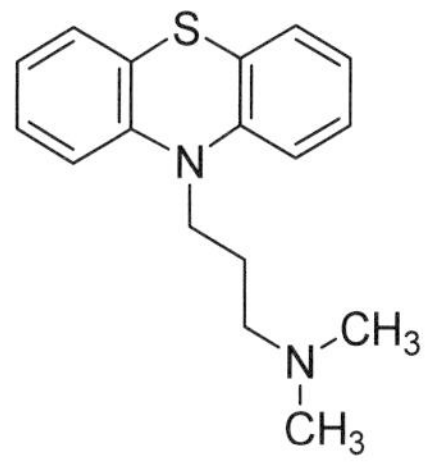

[58-40-2], 10-[3-(Dimethylamino)propyl]phenothiazine, $C_{17}H_{20}N_2S$, M_r 284.43; [53-60-1], hydrochloride salt, $C_{17}H_{21}ClN_2S$, Mr 320.89; [004701-69-3], maleate salt, $C_{21}H_{24}N_2O_4S$, Mr 400.50; [1508-27-6], phosphate salt, $C_{17}H_{23}N_2PO_4S$, Mr 382.42; embonate ester, $C_{40}H_{36}N_2O_6S$, Mr 957.20

Synthesis (Charpentier, 1950): The compound is obtained by the addition of 3-dimethylaminopropyl-chloride to phenotiazine, after heating under reflux in a mixture with xylene and sodamide.

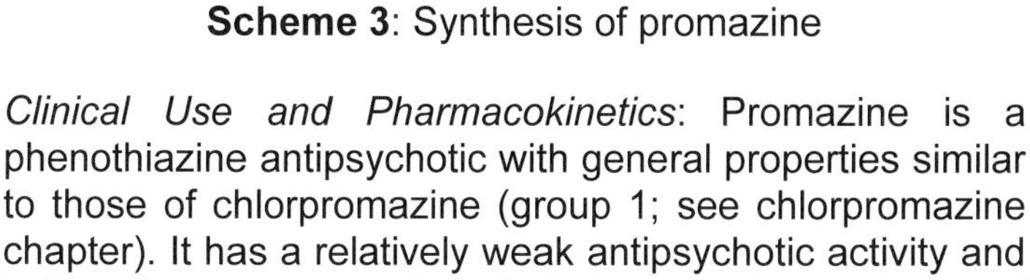

Scheme 3: Synthesis of promazine

Clinical Use and Pharmacokinetics: Promazine is a phenothiazine antipsychotic with general properties similar to those of chlorpromazine (group 1; see chlorpromazine chapter). It has a relatively weak antipsychotic activity and is therefore not generally a first choice in the treatment of psychoses. It is used in the short-term management of disturbed or agitated behavior, and has also been given for the alleviation of nausea and vomiting, similar to prochlorperazine. In the treatment of agitated behavior, promazine is usally given as hydrochloride salt in doses of 100–200 mg four times daily. It can also be applied intramuscularly, or by slow intravenous injection. Adverse effects (John, 1975), drug interactions and pharmacokinetics are very comparable to chlorpromazine.

Levomepromazine

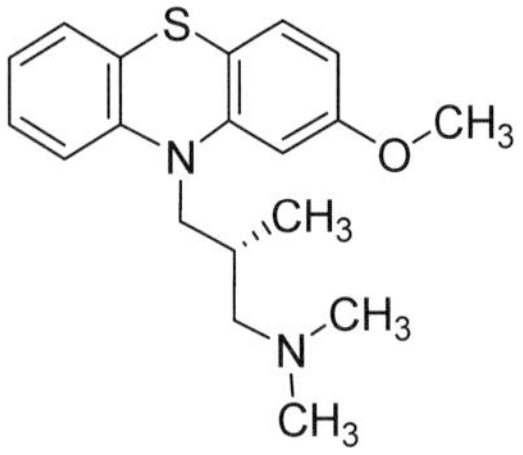

Levomepromazine

[60-99-1], (-)-(R)-N,N,β-Trimethyl-2-methoxy-10H-phenothiazine-10-propanamide, $C_{19}H_{24}N_2OS$, M_r 328.48;
[1236-99-3], hydrochloride salt, $C_{19}H_{25}ClN_2OS$, M_r 364.94;
[7104-38-3], maleate salt, $C_{23}H_{28}N_2O_5S$, M_r 444.55

- The compound patent was filed in 1954 by Rhone-Poulenc (Jacob and Robert, 1958)

- First launched in the USA by Wyeth for the treatment of schizophrenia in 1957

- Levomepromazine is a phenothiazine with pharmacological activity similar to that of chlorpromazine. It is used in the treatment of various psychoses including schizophrenia, as an analgesic for moderate to severe pain, for the control of symptoms such as restlessness, agitation and vomiting, and as an adjunct to opioid analgesia in terminally ill patients

Synthesis (Jacob and Robert, 1958): Levorotatory final compound is obtained by dissolving the racemate 3-(3 methoxy-10H-phenothiazinyl)-2-methyl-1-dimethylaminopropane with L-tartaric acid in boiling isopropanol, and the subsequent filtration of the solid. The racemate is obtained by condensation of 2-methoxy-phenotiazine and 3-dimethylamino-2-methylpropylchloride.

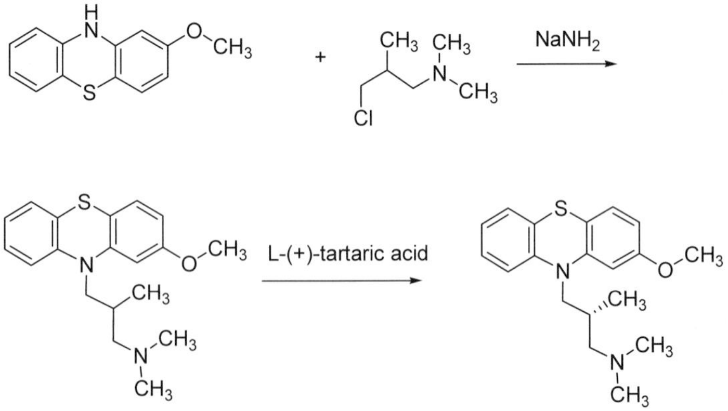

Scheme 4: Synthesis of levomepromazine

Clinical Use, Pharmacokinetics and Metabolism: Levomepromazine is a phenothiazine antipsychotic with general properties very similar to both chlorpromazine and promethazine. With regard to its antihistaminic and CNS effects, it resembles very much chlorpromazine. It is used widely in the treatment of psychoses like schizophrenia and due to its analgesic activity, as an adjunct to opioid analgesia in terminally ill patients and as a general analgesic for moderate to severe pain. In addition, it is used in the control of symptoms such as restlessness, agitation and vomiting. Levomepromazine is also used in veterinary medicine as well. Initial doses for levomepromazine schizophrenia therapy range from 25 to 50 mg daily p.o. (maleate or hydrochloride salt). In some countries, levomepromazine is licensed as anxiolytic, sedative and analgesic (Oliver, 1985; Stiell et al., 1991; Patt et al., 1994; O'Neill and Fountain, 1999; Skinner and Skinner, 1999). Adverse effects are comparable to those observed with chlorpromazine, although levomepromazine is usually more sedating. Levomepromazine may provoke severe orthostatic hypotension, particularly dangerous for elderly patients and children. Drug interactions are comparable to those with chlorpromazine treatment, especially in combination with antidepressants (Barsa and Saunders; 1964; McQueen, 1980). Peak plasma concentrations of levomepromazine are usually obtained 1–4 hours after p.o. administration. Oral bioavailability ranges around 50%. The main metabolite identified is levomepromazine sulfoxide, usually found in concentrations higher than the parent compound levomepromazine. Plasma half-lives determined show great variability, ranging from 16.5–77.8 hours, without good correlation to the dose applied (Dahl, 1976; Dahl et al., 1977).

Perphenazine

- The compound patent was filed in 1958 by Searle (Cusic and Hamilton, 1958)

- First launched by Schering-Plough for the treatment of nausea and psychosis in 1957

- Perphenazine is a phenothiazine with general properties similar to those of chlorpromazine. It is used in the treatment of various psychoses including schizophrenia and mania as well as disturbed behavior and in the short-term, adjunctive management of severe anxiety. It is also used for the management of post-operative or chemotherapy-induced nausea and vomiting and for the treatment of intractable hiccup

Perphenazine

[58-39-9], 4-[3-(2-Chloro-10H-phenothiazin-10-yl)propyl]-1-piperazineethanol, $C_{21}H_{26}ClN_3OS$, M_r 403.98; [17528-28-8], enantate ester, $C_{28}H_{38}ClN_3O_2S$, M_r 516.10; decanoate ester, $C_{31}H_{44}ClN_3O_2S$, M_r 558.20

- Long-acting decanoate or enantate esters, available in some countries, are given by intramuscular injection

Synthesis (Cusic and Hamilton, 1958): A mixture of 2-chloro-10-(3-chloropropyl)-10*H*-phenothiazine, sodium, iodide, piperazine and butanone is refluxed, concentrated and extracted with dilute hydrochloric acid, to yield 2-chloro-10-(3-(piperazin-1-yl)propyl)-10*H*-phenothiazine. A mixture of this compound with 2-bromoethanol, potassium carbonate and toluene is refluxed, and perphenazine is obtained as the final compound.

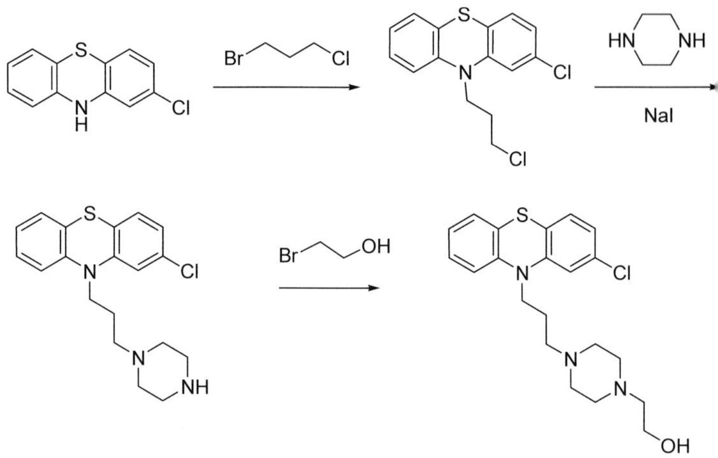

Scheme 5: Synthesis of perphenazine

Clinical Use, Pharmacokinetics and Metabolism: Perphenazine is a phenothiazine antipsychotic with general properties similar to those of chlorpromazine. It is used in the treatment of various psychoses including schizophrenia and mania as well as in disturbed behavior and in the short-term, adjunctive management of severe anxiety. Perphenazine is also used in the management of postoperative and chemotherapy-induced nausea and vomiting and for the treatment of intractable hiccup. Initial doses for the treatment of schizophrenia, mania and other psychoses range from around 4 mg three to four times a day and can be adjusted according to individual needs up to 24 mg to 64 mg daily. Long-acting decanoate or enantate esters are available for intramuscular injection (depot injections). In terms of its side-effect profile, perphenazine can be compared with chlorpromazine, however with less sedative and antimuscarinic, but with more pronounced extrapyramidal side-effects (group 3;

see chlorpromazine chapter). Perphenazine is well absorbed from the GI tract following oral administration and undergoes some first-pass metabolism, resulting in good oral bioavailabilities ranging from 60% to 80%. It is widely distributed and crosses the placenta (effects on breast feeding; see: American Academy of Pediatrics, 2001; Olesen et al., 1990). Extensive metabolism leads to 70% excretion via urine, mainly in the form of metabolites (Eggert et al., 1976). Interindividual differences in perphenazine disposition according to polymorphic debrisoquin hydroxylation have been reported (Dahl-Puustinen et al., 1989).

Triflupromazine

- The compound patent was filed in 1956 by Smith Kline & French (Ullyot, 1960)
- First launched in the USA by Bristol-Myers Squibb for the treatment of schizophrenia in 1957
- Triflupromazine is a phenothiazine with general properties similar to those of chlorpromazine. It is used mainly in the management of psychoses and the control of nausea and vomiting

Synthesis (Ullyot, 1960): 2-Trifluromethylphenothiazine is treated with sodamide in refluxing toluene and allowed to react with 3-chloro-N,N-dimethylpropan-1-amine.

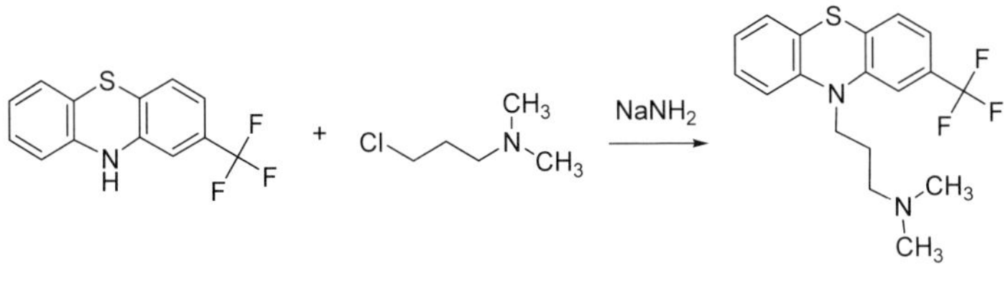

Triflupromazine

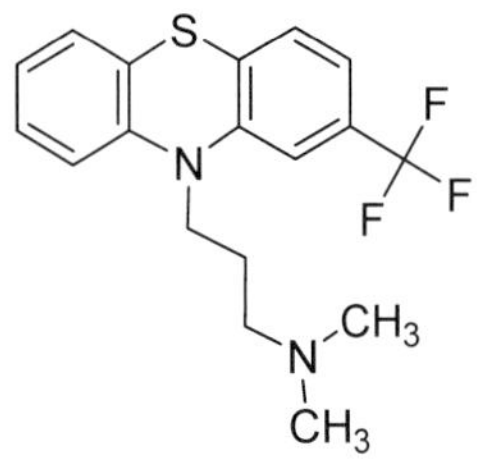

[146-54-3], N,N-Dimethyl-2-(trifluoromethyl)-10H-phenothiazine-10-propanamide, $C_{18}H_{19}F_3N_2S$; M_r 352.41;
[1098-60-8] hydrochloride salt, $C_{18}H_{20}ClF_3N_2S$, M_r 388.88

Scheme 6: Synthesis of triflupromazine

Clinical Use, Pharmacokinetics and Metabolism: Triflupromazine hydrochloride is a phenothiazine antipsychotic with general properties similar to those of chlorpromazine. Its predominant use is in the treatment of psychoses and the management of nausea and vomiting. It is usally given by injection, even though in some countries p.o. preparations are also available. The usual dose for the treatment of psychoses ranges from 60 mg to

150 mg daily (intramuscular injection of the HCl salt). Reduced doses should be given in elderly or debilitated patients. A comparative ADME study with the methiodides of triflupromazine, promazine and chlorpromazine has been described by Huang et al. (1970).

Thioridazine

Thioridazine

[50-52-2] 10-[2-(1-Methylpiperidin-2-yl)ethyl]-2-(methylsulfanyl)-10H-phenothiazine, $C_{21}H_{26}N_2S_2$, M_r 370.58;
[130-61-0] hydrochloride salt, $C_{21}H_{27}ClN_2S_2$, M_r 407.04;
[1257-76-7] tartrate salt, $C_{25}H_{32}N_2O_6S_2$, M_r 520.67

- The compound patent was filed in 1956 by Sandoz (Renz and Bourquin, 1966)

- First launched in the USA by Novartis for the treatment of schizophrenia in 1959

- Thioridazine is a phenothiazine with general properties similar to those of chlorpromazine. The distinctive features of thioridazine are its low propensity to cause extrapyramidal side-effects and its low antiemetic activity

- The manufacturer of thioridazine in the USA, Canada and Europe, Novartis/Sandoz/Wander, discontinued the drug worldwide in June 2005

- Thioridazine may still be available from other manufacturers as a generic drug with the precaution that it is used only in psychotic patients refractory to other forms of drug treatment. ECG-monitoring and frequent white blood cell counts are required before initiating therapy and in close intervals afterwards

Synthesis (Renz and Bourquin, 1966): N-(m-Methylmercapto-phenyl)-aniline is prepared by condensing m-methylmercapto-aniline with the potassium salt of o-chloro-benzoic acid and decarboxylating the resultant N-(m-methylmercapto-phenyl)-anthranilic acid by heating. Treatment with sulfur and powdered iodine produces 3-methylmercapto-phenotiazine. This compound is condensed with 2-(2-chloroethyl)-1-methylpiperidine by sodamide and xylene to produce thioridazine.

Clinical Use, Pharmacokinetics and Metabolism: The use of thioridazine is restricted to schizophrenia patients unresponsive towards treatment with other antipsychotics. Because of its cardiotoxic potential, thioridazine is not used any more in other psychiatric disorders. It is recommended for thioridazine treatment to be accompanied by baseline ECG and electrolyte screening. Thioridazine is administered in the form of its hydrochlo

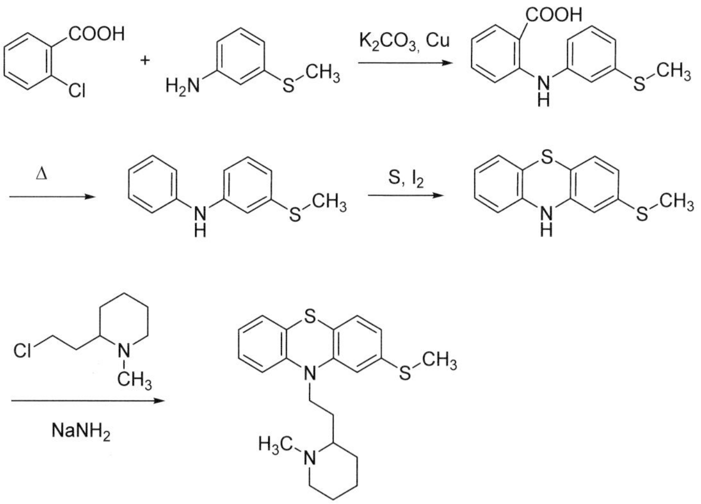

Scheme 7: Synthesis of thioridazine

ride salt or the free base in a dose of 50–300 mg HCl salt for outpatients as a treatment for psychoses. Thioridazine treatment should not be discontinued abruptly in order to avoid withdrawal symptoms e.g., GI disorders, dizziness, anxiety and insomnia. Instead, the dose should be gradually reduced over 1–2 weeks. Thioridazine has been associated with a higher incidence of antimuscarinic side-effects, but with a lower incidence of extrapyramidal symptoms, as compared to chlorpromazine (group 2: moderate sedation, marked antimuscarinic and few extrapyramidal side-effects; see chlorpromazine chapter). However, it is more likely to induce hypotension and has a cardiotoxic risk associated with Q–T interval prolongation (Torsade de Points and Sudden Death; Committee on Safety of Medicines/Medicines Control Agency, 2001). Sexual dysfunction is also more frequent with thioridazine than with chlorpromazine. In rare cases, pigmentary retinopathy and hypersensitivity (Sell, 1985) has been reported. Rhabdomyolysis has been reported in combination with overdosage (Nankivell et al., 1994). Thioridazine should not be used in patients with clinically significant cardiac disorders, such as e.g., uncorrected hypokalaemia, electrolyte imbalance, suspected Q–T prolongation or with a history of ventricular arrhythmias. As thioridazine is a strong inhibitor of the cytochrome P450 isoenzyme CYP2D6 and its metabolism is mainly mediated by this isoenzyme, it is contraindicated in patients with reduced CYP2D6. The therapeutic use of combinations of thioridazine with drugs that inhibit or are substrates of CYP2D6 is also contraindicated. Examples include antidepressants like the SSRIs or the TCAs, antiarrhytmics, β blockers or HIV-protease inhibitors and opiates. Pharmacokinetics of thioridazine are similar to those reported for chlorpromazine. Thioridazine is mainly metabolized to by the cytochrome P450 isoenzyme CYP2D6 to its main active metabolite mesoridazine (side-chain sulfoxide). Another metabolite, sulforidazine (side chain sulfone), also contributes to the activity of thioridazine. Thioridazine and its active metabolites are reported to have high plasma protein binding activity. The plasma half-life has been estimated to be about 4–10 hours (Mårtensson and Roos, 1973; Axelsson and Mårtensson, 1976; Axelsson, 1977). There seems to be some effect of debrisoquine hydroxylation phentotype on metabolism of thioridazine (von Bahr et al., 1991).

Trifluoperazine

Trifluoperazine

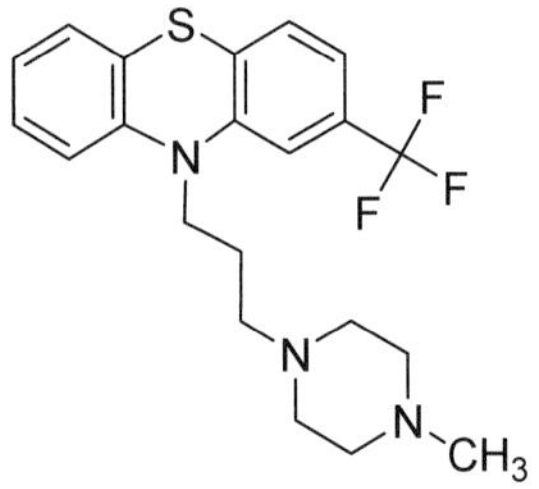

[117-89-5], 10-[3-(4-Methyl-1-piperazinyl)propyl]-2-(trifluoromethyl)-10H-phenothiazine, $C_{21}H_{24}F_3N_3S$, M_r 407.50;
[440-17-5] dihydrochloride salt, $C_{21}H_{26}Cl_2F_3N_3S$, M_r 480.42

- The first patent for trifluoperazine was filed in 1955 by Smith Kline & French Laboratories (Mathieson, 1960a)

- In 1959, the compound was launched as an antipsychotic and anxiolytic (see Chapter 3) by GlaxoSmithKline

- Trifluoperazine is a phenothiazine antipsychotic with general properties similar to those of chlorpromazine. Its use in many parts of the world has declined because of highly frequent and severe early and late (tardive dyskinesia) extrapyramidal symptoms

Synthesis (Mathieson, 1960a; 1960b): 2-Trifluoromethylphenothiazine is treated with sodamide in refluxing toluene and allowed to react with 1-(3-chloropropyl)-4-methylpiperazine chloride to give trifluoperazine.

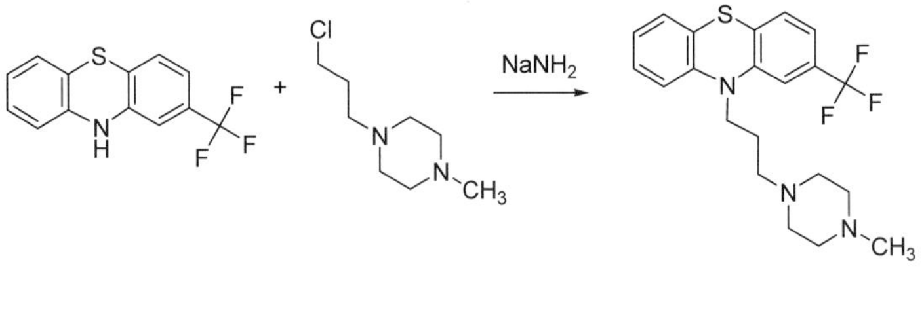

Scheme 8: Synthesis of trifluoperazine

Clinical Use and Pharmacokinetics: The antipsychotic use of trifluoperazine is similar to the one of chlorpromazine. Trifluoperazine is used in the treatment of psychiatric disorders including schizophrenia, severe anxiety, and disturbed behavior. It is also used for the short-term control of nausea and vomiting. Trifluoperazine is given p.o. in the form of its hydrochloride with initial doses ranging from 2–5 mg b.i.d. for schizophrenia treatment. The dose can gradually be increased up to 20, or even 40 mg per day. Depot application in form of deep intramuscular injection is also possible. In terms of its side-effect profile, trifluoperazine can be compared with chlorpromazine, though with less sedation, hypotension, hypothermia and antimuscarinic side-effects, but with more pronounced extrapyramidal side-effects, particularly when the daily dose exceeds 6 mg (group 3; see chlorpromazine chapter). Trifluoperazine should not be used when breast feeding (American Academy of Pedriatrics, 2001). Trifluoperazine reaches peak plasma concentrations from 1.5–4.5 hours after ingestion (p.o.). Elimination was detected to be multiphasic, with elimination half-lives being around 5 hours (Midha et al., 1983a).

Fluphenazine

- The compound patent was filed in 1956 by Smith Kline & French Laboratories (Mathieson, 1960b)

- Between 1959 and 1966, fluphenazine was launched as an antipsychotic and anxiolytic drug and for the treatment of bipolar disorder by Bristol-Myers Squibb, Sanofi-Aventis and Schering-Plough. (see Chapter 3.5)

- Fluphenazine is a phenothiazine with general properties similar to those of chlorpromazine. Fluphenazine is used in the treatment of a variety of psychiatric disorders including schizophrenia, mania, severe anxiety and behavioral disturbances

- The long-acting decanoate or enantate esters of fluphenazine are usually given by deep intramuscular injection and are used mainly for the maintenance treatment to patients with schizophrenia or other chronic psychoses

Synthesis (Mathieson, 1960a; 1960b): The *N*-alkylation of 2-trifluromethylphenothiazine with 4-(3-chloropropyl)-piperazine-1-carboxaldehyde in the presence of sodamide leads to 10-[3-(piperazinyl)propyl]-2-trifluoromethyl-

Fluphenazine

[69-23-8] 4-[3-[2-(Trifluoromethyl)phenothiazin-10-yl]propyl]-1-piperazineethanol, $C_{22}H_{26}F_3N_3OS$, M_r 437.53;
[146-56-5] dihydrochloride salt, $C_{22}H_{28}Cl_2F_3N_3OS$, M_r 473.99;
[5002-47-1] decanoate ester, $C_{32}H_{46}F_3N_3O_3S$, M_r 609.79;
[2746-81-8] eantate ester, $C_{29}H_{38}F_3N_3O_2S$, M_r 549.70

phenothiazine1-carboxaldehyde. Deprotection of the amino group with sodium hydroxide, followed by alkylation with 2-bromoethyl acetate and treatment with aqueous hydrochloric acid gives the target compound (see Chapter 3.5 for additional synthesis details).

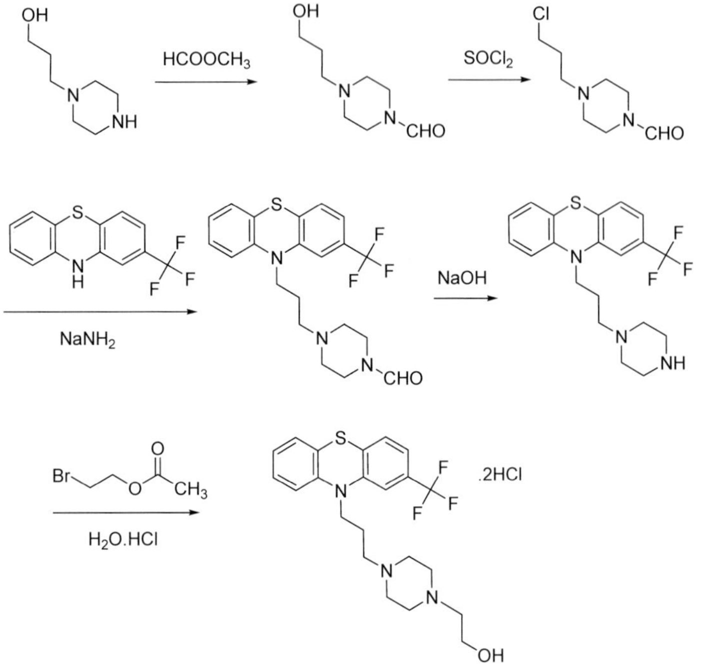

Scheme 9: Synthesis of fluphenazine

Clinical Use, Pharmacokinetics and Metabolism: Fluphenazine hydrochloride is a phenothiazine antipsychotic with general properties similar to those of chlorpromazine. It is used in the treatment of a variety of psychiatric disorders including schizophrenia, mania, severe anxiety and other behavioral disturbances. Initial doses of fluphenazine for antipsychotic treatment are in the range of 2.5–10 mg daily p.o., which can be increased up to 20 mg per day depending upon response. Decanoate and enantate salts are available for intramuscular or subcutaneous depot injection (maintenance treatment) (Kane et al., 1983; Marder et al., 1987; Hogarty et al., 1988; Carpenter et al., 1999; Adams and Eisenbruch, 2004). In terms of its side-effect profile, fluphenazine can be compared with chlorpromazine, however with less sedation, hypotension and antimuscarinic side-effects, but with more pronounced extrapyramidal side-effects (group 3; see chlorpromazine chapter). Effects on the liver (Kennedy, 1983) and on the CNS (Curry et al., 1979a) as well as experiences with overdosage (Ladhani, 1974; Cheung and Yu, 1983) have been described. Fluphenazine exerts depressant activity and should not be given to a depressed patients. Other contraindications are similar to those reported for chlorpromazine. Fluphenazine hydrochloride is absorbed following oral administration, to give reported plasma half-life of around 5 to 15 hours after ingestion The compound is widely distributed, crosses the blood-brain barrier and the placenta, and may be excreted into breast milk. It is more than 90% bound to proteins and highly metabolized by the liver. The plasma half-life of fluphenazine enanthate is around 4 days, and of fluphenazine decanoate, from 7–10 days (Curry et al., 1979b; Wistedt et al., 1981; Midha et al., 1983b; Marder et al., 1989).

Haloperidol

- Haloperidol was developed in 1958 by the Belgian company Janssen Pharmaceutica (Janssen, 1969a) and launched in 1959. After being rejected by the USA company Searle due to side-effects, it was later marketed in the USA by McNeil Laboratories

- Haloperidol is a butyrophenone antipsychotic with general properties similar to those of the phenothiazine antipsychotics with a piperazine chain and to chlorpromazine. Haloperidol is a strong D_2 receptor antagonist and approximately 50 times more potent than chlorpromazine. Haloperidol possesses a strong activity against delusions and hallucinations, most likely due to an effective dopaminergic receptor blockage. It has minor antihistaminic and anticholinergic properties

- Haloperidol is used in the treatment of various psychoses including schizophrenia and mania, and in behavioral disturbances, Tourette's syndrome, severe tics, intractable hiccup and severe anxiety. Haloperidol has also been used in the control of nausea and vomiting. Haloperidol also has sedative properties and is an effective treatment for mania and states of agitation. Additionally, it can be given as an adjuvant in the therapy of severe chronic pain

Haloperidol

[52-86-8], 4-[4-Acetoxy-4-(4-chlorophenyl)piperidin-1-yl]-1-(4-fluorophenyl)butan-1-one, $C_{21}H_{23}ClFNO_2$, M_r 375.87;
[74050-97-8], decanoat ester, $C_{31}H_{41}ClFNO_3$, M_r 530.11

- The decanoate ester of haloperidol, which has a greatly extended duration of effect, was launched in 1984

Synthesis (Janssen et al., 1959; Janssen, 1969a, 1970)
The compound haloperidol is obtained by condensation of 4-chloro-1-(4-fluorophenyl)butan-1-one with 4-(4-chlorophenyl)piperidin-4-ol in an apolar solvent, such as toluene. The 4-chloro-1-(4-fluorophenyl)butan-1-one is prepared by a Friedel-Craft reaction using 5-chloropentanoyl chloride and fluorobenzene. The 4-(4-chlorophenyl)piperidin-4-ol is obtained from 1-chloro-4-(prop-1-en-2-yl)benzene by condensation with formaldehyde and ammonium chloride to give a 6-(4-chlorophenyl)-6-methyl-1,3-oxazinane. This compound is converted with excess acid to the corresponding 4-(4-chlorophenyl)-1,2,3,6-tetrahydropyridine. Addition of hydrobromic acid gives an intermediate which hydrolyses easily to yield the desired 4-(4-chlorophenyl)piperidin-4-ol.

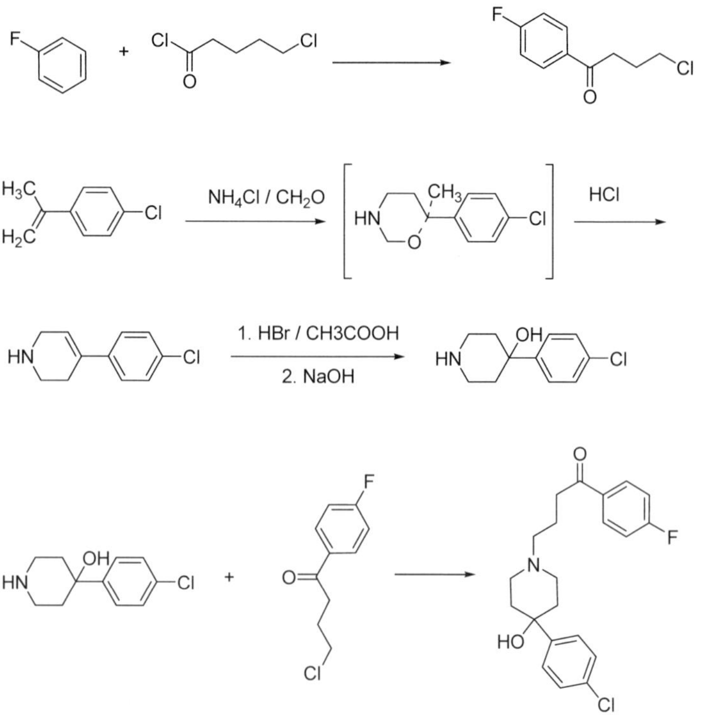

Scheme 10: Synthesis of Haloperidol

Clinical Use, Pharmacokinetics and Metabolism: Haloperidol is a butyrophenone antipsychotic with general properties similar to those of the phenothiazine antipsychotics with a piperazine chain and to chlorpromazine. It is used in the treatment of various psychoses including schizophrenia and mania, behavioral disturbances, Tourette's syndrome and severe tics, in intractable hiccup and in severe anxiety. Haloperidol has also been used in control of nausea and vomiting such as that resulting from chemotherapy due to its peripheral antidopaminergic effects. Initial doses for the treatment of psychoses range fromm between 0.5 mg and 5 mg two or three times daily p.o. In severe psychoses, doses of up to 100 mg daily may be required. For the control of acute psychotic conditions, intramuscular (decanoate ester as depot) or intravenous application of haloperidol is possible. As compared with chlorpromazine, haloperidol is less likely to cause sedation, hypotension or antimuscarinic effects, but shows a higher incidence of extrapyramidal side-effects. It should be used only with great caution in children and adolescents because of possible dystonic reactions. Adverse effects when breast feeding (American Academy of Pedriatrics, 2001; Stewart et al., 1980) have been described as well as effects on the liver (Dincsoy and Saelinger, 1982), side-effects associated with overdosage (Scialli and Thornton, 1978; Sinaniotis et al., 1978; Cummingham and Challapalli, 1979), porphyria, obstructive uropathy (Jeffries et al., 1982) and toxic encephalopathy (Maxa et al., 1997). Drug interactions are similar to the ones described for chlorpromazine. It must be used with extreme caution in patients receiving lithium treatment because of the risk of encephalopathic syndrome. Haloperidol is readily absorbed from the GI tract, metabolized in the liver and excreted in urine and – via bile – in faeces. After oral administration, it undergoes significant first-pass metabolism in the liver, decreasing oral bioavailability. There is wide interindividual variation in plasma concentrations, and – in addition – no strong correlation between plasma concentration and therapeutic effect has been found. Haloperidol was reported to have half-lives between 12 hours to 38 hours after oral administration. 92% of it is bound to plasma proteins. Haloperidol is widely distributed throughout the body, and is also distributed into breast milk (Kudo and Ishizaki, 1999). The significance of metabolites, especially of "reduced" haloperidol, has been discussed widely, especially in terms of the explanation of the occurrence of haloperidol non-responders. Pyridinium metabolites have

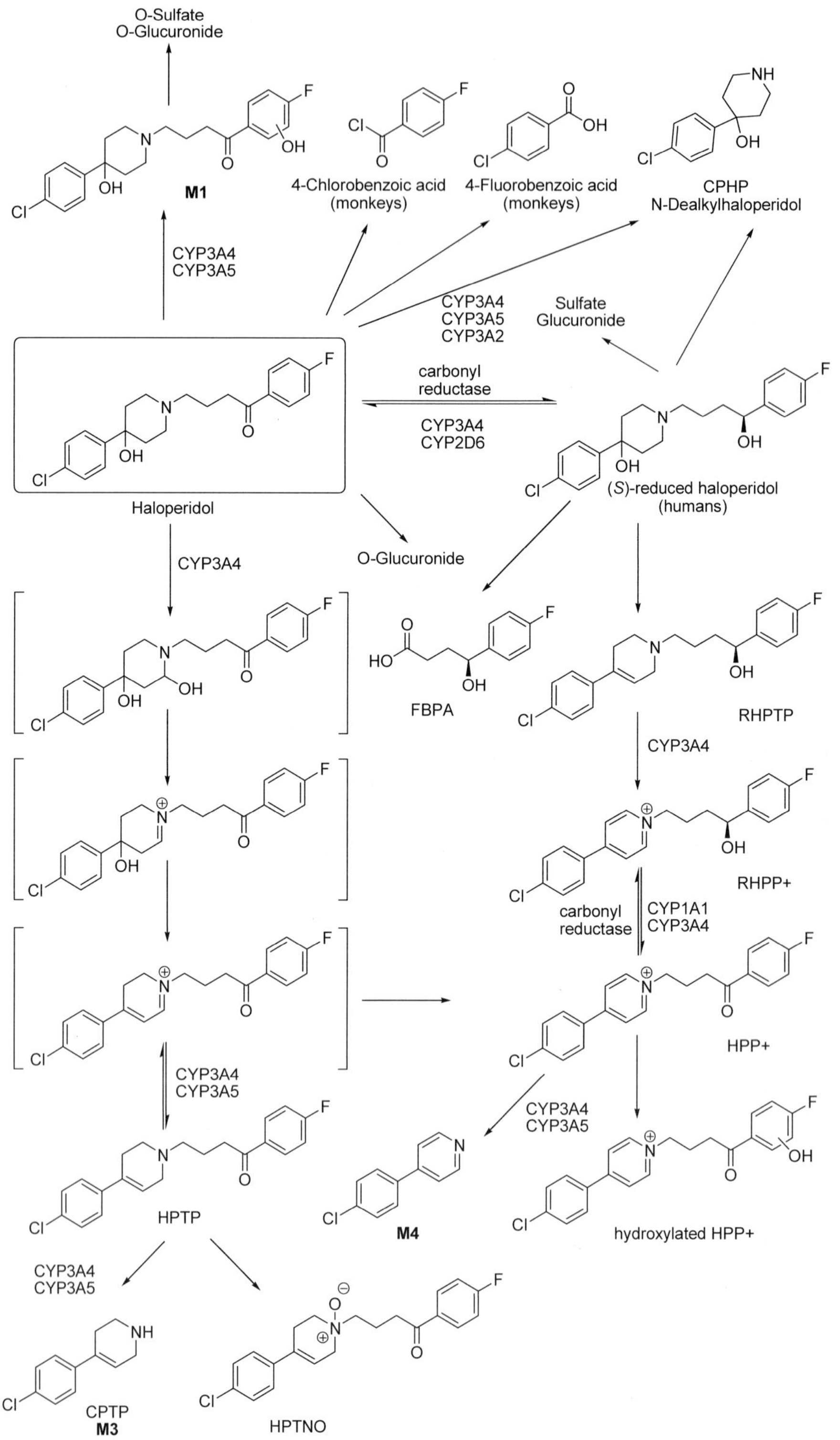

Scheme 11: Metabolism of haloperidol

been discussed in connection with neurotoxicity and irreversible Parkinsonism (MPTP-like behavior) (Sramek et al., 1988; Froemming et al., 1989; Chakraborty et al., 1989; Eyles et al., 1994). The compliance of haloperidol patients can be monitored by measurements of concentrations of haloperidol and/or its metabolite in scalp hair (Uematsu et al., 1989; Ulrich et al., 1989; Matsuno et al., 1990). At least 15 metabolites have been identified in humans and experimental animals *in vivo* or *in vitro*. The major metabolic pathways include oxidative N-dealkylation, oxidation to pyridinium-related metabolites, glucuronidation and carbonyl reduction. In humans, haloperidol is reduced only to the (*S*)-enantiomer (a major metabolite in blood and brain) via the regulations of a carbonyl reductase, CYP3A4 and CYP2D6. Both haloperidol and its reduced form undergo sequential oxidations via several intermediates forming HPTP and RHPTP, respectively. The latter two (both found *in vitro* but not *in vivo*) are further oxidized to the neurotoxic pyridinium derivatives HPP+ and RHPP+, detectable in human blood and urine. HPP+ can also be reversibly formed from RHPP+ via the mediation of carbonyl reductase, CYP1A1 and CYP3A4. HPP+ is also the major metabolite detected in human liver microsomes, while the second predominant metabolite is a fluorophenyl ring-hydroxylated derivative (M1) of haloperidol. Both *O*-sulfate and *O*-glucuronide conjugates of M1 are detected in human urine, whereas a fluorophenyl ring hydroxylated derivative of HPP+ is detected in mouse urine and brain tissue extracts. Haloperidol undergoes *N*-dealkylation forming CPHP via regulations of CYP3A4, CYP3A5 and CYP3A2. Similar cleavage products M3 and M4 (derived from HPTP and HPP+) are detected in incubates with recombinant CYP3A4 or CYP3A5. Two carboxylic acid derivatives FBPA and FBHP are formed from *N*-dealkylations of haloperidol and its reduced form under the main regulation of CYP3A4 (in rat liver CYP3A2 catalyzed the formation of FBPA). 4-chlorobenzoic acid and 4-fluorobenzoic acid are detected in baboon urine. The *N*-oxide of HPTP (HPTNO) has also been reported (Fang et al., 1995, 2001; Eyles et al., 1996; Kudo and Ishizaki, 1999; Shin et al., 2001; Kalgutkar et al., 2003).

Zuclopenthixol

Zuclopenthixol

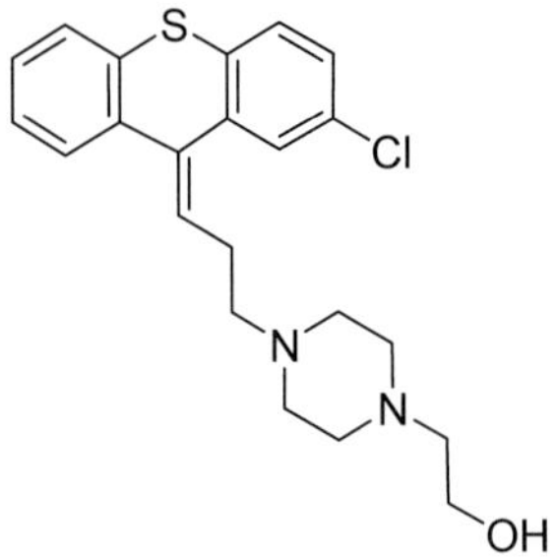

[53772-83-1], (Z)-4-[3-(2-Chlorothioxanthen-9-ylidene)propyl]-1-piperazineethanol, $C_{22}H_{25}ClN_2OS$, M_r 400.97; [85721-05-7], acetate salt, $C_{24}H_{27}ClN_2O_2S$, M_r 443.00; [58045-23-1], dihydrochloride salt, $C_{22}H_{27}Cl_3N_2OS$, M_r 473.89; [64053-00-5], decanoate ester, $C_{32}H_{43}ClN_2OS$, M_r 555.20

- The compound patent was filed in 1958 by Kefalas (Petersen et al., 1963)
- It has been launched by Lundbeck for the treatment of schizophrenia in 1962 in a number of countries
- Zuclopenthixol is a dopamine D_1/D_2 antagonist antipsychotic
- Zuclopenthixol is a thioxanthene of high potency with general properties similar to those of chlorpromazine. It is used for the treatment of psychoses like schizophrenia and mania. It is particularly suitable for agitated or aggressive patients who may become over-excited with flupentixol
- The longer-acting decanoate ester is used for the maintenance treatment of chronic psychoses

Synthesis (Petersen et al., 1963): Treatment of 9-allyl-2-chloro-9*H*-thioxanthen-9-ol with a dehydration agent gives the corresponding 9-allylidene compound which is mixed with 2-(piperazin-1-yl)ethanol to yield zuclopenthixol

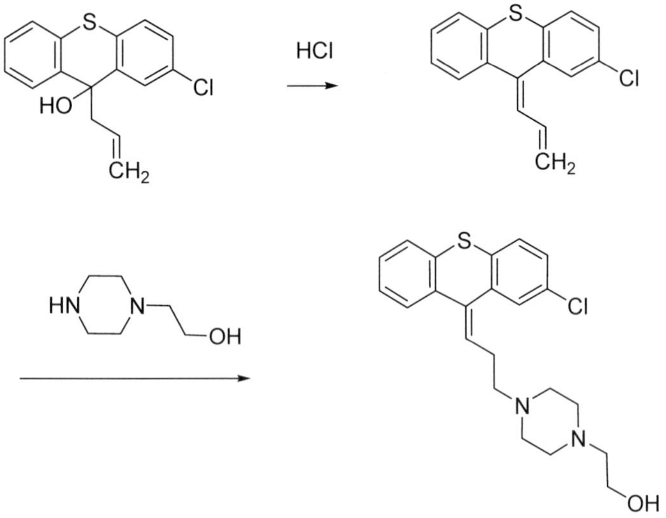

Scheme 12: Synthesis of zuclopenthixol

Clinical Use, Pharmacokinetics and Metabolism: Zuclopenthixol is used for the treatment of psychoses like schizophrenia (Coutinho et al., 2004; Gibson et al., 2004) and mania. Initial doses (hydrochloride salt) for antipsychotic therapy range between 20 mg to 30 mg per day, and can be increased according to response up to 150 mg daily. As compared with chlorpromazine, zuclopenthixol is less likely to cause sedation. It should not be used in apathetic or withdrawn states, and its dose needs adjustment for patients with hepatic or renal impairment (Coutinho et al., 2004; Fenton et al., 2004). Zuclopenthixol is absorbed from the GI tract after oral administration to give peak plasma concentrations 3–6 hours after ingestion. The biological half-life is reported to be around 1 day. The metabolism of zuclopenthixol is mainly by sulfoxidation, side chain N-dealkylation and glucuronic acid conjugation. The metabolites are devoid of pharmacological activity. Zuclopenthixol is excreted mainly in faeces with about 10% excreted in the urine. Approximately 0.1% of a dose is excreted unchanged in the urine. The systemic clearance is approximately 0.9%. Zuclopenthixol is distributed widely throughout the body, including in breast milk, and crosses the placenta. Depot injections have a relatively quick onset of action and a duration of up to 3 days (acetate). The effect of the decanoate salt injection lasts even much longer (maintenance treatment).

Droperidol

- The compound patent was filed in 1961 by Janssen (Janssen, 1965a)

- The compound was first introduced to the USA market in 1970 for anesthesia by Akorn. It was launched in France by Janssen-Cilag for the treatment of mania and psychosis in 1976

- Droperidol is a butyrophenone with general properties similar to those of haloperidol, and is a potent dopamine D_2 receptor antagonist with some histamine and serotonin antagonist activity

- Janssen-Cilag voluntarily withdrew it from the market worldwide in March 2001 following reports of Q–T prolongation, serious ventricular arrhythmias or sudden death. However, in the USA, droperidol remains available although its use is restricted to the management of nausea and vomiting following surgical or diagnostic procedures. In some other countries is still available as a premedicant, as an adjunct in anaesthesia and for the control of agitated patients in acute psychoses and in mania. It has also been used with an

Droperidol

[548-73-2], 1-[1-[4-(4-Fluorophenyl)-4-oxobutyl]-1,2,3,6-tetrahydro-4-pyridinyl]-2,3-dihydro-1H-benzimidazol-2-one, $C_{22}H_{22}FN_3O_2$, M_r 379.43

opioid analgesic such as fentanyl to maintain patients in state of neuroleptanalgesia

Synthesis (Janssen, 1964; Janssen, 1965a): 1-Benzyl-3-carbethoxy-2-piperidone is heated with 1,2-phenylenediamine in xylene. This gives the 1-(1-benzyl-1,2,3,6-tetrahydro-4-pyridyl)-2-benzimidazolinone. Catalytic hydrogenation removes the benzyl group to give the desired 1-(2,2,3,6-tetrahydro-4-pyridyl)-2-benzimidazolinone. Reaction of a mixture of 4-chloro-4-fluorobutyrophenone, 1-(2,2,3,6-tetrahydro-4-pyridyl)-2-benzimidazolinone, sodium carbonate and potassium iodide in 4-methyl-2-pentanone yields the final compound.

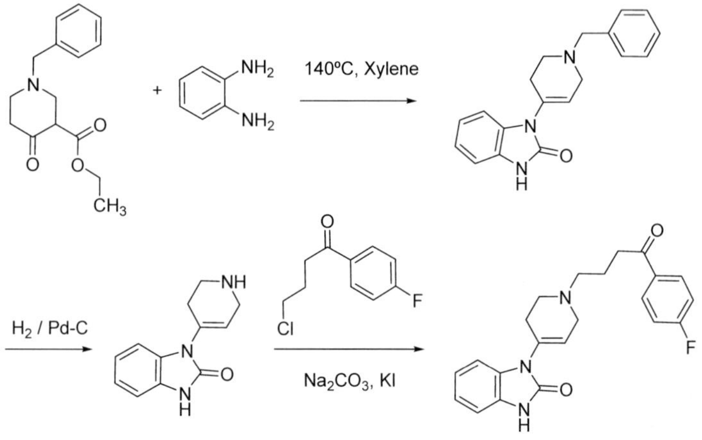

Scheme 13: Synthesis of droperidol

Clinical Use and Pharmacokinetics: Droperidol is a butyrophenone antipsychotic with general properties similar to haloperidol. It has been withdrawn from the market in some countries after reports of Q–T interval prolongation, serious ventricular arrhythmias and sudden death in association with its use. However, in some countries like the USA, droperidol still remains available, but with its use being restricted to the management of nausea and vomiting following surgical or diagnostic procedures. In other countries, it is used as premedicant, as adjunct in anaesthesia, and for the control of agitated patients in acute psychoses and mania. It also has been used (in combination with opioid analgesics like fentanyl) in neuroleptanalgesia in order to keep patients calm and indifferent to surroundings (e.g., along with surgery). Droperidol has an action of 2–4 hours. Alteration of alertness may last up to 12 hours or even longer. Side-effects are similar to chlorpromazine. Droperidol has an increased risk for cardiotoxicity along with a prolongation of the Q–T interval. Therefore it should be used with extreme caution in patients with arrhythmias, as well as in those with cardiac function impairment, hypokalaemia or electrolyte imbalance. It should not be used in patients with known or suspected Q–T interval prolongation.

Periciazine

- The compound patent was filed in 1957 by Rhone-Poulenc (Jacob and Robert, 1963)
- First launched by Sanofi-Aventis for the treatment of schizophrenia in 1964
- Periciazine is a phenothiazine with general properties similar to those of chlorpromazine. It is used in the treatment of various psychoses including schizophrenia and disturbed behaviors, and in the short-term management of severe anxiety

Synthesis (Jacob and Robert, 1963): 2-Cyano phenothiazine is condensed with 3-chloropropyl-4-methylbenzenesulfonate, to yield the corresponding 3-cyano-10-(3-toluene-*p*-sulpronyloxypropyl)phenothiazine. A solution of this compound and 4-piperidinol in toluene is heated under reflux to give periciazine.

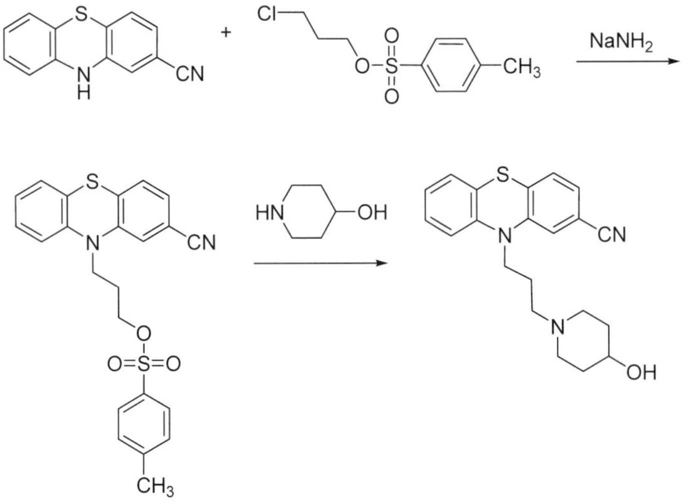

Periciazine

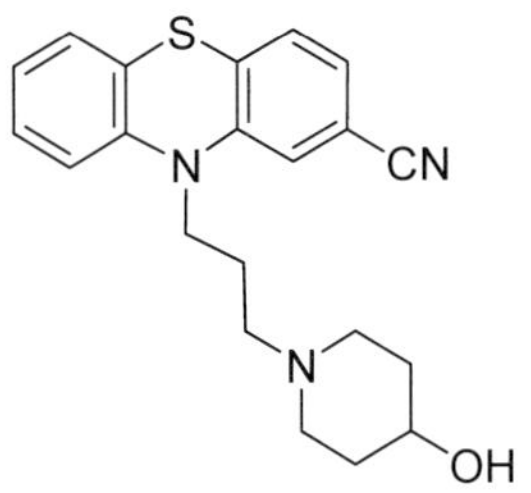

[2622-26-6], 10-[3-(4-Hydroxy-1-piperidinyl)propyl]-10*H*-phenothiazine-2-carbonitrile, $C_{21}H_{23}N_3OS$, M_r 365.49

Scheme 14: Synthesis of periciazine

Clinical Use: Periciazine is a phenothiazine antipsychotic with general properties similar to those of chlorpromazine. It has also adrenolytic, anticholinergic, metabolic and endocrine effects, and an action on the extra-pyramidal system. Like other phenothiazines, it is presumed to act principally in the subcortical areas, by producing a central adrenergic blockade. It is used in the treatment of different psychoses, including schizophrenia and disturbed behavior, and in the short-term management of severe anxiety. Usually it is applied p.o. in form of the base, even though mesilate and tartrate salts are available as well. Initial doses for antipsychotic treatment range around 75 mg per day (divided doses) and can be increased (in weekly steps of +25 mg) according to response up to a maximum of 300 mg daily.

Thiothixene

Thiothixene

[3313-26-6], (Z)-N,N-Dimethyl-9-[3-(4-methyl-1-piperazinyl)propylidene]thioxanthene-2-sulfonamide, $C_{23}H_{29}N_3O_2S_2$, M_r 443.63; [22189-31-7], dihydrochloride dihydrate salt, $C_{23}H_{35}Cl_2N_3O_4S_2$, M_r 552.59.

- The compound patent was filed in 1962 by Pfizer (Bloom and Muren, 1967)

- First launched by Pfizer for the treatment of schizophrenia in 1965

- Thiothixene is a thioxanthene with general properties similar to those of chlorpromazine. It is used in the treatment of various psychoses including schizophrenia

Synthesis (Bloom and Muren, 1967): Thioxanthene is converted in thioxanthene-2-sulfonate by successive treatment with chlorosulfonic acid and sodium chloride. The sulfonate group is converted to the corresponding chlorosulfonyl group and then transformed in a sulfamyl group by reaction with dimethylamine. The compound 2-dimethylsulfamyl thioxanthene is first treated with *n*-butyl lithium and then with methyl acetate, to yield 9-acetyl-2-dimethylsulfamyl thioxantene. This compound is first treated with formaldehyde and dimethylamine, and then with 1-methyl piperazine, to obtain 9-[3-(4-methyl-1-piperazinyl) propanoyll]-2-sulfamylthioxanthene. Treatment with sodium borohydride and subsequent dehydration, e.g. with phosphorus oxychloride in the presence of pyridine, yields the final compound thiothixene.

Clinical Use, Pharmacokinetics and Metabolism: Thiothixene is a thioxanthene antipsychotic with general properties similar to those of the phenothiazine chlorpromazine.

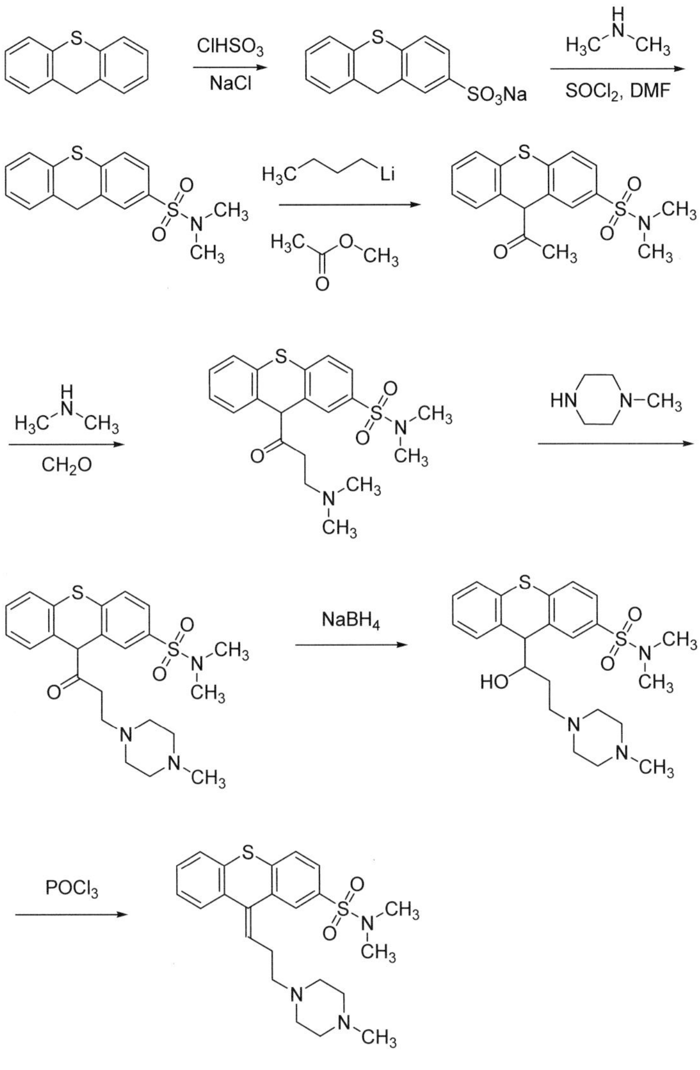

Scheme 15: Synthesis of thiothixene

It is used in the treatment of different psychoses including schizophrenia. Thiothixene is usually applied p.o. as the base or the hydrochloride salt, or via intramuscular injection (hydrochloride). Initial doses for antipsychotic treatment range around 2–5 mg three times a day, which can be increased according to response up to 20–30 mg daily. After oral ingestion, peak concentrations are reached after 1–3 hours, indicating a rapid absorption with absorption half-lives of around 30 minutes. Early plasma half-life ranges around 210 minutes (late half-life lies around 34 hours). Resurgence of drug concentration observed in some individuals may be due to enterohepatic recycling (Hobbs et al., 1974). Thiothixene is metabolized in a similar way as other related tricyclic psychotherapeutics to a number of metabolites, which are present in some tissues and are mainly secreted into the bile. In contrast, the brain contains only unchanged thiothixene (Hobbs, 1968). Thiothixene is inducing its own metabolism (autocatalysis) (Bergling et al., 1975).

Benperidol

Benperidol

[2062-84-2], 1-[1-[4-(4-Fluorophenyl)-4-oxobutyl]-4-piperidinyl]-1,3-dihydro-2H-benzimidazol-2-one, $C_{22}H_{24}FN_3O_2$, M_r 381.44

- The compound patent was filed in 1961 by Janssen (Janssen, 1964)

- First launched by Janssen for the treatment of psychosis in 1966

- Benperidol is a butyrophenone with general properties similar to those of haloperidol. In some countries it is prescribed for the treatment of psychotic conditions, and in the management of deviant sexual behavior

Synthesis (Janssen, 1964): 1-Benzyl-3-carbethoxy-2-piperidone is heated with 1,2-phenylenediamine in xylene to give 1-(1-benzyl-4-piperidyl)-2-benzimidazolinone. Catalytic hydrogenation removes the benzyl group to yield 1-(4-piperidyl)-2-benzimidazolinone. Reaction of a mixture of 4-chloro-4-fluorobutyrophenone, 1-(4-piperidyl)-2-benzimidazolinone, sodium carbonate and potassium iodide in 4-methyl-2-pentanone gives the final compound.

Clinical Use, Pharmacokinetics and Metabolism: Benperidol is a butyrophenone antipsychotic with general properties similar to those of haloperidol. It is used in the treatment of different psychoses in some countries (orally or parenterally). Another use lies in the treatment of deviant sexual behavior (divided doses of 0.25 mg to 1.5 mg daily), even though the effects exerted seem unlikely to be sufficient to control severe forms (Tennent et al., 1974). Benperidol is rapidly absorbed after oral administration, with peak plasma concentration reached at 2–3 hours, and largely distributed. The elimination half-life is around 5.5–9.5 hours. Urinary excretion represented only a minimal fraction of ingested dose. Variability of the

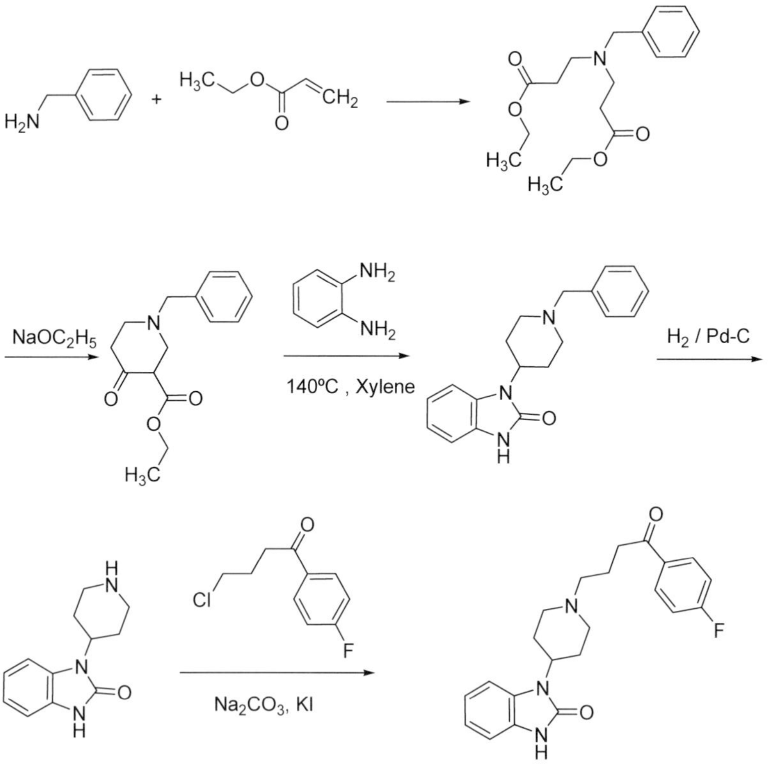

Scheme 16: Synthesis of benperidol

area under the curve makes a first-pass metabolism a reasonable possibility (Furlanut et al., 1988).

Flupentixol

- The compound patent was filed in 1958 by Kefalas (Petersen et al., 1963)
- First launched by Lundbeck for the treatment of psychosis in 1966
- Flupentixol is a thioxanthene antipsychotic with general properties similar to those of chlorpromazine. Flupentixol is used mainly in the treatment of schizophrenia and other psychoses. Unlike chlorpromazine, an activating effect has been attributed to flupentixol and it is not indicated in overactive or manic patients. It has also been used because of its antidepressant activities
- Flupentixol is administered orally as the hydrochloride and as the long-acting decanoate ester by deep intramuscular injection. The long-acting preparation

Flupentixol

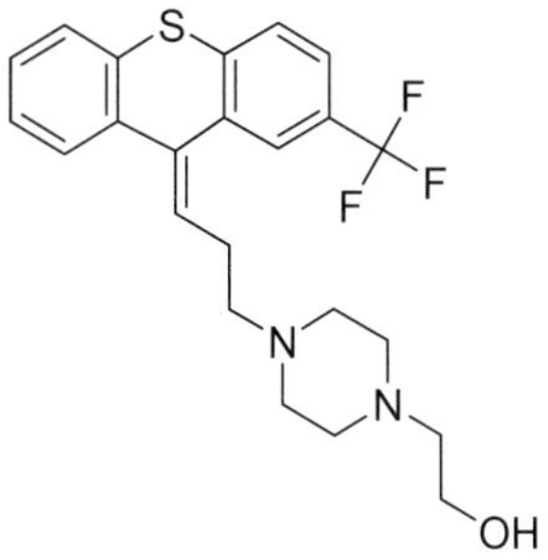

[2709-56-0], 2-Trifluoromethyl-9-[3-[4-(2-hydroxyethyl)piperazin-1-yl]propylidene]thioxanthene, $C_{23}H_{25}F_3N_2OS$, M_r 434.53; [2413-38-9] dihydrochloride salt, $C_{23}H_{27}Cl_2F_3N_2OS$, M_r 434.53; [30909-51-4] decanoate ester, $C_{33}H_{45}F_3N_2O_3S$, M_r 606.79; [53772-82-0] cis-isomer; $C_{23}H_{25}F_3N_2OS$, M_r 434.53; [51529-01-2] cis-isomer, dihydrochloride salt, $C_{23}H_{27}Cl_2F_3N_2OS$, M_r 434.53

available in the UK contains cis-(Z)-flupentixol decanoate

Synthesis (Petersen et al., 1963): 2-(Trifluoromethyl)-9H-thioxanthen-9-one is combined with 3-bromomagnesium-1-propene, to give 9-allyl-2-(trifluoromethyl)-9H-thioxanthen-9-ol. Treatment with a dehydration agent gives the corresponding 9-allylidene compound which is mixed with 2-(piperazin-1-yl)ethanol to yield flupentixol.

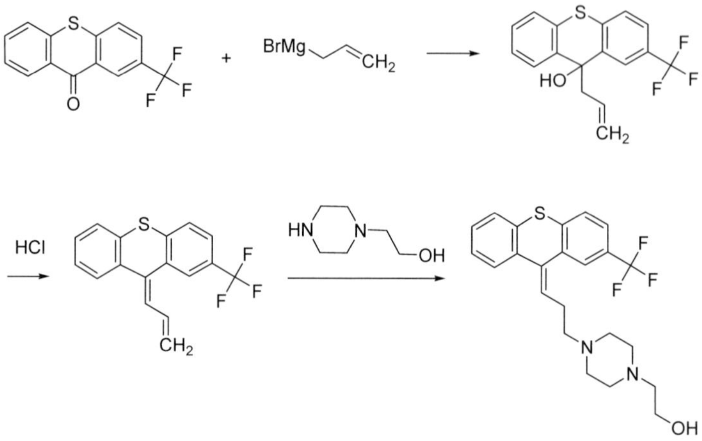

Scheme 17: Synthesis of flupentixol

An alternative synthesis is described by Craig and Zirkle (1966), in which 2-(trifluoromethyl)-9H-thioxanthen-9-one reacts with 1-(2-(benzyloxy)ethyl)-4-(3-chloropropyl)piperazine (obtained by condensation of 1-(2-(benzyloxy)ethyl)piperazine and 3-bromopropan-1-ol and subsequent treatment with $SOCl_2$) to give the intermediate 10-benzyloxymethylene-piperazinylpropyl-10-hydroxy-trifluoromethyl thioxanthene. Dehydration and removal of the protective benzyl group is accomplished by treatment with hydrochloric acid.

Clinical Use, Pharmacokinetics and Metabolism: Flupentixol is a thioxanthene antipsychotic with general properties similar to those of the phenothiazine chlorpromazine. It is mainly used in the treatment of different psychoses including schizophrenia. Unlike chlorpromazine, flupentixol exerts an activating effect and thus is not indicated for overactive or manic patients. Flupentixol has also been used for its antidepressant properties.

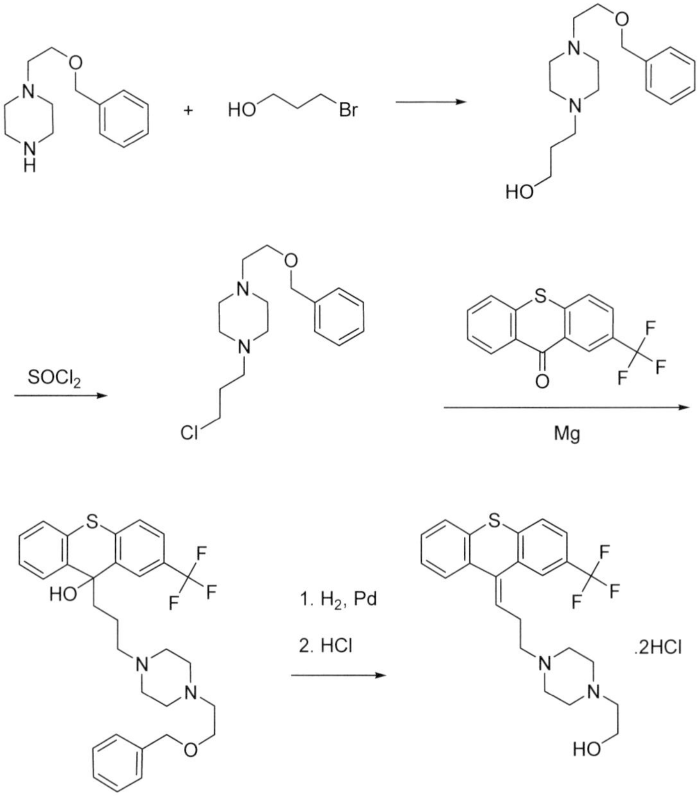

Scheme 18: Alternative synthesis of flupentixol

Flupentixol is given p.o. in form of its hydrochloride salt, or by deep intramuscular depot injection (in form of the decanoate ester). Long-acting preparations in the UK contain the active cis-(Z)-flupentixol decanoate isomer. Initial doses for antipsychotic treatment range from 3 mg to 9 mg of flupentixol hydrochloride p.o. They can be increased to a maximum of 18 mg daily, depending upon the response of the patient. Investigations on the 2 double bond isomers indicate a more favorable clinical profile for the cis-(Z)-isomer (α-flupentixol) as compared to β-flupentixol (trans-(E)-isomer) (Johnstone et al., 1978; Crow and Johnstone, 1978). Depot flupentixol decanoate has shown to be promising in reducing cocaine usage by inducing aversion (Gawin et al., 1996; Evans et al., 2001). Adverse effects are comparable to chlorpromazine, with less sedation, but a higher incidence of extrapyramidal side-effects. Sudden death following depot injections has been reported (Turbott and Smeeton, 1984). Flupentixol is not recommended for patients with states of excitement or overactivity, including mania. It is considered unsafe in patients with porphyria. Other adverse effects and drug interactions are similar to those reported for chlorpromazine. Flupentixol is readily absorbed from the GI tract and undergoes a first-pass metabolism in the gut wall, with a bioavailability of about 50%. The average time to peak plasma concentration is 4 hours (from 3–8 hours). It is extensively metabolized in the liver and is excreted in urine and faeces in form of numerous metabolites. Evidence for enterohepatic recycling has been found. Due to the first-pass effect, plasma concentrations after oral administration are much lower than those obtained via intramuscular application. Flupentixol decanoate is slowly released from the depot site, with a half-life of 3–8 days. The serum peak concentration for intramuscular flupentixol decanoate is 3–5 days (Jorgensen, 1980; Jorgensen et al., 1982). There is strong interindividual variation in plasma concentrations. No correlation has been found between plasma concentrations of flupentixol and/or its metabolites and the observed therapeutic effect. Metabolism includes sulfoxidation, side chain N-dealkylation and conjugation with glucuronic acid. Flupentixol is widely distributed throughout the body and crosses the blood-brain barrier.

Sulpiride

Sulpiride

[15676-16-1], N-(1-Ethylpyrrolidin-2-ylmethyl)-2-methoxy-5-sulfamoylbenzamide, $C_{15}H_{23}N_3O_4S$, M_r 341.43

- The compound patent was filed in 1964 by Société d'Etudes Scientifiques et Industrielles de L'Ile de France (Miller et al., 1967)

- It was launched in Japan by Astellas Pharma in 1968 and in Europe by Janssen in 1969, for the treatment of psychosis

- Sulpiride is more commonly used in Europe and Japan. So far it has not been approved in the USA and Canada

- Sulpiride is a substituted benzamide antipsychotic selective antagonist of central dopamine D_2, D_3 and D_4 receptors. It has mood elevating properties, and is mainly used in the treatment of psychoses such as schizophrenia. It has also been given in the management of Tourette's syndrome, anxiety disorders, vertigo and benign peptic ulceration

- Levosulpiride (see below), the L-isomer of sulpiride, has been used similarly to sulpiride

Synthesis (Miller et al., 1967): Carbonyldiimadazole is dissolved in tetrahydrofurane and reacted with 2-methoxy-5-sulfamoylbenzoic acid (obtained by treatment of 5-aminosulfonyl-salicylic acid with dimethylsulfate). The final compound is obtained by reacting the *N*-2-methoxy-5-sulfamoylbenzoyl)-imidazole with 1-ethyl-2-aminomethylpyrrolidine.

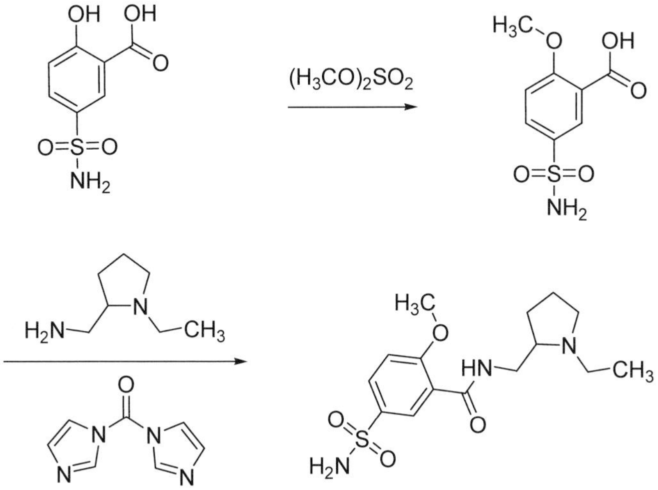

Scheme 19: Synthesis of sulpiride

Some alternative syntheses have been described, e.g. Kosary et al. (1977) describe that sulpiride is obtained by reacting 2-chlorobenzoic acid with chlorosulfonic acid, amidating the resulting 2-chloro-5-chlorosulfonylbenzoic with ammonia, and reacting the 2-chloro-5-sulfanoylbenzoic acid with 1-ethyl-2-aminomethylpyrrolidine. The chlorine atom is finally displaced by a methoxy group.

Clinical Use, Pharmacokinetics and Metabolism: Sulpiride is a substituted benzamide antipsychotic that is reported to be a selective antagonist of central dopamine D_2, D_3 and D_4 receptors. It is claimed to have mood elevating properties and is mainly used in the treatment of psychoses such as schizophrenia. It also has been used in the management of Tourette's syndrome, anxiety disorders, vertigo, and benign peptic ulceration. Levosulpiride, the L-isomer of sulpiride, has been developed as a follow-up drug of sulpiride (racemic switch; see chapter levosulpiride below). For the treatment of schizophrenia, initial p.o. doses of about 200–400 mg are given b.i.d. According to the response, the dose can be increased to a maximum of 2.4 g per day (Caley and Weber, 1995; Mauri et al., 1996). The antipsychotic potential of sulpiride was reviewed by Soares et al. (2004b) to be as good as the classic antipsychotics for schizophrenia and to have comparably less side-effects. When used in patients with renal impairment, the dose and frequency of intake needs adjustment, or treatment should be avoided at all (Bressolle et al., 1989). Sulpiride has been used in Tourette's syndrome (Robertson et al., 1990), chorea (Quinn and Marsden, 1984), gastrointestinal disorders (peptic ulcer disease) and occasionally to stimulate lactation in breast-feeding mothers (Aono et al., 1982; Ylikorkala et al., 1982; Ylikorkala et al., 1984; Pons et al., 1994). Side-effects are similar to chlorpromazine and comprise sleep disturbances, overstimulation and agitation, mild extrapyramidal effects and – very rarely – antimuscarinic effects. Sulpiride is distributed into breast milk in relatively large amounts (Polatti, 1982). Effects on the cardiovascular system (Corvol et al., 1974), porphyria and renal impairment are amongst other reported side-effects. Sulpiride is absorbed slowly from the GI tract, leading to low and variable bioavailabilities. It is widely distributed throughout the body, but passage through the blood-brain barrier is limited. It has low plasma-protein binding (40%) and has a plasma half-life of about 8 to 9 hours. It is excreted mainly in unchanged form in urine and faeces (Wiesel et al., 1980; Bressolle et al., 1984; Bressolle et al., 1992).

Pimozide

- The compound patent was filed in 1963 by Janssen (Janssen, 1965b)

- First launched in France by Janssen for the treatment of schizophrenia in 1969

- Pimozide is a diphenylbutylpiperidine antipsychotic and is structurally similar to the butyrophenones. It is a long acting antipsychotic with general properties similar to those of chlorpromazine

- Pimozide blocks the D_2, D_3, α_1 and 5-HT$_{2A}$ receptors. Also shows moderate affinity for D_1, D_4 and α_2 receptors. It also possesses some calcium blocking activity

- Pimozide is used in its oral preparation in schizophrenia and chronic psychosis (on-label indications in Europe only), Gilles de la Tourette syndrome and

Pimozide

[2062-78-4], 1-[1-[4,4-Bis(4-fluorophenyl)butyl]-4-piperidinyl]-1,3-dihydro-2H-benzimidazol-2-one, $C_{28}H_{29}F_2N_3O$, M_r 461.55

resistant tics (Europe, USA and Canada). In Germany, the 1mg tablet is indicated for the treatment of some forms of reactive depression

Synthesis (Janssen, 1965b): Pimozide is obtained by condensation of 4-chloro-1,1-bis(4-fluorophenyl)butane with 4-(2-oxo-1-benzimidazolinyl)-piperidine.

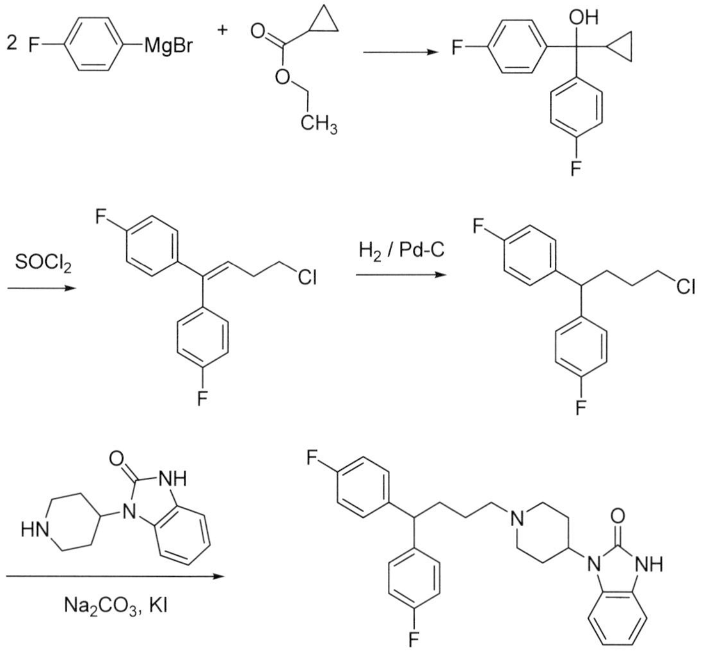

Scheme 20: Synthesis of pimozide

Clinical Use, Pharmacokinetics and Metabolism: Pimozide is a diphenylbutylpiperidine antipsychotic, chemically related to the butyrophenones. It has long-acting antipsychotic properties similar to those of the phenothiazine chlorpromazine. In addition, it has calcium-blocking activity. Pimozide can be used p.o. in the treatment of psychoses like schizophrenia, paranoid states, monosymptomatic hypochondria and in Tourette's syndrome. Because of its ability to prolong the Q–T interval, ECG measurement of the patient should be done before starting the therapy with pimozide. Starting dose for schizophrenia treatment is 2 mg daily, which can be gradually (in 2 to 4 mg steps and week intervals) increased to a maximum dose of 20 mg daily. Due to its long half-life, once-a-day application of pimozide is usually prescribed (Sultana and

Fluspirilene

[60-99-1], (-)-(R)-N,N,β-Trimethyl-2-methoxy-10H-phenothiazine-10-propanamide, $C_{19}H_{24}N_2OS$, M_r 328.48; [1236-99-3], hydrochloride salt, $C_{19}H_{25}ClN_2OS$, M_r 394.94; [7104-38-3], maleate salt, $C_{23}H_{28}N_2O_5S$, M_r 444.55

McMonagle, 2004). Pimozide has been used for the treatment of idiopathic dystonia (Marsden and Quinn, 1990), taste disorders, Tourette's syndrome (Shapiro et al., 1989; Sallee et al., 1997), and chorea (Shannon and Fenichel, 1990). As with respect to adverse effects, pimozide shows a similar profile as chlorpromazine with more pronounced extrapyramidal symptoms and less sedation, hypotension and antimuscarinic effects. Pimozide has an increased risk for cardiotoxicity along with a prolongation of the Q–T interval and T-wave changes. Therefore it should be avoided in patients with arrhythmias, cardiac function impairment, hypokalaemia, hypomagnesaemia or electrolyte imbalance. It should not be used in patients with known or suspected Q–T interval prolongation. Patients receiving higher doses of pimozide over a longer period should undergo periodic cardiac function assessment. Pimozide may not be combined with other drugs prolonging Q–T interval, such as e.g. other antipsychotics, antidepressants, the antihistaminics terfenadine and astemizole, antimalarials or cisapride. Use along with diuretics should be avoided. The concomitant use with drugs inhibiting cytochrome P450 isoenzyme CYP3A4 is contra-indicated, as these decrease the metabolism of pimozide. Following p.o. application, over 50% of the dose is absorbed from the GI tract, undergoing a significant first-pass metabolism. Peak plasma concentrations are observed 4–12 hours after ingestion, with considerable interindividual variation. Pimozide is mainly metabolized by N-dealkylation (CYP3A4, and to a lesser degree CYP2D6 and CYP1A2) and excreted in urine and faeces. Pimozide has a very long terminal half-life (about 55 hours).

Fluspirilene

- The compound patent was filed in 1962 by Janssen (Janssen, 1969b)

- First launched in France by Janssen for the treatment of psychosis in 1970

- Fluspirilene is a diphenylbutylpiperidine antipsychotic with general properties similar to those of chlorpromazine, though is less likely to cause sedation. Fluspirilene can be given by deep intramuscular injection for the treatment of psychoses including schizophrenia

Synthesis (Janssen, 1966; 1969a): Condensation of 4-hydroxy-piperidine, protected at the nitrogen with a benzyl group, with aniline and an alkali metal cyanide effects both introduction of the aniline and nitrile in the piperidyl ring. The nitrile function is converted to the amide by acid hydrolysis. Cyclization of the resulting compound is obtained by treatment with formamide. Debenzylation is afforded by means of hydrogen, activated by Pd-C. Finally, the resulting compound is condensed with 4-bromo-1,1-bis(4-fluorophenyl)butane.

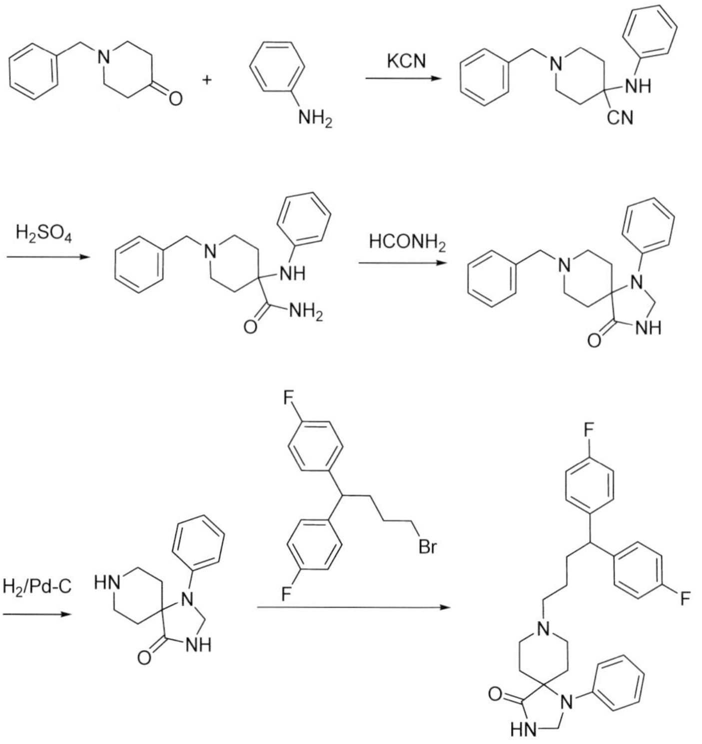

Scheme 21: Synthesis of fluspirilene

Clinical Use: Fluspirilene is a diphenylbutylpiperidine antipsychotic, chemically related to the butyrophenones. It has antipsychotic properties similar to those of the phenothiazine chlorpromazine. In comparison to chlorpromazine, it has a comparable side-effect profile, but is less likely to cause sedation. It is usually given as deep intramuscular injection (depot) in schizophrenia treatment, but no significant advantage over commonly used antipsychotics like chlorpromazine was found in a systematic review article (Quraishi and David, 2000).

Cyamemazine

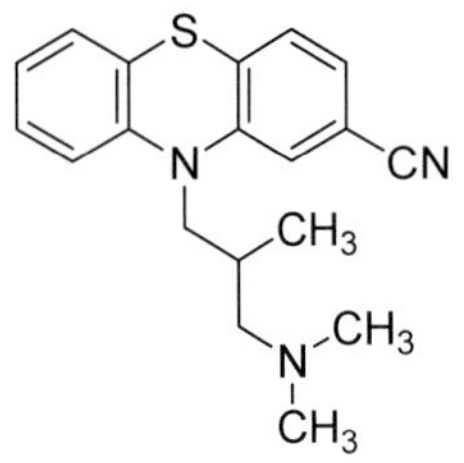

[3546-03-0], 10-[3-(Dimethylamino)-2-methylpropyl]-10H-phenothiazine-2-carbonitrile, $C_{19}H_{21}N_3S$, M_r 323.46

Usual initial dose for antipsychotic treatment is up to 2 mg weekly (intramuscularly). Usual maintenance doses range from 1–10 mg per week. Adverse effects are similar to those reported for chlorpromazine. Toxic necrosis has been described along with fluspirilene treatment (McCreadie et al., 1979).

Cyamemazine

- The compound patent was filed in 1955 by Rhone-Poulenc (Jacob and Robert, 1956)
- Launched in 1972 by Sanofi-Aventis for the treatment of symptomatic, short-term treatment of anxiety in adults unresponsive to other treatments and short-term treatment of aggressivity during acute and chronic psychotic syndromes including schizophrenia (see Chapter 3.5)
- Cyamemazine is a phenothiazine with general properties similar to those of chlorpromazine

Synthesis (Jacob and Robert, 1956; 1959). The 10H-Phenothiazine-2-carbonitrile used as starting material can be prepared by the action of cupric cyanide on 2-chloro-10H-phenothiazine in boiling quinoline. Treatment with NaNH$_2$ and condensation with 3-chloro-N,N,2-trimethylpropan-1-amine yields cyamemazine.

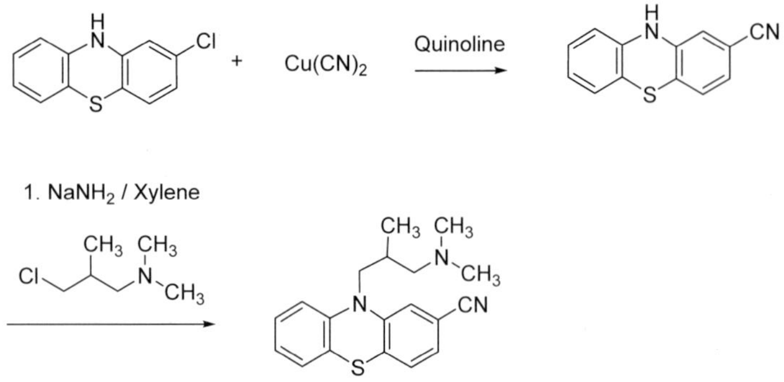

Scheme 22: Synthesis of cyamemazine

Clinical Use: Cyamemazine is a phenothiazine antipsychotic with general properties similar to those of chlorpromazine. It is used in the preparation for the management of different psychoses, including anxiety and aggressive behavior. Cyamemazine is given p.o. in form of its tartrate salt or as free base. Usual doses range between 25 mg to 600 mg daily p.o. (depending on individual and conditions), divided in 2–3 doses. Cyamemazine can also be given as intramuscular injection.

Pipotiazine

- The compound patent was filed in 1957 by Rhone-Poulenc (Jacob and Robert, 1963)
- First launched by Sanofi-Aventis for the treatment of schizophrenia in 1973
- Pipotiazine is a phenothiazine with general properties similar to those of chlorpromazine. It is used in the treatment of schizophrenia and other psychoses

Synthesis (Jacob and Robert, 1963): 2-(Dimethylsulfamyl) phenothiazine is condensed with 3-chloropropyl-4-methylbenzenesulfonate, to yield the corresponding 3-(10H-phenothiazin-10-yl) propyl 4-methylbenzenesulfonate. This compound is mixed with 2-(piperidin-4-yl) ethanol in toluene and heated under reflux to give pipotiazine.

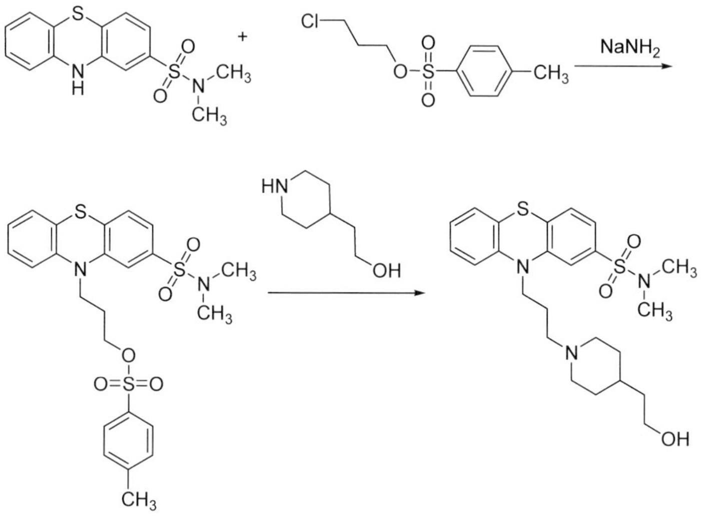

Pipotiazine

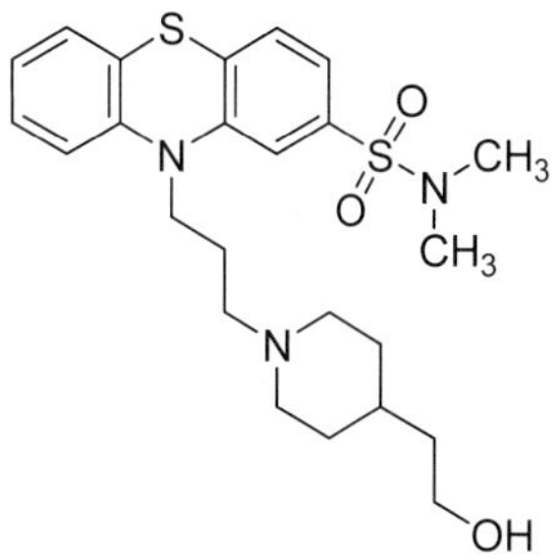

[39860-99-6], 10-[3-[4-(2-Hydroxyethyl)-1-piperidinyl]propyl]-N,N-dimethyl-10H-phenothiazine-2-sulfonamide, $C_{24}H_{33}N_3O_3S_2$, M_r 475.67; [37517-26-3], palmitate ester, $C_{40}H_{63}N_3O_4S_2$, M_r 714.10

Scheme 23: Synthesis of pipotiazine

Clinical Use, Pharmacokinetics and Metabolism: Pipotiazine is a phenothiazine antipsychotic with general properties similar to those of chlorpromazine. It is used in the management of different psychoses, including schizophrenia. Cyamemazine is given p.o. in form of the free base or by deep intramuscular injection (palmitate ester, depot injection). Usual starting doses for antipsychotic treatment are 5–25 mg daily (single dose), and can be slightly increased when treating severe psychoses for brief periods. A systematic analysis of the antipsychotic effects of pipotiazine concludes that pipotiazine shows no significant advantages over other antipsychotics (Quraishi and David, 2004). Pipotiazine has similar adverse effects and contraindications as those reported for chlorpromazine. Manic symptoms have been reported along with pipotiazine depot treatment (Singh and Maguire, 1984). Plasma kinetics of pipotiazine have been studied in chronic schizophrenic patients after both oral (25 mg) and i.v. (5 mg) administration of pipotiazine (De Schepper et al., 1979). Peak plasma concentrations of unchanged pipotiazine were reached 1–2 hour after oral administration and showed a five-fold interindividual variation. The mean terminal elimination half-life was 11.2 hours. After i.v. administration, plasma concentration declined bi-exponentially with mean half-life values of 2.7 hours and 8.8 hours. Data indicate that biotransformation of the drug was not dependent on the route of administration.

Loxapine

Loxapine

[*1977-10-2*], 2-Chloro-11-(4-methylpiperazin-1-yl)dibenz[*b,f*][1,4]oxazepine, $C_{18}H_{18}ClN_3O$, M_r 327.82; [*27833-64-3*] succinate salt, $C_{22}H_{24}ClN_3O_5$, M_r 445.91

- The compound patent was filed in 1965 by American Cyanamid (Coppola, 1968)

- First launched in USA by Wyeth for the treatment of schizophrenia in 1975

- Loxapine is a dibenzoxazepine with general properties similar to those of chlorpromazine

Synthesis (Coppola, 1968): Ethyl chloroformate is added to a mixture or *o-(p-chlorophenoxy)* aniline hydrochloride in pyridine. After the addition is complete, the mixture is heated under reflux to give ethyl *o-(p-chlorophenoxy)* carbanilate. The crude ethyl *o-(p-chlorophenoxy)* carbanilate is dissolved in benzene, and 1-methylpiperazine and sodium methylate are added, to obtain 4-methyl-2'-(*p*-chlorophenoxy)-1-piperazinecarboxanilide. A mixture of this compound and phosphorus pentoxide is heated under reflux to yield 2-chloro-11-(4-methylpiperazin-1-yl) dibenz[b,f][1,4]oxa-zepine as target compound.

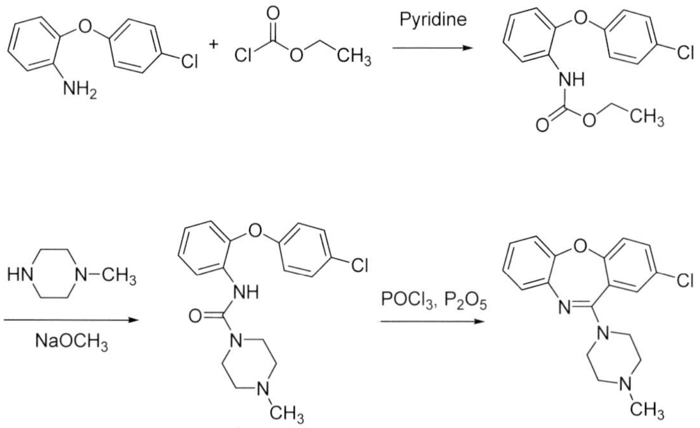

Scheme 24: Synthesis of loxapine

Clinical Use, Pharmacokinetics and Metabolism: Loxapine is a dibenzooxazepine with general properties similar to those of the phenothiazine chlorpromazine. It is likely that loxapine acts very similar to many other antipsychotics (Fenton et al., 2004). Loxapine has been used in the treatment of disturbed behavior (Carlyle et al., 1993). It can be given p.o. (succinate or hydrochloride salts) or as deep muscular injection (free base as depot) in the treatment of psychoses like schizophrenia, with usual doses ranging from 20mg to 50 mg daily initially, divided into 2 individual doses. Doses can be increased over the next 7–10 days to 60–100 mg per day, divided into 2–4 individual daily doses. Maximum recommended dose is 250 mg daily. Claims that loxapine could be especially effective in patients with paranoid schizophrenia could not be verified (Anonymous, 1991). Adverse effects are similar to those reported for chlorpromazine and comprise also effects on carbohydrate metabolism (Tollefson and Lesar, 1983) and mania (Gojer, 1992). Experience on loxapine abuse (Sperry et al., 1984) and overdosage (Tarricone, 1998) has been reported. Loxapine should not be given to patients with porphyria. Loxapine is readily absorbed from the GI tract and is rapidly and extensively metabolized, presumably undergoing a first-pass metabolism. It is mainly excreted in urine in the form of its conjugated metabolites. Major metabolites comprise active 7- and 8-hydroxyloxapine and hydroxyloxapine-N-oxide, loxapine-

N-oxide, and hydroxydesmethylloxapine (hydroxyamoxapine), which then – in most cases – undergo conjugate addition with glucuronic acid (phase II metabolism). Loxapine is widely distributed throughout the body and is distributed across the placenta and into breast milk (Cooper et al., 1979).

Sultopride

Sultopride

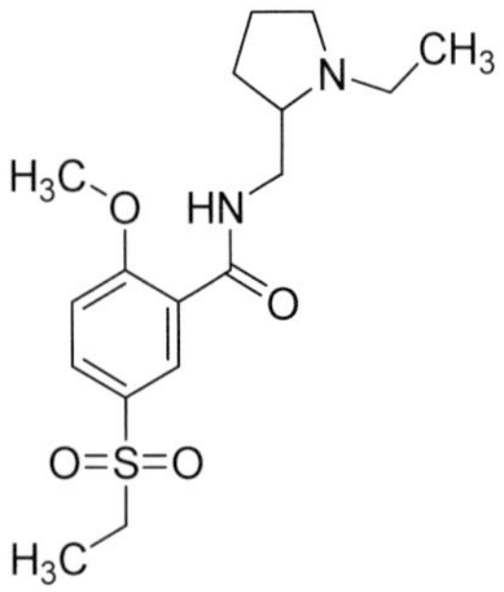

[*53583-79-2*], *N*-[(1-Methyl-2-pyrrolidinyl)methyl]-5-(ethylsulfonyl)-2-methoxybenzamide, $C_{17}H_{26}N_2O_4S$, M_r 354.47; [*523694-17-9*], hydrochloride salt, $C_{17}H_{27}ClN_2O_4S$, M_r 390.93

- The compound patent was filed in 1972 by Société d'Etudes Scientifiques et Industrielles de L'Ile de France (Bulteau and Acher, 1973)
- First launched in France by Sanofi-Aventis for the treatment of psychosis in 1976
- Sultopride is a substituted benzamide wih general properties similar to those of sulpiride. It is used in the emergency management of agitation in psychotic or aggressive patients and in psychoses such as schizophrenia

Synthesis (Bulteau and Acher, 1973; Bulteau et al., 1974): Prepared by reacting 2-methoxy-5-ethylsulfonylbenzoic acid with isobutyl chloroformate in dioxane to give the corresponding anhydride, which is finally condensed with 1-ethyl-2-amino methyl-pyrrolidine.

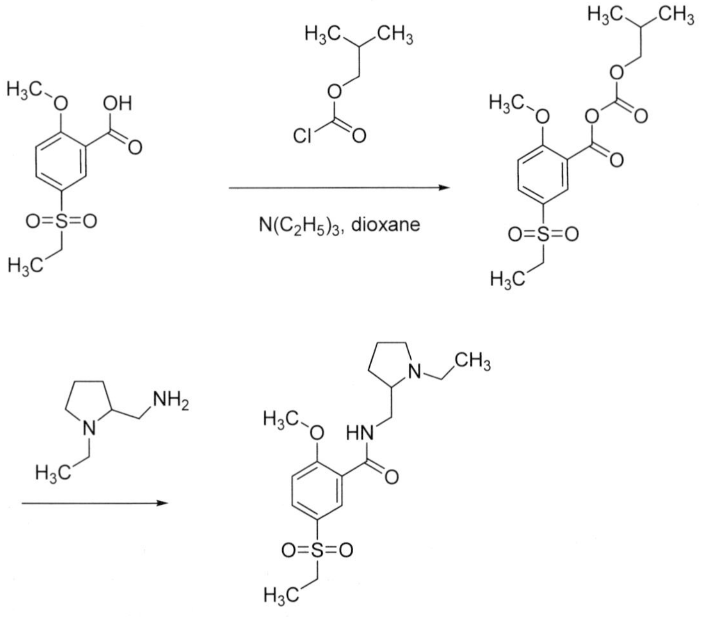

Scheme 25: Synthesis of sultopride

Clinical Use and Pharmacokinetics: Sultopride is a substituted benzamide antipsychotic with properties similar to those reported for sulpiride. It is used mainly in the emergency management of agitation in psychotic or aggressive patients, and in psychoses such as schizophrenia. It is given as hydrochloride salt in doses between 400 mg to 800 mg daily p.o. or intramuscularly (acute agitation). In psychoses, the dose may be increased up to 1200 mg p.o. Sultopride is not considered safe in patients with porphyria. As severe side-effects, ventricular arrhythmias, including torsade de pointes (due to Q–T interval prolongation) have been reported. It should not be used in patients with bradycardia. Half-life of sultopride determined from either plasma or red blood cell concentrations was 5 h (Bressolle and Bres, 1985).

Carpipramine

- The compound patent was filed in 1966 by Yoshitomi Pharmaceutical Industries (Nakanishi and Munakata, 1966)

- First launched in France by Pierre Fabre for the treatment of schizophrenia and anxiety in 1977

- Medical use of carpipramine as anxiolytic and hypnotic was patented by Rhone-Poulenc Sante in 1988 (see Chapter 3.5)

- Carpipramine is structurally related both to imipramine and to butyrophenones such as haloperidol

Synthesis (Nakanishi and Munakata, 1966; Nakanishi et al., 1970). Condensation of 10,11-dihydro-5H-dibenz[b,f]azepine with 1,3-dibromopropane yields 3-(10,11-dihydro-5H-dibenz[b,f]azepin-5-yl)propyl bromide. Reaction with 4-(piperidin-1-yl)piperidine-4-carboxamide affords carpipramine. (see Chapter 3.5 for additional synthesis details).

Carpipramine

[*5942-95-0*], 5-[3-[4-Carbamoyl-4-(1-piperidinyl)piperidin-1-yl]propyl]-10,11-dihydrodibenzo[b,f]azepine, $C_{28}H_{38}N_4O$, M_r 446.63; [*7075-03-8*], dihydrochloride monohydrate salt, $C_{28}H_{42}Cl_2N_4O_2$, M_r 537.57; [*100482-23-3*], maleate salt, $C_{32}H_{42}N_4O_5$, M_r 562.70

Clinical Use, Pharmacokinetics and Metabolism: Carpipramine is structurally related to both imipramine and to the butyrophenones such as haloperidol. It is used in the treatment of anxiety disorders and psychoses such as schizophrenia. It is usually given p.o. as hydrochloride salt, with usual doses around 50 mg three times a day. Carpipramine is not considered safe in patients with porphyria, although there is conflicting evidence from animal data on porphyrinogenicity. Carpipramine

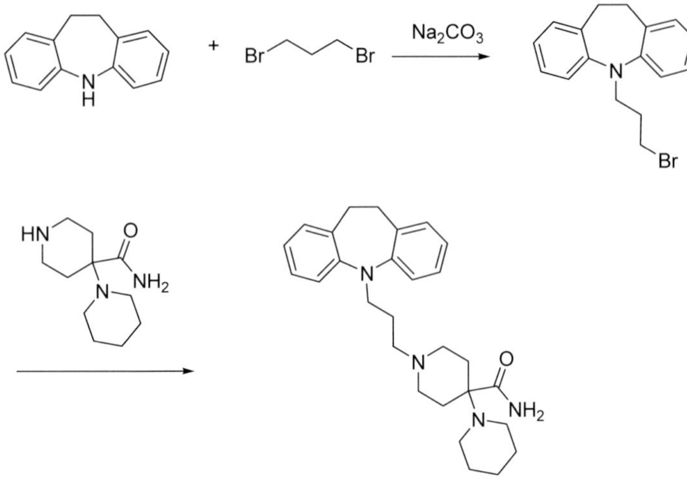

Scheme 26: Synthesis of carpipramine

administered orally is excreted via the urine and faeces in rat, rabbit, dog and man. Many metabolites are formed, including several conjugates in the urine. Three metabolic pathways were observed: hydroxylation of the iminodibenzyl ring to a phenol or alcohol without modification of the side-chain, hydroxylation of the terminal piperidine of the 2-piperidinol side-chain, and cyclization and dehydrogenation of the same 2-piperidinol group (Bieder et al., 1985).

Tiapride

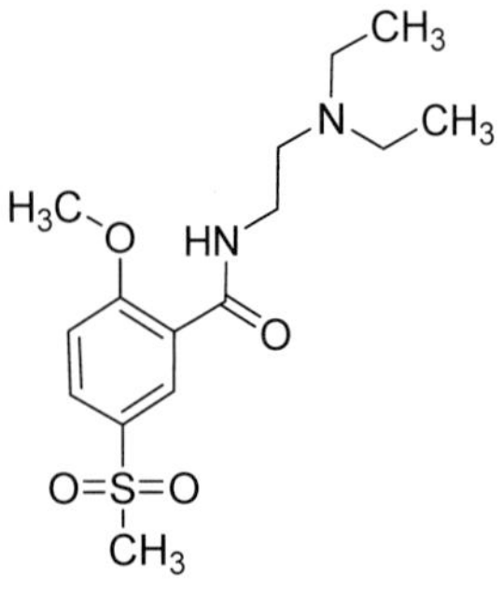

Tiapride

[51012-32-9], N-[2-(Diethylamino)ethyl]-2-methoxy-5-(methylsulfonyl)benzamide, $C_{15}H_{25}ClN_2O_4S$, M_r 364.89; [51012-33-0] hydrochloride salt, $C_{15}H_{26}Cl_2N_2O_4S$, M_r 401.35

- The compound patent was filed in 1976 by Société d'Etudes Scientifiques et Industrielles de L'Ile de France (Bulteau and Acher, 1973)

- First launched in France by Sanofi-Aventis in 1977 for the treatment of aggresiveness, delirium tremens, psychosis and Tourette's disease

- Tiapride is a substituted benzamide with general properties similar to those of sulpiride. It is also used for the management of behavior disorders and to treat dyskinesias

Synthesis (Bulteau and Acher, 1973; Bulteau et al., 1974): Prepared by reacting 2-methoxy-5-methylsulfonylbenzoic acid with isobutyl chloroformate in dioxane to give the corresponding anhydride, which is finally condensed with N,N-diethylethylenediamine.

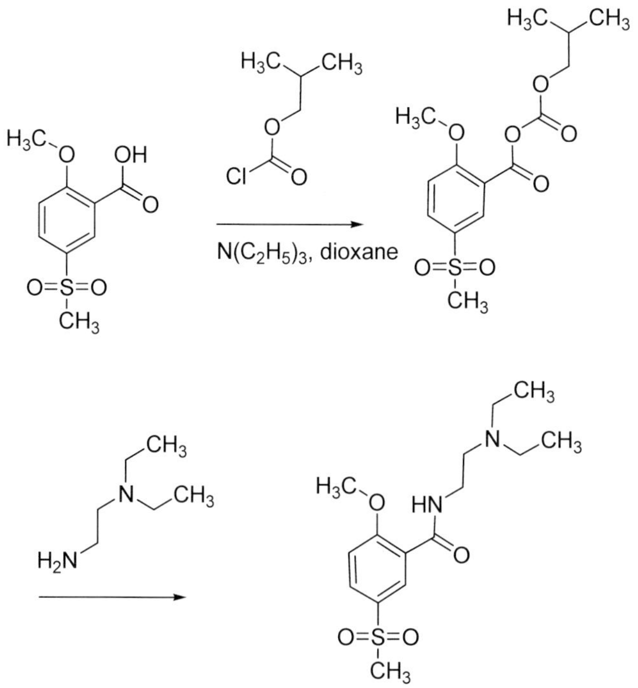

Scheme 27: Synthesis of tiapride

Clinical Use, Pharmacokinetics and Metabolism: Tiapride is a substituted benzamide antipsychotic with properties similar to those reported for sulpiride. It is usually given in form of the hydrochloride salt in the management of behavior disorders (Gutzmann et al., 1997) and to treat dyskinesias. It has also been used in the short-term management of extrapyramidal disorders, in the treatment of Tourette's syndrome and in Huntington's chorea (Roos et al., 1982; Deroover et al., 1984). Its use in the management of substance dependence is discussed controversically (Peters and Faulds, 1994; Shaw et al., 1994; Franz et al., 2001). Doses of 200–400 mg daily p.o. are usually given in dyskinesia treatment, which can be increased according to response. Tiapride hydrochloride can also be given i.v. or intramuscularly. Adverse effects and interactions with tiapride are similar to those reported with chlorpromazine. As severe side-effects, ventricular arrhythmias, including torsade de pointes (due to Q–T interval prolongation) have been reported (Iglesias et al., 2000). It should not be used in patients with a risk factor for this kind of arrythmias. Tiapride is readily absorbed from the GI tract after p.o. application and is excreted mainly in unchanged form in the urine. Plasma half-life is reported to range from 3–4 hours. One presumable metabolite is *N*-monodesethyl-tiapride (Roos et al., 1986).

Veralipride

- The compound patent was filed in 1976 by Société d'Etudes Scientifiques et Industrielles de L'Ile de France (Thominet and Perrot, 1985)
- First launched in France by Sanofi-Aventis for the treatment of psychosis in 1980
- Veralipride is a substituted benzamide antipsychotic. It has been used in the treatment of cardiovascular and psychological symptoms associated with menopause

Veralipride

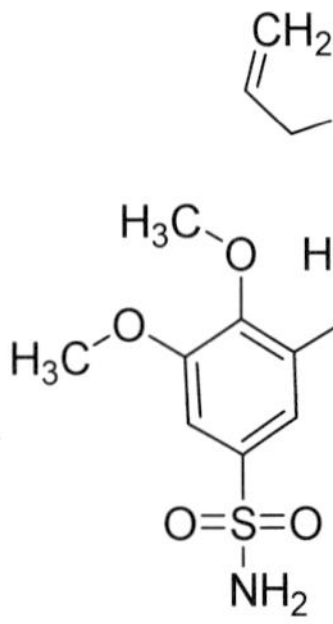

[66644-81-3], 5-(Aminosulfonyl)-2,3-dimethoxy-*N*-[[1-(2-propenyl)-2-pyrrolidinyl]methyl]benzamide, $C_{17}H_{25}N_3O_5S$, M_r 383.46; [66644-83-5] hhydrochloride salt, $C_{17}H_{25}ClN_3O_5S$, M_r 419.92

Synthesis (Thominet and Perrot, 1985): By reaction of 5-aminosulfonyl-2,3-dimethoxybenzoic acid with 1-allyl-2-aminomethylpyrrolidine.

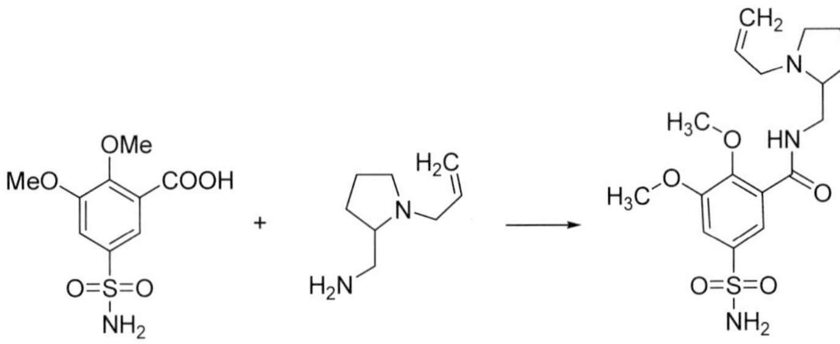

Scheme 28: Synthesis of veralipride

Clinical Use and Pharmacokinetics: Veralipride is a substituted benzamide antipsychotic with properties similar to those reported for sulpiride. It is used in the symptomatic treatment of cardiovascular and psychological problems associated with menopause (Young et al., 1990). Usual dose is 100 mg daily p.o. for 20 days, repeated at intervals of 7–10 days. Veralipride is not indicated to be safe for patients with porphyria, on the basis of *in vitro* data on porphyrinogenic potential. Pharmacokinetics of veralipride after chronic application in humans have been described by Staveris et al. (1978).

Bromperidol

- The compound patent was filed in 1959 by Janssen (Janssen, 1969c)
- First launched in The Netherlands by Janssen for the treatment of schizophrenia in 1981
- Bromperidol is a butyrophenone with general properties similar to those of haloperidol. It is given in the treatment of schizophrenia and other psychoses
- The long-acting decanoate ester may be used for patients requiring long-term therapy with bromperidol

Synthesis (Janssen, 1969c): 4-(4-Bromophenyl)-1,2,3,6-tetrahydropyridine is obtained by addition of formaldehyde to 1-bromo-4-(prop-1-en-2-yl)benzene. It is converted to 4-(4-bromophenyl)piperidin-4-ol in sodium hydroxide. The final compound is obtained by condensation of 4-(4-bromophenyl)piperidin-4-ol with 4-chloro-1-(4-fluorophenyl)butan-1-one.

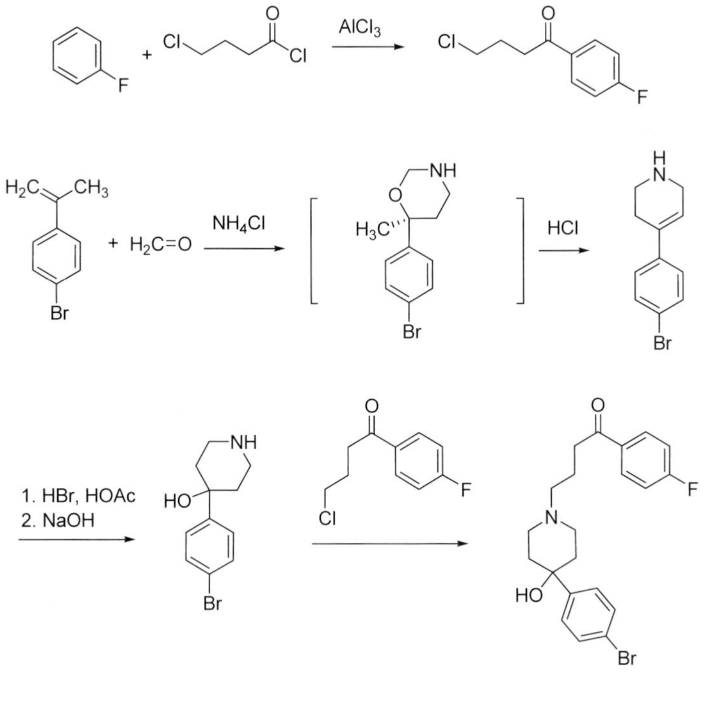

Bromperidol

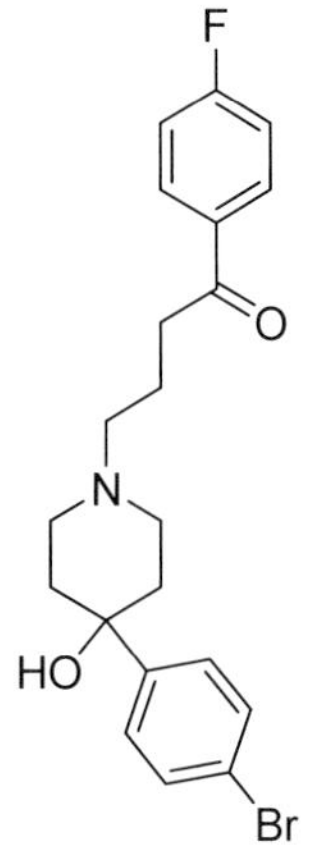

[10457-90-6], 4-[4-(4-Bromophenyl)-4-hydroxy-1-piperidinyl]-1-(4-fluorophenyl)-1-butanone, $C_{21}H_{23}BrFNO_2$, M_r 420.32;
[59453-24-6] hydrochloride salt, $C_{21}H_{24}BrClFNO_2$, M_r 456.78;
[75067-66-2] decanoate ester, $C_{31}H_{41}BrFNO_3$, M_r 574.60

Scheme 29: Synthesis of bromperidol

Clinical Use, Pharmacokinetics and Metabolism: Bromperidol is a butyrophenone antipsychotic with general properties similar to those of haloperidol. It is used in the treatment of psychoses like schizophrenia, either as decanoate ester (depot intramuscular injection) or as lactate salt (p.o.). Usual doses are between 1 mg to 15 mg per day p.o., and can be increased to a maximum of 50 mg daily according to treatment response. In schizophrenia, depot bromperidol has some advantages in side-effect profile as compared with haloperidol or fluphenazine, but is somewhat less potent (Wong et al., 2004). The pharmacokinetic and pharmacodynamic properties of bromperidol have been reviewed by Benfield et al. (1988). Bromperidol undergoes oxidative N-dealkylation, O-glucuronide conjugation, and carbonyl reduction, yielding reduced bromperidol (Wong et al., 1983). In addition, carbonyl reduction was confirmed to be mediated via carbonyl reductase, and reduced bromperidol was found to undergo oxidation back to bromperidol (Someya et al., 1991). CYP3A4 catalyzes the N-dealkylation of bromperidol and its reduced metabolite. CYP3A4 also catalyzes the dehydration of bromperidol to bromperidol 1,2,3,6-tetrahydropyridine, metabolizes bromperidol to bromperidol pyridinium, and catalyzes the oxidation of reduced bromperidol back to bromperidol (Sato et al., 2000).

Levosulpiride

Levosulpiride

[23672-07-3], (S)-5-(Aminosulfonyl)-N-[(1-ethyl-2-pyrrolidinyl)methyl]-2-methoxybenzamide, $C_{15}H_{23}N_3O_4S$, M_r 341.43; [77111-58-1] hydrochloride salt, $C_{15}H_{24}ClN_3O_4S$, M_r 377.89

- The compound patent was filed in 1978 by Ravizza (Mauri, 1979)

- First launched by Abbott for the treatment of nausea and schizophrenia in 1987

- Levosulpiride is the L-enantiomer of the substitute benzamide sulpiride and shows similar activities (see above), accounting for its binding to the D_2, D_3 and D_4 dopamine receptors

Synthesis (Mauri, 1979, 1982): Levosulpiride is obtained by condensation of 2(S)-(aminomethyl)-1-ethylpyrrolidine with 5-(aminosulfonyl)-2-methoxybenzoyl chloride by means of K_2CO_3 in acetone. The acylation of 2(S)-(hydroxymethyl)pyrrolidine with acetic anhydride in refluxing methanol gives 1-acetyl-2(S)-(hydroxymethyl)pyrrolidine, which by reaction with $SOCl_2$ in $CHCl_3$ is converted into 1-acetyl-2(S)-(chloromethyl)pyrrolidine. The reaction of 1-acetyl-2(S)-(chloromethyl)pyrrolidine with sodium azide in hot DMF affords the corresponding azide, which is reduced with LiAlH₄ in refluxing THF to yield the corresponding amine. This compound is condensed with 5-(aminosulfonyl)-2-methoxybenzoyl chloride to yield levosulpiride.

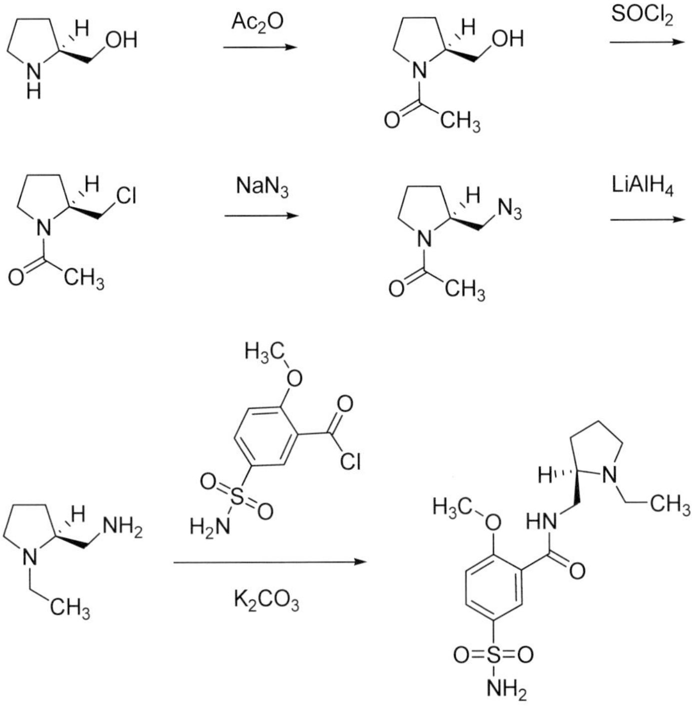

Scheme 30: Synthesis of levosulpiride

Clinical Use, Pharmacokinetics and Metabolism: Levosulpiride is a substituted benzamide antipsychotic that is reported to be a selective antagonist of central dopamine D_2, D_3 and D_4 receptors. It is claimed to have mood elevating properties and is mainly used in the treatment of psychoses such as schizophrenia. It also has been used in the management of Tourette's syndrome, anxiety disorders, vertigo, and benign peptic ulceration. Levosulpiride is a follow-up drug derived from sulpiride, by "chiral switch", i.e. it is the more active enantiomer (L-enantiomer) of the racemate sulpiride. With regard to its pharmacological actions, its side-effect profile, its adverse reactions and its pharmacokinetic behavior and its metabolism, it is very comparable to sulpiride (see chapter sulpiride above).

Remoxipride

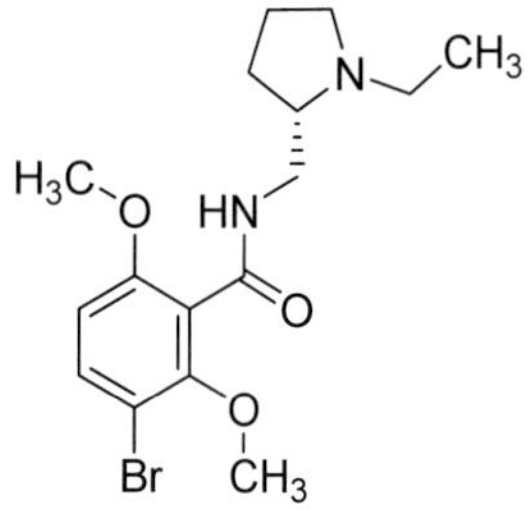

[80125-14-0], (-)-(S)-3-
Bromo-N-[(1-ethyl-2-
pyrrolidinyl)methyl]-2,6-
dimethoxybenzamide,
$C_{16}H_{23}BrN_2O_3$, M_r 371.27;
[82935-42-0] hydrochloride
salt, $C_{16}H_{24}BrClN_2O_3$,
M_r 407.73;
[117591-79-4] hydrochloride
monohydrate,
$C_{16}H_{26}BrClN_2O_4$, M_r 425.75

Remoxipride

- The compound patent was filed in 1978 by Astra Laekemedel (Florvall and Ögren, 1980)
- Remoxipride was first launched in 1990 in Denmark by Astra and was subsequently launched in most of Europe. Astra recommended restrictions on its use in 1993 due to reported harmful side-effects on bone marrow. Licensee Merck & Co suspended ongoing clinical trials in the USA in 1993
- Remoxipride has been suspended in all markets according to Astra, although it will still be available on a named patient basis for patients who are intolerant to standard antipsychotics
- Remoxipride is a substituted benzamide and acts as dopamine D_2/D_3 antagonist and sigma ligand

Synthesis (Florvall and Ögren, 1980, 1982): The bromination of 2,6-dimethoxybenzoic acid with Br_2 in dioxane gives 3-bromo-2,6-dimethoxy benzoic acid, which is treated with refluxing $SOCl_2$ to yield the corresponding acyl chloride. Finally, this compound is condensed with (-)-(S)-1-ethyl 2-(aminomethyl)pyrrolidine.

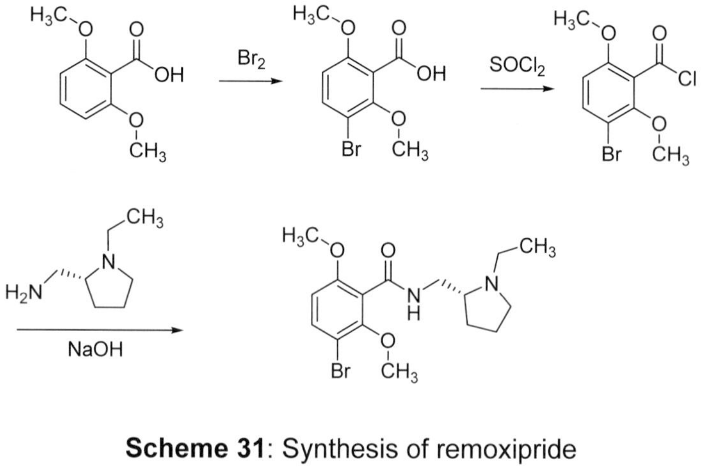

Scheme 31: Synthesis of remoxipride

Clinical Use, Pharmacokinetics and Metabolism: Remoxipride is a substituted benzamide antipsychotic with chemical similarity to sulpiride. It is a mixed dopamine D_2/D_3 antagonist, and also shows binding to the sigma receptor. Due to harmful side-effects on bone marrow, remoxipride has been suspended from most markets by its originator Astra Laekemedel and is only available for a limited number of named patients intolerant to other antipsychotic treatment. The clinical pharmacokinetics of remoxipride, a pure enantiomer, have been studied in healthy volunteers and patients. After oral administration the drug is rapidly and almost completely absorbed with a bioavailability above 90%. Thus remoxipride is a low clearance drug, without any first-pass metabolism. The apparent volume of distribution is 0.7 l/kg, about 80% being bound to plasma proteins (mainly alpha 1-acid glycoprotein). Remoxipride has a plasma half-life in the range of 4–7 hours and is eliminated by both hepatic metabolism and renal excretion. Slightly more than 70% of the dose is recovered as urinary metabolites and about 25% is excreted unchanged. Steady-state plasma levels are reached within 2 days, and they increase linearly with doses up to 600 mg daily. There is no evidence that active metabolites of remoxipride are present in the blood. Decreased renal function is associated with increased levels of remoxipride, whereas moderate cirrhosis of the liver only slightly affects elimination (von Bahr et al., 1990; Movin-Osswald and Hammarlund-Udenase, 1991).

Mosapramine

- The compound patent was filed in 1981 by Mitsubishi Pharma (Tashiro and Horii, 1982)

- First launched in Japan by Mitsubishi Pharma for the treatment of psychosis in 1991

- Mosapramine is a iminodibenzyl antipsychotic

Synthesis (Tashiro and Horii, 1982): By condensation of 3-chloro-5-(3-methylsulfonyloxypropyl)-10,11-dihydro-5*H*-di-benzo[*b,f*]azepine with 1,2,3,5,6,7,8,8a-octahydroimidazo[1,2-a]pyridine-3-spiro 4'-piperidin-2-one by means of K_2CO_3 in hot ethanol, and treatment with HCl to yield mosapramine hydrochloride salt.

The synthesis of the optical isomers of mosapramine has been reported (Tashiro et al., 1993): A solution of racemic spiro(octahydroimidazo[1,2-a]pyridine-3,4'-piperidin)-2-one with *N*-acetyl-L-tryptophan in hot isopropanol is allowed to crystallize. The precipitate collected is recrystallized from isopropanol and treated with aqueous NH_4OH, extracted with $CHCl_3$, evaporated and crystallized from isopropanol to obtain the R-(-)-enantiomer of spiro(octahydro-

Mosapramine

[89419-40-9], (±)-1'-[3-(3-Chloro-10,11-dihydro-5*H*-dibenz[*b,f*]azepin-5-yl)propyl]-1,2,3,5,6,7,8,8a-octahydrospiro[imidazo[1,2-a]pyridine-3,4'-piperidin]-2-one, $C_{28}H_{35}ClN_4O$, M_r 479.07;
[98043-60-8] dihydrochloride salt, $C_{28}H_{37}Cl_3N_4O$, M_r 551.99

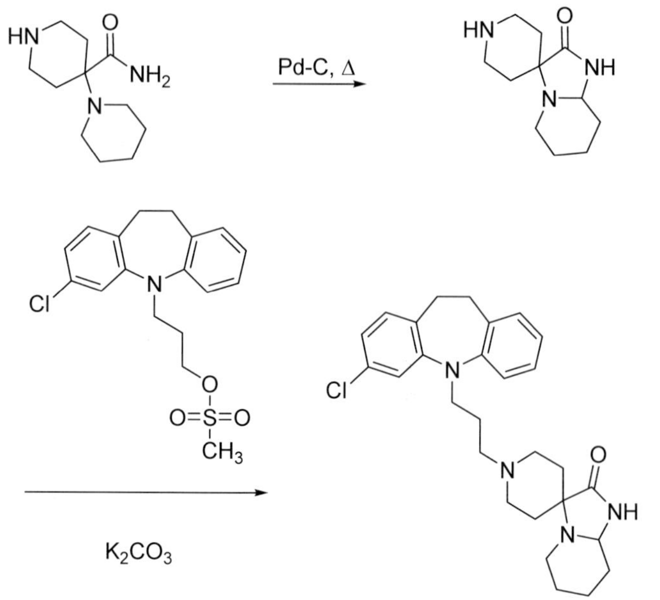

Scheme 32: Synthesis of mosapramine

imidazo[1,2-a]pyridine-3,4'-piperidin)-2-one. The condensation of this compound with 3-chloro-5-[3-(methanesulfonyloxy)propyl]-10,11-dihydro-5H-dibenzo[b,f] azepine by means of triethylamine in refluxing CHCl$_3$ affords the R-(−)-isomer of mosapramine. The mother liquors of the separation of R-(−)-spiro(octahydroimidazo[1,2-a]pyridine-3,4'-piperidin)-2-one are treated with aqueous NH$_4$OH and extracted with CHCl$_3$. This extract is purified by treatment with N-acetyl-D-tryptophan to obtain S-(+)-spiro(octahydroimidazo[1,2-a]pyridine-3,4'-piperidin)-2-one, which is then condensed with 3-chloro-5-[3-(methanesulfonyloxy)propyl]-10,11-dihydro-5H-dibenzo[b,f] azepine as before, yielding the S-(+)-isomer of mosapramine.

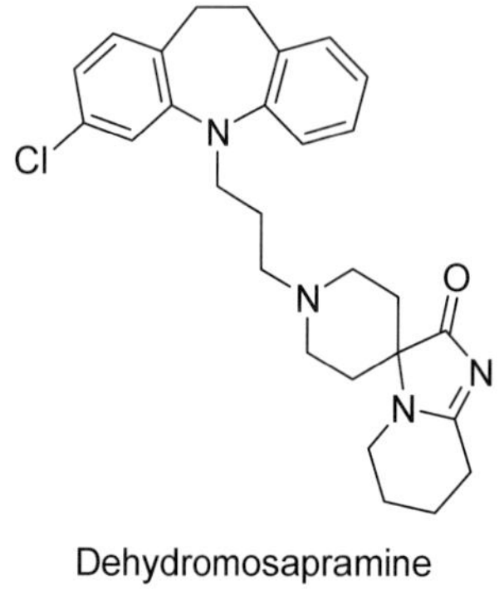

Clinical Use, Pharmacokinetics and Metabolism: Mosapramine is an iminodibenzyl antipsychotic launched for the neuroleptic treatment of schizophrenia (Takahashi et al., 1999). Data on pharmacokinetic behavior and metabolism shows that plasma concentrations of mosapramine are usually obtained 5–7 hours after administration, and plasma half-life ranges from 13–17 hours (Ishigooka et al., 1994; Ishigooka et al., 1989). The main metabolite detected after oral administration is dehydromosapramine in plasma and urine (Ishigooka et al., 1989).

Dehydromosapramine

Nemonapride

- The compound patent was filed in 1978 by Yamanouchi Pharma (Iwanami et al., 1980)
- The psycholeptic nemonapride was launched by Yamanouchi in Japan in 1991 as treatment for schizophrenia
- Nemonapride is a substituted benzamide with similar properties to those of sulpiride

Nemonapride

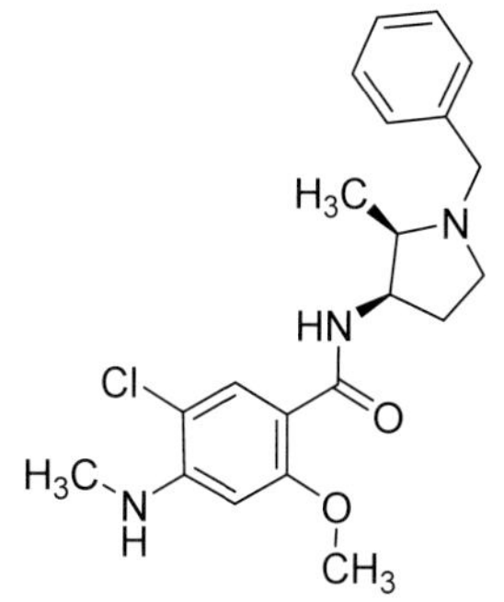

[75272-39-8], (±)-cis-*N*-(1-Benzyl-2-methylpyrrolidin-3-yl)-5-chloro-2-methoxy-4-methylaminobenzamide, $C_{21}H_{26}ClN_3O_2$, M_r 387.91

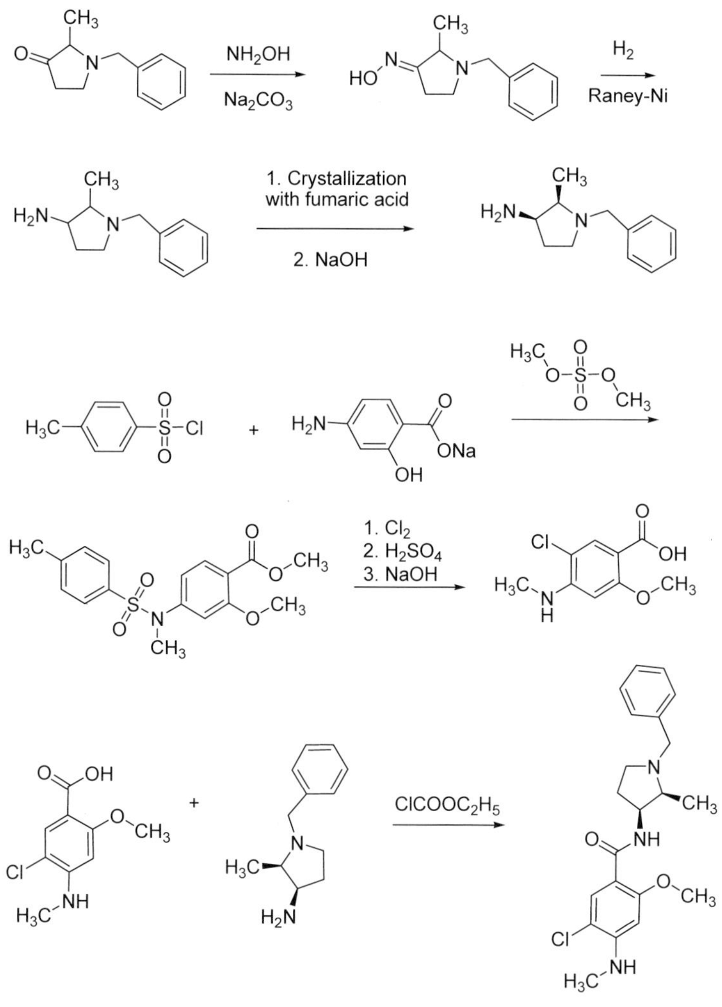

Synthesis (Iwanami et al., 1980): The reaction of 1-benzyl-2-methyl-3-pyrrolidone with hydroxylamine by means of Na_2CO_3 in hot ethanol gives the corresponding

Scheme 33: Synthesis of nemonapride

oxime, which is reduced with H₂ over Raney-Ni in methanol containing ammonia to afford 1-benzyl-2-methyl-3-amino-pyrrolidine. Finally, this compound is condensed with 2-methoxy-4-methylamino-5-chlorobenzoic acid by means of ethyl chlorocarbonate and triethylamine in methylene chloride.

Clinical Use: Nemonapride is a substituted benzamide antipsychotic with similar general properties as sulpiride. For the treatment of schizophrenia, it is given in divided doses ranging from 9–36 mg daily p.o. (Satoh et al., 1996). This dose can be increased up to a maximum dose of 60 mg daily, depending on the patient's response.

Tandospirone

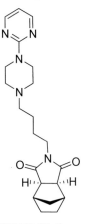

Tandospirone

[*87760-53-0*],
3aα,4α,5,6,7α,7aα-Hexahydro-2-[4-[4-(2-pyrimidinyl)-1-piperazinyl]butyl]-4,7-methano-1H-isoindole-1,3(2H)-dione dihydrogen, $C_{21}H_{29}N_5O_2$, M_r 383.50; [*112457-95-1*], citrate salt, $C_{27}H_{37}N_5O_9$, M_r 575.62

- The compound patent was filed in 1982 by Sumimoto Chemical Co (Ishizumi et al., 1983)

- Launched in 1996 for the treatment of anxiety, depression and psychosis by Dainippon Sumimoto Pharma (see Chapter 3.5)

- Tandospirone is structurally related to buspirone. It is a partial agonist at the serotonin 5-HT$_{1A}$ receptor

Synthesis (Ishizumi et al., 1983; Prous and Castaner, 1986). Condensation of 2-(piperazin-1-yl)pyrimidine with 4-chlorobutyronitrile by means of NaOH in acetone gives 4-[4-(2-pyrimidinyl)piperazin-1-yl]butyronitrile, which is reduced with LiAlH₄ in ether yielding 1-(4-aminobutyl)-4-(2-pyrimidinyl)piperazine. Finally, this compound is condensed with bicyclo[2.2.1]heptane-2,3-di-exocarboxylic acid anhydride in refluxing pyridine to give tandospirone. The acid anhydride can be obtained by hydrogenation of bicyclo[2.2.1]hept-5-ene-2,3-di-exo-carboxylic acid anhydride with H₂ over Pd/C. (see Chapter 3.5 for additional synthesis details).

Clinical Use: Tandospirone is a pyrimidinyl-piperazine antipsychotic, related to buspirone. It was launched for treatment of anxiety, depression and psychoses, and is used – in form of its citrate salt – in the management of schizophrenia for memory enhancement (Sumiyoshi et al., 2001) and also, due to its antidepressant effects, in the treatment of major depressive disorder (Yamada et al., 2003). Pharmacological action comprises partial agonism with respect to the serotonin 5-HT$_{1A}$ receptor. For the treatment of psychoses, it is given in divided doses ranging around 30 mg daily p.o.

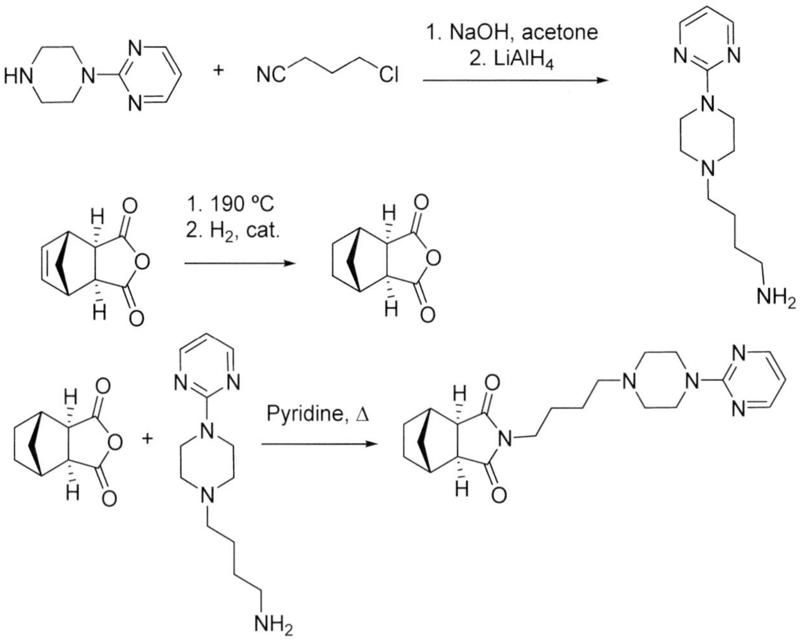

Scheme 34: Synthesis of tandospirone

2.5.3 Atypical antipsychotics

2.5.3.1 Atypical antipsychotics with dopaminergic activity only (D_4 antagonists)

Clozapine

- Clozapine was discovered in 1952 (Dr. A. Wander S.A, 1965), developed by Sandoz (Novartis), and introduced in Europe in 1972
- Clozapine is a dibenzodiazepine derivative that was the first of the atypical antipsychotics to be developed
- It has high affinity for dopamine D_4 receptor and relatively week activity at D_1, D_2, D_3, and D_5 receptors. Clozapine possesses α-adrenergic blocking, antimuscarinic, antihistaminic, antiserotoninergic, and sedative properties
- From a pharmacological point of view, the arrival of clozapine revolutionized schizophrenia treatment as

Clozapine

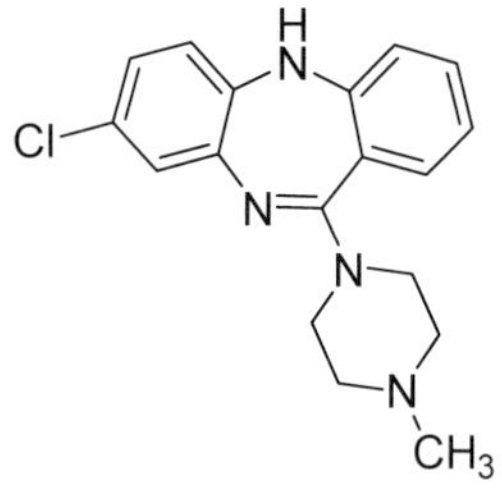

[5786-21-0], 8-Chloro-11-(4-methylpiperazin-1-yl)-5H-dibenzo[b,e]-[1,4]diazepine, $C_{18}H_{19}ClN_4$, M_r 326.82

much as the release of chlorpromazine did years earlier

- In 1975, after reports of agranulocytosis leading to death in some patients, clozapine was voluntarily withdrawn by the manufacturer
- After demonstrating that clozapine was more effective against treatment-resistant schizophrenia than other antipsychotics, in 1989 the FDA and health authorities in most other countries approved its use only for treatment-resistant schizophrenia. In 2002, it was also approved for reducing the risk of suicidal behavior in patients with schizophrenia

Synthesis (Dr. A. Wander S.A, 1965): 4-Chloro-2-nitrobenzenamine is combined with methyl 2-chlorobenzoate, and further addition of 1-methylpiperazine produces the intermediate (2-(4-chloro-2-nitrophenyl-amino)phenyl)(4-methylpiperazin-1-yl)methanone. Hydrogenation of the intermediate and ring closure with with a dehydrating agent result in the final compound clozapine.

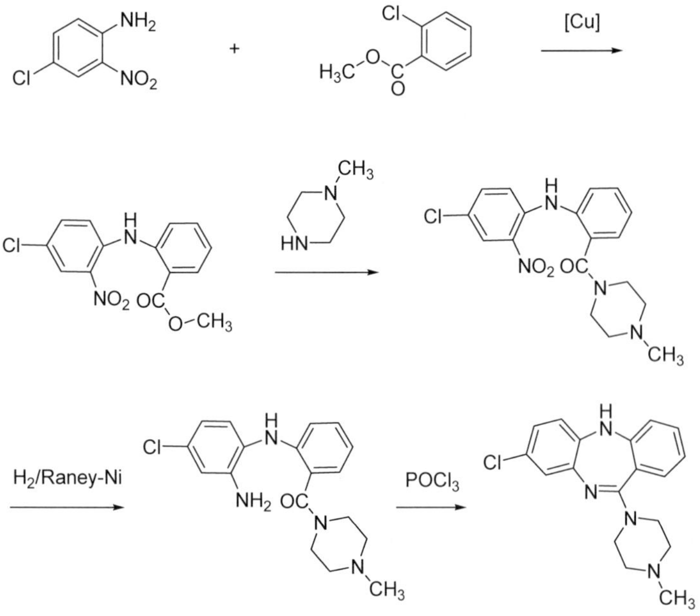

Scheme 35: Synthesis of Clozapine

Clinical Use, Pharmacokinetics and Metabolism: Clozapine is a dibenzodiazepine derivative described to be an atypical antipsychotic. It has only weak affinity towards dopamine D_1, D_2, D_3 and D_5 receptors, but is a potent D_4 receptor antagonist. In addition, it possesses α-adrenergic blocking, antimuscarinic, antihistaminic, antiserotoninergic, and sedative properties. Its exact antipsychotic mechanism of action is not known, but it seems not to excert its action via dopamine D_2 receptor blockade, and also lacks extrapyramidal side-effects (Kerwin, 1994). Clozapine is used for the management of schizophrenia, however, due to the risk of agranulocytosis, there are restrictions on its use (Fiston and Heel, 1990; Anonymous, 1991; Baldessarini and Frankenburg, 1991; Hirsch and Puri, 1993; Taylor, 1995; Kerwin, 1995). Thus, it should be used only in patients who fail to respond to other antipsychotics (including atypical ones) or who experience severe neurological side-effects with other antipsychotic treatments. Clozapine is especially used in different therapeutic fields related to schizophrenia (Kane et al., 1988; Buckley, 1998; Wahlbeck et al., 2004; Meltzer et al., 1989; Conley et al., 1997), e.g. for reducing the risk of recurrent suicidal behavior associated with schizophrenia (Meltzer and Okayli, 1995; Kerwin, 1995; Meltzer et al., 2003), and in treatment-resistant psychoses associated with Parkinson's disease, though there are controversial reports on therapeutic success (Klein et al., 2003; Pfeiffer and Wagner, 1994; The Parkinson Study Group, 1990; The French Clozapine Parkinson Study Group, 1999; Auzou et al., 1994; Green, 1995). It also has been used in bipolar disorder, dementia, and to treat tardive dyskinesia (Tamminga et al., 1994; Nair et al., 1996). Its use must be accompanied by strict monitoring procedures (e.g., white blood cell count). Initial doses for antipsychotic treatment range from once or twice daily 12.5 mg p.o., followed by a 25 mg dose once or twice daily on the second day. Then, the dose can be increased according to response in increments of 25–50 mg up to a maximum dose of 300–450 mg per day (within 14–21 days). A maximum dose of 900 mg should not be exceeded. Due to the possibility of serious adverse effects (agranulocytosis, see above), it is important to adjust the minimum necessary dose properly and to monitor white blood cell count. In terms of its adverse effects profile, clozapine shares some of the side-effects of classic (typical) antipsychotics, but incidence and severity may vary. Antimuscarinic effects as well as weight gain and sedation may be more pronounced along with clozapine treatment as with other antipsychotics. As described above, clozapine can cause reversible neutropenia, possibly leading to fatal agranulocytosis

(Idänpään-Heikkilä et al., 1977; Anderman and Griffith, 1977; Lieberman et al., 1990; Gerson et al., 1991; Gerson and Meltzer, 1992; Safferman et al., 1992; Committee on Safety of Medicines/Medicines Control Agency, 1993a; Alvir et al., 1993; Finkel and Arellano, 1995; Atkin et al., 1996; Fisher and Baigent, 1996; Honigfeld et al., 1998). Eosinophilia may also occur. Extrapyramidal side-effects, including tardive dyskinesia, are rarely observed along with clozapine treatment. Clozapine has little effect on prolactin secretion. It appears to have a greater epileptogenic potential as e.g. chlorpromazine, but is comparable to chlorpromazine with respect to cardiovascular effects such as tachycardia and orthostatic hypotension. Typical minor adverse effects comprise dizziness, hypersalivation (Davydov and Botts, 2000; Calderon et al., 2000), headache, nausea, vomiting, constipation, urinary incontinence and retention, anxiety, confusion, fatigue, and transient fever. Seizures and convulsions have been observed along with clozapine therapy (Devinski et al., 1991; Committee on Safety of Medicines, 1991), also effects on carbohydrate metabolism (Griffith and Springuel, 2001; Hedenmalm et al., 2002; Citrome and Jaffe, 2003; American Diabetes Association/American Psychiatric Association/American Association of Clinical Endocrinologists/North American Association for the Study of Obesity, 2004), on the cardiovascular system (Committee on Safety of Medicines/Medicines Control Agency, 1993b; Adverse Drug Reactions Advisory Committee, 1994; Gupta, 1994; Pokorny et al., 1994; Ghaeli and Dufresne, 1996; Ennis and Parker, 1997; Killian et al., 1999; Hägg et al., 2000; Wolstein et al., 2000; La Grenade et al., 2001; Committee on Safety of Medicines/Medicines Control Agency, 2002a), effects on electrolyte homeostasis (Ogilvie and Croy, 1992), on the gastrointestinal tract (Committee on Safety of Medicines/Medicines Control Agency, 1999a), on the kidneys (Elias et al., 1999), on the pancreas (Martin, 1992; Frankenburg and Kando, 1992; Jubert et al., 1994), neuroleptic malignant syndrome (Sachdev et al., 1995; Kargianis et al., 1999) and side-effects along with withdrawal (Stanilla et al., 1997), amongst others. Clozapine should not be given to patients with uncontrolled epilepsy, alcoholic or toxic psychoses, drug intoxication, or a history of circulatory collapse. It is amongst other conditions contraindicated in patients with bone-marrow suppression, myeloproliferative disorders or any abnormalities with white blood cell count, and when breast feeding (Barnas et al., 1994; American Academy of

Pediatrics, 2001). Clozapine may enhance central effects of CNS depressants including alcohol, benzodiazepins, and MAOI's (Sassim and Grohmann, 1988; Cobb et al., 1991; Friedman et al., 1991; Finkel and Schwimmer, 1991; Martin, 1993; Centorrino et al., 1994; Jerling et al., 1994; Jackson et al., 1995; Kingsbury and Puckett, 1995; Centorrino et al., 1996; Zerjav-Lacombe and Dewna, 2001). It should not be taken together with inhibitors or substrates of cytochrome P450 isoenzyme CYP1A2 (Taylor, 1997). Clozapine is well absorbed from the GI tract after oral administration. It undergoes extensive first-pass metabolism leading to an oral bioavailability of about 50%. Peak plasma concentrations are reached about 2.5 hours after ingestion. Clozapine has a plasma protein binding of 95% and a terminal elimination half-life of about 12 hours. It is almost completely metabolized via *N*-desmethylation, hydroxylation and *N*-oxidation phase I metabolism reactions. Desmethyl-clozapine (norclozapine) has only limited antipsychotic activity. Metabolism of clozapine is mainly CYP1A2-mediated. There is wide interindividual variation found in plasma concentrations, and no simple correlation of plasma concentration and therapeutic effect was found (Jann et al., 1993; Lin et al., 1994; Freeman and Oyewumi, 1997; Olesen, 1998; Guitton et al., 1999).

2.5.3.2 Atypical antipsychotics with dopaminergic and serotonergic activity (D_2 and 5-HT_{2A} antagonists)

Zotepine

- The compound patent was filed in 1968 by Fujisawa Pharmaceuticals (Umio et al., 1969)

- Zotepine has been launched in Japan by Fujisawa in 1982 and in Germany by Sanofi-Aventis in 1990, for treatment of schizophrenia

- Knoll has marketing rights for zotepine in the USA and Europe, except Austria and Germany

- Zotepine is a substituted dibenzothiepine tricyclic antipsychotic

- Zotepine has a high affinity for the dopamine D_1 and D_2 receptors. It also affects the serotonin $5HT_{2A}$, $5HT_{2C}$, $5HT_6$, and $5HT_7$, the adrenergic α_1 and the histamine H_1 receptors. In addition, it inhibits the

Zotepine

[*026615-21-4*], 8-Chloro-10-(2-dimethylaminoeth-oxy)dibenzo[*b,f*]thiepine, $C_{18}H_{18}ClNOS$, M_r 331.86.

reuptake of noradrenaline. It is thought that by hitting multiple targets, zotepine has a high efficacy for the negative symptoms of schizophrenia

Synthesis (Umio et al., 1969): The compound can be prepared in two different ways:

1. By condensation of 8-chlorodibenzo[*b,f*]thiepin-10(11*H*)-one with 2-dimethylaminoethyl chloride by means of NaH in benzene–DMF at 60°C.

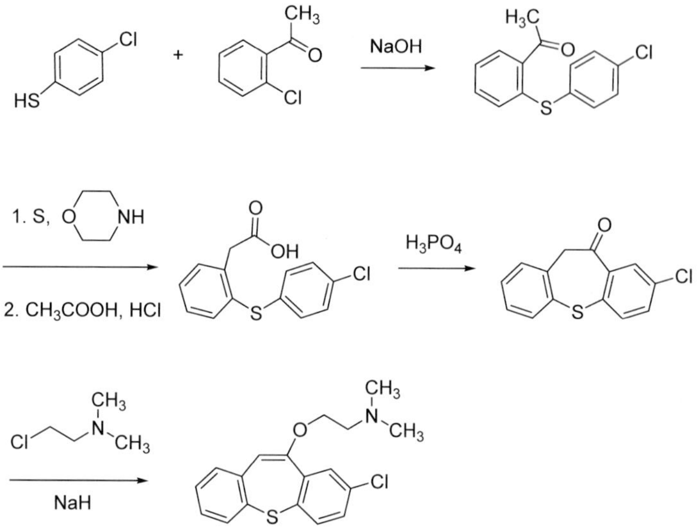

Scheme 36: Synthesis of zotepine

2. By condensation of 8-chloro-10-(2-chloroethoxy)dibenzo[*b,f*]thiepin with dimethylamine.

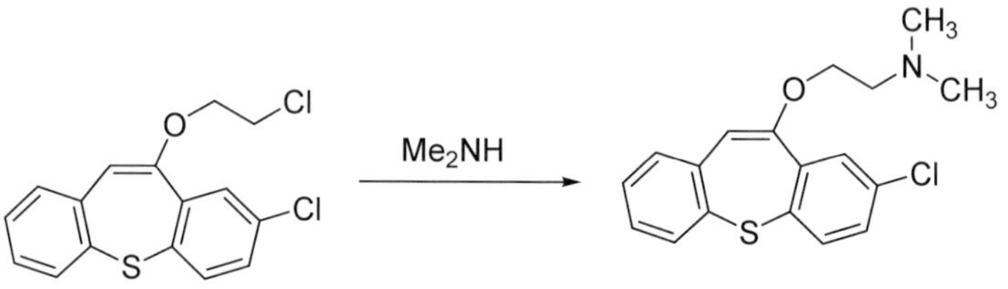

Scheme 37: Alternative synthesis of zotepine

Clinical Use, Pharmacokinetics and Metabolism: Zotepine is an atypical antipsychotic with a broader spectrum of receptor affinities, ranging from central dopamine D_1 and D_2 inhibition to serotonin 5-HT_2, α_1-adrenergic, and histamine H_1 receptor binding, and noradrenaline reuptake inhibition. It is used for the treatment of schizophrenia in initial doses of around 25 mg three times a day, which can be increased to a maximum dose of three times 100 mg a day, depending on the response of the patient. Zotepine may be especially valuable in the treatment of schizophrenia in patients with negative symptoms. In addition, it seems less likely to provoke extrapyramidal effects in comparison to classic antipsychotics (Morris et al., 2004). Adverse effects are similar to those described for classic antipsychotics, but incidence and severity may vary. Common adverse effects include asthenia, headache, hypotension, and less commonly orthostatic hypotension, tachycardia, gastrointestinal disturbances, elevated liver-enzyme values, leucopenia, agitation, anxiety, dizziness, insomnia, somnolence, rhinitis, sweating, and blurred vision. Zotepine can prolong the Q–T interval, leading to a potential for cardiotoxicity (torsade de pointes and sudden death). Patients with cardiovascular risk factors should be monitored (ECG and electrolytes) during therapy. Zotepine may – similarly to clozapine – enhance the effect of other CNS depressants including alcohol, benzodiazepines, and MAOI's (see clozapine Section above). Zotepine is absorbed from the GI tract after oral application to reach peak plasma concentrations 2–3 hours after ingestion. It undergoes extensive first-pass metabolism to equipotent metabolite norzotepine (*N*-desmethylzotepine) and other inactive metabolites. Metabolism is mainly mediated by cytochrome P450 isoenzymes CYP1A2 and CYP3A4. Zotepine shows a plasma protein binding of 97% and is excreted mainly in urine and faeces in the form of its metabolites. Elimination half-life is about 14 hours.

Risperidone

- The compound patent was filed in 1985 by Janssen (Kennis and Vandenberk, 1986)
- Developed by Janssen, risperidone was initially launched in 1993 for the treatment of schizophrenia
- It has since been marketed for the treatment of mania and psychosis and is currently awaiting registration for the treatment of autism

Risperidone

[106266-06-2], 4-[2-[4-(6-fluorobenzo[d]isoxazol-3-yl)-1-piperidyl]ethyl]-3-methyl-2,6-diazabicyclo[4.4.0]deca-1,3-dien-5-one, $C_{23}H_{27}FN_4O_2$, M_r 410.48.

- In 1997, Janssen signed a supply agreement with Alkermes for the production of a long-acting injectable form of risperidone
- Risperidone is a benzisoxazole atypical antipsychotic, reported to be a serotonin 5-HT$_2$, dopamine D$_2$, adrenergic α_1 and α_2, and histamine H$_1$ receptor antagonist
- Risperidone is also being evaluated in phase III clinical trials for the treatment of major depression and as maintenance therapy for bipolar disorder

Synthesis (Kennis and Vandenberk, 1986): The Friedel–Crafts condensation of 1,3-difluorobenzene with 1-acetylpiperidine-4-carbonyl chloride by means of AlCl$_3$ in dichloromethane gives 1-acetyl-4-(2,4-difluorobenzoyl)piperidine, which is hydrolyzed with refluxing 6N HCl to yield 4-(2,4-difluorobenzoyl)piperidine. The reaction of 4-(2,4-difluorobenzoyl)piperidine with hydroxylamine in refluxing ethanol affords the corresponding oxime, which is cyclized by means of KOH in boiling water to give 6-fluoro-3-(4-piperidinyl)-1,2-benzisoxazole. Finally, this compound is condensed with 3-(2-chloroethyl)-2-methyl-6,7,8,9-tetrahydro-4*H*-pyrido[1,2-a]pyrimidin-4-one by means of K$_2$CO$_3$ and KI in a variety of solvents to give risperidone.

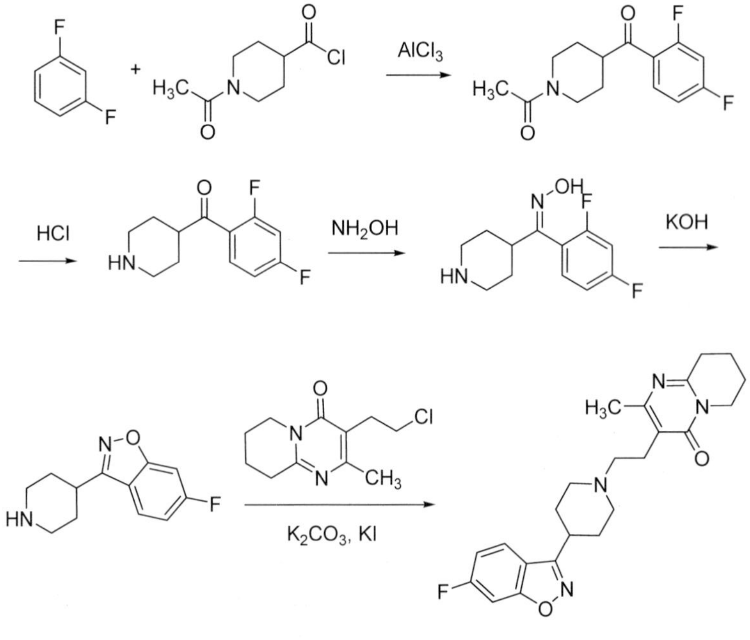

Scheme 38: Synthesis of risperidone

An additional process has been described (Marquillas et al., 1994) in which condensation of 4-(2,4-difluorobenzoyl)piperidine with 3-(2-chloroethyl)-2-methyl-6,7,8,9-tetrahydro-4H-pyrido[1,2-a]pyrimidin-4-one by means of KI and $NaHCO_3$ in refluxing acetonitrile affords an adduct which is treated with hydroxylamine hydrochloride and KOH in refluxing pyridine/ethanol to provide the corresponding oxime. Finally, this compound is cyclized by means of KOH in refluxing water or with NaH in refluxing THF to afford in both cases the target 1,2-benzisoxazole.

The antipsychotic compound risperidone is also prepared through a variant of the original method (Srinivasa et al., 2005) involving the condensation of 6-fluoro-3-(4-piperidinyl)-1,2-benzisoxazole and 3-(2-chloroethyl)-2-methyl-6,7,8,9-tetrahydro-4H-pyrido[1,2-a]pyrimidin-4-one, both as their hydrochloride salts, in the presence of Na_2CO_3 in water and/or a water-miscible solvent.

Clinical Use, Pharmacokinetics and Metabolism: Risperidone is a benzisoxazole atypical antipsychotic, which is an antagonist at dopaminergic (D_2), serotoninergic (5-HT_2), adrenergic (α_1 and α_2) and histaminergic (H_1) receptors. It has been proposed that the atypical properties of risperidone are based on the ratio between 5-HT_2 and D_2 affinity, leading to less extrapyramidal adverse effects, as compared e.g. with haloperidol (Kerwin, 1994). It is used p.o. in the treatment of schizophrenia and other psychoses and in the short-term treatment of mania associated with bipolar disorder. It can also be applied by intramuscular injection (depot) for maintenance therapy. Initial dose for schizophrenia treatment is 2 mg p.o., which is increased to 4 mg on the 2nd day. Subsequent increases in 1–2 mg intervals per week can be realized until the desired effect is obtained. Usual doses are between 4 mg to 6 mg per day. Risperidone has been used in HIV-related psychosis (Singh et al., 1997), anxiety (Jacobsen, 1995), bipolar disorders (Segal et al., 1998; Sachs et al., 2002; Yatham et al., 2003), disturbed behavior (Allen et al., 1995; McKeith et al., 1995; De Deyn et al., 1999; Falsetti, 2000; Research Units on Pediatric Psychopharmacology Autism Network, 2002; Valiquette, 2002), dystonias (Zuddas and Cianchetti, 1996), Parkinsonism (Ford et al., 1994; Meco et al., 1994; Leopold, 2000), schizophrenia (Chouinard et al., 1993; Livingstone, 1994; Marder and Meibach, 1994; Owens, 1994; Klieser et al., 1995; Peuskens, 1995; Musser and Kirisci, 1995; Csernansky et al., 2002; Hunter et al., 2004), stuttering (Maquire et al., 2000; Lee et al., 2001), and Tourette's syndrome (Bruun and Budman, 1996; Bruggeman et al., 2001; Scahill et al., 2003). Adverse effects are similar to those described for classic antipsychotics, but incidence and severity may vary. Riperidone is reported to less likely cause sedation or extrapyramidal effects, as compared e.g. with chlorpromazine, but may induce agitation more frequently. Common adverse effects include insomnia, anxiety, and headache, and – less commonly – dyspepsia, nausea, abdominal pain, constipation, blurred vision, sexual dysfunction including priapism, urinary incontinence, rash and other allergic reactions, drowsiness, concentration difficulties, dizziness, fatigue, and rhinitis. Risperidone should be used with care in patients with cardiovascular diseases, including conditions associated with Q–T prolongation, as well as in patients with Parkinson's disease or epilepsy. Experiences when breast feeding (Hill et al., 2000) and adverse effects on carbohydrate metabolism (Janssen-Ortho Inc./Health Canada, 2002), on the liver (Fuller et al., 1996; Philips et al., 1998), on the skin (Cooney and Nagy, 1995), extrapyramidal disorders (Vercueil and Foucher, 1999; Krebs and Olie, 1999; Miller, 2000), mania (Dwight et al., 1994; Zolezzi and Badr, 1999) neuroleptic malignant syndrome (Sharma et al., 1996; Tarsy, 1996; Reeves et al., 2001) and experiences with overdosage (Cheslik and Erramouspe, 1996; Himstreet and Daya, 1998) have been described. Risperidone may – similarly to clozapine – enhance the effect of other CNS depressants including alcohol, benzodiazepines, and MAOI's (see clozapine chapter above). Risperidone may also enhance the effect of antihypertensives. There may be an increased risk of Q–T prolongation when risperidone is given together with a drug known to cause this effect. Extrapyramidal symptoms have been described for risperidone in combination with HIV treatment with ritonavir/indinavir (Kelly et al., 2002). Risperidone is readily absorbed after oral ingestion and reaches plasma peak concentrations after 1–2 hours. It is extensively metabolized in the liver, mainly by hydroxylation to its main active metabolite, 9-hydroxyrisperidone (by CYP2D6; subject to genetic polymorphism). Oxidative dealkylation is a minor metabolic pathway. Excretion occurs mainly in urine, to a lesser degree in faeces. Risperidone and 9-hydroxyrisperidone are about 90% and 77% plasma protein bound, respectively. Both compounds are distributed widely throughout the body, and also into breast milk (Huang et al., 1993).

Olanzapine

- The compound patent was filed in 1990 (Chakrabarti et al., 1991). Olanzapine was developed by Eli Lilly and Company and was the second atypical antipsychotic to gain FDA approval. It was initially launched in 1996 in the USA for the treatment of schizophrenia

- Olanzapine has been FDA approved for the treatment of schizophrenia, acute mania in bipolar disorder, agitation associated with schizophrenia and bipolar disorder, and as maintenance treatment in bipolar disorder

- Olanzapine is structurally similar to clozapine, and has a high affinity for dopamine and serotonin receptors, and lower affinity for histamine H_1, cholinergic, muscarinic and α_1 adrenergic receptors. The mechanism of action of olanzapine is theorized to be mediated primarily by antagonism at dopamine D_2 and serotonin 5-HT_{2A} receptors

- Lilly has developed and launched a combination of fluoxetine and olanzapine for the treatment of bipolar depression in April 2004 (USA) (Bymaster et al., 1998; Tollefson, 1999)

Olanzapine

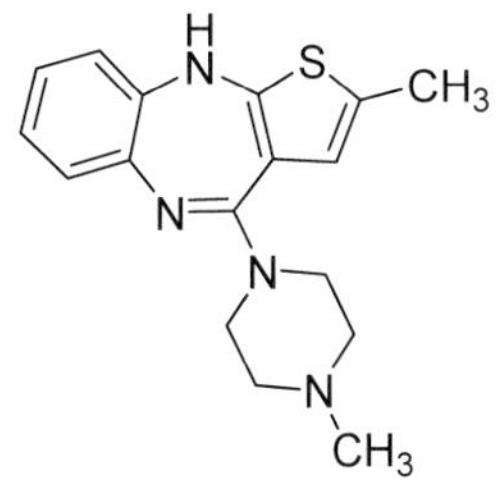

[*132539-06-1*], 2-Methyl-4-(4-methylpiperazin-1-yl)-10H-thieno[2,3-b][1,5]benzodiazepine, $C_{17}H_{20}N_4S$, M_r 312.44; [*221373-18-8*] pamoate salt, $C_{40}H_{36}N_4O_6S$, M_r 700.82

Synthesis (Chakrabarti et al., 1991): The condensation of o-chloronitrobenzene with 2-amino-5-methylthiophene-3-carbonitrile by means of lithium hydroxide in DMSO gives 2-(2-nitroanilino)-5-methylthiophene-3-carbonitrile. The reductive cyclization of this compound using stannous chloride in aqueous ethanolic hydrochloric acid gives the primary amidine hydrochloride, which is condensed with N-methylpiperazine in a mixture of 4:1 toluene:DMSO.

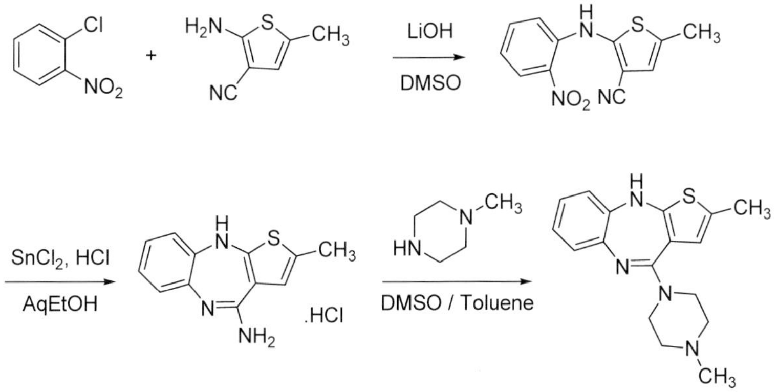

Scheme 39: Synthesis of olanzapine

Improved methods are disclosed for the last step of the synthesis of olanzapine (Dolitzky and Diller, 2005) avoiding the use of high boiling organic solvents. The condensation between 4-amino-2-methylthieno[2,3-b][1,5]benzodiazepine·HCl and N-methylpiperazine can be effected in the absence of solvents at 110-145 °C, followed by cooling and precipitation of the product upon addition of a suitable solvent. Alternatively, the condensation can be carried out by heating in a low boiling point solvent such as acetone or acetonitrile.

The antipsychotic compound olanzapine is also synthesized by condensation of 4-amino-2-methyl-10H-thieno[2,3-b][1,5]benzodiazepine with N-formylpiperazine (Keltjens and Peters, 2005), followed by reduction of the resultant N-formyl olanzapine by means of catalytic hydrogenation or using an aluminium hydride, especially sodium bis(2-metoxyethoxy)aluminum hydride (RedAl).

Clinical Use, Pharmacokinetics and Metabolism: Olanzapine is a thienobenzodiazepine atypical antipsychotic. It shows high affinity for dopaminergic and serotonin receptors, and moderate affinity for histamine H_1, muscarinic, cholinergic, and α_1-adrenergic receptors. The presumable antipsychotic mechanism of action for olanzapine is via inhibition (antagonism) of dopamine D_2 and serotonin $5-HT_2$ receptors. Olanzapine is used in the management of schizophrenia and for the treatment of moderate to severe mania associated with polar disorder (Shelton et al., 2001; Rendell et al., 2004). Studies suggest that olanzapine is as effective as haloperidol against positive symptoms of schizophrenia, and more effective against negative symptoms in the short-term and possibly also in the long-term treatment (Beasley et al., 1996a, b, 1997; Tollefson et al., 1997; Tran et al., 1997; Hamilton et al., 1998; Bhana et al., 2001; Duggan et al., 2004). Olanzapine has also been used in the treatment of Parkinsonism (Wolters et al., 1996), stuttering (Lavid et al., 1999), and Tourette's syndrome (Stamenkovic et al., 2000; Onofrj et al., 2000; Budman et al., 2001). Initial and maintenance dose for schizophrenia treatment with olanzapine is 10 mg daily (single dose, p.o.). According to the response, in the clinical situation the dose may be increased in 5 mg intervals up to a maximum dose of 20 mg daily. For rapid control of agitation and disturbed behavior in patients with mania or schizophrenia, olanzapine can be given intramuscularly (initial dose of 5–10 mg). Adverse effects are similar to those described for classic antipsychotics like chlorpromazine, but incidence and severity may vary. Most frequent adverse effects comprise somnolence, (usually) asymptomatic hyperprolactinaemia and weight gain. Other adverse effects include increased appetite, dizziness, elevated plasma glucose, triglyceride, and liver enzyme values, eosinophilia, oedema, orthostatic hypotension, and mild transient antimuscarinic effects like constipation and dry mouth. In addition, cerebrovascular disorders (stroke), dementia, effects on the blood (Flynn et al., 1997; Bogunovic and Viswanathan, 2000; Tolosa-Vilella et al., 2002), carbohydrate metabolism (Bettinger et al., 2000; Roefaro and Mukherjee, 2001; Bonanno et al., 2001; Ragucci and Wells, 2001; Koller et al., 2001; Anonymous, 2002; Koro et al., 2002), on the liver (Jadallah et al., 2003), on the nervous system (Lee et al., 1999; Bonelli, 2003), on the pancreas (Doucette et al., 2000; Hagger et al., 2000), on sexual function (Deirmenjian et al., 1998), extrapyramidal disorders (Herrán and Vázquez-Barquero, 1999), mania (Lindenmayer and Klebanov, 1998; Fitz-Gerald et al., 1999; Aubry et al., 2000; Henry and Demotes-Mainard, 2002; Baker et al., 2003), neuroleptic malignant syndrome (Filice et al., 1998; Nyfort-Hansen and Alderman, 2000; Kogoj and Velikonja, 2003) and worsening of motor function in Parkinsonism (Graham et al., 1998; Molho and Factor, 1999; Goetz et al., 2000; Manson et al., 2000) have been described amongst others. Experience with overdosage has been described (Yip et al., 1998; Chue and Singer, 2003). Olanzapine may – similarly to clozapine – enhance the effect of other CNS depressants including alcohol, benzodiazepins, and MAOI's (see clozapine Chapter above). CYP1A2 inhibitor or substrate drugs like e.g. fluvoxamine should not be combined with olanzapine. Olanzapine is well absorbed from the GI tract after oral administration and is subject to considerable first-pass metabolism. Peak plasma concentrations are reached 5–8 hours after ingestion and about 15–45 minutes after intramuscular injection. Olanzapine is up to 93% plasma protein bound and is metabolized in the liver primarily by oxidation (CYP1A2, and to a lesser degree CYP2D6) and by (direct and secondary) glucuronidation. Two major metabolites identified are inactive 10-*N*-glucuronide and 4'-*N*-desmethyl-olanzapine (see below). About 57% of olanzapine is excreted in urine, mainly in the form of metabolites, and about 30% in faeces. Plasma elimination half-lives are around 30–38 hours, with gender differences. Olanzapine is distributed into breast milk

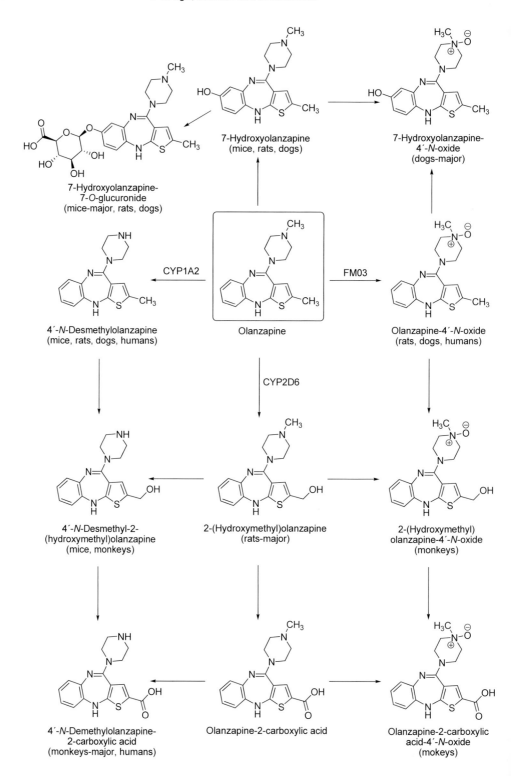

Scheme 40: Metabolism of olanzapine

(Callaghan et al., 1999). Olanzapine undergoes extensive metabolism in animal models (mice, rats, dogs and monkeys) and humans after oral administration. A total of 13 metabolites have been identified, which are formed via N-glucuronidation, allylic hydroxylation, hydroxylation on the aromatic ring, N-oxidation, N-dealkylation and a combination thereof. Allylic oxidation, forming 2-hydroxymethyl olanzapine and the carboxylic acid derivative, is a common pathway in all five species and a major route in rats. 4'-N-Desmethylation followed by allylic oxidation results in 4'-N-desmethyl olanzapine, 4'-N-desmethyl-2-hydroxymethyl olanzapine and 4'-N-desmethyl olanzapine 2-carboxylic acid, the latter is the major metabolite in monkeys. 4'-N-oxidation followed by allylic oxidation is another important metabolic route in monkeys. 7-Hydroxylation on the aromatic ring followed by 4'-N-oxidation is the major metabolic pathway in dogs, while the O-glucuronide of 7-hydroxylated olanzapine is the major metabolite in mice. 4'-N- and 10-N-glucuronidation of olanzapine are the metabolic routes identified in humans, with the 10-N-glucuronide being the major metabolite in humans although being detectable in trace amounts from dog urine. In human liver microsomes or other liver preparations, CYP1A2 catalyzes N-desmethylation and 7-hydroxylation, CYP2D6 catalyzes 2-hydroxylation and FMO3 catalyzes 4'-N-oxidation, while N-glucuronidations are mediated by UGT1A4 (Ring et al., 1996; Kassahun et al., 1997, 1998; Mattiuz et al., 1997; Linnet, 2002).

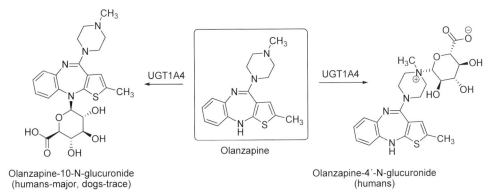

Olanzapine-10-N-glucuronide
(humans-major, dogs-trace)

Olanzapine

Olanzapine-4'-N-glucuronide
(humans)

Scheme 41: *N*-Glucuronidation of olanzapine mediated by UGT1A4

Quetiapine

Quetiapine

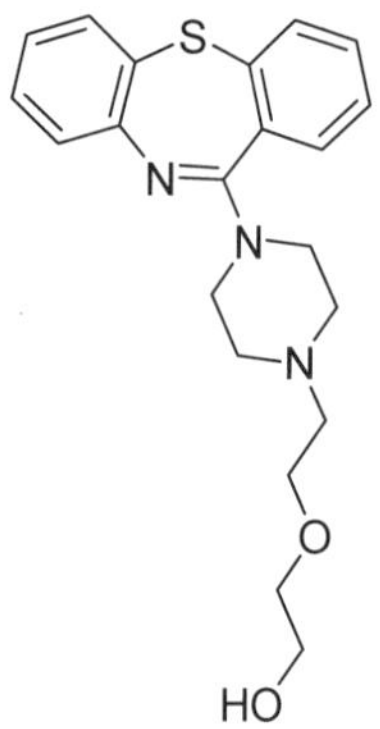

[111974-69-7], 11-[4-[2-(2-Hydroxyethoxy)ethyl]piperazin-1-yl]dibenzo[b,f][1,4]thiazepine $C_{21}H_{25}N_3O_2S$, M_r 383.51; [082935-42-0] hydrochloride salt, $C_{21}H_{26}ClN_3O_2S$, M_r 419.97; [111974-72-2] fumarate salt, $C_{25}H_{29}N_3O_6S$, M_r 499.58

- The compound patent was filed in 1986 by AstraZeneca (Warawa and Migler, 1987)
- Quetiapine fumarate was launched in 1997 by AstraZeneca for the treatment of schizophrenia. In 2004, the product became available for the treatment of acute manic episodes associated with bipolar disorder, as either monotherapy or as adjunct therapy
- Astellas Pharma obtained exclusive marketing rights to the product in Japan in January 1999
- The compound was recently approved from the FDA for bipolar disorders and is being evaluated in phase III clinical trials for the treatment of generalized anxiety, major depression and obsessive-compulsive disorder
- Quetiapine, a dibenzothiazepine antipsychotic, is a dopamine D_2 and 5-HT$_{2A}$ receptor antagonist, as well as histamine H_1 and adrenergic α_1 and α_2 receptors antagonist

Synthesis (Warawa and Migler, 1987): By reaction of dibenzo [b,f][1,4]thiazepine-11(10H)one with refluxing POCl$_3$ and N,N-dimethylaniline the corresponding 11-chlorodibenzo[b,f][1,4]thiazepine compound is obtained. This imino chloride is finally condensed with 4-[2-(2-hydroxyethoxy)ethyl]piperazine by means of NaOH in refluxing xylene to yield quetiapine.

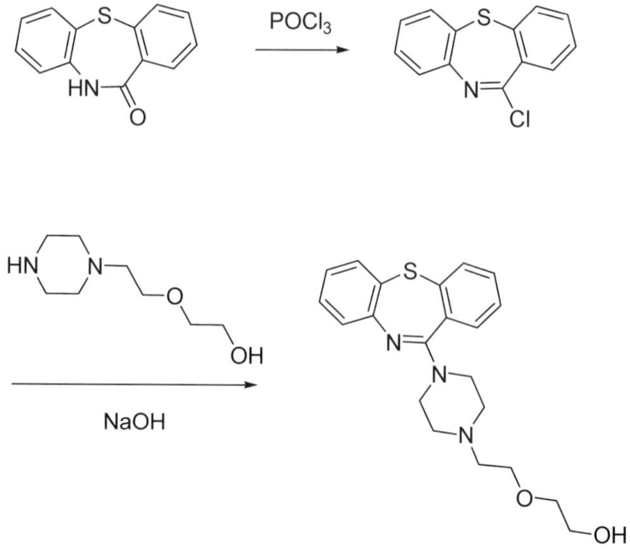

Scheme 42: Synthesis of quetiapine

Clinical Use, Pharmacokinetics and Metabolism: Quetiapine fumarate is a dibenzothiazepine atypical antipsychotic. It is reported to be a serotonin 5-HT_2 and dopamine D_2 receptor antagonist, as well as a histamine H_1 and adrenergic (α_1 and α_2) receptor antagonist. It is used in the treatment of schizophrenia and of mania associated with bipolar disorder (Sajatovic et al., 2001; Vieta et al., 2002; Delbello et al., 2002; Altamura et al., 2003). Recently (in 2006), it received FDA approval as treatment for bipolar disorder as well. For the treatment of schizophrenia, quetiapine fumarate is given p.o. in initial doses of 25 mg twice daily on day one, 50 mg twice daily on day two, 100 mg twice daily on day three, and 150 mg twice daily on day four. The dosage is then adjusted according to the response up to a usual range of 300–450 mg daily, divided into three to four individual doses. Maximum recommended dose is 750 mg per day. A systematic review concluded that the short-term benefits of quetiapine for positive and negative symptoms of schizophrenia quetiapine were comparable to classic antidepressants, but side-effects such as extrapyramidal ones were low (comparable to placebo) (Srisurapanont et al., 2004). Quetiapine has also been used as antipsychotic treatment in patients with Parkinsonism. Quetiapine shares some of the side-effects of classic antipsychotics, but intensity and severity may vary and seem less pronounced. Most frequent adverse effects include non-severe, minor ones such as somnolence, dizziness, constipation, orthostatic hypotension, dry mouth, and elevated liver enzyme values. Quetiapine has been associated with a very low incidence of extrapyramidal adverse effects. In addition, as for many other antipsychotics, effects on carbohydrate metabolism (risk of glucose intolerance and diabetes mellitus requires monitoring of hyperglycaemia during long-term therapy of risk patients) and experiences concerning overdosage (Pollak and Zbuk, 2000; Beelen et al., 2001; Balit et al., 2003) have been reported. Quetiapine may – similarly to clozapine – enhance the effect of other CNS depressants including alcohol. Quetiapine dose needs adjustment (lower dose) when given together with drugs which are inhibitors of cytochrome P450 isoenzyme CYP3A4, e.g. erythromycin, fluconazole or ketoconazole. Quetiapine should not be combined with drugs prolonging the Q–T interval. Quetiapine has a favorable PK profile and is well absorbed after oral administration and distributed throughout the body. Peak concentrations are reached after 1.5 hours. It is about 83% plasma protein bound. Quetiapine is

Ziprasidone

Ziprasidone

[146939-27-7], 5-[2-[4-(1,2-Benzisothiazol-3-yl)piperazin-1-yl]ethyl]-6-chloroindolin-2-one,
$C_{21}H_{21}ClN_4OS$, M_r 412.94;
[138982-67-9], hydrochloride hydrate,
$C_{21}H_{22}Cl_2N_4OS$, M_r 467.42;
[118289-78-4], hemihydrate,
$C_{21}H_{22}ClN_4O_{1.5}S$, M_r 421.95;
[122883-93-6] monohydrate,
$C_{21}H_{23}ClN_4O_2S$, M_r 430.96;
[199191-69-0] methanesulfonate trihydrate
$C_{22}H_{31}ClN_4O_7S_2$, M_r 563.11.

- The compound patent was filed in 1987 by Pfizer (Lowe and Nagel, 1988)
- Ziprasidone hydrochloride was initially launched in Sweden by Pfizer in 2000 for the treatment of schizophrenia. In 2003, the drug was commercialized for the additional indication of treatment of acute bipolar mania, including manic and mixed episodes
- Ziprasidone is a substituted benzothiazolyl-piperazine
- Ziprasidone is a dopamine D_2, 5-HT_{2A}, 5-HT_{2C} and 5-HT_{1D} receptor antagonist
- An injectable form of ziprasidone, as a mesilate salt, was approved by FDA in 2002 for the rapid control of agitated behavior and psychotic symptoms

Synthesis (Lowe and Nagel, 1988; Bowles, 1993): Wolff–Kishner reduction of 6-chloroisatin gives 6-chlorooxindole, which is treated with chloroacetyl chloride under Friedel-Crafts conditions to yield 5-chloroacetyl-6-chlorooxindole. This ketone is reduced using triethylsilane in trifluoroacetic acid to produce 6-chloro-5-(2-chloroethyl)oxindole. 1,2-Benzisothiazolin-3-one is converted to 3-chloro-1,2-benzisothiazole using phosphorus oxychloride and is then condensed with piperazine to provide 1-(1,2-benzisothiazol-3-yl)piperazine. Finally, intermediate 1-(1,2-benzisothiazol-3-yl)piperazine is alkylated by 6-chloro-5-(2-chloroethyl)oxindole in the presence of sodium carbonate in water and is converted into its salt with aqueous hydrochloric acid.

Clinical Use, Pharmacokinetics and Metabolism: Ziprasidone has affinity for adrenergic (α_1), histamine (H_1), serotonin (5-HT_{2A}, 5-HT_{2C}, and 5-HT_{1D} antagonist) and dopamine (D_2 antagonist) receptors. Ziprasidone is given as hydrochloride salt p.o. in initial doses of 20 mg b.i.d. with food (antipsychotic treatment). Doses may be increased stepwise in intervals of no less than 2 days up

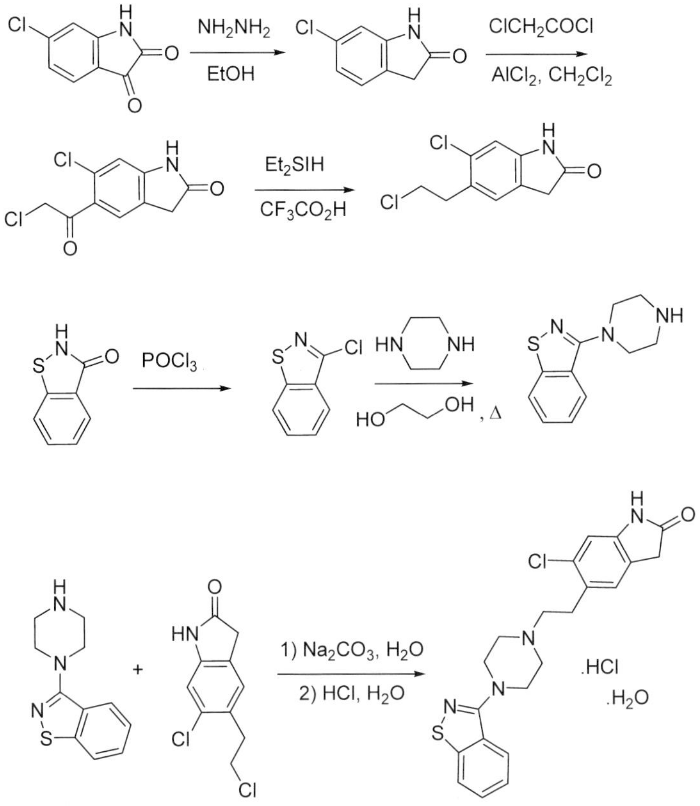

Scheme 43: Synthesis of ziprasidone

to 80 mg twice daily, if necessary. The mesylate salt of ziprasidone is used for intramuscular injection. Ziprasidone has shown efficacy in bipolar disorder (but may be associated with the induction of mania and hypomania) (Keck et al., 2003a; Baldassano et al., 2003), schizophrenia [as effective as haloperidol (p.o.) and superior to haloperidol (intramuscular application), with low incidence of extrapyramidal effects, but it seems to cause more nausea and vomiting] (Brook et al., 2000; Bagnall et al., 2004), and some efficacy in Tourette's syndrome (Sallee et al., 2000). Adverse effects are similar to those described for classic antipsychotics, but incidence and severity may vary. Frequent adverse effects include somnolence, rash or urticaria, gastrointestinal disturbances, dizziness, flu-like symptoms, hypertension, headache, agitation, confusion and dyspnoea. Orthostatic hypotension may be a problem, especially in the beginning of the therapy. Ziprasidone has been infrequently associated with sexual dysfunction and sometimes with weight gain. It is known to prolong the Q–T interval in a dose-dependent manner, possibly resulting in life-threatening arrhythmias like torsade de pointes and sudden death, so that it is contraindicated in patients with a history of Q–T prolongation or with arrhythmias, with acute myocardial infarction, or with decompensated heart failure. Concomitant use with drugs known to interfere with the metabolism of ziprasidone (cytochrome P450 isoenzyme CYP3A4 substrates and/or inhibitors) and drugs known to prolong the Q–T interval or to cause electrolyte imbalance should be avoided. Baseline serum potassium and magnesium screening should be performed for all patients starting ziprasidone treatment. Ziprasidone has, as quetiapine and other antipsychotics, an increased risk of glucose tolerance and diabetes mellitus. Ziprasidone is well absorbed following oral administration, with peak plasma concentrations being reached after 6–8 hours (1 hour after intramuscular injection). The presence of food doubles absorption (food effect). Plasma protein binding is high (99%). Terminal half-life is reported to be around 7 hours. Ziprasidone is excreted predominantly in form of metabolites in faeces (66%) and urine (20%) (Various, 2000). Ziprasidone undergoes extensive metabolism in both rats and humans after oral administration. At least twelve metabolites have been identified from serum, urine, bile and faeces of rats and humans. Four major and two minor routes of metabolism were identified. The four major routes include: (1) *N*-dealkylation of the ethyl side chain attached to the piperazinyl nitrogen (forming cleavage products M5 and OX-CHO), (2) oxidation at sulfur resulting in the formation of sulfoxide and sulfone (M10 and M8), (3) reductive cleavage of the benzisothiazole moiety (the formation of the intermediate dihydroziprasidone), and (4) hydration of the C=N bond and subsequent ring opening or *N*-dearylation of the benzisothiazole moiety (formation of M4A and M7 from M10). M5 undergoes further *S*- and *O*-oxidation forming metabolites M2, M1 and M3, while OX-CHO undertakes further oxidation and conjugation forming oxindole-acetic acid M4 and its glucuronide M3A. The intermediate dihydroziprasidone undergoes *S*-methylation and *S*-oxidation forming M9 and M6. Two minor routes only detected in rats involve *N*-oxidation on the piperazine ring (forming ziprasidone *N*-oxide in faeces) and hydrolysis of the oxindole moiety (forming M12 in rat bile). M1 and M9 are major serum metabolites in humans, whereas the main urinary metabolites were identified as

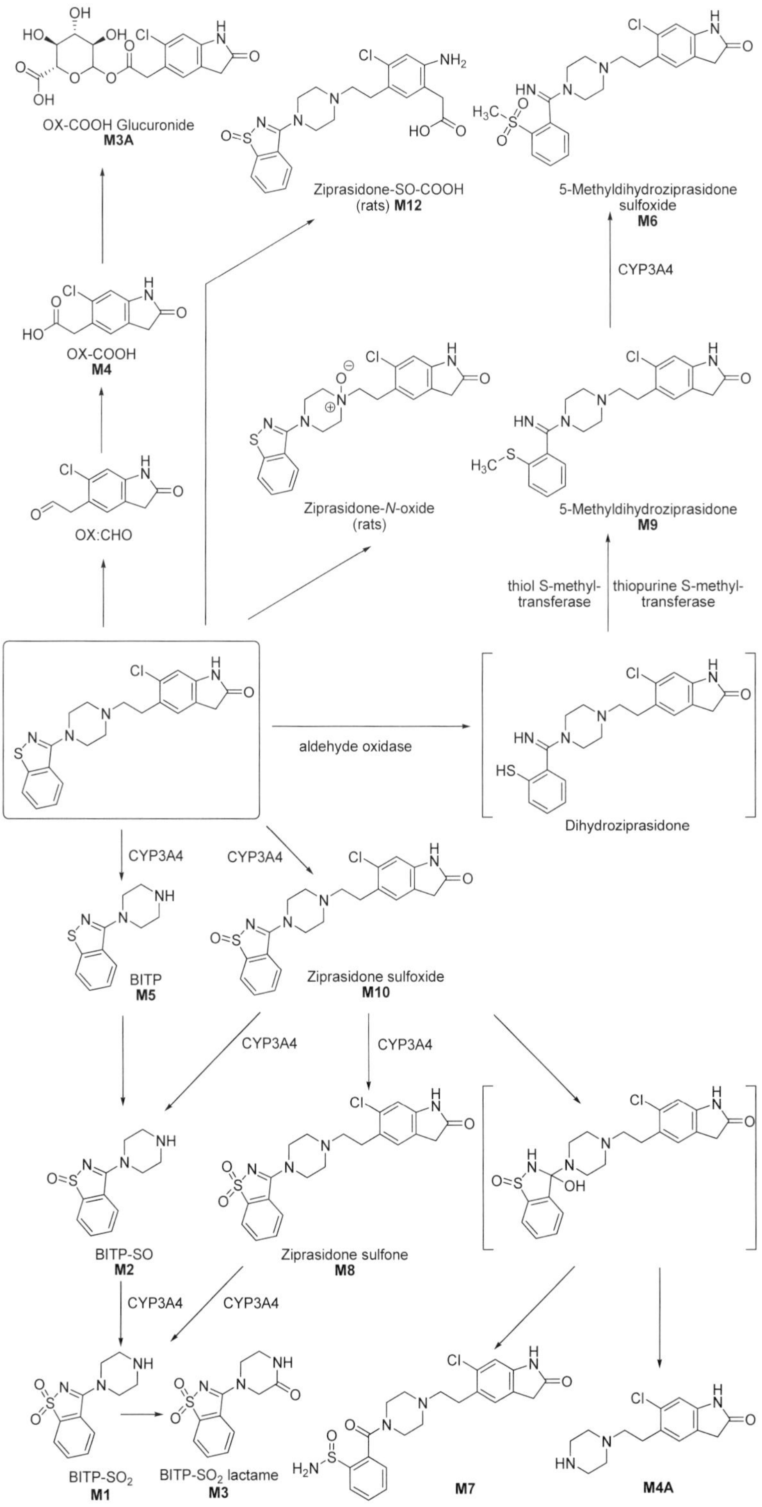

Scheme 44: Metabolism of ziprasidone

M9, M6, M4, M3A and M1. A similar metabolic profile is observed in rat urine, while in rat bile M6, M9 and M7 are the main components, and in rat faeces unchanged ziprasidone is the principal component detected. Ziprasidone is excreted through urine and faeces. Studies in human liver homogenate (S-9) and liver microsomes indicate that cytosolic aldehyde oxidase and thio S-methyltransferase are responsible for the reduction and S-methylation from ziprasidone to M9, whereas CYP3A4 is responsible for the S-oxidation of ziprasidone and its metabolites (Prakash et al., 1997a, b, c; Beedham et al., 2003; Miao et al., 2005).

Perospirone

Perospirone

[150915-41-6], cis-2-[4-[4-(1,2-Benzisothiazol-3-yl)-1-piperazinyl]butyl]octahydro-1H-isoindole-1,3-dione, $C_{23}H_{30}N_4O_2S$, M_r 463.04; [129273-38-7] hydrochloride salt, $C_{23}H_{31}ClN_4O_2S$, M_r 499.50

- The compound patent was filed in 1985 by Sumitomo Pharmaceuticals Co., Ltd. (Ishizumi et al., 1986)

- The compound was launched in Japan on February 2001 for the treatment of schizophrenia

- Perospirone is a D_2 and 5-HT_{2A} antagonist

Synthesis (Ishizumi et al., 1986): The reaction of cis-octahydroisobenzofuran-1,3-dione with ammonia in water at 180–190°C gives cis-octahydro-1H-isoindole-1,3-dione, which is alkylated with 1,4-dibromobutane and K_2CO_3 in refluxing acetone to yield cis-2-(4-bromobutyl)octahydro-1H-isoindole-1,3-dione. The condensation of cis-2-(4-bromobutyl)octahydro-1H-isoindole-1,3-dione with piperazine in refluxing toluene affords cis-2-(4-piperazinobutyl)octahydro-1H-isoindole-1,3-dione, which is finally condensed with 3-chloro-1,2-benzisothiazole and K_2CO_3 in refluxing toluene.

This compound can also be obtained by a related way, in which the condensation of 3-chloro-1,2-benzisothiazole with piperazine by heating at 120°C gives 3-piperazino-1,2-benzisothiazole, which is then condensed with cis-2-(4-bromobutyl)octahydro-1H-isoindole-1,3-dione by means of K_2CO_3 and KI in hot DMF.

Clinical Use, Pharmacokinetics and Metabolism: Perospirone is an atypical antipsychotic used in the treatment of schizophrenia. It is a D_2 and a 5-HT_{2A} antagonist. It is given in form of its hydrochloride salt p.o. in usual doses ranging from 12–48 mg daily, given in 3 divided doses (Onrust and McClellan, 2001). Steady-state

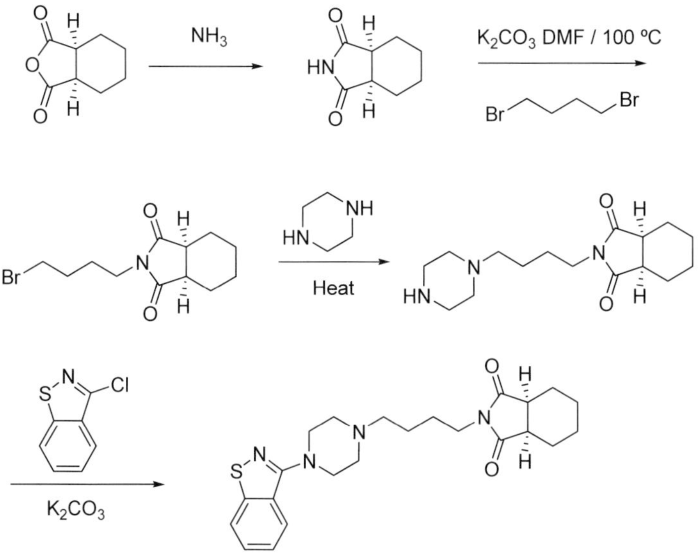

Scheme 45: Synthesis of perospirone

pharmacokinetics of perospirone and its active metabolite hydroxyperospirone was evaluated in schizophrenic patients receiving 16 mg twice daily. The peak plasma concentration of perospirone occurs at 0.8 hours after administration, and the half-life is 1.9 hours. The plasma concentration of hydroxyperospirone is higher than that of the parent compound, and it reaches maximum concentration 1.1 hours after administration (Yasui-Furukori et al., 2004).

Aripiprazole

- The compound patent was filed in 1988 by Otsuka Pharmaceutical Co., Ltd. (Oshiro et al., 1990)
- The product was licensed to Bristol-Myers Squibb as a result of a development, commercialization and collaboration agreement signed in 1999
- The compound was launched in 2002 in USA for the treatment of schizophrenia. The drug has since been marketed by the companies for the treatment of acute bipolar disorder and mania, including manic and mixed episodes associated with bipolar disorder
- Aripiprazole is a quinolinone derivative and the first of a new class of atypical antipsychotics. It is a dopamine

Aripiprazole

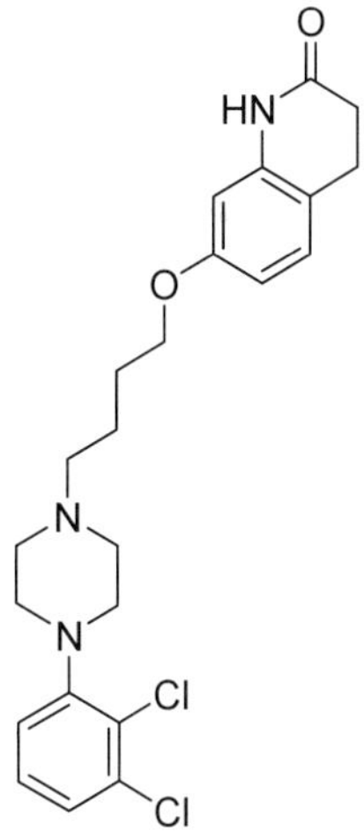

[129722-12-9], 7-[4-[4-(2,3-dichlorophenyl)piperazin-1-yl]butoxy]-3,4-dihydro-1H-quinolin-2-one, $C_{23}H_{27}Cl_2N_3O_2$, M_r 448.38; [129722-13-0] hydrochloride salt, $C_{23}H_{28}Cl_3N_3O_2$, M_r 484.84; [129722-15-2] fumarate salt, $C_{27}H_{31}Cl_2N_3O_6$, M_r 564.45; [129722-16-3] maleate salt, $C_{27}H_{31}Cl_2N_3O_6$, M_r 564.45

D$_2$ and serotonin 5HT$_{1A}$ partial agonist and a serotonin 5HT$_{2A}$ antagonist

Synthesis (Oshiro et al., 1990; Oshiro et al., 1998): The condensation of 7-hydroxy-3,4-dihydro-1*H*-quinolin-2-one with 1,4-dibromobutane by means of K$_2$CO$_3$ in hot DMF gives 7-(4-bromobutoxy)-3,4-dihydro-1*H*-quinolin-2-one, which is then condensed with 1-(2,3-dichlorophenyl)piperazine by means of NaI and triethylamine in refluxing acetonitrile. The final product is extracted from the residue with ethyl acetate.

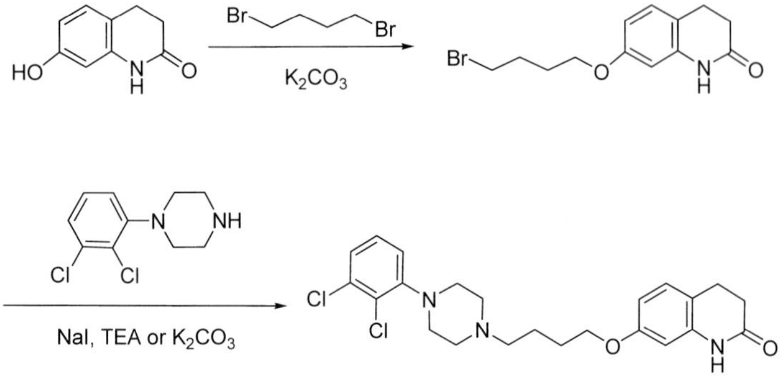

Scheme 46: Synthesis of aripiprazole

Clinical Use, Pharmacokinetics and Metabolism: Aripiprazole is a recently developed atypical antipsychotic used in the treatment of schizophrenia and bipolar disorder. It is a serotonin 5-HT$_{1A}$ receptor partial agonist and a 5-HT$_{2A}$ antagonist, as well as a partial agonist at the dopamine D$_2$ receptor. It is given p.o. in initial doses of 10 or 15 mg once daily. The dose can be adjusted at intervals of no less than 2 weeks up to a maximum of 30 mg daily (McGavin and Goa, 2002; Goodnick and Jerry, 2002; Taylor, 2003; Keck and McElroy, 2003; Bowles and Levin, 2003; Keck et al., 2003b). Adverse effects are similar to those described for classic antipsychotics, but incidence and severity may vary. Common adverse effects along with aripiprazole treatment comprise headache, gastrointestinal disorders such as constipation, nausea and vomiting, anxiety, insomnia, lightheadedness, and drowsiness. Slight weight gain has been reported. The incidence of extrapyramidal adverse effects is low (Marder et al., 2003). Aripiprazole has, like other antipsychotics, an increased risk of glucose tolerance and diabetes mellitus. Aripiprazol may – similarly to clozapine – enhance the effect of other CNS depressants including alcohol. As aripiprazole is mainly metabolized by cytochrome P450 isoenzymes CYP3A4 and CYP2D6, it should not be combined with drugs known to be substrates and/or inhibitors of these isoenzymes, like e.g. ketoconazole, unless the dose is adjusted (lowered). Aripiprazole is well absorbed from the GI tract after oral administration, leading to plasma peak concentrations after 3 to 5 hours. It is metabolized in the liver using dehydrogenation and hydroxylation phase I metabolic reactions (CYP3A4 and CYP2D6). Further metabolic mechanisms include N-dealkylation (CYP3A4 mediated). The major metabolite is dehydro-aripiprazole, which is active and represents about 40% of the plasma levels of aripiprazole. Elimination half lives of aripiprazole and dehydro-aripiprazole are very long, about 75 and 95 hours, respectively. Protein binding is high for aripiprazole and dehydro-aripiprazole (99%). Excretion occurs mainly in faeces (55%), and to a lesser degree (25%) in urine (mainly in the form of metabolites).

Sertindole

- The compound patent was filed by Lundbeck in 1985. Sertindole was first launched in the U.K. in 1996 for the treatment of schizophrenia (Perregaard, 1987)

- Sertindole was withdrawn from the market on December 1998 due to concerns over the risk of cardiac arrhythmia and sudden death

- Following the suspension of the marketing authorization, Lundbeck included an additional 5,000 patients in a study confirming that the product could be prescribed safely. In April 2005, the European Committee for Medicinal Products for Human Use (CHMP) lifted the marketing restrictions

- In January 2006, sertindole was approved in the E.U. for the treatment of schizophrenia and launched in Estonia in early 2006

- Sertindole is a phenylindole derivative. The compound is active on central dopamine D_2 and serotonin $5HT_2$ receptors. In addition it has an inhibitory effect on α-adrenergic receptors

Synthesis (Perregaard, 1987): Acid-catalyzed condensation of 4-piperidone with 5-chloro-1-(4-fluorophenyl)-1H-indole produces the tetrahydropyridyl indole. This is

Sertindole

[106516-24-9], 5-Chloro-1-(4-fluorophenyl)-3-[1-(2-oxoimidazolidin-1-ylethyl)piperidin-4-yl]-1H-indole 1-[2-[4-[5-Chloro-1-(4-fluorophenyl)-1H-indol-3-yl]piperidin-1-yl]ethyl]imidazolidin-2-one, $C_{24}H_{26}ClFN_4O$, M_r 440.94; monomaleate [106516-25-0], $C_{28}H_{30}ClFN_4O_5$, M_r 557.01

alkylated with 1-(2-chloroethyl)imidazo-linone (prepared by chlorination of the corresponding hydroxyethyl imidazolinone) to afford the *N*-substituted tetrahydropyridine. The target piperidine derivative is obtained by catalytic hydrogenation of tetrahydropyridine.

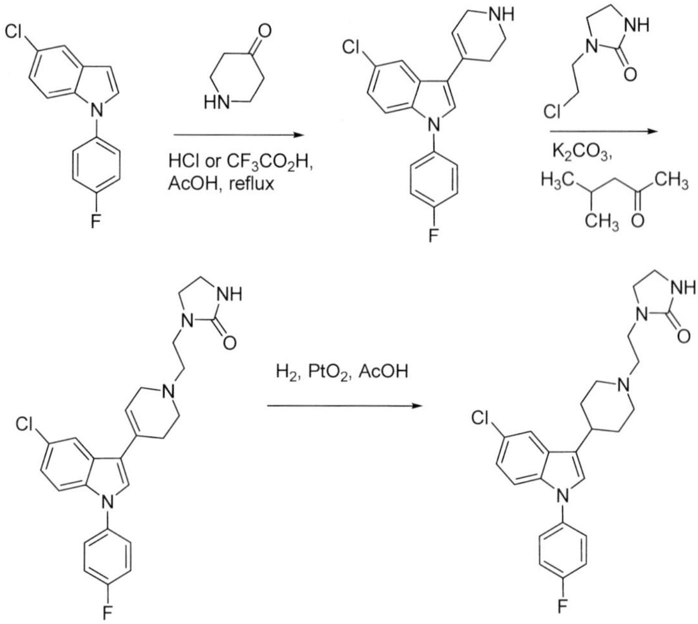

Scheme 47: Synthesis of sertindole

Clinical Use, Pharmacokinetics and Metabolism: Sertindole is an atypical antipsychotic with central dopaminergic (D_2 antagonist), serotoninergic (5-HT_2 antagonist) and adrenergic (α_1 antagonist) activity. It is used in the treatment of schizophrenia in patients who are intolerant towards other antipsychotics. Sertindole should only be prescribed along with adequate cardiac monitoring, especially by regular ECG measurements (Barnett, 1996; Committee on Safety of Medicines/Medicines Control Agency, 1999b; Committee on Safety of Medicines/Medicines Control Agency, 2002b). It is given p.o. in an initial dose of 4 mg once daily, which then is gradually increased in 4 mg intervals until about 12–20 mg once daily as usual maintenance dose. Maximum p.o. dose recommended is 24 mg per day. Adverse effects are similar to those described for classic antipsychotics, but incidence and severity may vary. Thus, in comparison to e.g. chlorpromazine, extrapyramidal effects and sedation, as well as prolactin elevation seem to occur less frequently. Adverse effects along with sertindole treatment comprise peripheral oedemia, rhinitis, dyspnoea, sexual dysfunction, dizziness, dry mouth, orthostatic hypotension, weight gain, and parasthesia. The marketing and distribution of sertindole has been restricted because of cardiac arrhythmias and sudden death associated with its use. Sertindole is associated with a dose-dependent prolongation of the Q–T interval, possibly leading to life-threatening arrhythmias like torsade de pointes and sudden death, especially within the first 3 to 6 weeks of treatment. ECG measurement should be performed before the start of the therapy, and periodically during treatment. Sertindole is contraindicated for patients with a history of Q–T interval prolongation, arrhythmias, cardiovascular diseases, heart failure, cardiac hypertrophy, bradycardia, or a family history of congenital Q–T prolongation. The risk of arrhythmias may be increased by concomitant use of sertindole with other drugs which prolong the Q–T interval or by concomitant use of sertindole with drugs which are inhibitors and/or substrates of the cytochrome P450 isoenzymes CYP3A4 and CYP2D6 (e.g., macrolide antibiotics, calcium-channel blockers, fluoxetine or paroxetine), the main metabolic enzymes of sertindole. Sertindole is slowly absorbed from the GI tract and reaches peak concentrations about 10 hours after ingestion. It is about 99.5% plasma protein bound and readily crosses the placenta. Poor metabolizers (lack of CYP2D6) may have plasma concentrations 2 to 3 times higher than other patients. Sertindole and its metabolites are excreted slowly mainly in faeces, and to a lesser degree in urine. Mean terminal half-life is about 3 days. A total of 13 metabolites of sertindole have been identified in the rat, monkey, dog and man. Sertindole is mainly metabolized by hydroxylation at the 4- and 5-positions on the imidazolidinone ring, *N*-dealkylation, and a NIH shift at the fluorophenyl group forming metabolites 4-hydroxysertindole (Lu-30-148), 5-hydroxysertindole (Lu-30-131), norsertindole (Lu-25-073), and 3'-fluoro-4'-hydroxysertindole (Lu-31-096), respectively. Both Lu-30-131 and Lu-30-148 undertake further oxidative reactions (dehydration, oxidation and hydroxylation) to give dehydrosertindole (Lu-28-092), 5-oxosertindole (Lu-25-126), 4-oxosertindole (Lu-28-115), 5-hydroxy-4-oxosertindole (Lu-31-024) and 4,5-dihydroxysertindole (Lu-32-035). Dehydrosertindole also undergoes a NIH shift at the fluorophenyl group forming 3'-fluoro-4'-hydroxydehydrosertindole (Lu-31-154). Glucuronidation and/or sulfation post the NIH shift was observed in Lu-31-

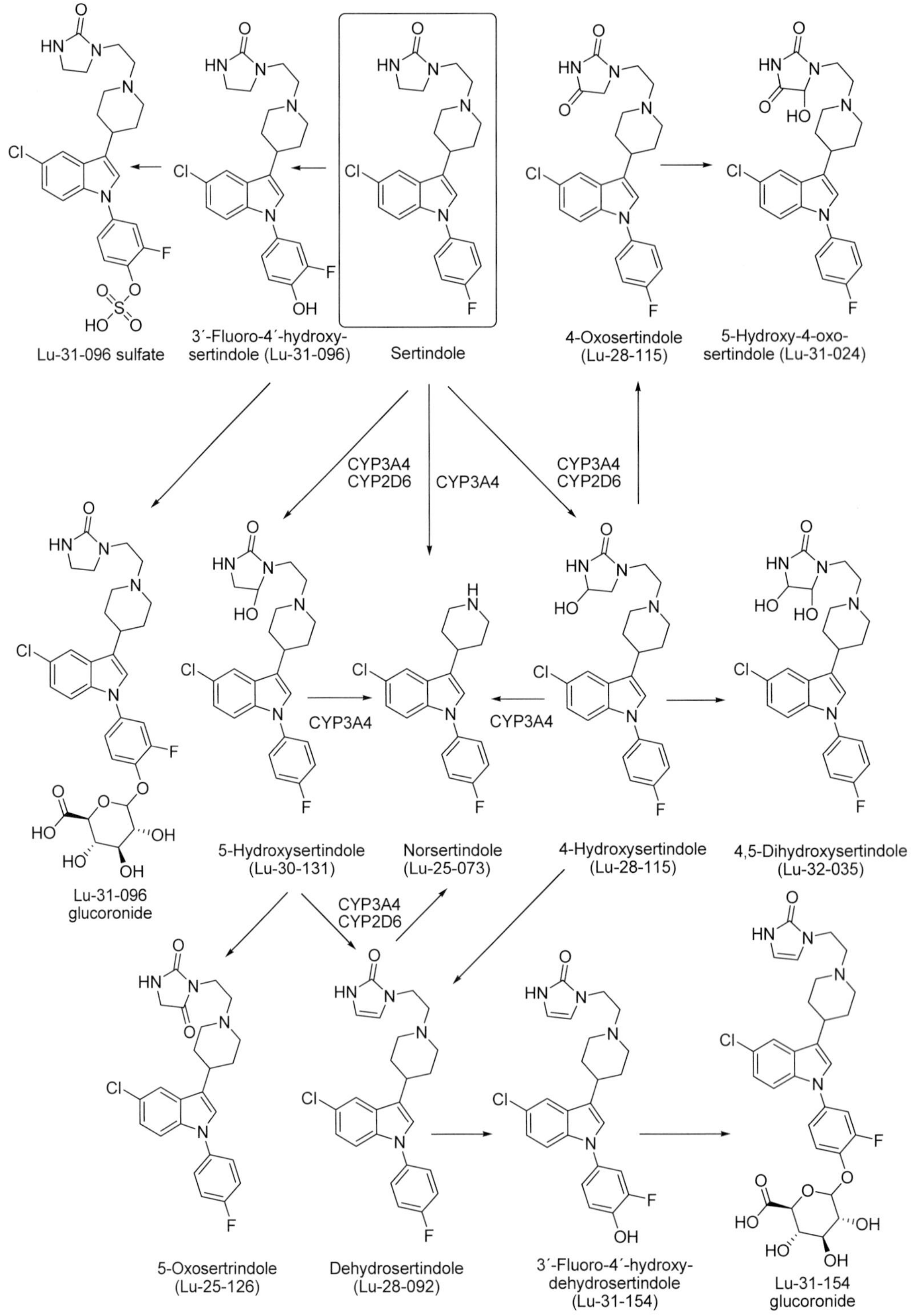

Scheme 48: Metabolism of sertindole

096 and Lu-31-154. Similar metabolic patterns were found in rat, monkey and man, with 4- and 5- hydroxylation and N-dealkylation being the main metabolic pathways. However, in dog, 4-hydroxylation and the NIH shift at the fluorophenyl group followed by conjugation are the main metabolic reactions. Inhibition studies and studies with characterized microsomes showed that the formations of dehydrosertindole and its transiently present precursor hydroxy-metabolites are mediated by both CYP2D6 and CYP3A4, whereas N-dealkylations from sertindole or its metabolites are catalyzed only by CYP3A4 (Sakamoto et al., 1995; Wong et al., 1997).

2.5.3.3 Atypical antipsychotics with dopaminergic activity only (D_2 and D_3 antagonists)

Amisulpride

- The compound patent was filed in 1978 by Société d'Etudes Scientifiques et Industrielles de L'Ile de France (Thominet et al., 1981), and has been developed and launched by Synthelabo (now Sanofi-Aventis) for the treatment of schizophrenia and dysthymia

- The compound has been launched in 1986 for schizophrenia in France, Portugal and Italy, and is marketed in 43 countries in Europe as well as in Asia, Africa, and Latin America

- By 1999 a clinical development program was planned in the USA and phase III clinical trials of the compound had been initiated. In 2000, the development of amisulpride in the USA was discontinued because patent coverage expired in 2002/3 and the FDA had requested further filing information

- Amisulpride is a substituted benzamide antipsychotic, structurally related to sulpiride. It is a dopamine D_2 and D_3 receptor antagonist

Amisulpride

[071675-85-9], 4-Amino-N-(1-ethyl-2-pyrrolidinylmethyl)-5-(ethylsulfonyl)-2-methoxybenzamide, $C_{17}H_{27}N_3O_4S$, M_r 369.48

Synthesis (Thominet et al., 1981; Acher and Monier, 1981): 2-Methoxy-4-amino-5-mercaptobenzoic acid is treated with diethyl sulfate and Na_2CO_3 to give 2-methoxy-4-amino-5-ethylthiobenzoic acid, which is oxidized with H_2O_2 in acetic acid to provide 2-methoxy-4-amino-5-(ethylsulfonyl)benzoic acid. Finally, this compound is condensed with *N*-ethyl-2-aminomethylpyrrolidine by means of ethyl chloroformate.

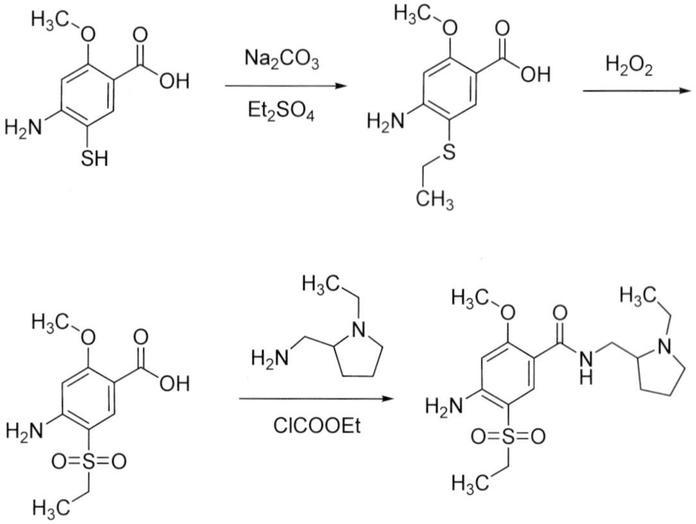

Scheme 49: Synthesis of amisulpride

Clinical Use, Pharmacokinetics and Metabolism: Amisulpride is a substituted benzamide atypical antipsychotic, structurally related to sulpiride. It has a high affinity to dopamine D_2 and D_3 receptors (antagonist at both receptors). It is used predominantly in the management of psychoses like schizophrenia, and in some countries in depression. Initial doses for antipsychotic treatment are between 400 mg and 800 mg daily (given in 2 divided doses), which can be increased according to response up to a maximum of 1200 mg daily (Boyer et al., 1995; Möller et al., 1997; Loo et al., 1997; Lecrubier et al., 1997; Smeraldi, 1998; Boyer et al., 1999; Curran and Perry, 2001). Adverse effects are similar to those described for classic antipsychotics, but incidence and severity may vary. Common adverse effects along with amisulpride treatment comprise insomnia, anxiety and agitation. Less common side-effects include drowsiness, gastrointestinal disorders such as constipation, nausea, vomiting, and dry mouth. Allergic reactions and abnormal liver function tests have been reported rarely. Amisulpride has been associated with Q–T interval prolongation, possibly resulting in life-threatening arrhythmias like torsade de pointes and sudden death, so that it is not recommended in patients with a history of Q–T prolongation or with arrhythmias, with acute myocardial infarction, or with decompensated heart failure. Co-administration with drugs known to prolong the Q–T interval (e.g., cisapride, erythromycin, halofantrin, and thioridazine) or to cause arrhythmias or to produce bradycardia or hypokalaemia (e.g., β-blockers, some calcium-channel blockers, clonidine, digoxin, guanfacin, haloperidol, TCAs etc.) should be avoided. Experiences with overdosage have been described (Tracqui et al., 1995). Amisulpride may – similarly to clozapine and other antipsychotics – enhance the effect of other CNS depressants including alcohol. Amisulpride is absorbed from the GI tract after oral ingestion with reported bioavailabilities of around 48%. An initial peak concentration has been reported to occur 1 hour after application (p.o.), and a second after 3–4 hours. Plasma protein binding is low (16%) and metabolism is limited, with most of the dose being excreted in unchanged form in urine and faeces. Terminal half-life is reported to be about 12 hours.

2.5.4 Drugs in development

A large number of compounds are currently under different development phases. In Table 3 is detailed the code, structures (only compounds with known structure are depicted in this summary) and proposed mechanisms described in different drug databases (Integrity: http://integrity.prous.com, IDDB3: http://www.iddb3.com). Some mechanisms are already known from marketed compounds as in the case of dopamine and/or serotonin receptors antagonist, but others are novel as for instance action on metabotropic glutamate or nicotinic acetylcholine receptors.

Several dual-acting dopamine D_2 and $5HT_{2A}$ antagonists are in advanced stages of development: two in pre-registration (blonanserin, Dainippon; paliperidone, Johnson & Johnson), four in phase III (asenapine, Organon/Pfizer; bis(olanzapine) pamoate, Lilly; iloperidone, Titan/Vanda; paliperidone palmitate, Johnson & Johnson), two in phase II (lurasidone, Dainippon; ocaperidone, Neuro3d) and one in phase I (abaperidone, Ferrer).

Other agents with activity on only dopamine or serotonin receptors, as well as combinations of those receptors, have also been described to be in clinical phases: dopamine receptors (ACR-16, Carlsson Research, Phase II); dopamine D_2 antagonists (YKP-1358, SK Bio-

Pharmaceuticals, phase I); dopamine D_2/D_3 antagonists (RGH-188, Forest/Gedeon, phase II); dopamine/serotonin antagonists (SB-773812, GlaxoSmithKline, phase II); dopamine D_2 antagonists/5-HT reuptake inhibitors (SLV-314, Solvay/Wyeth, phase I); dopamine D_2 antagonists/5-HT_{1A} agonists (bifeprunox, Lundbeck/Solvay/Wyeth, phase III); 5-HT_{2A} inverse agonists (ACP-103, Acadia, phase II); 5-HT_{2C} agonists (vabicaserin, Wyeth, phase II); 5-HT_3 antagonists (ondansetron, National Institute of Mental Health, phase II); 5HT_6 antagonists (SGS-518, Lilly/Saegis, phase II, cognitive enhancement associated with schizophrenia).

Talnetant is the unique representative in clinical trials as member of the NK_3 antagonist family (GlaxoSmithKline, phase II).

Drugs acting as agonists, antagonists and modulators of metabotropic glutamate receptors have also been developed: NMDA glycine-site modulators (XY-2401, Xytis, phase I); NMDA glycine-site modulators/norepinephrine reuptake inhibitos (neboglamine, Rotapharm, phase I); GlyT-1 inhibitors (SSR-504734, Sanofi-Aventis, phase I; ORG-25935, Organon, phase I); AMPA receptor modulators (farampator, Organon, phase II); $mgluR_2/mgluR_3$ agonists (LY-2140023 and LY-404039, Lilly, phase II); glutamate release inhibitors/sodium channel blockers (lamotrigine, GlaxoSmithKline, phase III).

Drugs acting on nicotininc acetylcholine receptors have also been proposed for the treatment of schizophrenia and/or associated cognitive dysfunction, as DMXB-anabaseine, a nicotinic $α_7$ agonist (Athenage, phase II) and MEM-3454, a nicotinic $α_7$ partial agonist (Memory Pharmaceuticals, phase I).

A group of miscellaneous agents has been described with alternative mechanisms as: cannabinoid receptor agonists (cannabidiol, GW Pharmaceuticals, phase I), CB_1 antagonists (AVE-1625, Sanofi-Aventis, phase II), COX-2 inhibitors (644784, GlaxoSmithKline, phase I), antioxidants/AGE inhibitors (carnosine, University of Pittsburgh, Phase II), muscarinic M1 agonist (norclozapine/N-desmethylclozapine, Acadia, phase II) and $α_1$ adrenoceptor agonists (modafinil, National Institute of Mental Health, phase II).

Finally, other compounds without a clear defined mechanism of action in clinical trials are PF-00184562 (Pfizer, phase I), BL-1020 (Bar-Llan University/BiolineRx/Ramot, phase I), Lu-31-130 (Lundbeck, phase I), PW-4123 (Penwest, Phase I), R-1678 (Roche, phase I), CRD-101 (Curidium Ltd., phase I); and mifepristone (Corcept, phase III).

Lamotrigine, ondansetron, modafinil and mifepristone (described above) were previously launched for different indications.

Table 3. Antipsychotic drugs in development by clinical phase

Code or name	Mechanism of action	Structure
Phase I		
Abaperidone	5-HT$_{2A}$ / Dopamine D$_2$ antagonists	
YKP-1358	Dopamine D2 antagonists	Not Known
SLV-314	5-HT reuptake inhibitors; Dopamine D$_2$ antagonists	
SSR-504734	GlyT-1 inhibitors	
ORG-25935	GlyT-1 inhibitors	

Neboglamine	NMDA Glycine-site Modulators / Norepinephrine reuptake inhibitors	*(structure: glutamic acid amide with 4,4-dimethylcyclohexylamine)*
PF-00184562	Not known	*(structure: 5-chloro-1-(4-fluorophenyl)indole with piperidine-ethyl-imidazolidinone substituent)*
XY-2401	NMDA Glycine site modulators	Not known
MEM-3454	Nicotinic α$_7$ partial agonists	Not known
644784	Cyclooxygenase-2 inhibitors	Not known
CRD-101	CNS modulator	Not known
Lu-31-130	Not known	Not known
BL-1020	Not known	Not known
PW-4123	Not known	Not known
R-1678	Not known	Not known
Phase II		
Ocaperidone	5-HT$_{2A}$ / Dopamine D$_2$ antagonists	*(structure: pyrido-pyrimidinone linked via ethyl to piperidine-benzisoxazole, fluorinated)*
Lurasidone	5-HT$_{2A}$ / Dopamine D$_2$ antagonists	*(structure: norbornane-dicarboximide-methyl cyclohexyl-methyl piperazinyl benzisothiazole)*
ACR-16	Dopamine receptors	*(structure: 1-propyl-4-[3-(methylsulfonyl)phenyl]piperidine)*

RGH-188	Dopamine D$_2$/D$_3$ antagonists	Not known
SB-773812	Dopamine/Serotonin receptor antagonists	Not known
SLV-313	5-HT$_{1A}$ agonists / Dopamine D$_2$ antagonists	
ACP-103	5-HT$_{2A}$ inverse agonists	
Vabicaserin	5-HT$_{2C}$ agonists	
Ondansetron	5-HT$_3$ antagonists	
SGS-518	5-HT$_6$	
Talnetant	Tachykinin NK$_3$ antagonists	
Farampator	AMPA receptor modulators	

LY-2140023	mgluR$_2$ / mgluR$_3$ agonists	
LY-404039	mgluR$_2$ / mgluR$_3$ agonists	
DMXB-Anabaseine	Nicotinic α$_7$ agonists	
Cannabidiol	Cannabinoid receptor agonists	
AVE-1625	Cannabinoid CB$_1$ antagonists	
Carnosine	AGE inhibitors / Antioxidants	
Norclozapine	Muscarinic M$_1$ agonists	

Modafinil	alpha₁-adrenoceptor agonists	

Phase III

Asenapine	5-HT₂ / Dopamine D₁ / Dopamine D₂ antagonists	
Iloperidone	5-HT₂ₐ / Dopamine D₂ antagonists	
Bis(olanzapine) pamoate monohydrate	5-HT₂ₐ / Dopamine D₂ antagonists	
Paliperidone palmitate	5-HT₂ₐ / Dopamine D₂ antagonists	
Bifeprunox	5-HT₁ₐ agonists / Dopamine D₂ partial agonists	

Lamotrigine	Glutamate Release Inhibitors / Sodium channel blockers	
Mifepristone	Glucocorticoid receptor antagonists	

Pre-registration

Paliperidone	5-HT$_{2A}$ / Dopamine D$_2$ antagonists	
Blonanserin	5-HT$_{2A}$ / Dopamine D$_2$ antagonists	

References

Acher J and Monier J-C. *Nouveau Procédé de préparation de dérivés de 4-amino 5-alkylsulfonyl ortho-anisamides et nouveaux dérivés de 4-nitro 5-alkylsulfonyl ortho-anisamides utiles comme intermédiaires de synthése.* FR2460930 (**1981**); priority: 1979.

Adams CE, Eisenbruch M. *Depot fluphenazine for schizophrenia.* Available in The Cochrane Library; Issue 2. Chichester: John Wiley. **2004**.

Adler LA, Edson R, Lavori P, Peselow E, Duncan E, Rosenthal M, Rotrosen J. *Long-term treatment effects of vitamin E for tardive dyskinesia.* Biol Psychiatry. **1998**, 43:868-872.

Adnet P, Lestavel P, Krivosic-Horber R. *Neuroleptic malignant syndrome.* Br J Anaesth. **2000**, 85:129-135.

Adverse Drug Reactions Advisory Committee. *Clozapine and myocarditis.* Aust Adverse Drug React Bull **1994**, 13(Nov):14-15.

Allen RL, Walker Z, D'Ath PJ, Katona CL. *Risperidone for psychotic and behavioural symptoms in Lewy body dementia.* Lancet. **1995**, 346:185.

Allison DB, Mentore JL, Heo M, Chandler LP, Cappelleri JC, Infante MC, Weiden PJ. *Antipsychotic-induced weight gain: a comprehensive research synthesis.* Am J Psychiatry. **1999**, 156:1686-1696.

Altamura AC, Salvadori D, Madaro D, Santini A, Mundo E. *Efficacy and tolerability of quetiapine in the treatment of bipolar disorder: preliminary evidence from a 12-month open-label study.* J Affect Disord. **2003**, 76:267-271.

Alvir JM, Lieberman JA, Safferman AZ, Schwimmer JL, Schaaf JA. *Clozapine-induced agranulocytosis. Incidence and risk factors in the United States.* N Engl J Med. **1993**, 329:162-167.

Amdisen A. *Clinical features and management of lithium poisoning.* Med Toxicol Adverse Drug Exp. **1988**, 3:18-32.

American Academy of Pediatrics. *The transfer of drugs and other chemicals into humans milk.* Pediatrics. **2001**; 108:776-89. Correction. ibid.;1029.

American Diabetes Association; American Psychiatric Association; American Association of Clinical Endocrinologists; North American Association for the Study of Obesity. *Concensus development conference on antipsychotic drugs and obesity and diabetes.* Diabetes Care **2004**, 27:596-601.

Anderman B, Griffith RW. *Clozapine-induced agranulocytosis: a situation report up to Agust 1976.* Eur J Clin Pharmacol **1977**, 11:199-201.

Anonymous. *Now we understand antipsychotics?* Lancet. **1990**, 336:1222-1223.

Anonymous. *Clozapine and loxapine for schizophrenia.* Drug Ther Bull **1991**, 29:41-42. Correction. ibid.; 52.

Anonymous. *Olanzapine (Zyprexa) and diabetes.* Current Problems. **2002**, 28:3.

Aono T, Aki T, Koike K, Kurachi K. *Effect of sulpiride on poor puerperal lactation.* Am J Obstet Gynecol. **1982**, 143:927-932.

Aronson JK, Reynolds DJ. *ABC of monitoring drug therapy. Lithium.* BMJ. **1992**, 305:1273-1274.

Asplund K, Hagg E, Lindqvist M, Rapp W. *Phenothiazine drugs and pituitary tumors.* Ann Intern Med. **1982**, 96:533.

Atkin K, Kendall F, Gould D, Freeman H, Liberman J, O'Sullivan D. *Neutropenia and agranulocytosis in patients receiving clozapine in the UK and Ireland.* Br J Psychiatry. **1996**, 169:483-488.

Aubry JM, Simon AE, Bertschy G. *Possible induction of mania and hypomania by olanzapine or risperidone: a critical review of reported cases.* J Clin Psychiatry. **2000**, 61:649-655.

Auzou P, Hannequin D, Landrin I, Cochin JP, Moore N. *Worsening of psychotic symptoms by clozapine in Parkinson's disease.* Lancet. **1994**, 344:955.

Axelsson R, Mårtensson E. *Serum concentration and elimination from serum of thioridazine in psychiatric patients*. Curr Ther Res. **1976**, 19:242-265.

Axelsson R. *On the serum concentrations and antipsychotics effects of thioridazine, thioridazine side-chain sulfoxide and thioridazine side-chain sulfone, in chronic psychotic patients*. Curr Ther Res. **1977**, 21:587-605.

Baandrup U. Muscle. In: Johnson FN, ed. *Depression & mania: modern lithium therapy*. IRL Press, Oxford, **1987**, pp 236-238.

Bagnall A-M, Lewis RA, Leitner ML. Ziprasidone for schizophrenia and severe mental illness. Available in The Cochrane Library; Issue 2. Chichester: John Wiley. **2004**.

Baker RW, Milton DR, Stauffer VL, Gelenberg A, Tohen M. *Placebo-controlled trials do not find association of olanzapine with exacerbation of bipolar mania*. J Affect Disord. **2003**, 73:147-153.

Baldassano CF, Ballas C, Datto SM, Kim D, Littman L, O'Reardon J, Rynn MA. *Ziprasidone-associated mania: a case series and review of the mechanism*. Bipolar Disord. **2003**, 5:72-75.

Baldessarini RJ, Cohen BM, Teicher MH. *Significance of neuroleptic dose and plasma level in the pharmacological treatment of psychoses*. Arch Gen Psychiatry. **1988**, 45:77-91.

Baldessarini RJ, Frankenburg F. *Clozapine: a novel antipsychotic agent*. N Engl J Med **1991**, 324:746-754.

Balit CR, Isbister GK, Hackett LP, Whyte IM. *Quetiapine poisoning: a case series*. Ann Emerg Med. **2003**, 42:751-758.

Banos JE, Bosch F, Farre M. *Drug-induced priapism. Its aetiology, incidence and treatment*. Med Toxicol Adverse Drug Exp. **1989**, 4:46-58.

Barnas C, Bergant A, Hummer M, Saria A, Fleischhacker WW. *Clozapine concentrations in maternal and fetal plasma, amniotic fluid, and breast milk*. Am J Psychiatry. **1994**, 151:945.

Barnes TRE. *Comment on the WHO consensus statement*. Br J Psychiatry. **1990**, 156:413-414.

Barnett AA. *Safety concerns over antipsychotic drug, sertindole*. Lancet **1996**, 348:256.

Barsa JA, Saunders JC. *A comparative study of tranylcypromine and pargyline*. Psychopharmacologia. **1964**, 6:295-298.

Bateman DN, Darling WM, Boys R, Rawlins MD. *Extrapyramidal reactions to metoclopramide and prochlorperazine*. Q J Med. **1989**, 71:307-311.

Beasley C, Tran P, Beuzen JN, Tamura R, Dellva MA, Bailey J, Krueger J, Tollefson G. *Olanzapine versus haloperidol: long-term results of the multi-center international trial*. Eur Neuropsychopharmacol. **1996a**, 6(suppl 3):59.

Beasley CM, Jr., Tollefson G, Tran P, Satterlee W, Sanger T, Hamilton S. *Olanzapine versus placebo and haloperidol: acute phase results of the North American double-blind olanzapine trial*. Neuropsychopharmacology. **1996b**, 14:111-123.

Beasley CM, Jr., Hamilton SH, Crawford AM, Dellva MA, Tollefson GD, Tran PV, Blin O, Beuzen JN. *Olanzapine versus haloperidol: acute phase results of the international double-blind olanzapine trial*. Eur Neuropsychopharmacol. **1997**, 7:125-137.

Beedham C, Miceli JJ, Obach RS. *Ziprasidone metabolism, aldehyde oxidase, and clinical implications*. J Clin Psychopharmacol. **2003**, 23:229-232.

Beelen AP, Yeo KT, Lewis LD. *Asymptomatic QTc prolongation associated with quetiapine fumarate overdose in a patient being treated with risperidone*. Hum Exp Toxicol. **2001**, 20:215-219.

Beeley L. *Drug-induced sexual dysfunction and infertility*. Adverse Drug React Acute Poisoning Rev. **1984**, 3:23-42.

Benfield P, Ward A, Clark BG, Jue SG. *Bromperidol. A preliminary review of its pharmacodynamic and pharmacokinetic properties, and therapeutic efficacy in psychoses*. Drugs. **1988**, 35:670-684.

Bergling R, Mjorndal T, Oreland L, Rapp W, Wold S. *Plasma levels and clinical effects of thioridazine and thiothixene*. J Clin Pharmacol. **1975**, 15:178-186.

Bett JHN, Holt GW. *Malignant ventricular tachyarrhythmia and haloperidol*. BMJ. **1983**, 287:1264.

Bettinger TL, Mendelson SC, Dorson PG, Crismon ML. *Olanzapine-induced glucose dysregulation*. Ann Pharmacother. **2000**, 34:865-867.

Bhana N, Foster RH, Olney R, Plosker GL. *Olanzapine: an updated review of its use in the management of schizophrenia*. Drugs. **2001**, 61:111-161.

Bieder A, Decouvelaere B, Gaillard C, Gaillot J, Depaire H, Raynaud L, Snozzi C. *Carpipramine metabolism in the rat, rabbit and dog and in man after oral administration*. Xenobiotica. **1985**, 15:421-435.

Birch NJ, Grof P, Hullin RP, Kehoe RF, Schou M, Srinivasan DP. *Lithium prophylaxis: proposed guidelines for good clinical practice*. Lithium. **1993**, 4:225-230.

Bloom BM and Muren JF. *Alkylated thioxathenesulfonamides*. US3310553 (**1967**); priority: 1962

Bogunovic O, Viswanathan R. *Thrombocytopenia possibly associated with olanzapine and subsequently with benztropine mesylate*. Psychosomatics. **2000**, 41:277-288.

Bonanno DG, Davydov L, Botts SR. *Olanzapine-induced diabetes mellitus*. Ann Pharmacother. **2001**, 35:563-565.

Bonelli RM. *Olanzapine associated seizure*. Ann Pharmacother. **2003**, 37:149-150.

Bowles P. *Process for preparing aryl piperazinyl-heterocyclic compounds*. US5206366 (**1993**); priority: 1992

Bowles TM, Levin GM. *Aripiprazole: a new atypical antipsychotic drug*. Ann Pharmacother. **2003**, 37:687-694.

Boyer P, Lecrubier Y, Puech AJ, Dewailly J, Aubin F. *Treatment of negative symptoms in schizophrenia with amisulpride*. Br J Psychiatry. **1995**, 166:68-72.

Boyer P, Lecrubier Y, Stalla-Bourdillon A, Fleurot O. *Amisulpride versus amineptine and placebo for the treatment of dysthymia*. Neuropsychobiology. **1999**, 39:25-32.

Bressolle F, Bres J. *Determination of sulpiride and sultopride by high-performance liquid chromatography for pharmacokinetic studies*. J Chromatogr. **1985**, 341:391-399.

Bressolle F, Bres J, Blanchin MD, Gomeni R. *Sulpiride pharmacokinetics in humans after intramuscular administration at three dose levels.* J Pharm Sci. **1984**, 73:1128-1136.

Bressolle F, Bres J, Mourad G. *Pharmacokinetics of sulpiride after intravenous administration in patients with impaired renal function.* Clin Pharmacokinet. **1989**, 17:367-373.

Bressolle F, Bres J, Faure-Jeantis A. *Absolute bioavailability, rate of absorption, and dose proportionality of sulpiride in humans.* J Pharm Sci. **1992**, 81:26-32.

Bristow MF, Kohen D. *How "malignant" is the neuroleptic malignant syndrome?* BMJ.**1993**, 307:1223-1224.

Bristow MF, Kohen D. *Neuroleptic malignant syndrome.* Br J Hosp Med. **1996**, 55 :517-520.

Brook S, Lucey JV, Gunn KP. *Intramuscular ziprasidone compared with intramuscular haloperidol in the treatment of acute psychosis. Ziprasidone I.M. Study Group.* J Clin Psychiatry. **2000**, 61:933-941.

Brown WA, Laughren TP, Williams B. *Differential effects of neuroleptic agents on the pituitary-gonadal axis in men.* Arch Gen Psychiatry. **1981**, 38:1270-1272.

Bruggeman R, van der LC, Buitelaar JK, Gericke GS, Hawkridge SM, Temlett JA. *Risperidone versus pimozide in Tourette's disorder: a comparative double-blind parallel-group study.* J Clin Psychiatry. **2001**, 62:50-56.

Bruun RD, Budman CL. *Risperidone as a treatment for Tourette's syndrome.* J Clin Psychiatry **1996**, 57:29-31.

Buckey PF. *New dimensions in the pharmacologic treatment of schizophrenia and related psychoses.* J Clin Pharmacol **1997**, 37:363-378. Correction. ibid.;1998, 38:27.

Budman CL, Gayer A, Lesser M, Shi Q, Bruun RD. *An open-label study of the treatment efficacy of olanzapine for Tourette's disorder.* J Clin Psychiatry. **2001**, 62:290-294.

Bulteau G and Acher J. *Process for preparing benzamides.* DE2327192 (**1973**); priority: 1972

Bulteau G, Acher J, and Monier J-C. *Verfahren zur herstellung von 2-methoxy-5-alkylsulfonylbenzamiden.* DE2327193 (**1974**); priority: 1972

Bymaster FP, Perry KW, and Tollefson GD. *Combination therapy for the treatment of psychoses.* WO9811897 (**1998**); priority: 1997

Caccia S, Garattini S. *Formation of active metabolites of psychotropic drugs: an updated review of their significance.* Clin Pharmacokinet. **1990**, 18:434-459.

Calderon J, Rubin E, Sobota WL. *Potential use of ipatropium bromide for the treatment of clozapine-induced hypersalivation: a preliminary report.* Int Clin Psychopharmacol. **2000**, 15:49-52.

Caley CF, Weber SS. *Sulpiride: an antipsychotics with selective dopaminergic antagonist properties.* Ann Pharmacother. **1995**, 29:152-160.

Callaghan JT, Bergstrom RF, Ptak LR, Beasley CM. *Olanzapine. Pharmacokinetic and pharmacodynamic profile.* Clin Pharmacokinet. **1999**, 37:177-193.

Carlyle W, Ancill RJ, Sheldon L. *Aggression in the demented patient: a double-blind study of loxapine versus haloperidol.* Int Clin Psychopharmacol. **1993**, 8:103-108.

Carpenter WT, Jr., Buchanan RW, Kirkpatrick B, Lann HD, Breier AF, Summerfelt AT. *Comparative effectiveness of fluphenazine decanoate injections every 2 weeks versus every 6 weeks.* Am J Psychiatry. **1999**, 156:412-418.

Casey DE. *Tardive dyskinesia and atypical antipsychotic drugs.* Schizophr Res. **1999**, 35 (suppl):S61-S66

Centorrino F, Baldessarini RJ, Kando J, Frankenburg FR, Volpicelli SA, Puopolo PR, Flood JG. *Serum concentrations of clozapine and its major metabolites: effects of cotreatment with fluoxetine or valproate.* Am J Psychiatry. **1994**, 151:123-125.

Centorrino F, Baldessarini RJ, Frankenburg FR, Kando J, Volpicelli SA, Flood JG. *Serum levels of clozapine and norclozapine in patients treated with selective serotonin reuptake inhibitors.* Am J Psychiatry. **1996**, 153:820-822.

Chakrabarti JK, Hotten TM, and Tupper DE. *Pharmaceutical compounds.* EP0454436 (**1991**); priority: 1990

Chakraborty BS, Hubbard JW, Hawes EM, McKay G, Cooper JK, Gurnsey T, Korchinski ED, Midha KK. *Interconversion between haloperidol and reduced haloperidol in healthy volunteers.* Eur J Clin Pharmacol. **1989**, 37:45-48.

Chan J, Alldredge BK, Baskin LS. *Perphenazine-induced priapism.* DICP. **1990**, 24:246-249.

Charpentier P. *Beta-dimethylaminoethylphenothiazines and their production.* US2519886 (**1950**); priority: 1945

Charpentier P. *Phenthiazine derivatives.* US2645640 (**1953**); priority: 1950

Cheslik TA. Erramouspe J. Extrapyramidal *symptoms following accidental ingestion of risperidone in a child.* Ann Pharmacother **1996**, 30:360-363.

Cheung HK, Yu ECS. *Effect of 1050 mg fluphenazine decanoate given intramusculary over six days.* BMJ. **1983**, 286:1016-1017.

Chouinard G, Jones B, Remington G, Bloom D, Addington D, MacEwan GW, Labelle A, Beauclair L, Arnott W. *A Canadian multicenter placebo-controlled study of fixed doses of risperidone and haloperidol in the treatment of chronic schizophrenic patients.* J Clin Psychopharmacol. **1993**, 13:25-40.

Chue P, Singer P. *A review of olanzapine-associated toxicity and fatality in overdose.* J Psychiatry Neurosci. **2003**, 28:253-261.

Citrome LL, Jaffe AB. *Relationship of atypical antipsychotics with development of diabetes mellitus.* Ann Pharmacother **2003**, 37:1849-1857.

Cobb CD, Anderson CB, Seidel DR. *Possible interaction between clozapine and lorazepam.* Am J Psychiatry. **1991**, 148:1606-1607.

Cold JA, Wells BG, Froemming JH. *Seizure activity associated with antipsychotic therapy.* DICP Ann Pharmacother. **1990**, 24:601-606.

Collins S. *Thrombocytopenia associated with lithium carbonate.* BMJ. **1992**, 305:159.

Committee on Safety of Medicines. *Convulsions may occur in patients receiving clozapine(Clozaril®), Sandoz.* Current Problems. **1991**, 31.

Committee on Safety of Medicines/Medicines Control Agency. *Drug-induced neutropenia and agranulocytosis*. Current Problems. **1993a**, 19:10 -11.

Committee on Safety of Medicines/Medicines Control Agency. *Myocarditis with antipsychotics: recent cases with clozapine. (Clozaril)* Current Problems. **1993b**, 19:9 -10.

Committee on Safety of Medicines/Medicines Control Agency. *Drug-induced extrapyramidal reactions.* Current Problems. **1994**, 20:15 -16.

Committee on Safety of Medicines/Medicines Control Agency. *Clozapine (Clozaril) and gastrointestinal obstruction.* Current Problems. **1999a**, 25:5.

Committee on Safety of Medicines/Medicines Control Agency. *Supension of availability of sertindole (Serdolect).* Current Problems **1999b**, 25:1.

Committee on Safety of Medicines/Medicines Control Agency. *QT interval prolongation with antipsychotics.* Current Problems. **2001**, 27:4.

Committee on Safety of Medicines/Medicines Control Agency. *Clozapine and cardiac safety: updated advice for prescribers.* Current Problems. **2002a**, 28:8.

Committee on Safety of Medicines/Medicines Control Agency. *Restricted re-introduction of the atypical antipsychotic sertindole (Serdolect)* **2002b**. Available at: http//www.mca.gov.uk/ourwork/monitorsafequalmed/safetymessages/serdolect3.htm

Conley RR, Carpenter WT, Jr., Tamminga CA. *Time to clozapine response in a standardized trial.* Am J Psychiatry. **1997**, 154:1243-1247.

Cooney C, Nagy A. *Angio-oedema associated with risperidone.* BMJ **1995**, 311:1204.

Cooper SF, Dugal R, Bertrand MJ. *Determination of loxapine in human plasma and urine and identification of three urinary metabolites.* Xenobiotica. **1979**, 9:405-414.

Coppola JA. *11-(4-methyl-1-piperazinyl)dibenz[b, f][1, 4]oxazepines or thiazepines for controlling fertility.* US3412193 (**1968**); priority: 1965

Corvol P, Bisseliches F, Alexandre JM, Bohuon C, Heurtault JP, Menard J, Tcherdakoff P, Vaysse JC, Milliez P. *[Hypertensive episodes triggered by sulpiride].* Sem Hop. **1974**, 50:1265-1269.

Coutinho E, Fenton M, Quraishi S. *Zuclopenthixol decanoate for schizophrenia and other serious mental illnesses.* Available in The Cochrane Library; Issue 3. Chichester: John Wiley. **2004**.

Craig PN and Zirkle CL. *Hydroxyalkylenepiperazine derivatives and analogs thereof.* US3282930 (**1966**); priority: 1960

Crombie AL. *Drugs causing eye problems.* Prescribers' J. **1981**, 21:222-227.

Crow TJ, Johnstone EC. *Mechanism of action of neuroleptic drugs.* Lancet. **1978**, 1:1050.

Csernansky JG, Mahmoud R, Brenner R. *A comparison of risperidone and haloperidol for the prevention of relapse in patients with schizophrenia.* N Engl J Med. **2002**, 346:16-22.

Cummingham DG, Challapalli M. *Hypertension in acute haloperidol poisoning.* J Pediatr. **1979**, 95:489-490.

Curran MP, Perry CM. *Amilsupride: a review of its use in the management of schizophrenia*. Drugs. **2001**, 61: 2123-2150.

Curry SH, Altamura AC, Montgomery S. *Unwanted effects of fluphenazine enanthate and decanoate*. Lancet. **1979a**, 1:331-332.

Curry SH, Whelpton R, de Schepper PJ, Vranckx S, Schiff AA. *Kinetics of fluphenazine after fluphenazine dihydrochloride, enanthate and decanoate administration to man*. Br J Clin Pharmacol. **1979b**, 7:325-331.

Cusic JW and Hamilton RW. *Hydroxyalkylpiperazinoalkylhalophenothiazines*. US2838507 (**1958**); priority: 1954

Dahl SG, Strandjord RE, Sigfusson S. *Pharmacokinetics and relative bioavailability of levomepromazine after repeated administration of tablets and syrup*. Eur J Clin Pharmacol. **1977**, 11:305-310.

Dahl SG. *Pharmacokinetics of methotrimeprazine after single and multiple doses*. Clin Pharmacol Ther. **1976**, 19:435-442.

Dahl-Puustinen ML, Liden A, Alm C, Nordin C, Bertilsson L. *Disposition of perphenazine is related to polymorphic debrisoquin hydroxylation in human beings*. Clin Pharmacol Ther. **1989**, 46:78-81.

Davydov L, Botts SR. *Clozapine-induced hypersalivation*. Ann Pharmacother **2000**, 34:662-365.

De Deyn PP, Rabheru K, Rasmussen A, Bocksberger JP, Dautzenberg PL, Eriksson S, Lawlor BA. *A randomized trial of risperidone, placebo, and haloperidol for behavioral symptoms of dementia*. Neurology. **1999**, 53:946-955.

De Schepper PJ, Vranckx C, Verbeeck R, Van den Berghe ML. *Pipotiazine pharmacokinetics after p.o. and i.v. administration in man. Correlation between blood levels and effect on the handwriting area*. Arzneimittelforschung. **1979**, 29:1056-1062.

Deirmenjian JM, Erhart SM, Wirshing DA, Spellberg BJ, Wirshing WC. *Olanzapine-induced reversible priaprism: a case report*. J Clin Psychopharmacol. **1998**, 18:351-353.

Delbello MP, Schwiers ML, Rosenberg HL, Strakowski SM. *A double-blind, randomized, placebo-controlled study of quetiapine as adjunctive treatment for adolescent mania*. J Am Acad Child Adolesc Psychiatry. **2002**, 41:1216-1223.

Demers R, Lukesh R, Prichard J. *Convulsion during lithium therapy*. Lancet. **1970**, 2:315-316.

Demers RG, Heninger GR. *Electrocardiographic T-wave changes during lithium carbonate treatment*. JAMA. **1971**, 218:381-386.

Deroover J, Baro F, Bourguignon RP, Smets P. *Tiapride versus placebo: a double-blind comparative study in the management of Huntington's chorea*. Curr Med Res Opin. **1984**, 9:329-338.

DeVane CL, Nemeroff CB. *Clinical pharmacokinetics of quetiapine : an atypical antipsychotic*. Clin Pharmacokinet. **2001**, 40:509-522.

Devinsky O, Honigfeld G, Patin J. *Clozapine-related seizures*. Neurology. **1991**, 41:369-371.

DiGiacomo J. *Cardiovascular effects of psychotropic drugs*. Cardiovasc Rev Rep **1989**, 10:31-32, 39-41, and 47.

Dilsaver SC. *Withdrawal phenomena associated with antidepressant and antipsychotic agents.* Drug Safety. **1994**, 10:103-114.

Dincsoy HP, Saelinger DA. *Haloperidol-induced chronic cholestatic liver disease.* Gastroenterology **1982**, 83: 694-700.

Dolitzky B and Diller D. *Methods for preparing olanzapine.* WO2005063771 (**2005**); priority: 2003.

Doucette DE, Grenier JP, Robertson PS. *Olanzapine-induced acute pancreatitis.* Ann Pharmacother. **2000**, 34:1128-1131.

Dr. A. Wander S.A. *Diazepine and thiazepine compounds.* GB980853 (**1965**); priority: 1960

Duggan L, Fenton M, Dardennes RM. *Olanzapine for schizophrenia.* Available in The Cochrane Library; Issue 2. Chichester: John Wiley. **2004**.

Duxbury AJ, Ead RD, Turner EP. *Erosive cheilitis related to prochlorperazine maleate.* Br Dent J. **1982**, 153:271-272.

Dwight MM, Keck PE, Jr., Stanton SP, Strakowski SM, McElroy SL. *Antidepressant activity and mania associated with risperidone treatment of schizoaffective disorder.* Lancet. **1994**, 344:554-555.

Ebadi M, Srinivasan SK. *Pathogenesis, prevention, and treatment of neuroleptic-induced movement disorders.* Pharmacol Rev. **1995**, 47:575-604.

Egan MF, Apud J, Wyatt RJ. *Treatment of tardive dyskinesia.* Schizophr Bull. **1997**, 23:583-609.

Eggert HC, Rosted CT, Elley J, Bolvig HL, Kragh-Sorensen P, Larsen NE, Naestoft J, Hvidberg EF. *Clinical pharmacokinetic studies of perphenazine.* Br J Clin Pharmacol. **1976**, 3:915-923.

Elias TJ, Bannister KM, Clarkson AR, Faull D, Faull RJ. *Clozapine-induced acute interstitial nephritis.* Lancet. **1999**, 354:1180-1181.

Ennis LM, Parker RM. *Paradoxical hypertension associated with clozapine.* Med J Aus **1997**, 166:278.

Ereshefsky L, Tran-Johnson TK, Watanabe MD. *Pathophysiologic basis for schizophrenia and the efficacy of antipsychotics.* Clin Pharm. **1990**, 9:682-707.

Evans SM, Walsh SL, Levin FR, Foltin RW, Fischman MW, Bigelow GE. *Effect of flupenthixol on subjective and cardiovascular responses to intravenous cocaine in humans.* Drug Alcohol Depend. **2001**, 64:271-283.

Eyles DW, McLennan HR, Jones A, McGrath JJ, Stedman TJ, Pond SM. *Quantitative analysis of two pyridinium metabolites of haloperidol in patients with schizophrenia.* Clin Pharmacol Ther. **1994**, 56:512-520.

Eyles DW, McGrath JJ, Pond SM. *Formation of pyridinium species of haloperidol in human liver and brain.* Psychopharmacology (Berl). **1996**, 125:214-219.

Fabian J-L. *Psychotrophic medications and priapism.* Am J Psychiatry. **1993**, 150:349-350.

Falsetti AE. *Risperidone for control of agitation in dementia patients.* Am J Health-Syst Pharm **2000**, 57:862-870.

Fang J, Yu PH, Gorrod JW, Boulton AA. *Inhibition of monoamine oxidases by haloperidol and its metabolites: pharmacological implications for the chemotherapy of schizophrenia*. Psychopharmacology (Berl). **1995**, 118:206-212.

Fang J, McKay G, Song J, Remillrd A, Li X, Midha K. *In vitro characterization of the metabolism of haloperidol using recombinant cytochrome p450 enzymes and human liver microsomes*. Drug Metab Dispos. **2001**, 29:1638-1643.

Fenton M, Murphy B, Wood J. *Loxapine for schizophrenia*. Available in The Cochrane Library; Issue 3. Chichester: John Wiley. **2004**.

Ferguson J, Walker EM, Johnson BE. *Further clinical and investigative studies of chlorpromazine phototoxicity*. Br J Dermatol. **1986**, 115 (suppl 30):35.

Filice GA, McDougall BC, Ercan-Fang N, Billington CJ. *Neuroleptic malignant syndrome associated with olanzapine*. Ann Pharmacother. **1998**, 32:1158-1159.

Finkel MJ, Arellano F. *White-blood-cell monitoring and clozapine*. Lancet. **1995**, 346:849.

Finkel MJ, Schwimmer JL, *Clozapine- a novel antipsychotic agent*. N Engl J Med. **1991**, 325:518-19.

Fisher N, Baigent B. *Treatment with clozapine: black patients' low white cell counts currently mean that they cannot be treated*. BMJ. **1996**, 313:1262.

Fiston A, Heel RC. *Clozapine: a review of its pharmacological properties, and therapeutic use in schizophrenia*. Drugs **1990**, 722-747.

Fitz-Gerald MJ, Pinkofsky HB, Brannon G, Dandridge E, Calhoun A. *Olanzapine-induced mania*. Am J Psychiatry. **1999**, 156:1114.

Florvall L and Ögren S-O. *2,6-Dialkoxybenzamides, intermediates, pharmaceutical compositions and methods for treatment of psychotic disorders*. US4232037 (**1980**); priority: 1978

Florvall L, Ögren S-O. *Potential neuroleptic agents. 2,6-Dialkoxybenzamide derivatives with potent dopamine receptor blocking activities*. J Med Chem. **1982**, 25:1280-1286.

Flynn SW, Altman S, MacEwan GW, Black LL, Greenidge LL, Honer WG. *Prolongation of clozapine-induced granulocytopenia associated with olanzapine*. J Clin Psychopharmacol. **1997**, 17:494-495.

Ford B, Lynch T, Greene P. *Risperidone in Parkinson's disease*. Lancet. **1994**, 344:681.

Frankenburg FR, Kando J. *Eosinophilia, clozapine, and pancreatitis*. Lancet. **1992**, 340:251.

Franz M, Dlabal H, Kunz S, Ulferts J, Gruppe H, Gallhofer B. *Treatment of alcohol withdrawal: tiapride and carbamazepine versus clomethiazole. A pilot study*. Eur Arch Psychiatry Clin Neurosci. **2001**, 251:185-192.

Fraunfelder FT, Fraunfelder FW, Jefferson JW. *The effects of lithium on the human visual system*. J Toxicol Cutan Ocul Toxicol. **1992**, 11:97-169.

Freeman DJ, Oyewumi LK. *Will routine therapeutic drug monitoring have a place in clozapine therapy?* Clin Pharmacokinet. **1997**, 32:93-100.

Friedman LJ, Tabb SE, Worthington JJ, Sanchez CJ, Sved M. *Clozapine--a novel antipsychotic agent*. N Engl J Med. **1991**, 325:518-519.

Froemming JS, Lam YW, Jann MW, Davis CM. *Pharmacokinetics of haloperidol.* Clin Pharmacokinet. **1989**, 17:396-423.

Fukuzako H, Hashiguchi T, Nomaguchi M, Nagatomo I, Matsumoto K. *Ultrasonography detected a higher incidence of gallstones in psychiatric inpatients.* Acta Psychiatr Scand. **1991**, 84:83-85.

Fuller MA, Simon MR, Freedman L. *Risperidone-associated hepatotoxicity.* J Clin Psychopharmacol. **1996**, 16:84-85.

Furlanut M, Benetello P, Perosa A, Colombo G, Gallo F, Forgione A. *Pharmacokinetics of benperidol in volunteers after oral administration.* Int J Clin Pharmacol Res. **1988**, 8:13-16.

Furlanut M, Benetello P, Baraldo M, Zara G, Montanari G, Donzelli F. *Chlorpromazine disposition in relation to age in children.* Clin Pharmacokinet. **1990**, 18:329-331.

Gawin FH, Khalsa-Denison ME, Jatlow P. *Flupentixol-induced aversion to crack cocaine.* N Engl J Med. **1996**, 334:1340-1341.

George CR. *Renal aspects of lithium toxicity.* Med J Aust. **1989**, 150:291-292.

Gerson SL, Lieberman JA, Friedenberg WR, Lee D, Marx JJ, Jr., Meltzer H. *Polypharmacy in fatal clozapine-associated agranulocytosis.* Lancet. **1991**, 338:262-263.

Gerson SL, Meltzer H. *Mechanisms of clozapine-induced agranulocytosis.* Drug Saf. **1992**, 7 Suppl 1:17-25.

Ghaeli P, Dufresne RL. *Serum triglyceride levels in patients treated with clozapine.* Am J Health-Syst Pharm Res. **1996**, 53:2079-2081.

Gibson RC, Fenton M, Coutinho ES, Campbell C. *Zuclopenthixol acetate in the treatment of acute schizophrenia and similar serious mental illnesses.* Available in The Cochrane Library; Issue 2. Chichester: John Wiley. **2004**.

Goetz CG, Blasucci LM, Leurgans S, Pappert EJ. *Olanzapine and clozapine: comparative effects on motor function in hallucinating PD patients.* Neurology. **2000**, 55:789-794.

Gojer JAC. *Possible manic side-effects of loxapine.* Can J Psychiatry. **1992**, 37:669-670.

Gomolin IH. Coping with excessive doses. In: Johnson FN, ed. *Depression & mania: modern lithium therapy.* IRL Press, Oxford, **1987**, pp 154-157.

Goodnick PJ, Jerry JM. *Aripiprazole: profile on efficacy and safety.* Expert Opin Pharmacother. **2002**, 3:1773-1781.

Graham JM, Sussman JD, Ford KS, Sagar HJ. *Olanzapine in the treatment of hallucinosis in idiopathic Parkinson's disease: a cautionary note.* J Neurol Neurosurg Psychiatry. **1998**, 65:774-777.

Greene P. *Clozapine therapeutic plunge in patient with Parkinson's disease.* Lancet. **1995**, 345:1172-1173.

Griffiths J, Springuel P. *Atypical antipsychotics: impaired glucose metabolism.* Can Adverse Drug React News. **2001**, 11:3-6.

Grupta S. *Paradoxical hypertension associated with clozapine.* Am J Psychiatry. **1994**,151:148.

Guitton C, Kinowski JM, Abbar M, Chabrand P, Bressolle F. *Clozapine and metabolite concentrations during treatment of patients with chronic schizophrenia*. J Clin Pharmacol. **1999**, 39:721-728.

Gunnet JW, Moore KE. *Neuroleptics and neuroendocrine function*. Ann Rev Pharmacol Toxicol. **1988**, 28:347-366.

Gutzmann H, Kuhl KP, Kanowski S, Khan-Boluki J. *Measuring the efficacy of psychopharmacological treatment of psychomotoric restlessness in dementia: clinical evaluation of tiapride*. Pharmacopsychiatry. **1997**, 30:6-11.

Haag H, Ruther E, Hippius H eds. *Tardive Dyskinesia. WHO Expert Series on Biological Psychiatry Volume 1. Seattle:* Hogrefe & Huber. **1992**.

Haddad PM, Anderson IM. *Antipsychotics-related QTc prolongation, torsade de pointes and sudden death*. Drugs. **2002**, 62:1649-1671.

Hagg S, Spigset O, Soderstrom TG. *Association of venous thromboembolism and clozapine*. Lancet. **2000**, 355:1155-1156.

Hagger R, Brown C, Hurley P. *Olanzapine and pancreatitis*. Br J Psychiatry. **2000**, 177:567.

Hamilton SH, Revicki DA, Genduso LA, Beasley CM, Jr. *Olanzapine versus placebo and haloperidol: quality of life and efficacy results of the North American double-blind trial*. Neuropsychopharmacology. **1998**, 18:41-49.

Harth Y, Rappoport M. *Photosensitivity associated with antipsychotics, antidepressants and anxiolytics*. Drug Safety. **1996**, 14: 252-259.

Hay J. *Complications at site of injection of depot neuroleptics*. BMJ. **1995**, 311:421.

Hedenmalm K, Hagg S, Stahl M, Mortimer O, Spigset O. *Glucose intolerance with atypical antipsychotics*. Drug Saf. **2002**, 25:1107-1116.

Henderson RA, Lane S, Henry JA. *Life-threatening ventricular arrhythmia (torsades de pointes) after haloperidol overdose*. Hum Exp Toxicol. **1991**, 10:59-62.

Henkin Y, Como JA, Oberman A. *Secondary dyslipidemia. Inadvertent effects of drugs in clinical practice*. JAMA. **1992**, 267:961-968.

Henry C, Demotes-Mainard J. *Olanzapine-induced mania and bipolar disorders*. J Psychiatry Neurosci. **2002**, 27:200-201.

Herrán A, Vázquez Baquero JL. *Tardive dyskinesia associated with Olanzapine*. Ann Intern Med. **1999**, 131:72.

Hessell PG, Lloyd-Jones JG, Muir NC, Parr GD, Sugden K. *A comparison of the availability of prochlorperazine following im buccal and oral administration*. Int J Pharmaceutics. **1989**, 52:159-164.

Hill RC, McIvor RJ, Wojnar-Horton RE, Hackett LP, Ilett KF. *Risperidone distribution and excretion into human milk: case report and estimated infant exposure during breast-feeding*. J Clin Psychopharmacol. **2000**, 20:285-286.

Himmelhoch JM, Hanin I. *Letter: Side effects of lithium carbonate*. Br Med J. **1974**, 4:233.

Himstreet JE. Daya M. Hypotension *and orthostasis following a risperidone overdose.* Ann Pharmacother. **1998**, 32:267.

Hirsch SR, Puri BK. *Clozapine: progress in treating refractory schizophrenia.* BMJ. **1993**, 306:1427-8.

Hobbs DC. *Metabolism of thiothixene.* J Pharm Sci. **1968**, 57:105-111.

Hobbs DC, Welch WM, Short MJ, Moody WA, Van der Velde CD. *Pharmacokinetics of thiothixene in man.* Clin Pharmacol Ther. **1974**, 16:473-478.

Hogarty GE, McEvoy JP, Munetz M, DiBarry AL, Bartone P, Cather R, Cooley SJ, Ulrich RF, Carter M, Madonia MJ. *Dose of fluphenazine, familial expressed emotion, and outcome in schizophrenia. Results of a two-year controlled study.* Arch Gen Psychiatry. **1988**, 45:797-805.

Holloman LC, Marder SR. *Management of acute extrapyramidal effects induced by antipsychotics drugs.* Am J Health-Syst Pharm. **1997**, 54:2461-2477.

Honigfeld G, Arellano F, Sethi J, Bianchini A, Schein J. *Reducing clozapine-related morbidity and mortality: 5 years of experience with the Clozaril National Registry.* J Clin Psychiatry. **1998**, 59 Suppl 3:3-7.

Horclois RJ. *Phenthiazine derivatives.* US2902485 (**1958**); priority: 1955

Horclois RJ. *Phenthiazine derivatives and processes for their preparation.* US2902484 (**1959**); priority: 1957

Howes O, Smith S. *Hyperprolactinaemia caused by antipsychotic drugs: endocrine antipsychotic side effects must be systematically assessed.* BMJ. **2002**, 324:1278.

Huang CL, Sands FL. *Effect of ultraviolet irradiation on chlorpromazine. II: anaerobic condition.* J Pharm Sci. **1967**, 56:259-264.

Huang CL, Yeh JZ, Muni IA. *Comparative distribution, excretion, and metabolism of quaternary ammonium compounds of promazine, chlorpromazine, and triflupromazine.* J Pharm Sci. **1970**, 59:1114-1118.

Huang ML, Van Peer A, Woestenborghs R, De Coster R, Heykants J, Jansen AA, Zylicz Z, Visscher HW, Jonkman JH. *Pharmacokinetics of the novel antipsychotic agent risperidone and the prolactin response in healthy subjects.* Clin Pharmacol Ther. **1993**, 54:257-268.

Hui WK, Mitchell LB, Kavanagh KM, Gillis AM, Wyse DG, Manyari DE, Duff HJ. *Melperone: electrophysiologic and antiarrhythmic activity in humans.* J Cardiovasc Pharmacol. **1990**, 15:144-149.

Hunter RH, Joy CB, Keennedy E, Gilbody SM, Song F. Risperidone versus typical antipsychotic medication for schizophrenia. Available in The Cochrane Library; Issue 2. Chichester: John Wiley. **2004**.

Idanpaan-Heikkila J, Alhava E, Olkinuora M, Palva IP. *Agranulocytosis during treatment with chlozapine.* Eur J Clin Pharmacol. **1977**, 11:193-198.

Iglesias E, Esteban E, Zabala S, Gascon A. *Tiapride-induced torsade de pointes.* Am J Med. **2000**, 109:509.

Ishigooka J, Murasaki M, Wakatabe H, Miura S, Hikida K, Shibata M, Nobunaga H. *Pharmacokinetic study of iminodibenzyl antipsychotic drugs, clocapramine and Y-516 in dog and man.* Psychopharmacology (Berl). **1989**, 97:303-308.

Ishigooka J, Murasaki M, Miura S, Shibata M, Takamatsu R. *Pilot study of plasma concentrations of mosapramine, a new iminodibenzyl antipsychotic agent, after multiple oral administration in schizophrenic patients.* Curr Ther Res. **1994**, 55:331-342.

Ishizumi K, Antoku F, and Asami Y. *Succinimide derivates and process for preparation thereof.* EP82402 (**1983**); priority: 1982

Ishizumi K, Antoku F, Maruyama I, and Kojima A. *Imide derivatives, their production and use.* EP0196096 (**1986**); priority: 1985

Iwanami S, Takashima M, and Usuda S. *Benzamide derivatives.* US4210660 (**1980**); priority: 1978

Jackson CW, Markowitz JS, Brewerton TD. *Delirium associated with clozapine and benzodiazepine combinations.* Ann Clin Psychiatry. **1995**, 7:139-141.

Jacob RM and Robert JG. *Phenothiazine derivatives.* DE1056611 (**1956**); priority: 1955

Jacob RM and Robert JG. *Phenthiazine compounds.* US2837518 (**1958**); priority: 1954

Jacob RM and Robert JG. *3-cyano substituted phenothiazines.* US2877224 (**1959**); priority: 1955

Jacob RM and Robert JG. *Phenthiazine derivatives.* US3075976 (**1963**); priority: 1957

Jacobsen FM. *Risperidone in the treatment of affective illness and obsessive-compulsive disorder.* J Clin Psychiatry. **1995**, 56:423-429.

Jadallah KA, Limauro DL, Colatrella AM. *Acute hepatocellular-cholestatic liver injury after olanzapine therapy.* Ann Intern Med. **2003**, 138:357-358.

Jann MW, Grimsley SR, Gray EC, Chang WH. *Pharmacokinetics and pharmacodynamics of clozapine.* Clin Pharmacokinet. **1993**, 24:161-176.

Janssen PA, Van De West, Jageneau AH, Demoen PJ, Hermans BK, Van Daele GH, Schellekens KH, Van Der Eycken CA. *Chemistry and pharmacology of CNS depressants related to 4-(4-hydroxy-phenylpiperidino)butyrophenone. I. Synthesis and screening data in mice.* J Med Pharm Chem. **1959**, 1:281-297.

Janssen PA. *1-(1-aroylpropyl-4-piperidyl)-2-benzimidazolinones and related compounds.* US3161645 (**1964**); priority: 1961

Janssen PA. *1-(1-aroylpropyl-4-piperidyl)-2-benzimidazolinones and related compounds.* GB989755 (**1965a**); priority: 1961

Janssen PA. *Benzimidazolinyl piperidines.* US3196157 (**1965b**); priority: 1963

Janssen PA. *Substituted 1,3,8-triaza-spiro (4,5) decanes.* US3238216 (**1966**); priority: 1963

Janssen PA. *Arylpiperidine derivatives.* DE1289845 (**1969a**); priority: 1958

Janssen PA. *Verfahren zur Herstellung von neuen Triazaspirodecanen und ihren Saeureadditionssalzen.* DE1470125 (**1969b**); priority: 1962

Janssen PA. *1-Aroylalkyl derivatives of arylhydroxypyrrolidines and arylhydroxy-piperidines.* US3438991 (**1969c**); priority: 1959

Janssen PA. *Substituted 1-Benzoylpropyl-4-Hydroxy-4-Phenyl Piperidine Derivatives.* US3518276 (**1970**); priority: 1965

Janssen-Ortho Inc./Health Canada. *Important drug safety information: Risperdal (risperidone) and cerebrovascular adverse events in placebo-controlled dementia trials* (issued 11/10/02). Available at: http://www.hc-sc.gc.ca/hpfb-dgpsa/tpd-dpt/risperdal1_e.html

Jeffries JJ, Lyall WA, Bezchlibnyk K, Papoff PM, Newman F. *Retroperitoneal fibrosis and haloperidol.* Am J Psychiatry. **1982**, 139:1524-1525.

Jerling M, Lindstrom L, Bondesson U, Bertilsson L. *Fluvoxamine inhibition and carbamazepine induction of the metabolism of clozapine: evidence from a therapeutic drug monitoring service.* Ther Drug Monit. **1994**, 16:368-374.

Jiménez-Jiménez FJ, García-Ruíz PJ, Molina JA. *Drug-induced movement disorders.* Drug Saf. **1997**, 16:180-204.

John E. *Promazine and neonatal hyperbilirubinaemia.* Med J Aust. **1975**, 2:342-344.

Johns S, Harris B. *Tremor.* Br Med J (Clin Res Ed). **1984**, 288:1309.

Johnson FNed. *Depression & mania: modern lithium therapy.* Oxford: IRL Press; Oxford, **1987**.

Johnstone EC, Whaley K. *Antinuclear antibodies in psychiatric illness: their relationship to diagnosis and drug treatment.* Br Med J. **1975**, 2:724-725.

Johnstone EC, Crow TJ, Frith CD, Carney MW, Price JS. *Mechanism of the antipsychotic effect in the treatment of acute schizophrenia.* Lancet. **1978**, 1:848-851.

Jones HM, Pilowsky LS. *Dopamine and antipsychotic drug action revisited.* Br J Psychiatry. **2002**, 181:271-275.

Jorgensen A. *Pharmacokinetic studies in volunteers of intravenous and oral cis (Z)-flupentixol and intramuscular cis (Z)-flupentixol decanoate in Viscoleo.* Eur J Clin Pharmacol. **1980**, 18:355-360.

Jorgensen A, Andersen J, Bjorndal N, Dencker SJ, Lundin L, Malm U. *Serum concentrations of cis(Z)-flupentixol and prolactin in chronic schizophrenic patients treated with flupentixol and cis(Z)-flupentixol decanoate.* Psychopharmacology (Berl). **1982**, 77:58-65.

Jubert P, Fernandez R, Ruiz A. *Clozapine-related pancreatitis.* Ann Intern Med. **1994**, 121:722-723.

Kalgutkar AS, Taylor TJ, Venkatakrishnan K, Isin EM. *Assessment of the contributions of CYP3A4 and CYP3A5 in the metabolism of the antipsychotic agent haloperidol to its potentially neurotoxic pyridinium metabolite and effect of antidepressants on the bioactivation pathway.* Drug Metab Dispos. **2003**, 31:243-249.

Kane JM, Rifkin A, Woerner M, Reardon G, Sarantakos S, Schiebel D, Ramos-Lorenzi J. *Low-dose neuroleptic treatment of outpatient schizophrenics. I. Preliminary results for relapse rates.* Arch Gen Psychiatry. **1983**, 40:893-896.

Kane J, Honigfeld G, Singer J, Meltzer H. *Clozapine for the treatment-resistant schizophrenic. A double-blind comparison with chlorpromazine.* Arch Gen Psychiatry. **1988**, 45:789-796.

Kane JM. *Tardive dyskinesia in affective disorders.* J Clin Psychiatry. **1999**, 60(suppl 5):43-47.

Karagianis JL, Phillips LC, Hogan KP, LeDrew KK. *Clozapine-associated neuroleptic malignant syndrome: two new cases and a review of the literature.* Ann Pharmacother **1999**, 33:623-630. Correction. ibid.; 1011.

Kassahun K, Mattiuz E, Nyhart E Jr, Obermeyer B, Gillespie T, Murphy A, Goodwin RM, Tupper D, Callaghan JT, Lemberger L. *Disposition and biotransformation of the antipsychotic agent olanzapine in humans.* Drug Metab Dispos. **1997**, 25:81-93.

Kassahun K, Mattiuz E, Franklin R, Gillespie T. *Olanzapine 10-N-glucuronide. A tertiary N-glucuronide unique to humans.* Drug Metab Dispos. **1998**, 26:848-855.

Keck PE, Jr., Versiani M, Potkin S, West SA, Giller E, Ice K. *Ziprasidone in the treatment of acute bipolar mania: a three-week, placebo-controlled, double-blind, randomized trial.* Am J Psychiatry. **2003a**, 160:741-748.

Keck PE, Jr., Marcus R, Tourkodimitris S, Ali M, Liebeskind A, Saha A, Ingenito G. *A placebo-controlled, double-blind study of the efficacy and safety of aripiprazole in patients with acute bipolar mania.* Am J Psychiatry. **2003b**, 160:1651-1658.

Keck PE, McElroy SL. *Aripiprazole: a partial dopamine D2 receptor agonist antipsychotic.* Expert Opin Invest Drugs. **2003**, 12:655-662.

Kelly DV, Beique LC, Bowmer MI. *Extrapyramidal symptoms with ritonavir/indinavir plus risperidone.* Ann Pharmacother. **2002**, 36:827-830.

Keltjens R and Peters THA. *Synthesis of olanzapine and intermediates thereof.* WO2005070939 (**2005**); priority: 2003

Kennedy P. *Liver cross-sensitivity to antipsychotic drugs.* Br J Psychiatry. **1983**, 143:312.

Kennis LEJ and Vanderberk J. *Novel 1,2-benzisoxazol-3-yl and 1,2-benzisothiazol-3-yl derivatives.* GR860800 (**1986**); priority: 1985

Kerwin RW. *The new atypical antipsychotics. A lack of extrapyramidal side-effects and new routes in schizophrenia research.* Br J Psychiatry. **1994**, 164:141-148.

Kerwin RW. *Clozapine: back to the future for schizophrenia research.* Lancet. **1995**, 345:1063-1064.

Killian JG, Kerr K, Lawrence C, Celermajer DS. *Myocarditis and cardiomyopathy associated with clozapine.* Lancet. **1999**, 354:1841-1845.

Kingsbury SJ, Puckett KM. *Effects of fluoxetine on serum clozapine levels.* Am J Psychiatry. **1995**, 152:473-474.

Kleeman A, Engel J, Kutscher B, Reichert D. *Pharmaceutical Substances: Syntheses, Patents Applications.* 4th Ed. (2 Volumes). New York: Thieme, **2001**.

Klein C, Gordon J, Pollak L, Rabey JM. *Clozapine in Parkinson's disease psychosis: 5-year follow-up review.* Clin Neuropharmacol. **2003**, 26:8-11.

Klieser E, Lehmann E, Kinzler E, Wurthmann C, Heinrich K. *Randomized, double-blind, controlled trial of risperidone versus clozapine in patients with chronic schizophrenia.* J Clin Psychopharmacol. **1995**, 15:45S-51S.

Koek RJ, Pi EH. Acute *laryngeal dystonic reactions to neuroleptics*. Psychosomatics. **1989**, 30:359-364.

Kogoj A, Velikonja L. *Olanzapine-induced neuroleptic malignant syndrome- a case review*. Hum Psychopharmacol. **2003**, 18:301-309.

Koller E, Malozowski S, Doraiswamy PM. *Atypical antipsychotic drugs and hyperglycemia in adolescents*. JAMA. **2001**, 286:2547-2548.

Kornhuber J, Weller M. *Neuroleptic malignant syndrome*. Curr Opin Neurol. **1994**, 7:353-357.

Koro CE, Fedder DO, L'Italien GJ, Weiss SS, Magder LS, Kreyenbuhl J, Revicki DA, Buchanan RW. *Assessment of independent effect of olanzapine and risperidone on risk of diabetes among patients with schizophrenia: population based nested case-control study*. BMJ. **2002**, 325:243.

Kosary J, Kasztriner E, Farkas L, Borvendeg J, Eggenhofer J, Pap V, Balogh T, Somogyi G, Orban E, Koczka E, and Bursich E. *Methoxybenzene sulphonamide derivatives*. GB1492166 (**1977**); priority: 1975

Kotin J, Wilbert DE, Verburg D, Soldinger SM. *Thioridazine and sexual dysfunction*. Am J Psychiatry. **1976**, 133:82-85.

Krebs MO, Olie JP. *Tardive dystonia induced by risperidone*. Can J Psychiatry. **1999**, 44:507-508.

Kudo S, Ishizaki T. *Pharmacokinetics of haloperidol: an update*. Clin Pharmacokinet. **1999**, 37:435-456.

La Grenade L, Graham D, Trontell A. *Myocarditis and cardiomyopathy associated with clozapine use in the United States*. N Engl J Med. **2001**, 345:224-225.

Ladhani FM. *Severe extrapyramidal manifestations following fluphenazine overdose*. Med J Aust. **1974**, 2:26.

Lam RW, Remick RA. *Pigmentary retinopathy associated with low-dose thioridazine treatment*. CAN Med Assoc J. **1985**, 132:737.

Lambert D, Dalac S. Skin, hair and nails. In: Johnson FN, ed. *Depression & mania: modern lithium therapy*. IRL Press, Oxford, **1987**, pp 232-234.

Lavid N, Franklin DL, Maguire GA. *Management of child and adolescent stuttering with olanzapine: three case reports*. Ann Clin Psychiatry. **1999**, 11:233-236.

Lecrubier Y, Boyer P, Turjanski S, Rein W. *Amisulpride versus imipramine and placebo in dysthymia and major depression. Amisulpride Study Group*. J Affect Disord. **1997**, 43:95-103.

Lee HJ, Lee HS, Kim L, Lee MS, Suh KY, Kwak DI. *A case of risperidone-induced stuttering*. J Clin Psychopharmacol. **2001**, 21:115-116.

Lee JW, Crismon ML, Dorson PG. *Seizure associated with olanzapine*. Ann Pharmacother. **1999**, 33:554-556.

Leopold NA. *Risperidone treatment of drug-related phychosis in patients with parkinsonism*. Mov Disord. **2000**, 15:301-304.

Lidenmayer J-P, Klebanov R. *Olanzapine-induced manic-like syndrome.* J Clin Psychiatry. **1998**, 318-319.

Lieberman JA, Yunis J, Egea E, Canoso RT, Kane JM, Yunis EJ. *HLA-B38, DR4, DQw3 and clozapine-induced agranulocytosis in Jewish patients with schizophrenia.* Arch Gen Psychiatry. **1990**, 47:945-948.

Lilford VA, Lilford RJ, Dacie JE, Rees LA, Browne PD, Chard T. *Long-term phenothiazine treatment does not cause pituitary tumours.* Br J Psychiatry. **1984**, 144:421-424.

Lin SK, Chang WH, Chung MC, Lam YW, Jann MW. *Disposition of clozapine and desmethylclozapine in schizophrenic patients.* J Clin Pharmacol. **1994**, 34:318-324.

Linnet K, Wiborg O. *Influence of Cyp2D6 genetic polymorphism on ratios of steady-state serum concentration to dose of the neuroleptic zuclopenthixol.* Ther Drug Monit. **1996**, 18:629-634.

Linnet K. *Glucuronidation of olanzapine by cDNA-expressed human UDP-glucuronosyltransferases and human liver microsomes.* Hum Psychopharmacol. **2002**, 17:233-238.

Livingstone MG, *Risperidone.* Lancet. **1994**, 343:457-60.

Loo H, Poirier-Littre MF, Theron M, Rein W, Fleurot O. *Amisulpride versus placebo in the medium-term treatment of the negative symptoms of schizophrenia.* Br J Psychiatry. **1997**, 170:18-22.

Lowe JAI and Nagel AA. *Piperazinyl-heterocyclic compounds.* EP0281309 (**1988**); priority: 1987

Maguire GA, Riley GD, Franklin DL, Gottschalk LA. *Risperidone for the treatment of stuttering.* J Clin Psychopharmacol. **2000**, 20:479-482.

Mamo DC, Sweet RA, Keshavan MS. *Managing antipsychotic-induced parkinsonism.* Drug Saf. **1999**, 20:269-275.

Manson AJ, Schrag A, Lees AJ. *Low-dose olanzapine for levodopa induced dyskinesias.* Neurology. **2000**, 55:795-799.

Marder SR, Van Putten T, Mintz J, Lebell M, McKenzie J, May PR. *Low- and conventional-dose maintenance therapy with fluphenazine decanoate. Two-year outcome.* Arch Gen Psychiatry. **1987**, 44:518-521.

Marder SR, Van Putten T, Aravagiri M, Hubbard JW, Hawes EM, McKay G, Midha KK. *Plasma levels of parent drug and metabolites in patients receiving oral and depot fluphenazine.* Psychopharmacol Bull. **1989**, 25:479-482.

Marder SR, Meibach RC. *Risperidone in the treatment of schizophrenia.* Am J Psychiatry **1994**, 151:825-835.

Marder SR, McQuade RD, Stock E, Kaplita S, Marcus R, Safferman AZ, Saha A, Ali M, Iwamoto T. *Aripiprazole in the treatment of schizophrenia: safety and tolerability in short-term, placebo-controlled trials.* Schizophr Res. **2003**, 61:123-136.

Marmor MF. *Is thioridazine retinopathy progressive? Relationship of pigmentary changes to visual function.* Br J Ophthalmol **1990**, 74:739-742.

Marquillas Olondriz F, Bosch Rovira A, Dalmases Barjoan P, and Caldero Ges JM. *Procedure for obtaining 3-[2-[4-(6-fluoro-1,2-benzisoxazol-3-yl)piperidin]ethyl]-2-methyl-6,7,8,9-tetrahydro-4H-pyrido[1,2-a]pyrimidin-4-one*. ES2050069 (**1994**); priority: 1992

Marsden CD, Quinn NP. *The dystonias*. BMJ. **1990**, 300:139-144.

Mårtensson E, Roos B-E. *Serum levels of thioridazine in psychiatric patients and healthy volunteers*. Eur J Clin Pharmacol. **1973**, 6:181-186

Martin A. *Acute pancreatitis associated with clozapine use*. Am J Psychiatry. **1992**, 149:714.

Martin CA, Piascik MT. *First degree A-V block in patients on lithium carbonate*. Can J Psychiatry. **1985**, 30:114-116.

Martin SD. *Drug-induced parotid swelling*. Br J Hosp Med. **1993**, 50:426.

Massey EW, Folger WN. *Seizures activated by therapeutic levels of lithium carbonate*. South Med J. **1984**, 77:1173-1175.

Mathieson O. *Trifluoromethyl-phenothiazine derivatives*. GB857546 (**1960a**); priority: 1955

Mathieson O. *Trifluoromethyl-phenothiazine derivatives*. GB857547 (**1960b**); priority: 1956

Matsuno H, Uematsu T, Nakashima M. *The measurement of haloperidol and reduced haloperidol in hair as an index of dosage history*. Br J Clin Pharmacol. **1990**, 29:187-194.

Mattiuz E, Franklin R, Gillespie T, Murphy A, Bernstein J, Chiu A, Hotten T, Kassahun K. *Disposition and metabolism of olanzapine in mice, dogs, and rhesus monkeys*. Drug Metab Dispos. **1997**, 25:573-583.

Mauri F. *Optically active benzamide, process for their producton and their application*. DE2903891 (**1979**); priority: 1978

Mauri F. *Raceme sulpiride resolution process*. EP0053584 (**1982**); priority: 1980

Mauri MC, Bravin S, Bitetto A, Rudelli R, Invernizzi G. *A risk-benefit assessment of sulpiride in the treatment of schizophrenia*. Drug Saf. **1996**, 14:288-298.

Maxa JL, Taleghani AM, Ogu CC, Tanzi M. *Possible toxic encephalopathy following high-dose intravenous haloperidol*. Ann Pharmacother. **1997**, 31:736-737.

Mc Gavin JK, Goa KL. *Aripiprazole*. CNS. Drugs. **2002**, 16:779-786.

McCreadie RG, Kiernan WE, Venner RM, Denholm RB. *Probable toxic necrosis after prolonged fluspirilene administration*. Br Med J. **1979**, 1:523-524.

McGovern GP. *Lithium induced constructional dyspraxia*. Br Med J (Clin Res Ed). **1983**, 286:646.

McKeith IG, Ballard CG, Harrison RW. *Neuroleptic sensitivity to risperidone in Lewy body dementia*. Lancet. **1995**, 346:699.

McQueen EG. *New Zealand committee on adverse drug reactions: fourteenth annual repot*. **1979**. N Z Med J. **1980**, 91:226-229.

Meco G, Alessandria A, Bonifati V, Giustini P. *Risperidone for hallucinations in levodopa-treated Parkinson's disease patients*. Lancet. **1994**, 343:1370-1371.

Mehta D, Mehta S, Petit J, Shriner W. *Cardiac arrhythmia and haloperidol*. Am J Psychiatry. **1979**, 136:1468-1469.

Meltzer HY, Bastani B, Kwon KY, Ramirez LF, Burnett S, Sharpe J. *A prospective study of clozapine in treatment-resistant schizophrenic patients. I. Preliminary report*. Psychopharmacology (Berl). **1989**, 99 Suppl:S68-S72.

Meltzer HY, Okayli G. *Reduction of suicidality during clozapine treatment of neuroleptic-resistant schizophrenia: impact on risk-benefit assessment*. Am J Psychiatry. **1995**; 152:183-190.

Meltzer HY, Alphs L, Green AI, Altamura AC, Anand R, Bertoldi A, Bourgeois M, Chouinard G, Islam MZ, Kane J, Krishnan R, Lindenmayer JP, Potkin S. *Clozapine treatment for suicidality in schizophrenia: International Suicide Prevention Trial (InterSePT)*. Arch Gen Psychiatry. **2003**, 60:82-91.

Miao Z, Kamel A, Prakash C. *Characterization of a novel metabolite intermediate of ziprasidone in hepatic cytosolic fractions of rat, dog, and human by ESI-MS/MS, hydrogen/deuterium exchange, and chemical derivatization*. Drug Metab Dispos. **2005**, 33:879-883.

Midha KK, Korchinski ED, Verbeeck RK, Roscoe RM, Hawes EM, Cooper JK, McKay G. *Kinetics of oral trifluoperazine disposition in man*. Br J Clin Pharmacol. **1983a**, 15:380-382.

Midha KK, McKay G, Edom R, Korchinski ED, Hawes EM, Hall K. *Kinetics of oral fluphenazine disposition in humans by GC-MS*. Eur J Clin Pharmacol. **1983b**, 25:709-711.

Miller CH, Fleischhacker WW. *Managing antipsychotic-induced acute and chronic akathisia*. Drug Safety. **2000**, 22:73-81.

Miller CS, Engelhardt EL, and Thominet ML. *Heterocyclic aminoalkyl benzamides*. US3342826 (**1967**); priority: 1967

Miller LJ. *Withdrawal-emergent dyskinesia in a patient taking risperidone/citalopram*. Ann Pharmacother. **2000**, 34:269.

Mogelvang JC, Petersen EN, Folke PE, Ovesen L. *Antiarrhythmic properties of a neuroleptic butyrophenone, melperone, in acute myocardial infarction. A double-blind trial*. Acta Med Scand. **1980**, 208:61-64.

Moleman P, Janzen G, von Bargen BA, Kappers EJ, Pepplinkhuizen L, Schmitz PI. *Relationship between age and incidence of parkinsonism in psychiatric patients treated with haloperidol*. Am J Psychiatry. **1986**, 143:232-234.

Molho ES, Factor SA. *Worsening of motor features of parkinsonism with olanzapine*. Mov Disord. **1999**, 14:1014-1016.

Möller HJ, Boyer P, Fleurot O, Rein W. *Improvement of acute exacerbations of schizophrenia with amisulpride: a comparison with haloperidol. PROD-ASLP Study Group*. Psychopharmacology (Berl). **1997**, 132:396-401.

Montalescot G, Levy Y, Hatt PY. *Serious sinus node dysfunction caused by therapeutic doses of lithium*. Int J Cardiol. **1984**, 5:94-96.

Morris S, Bagnall A, Cooper SJ, DeSilva P, Fenton M, Gammelin GO, Leitner M. *Zotepine for schizophrenia*. Available in The Cochrane Library; Issue 2. Chichester: John Wiley. **2004**.

Mortensen PB. *The incidence of cancer in schizophrenic patients*. J Epidemiol Community Health. **1989**, 43:43-47.

Mortensen PB. *The ocurrence of cancer in first admitted schizophrenic patients*. Schizophr Res. **1994**, 12:185-194.

Movin-Osswald G, Hammarlund-Udenaes M. *Remoxipride: pharmacokinetics and effect on plasma prolactin*. Br J Clin Pharmacol. **1991**, 32:355-360.

Musser WS, Kirisci L. *Critique of the Canadian multicenter placebo-controlled study of risperidone and haloperidol*. J Clin Psychopharmacol. **1995**, 15:226-228.

Myers DH, West TET. Hormone systems. In: Johnson FN, ed. *Depression & mania: modern lithium therapy*. Oxford: IRL Press, Oxford, **1987**, pp 220-226.

Nair C, Abraham G, de Leon J, Stanilla JK, Josiassen RC, Simpson G. *Dose-related effects of clozapine on tardive dyskinesia among "treatment-refractory" patients with schizophrenia*. Biol Psychiatry **1996**, 39:529-530.

Najib J. *Tardive dyskinesia: a review and current treatment options*. Am J Ther.**1999**, 6:51-60.

Nakanishi M and Munakata T. *Azepine derivatives*. JP41006752 (**1966**); priority: 1963

Nakanishi M, Tashiro C, Munakata T, Araki K, Tsumagari T, Imamura H. *Studies of piperidine derivatives. I*. J Med Chem. **1970**, 13:644-648.

Nankivell BJ, Bhandari PK, Koller LJ. *Rhabdomyolysis induced by thioridazine*. BMJ. **1994**, 309:378.

Ngen CC, Singh P. *Long-term phenothiazine administration and the eye in 100 Malaysians*. Br J Psychiatry **1988**,152:278-281.

Nordenstrom J, Strigard K, Perbeck L, Willems J, Bagedahl-Strindlund M, Linder J. *Hyperparathyroidism associated with treatment of manic-depressive disorders by lithium*. Eur J Surg. **1992**, 158:207-211.

Nyfort-Hansen K, Alderman CP. *Possible neuroleptic malignant syndrome associated with olanzapine*. Ann Pharmacother. **2000**, 34:667.

O'Neill J, Fountain A. *Levomepromazine (methotrimeprazine) and the last 48 hours*. Hosp Med. **1999**, 60:564-567.

O'Brien JM, Rockwood RP, Suh KI. *Haloperidol-induced torsade de pointes*. Ann Pharmacother. **1999**, 33:1046-1050.

Ogilvie AD, Croy MF. *Clozapine and hyponatraemia*. Lancet. **1992**, 340:672.

Okusa MD, Crystal LJ. *Clinical manifestations and management of acute lithium intoxication*. Am J Med. **1994**, 97:383-389.

Olesen OV. *Therapeutic drug monitoring of clozapine treatment therapeutic threshold value for serum clozapine concentratios*. Clin Pharmacokinet. **1998**, 34:497-502.

Olesen OV, Bartels U, Poulsen JH. *Perphenazine in breast milk and serum*. Am J Psychiatry. **1990**, 147:1378-1379.

Oliver DJ. The *use of methotrimeprazine in terminal care*. Br J Clin Pract. **1985**, 39:339-340.

Omrust SV, McClellan K. *Perospirone*. CNS Drugs. **2001**, 15:329-337.

Onofrj M, Paci C, D'Andreamatteo G, Toma L. *Olanzapine in severe Gilles de la Tourette syndrome: a 52-week double-blind cross-over study vs. low-dose pimozide*. J Neurol. **2000**, 247:443-446.

Oshiro Y, Sato S, and Kurahashi N. *Carbostyril derivatives*. EP0367141 (**1990**); priority: 1988

Oshiro Y, Sato S, Kurahashi N, Tanaka T, Kikuchi T, Tottori K, Uwahodo Y, Nishi T. *Novel antipsychotic agents with dopamine autoreceptor agonist properties: synthesis and pharmacology of 7-[4-(4-phenyl-1-piperazinyl)butoxy]-3,4-dihydro-2(1H)-quinolinone derivatives*. J Med Chem. **1998**, 41:658-667.

Owens DGH. *Extrapyramidal side effects and tolerability of risperidone: a review*. J Clin Psychiatry. **1994**, 55(supp 5): 29-35.

Palileo EV, Coelho A, Westveer D, Dhingra R, Rosen KM. *Persistent sinus node dysfunction secondary to lithium therapy*. Am Heart J. **1983**, 106:1443-1444.

Pandit MK, Burke J, Gustafson AB, Minocha A, Peiris AN. *Drug-induced disorders of glucose tolerance*. Ann Intern Med. **1993**, 118:529-539.

Patel AG, Mukherji K, Lee A. *Priapism associated with psychotropic drugs*. Br J Hosp Med. **1996**, 55:315-319.

Patt RB, Proper G, Reddy S. *The neuroleptics as adjuvant analgesics*. J Pain Symptom Manage. **1994**, 9:446-453.

Peet M, Pratt JP. *Lithium. Current status in psychiatric disorders*. Drugs. **1993**, 46:7-17.

Perregaard JK. *1-(4'-fluorophenyl)-3,5-substituted indoles useful in the treatment of psychic disorders and pharmaceutical compositions thereof*. US4710500 (**1987**); priority: 1985

Peters DH, Faudls D. *Tiapride: a review of its pharmacology and therapeutic potential in the management of alcohol dependence syndrome*. Drugs. **1994**, 47:1010-1032.

Petersen PV, Lassen NO, and Holm TO. *9-(propene-3-ylidene-1), and 9-[3'-(nu-hydroxyalkylpiperazino-nu)-propylidene], xanthenes and thiaxanthenes, and processes for their preparation*. US3116291 (**1963**); priority: 1958

Peuskens J. *Risperidone in the treatment of patients with chronic schizophrenia: a multi-national, multi-centre, double-blind, parallel-group study versus haloperidol. Risperidone Study Group*. Br J Psychiatry. **1995**, 166:712-726.

Pfeiffer C, Wagner ML. *Clozapine therapy for Parkinson's disease and other movement disorders*. Am J Hosp Pharm. **1994**, 51:3047-3053.

Phillips EJ, Liu BA, Knowles SR. *Rapid onset of risperidone-induced hepatotoxicity*. Ann Pharmacother. **1998**, 32:843.

Pokorny R, Finkel MJ, Robinson WT. *Normal volunteers should not be used for bioavailability or bioequivalence studies of clozapine*. Pharm Res. **1994**, 11:1221.

Polatti F. *Sulpiride isomers and milk secretion in Puerperium*. Clin Exp Obstet Gynecol. **1982**, 9:144-147.

Pollak PT, Zbuk K. *Quetiapine fumarate overdose: clinical and pharmacokinetic lessons from extreme conditions*. Clin Pharmacol Ther. **2000**, 68:92-97.

Pons G, Rey E, Matheson I. *Excretion of psychoactive drugs into breast milk. Pharmacokinetic principles and recommendations*. Clin Pharmacokinet. **1994**, 27:270-289.

Power WJ, Travers SP, Mooney DJ. *Welding arc maculopathy and fluphenazine*. Br J Ophthalmol. **1991**, 75:433-435.

Prakash C, Kamel A, Anderson W, Howard H. *Metabolism and excretion of the novel antipsychotic drug ziprasidone in rats after oral administration of a mixture of 14C- and 3H-labeled ziprasidone*. Drug Metab Dispos. **1997a**, 25:206-218.

Prakash C, Kamel A, Gummerus J, Wilner K. *Metabolism and excretion of a new antipsychotic drug, ziprasidone, in humans*. Drug Metab Dispos. **1997b**, 25:863-872.

Prakash C, Kamel A, Cui D. *Characterization of the novel benzisothiazole ring-cleaved products of the antipsychotic drug ziprasidone*. Drug Metab Dispos. **1997c**, 25:897-901.

Presley AP, Kahn A, Williamson N. *Antinuclear antibodies in patients on lithium carbonate*. Br Med J. **1976**, 2:280-281.

Price LH, Heninger GR. *Lithium in the treatment of mood disorders*. N Engl J Med. **1994**, 331:591-598.

Prous J, Castaner J. *SM-3997*. Drugs of the future. **1986**, 11:949.

Quinn N, Marsden CD. *A double blind trial of sulpiride in Huntington's disease and tardive dyskinesia*. J Neurol Neurosurg Psychiatry. **1984**, 47:844-847.

Quraishi S, David A. *Depot fluspirilene for schizophrenia*. Available in The Cochrane Library; Issue 2. Chichester: John Wiley. **2000**.

Quraishi S., David A. *Depot pipothiazine palmitate and undecylenate for schizophrenia*. Available in The Cochrane Library; Issue 2. Chichester: John Wiley. **2004**.

Ragucci KR, Wells BJ. *Olanzapine-induced diabetic ketoacidosis*. Ann Pharmacother. **2001**, 35:1556-1558.

Raja M. *Managing antipsychotic-induced acute and tardive dystonia*. Drug Safety. **1998**, 19:57-72.

Raoof NT, Pearson RM, Turner P. *Lithium inhibits human sperm motility in vitro*. Br J Clin Pharmacol. **1989**, 28:715-717.

Ray WA, Meredith S, Thapa PB, Meador KG, Hall K, Murray KT. *Antipsychotics and the risk of sudden cardiac death*. Arch Gen Psychiatry. **2001**, 58:1161-1167.

Reeves RR, Mack JE, Torres RA. *Neuroleptic malignant syndrome during a change from haloperidol to risperidone*. Ann Pharmacother. **2001**, 35:698-701.

Reilly GD, Wood ML. *Prochlorperazine-an unusual cause of lip ulceration*. Acta Derm Venereol (Stockh) **1984**, 64:270-271.

Reilly JG, Ayis SA, Ferrier IN, Jones SJ, Thomas SH. *QTc-interval abnormalities and psychotropic drug therapy in psychiatric patients*. Lancet. **2000**, 355:1048-1052.

Reilly JG, Ayis SA, Ferrier IN, Jones SJ, Thomas SH. *Thioridazine and sudden unexplained death in psychiatric in-patients*. Br J Psychiatry. **2002**, 180:515-522.

Reiss RA, Haas CE, Karki SD, Gumbiner B, Welle SL, Carson SW. *Lithium pharmacokinetics in the obese*. Clin Pharmacol Ther. **1994**, 55:392-398.

Remington G. *Understanding antipsychotic 'atypicality': a clinical and pharmacological moving target*. J Psychiatry Neurosci. **2003**, 28:275-284.

Rendell JM, Gijsman HJ, Keck P, Goodwin GM, Geddes JR. *Olanzapine alone or in combination for acute mania*. Available in The Cochrane Library; Issue 2. Chichester: John Wiley. **2004**.

Renz J and Bourquin JP. *Phenothiazine derivatives substituted by a monovalent sulfur function in 3-position*. US3239514 (**1966**); priority: 1956

Research Units on Pediatric Psychopharmacology Autism Network. *Risperidone in children with autism and serious behavioral problems*. N Engl J Med. **2002**, 347:314-321.

Rider JM, Mauger TF, Jameson JP, Notman DD. *Water handling in patients receiving haloperidol decanoate*. Ann Pharmacother. **1995**, 29:663-666.

Ring BJ, Catlow J, Lindsay TJ, Gillespie T, Roskos LK, Cerimele BJ, Swanson SP, Hamman MA, Wrighton SA. *Identification of the human cytochromes P450 responsible for the in vitro formation of the major oxidative metabolites of the antipsychotic agent olanzapine*. J Pharmacol Exp Ther. **1996**, 276:658-666.

Rivera-Calimlim L, Griesbach PH, Perlmutter R. *Plasma chlorpromazine concentrations in children with behavioral disorders and mental illness*. Clin Pharmacol Ther. **1979**, 26:114-121.

Roche H, Hyman G, Nahas G. *Hypertension and intravenous antidopaminergic drugs*. N Engl J Med. **1985**, 312:1125-1126.

Roefaro J, Mukherjee SM. *Olanzapine-induced hyperglicemic nonketonic coma*. Ann Pharmacother. **2001**, 35:300-302.

Roos RA, Buruma OJ, Bruyn GW, Kemp B, van der Velde EA. *Tiapride in the treatment of Huntington's chorea*. Acta Neurol Scand. **1982**, 65:45-50.

Roos RA, de Haas EJ, Buruma OJ, de Wolff FA. *Pharmacokinetics of tiapride in patients with tardive dyskinesia and Huntington's disease*. Eur J Clin Pharmacol. **1986**, 31:191-194.

Rosenblatt S, Hershman JM, Marder S, Garai T. *Chronic phenothiazine therapy does not increase sellar size*. Lancet. **1978**, 2:319-320.

Sachdev P, Kruk J, Kneebone M, Kissane D. *Clozapine-induced neuroleptic malignant syndrome: review and report of new cases*. J Clin Psychopharmacol. **1995**, 15:365-371.

Sachs GS, Grossman F, Ghaemi SN, Okamoto A, Bowden CL. *Combination of a mood stabilizer with risperidone or haloperidol for treatment of acute mania: a double-blind, placebo-controlled comparison of efficacy and safety*. Am J Psychiatry. **2002**, 159:1146-1154.

Safferman AZ, Lieberman JA, Alvir JM, Howard A. *Rechallenge in clozapine-induced agranulocytosis*. Lancet. **1992**, 339:1296-1297.

Sajatovic M, Brescan DW, Perez DE, DiGiovanni SK, Hattab H, Ray JB, Bingham CR. *Quetiapine alone and added to a mood stabilizer for serious mood disorders*. J Clin Psychiatry. **2001**, 62:728-732.

Sakamoto K, Nakamura Y, Aikoh S, Baba T, Perregaard J, Pedersen H, Moltzen EK, Mulford DJ, Yamaguchi T. *Metabolism of sertindole: identification of the metabolites in the rat and dog, and species comparison of liver microsomal metabolism.* Xenobiotica. **1995**, 25:1327-1343.

Salado J, Blazquez A, Diaz-Simon R, Lopez-Munoz F, Alamo C, Rubio G. *Priapism associated with zuclopenthixol.* Ann Pharmacother. **2002**, 36:1016-1018.

Salas IG, Pearson RM, Lawson M. *Lithium carbonate concentration in cervicovaginal mucus and serum after repeated oral dose administration.* Br J Clin Pharmacol. **1989**, 28:751P.

Sallee FR, Nesbitt L, Jackson C, Sine L, Sethuraman G. *Relative efficacy of haloperidol and pimozide in children and adolescents with Tourette's disorder.* Am J Psychiatry. **1997**, 154:1057-1062.

Sallee FR, Kurlan R, Goetz CG, Singer H, Scahill L, Law G, Dittman VM, Chappell PB. *Ziprasidone treatment of children and adolescents with Tourette's syndrome: a pilot study.* J Am Acad Child Adolesc Psychiatry. **2000**, 39:292-299.

Sansone ME, Ziegler DK. Brain and nervous system. In: Johnson FN, ed. *Depression & mania: modern lithium therapy.* Oxford: IRL Press, Oxford, **1987**, pp 240-245.

Sassim N, Grohmann R. *Adverse drug reactions with clozapine and simultaneous application of benzodiazepines.* Pharmacopsychiatry. **1988**, 21:306-307.

Sato S, Someya T, Shioiri O, Koitabashi T, Inoue Y. *Involvement of CYP3A4 in the metabolism of bromperidol in vitro.* Pharmacol Toxicol. **2000**, 86:145-148.

Satoh K, Someya T, Shibasaki M. *Effects of nemonapride on positive and negative symptoms of schizophrenia.* Int Clin Psychopharmacol. **1996**, 11:279-281.

Scahill L, Leckman JF, Schultz RT, Katsovich L, Peterson BS. *A placebo-controlled trial of risperidone in Tourette syndrome .* Neurology. **2003**, 60:1130-1135.

Schou M. Use in non-psychiatric conditions. In: Johnson FN, ed. *Depression & mania: modern lithium therapy.* IRL Press, Oxford, **1987**, pp 46-50.

Schou M. *Lithium treatment of manic-depressive illness. Past, present, and perspectives.* JAMA. **1988a**, 259:1834-1836.

Schou M. *Serum lithium monitoring of prophylactic treatment. Critical review and updated recommendations.* Clin Pharmacokinet. **1988b**, 15:283-286.

Schou M. *Lithium treatment of manic-depressive illness: a practical guide.* 4th rev.ed. ed. Karger AG; Basel, **1989**.

Scialli JVK, Thornton WE. *Toxic reactions from a haloperidol overdose in two children: thermal and cardiac manifestations.* JAMA. **1978**, 239:48-49.

Segal J, Berk M, Brook S. *Risperidone compared with both lithium and haloperidol in mania: a double-blind randomized controlled trial.* Clin Neuropharmacol. **1998**, 21:176-180.

Segraves RT. *Psychiatric drugs and inhibited female orgasm.* J Sex Marital Ther. **1988**,14:202-207.

Sell MB. *Sensitization to thioridazine through sexual intercourse.* Am J Psychiatry. **1985**, 142:271-272.

Shannon KM, Fenichel GM. *Pimozide treatment of Sydenham's chorea.* Neurology. **1990**, 40:186.

Shapiro E, Shapiro AK, Fulop G, Hubbard M, Mandeli J, Nordlie J, Phillips RA. *Controlled study of haloperidol, pimozide and placebo for the treatment of Gilles de la Tourette's syndrome.* Arch Gen Psychiatry. **1989**, 46:722-730.

Sharma R, Trappler B, Ng YK, Leeman CP. *Risperidone-induced neuroleptic malignant syndrome.* Ann Pharmacother. **1996**, 30:775-778.

Shaw GK, Waller S, Majumdar SK, Alberts JL, Latham CJ, Dunn G. *Tiapride in the prevention of relapse in recently detoxified alcoholics.* Br J Psychiatry. **1994**, 165:515-523.

Shelton RC, Tollefson GD, Tohen M, Stahl S, Gannon KS, Jacobs TG, Buras WR, Bymaster FP, Zhang W, Spencer KA, Feldman PD, Meltzer HY. *A novel augmentation strategy for treating resistant major depression.* Am J Psychiatry. **2001**, 158:131-134.

Sherlock S. *The spectrum of hepatotoxicity due to drugs.* Lancet. **1986**, 2:440-444.

Shin JG, Kane K, Flockhart DA. *Potent inhibition of CYP2D6 by haloperidol metabolites: stereoselective inhibition by reduced haloperidol.* Br J Clin Pharmacol. **2001**, 51:45-52.

Shukla VR, Borison RL. *Lithium and lupuslike syndrome.* JAMA. **1982**, 248:921-922.

Sinaniotis CA, Spyrides P, Vlachos P, Papadatos C. *Acute haloperidol poisoning in children.* J Pediatr. **1978**, 93:1038-1039.

Singh AN, Golledge H, Catalan J. *Treatment of HIV-related psychotic disorders with risperidone: a series of 21 cases.* J Psychosom Res. **1997**, 42:489-493.

Skinner J, Skinner A. *Levomepromazine for nausea and vomiting in advanced cancer.* Hosp Med. **1999**, 60:568-70.

Smeraldi E. *Amisulpride versus fluoxetine in patients with dysthymia or major depression in partial remission: a double-blind, comparative study.* J Affect Disord. **1998**, 48:47-56.

Snyder SH. *Drug and neurotransmitter receptors: new perspectives with clinical relevance.* JAMA. **1989**, 261:3126-3129.

Soares BGO, Fenton M, Chue P. *Sulpiride for schizophrenia.* Available in The Cochrane Library; Issue 2. Chichester: John Wiley. **2004a**.

Soares KVS, McGrath JJ, Deeks JJ. *Gamma-aminobutyric acid agonists for neuroleptic-induced tardive dyskinesia.* Available in The Cochrane Library; Issue 1. Chichester: John Wiley. **2004b**.

Soares KVS, McGrath JJ. *Calcium channel blockers for neuroleptic-induced tardive dyskinesia.* Available in The Cochrane Library; Isue 2. Chichester: John Wiley. **2004**.

Solomon K, Vickers R. *Dysarthria resulting from lithium carbonate. A case report.* JAMA. **1975**, 231:280.

Someya T, Inaba T, Tyndale RF, Tang SW, Takahashi S. *Conversion of bromperidol to reduced bromperidol in human liver.* Neuropsychopharmacology. **1991**, 5:177-182.

Sperry L, Hudson B, Chan CH. *Loxapine abuse.* N Engl J Med. **1984**, 310:598.

Spigset O, Hedenmalm K. *Hyponatraemia and the syndrome of inappropriate antidiuretic hormone secretion (SIADH) induced by psychotropic drugs.* Drug Safety. **1995**, 12:209-225.

Spiteri MA, James DG. *Adverse ocular reactions to drugs.* Post-grad Med J. **1983**, 59:343-349.

Sramek JJ, Potkin SG, Hahn R. *Neuroleptic plasma concentrations and clinical response: in search of a therapeutic window.* Drug Intell Clin Pharm. **1988**, 22:373-380.

Srinivasa Rao G, Prasanna Kumar BN, Manjunatha SG, and Kulkarni AK. *Process for the preparation of risperidone.* WO2005030772 (**2005**); priority: 2003

Srisurapanont M, Maneeton B, Maneeton N. *Quetiapine for schizophrenia.* Available in The Cochrane Library; Issue 2. Chichester: John Wiley. **2004**.

Stamenkovic M, Schindler SD, Aschauer HN, De Zwaan M, Willinger U, Resinger E, Kasper S. *Effective open-label treatment of tourette's disorder with olanzapine.* Int Clin Psychopharmacol. **2000**, 15:23-28.

Stanilla JK, de Leon J, Simpson GM. *Clozapine withdrawal resulting in delirium with psychosis: a report of three cases.* J Clin Psychiatry. **1997**, 58:252-255.

Staveris S, Plusquellec Y, Campistron G, Barre J, Rochas MA, Jung L, Tillement JP, Koffell JC, Houin G. *Pharmacokinetics of veralipride after chronic administration in humans.* J Pharm Sci. **1978**, 77:64-67.

Stewart RB, Karas B, Springer PK. *Haloperidol excretion in human milk.* Am J Psychiatry. **1980**, 137:849-850.

Stiell IG, Dufour DG, Moher D, Yen M, Beilby WJ, Smith NA. *Methotrimeprazine versus meperidine and dimenhydrinate in the treatment of severe migraine: a randomized, controlled trial.* Ann Emerg Med. **1991**, 20:1201-1205.

Sultana A, McMonagle T. *Pimozine for schizophrenia or related psychoses.* Available in The Cochrane Library; Issue 2. Chichester: John Wiley. **2004**.

Sumiyoshi T, Matsui M, Yamashita I, Nohara S, Kurachi M, Uehara T, Sumiyoshi S, Sumiyoshi C, Meltzer HY. *The effect of tandospirone, a serotonin(1A) agonist, on memory function in schizophrenia.* Biol Psychiatry. **2001**, 49:861-868.

Swartz CM, Jones P. *Hyperlithemia correction and persistent delirium.* J Clin Pharmacol. **1994**, 34:865-870.

Sweetman SC. *MARTINDALE: The Complete drug reference.* 34th ed. London: Pharmaceutical Press, **2004**.

Takahashi N, Terao T, Oga T, Okada M. *Comparison of risperidone and mosapramine addition to neuroleptic treatment in chronic schizophrenia.* Neuropsychobiology. **1999**, 39:81-85.

Tamminga CA, Thaker GK, Moran M, Kakigi T, Gao XM. *Clozapine in tardive dyskinesia: observations from human and animal model studies.* J Clin Psychiatry. **1994**, 55 Suppl B:102-106.

Tangedahl TN, Gau GT. *Myocardial irritability associated with lithium carbonate therapy.* N Engl J Med. **1972**, 287:867-869.

Tarricone NW. *Loxitane overdose.* Pediatrics. **1998**, 101:496.

Tarsy D. *Risperidone and neuroleptic malignant syndrome.* JAMA. **1996**, 275:446.

Tashiro C and Horii I. *Imidazopyridine-spiro-piperidine compounds.* US4337260 (**1982**); priority: 1981

Tashiro C, Setoguchi S, Fukuda T, Marubayashi N. *Syntheses and biological activities of optical isomers of 3-chloro-5-[3-(2-oxo-1,2,3,5,6,7,8,8a-octahydroimidazo[1,2-a]pyridine- 3-spiro-4'-piperidino)propyl]-10,11-dihydro-5H-dibenz[b,f]azepine (mosapramine) dihydrochloride.* Chem Pharm Bull (Tokyo). **1993**, 41:1074-1078.

Taylor D. *Clozapine- five years on.* Pharm J. **1995**, 254:260-263.

Taylor D. *Pharmacokinetic interaction involving clozapine.* Br J Psychiatry. **1997**, 171:109-12.

Taylor DM. *Aripiprazole: review of its pharmacology and clinical use.* Int J Clin Pract. **2003**, 57:49-54.

Taylor JW, Bell AJ. *Lithium-induced parathyroid dysfunction: a case report and review of the literature.* Ann Pharmacother. **1993**, 27:1040-1043.

Taylor WB, Bateman DN. *Preliminary studies of the pharmacokinetics and pharmacodynamics of prochlorperazine in healthy volunteers.* Br J Clin Pharmacol. **1987**, 23:137-142.

Tennent G, Bancroft J, Cass J. *The control of deviant sexual behavior by drugs: a double-blind controlled study of benperidol, chlorpromazine, and placebo.* Arch Sex Behav. **1974**, 3:261-271.

The French Clozapine Parkinson Study Group. *Clozapine in drug-induced psychosis in Parkinson's disease.* Lancet. **1999**, 353:2041-2042.

The Parkinson Study Group. *Low-dose clozapine for the treatment of drug-induced psychosis in Parkinson's disease.* N Engl J Med. **1999**, 340:757-763.

Thominet M, Acher J, and Monier J-C. *New derivatives of 4-amino-5-alkyl sulphonyl orthoanisamides, methods of preparing them and their application as psychotropic agents.* US4294828 (**1981**); priority: 1978

Thominet ML and Perrot JJ. *N-(1-allyl-2-pyrrolidylmethyl)-2,3-dimethoxy-5-sulfamoylbenzamide and derivatives thereof and method for treating hot flushes associated with natural or surgical menopause.* US4499019 (**1985**); priority: 1976

Thompson C. *Royal College of Psychiatrists'. Consensus Panel. The use of high-dose antipsychotic medication.* Br J Psychiatry. **1994**, 164:448-458.

Tollefson G, Lesar T. *Nonketotic hyperglycemia associated with loxapine and amoxapine: case report.* J Clin Psychiatry. **1983**, 44:347-348.

Tollefson GD, Beasley CM, Jr., Tran PV, Street JS, Krueger JA, Tamura RN, Graffeo KA, Thieme ME. *Olanzapine versus haloperidol in the treatment of schizophrenia and schizoaffective and schizophreniform disorders: results of an international collaborative trial.* Am J Psychiatry. **1997**, 154:457-465.

Tollefson GD. *Combination therapy for treatment of bipolar disorders.* WO9962522 (**1999**); priority: 1998

Tolosa-Vilella C, Ruiz-Ripoll A, Mari-Alfonso B, Naval-Sendra E. *Olanzapine-induced agranulocytosis: a case report and review of the literature.* Prog Neuropsychopharmacol Biol Psychiatry. **2002**, 26:411-414.

Tracqui A, Mutter-Schmidt C, Kintz P, Berton C, Mangin P. *Amisulpride poisoning: a report on two cases*. Hum Exp Toxicol. **1995**, 14:294-298.

Tran PV, Hamilton SH, Kuntz AJ, Potvin JH, Andersen SW, Beasley C, Jr., Tollefson GD. *Double-blind comparison of olanzapine versus risperidone in the treatment of schizophrenia and other psychotic disorders*. J Clin Psychopharmacol. **1997**, 17:407-418.

Turbott J, Smeeton WMI. *Sudden death and flupentixol decanoate*. Aust N Z J Psychiatry. **1984**, 18:91-94.

Uematsu T, Sato R, Suzuki K, Yamaguchi S, Nakashima M. *Human scalp hair as evidence of individual dosage history of haloperidol: method and retrospective study*. Eur J Clin Pharmacol. **1989**, 37:239-244.

Ullyot GE. *Substituted triflurormethylpheno-thiazine derivatives*. US2921069 (**1960**); priority: 1956

Ulrich S, Wurthmann C, Brosz M, Meyer FP. *The relationship between serum concentration and therapeutic effect of haloperidol in patients with acute schizophrenia*. Clin Pharmacokinet. **1998**, 34:227-263.

Umio S, Ueda I, Maeno S, and Sato Y. *Tricyclic enol ether compounds*. DE1907670 (**1969**); priority: 1968

Valiquette G. *Risperidone in children with autism and serious behavioral problems*. N Engl J Med. **2002**, 347:1890-1891.

van Harten PN, Hoek HW, Kahn RS. *Acute dystonia induced by drug treatment*. BMJ. **1999**, 319:623-626.

Various. *The Pharmacokinetics of ziprasidone*. Br J Clin Pharmacol. **2000**, 49(suppl 1):1S-76S.

Velamoor VR, Swamy GN, Parmar RS, Williamson P, Caroff SN. *Management of suspected neuroleptic malignant syndrome*. Can J Psychiatry. **1995**, 40:545-550.

Velamoor VR. *Neuroleptic malignant syndrome: recognition, prevention and management*. Drug Safety. **1998**, 19 :73-82.

Vercueil L, Foucher J. *Risperidone-induced tardive dystonia and psychosis*. Lancet. **1999**, 353:981

Vieta E, Parramon G, Padrell E, Nieto E, Martinez-Aran A, Corbella B, Colom F, Reinares M, Goikolea JM, Torrent C. *Quetiapine in the treatment of rapid cycling bipolar disorder*. Bipolar Disord. **2002**, 4:335-340.

von Bahr C, Movin G, Yisak WA, Jostell KG, Widman M. *Clinical pharmacokinetics of remoxipride*. Acta Psychiatr Scand Suppl. **1990**, 358:41-44.

von Bahr C, Movin G, Nordin C, Liden A, Hammarlund-Udenaes M, Hedberg A, Ring H, Sjoqvist F. *Plasma levels of thioridazine and metabolites are influenced by the debrisoquin hydroxylation phenotype*. Clin Pharmacol Ther. **1991**, 49:234-240.

Wahlbeck K, Cheine M, Essali MA. *Clozapine versus typical neuroleptic medication for schizophrenia*. Available in The Cochrane Library; Issue 2. Chichester: John Wiley. **2004**.

Walker RG, Kincaid-Smith P. *Kidneys and the fluid regulatory system*. In: Johnson FNed, ed. *Depression & mania: modern lithium therapy*. Oxford: IRL Press, Oxford, **1987**, pp 206-213.

Wang PS, Walker AM, Tsuang MT, Orav EJ, Glynn RJ, Levin R, Avorn J. *Dopamine antagonists and the development of breast cancer.* Arch Gen Psychiatry. **2002**, 59:1147-1154.

Warawa EJ and Migler BM. *Thiazepine compounds.* EP0240228 (**1987**); priority: 1986

Ward ME, Musa MN, Bailey L. *Clinical pharmacokinetics of lithium.* J Clin Pharmacol. **1994**, 34:280-285.

Watson RG, Olomu A, Clements D, Waring RH, Mitchell S, Elias E. *A proposed mechanism for chlorpromazine jaundice--defective hepatic sulphoxidation combined with rapid hydroxylation.* J Hepatol. **1988**, 7:72-78.

Weiner M, Chausow A, Wolpert E, Addington W, Szidon P. *Effect of lithium on the responses to added respiratory resistances.* N Engl J Med. **1983**, 308:319-322.

Wells AJ, Sommi RW, Crismon ML. *Neuroleptic rechallenge after neuroleptic malignant syndrome: case report and literature review.* Drug Intell Clin Pharm. **1988**, 22:475-480.

Wells BG. *Amiloride in lithium-induced polyuria.* Ann Pharmacother. **1994**, 28:888-889.

Wetterling T. *Bodyweight gain with atypical antipsychotics: a comparative review.* Drug Safety. **2001**, 24:59-73.

WHO. *Prophylactic use of anticholinergics in patients on long-term neuroleptics treatment: a consensus statement.* Br J Psychiatry. **1990**, 156:412.

Wieck A, Haddad P. *Hyperprolactinaemia caused by antipsychotic drugs.* BMJ. **2002**, 324:250-252.

Wiesel FA, Alfredsson G, Ehrnebo M, Sedvall G. *The pharmacokinetics of intravenous and oral sulpiride in healthy human subjects.* Eur J Clin Pharmacol. **1980**, 17:385-391.

Wilt JL, Minnema AM, Johnson RF, Rosenblum AM. *Torsade de pointes associated with the use of intravenous haloperidol.* Ann Intern Med. **1993**, 119:391-394.

Wistedt B, Wiles D, Kolakowska T. *Slow decline of plasma drug and prolactin levels after discontinuation of chronic treatment with depot neuroleptics.* Lancet. **1981**, 1:1163.

Witz L, Shapiro MS, Shenkman L. *Chlorpromazine induced fluid retention masquerading as idiopathic oedema.* Br Med J (Clin Res Ed). **1987**, 294:807-808.

Wolstein J, Grohmann R, Ruther E, Hippius H. *Antipsychotic drugs and venous thromboembolism.* Lancet. **2000**, 356:252.

Wolters EC, Jansen EN, Tuynman-Qua HG, Bergmans PL. *Olanzapine in the treatment of dopaminomimetic psychosis in patients with Parkinson's disease.* Neurology. **1996**, 47:1085-1087.

Wong FA, Bateman CP, Shaw CJ, Patrick JE. *Biotransformation of bromperidol in rat, dog, and man.* Drug Metab Dispos. **1983**, 11:301-307.

Wong SL, Cao G, Mack RJ, Granneman GR. *The effect of erythromycin on the CYP3A component of sertindole clearance in healthy volunteers.* J Clin Pharmacol. **1997**, 37:1056-1061.

Wong D, Adams CE, David A, Quraishi SN. *Depot bromperidol decanoate for schizophrenia.* Available in The Cochrane Library; Issue 2. Chichester: John Wiley. **2004**.

Worrall EP, Gillham RA. *Lithium-induced constructional dyspraxia.* Br Med J (Clin Res Ed). **1983**, 286:189.

Yamada K, Yagi G, Kanba S. *Clinical efficacy of tandospirone augmentation in patients with major depressive disorder: a randomized controlled trial.* Psychiatry Clin Neurosci. **2003**, 57:183-187.

Yasui-Furukori N, Furukori H, Nakagami T, Saito M, Inoue Y, Kaneko S, Tateishi T. *Steady-state pharmacokinetics of a new antipsychotic agent perospirone and its active metabolite, and its relationship with prolactin response.* Ther Drug Monit. **2004**, 26:361-365.

Yatham LN, Grossman F, Augustyns I, Vieta E, Ravindran A. *Mood stabilisers plus risperidone or placebo in the treatment of acute mania. International, double-blind, randomised controlled trial.* Br J Psychiatry. **2003**, 182:141-147. Correction. Ibi.;369.

Yeung PK, Hubbard JW, Korchinski ED, Midha KK. *Pharmacokinetics of chlorpromazine and key metabolites.* Eur J Clin Pharmacol. **1993**, 45:563-569.

Yin L, Wang J, Klein PS, Lazar MA. *Nuclear receptor Rev-erbalpha is a critical lithium-sensitive component of the circadian clock.* Science. **2006**, 311:1002-1005.

Yip L, Dart RC, Graham K. *Olanzapine toxicity in a toddler.* Pediatrics. **1998**, 102:1494.

Ylikorkala O, Kauppila A, Kivinen S, Viinikka L. *Sulpiride improves inadequate lactation.* Br Med J (Clin Res Ed). **1982**, 285:249-251.

Ylikorkala O, Kauppila A, Kivinen S, Viinikka L. *Treatment of inadequate lactation with oral sulpiride and buccal oxytocin.* Obstet Gynecol. **1984**, 63:57-60.

Young RL, Kumar NS, Goldzieher JW. *Management of menopause when estrogen cannot be used.* Drugs. **1990**, 40:220-230.

Zaccara G, Muscas GC, Messori A. *Clinical features, pathogenesis and management of drug-induced seizures.* Drug Safety. **1990**, 5:109-151.

Zee-Cheng CS, Mueller CE, Seifert CF, Gibbs HR. *Haloperidol and torsades de pointes.* Ann Intern Med. **1985**, 102:418.

Zerjav-Lacombe S, Dewan V. *Possible serotonin syndrome associated with clomipramine after withdrawl of clozapine.* Ann Pharmacother. **2001**, 35:180-182.

Zolezzi M, Badr MG. *Risperidone-induced mania.* Ann Pharmacother. **1999**, 33:380-381.

Zornberg GL, Jick H. *Antipsychotic drugs use and risk of firstime idiopathic venous thromboembolism: a case-control study.* Lancet. **2000**, 356:1219-23.

Zuddas A, Chianchetti C. *Efficacy of Risperidone in idiopathic segmental dystonia.* Lancet. **1996**, 347:127-128.

Related Titles

J. Fischer, C. R. Ganellin

Analogue-based Drug Discovery

2006
ISBN 3-527-31257-9

J. Licinio, M.-L. Wong (eds.)

Biology of Depression

From Novel Insights to Therapeutic Strategies

2005
ISBN 3-527-30785-0

R. Spiegel

Psychopharmacology - An Introduction 4e

2003
ISBN 0-471-56039-1

H. Buschmann, T. Christoph, E. Friderichs, C. Maul, B. Sundermann (eds.)

Analgesics

From Chemistry and Pharmacology to Clinical Application

2002
ISBN 3-527-30403-7